$\mathbf{x}_h$, solution of the homogeneous system $A\mathbf{x} = \mathbf{0}$

$[\mathbf{x}]_B$, coordinate matrix (or coordinate vector) of $\mathbf{x}$ relative to the basis B

$[\mathbf{x}]_{B'}$, coordinate matrix (or coordinate vector) of $\mathbf{x}$ relative to the basis B'

P, transition matrix from B' to B

P^{-1}, transition matrix from B to B'

W, Wronskian of a set of functions

$V \cup W$, union of V and W

$V \oplus W$, direct sum of V and W

Chapter 5

$\|\mathbf{v}\|$, norm (length, magnitude) of the vector $\mathbf{v}$

$\{\mathbf{i}, \mathbf{j}\}$, standard unit vector in R^2

$\{\mathbf{i}, \mathbf{j}, \mathbf{k}\}$, standard unit vector in R^3

$d(\mathbf{u}, \mathbf{v})$, distance between two vectors $\mathbf{u}$ and $\mathbf{v}$

$\|\mathbf{u} - \mathbf{v}\|$, distance between two vectors $\mathbf{u}$ and $\mathbf{v}$

$\mathbf{u} \cdot \mathbf{v}$, dot product of $\mathbf{u}$ and $\mathbf{v}$, or the Euclidean inner product for R^n

$|\mathbf{u} \cdot \mathbf{v}|$, absolute value of $\mathbf{u} \cdot \mathbf{v}$

$\mathbf{u}^T\mathbf{v}$, matrix product of the transpose of $\mathbf{u}$ times $\mathbf{v}$ (equal to the dot product)

$\langle \mathbf{u}, \mathbf{v} \rangle$, general inner product

$\text{proj}_{\mathbf{v}}\mathbf{u}$, orthogonal projection of $\mathbf{u}$ onto $\mathbf{v}$

$S^{\perp}$, orthogonal complement of S

$V \oplus W$, direct sum of V and W

$\text{proj}_S\mathbf{v}$, projection of $\mathbf{v}$ onto the subspace S

$R(A)$, column space of A, or row space of A

$A^TA\mathbf{x} = A^T\mathbf{b}$, normal equations of the least squares problem $A\mathbf{x} = \mathbf{b}$

$\mathbf{u} \times \mathbf{v}$, cross product of $\mathbf{u}$ and $\mathbf{v}$

Chapter 6

T, often denotes a linear transformation

$T: V \to W$, functions that map a vector space V into a vector space W

$T(\mathbf{v}) = \mathbf{w}$, $\mathbf{w}$ is the image of $\mathbf{v}$ under T

$T(\mathbf{v}) = \mathbf{0}$, zero transformation

$T(\mathbf{v}) = \mathbf{v}$, identity transformation

$T(\mathbf{v}) = A\mathbf{v}$, linear transformation from R^n into R^m

D_x, differential operator

$\ker(T)$, kernel of T

$\text{range}(T)$, range of T

$\text{nullity}(T)$, nullity of T

$\text{rank}(T)$, rank of T

$T_2 \circ T_1$, composition of T_1 and T_2

$[T(\mathbf{v})]_{B'} = A[\mathbf{v}]_B$, A is the matrix of T relative to the bases B and B'

$A' = P^{-1}AP$, square matrices A and A' are similar

$T(x, y) = (kx, y)$, horizontal contractions and expansions

$T(x, y) = (x, ky)$, vertical contractions and expansions

Chapter 7

λ, eigenvalue of an $n \times n$ matrix

$A\mathbf{x} = \lambda\mathbf{x}$, fundamental equation for the eigenvalue problem

$|\lambda I - A| = 0$, characteristic equation

k, multiplicity of an eigenvalue

D, often denotes a diagonal matrix

$A = A^T$, square matrix A is symmetric

$P^{-1} = P^T$, square matrix P is orthogonal

ELEMENTARY LINEAR ALGEBRA

FIFTH EDITION

RON LARSON
The Pennsylvania State University
The Behrend College

BRUCE H. EDWARDS
The University of Florida

DAVID C. FALVO
The Pennsylvania State University
The Behrend College

HOUGHTON MIFFLIN COMPANY   *Boston   New York*

Vice President and Publisher: Jack Shira
Associate Sponsoring Editor: Cathy Cantin
Development Manager: Maureen Ross
Development Editor: Laura Wheel
Assistant Editor: Marika Hoe
Assistant Editor: James Cohen
Supervising Editor: Karen Carter
Senior Project Editor: Patty Bergin
Editorial Assistant: Allison Seymour
Production Technology Supervisor: Gary Crespo
Executive Marketing Manager: Michael Busnach
Marketing Associate: Nicole Mollica
Senior Manufacturing Coordinator: Jane Spelman
Composition and Art: Meridian Creative Group
Cover Design Manager: Diana Coe

Cover Image: Multicolored background © EyeWire/EyeWire Collection
Photo Credits: page 1, Corbis-Bettmann; pages 45, 119, 174, 270, 352, 410, The Granger Collection.

Trademark Acknowledgments: TI is a registered trademark of Texas Instruments Incorporated. Maple is a registered trademark of Waterloo Maple Software. *Mathematica* is a registered trademark of Wolfram Research, Inc. Derive is a registered trademark of Texas Instruments Incorporated. IBM PC is a registered trademark of International Business Machines Corporation. Macintosh is a registered trademark of Apple Computer, Inc. MATLAB is a registered trademark of The MathWorks, Inc. For information on MATLAB, contact The MathWorks, Inc., 3 Apple Hill Drive, Natick, Massachusetts, 01760-2098.

Copyright © 2004 by Houghton Mifflin Company. All rights reserved.

No part of this work may be reproduced or transmitted in any form or by any means, electronic or mechanical, including photocopying and recording, or by any information storage or retrieval system, without the prior written permission of Houghton Mifflin Company unless such copying is expressly permitted by federal copyright law. Address inquiries to College Permissions, Houghton Mifflin Company, 222 Berkeley Street, Boston, MA 02116-3764.

Printed in the United States of America.

Library of Congress Catalogue Card Number: 2002117256

ISBN: 0-618-33567-6

123456789–DOC–08 07 06 05 04 03

CONTENTS

Contents

iii

* Chapters 8, 9, and 10 are available on the Learning Tools Student CD-ROM and the textbook website. To access the chapters online, go to math.college.hmco.com/students and link to Elementary Linear Algebra, *Fifth Edition.*

Elementary Linear Algebra, FIFTH EDITION
Learning Tools Student CD-ROM

The Learning Tools Student CD-ROM that accompanies the text provides students with the electronic tools described below. These tools are designed to offer students additional practice while also giving them opportunities to further explore selected topics.

The symbol *Data* next to an exercise in the text signifies an **Electronic Data Set.** The data sets for these exercises are supplied on the CD-ROM to help students reinforce and broaden their technology skills using MATLAB, *Mathematica*, Maple, Derive, and the TI-83/TI-83 Plus, TI-86, TI-89, and TI-92 graphing calculators.

The **Graphing Calculator Keystroke Guide** offers commands and instructions for various graphing calculators and includes examples with step-by-step solutions, technology tips, and programs.

MATLAB Exercises are included to enhance students' understanding of concepts using MATLAB software. These optional exercises correlate to chapters in the text.

Chapters 8, 9, and 10 are provided for those students who want to explore additional topics such as complex vector spaces, linear programming, and numerical methods. The answers to the odd-numbered exercises for these chapters are also included on the CD-ROM.

The symbol next to an example in the text signifies a **Simulation.** The simulations provide hands-on experimentation by giving students the opportunity to change variables and observe the outcomes of these changes. Each chapter contains a simulation so that students can explore key concepts throughout the text.

PREFACE

Elementary Linear Algebra, Fifth Edition, is designed for the introductory linear algebra course generally taken by sophomores and juniors majoring in engineering, computer science, mathematics, economics, statistics, science, or operations research. The primary prerequisite for this course is algebra, but we also assume familiarity with analytic geometry and trigonometry. Calculus is not a prerequisite for this course. Although several examples and exercises require calculus, these are clearly labeled and can be omitted without loss of continuity.

Many students will encounter mathematical formalism for the first time in this course. Thus, our primary goal is to present the major concepts of linear algebra clearly and concisely. To this end, we have carefully selected the examples and exercises to balance theory with applications and geometric intuition. In addition, the instructor may choose to cover the optional applications sections, which contain a wide variety of engaging and motivational material.

Coverage of Linear Algebra Topics

The order and coverage of topics were chosen for maximum efficiency, effectiveness, and balance. For example, in Chapter 4 we present the main ideas of vector spaces and bases, beginning with a brief look at R^2 and R^n and leading into the vector space concept as a natural extension of these familiar examples. This material is often the most difficult for students, but our approach to linear independence, span, basis, and dimension is carefully explained and illustrated by examples. The eigenvalue problem is developed in detail in Chapter 7, but we lay an intuitive foundation for students earlier in exercises in Section 1.2, Section 3.1, and Chapter 4. In addition, the newly revised Section 3.4 provides an early optional introduction to eigenvalues that instructors may choose to include or omit at their discretion.

The Fifth Edition accommodates different approaches to teaching this course. In response to feedback from linear algebra instructors, we have condensed our coverage of topics to include systems of linear equations, matrices, determinants, vector spaces, inner product spaces, linear transformations, and eigenvalues and eigenvectors. Additional topics such as complex vector spaces, linear programming, and numerical methods can be found on the *Elementary Linear Algebra* Learning Tools Student CD-ROM and on our website *math.college.hmco.com/students*. For those instructors who emphasize applications, there are numerous examples and exercises throughout the text. Each chapter also concludes with a section of applications. (A list of these applications appears on page xiii.) For the proof-oriented course, the text is sufficiently rigorous. In particular, the presentation of theorems focuses on developing the skills needed to construct accurate proofs. For the computer-oriented course, Chapter 6 covers computer graphics. Brief introductions to the latest computer algebra systems and graphing calculators are offered in Appendix B. In addition, throughout the text there are new and updated opportunities for using technology to analyze examples and solutions, as well as work the exercises.

Features

Theorems and Proofs

Every effort has been made to present theorems and their proofs using language that is both clear to students taking this course *and* mathematically precise. Proofs of theorems are treated in several ways in the Fifth Edition. When it is pedagogically sound to do so, proofs are presented in complete detail. Alternatively, some proofs are presented in outline form, omitting unwieldy calculations. Specialized exercises labeled Guided Proofs lead students through the initial steps of constructing proofs and then utilizing the results. Finally, to give students additional practice, the proofs of several theorems are left as exercises.

Graphics and Geometric Emphasis

Visualization is a problem-solving skill that is necessary for the understanding of mathematical concepts and theory. In the Fifth Edition, we have added several new graphs to examples, particularly when presenting vector spaces and inner product spaces. Computer-generated illustrations are used frequently to help students grasp complex concepts and mathematical relationships. For example, graphics offer geometric interpretations of problems, show how linear transformations can be used to rotate figures in two- and three-dimensional space, and illustrate the geometric aspects of projections and reflections.

Applications

Each chapter ends with a section on real-life applications of linear algebra concepts. These interesting and substantive problems address a wide variety of topics, including computer graphics, cryptography, differential equations, population growth, least squares approximations, and Fourier approximations. (A list of these applications appears on page xiii.) The material covered in the applications section is not required for the work in subsequent chapters and can be omitted without loss of continuity.

Technology and Technology Notes

In the Fifth Edition, we continue to take advantage of opportunities to use graphing calculators and computer algebra systems where appropriate. However, the use of such graphing utilities remains optional. Exercises that require the use of technology are clearly labeled. For students who do have access to graphing utilities in conjunction with this course, there are ample opportunities to use technology as an efficient problem-solving tool, as well as a means of enhancing or reinforcing understanding through discovery and alternative approaches. Through Technology Notes, we ensure that students use technology appropriately in the problem-solving process. Appendix B, "Computer Algebra Systems and Graphing Calculators," has been updated to reflect the most current systems and software. The newly updated MATLAB exercises are also available on the Learning Tools Student CD-ROM and on our website at *math.college.hmco.com/students*.

Examples

Numerous carefully chosen Examples illustrate concepts, computational techniques, and problem-solving approaches. All those that use real data have been updated.

Discovery

The Discovery features are designed to help students develop an intuitive understanding of mathematical concepts and relationships.

Historical Notes

Each chapter opens with a brief biographical sketch of a mathematician who played a role in the development of linear algebra.

Exercises

The Fifth Edition contains nearly 3000 problems. Each exercise set is carefully designed to give students practice in problem-solving techniques and to test their understanding of theoretical concepts. To help instructors plan assignments, particular types of problems are clearly identified. The icon ⬛ indicates that technology is required to work the problem. *Guided Proof* and *Writing* exercises are identified by these specific headings. For the Fifth Edition, all real-data exercises have been updated. In addition, new exercises have been added, including technology problems, data-driven exercises, matrix problems, and writing exercises.

We have also created several electronic data sets to accompany many problems in the Fifth Edition. These data sets, located on the Learning Tools Student CD-ROM and on our website at *math.college.hmco.com/students*, may be used with MATLAB, *Mathematica*, Maple, Derive, and the TI-83/TI-83 Plus, TI-86, TI-89, or TI-92 graphing calculators. The icon ⬛ appears next to those exercises that utilize these data sets.

Guided Proofs

Guided Proof exercises are designed to help students successfully complete theoretical proofs by leading them through the logical sequence of statements necessary to reach the correct conclusion.

True or False?

Exercises labeled *True or False?* test a student's understanding of mathematical concepts. For each exercise, students are required to give examples or justifications to support their conclusions.

Chapter Summaries

A Chapter Summary is provided at the end of each chapter. These Summaries highlight the key concepts covered in the chapter and function as a useful organizational tool for student study and review.

Chapter Projects

Each chapter concludes with two Chapter Projects. These Projects offer the opportunity for group activities or more extensive homework assignments. Some focus on theoretical concepts, whereas others are applications-oriented. Many encourage the use of technology. See the list of Chapter Projects on page xiv.

Cumulative Tests

Following Chapters 3, 5, and 7, Cumulative Tests help students synthesize the knowledge they have accumulated throughout the text, as well as prepare for exams and future mathematics courses.

Supplements

Learning Tools Student CD-ROM (0-618-33570-6)

- Simulations to help illustrate key concepts presented throughout the text
- Electronic data sets for use with MATLAB, *Mathematica*, Maple, Derive, and the TI-83/TI-83 Plus, TI-86, TI-89, and TI-92 graphing calculators.
- Graphing Calculator Keystroke Guide
- MATLAB exercises
- Chapters 8, 9, and 10

HMClass Prep™ with HM Testing 6.1 (0-618-33571-4)

- *HMClassPrep* contains a variety of instructor resources, including the textbook solutions and online Chapters 8, 9, and 10.
- *HM Testing* software provides instructors with an array of test items that can be used to create tests quickly and efficiently. This computerized software includes gradebook features and offers interested instructors the ability to deliver testing online via a Local Area Network (LAN) or the Internet.

Student Solutions Guide (0-618-33568-4)

- Step-by-step solutions to all odd-numbered text exercises
- Step-by-step solutions to all Cumulative Test items

Instructor's Solutions Guide and Test Item File (0-618-33569-2)

- Step-by-step solutions to all even-numbered text exercises
- Step-by-step solutions to all Chapter Projects
- Multiple-choice and open-ended test items with answers

Website for Elementary Linear Algebra, *Fifth Edition (math.college.hmco.com)*

- Student resources
- Instructor resources

Acknowledgments

We would like to thank the many people who have helped us during various stages of this project. In particular, we appreciate the efforts of the following colleagues who made many helpful suggestions along the way: Elena Black, University of Oklahoma; Roxanne M. Byrne, University of Colorado at Denver; Minerva Cordero-Epperson, University of Texas at Arlington; Elwyn H. Davis, Pittsburg State University; William D. Etling, Edinboro University of Pennsylvania; Guy Martin Forrest, Louisiana State University; Michael Y. Li, Mississippi State University; Karl P. Reitz, Chapman University; Rick Vaughn, Paradise Valley Community College; Charles Waters, Minnesota State University, Mankato; and John Woods, Southwestern Oklahoma State University. In addition, we wish to thank Donna Flint from South Dakota State University and Helen Medley for their careful accuracy checking of the textbook.

We would like to extend a special thanks to Elena Black for her contributions to this textbook.

On a personal level, we are grateful to our wives, Deanna Gilbert Larson, Consuelo Edwards, and Susan Falvo, for their love, patience, and support. Also, special thanks go to R. Scott O'Neil.

If you have suggestions for improving this text, please feel free to write to us.

Ron Larson
Bruce H. Edwards
David C. Falvo

INDEX OF APPLICATIONS

CHAPTER PROJECTS

WHAT IS LINEAR ALGEBRA?

To answer the question "What is linear algebra?," take a closer look at what you will study in this course. The most fundamental theme of linear algebra, and the first topic covered in this textbook, is the theory of systems of linear equations. You have probably encountered small systems of linear equations in your previous mathematics courses. For example, suppose you travel on an airplane between two cities that are 5000 kilometers apart. If the trip one way against a headwind takes $6\frac{1}{4}$ hours and the return trip the same day in the direction of the wind takes only 5 hours, can you find the ground speed of the plane and the speed of the wind, assuming that both remain constant?

If you let x represent the speed of the plane and y the speed of the wind, then the following system models the problem.

$$6.25(x - y) = 5000$$
$$5(x + y) = 5000$$

This system of two equations and two unknowns simplifies to

$$x - y = 800$$
$$x + y = 1000,$$

and the solution is $x = 900$ kilometers per hour and $y = 100$ kilometers per hour. Geometrically, this system represents two lines in the xy–plane. You can see in the figure that these lines intersect at the point $(900, 100)$, which verifies the answer that was obtained.

Solving systems of linear equations is one of the most important applications of linear algebra. It has been argued that the majority of all mathematical problems encountered in scientific and industrial applications involve solving a linear system at some point. Linear applications arise in such diverse areas as engineering, chemistry, economics, business, ecology, biology, and psychology.

Of course, the small system presented in the airplane example above is very easy to solve. In real-world situations, it is not unusual to have to solve systems of hundreds or even thousands of equations. One of the early goals of this course is to develop an algorithm that helps solve larger systems in an orderly manner and is amenable to computer implementation.

Original Flight

$x - y$

Return Flight

$x + y$

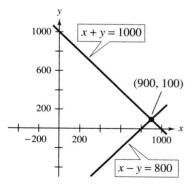

The lines intersect at (900, 100).

Vectors in the Plane

LINEAR ALGEBRA The branch of algebra in which one studies vector (linear) spaces, linear operators (linear mappings), and linear, bilinear, and quadratic functions (functionals and forms) on vector spaces. *(Encyclopedia of Mathematics, Kluwer Academic Press, 1990)*

The first three chapters of this textbook cover linear systems and two other computational areas you may have studied before: matrices and determinants. These discussions prepare the way for the central theoretical topic of linear algebra: the concept of a vector space. Vector spaces generalize the familiar properties of vectors in the plane. It is at this point in the text that you will begin to write proofs and learn to verify theoretical properties of vector spaces.

The concept of a vector space permits you to develop an entire theory of its properties. The theorems you prove will apply to all vector spaces. For example, in Chapter 6 you will study linear transformations, which are special functions between vector spaces. The applications of linear transformations appear almost everywhere—computer graphics, differential equations, and satellite data transmission, to name just a few examples.

Another major focus of linear algebra is the so-called eigenvalue $\left(\bar{I}\text{–g}\partial\text{n–value}\right)$ problem. Eigenvalues are certain numbers associated with square matrices and are fundamental in applications as diverse as population dynamics, electrical networks, chemical reactions, differential equations, and economics.

Linear algebra strikes a wonderful balance between computation and theory. As you proceed, you will become adept at matrix computations and will simultaneously develop abstract reasoning skills. Furthermore, you will see immediately that the applications of linear algebra to other disciplines are plentiful. In fact, you will notice that each chapter of this textbook closes with a section of applications. You might want to peruse some of these sections to see the many diverse areas to which linear algebra can be applied. (An index of these applications is given on page xiii.)

Linear algebra has become a central course for mathematics majors as well as students of science, business, and engineering. Its balance of computation, theory, and applications to real life, geometry, and other areas makes linear algebra unique among mathematics courses. For the many people who make use of pure and applied mathematics in their professional careers, an understanding and appreciation of linear algebra is indispensable.

CARL FRIEDRICH GAUSS is often ranked—along with Archimedes and Newton—as one of the greatest mathematicians in history.

Gauss earned his reputation as "prince of mathematics" at an early age. At 18 he invented the method of least squares and soon after proved that a 17-sided polygon is constructible using only a straightedge and a compass. In his doctoral thesis, Gauss proved the Fundamental Theorem of Algebra.

Gauss's landmark work in differential geometry derived from his interest in surveying and mapmaking. He also made many other contributions to algebra, complex functions, and potential theory.

Gauss wrote much more than he published. He kept files of many unpublished papers that he considered to be unfinished, inelegant, nonrigorous, or inconcise. Later readers of these papers have generally felt that Gauss was too critical of his own work. His unpublished papers covered work in two major fields: elliptic function theory and non-Euclidean geometry.

In addition to mathematics, Gauss had a keen interest in the physical sciences, especially astronomy; he held an appointment as director of the Göttingen observatory for nearly 50 years. One of his best-known publications in this area is *Theory of Motion of the Heavenly Bodies*, published in 1809. Gauss also earned great distinction for his research on theoretical and experimental magnetism. He helped improve existing techniques in telegraphy and created a new model for handling optic problems.

CARL FRIEDRICH GAUSS

1 7 7 7 – 1 8 5 5

SYSTEMS OF LINEAR EQUATIONS

CHAPTER 1

1.1 | Introduction to Systems of Linear Equations

Linear algebra is a branch of mathematics rich in theory and applications. This text strikes a balance between the theoretical and the practical. Because linear algebra arose from the study of systems of linear equations, you shall begin with linear equations. Although some material in this first chapter will be familiar to you, it is suggested that you study carefully the methods presented here. Doing so will cultivate and clarify your intuition for the more abstract material that follows.

The study of linear algebra demands familiarity with algebra, analytic geometry, and trigonometry. Occasionally you will find examples and exercises requiring a knowledge of calculus; these are clearly marked in the text.

Early in your study of linear algebra you will discover that many of the solution methods involve dozens of arithmetic steps; it is therefore essential to strive to avoid careless errors. A computer or calculator can be very useful in checking your work, as well as in performing many of the routine computations in linear algebra.

Linear Equations in n Variables

Recall from analytic geometry that the equation of a line in two-dimensional space has the form

$$a_1 x + a_2 y = b, \quad a_1, a_2, \text{ and } b \text{ are constants.}$$

This is a **linear equation in two variables** x and y. Similarly, the equation of a plane in three-dimensional space has the form

$$a_1 x + a_2 y + a_3 z = b, \quad a_1, a_2, a_3, \text{ and } b \text{ are constants.}$$

Such an equation is called a **linear equation in three variables** x, y, and z. In general, a linear equation in n variables is defined as follows.

Definition of a Linear Equation in n Variables

A **linear equation in n variables** $x_1, x_2, x_3, \ldots, x_n$ has the form

$$a_1 x_1 + a_2 x_2 + a_3 x_3 + \cdots + a_n x_n = b.$$

The **coefficients** $a_1, a_2, a_3, \ldots, a_n$ are real numbers, and the **constant term** b is a real number. The number a_1 is the **leading coefficient,** and x_1 is the **leading variable.**

REMARK: Letters that occur early in the alphabet are used to represent constants, and letters that occur late in the alphabet are used to represent variables.

Linear equations have no products or roots of variables and no variables involved in trigonometric, exponential, or logarithmic functions. Variables appear only to the first power. Example 1 lists some equations that are linear and some that are not linear.

EXAMPLE 1 *Examples of Linear Equations and Nonlinear Equations*

Each equation is linear.

(a) $3x + 2y = 7$

(b) $\frac{1}{2}x + y - \pi z = \sqrt{2}$

(c) $x_1 - 2x_2 + 10x_3 + x_4 = 0$

(d) $\left(\sin \frac{\pi}{2} \right) x_1 - 4x_2 = e^2$

Each equation is not linear.

(a) $xy + z = 2$

(b) $e^x - 2y = 4$

(c) $\sin x_1 + 2x_2 - 3x_3 = 0$

(d) $\frac{1}{x} + \frac{1}{y} = 4$

A **solution** of a linear equation in n variables is a sequence of n real numbers $s_1, s_2, s_3, \ldots, s_n$ arranged so that the equation is satisfied when the values

$$x_1 = s_1, \qquad x_2 = s_2, \qquad x_3 = s_3, \qquad \ldots, \qquad x_n = s_n$$

are substituted into the equation. For example, the equation

$$x_1 + 2x_2 = 4$$

is satisfied when $x_1 = 2$ and $x_2 = 1$. Some other solutions are $x_1 = -4$ and $x_2 = 4$, $x_1 = 0$ and $x_2 = 2$, and $x_1 = -2$ and $x_2 = 3$.

The set of *all* solutions of a linear equation is called its **solution set,** and when this set is found, the equation is said to have been **solved.** To describe the entire solution set of a linear equation, a **parametric representation** is often used, as illustrated in Examples 2 and 3.

EXAMPLE 2   *Parametric Representation of a Solution Set*

Solve the linear equation $x_1 + 2x_2 = 4$.

Solution   To find the solution set of an equation involving two variables, solve for one of the variables in terms of the other variable. For instance, if you solve for x_1 in terms of x_2, you obtain

$$x_1 = 4 - 2x_2.$$

In this form, the variable x_2 is **free,** which means that it can take on any real value. The variable x_1 is not free because its value depends on the value assigned to x_2. To represent the infinite number of solutions of this equation, it is convenient to introduce a third variable, t, called a **parameter.** Thus, by letting $x_2 = t$, you can represent the solution set as

$$x_1 = 4 - 2t, \qquad x_2 = t, \quad t \text{ is any real number.}$$

Particular solutions can be obtained by assigning values to the parameter t. For instance, $t = 1$ yields the solution $x_1 = 2$ and $x_2 = 1$, and $t = 4$ yields the solution $x_1 = -4$ and $x_2 = 4$.

_____

The solution set of a linear equation can be represented parametrically in more than one way. For instance, in Example 2 you could have chosen x_1 to be the free variable. The parametric representation of the solution set would then have taken the form

$$x_1 = s, \qquad x_2 = 2 - \tfrac{1}{2}s, \quad s \text{ is any real number.}$$

For convenience, choose the variables that occur last in a given equation to be free variables.

EXAMPLE 3   *Parametric Representation of a Solution Set*

Solve the linear equation $3x + 2y - z = 3$.

Solution   Choosing y and z to be the free variables, begin by solving for x to obtain

$$3x = 3 - 2y + z$$
$$x = 1 - \tfrac{2}{3}y + \tfrac{1}{3}z.$$

Letting $y = s$ and $z = t$, you obtain the parametric representation

$$x = 1 - \tfrac{2}{3}s + \tfrac{1}{3}t, \qquad y = s, \qquad z = t$$

where s and t are any real numbers. Two particular solutions are

$$x = 1, y = 0, z = 0 \qquad \text{and} \qquad x = 1, y = 1, z = 2.$$

Systems of Linear Equations

A **system of m linear equations in n variables** is a set of m equations, each of which is linear in the same n variables:

$$
\begin{aligned}
a_{11}x_1 + a_{12}x_2 + a_{13}x_3 + \cdots + a_{1n}x_n &= b_1 \\
a_{21}x_1 + a_{22}x_2 + a_{23}x_3 + \cdots + a_{2n}x_n &= b_2 \\
a_{31}x_1 + a_{32}x_2 + a_{33}x_3 + \cdots + a_{3n}x_n &= b_3 \\
&\ \vdots \\
a_{m1}x_1 + a_{m2}x_2 + a_{m3}x_3 + \cdots + a_{mn}x_n &= b_m.
\end{aligned}
$$

REMARK: The double-subscript notation indicates that a_{ij} is the coefficient of x_j in the ith equation.

A **solution** of a system of linear equations is a sequence of numbers $s_1, s_2, s_3, \ldots, s_n$ that is a solution of each of the linear equations in the system. For example, the system

$$
\begin{aligned}
3x_1 + 2x_2 &= 3 \\
-x_1 + x_2 &= 4
\end{aligned}
$$

has $x_1 = -1$ and $x_2 = 3$ as a solution because *both* equations are satisfied when $x_1 = -1$ and $x_2 = 3$. On the other hand, $x_1 = 1$ and $x_2 = 0$ is not a solution of the system because these values satisfy only the first equation in the system.

DISCOVERY

Graph the two lines

$$
\begin{aligned}
3x - y &= 1 \\
2x - y &= 0
\end{aligned}
$$

in the xy-plane. Where do they intersect? How many solutions does this system of linear equations have?
  Repeat this analysis for the pairs of lines

$$
\begin{array}{ll}
3x - y = 1 & \qquad 3x - y = 1 \\
3x - y = 0 & \qquad 6x - 2y = 2.
\end{array}
$$

In general, what basic types of solution sets are possible for a system of two equations in two unknowns?

It can happen that a system of linear equations has exactly one solution, an infinite number of solutions, or no solution. A system of linear equations is called **consistent** if it has at least one solution and **inconsistent** if it has no solution.

EXAMPLE 4   *Systems of Two Equations in Two Variables*

Solve each system of linear equations, and graph each system as a pair of straight lines.

(a) $x + y = 3$ (b) $x + y = 3$ (c) $x + y = 3$
$x - y = -1$ $2x + 2y = 6$ $x + y = 1$

Solution   (a) This system has exactly one solution, $x = 1$ and $y = 2$. This solution can be obtained by adding the two equations to give $2x = 2$, which implies that $x = 1$ and hence $y = 2$. The graph of this system is represented by two *intersecting* lines, as shown in Figure 1.1(a).

(b) This system has an infinite number of solutions because the second equation is the result of multiplying both sides of the first equation by 2. A parametric representation of the solution set is given by

$$x = 3 - t, \qquad y = t, \quad t \text{ is any real number.}$$

The graph of this system is represented by two *coincident* lines, as shown in Figure 1.1(b).

(c) This system has no solution because it is impossible for the sum of two numbers to be 3 and 1 simultaneously. The graph of this system is represented by two *parallel* lines, as shown in Figure 1.1(c).

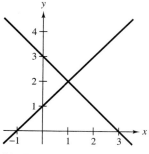

(a) Two intersecting lines:

$$x + y = 3$$
$$x - y = -1$$

(b) Two coincident lines:

$$x + y = 3$$
$$2x + 2y = 6$$

(c) Two parallel lines:

$$x + y = 3$$
$$x + y = 1$$

Figure 1.1

Example 4 illustrates the three basic types of solution sets that are possible for a system of linear equations. This result is stated here without proof. (The proof is given later in Theorem 2.5.)

Number of Solutions of a System of Linear Equations	For a system of linear equations in n variables, precisely one of the following is true. 1. The system has exactly one solution (consistent system). 2. The system has an infinite number of solutions (consistent system). 3. The system has no solution (inconsistent system).

Solving a System of Linear Equations

Which system is easier to solve algebraically?

$$\begin{aligned} x - 2y + 3z &= 9 \\ -x + 3y &= -4 \\ 2x - 5y + 5z &= 17 \end{aligned} \qquad\qquad \begin{aligned} x - 2y + 3z &= 9 \\ y + 3z &= 5 \\ z &= 2 \end{aligned}$$

The system on the right is clearly easier to solve. This system is in **row-echelon form,** which means that it follows a stair-step pattern and has leading coefficients of 1. To solve such a system, use a procedure called **back-substitution.**

EXAMPLE 5   *Using Back-Substitution to Solve a System in Row-Echelon Form*

Use back-substitution to solve the system

$$\begin{aligned} x - 2y &= 5 &&\text{Equation 1}\\ y &= -2. &&\text{Equation 2}\end{aligned}$$

Solution   From Equation 2 you know that $y = -2$. By substituting this value of y into Equation 1, you obtain

$$\begin{aligned} x - 2(-2) &= 5 &&\text{Substitute } y = -2.\\ x &= 1. &&\text{Solve for } x.\end{aligned}$$

Thus the system has exactly one solution: $x = 1$ and $y = -2$.

The term "back-substitution" implies that you work *backward*. For instance, in Example 5, the second equation gave you the value for y. Then you substituted that value into the first equation to solve for x. Example 6 further demonstrates this procedure.

EXAMPLE 6   *Using Back-Substitution to Solve a System in Row-Echelon Form*

Solve the system.

$$\begin{aligned} x - 2y + 3z &= 9 &&\text{Equation 1}\\ y + 3z &= 5 &&\text{Equation 2}\\ z &= 2 &&\text{Equation 3}\end{aligned}$$

Solution   From Equation 3 you already know the value of z. To solve for y, substitute $z = 2$ into Equation 2 to obtain

$$y + 3(2) = 5 \qquad \text{Substitute } z = 2.$$
$$y = -1. \qquad \text{Solve for } y.$$

Finally, substitute $y = -1$ and $z = 2$ in Equation 1 to obtain

$$x - 2(-1) + 3(2) = 9 \qquad \text{Substitute } y = -1, z = 2.$$
$$x = 1. \qquad \text{Solve for } x.$$

Thus the solution is $x = 1$, $y = -1$, and $z = 2$.

Two systems of linear equations are called **equivalent** if they have precisely the same solution set. To solve a system that is not in row-echelon form, first change it to an *equivalent* system that is in row-echelon form by using the following operations.

Operations That Lead to Equivalent Systems of Equations

Each of the following operations on a system of linear equations produces an *equivalent* system.
1. Interchange two equations.
2. Multiply an equation by a nonzero constant.
3. Add a multiple of an equation to another equation.

Rewriting a system of linear equations in row-echelon form usually involves a *chain* of equivalent systems, each of which is obtained by using one of the three basic operations. This process is called **Gaussian elimination,** after the German mathematician Carl Friedrich Gauss (1777–1855).

EXAMPLE 7   *Using Elimination to Rewrite a System in Row-Echelon Form*

Solve the system.

$$\begin{array}{rcrcrcr} x & - & 2y & + & 3z & = & 9 \\ -x & + & 3y & & & = & -4 \\ 2x & - & 5y & + & 5z & = & 17 \end{array}$$

Solution   Although there are several ways to begin, you want to use a systematic procedure that can be applied easily to large systems. Work from the upper left corner of the system, saving the x in the upper left position and eliminating the other x's from the first column.

$$\begin{array}{rcrcrcr} x & - & 2y & + & 3z & = & 9 \\ & & y & + & 3z & = & 5 \\ 2x & - & 5y & + & 5z & = & 17 \end{array} \qquad \longleftarrow \qquad \begin{array}{l} \text{Adding the first equation to} \\ \text{the second equation produces} \\ \text{a new second equation.} \end{array}$$

$$\begin{array}{rcrcrcr} x & - & 2y & + & 3z & = & 9 \\ & & y & + & 3z & = & 5 \\ & - & y & - & z & = & -1 \end{array} \qquad \longleftarrow \qquad \begin{array}{l} \text{Adding } -2 \text{ times the first} \\ \text{equation to the third equation} \\ \text{produces a new third equation.} \end{array}$$

Now that everything but the first x has been eliminated from the first column, go to work on the second column.

$$\begin{aligned} x - 2y + 3z &= 9 \\ y + 3z &= 5 \\ 2z &= 4 \end{aligned}$$ ⟵   Adding the second equation to the third equation produces a new third equation.

$$\begin{aligned} x - 2y + 3z &= 9 \\ y + 3z &= 5 \\ z &= 2 \end{aligned}$$ ⟵   Multiplying the third equation by $\frac{1}{2}$ produces a new third equation.

This is the same system you solved in Example 6, and, as in that example, the solution is

$$x = 1, \qquad y = -1, \qquad z = 2.$$

Each of the three equations in Example 7 is represented in a three-dimensional coordinate system by a plane. Because the unique solution of the system is the point

$$(x, y, z) = (1, -1, 2),$$

the three planes intersect at the point represented by these coordinates, as shown in Figure 1.2.

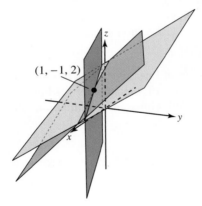

Figure 1.2

Technology Note

Many graphing calculators and computer software programs can solve a system of m linear equations in n variables. Try solving the system in Example 7. If you are using a TI-86, you can use the *simult* feature. Enter 3 as the number of equations, and then enter the coefficients and constant term for each equation.

Because many steps are required to solve a system of linear equations, it is very easy to make errors in arithmetic; hence it is suggested that you develop the habit of *checking your solution by substituting it into each equation of the original system.* For instance, in Example 7, you can check the solution $x = 1$, $y = -1$, and $z = 2$ as follows.

Equation 1:	$(1) - 2(-1) + 3(2) = \quad 9$	Substitute solution in
Equation 2:	$-(1) + 3(-1) \qquad\quad = -4$	each equation of
Equation 3:	$2(1) - 5(-1) + 5(2) = \quad 17$	original system.

Each of the systems in Examples 5, 6, and 7 has exactly one solution. You will now look at an inconsistent system—one that has no solution. The key to recognizing an inconsistent system is reaching a false statement such as $0 = 7$ at some stage of the elimination process. This is demonstrated in Example 8.

EXAMPLE 8   *An Inconsistent System*

Solve the system.

$$
\begin{aligned}
x_1 - 3x_2 + x_3 &= 1 \\
2x_1 - x_2 - 2x_3 &= 2 \\
x_1 + 2x_2 - 3x_3 &= -1
\end{aligned}
$$

Solution

$$
\begin{aligned}
x_1 - 3x_2 + x_3 &= 1 \\
5x_2 - 4x_3 &= 0 \\
x_1 + 2x_2 - 3x_3 &= -1
\end{aligned}
$$

⟵ Adding -2 times the first equation to the second equation produces a new second equation.

$$
\begin{aligned}
x_1 - 3x_2 + x_3 &= 1 \\
5x_2 - 4x_3 &= 0 \\
5x_2 - 4x_3 &= -2
\end{aligned}
$$

⟵ Adding -1 times the first equation to the third equation produces a new third equation.

(Another way of describing this operation is to say that you *subtracted* the first equation from the third equation to produce a new third equation.) Now, continuing the elimination process, add -1 times the second equation to the third equation to produce a new third equation.

$$
\begin{aligned}
x_1 - 3x_2 + x_3 &= 1 \\
5x_2 - 4x_3 &= 0 \\
0 &= -2
\end{aligned}
$$

⟵ Adding -1 times the second equation to the third equation produces a new third equation.

Because the third "equation" is a false statement, this system has no solution. Moreover, because this system is equivalent to the original system, you can conclude that the original system also has no solution.

As in Example 7, the three equations in Example 8 represent planes in a three-dimensional coordinate system. In this example, however, the system is inconsistent; hence the planes do not have a point in common, as shown in Figure 1.3 on the next page.

Figure 1.3

This section ends with an example of a system of linear equations that has an infinite number of solutions. You can represent the solution set for such a system in parametric form, as you did in Examples 2 and 3.

EXAMPLE 9 *A System with an Infinite Number of Solutions*

Solve the system.

$$\begin{aligned} x_2 - x_3 &= 0 \\ x_1 \qquad - 3x_3 &= -1 \\ -x_1 + 3x_2 \qquad &= 1 \end{aligned}$$

Solution Begin by rewriting the system in row-echelon form as follows.

$$\begin{aligned} x_1 \qquad - 3x_3 &= -1 \\ x_2 - x_3 &= 0 \\ -x_1 + 3x_2 \qquad &= 1 \end{aligned}$$ $\longleftarrow$ The first two equations are interchanged.

$$\begin{aligned} x_1 \qquad - 3x_3 &= -1 \\ x_2 - x_3 &= 0 \\ 3x_2 - 3x_3 &= 0 \end{aligned}$$ $\longleftarrow$ Adding the first equation to the third equation produces a new third equation.

$$\begin{aligned} x_1 \qquad - 3x_3 &= -1 \\ x_2 - x_3 &= 0 \\ 0 &= 0 \end{aligned}$$ $\longleftarrow$ Adding -3 times the second equation to the third equation eliminates the third equation.

Because the third equation is unnecessary, drop it to obtain the following system.

$$\begin{aligned} x_1 \qquad - 3x_3 &= -1 \\ x_2 - x_3 &= 0 \end{aligned}$$

To represent the solutions, choose x_3 to be the free variable and represent it by the parameter t. Because $x_2 = x_3$ and $x_1 = 3x_3 - 1$, you can describe the solution set as

$$x_1 = 3t - 1, \qquad x_2 = t, \qquad x_3 = t, \quad t \text{ is any real number.}$$

┌─ **DISCOVERY**

Graph the two lines represented by the system of equations.

$$x - 2y = 1$$
$$-2x + 3y = -3$$

You can use Gaussian elimination to solve this system as follows.

$x - 2y = 1$	$x - 2y = 1$	$x = 3$
$-1y = -1$	$y = 1$	$y = 1$

Graph the system of equations you obtain at each step of this process. What do you observe about the lines? You are asked to repeat this graphical analysis for other systems in Exercises 79 and 80.

SECTION 1.1 Exercises

In Exercises 1–6, determine whether the equation is linear in the variables x and y.

1. $2x - 3y = 4$

2. $3x - 4xy = 0$

3. $\dfrac{3}{y} + \dfrac{2}{x} - 1 = 0$

4. $x^2 + y^2 = 4$

5. $2\sin x - y = 14$

6. $(\sin 2)x - y = 14$

In Exercises 7–10, find a parametric representation of the solution set of the linear equation.

7. $2x - 4y = 0$

8. $3x - \frac{1}{2}y = 9$

9. $x + y + z = 1$

10. $13x_1 - 26x_2 + 39x_3 = 13$

In Exercises 11–16, use back-substitution to solve the system.

11. $x_1 - x_2 = 2$
$ x_2 = 3$

12. $2x_1 - 4x_2 = 6$
$ 3x_2 = 9$

13. $-x + y - z = 0$
$ 2y + z = 3$
$ \frac{1}{2}z = 0$

14. $x - y = 4$
$ 2y + z = 6$
$ 3z = 6$

15. $5x_1 + 2x_2 + x_3 = 0$
$2x_1 + x_2 = 0$

16. $x_1 + x_2 + x_3 = 0$
$ x_2 = 0$

In Exercises 17–20, graph each system of equations as a pair of lines in the xy-plane. Solve each system and interpret your answer.

17. $2x + y = 4$
$x - y = 2$

18. $x + 3y = 2$
$-x + 2y = 3$

19. $x - y = 1$
$-2x + 2y = 5$

20. $\frac{1}{2}x - \frac{1}{3}y = 1$
$-2x + \frac{4}{3}y = -4$

In Exercises 21–26, complete the following for each system of equations.

(a) Use a graphing utility to graph the equations in the system.
(b) Use the graphs to determine whether the system is consistent or inconsistent.
(c) If the system is consistent, approximate the solution.
(d) Solve the system algebraically.
(e) Compare the solution in part (d) with the approximation in part (c). What can you conclude?

21. $-3x - y = 3$
$6x + 2y = 1$

22. $4x - 5y = 3$
$-8x + 10y = 14$

23. $2x - 8y = 3$
$\frac{1}{2}x + y = 0$

24. $9x - 4y = 5$
$\frac{1}{2}x + \frac{1}{3}y = 0$

25. $4x - 8y = 9$
$0.8x - 1.6y = 1.8$

26. $-5.3x + 2.1y = 1.25$
$15.9x - 6.3y = -3.75$

In Exercises 27–46, solve the system of linear equations.

27. $x_1 - x_2 = 0$
$3x_1 - 2x_2 = -1$

28. $3x + 2y = 2$
$6x + 4y = 14$

29. $2u + v = 120$
$u + 2v = 120$

30. $x_1 - 2x_2 = 0$
$6x_1 + 2x_2 = 0$

31. $9x - 3y = -1$
$\frac{1}{5}x + \frac{2}{5}y = -\frac{1}{3}$

32. $\frac{2}{3}x_1 + \frac{1}{6}x_2 = 0$
$4x_1 + \phantom{\frac{1}{6}}x_2 = 0$

33. $\dfrac{x-1}{2} + \dfrac{y+2}{3} = 4$
$\phantom{\dfrac{x-1}{2}} x - 2y = 5$

34. $\dfrac{x_1+3}{4} + \dfrac{x_2-1}{3} = 1$
$\phantom{\dfrac{x_1+3}{4}} 2x_1 - x_2 = 12$

The symbol ⬚ indicates an exercise in which you are instructed to use graphing technology or a symbolic computer algebra system.

35. $0.02x_1 - 0.05x_2 = -0.19$
$0.03x_1 + 0.04x_2 = 0.52$

36. $0.05x_1 - 0.03x_2 = 0.21$
$0.07x_1 + 0.02x_2 = 0.17$

37. $x + y + z = 6$
$2x - y + z = 3$
$3x \quad - z = 0$

38. $x + y + z = 2$
$-x + 3y + 2z = 8$
$4x + y \quad = 4$

39. $3x_1 - 2x_2 + 4x_3 = 1$
$x_1 + x_2 - 2x_3 = 3$
$2x_1 - 3x_2 + 6x_3 = 8$

40. $5x_1 - 3x_2 + 2x_3 = 3$
$2x_1 + 4x_2 - x_3 = 7$
$x_1 - 11x_2 + 4x_3 = 3$

41. $2x_1 + x_2 - 3x_3 = 4$
$4x_1 \quad + 2x_3 = 10$
$-2x_1 + 3x_2 - 13x_3 = -8$

42. $x_1 \quad + 4x_3 = 13$
$4x_1 - 2x_2 + x_3 = 7$
$2x_1 - 2x_2 - 7x_3 = -19$

43. $x - 3y + 2z = 18$
$5x - 15y + 10z = 18$

44. $x_1 - 2x_2 + 5x_3 = 2$
$3x_1 + 2x_2 - x_3 = -2$

45. $x + y + z + w = 6$
$2x + 3y \quad - w = 0$
$-3x + 4y + z + 2w = 4$
$x + 2y - z + w = 0$

46. $x_1 \quad + 3x_4 = 4$
$2x_2 - x_3 - x_4 = 0$
$3x_2 \quad - 2x_4 = 1$
$2x_1 - x_2 + 4x_3 \quad = 5$

In Exercises 47–54, use a computer or graphing calculator to solve the system of linear equations.

Data

47. $x_1 + 0.5x_2 + 0.33x_3 + 0.25x_4 = 1.1$
$0.5x_1 + 0.33x_2 + 0.25x_3 + 0.21x_4 = 1.2$
$0.33x_1 + 0.25x_2 + 0.2x_3 + 0.17x_4 = 1.3$
$0.25x_1 + 0.2x_2 + 0.17x_3 + 0.14x_4 = 1.4$

48. $0.1x - 2.5y + 1.2z - 0.75w = 108$
$2.4x + 1.5y - 1.8z + 0.25w = -81$
$0.4x - 3.2y + 1.6z - 1.4w = 148.8$
$1.6x + 1.2y - 3.2z + 0.6w = -143.2$

49. $123.5x + 61.3y - 32.4z = -262.74$
$54.7x - 45.6y + 98.2z = 197.4$
$42.4x - 89.3y + 12.9z = 33.66$

50. $120.2x + 62.4y - 36.5z = 258.64$
$56.8x - 42.8y + 27.3z = -71.44$
$88.1x + 72.5y - 28.5z = 225.88$

The symbol (icon) indicates that electronic data sets for these exercises are available on the *Elementary Linear Algebra* **Data** Learning Tools Student CD-ROM. These data sets are compatible with each of the following technologies: MATLAB, *Mathematica*, Maple, Derive, TI-83/TI-83 Plus, TI-86, TI-89, and TI-92.

51. $\frac{1}{2}x_1 - \frac{3}{7}x_2 + \frac{2}{9}x_3 = \frac{349}{630}$
$\frac{2}{3}x_1 + \frac{4}{9}x_2 - \frac{2}{5}x_3 = -\frac{19}{45}$
$\frac{4}{5}x_1 - \frac{1}{8}x_2 + \frac{4}{3}x_3 = \frac{139}{150}$

52. $\frac{1}{4}x_1 - \frac{3}{5}x_2 + \frac{1}{3}x_3 = \frac{43}{60}$
$\frac{2}{5}x_1 + \frac{1}{4}x_2 - \frac{5}{6}x_3 = -\frac{331}{600}$
$\frac{3}{4}x_1 - \frac{2}{5}x_2 + \frac{1}{5}x_3 = \frac{81}{100}$

53. $\frac{1}{8}x - \frac{1}{7}y + \frac{1}{6}z - \frac{1}{5}w = 1$
$\frac{1}{7}x + \frac{1}{6}y - \frac{1}{5}z + \frac{1}{4}w = 1$
$\frac{1}{6}x - \frac{1}{5}y + \frac{1}{4}z - \frac{1}{3}w = 1$
$\frac{1}{5}x + \frac{1}{4}y - \frac{1}{3}z + \frac{1}{2}w = 1$

54. $\frac{1}{8}x + \frac{1}{7}y - \frac{1}{6}z + \frac{1}{5}w = 1$
$\frac{1}{7}x - \frac{1}{6}y + \frac{1}{5}z - \frac{1}{4}w = 1$
$\frac{1}{6}x + \frac{1}{5}y - \frac{1}{4}z + \frac{1}{3}w = 1$
$\frac{1}{5}x - \frac{1}{4}y + \frac{1}{3}z - \frac{1}{2}w = 1$

In Exercises 55–58, state why each system of equations must have at least one solution. Then solve the system and determine if it has exactly one solution or an infinite number of solutions.

55. $4x + 3y + 17z = 0$
$5x + 4y + 22z = 0$
$4x + 2y + 19z = 0$

56. $2x + 3y = 0$
$4x + 3y - z = 0$
$8x + 3y + 3z = 0$

57. $5x + 5y - z = 0$
$10x + 5y + 2z = 0$
$5x + 15y - 9z = 0$

58. $12x + 5y + z = 0$
$12x + 4y - z = 0$

True or False? In Exercises 59 and 60, determine whether each statement is true or false. If a statement is true, give a reason or cite an appropriate statement in the text. If a statement is false, provide an example that shows that the statement is not true in all cases or cite an appropriate statement in the text.

59. (a) A system of one linear equation in two variables is always consistent.
(b) A system of two linear equations in three unknowns is always consistent.
(c) If a linear system is consistent, then it has an infinite number of solutions.

60. (a) A system of linear equations can have exactly two solutions.
(b) Two systems of linear equations are equivalent if they have the same solution set.
(c) A system of three linear equations in two unknowns is always inconsistent.

61. Find a system of two equations in two unknowns, x_1 and x_2, that has the solution set given by the parametric representation $x_1 = t$ and $x_2 = 3t - 4$, where t is any real number. Then show that the solutions to your system can also be written as
$$x_1 = \frac{4}{3} + \frac{t}{3} \text{ and } x_2 = t.$$

62. Find a system of two equations in three unknowns, x_1, x_2, and x_3, that has the solution set given by the parametric representation

$$x_1 = t, \quad x_2 = s, \quad \text{and} \quad x_3 = 3 + s - t,$$

where s and t are any real numbers. Then show that the solutions to your system can also be written as

$$x_1 = 3 + s - t, \quad x_2 = s, \quad \text{and} \quad x_3 = t.$$

In Exercises 63 and 64, solve the system of equations by letting

$X = 1/x$,

$Y = 1/y$, and

$Z = 1/z$.

63.
$$\frac{12}{x} - \frac{12}{y} = 7$$
$$\frac{3}{x} + \frac{4}{y} = 0$$

64.
$$\frac{2}{x} + \frac{1}{y} - \frac{3}{z} = 4$$
$$\frac{4}{x} \qquad + \frac{2}{z} = 10$$
$$-\frac{2}{x} + \frac{3}{y} - \frac{13}{z} = -8$$

In Exercises 65 and 66, solve the system of linear equations for x and y.

65. $(\cos \theta)x + (\sin \theta)y = 1$
$(-\sin \theta)x + (\cos \theta)y = 0$

66. $(\cos \theta)x + (\sin \theta)y = 1$
$(-\sin \theta)x + (\cos \theta)y = 1$

In Exercises 67–72, determine the value(s) of k such that the system of linear equations has the indicated number of solutions.

67. An infinite number of solutions
$$4x + ky = 6$$
$$kx + y = -3$$

68. An infinite number of solutions
$$kx + y = 4$$
$$2x - 3y = -12$$

69. Exactly one solution
$$x + ky = 0$$
$$kx + y = 0$$

70. No solution
$$x + ky = 2$$
$$kx + y = 4$$

71. No solution
$$x + 2y + kz = 6$$
$$3x + 6y + 8z = 4$$

72. Exactly one solution
$$kx + 2ky + 3kz = 4k$$
$$x + y + z = 0$$
$$2x - y + z = 1$$

73. Determine the values of k such that the system of linear equations does not have a unique solution.
$$x + y + kz = 3$$
$$x + ky + z = 2$$
$$kx + y + z = 1$$

74. Find values of a, b, and c such that the system of linear equations has (a) exactly one solution, (b) an infinite number of solutions, and (c) no solution.
$$x + 5y + z = 0$$
$$x + 6y - z = 0$$
$$2x + ay + bz = c$$

75. *Writing* Consider the system of linear equations in x and y.
$$a_1x + b_1y = c_1$$
$$a_2x + b_2y = c_2$$
$$a_3x + b_3y = c_3$$
Describe the graphs of these three equations in the xy-plane when the system has (a) exactly one solution, (b) an infinite number of solutions, and (c) no solution.

76. *Writing* Explain why the system of linear equations in Exercise 75 must be consistent if the constant terms c_1, c_2, and c_3 are all zero.

77. Show that if
$$ax^2 + bx + c = 0$$
for all x, then
$$a = b = c = 0.$$

78. Consider the system of linear equations in x and y.
$$ax + by = e$$
$$cx + dy = f$$
Under what conditions will the system have exactly one solution?

In Exercises 79 and 80, graph the lines determined by the system of linear equations. Then use Gaussian elimination to solve the system. At each step of the elimination process, graph the corresponding lines. What do you observe about these lines?

79. $x - 4y = -3$
$5x - 6y = 13$

80. $2x - 3y = 7$
$-4x + 6y = -14$

1.2 | Gaussian Elimination and Gauss-Jordan Elimination

In Section 1.1, Gaussian elimination was introduced as a procedure for solving a system of linear equations. In this section you will study this procedure more thoroughly, beginning with some definitions. The first is the definition of a **matrix.**

Definition of a Matrix	If m and n are positive integers, then an $m \times n$ **matrix** is a rectangular array

$$\begin{bmatrix} a_{11} & a_{12} & a_{13} & \cdots & a_{1n} \\ a_{21} & a_{22} & a_{23} & \cdots & a_{2n} \\ a_{31} & a_{32} & a_{33} & \cdots & a_{3n} \\ \vdots & \vdots & \vdots & & \vdots \\ a_{m1} & a_{m2} & a_{m3} & \cdots & a_{mn} \end{bmatrix} \Big\} \ m \ \text{rows}$$

$$\underbrace{\phantom{a_{11} \quad a_{12} \quad a_{13}}}_{n \ \text{columns}}$$

in which each **entry,** a_{ij}, of the matrix is a number. An $m \times n$ matrix (read "m by n") has m **rows** (horizontal lines) and n **columns** (vertical lines).

REMARK: The plural of matrix is *matrices.* If each entry of a matrix is a *real* number, then the matrix is called a **real matrix.** Unless stated otherwise, all matrices in this text are assumed to be real matrices.

The entry a_{ij} is located in the ith row and the jth column. The index i is called the **row subscript** because it identifies the row in which the entry lies, and the index j is called the **column subscript** because it identifies the column in which the entry lies.

A matrix with m rows and n columns (an $m \times n$ matrix) is said to be of **size** $m \times n$. If $m = n$, the matrix is called **square** of **order** n. For a square matrix, the entries a_{11}, a_{22}, a_{33}, . . . are called the **main diagonal** entries.

EXAMPLE 1 *Examples of Matrices*

Each matrix has the indicated size.

(a) Size: 1×1

$$[2]$$

(b) Size: 2×2

$$\begin{bmatrix} 0 & 0 \\ 0 & 0 \end{bmatrix}$$

(c) Size: 1×4

$$\begin{bmatrix} 1 & -3 & 0 & \frac{1}{2} \end{bmatrix}$$

(d) Size: 3×2

$$\begin{bmatrix} e & \pi \\ 2 & \sqrt{2} \\ -7 & 4 \end{bmatrix}$$

One very common use of matrices is to represent a system of linear equations. The matrix derived from the coefficients and constant terms of a system of linear equations is called the **augmented matrix** of the system. The matrix containing only the coefficients of the system is called the **coefficient matrix** of the system. Here is an example.

$$
\begin{array}{ccc}
\textit{System} & \textit{Augmented Matrix} & \textit{Coefficient Matrix} \\[4pt]
\begin{aligned}
x - 4y + 3z &= 5 \\
-x + 3y - z &= -3 \\
2x \qquad - 4z &= 6
\end{aligned}
&
\begin{bmatrix}
1 & -4 & 3 & 5 \\
-1 & 3 & -1 & -3 \\
2 & 0 & -4 & 6
\end{bmatrix}
&
\begin{bmatrix}
1 & -4 & 3 \\
-1 & 3 & -1 \\
2 & 0 & -4
\end{bmatrix}
\end{array}
$$

R E M A R K : Note the use of 0 to indicate that the coefficient of y in the third equation is zero. Also note the fourth column of constant terms in the augmented matrix.

When forming either the coefficient matrix or the augmented matrix of a system, you should begin by aligning the variables in the equations vertically.

$$
\begin{array}{ccc}
\textit{Given System} & \textit{Line Up Variables} & \textit{Form Augmented Matrix} \\[4pt]
\begin{aligned}
x_1 + 3x_2 &= 9 \\
-x_2 + 4x_3 &= -2 \\
x_1 - 5x_3 &= 0
\end{aligned}
&
\begin{aligned}
x_1 + 3x_2 \qquad &= 9 \\
-x_2 + 4x_3 &= -2 \\
x_1 \qquad - 5x_3 &= 0
\end{aligned}
&
\begin{bmatrix}
1 & 3 & 0 & 9 \\
0 & -1 & 4 & -2 \\
1 & 0 & -5 & 0
\end{bmatrix}
\end{array}
$$

Elementary Row Operations

In the previous section you studied three operations that can be used on a system of linear equations to produce equivalent systems.

1. Interchange two equations.
2. Multiply an equation by a nonzero constant.
3. Add a multiple of an equation to another equation.

In matrix terminology these three operations correspond to **elementary row operations.** An elementary row operation on an augmented matrix corresponding to a given system of linear equations produces a new augmented matrix corresponding to a new (but equivalent) system of linear equations. Two matrices are said to be **row-equivalent** if one can be obtained from the other by a finite sequence of elementary row operations.

Elementary Row Operations	1. Interchange two rows. 2. Multiply a row by a nonzero constant. 3. Add a multiple of a row to another row.

Although elementary row operations are simple to perform, they involve a lot of arithmetic. Because it is easy to make a mistake, you should get in the habit of noting the elementary row operation performed in each step so that you can go back to check your work.

Because solving some systems involves several steps, it is helpful to use a shorthand method of notation to keep track of each elementary row operation you perform. This notation is introduced in the next example.

EXAMPLE 2   *Elementary Row Operations*

(a) Interchange the first and second rows.

Original Matrix	*New Row Equivalent Matrix*	*Notation*
$\begin{bmatrix} 0 & 1 & 3 & 4 \\ -1 & 2 & 0 & 3 \\ 2 & -3 & 4 & 1 \end{bmatrix}$	$\begin{bmatrix} -1 & 2 & 0 & 3 \\ 0 & 1 & 3 & 4 \\ 2 & -3 & 4 & 1 \end{bmatrix}$	$R_1 \leftrightarrow R_2$

(b) Multiply the first row by $\frac{1}{2}$ to produce a new first row.

Original Matrix	*New Row Equivalent Matrix*	*Notation*
$\begin{bmatrix} 2 & -4 & 6 & -2 \\ 1 & 3 & -3 & 0 \\ 5 & -2 & 1 & 2 \end{bmatrix}$	$\begin{bmatrix} 1 & -2 & 3 & -1 \\ 1 & 3 & -3 & 0 \\ 5 & -2 & 1 & 2 \end{bmatrix}$	$\left(\frac{1}{2}\right)R_1 \rightarrow R_1$

(c) Add -2 times the first row to the third row to produce a new third row.

Original Matrix	*New Row Equivalent Matrix*	*Notation*
$\begin{bmatrix} 1 & 2 & -4 & 3 \\ 0 & 3 & -2 & -1 \\ 2 & 1 & 5 & -2 \end{bmatrix}$	$\begin{bmatrix} 1 & 2 & -4 & 3 \\ 0 & 3 & -2 & -1 \\ 0 & -3 & 13 & -8 \end{bmatrix}$	$R_3 + (-2)R_1 \rightarrow R_3$

REMARK: Notice in Example 2(c) that adding -2 times row 1 to row 3 does not change row 1.

Technology   Note

Many graphing calculators and computer software programs can perform elementary row operations on matrices. If you are using a TI-86, your screens for Example 2(c) will look like the following:

In Example 7 of Section 1.1, you used Gaussian elimination with back-substitution to solve a system of linear equations. You will now learn the matrix version of Gaussian elimination. The two methods used in the following example are essentially the same. The basic difference is that with the matrix method, you do not need to keep writing the variables.

EXAMPLE 3   *Using Elementary Row Operations to Solve a System*

Linear System

$$x - 2y + 3z = 9$$
$$-x + 3y \quad\quad = -4$$
$$2x - 5y + 5z = 17$$

Associated Augmented Matrix

$$\begin{bmatrix} 1 & -2 & 3 & 9 \\ -1 & 3 & 0 & -4 \\ 2 & -5 & 5 & 17 \end{bmatrix}$$

Add the first equation to the second equation.

$$x - 2y + 3z = 9$$
$$y + 3z = 5$$
$$2x - 5y + 5z = 17$$

Add the first row to the second row to produce a new second row.

$$\begin{bmatrix} 1 & -2 & 3 & 9 \\ 0 & 1 & 3 & 5 \\ 2 & -5 & 5 & 17 \end{bmatrix} \quad R_2 + R_1 \rightarrow R_2$$

Add -2 times the first equation to the third equation.

$$x - 2y + 3z = 9$$
$$y + 3z = 5$$
$$-y - z = -1$$

Add -2 times the first row to the third row to produce a new third row.

$$\begin{bmatrix} 1 & -2 & 3 & 9 \\ 0 & 1 & 3 & 5 \\ 0 & -1 & -1 & -1 \end{bmatrix} \quad R_3 + (-2)R_1 \rightarrow R_3$$

Add the second equation to the third equation.

$$x - 2y + 3z = 9$$
$$y + 3z = 5$$
$$2z = 4$$

Add the second row to the third row to produce a new third row.

$$\begin{bmatrix} 1 & -2 & 3 & 9 \\ 0 & 1 & 3 & 5 \\ 0 & 0 & 2 & 4 \end{bmatrix} \quad R_3 + R_2 \rightarrow R_3$$

Multiply the third equation by $\frac{1}{2}$.

$$x - 2y + 3z = 9$$
$$y + 3z = 5$$
$$z = 2$$

Multiply the third row by $\frac{1}{2}$ to produce a new third row.

$$\begin{bmatrix} 1 & -2 & 3 & 9 \\ 0 & 1 & 3 & 5 \\ 0 & 0 & 1 & 2 \end{bmatrix} \quad \left(\tfrac{1}{2}\right)R_3 \rightarrow R_3$$

Now you can use back-substitution to find the solution, as in Example 6 of Section 1.1.

The last matrix in Example 3 is said to be in **row-echelon form.** The term *echelon* refers to the stair-step pattern formed by the nonzero elements of the matrix. To be in row-echelon form, a matrix must have the following properties.

Definition of Row-Echelon *Form of a Matrix*	A matrix in **row-echelon form** has the following properties.

A matrix in **row-echelon form** has the following properties.
1. All rows consisting entirely of zeros occur at the bottom of the matrix.
2. For each row that does not consist entirely of zeros, the first nonzero entry is 1 (called a **leading 1**).
3. For two successive (nonzero) rows, the leading 1 in the higher row is farther to the left than the leading 1 in the lower row.

REMARK: A matrix in row-echelon form is in **reduced row-echelon form** if every column that has a leading 1 has zeros in every position above and below its leading 1.

EXAMPLE 4   *Row-Echelon Form*

The following matrices are in row-echelon form.

(a) $\begin{bmatrix} 1 & 2 & -1 & 4 \\ 0 & 1 & 0 & 3 \\ 0 & 0 & 1 & -2 \end{bmatrix}$ (b) $\begin{bmatrix} 0 & 1 & 0 & 5 \\ 0 & 0 & 1 & 3 \\ 0 & 0 & 0 & 0 \end{bmatrix}$

(c) $\begin{bmatrix} 1 & -5 & 2 & -1 & 3 \\ 0 & 0 & 1 & 3 & -2 \\ 0 & 0 & 0 & 1 & 4 \\ 0 & 0 & 0 & 0 & 1 \end{bmatrix}$ (d) $\begin{bmatrix} 1 & 0 & 0 & -1 \\ 0 & 1 & 0 & 2 \\ 0 & 0 & 1 & 3 \\ 0 & 0 & 0 & 0 \end{bmatrix}$

The matrices shown in parts (b) and (d) are in *reduced* row-echelon form. The following matrices are not in row-echelon form.

(e) $\begin{bmatrix} 1 & 2 & -3 & 4 \\ 0 & 2 & 1 & -1 \\ 0 & 0 & 1 & -3 \end{bmatrix}$ (f) $\begin{bmatrix} 1 & 2 & -1 & 2 \\ 0 & 0 & 0 & 0 \\ 0 & 1 & 2 & -4 \end{bmatrix}$

Technology Note

Some graphing calculators and computer software programs will produce the reduced row-echelon form of a matrix for you. If your tool has this capability, verify that the reduced row-echelon form of the matrix in part (f) of Example 4 is

$\begin{bmatrix} 1 & 0 & -5 & 10 \\ 0 & 1 & 2 & -4 \\ 0 & 0 & 0 & 0 \end{bmatrix}$.

It can be shown that every matrix is row-equivalent to a matrix in row-echelon form. For instance, in Example 4 you could change the matrix in part (e) to row-echelon form by multiplying the second row in the matrix by $\frac{1}{2}$.

The method of using Gaussian elimination with back-substitution to solve a system is as follows.

Gaussian Elimination with *Back-Substitution*	

1. Write the augmented matrix of the system of linear equations.
2. Use elementary row operations to rewrite the augmented matrix in row-echelon form.
3. Write the system of linear equations corresponding to the matrix in row-echelon form, and use back-substitution to find the solution.

Gaussian elimination with back-substitution works well as an algorithmic method for solving systems of linear equations with a computer. For this algorithm, the order in which the elementary row operations are performed is important. Move from *left to right by columns*, changing all entries directly below the leading 1's to zeros.

EXAMPLE 5   *Gaussian Elimination with Back-Substitution*

Solve the system.

$$\begin{aligned} x_2 + x_3 - 2x_4 &= -3 \\ x_1 + 2x_2 - x_3 &= 2 \\ 2x_1 + 4x_2 + x_3 - 3x_4 &= -2 \\ x_1 - 4x_2 - 7x_3 - x_4 &= -19 \end{aligned}$$

Solution   The augmented matrix for this system is

$$\begin{bmatrix} 0 & 1 & 1 & -2 & -3 \\ 1 & 2 & -1 & 0 & 2 \\ 2 & 4 & 1 & -3 & -2 \\ 1 & -4 & -7 & -1 & -19 \end{bmatrix}.$$

Obtain a leading 1 in the upper left corner and zeros elsewhere in the first column.

$$\begin{bmatrix} 1 & 2 & -1 & 0 & 2 \\ 0 & 1 & 1 & -2 & -3 \\ 2 & 4 & 1 & -3 & -2 \\ 1 & -4 & -7 & -1 & -19 \end{bmatrix}$$

⟵ The first two rows $R_1 \leftrightarrow R_2$
⟵ are interchanged.

$$\begin{bmatrix} 1 & 2 & -1 & 0 & 2 \\ 0 & 1 & 1 & -2 & -3 \\ 0 & 0 & 3 & -3 & -6 \\ 1 & -4 & -7 & -1 & -19 \end{bmatrix}$$

Adding -2 times the first row to the third row
⟵ produces a new third row. $R_3 + (-2)R_1 \rightarrow R_3$

$$\begin{bmatrix} 1 & 2 & -1 & 0 & 2 \\ 0 & 1 & 1 & -2 & -3 \\ 0 & 0 & 3 & -3 & -6 \\ 0 & -6 & -6 & -1 & -21 \end{bmatrix}$$

Adding -1 times the first row to the fourth row
⟵ produces a new fourth row. $R_4 + (-1)R_1 \rightarrow R_4$

Now that the first column is in the desired form, you should change the second column as follows.

$$\begin{bmatrix} 1 & 2 & -1 & 0 & 2 \\ 0 & 1 & 1 & -2 & -3 \\ 0 & 0 & 3 & -3 & -6 \\ 0 & 0 & 0 & -13 & -39 \end{bmatrix}$$

Adding 6 times the second row to the fourth row
⟵ produces a new fourth row. $R_4 + (6)R_2 \rightarrow R_4$

To write the third column in proper form, multiply the third row by $\frac{1}{3}$.

$$\begin{bmatrix} 1 & 2 & -1 & 0 & 2 \\ 0 & 1 & 1 & -2 & -3 \\ 0 & 0 & 1 & -1 & -2 \\ 0 & 0 & 0 & -13 & -39 \end{bmatrix}$$

⟵ Multiplying the third row by $\frac{1}{3}$ produces a new third row. $\left(\frac{1}{3}\right)R_3 \to R_3$

Similarly, to write the fourth column in proper form, you should multiply the fourth row by $-\frac{1}{13}$.

$$\begin{bmatrix} 1 & 2 & -1 & 0 & 2 \\ 0 & 1 & 1 & -2 & -3 \\ 0 & 0 & 1 & -1 & -2 \\ 0 & 0 & 0 & 1 & 3 \end{bmatrix}$$

⟵ Multiplying the fourth row by $-\frac{1}{13}$ produces a new fourth row. $\left(-\frac{1}{13}\right)R_4 \to R_4$

The matrix is now in row-echelon form, and the corresponding system of linear equations is as follows.

$$\begin{aligned} x_1 + 2x_2 - x_3 \quad\quad &= 2 \\ x_2 + x_3 - 2x_4 &= -3 \\ x_3 - x_4 &= -2 \\ x_4 &= 3 \end{aligned}$$

Using back-substitution, you can determine that the solution is

$$x_1 = -1, \quad x_2 = 2, \quad x_3 = 1, \quad x_4 = 3.$$

When solving a system of linear equations, remember that it is possible for the system to have no solution. If during the elimination process you obtain a row with zeros except for the last entry, it is unnecessary to continue the elimination process. You can simply conclude that the system is inconsistent and therefore has no solution.

EXAMPLE 6   *A System with No Solution*

Solve the system.

$$\begin{aligned} x_1 - x_2 + 2x_3 &= 4 \\ x_1 \quad\quad + x_3 &= 6 \\ 2x_1 - 3x_2 + 5x_3 &= 4 \\ 3x_1 + 2x_2 - x_3 &= 1 \end{aligned}$$

Solution   The augmented matrix for this system is

$$\begin{bmatrix} 1 & -1 & 2 & 4 \\ 1 & 0 & 1 & 6 \\ 2 & -3 & 5 & 4 \\ 3 & 2 & -1 & 1 \end{bmatrix}.$$

Apply Gaussian elimination to the augmented matrix.

$$\begin{bmatrix} 1 & -1 & 2 & 4 \\ 0 & 1 & -1 & 2 \\ 2 & -3 & 5 & 4 \\ 3 & 2 & -1 & 1 \end{bmatrix} \qquad R_2 + (-1)R_1 \rightarrow R_2$$

$$\begin{bmatrix} 1 & -1 & 2 & 4 \\ 0 & 1 & -1 & 2 \\ 0 & -1 & 1 & -4 \\ 3 & 2 & -1 & 1 \end{bmatrix} \qquad R_3 + (-2)R_1 \rightarrow R_3$$

$$\begin{bmatrix} 1 & -1 & 2 & 4 \\ 0 & 1 & -1 & 2 \\ 0 & -1 & 1 & -4 \\ 0 & 5 & -7 & -11 \end{bmatrix} \qquad R_4 + (-3)R_1 \rightarrow R_4$$

$$\begin{bmatrix} 1 & -1 & 2 & 4 \\ 0 & 1 & -1 & 2 \\ 0 & 0 & 0 & -2 \\ 0 & 5 & -7 & -11 \end{bmatrix} \qquad R_3 + R_2 \rightarrow R_3$$

Note that the third row of this matrix consists of zeros except for the last entry. This means that the original system of linear equations is *inconsistent*. You can see why this is true by converting back to a system of linear equations.

$$\begin{aligned} x_1 - x_2 + 2x_3 &= 4 \\ x_2 - x_3 &= 2 \\ 0 &= -2 \\ 5x_2 - 7x_3 &= -11 \end{aligned}$$

Because the third "equation" is a false statement, it follows that the system has no solution.

DISCOVERY

Consider the system of linear equations.

$$\begin{aligned} 2x_1 + 3x_2 + 5x_3 &= 0 \\ -5x_1 + 6x_2 - 17x_3 &= 0 \\ 7x_1 - 4x_2 + 3x_3 &= 0 \end{aligned}$$

Without doing any row operations, explain why this system is consistent.
The following system has more variables than equations. Why does it have an infinite number of solutions?

$$\begin{aligned} 2x_1 + 3x_2 + 5x_3 + 2x_4 &= 0 \\ -5x_1 + 6x_2 - 17x_3 - 3x_4 &= 0 \\ 7x_1 - 4x_2 + 3x_3 + 13x_4 &= 0 \end{aligned}$$

Gauss-Jordan Elimination

With Gaussian elimination, you apply elementary row operations to a matrix to obtain a (row-equivalent) row-echelon form. A second method of elimination, called **Gauss-Jordan elimination** after Carl Gauss and Wilhelm Jordan (1842–1899), continues the reduction process until a *reduced* row-echelon form is obtained. This procedure is demonstrated in the following example.

EXAMPLE 7   *Gauss-Jordan Elimination*

Use Gauss-Jordan elimination to solve the system

$$\begin{aligned} x - 2y + 3z &= 9 \\ -x + 3y &= -4 \\ 2x - 5y + 5z &= 17. \end{aligned}$$

Solution   In Example 3, Gaussian elimination was used to obtain the row-echelon form.

$$\begin{bmatrix} 1 & -2 & 3 & 9 \\ 0 & 1 & 3 & 5 \\ 0 & 0 & 1 & 2 \end{bmatrix}$$

Now, rather than using back-substitution, apply elementary row operations until you obtain a matrix in reduced row-echelon form. To do this, you must produce zeros above each of the leading 1's, as follows.

$$\begin{bmatrix} 1 & 0 & 9 & 19 \\ 0 & 1 & 3 & 5 \\ 0 & 0 & 1 & 2 \end{bmatrix} \qquad R_1 + (2)R_2 \rightarrow R_1$$

$$\begin{bmatrix} 1 & 0 & 9 & 19 \\ 0 & 1 & 0 & -1 \\ 0 & 0 & 1 & 2 \end{bmatrix} \qquad R_2 + (-3)R_3 \rightarrow R_2$$

$$\begin{bmatrix} 1 & 0 & 0 & 1 \\ 0 & 1 & 0 & -1 \\ 0 & 0 & 1 & 2 \end{bmatrix} \qquad R_1 + (-9)R_3 \rightarrow R_1$$

Now, converting back to a system of linear equations, you have

$$\begin{aligned} x &= 1 \\ y &= -1 \\ z &= 2. \end{aligned}$$

The Gaussian and Gauss-Jordan elimination procedures employ an algorithmic approach that is easily adapted to computer use. However, these elimination procedures make no effort to avoid fractional coefficients. For instance, if the system in Example 7 had been listed as

$$\begin{aligned} 2x - 5y + 5z &= 17 \\ x - 2y + 3z &= 9 \\ -x + 3y \quad\quad &= -4 \end{aligned}$$

both procedures would have required multiplying the first row by $\frac{1}{2}$, which would have introduced fractions in the first row. For hand computations, fractions can sometimes be avoided by judiciously choosing the order in which elementary row operations are applied. Moreover, it can be shown that no matter which order you use, the reduced row-echelon form will be the same.

The next example demonstrates how Gauss-Jordan elimination can be used to solve a system with an infinite number of solutions.

EXAMPLE 8   *A System with an Infinite Number of Solutions*

Solve the system of linear equations.

$$\begin{aligned} 2x_1 + 4x_2 - 2x_3 &= 0 \\ 3x_1 + 5x_2 \quad\quad &= 1 \end{aligned}$$

Solution   The augmented matrix of the system of linear equations is

$$\begin{bmatrix} 2 & 4 & -2 & 0 \\ 3 & 5 & 0 & 1 \end{bmatrix}.$$

Using a graphing calculator, a computer software program, or Gauss-Jordan elimination, you can verify that the reduced row-echelon form of the matrix is

$$\begin{bmatrix} 1 & 0 & 5 & 2 \\ 0 & 1 & -3 & -1 \end{bmatrix}.$$

The corresponding system of equations is

$$\begin{aligned} x_1 \quad\quad + 5x_3 &= 2 \\ x_2 - 3x_3 &= -1. \end{aligned}$$

Now, using the parameter t to represent the *nonleading* variable x_3, you have

$$x_1 = 2 - 5t, \qquad x_2 = -1 + 3t, \qquad x_3 = t$$

where t is any real number.

REMARK: Note that in Example 8 an arbitrary parameter was assigned to the nonleading variable x_3. You subsequently solved for the leading variables x_1 and x_2 as functions of t.

You now have looked at two elimination methods for solving a system of linear equations. Which is better? To some degree the answer depends on personal preference. In real-life applications of linear algebra, systems of linear equations are usually solved by computer. Most computer programs use a form of Gaussian elimination, with special emphasis on ways to reduce rounding errors and minimize storage of data. Because the examples and exercises in this text are generally much simpler and focus on the underlying concepts, you will need to know both methods.

Homogeneous Systems of Linear Equations

As the final topic of this section, you will look at systems of linear equations in which each of the constant terms is zero. We call such systems **homogeneous.** For example, a homogeneous system of m equations in n variables has the form

$$
\begin{aligned}
a_{11}x_1 + a_{12}x_2 + a_{13}x_3 + \cdots + a_{1n}x_n &= 0 \\
a_{21}x_1 + a_{22}x_2 + a_{23}x_3 + \cdots + a_{2n}x_n &= 0 \\
a_{31}x_1 + a_{32}x_2 + a_{33}x_3 + \cdots + a_{3n}x_n &= 0 \\
&\;\;\vdots \\
a_{m1}x_1 + a_{m2}x_2 + a_{m3}x_3 + \cdots + a_{mn}x_n &= 0.
\end{aligned}
$$

It is easy to see that a homogeneous system must have at least one solution. Specifically, if all variables in a homogeneous system have the value zero, then each of the equations must be satisfied. Such a solution is called **trivial** (or **obvious**). For instance, a homogeneous system of three equations in the three variables x_1, x_2, and x_3 must have $x_1 = 0$, $x_2 = 0$, and $x_3 = 0$ as a trivial solution.

EXAMPLE 9   *Solving a Homogeneous System of Linear Equations*

Solve the system of linear equations.

$$
\begin{aligned}
x_1 - x_2 + 3x_3 &= 0 \\
2x_1 + x_2 + 3x_3 &= 0
\end{aligned}
$$

Solution   Applying Gauss-Jordan elimination to the augmented matrix

$$
\begin{bmatrix} 1 & -1 & 3 & 0 \\ 2 & 1 & 3 & 0 \end{bmatrix}
$$

yields the following.

$$
\begin{bmatrix} 1 & -1 & 3 & 0 \\ 0 & 3 & -3 & 0 \end{bmatrix} \qquad R_2 + (-2)R_1 \rightarrow R_2
$$

$$
\begin{bmatrix} 1 & -1 & 3 & 0 \\ 0 & 1 & -1 & 0 \end{bmatrix} \qquad \left(\tfrac{1}{3}\right)R_2 \rightarrow R_2
$$

$$\begin{bmatrix} 1 & 0 & 2 & 0 \\ 0 & 1 & -1 & 0 \end{bmatrix} \qquad R_1 + R_2 \to R_1$$

The system of equations corresponding to this matrix is

$$\begin{aligned} x_1 \quad + 2x_3 &= 0 \\ x_2 - \quad x_3 &= 0. \end{aligned}$$

Using the parameter $t = x_3$, the solution set is given by

$$x_1 = -2t, \qquad x_2 = t, \qquad x_3 = t, \quad t \text{ is any real number.}$$

Therefore, this system of equations has an infinite number of solutions, one of which is the trivial solution (given by $t = 0$).

Example 9 illustrates an important point about homogeneous systems of linear equations. You began with two equations in three variables and discovered that the system has an infinite number of solutions. In general, a homogeneous system with fewer equations than variables has an infinite number of solutions.

Theorem 1.1

The Number of Solutions of a Homogeneous System

Every homogeneous system of linear equations is consistent. Moreover, if the system has fewer equations than variables, then it must have an infinite number of solutions.

SECTION 1.2 Exercises

In Exercises 1–6, determine the size of the matrix.

1. $\begin{bmatrix} 1 & 2 & -4 \\ 3 & -4 & 6 \\ 0 & 1 & 2 \end{bmatrix}$

2. $\begin{bmatrix} 2 & -1 & -1 & 1 \\ -6 & 2 & 0 & 1 \end{bmatrix}$

3. $\begin{bmatrix} 1 & 2 & 3 & 4 & -10 \end{bmatrix}$ **4.** $\begin{bmatrix} -1 \end{bmatrix}$

5. $\begin{bmatrix} 8 & 6 & 4 & 1 & 3 \\ 2 & 1 & -7 & 4 & 1 \\ 1 & 1 & -1 & 2 & 1 \\ 1 & -1 & 2 & 0 & 0 \end{bmatrix}$ **6.** $\begin{bmatrix} 1 \\ 2 \\ -1 \\ -2 \end{bmatrix}$

In Exercises 7–12, determine whether the matrix is in row-echelon form. If it is, determine whether it is also in reduced row-echelon form.

7. $\begin{bmatrix} 1 & 0 & 0 & 0 \\ 0 & 1 & 1 & 2 \\ 0 & 0 & 0 & 0 \end{bmatrix}$ **8.** $\begin{bmatrix} 0 & 1 & 0 & 0 \\ 1 & 0 & 2 & 1 \end{bmatrix}$

9. $\begin{bmatrix} 2 & 0 & 1 & 3 \\ 0 & -1 & 1 & 4 \\ 0 & 0 & 0 & 1 \end{bmatrix}$ **10.** $\begin{bmatrix} 1 & 0 & 2 & 1 \\ 0 & 1 & 3 & 4 \\ 0 & 0 & 1 & 0 \end{bmatrix}$

11. $\begin{bmatrix} 0 & 0 & 1 & 0 & 0 \\ 0 & 0 & 0 & 1 & 0 \\ 0 & 0 & 0 & 2 & 0 \end{bmatrix}$ **12.** $\begin{bmatrix} 1 & 0 & 0 & 0 \\ 0 & 0 & 0 & 1 \\ 0 & 0 & 0 & 0 \end{bmatrix}$

In Exercises 13–18, find the solution set of the system of linear equations represented by the augmented matrix.

13. $\begin{bmatrix} 1 & 0 & 0 \\ 0 & 1 & 2 \end{bmatrix}$

14. $\begin{bmatrix} 1 & 0 & 2 \\ 0 & 1 & 3 \end{bmatrix}$

15. $\begin{bmatrix} 1 & -1 & 0 & 3 \\ 0 & 1 & -2 & 1 \\ 0 & 0 & 1 & -1 \end{bmatrix}$

16. $\begin{bmatrix} 1 & 2 & 1 & 0 \\ 0 & 0 & 1 & -1 \\ 0 & 0 & 0 & 0 \end{bmatrix}$

17. $\begin{bmatrix} 1 & 2 & 0 & 1 & 4 \\ 0 & 1 & 2 & 1 & 3 \\ 0 & 0 & 1 & 2 & 1 \\ 0 & 0 & 0 & 1 & 4 \end{bmatrix}$

18. $\begin{bmatrix} 1 & 2 & 0 & 1 & 3 \\ 0 & 1 & 3 & 0 & 1 \\ 0 & 0 & 1 & 2 & 0 \\ 0 & 0 & 0 & 0 & 2 \end{bmatrix}$

In Exercises 19–32, solve the system using either Gaussian elimination with back-substitution or Gauss-Jordan elimination.

19. $x + 2y = 7$
$2x + y = 8$

20. $2x + 6y = 16$
$-2x - 6y = -16$

21. $-x + 2y = 1.5$
$2x - 4y = 3$

22. $2x - y = -0.1$
$3x + 2y = 1.6$

23. $-3x + 5y = -22$
$3x + 4y = 4$
$4x - 8y = 32$

24. $x + 2y = 0$
$x + y = 6$
$3x - 2y = 8$

25. $x_1 \quad\quad - 3x_3 = -2$
$3x_1 + x_2 - 2x_3 = 5$
$2x_1 + 2x_2 + x_3 = 4$

26. $2x_1 - x_2 + 3x_3 = 24$
$2x_2 - x_3 = 14$
$7x_1 - 5x_2 \quad\quad = 6$

27. $x_1 + x_2 - 5x_3 = 3$
$x_1 \quad\quad - 2x_3 = 1$
$2x_1 - x_2 - x_3 = 0$

28. $2x_1 + \quad\quad 3x_3 = 3$
$4x_1 - 3x_2 + 7x_3 = 5$
$8x_1 - 9x_2 + 15x_3 = 10$

29. $4x + 12y - 7z - 20w = 22$
$3x + 9y - 5z - 28w = 30$

30. $x + 2y + z = 8$
$-3x - 6y - 3z = -21$

31. $3x + 3y + 12z = 6$
$x + y + 4z = 2$
$2x + 5y + 20z = 10$
$-x + 2y + 8z = 4$

32. $2x + y - z + 2w = -6$
$3x + 4y \quad\quad + w = 1$
$x + 5y + 2z + 6w = -3$
$5x + 2y - z - w = 3$

In Exercises 33–38, use a computer or graphing calculator to solve the system of linear equations.

Data

33. $x_1 - 2x_2 + 5x_3 - 3x_4 = 23.6$
$x_1 + 4x_2 - 7x_3 - 2x_4 = 45.7$
$3x_1 - 5x_2 + 7x_3 + 4x_4 = 29.9$

34. $23.4x - 45.8y + 43.7z = 87.2$
$86.4x + 12.3y - 56.9z = 14.5$
$93.6x - 50.7y + 12.6z = 44.4$

35. $x_1 - x_2 + 2x_3 + 2x_4 + 6x_5 = 6$
$3x_1 - 2x_2 + 4x_3 + 4x_4 + 12x_5 = 14$
$x_2 - x_3 - x_4 - 3x_5 = -3$
$2x_1 - 2x_2 + 4x_3 + 5x_4 + 15x_5 = 10$
$2x_1 - 2x_2 + 4x_3 + 4x_4 + 13x_5 = 13$

36. $x_1 + x_2 - 2x_3 + 3x_4 + 2x_5 = 9$
$3x_1 + 3x_2 - x_3 + x_4 + x_5 = 5$
$2x_1 + 2x_2 - x_3 + x_4 - 2x_5 = 1$
$4x_1 + 4x_2 + x_3 \quad\quad - 3x_5 = 4$
$8x_1 + 5x_2 - 2x_3 - x_4 + 2x_5 = 3$

37. $4x_1 - 3x_2 + x_3 - x_4 + 2x_5 - x_6 = 8$
$x_1 - 2x_2 + x_3 - 3x_4 + x_5 - 4x_6 = 4$
$2x_1 + x_2 - 3x_3 + x_4 - 2x_5 + 5x_6 = 2$
$-2x_1 + 3x_2 - x_3 + x_4 - x_5 + 2x_6 = -7$
$x_1 - 3x_2 + x_3 - 2x_4 + x_5 - 2x_6 = 9$
$5x_1 - 4x_2 - x_3 - x_4 + 4x_5 + 5x_6 = 9$

38. $x_1 + 2x_2 - 2x_3 + 2x_4 - x_5 + 3x_6 = 0$
$2x_1 - x_2 + 3x_3 + x_4 - 3x_5 + 2x_6 = 17$
$x_1 + 3x_2 - 2x_3 + x_4 - 2x_5 - 3x_6 = -5$
$3x_1 - 2x_2 + x_3 - x_4 + 3x_5 - 2x_6 = -1$
$-x_1 - 2x_2 + x_3 + 2x_4 - 2x_5 + 3x_6 = 10$
$x_1 - 3x_2 + x_3 + 3x_4 - 2x_5 + x_6 = 11$

In Exercises 39–42, solve the homogeneous linear system corresponding to the given coefficient matrix.

39. $\begin{bmatrix} 1 & 0 & 0 \\ 0 & 1 & 1 \\ 0 & 0 & 0 \end{bmatrix}$

40. $\begin{bmatrix} 1 & 0 & 0 & 0 \\ 0 & 1 & 1 & 0 \end{bmatrix}$

41. $\begin{bmatrix} 1 & 0 & 0 & 1 \\ 0 & 0 & 1 & 0 \\ 0 & 0 & 0 & 0 \end{bmatrix}$

42. $\begin{bmatrix} 0 & 0 & 0 \\ 0 & 0 & 0 \\ 0 & 0 & 0 \end{bmatrix}$

43. Consider the matrix $A = \begin{bmatrix} 1 & k & 2 \\ -3 & 4 & 1 \end{bmatrix}$.

(a) If A is the *augmented* matrix of a system of linear equations, determine the number of equations and the number of unknowns.

(b) If A is the *augmented* matrix of a system of linear equations, find the value(s) of k such that the system is consistent.

(c) If A is the *coefficient* matrix of a *homogeneous* system of linear equations, determine the number of equations and the number of unknowns.

(d) If A is the *coefficient* matrix of a *homogeneous* system of linear equations, find the value(s) of k such that the system is consistent.

44. Consider the matrix $A = \begin{bmatrix} 2 & -1 & 3 \\ -4 & 2 & k \\ 4 & -2 & 6 \end{bmatrix}$.

(a) If A is the *augmented* matrix of a system of linear equations, determine the number of equations and the number of unknowns.

(b) If A is the *augmented* matrix of a system of linear equations, find the value(s) of k such that the system is consistent.

(c) If A is the *coefficient* matrix of a *homogeneous* system of linear equations, determine the number of equations and the number of unknowns.

(d) If A is the *coefficient* matrix of a *homogeneous* system of linear equations, find the value(s) of k such that the system is consistent.

In Exercises 45 and 46, find values of a, b, and c (if possible) such that the system of linear equations has (a) a unique solution, (b) no solution, and (c) an infinite number of solutions.

45. $\begin{aligned} x + y \quad &= 2 \\ y + z &= 2 \\ x \quad + z &= 2 \\ ax + by + cz &= 0 \end{aligned}$

46. $\begin{aligned} x + y \quad &= 0 \\ y + z &= 0 \\ x \quad + z &= 0 \\ ax + by + cz &= 0 \end{aligned}$

47. The following system has one solution: $x = 1$, $y = -1$, and $z = 2$.

$\begin{aligned} 4x - 2y + 5z &= 16 \qquad &\text{Equation 1} \\ x + y \quad &= 0 \qquad &\text{Equation 2} \\ -x - 3y + 2z &= 6 \qquad &\text{Equation 3} \end{aligned}$

Solve the systems given by (a) Equations 1 and 2, (b) Equations 1 and 3, and (c) Equations 2 and 3. (d) How many solutions does each of these systems have?

48. Assume that the following system has a unique solution.

$\begin{aligned} a_{11}x_1 + a_{12}x_2 + a_{13}x_3 &= b_1 \qquad &\text{Equation 1} \\ a_{21}x_1 + a_{22}x_2 + a_{23}x_3 &= b_2 \qquad &\text{Equation 2} \\ a_{31}x_1 + a_{32}x_2 + a_{33}x_3 &= b_3 \qquad &\text{Equation 3} \end{aligned}$

Does the system composed of Equations 1 and 2 have (a) a unique solution, (b) no solution, or (c) an infinite number of solutions?

In Exercises 49 and 50, find the unique reduced row-echelon matrix that is row-equivalent to the given matrix.

49. $\begin{bmatrix} 1 & 2 \\ -1 & 2 \end{bmatrix}$

50. $\begin{bmatrix} 1 & 2 & 3 \\ 4 & 5 & 6 \\ 7 & 8 & 9 \end{bmatrix}$

51. *Writing* Describe all possible 2×2 reduced row-echelon matrices. Support your answer with examples.

52. *Writing* Describe all possible 3×3 reduced row-echelon matrices. Support your answer with examples.

True or False? In Exercises 53 and 54, determine whether each statement is true or false. If a statement is true, give a reason or cite an appropriate statement in the text. If a statement is false, provide an example that shows that the statement is not true in all cases or cite an appropriate statement in the text.

53. (a) A 6×3 matrix has six rows.

(b) Every matrix is row-equivalent to a matrix in row-echelon form.

(c) If the row-echelon form of the augmented matrix of a system of linear equations contains the row $[1 \ 0 \ 0 \ 0 \ 0]$, then the original system is inconsistent.

(d) A homogeneous system of four linear equations in six unknowns has an infinite number of solutions.

54. (a) A 4×7 matrix has four columns.

(b) Every matrix has a unique reduced row-echelon form.

(c) A homogeneous system of four linear equations in four unknowns is always consistent.

(d) Multiplying a row of a matrix by a constant is one of the elementary row operations.

In Exercises 55 and 56, determine conditions on a, b, c, and d such that the matrix $\begin{bmatrix} a & b \\ c & d \end{bmatrix}$ will be row-equivalent to the given matrix.

55. $\begin{bmatrix} 1 & 0 \\ 0 & 1 \end{bmatrix}$

56. $\begin{bmatrix} 1 & 0 \\ 0 & 0 \end{bmatrix}$

In Exercises 57 and 58, find all values of λ (the Greek letter lambda) such that the homogeneous system of linear equations will have nontrivial solutions.

57. $(\lambda - 2)x + \qquad y = 0$
$\qquad x + (\lambda - 2)y = 0$

58. $(\lambda - 1)x + 2y = 0$
$\qquad x + \lambda y = 0$

59. *Writing* Is it possible that a system of linear equations with fewer equations than variables may have no solution? If so, give an example.

60. *Writing* Does a matrix have a unique row-echelon form? Illustrate your answer with examples. Is the reduced row-echelon form unique?

61. Solve the system for α and β, $0 \le \alpha, \beta \le 2\pi$.
$2 \cos \alpha - \quad \sin \beta = 0$
$4 \cos \alpha + 2 \sin \beta = 4$

62. Solve the system of equations for x and y.
$x^2 + 2y^3 = 2$
$3x^2 - y^3 = 13$

63. *Writing* Consider the 2×2 matrix $\begin{bmatrix} a & b \\ c & d \end{bmatrix}$. Perform the following sequence of row operations.

(a) Add (-1) times the second row to the first row.
(b) Add 1 times the first row to the second row.
(c) Add (-1) times the second row to the first row.
(d) Multiply the first row by (-1).

What has happened to the original matrix? Describe, in general, how to interchange two rows of a matrix using only the second and third elementary row operations.

1.3 | Applications of Systems of Linear Equations

Systems of linear equations arise in a wide variety of applications and are one of the central themes of linear algebra. In this section you will look at two such applications, and you will see many more in subsequent chapters. The first application shows how to fit a polynomial function to a set of data points in the plane. The second application focuses on networks and Kirchhoff's Laws for electricity.

Polynomial Curve Fitting

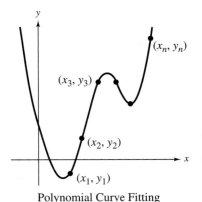

Polynomial Curve Fitting

Figure 1.4

Suppose a collection of data is represented by n points

$$(x_1, y_1), (x_2, y_2), \ldots, (x_n, y_n)$$

in the xy-plane, and you are asked to find a polynomial function

$$p(x) = a_0 + a_1 x + a_2 x^2 + \cdots + a_{n-1} x^{n-1}$$

of degree $n - 1$ whose graph passes through the given points. This procedure is called **polynomial curve fitting.** It can be shown that if all x-coordinates of the points are distinct, then there is precisely one polynomial function of degree $n - 1$ (or less) that fits the n points, as shown in Figure 1.4.

To solve for the n coefficients of $p(x)$, substitute each of the n points into the polynomial and obtain n linear equations in n variables $a_0, a_1, a_2, \ldots, a_{n-1}$:

$$a_0 + a_1 x_1 + a_2 x_1^2 + \cdots + a_{n-1} x_1^{n-1} = y_1$$
$$a_0 + a_1 x_2 + a_2 x_2^2 + \cdots + a_{n-1} x_2^{n-1} = y_2$$
$$\vdots$$
$$a_0 + a_1 x_n + a_2 x_n^2 + \cdots + a_{n-1} x_n^{n-1} = y_n.$$

This procedure is demonstrated with a second-degree polynomial in Example 1.

EXAMPLE 1   *Polynomial Curve Fitting*

Determine the polynomial $p(x) = a_0 + a_1x + a_2x^2$ whose graph passes through the points $(1, 4)$, $(2, 0)$, and $(3, 12)$.

Solution   Substituting $x = 1, 2,$ and 3 into $p(x)$ and equating the results to the respective y-values produces the following system of linear equations in the variables a_0, a_1, and a_2.

$$p(1) = a_0 + a_1(1) + a_2(1)^2 = a_0 + a_1 + a_2 = 4$$
$$p(2) = a_0 + a_1(2) + a_2(2)^2 = a_0 + 2a_1 + 4a_2 = 0$$
$$p(3) = a_0 + a_1(3) + a_2(3)^2 = a_0 + 3a_1 + 9a_2 = 12$$

The solution of this system is $a_0 = 24$, $a_1 = -28$, and $a_2 = 8$, so the polynomial is

$$p(x) = 24 - 28x + 8x^2.$$

The graph of p is shown in Figure 1.5.

Simulation

Explore this concept further with an electronic simulation available on the *Elementary Linear Algebra* Learning Tools Student CD-ROM.

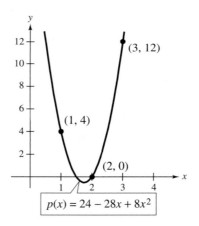

$$p(x) = 24 - 28x + 8x^2$$

Figure 1.5

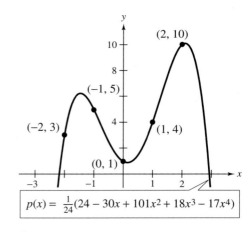

$$p(x) = \tfrac{1}{24}(24 - 30x + 101x^2 + 18x^3 - 17x^4)$$

Figure 1.6

EXAMPLE 2   *Polynomial Curve Fitting*

Find a polynomial that fits the points $(-2, 3)$, $(-1, 5)$, $(0, 1)$, $(1, 4)$, and $(2, 10)$.

Solution   Because you are given five points, choose a fourth-degree polynomial

$$p(x) = a_0 + a_1x + a_2x^2 + a_3x^3 + a_4x^4.$$

Substituting the given points into $p(x)$ produces the following system of linear equations.

$$\begin{aligned}
a_0 - 2a_1 + 4a_2 - 8a_3 + 16a_4 &= 3 \\
a_0 - a_1 + a_2 - a_3 + a_4 &= 5 \\
a_0 &= 1 \\
a_0 + a_1 + a_2 + a_3 + a_4 &= 4 \\
a_0 + 2a_1 + 4a_2 + 8a_3 + 16a_4 &= 10
\end{aligned}$$

The solution of these equations is

$$a_0 = 1, \qquad a_1 = -\tfrac{30}{24}, \qquad a_2 = \tfrac{101}{24}, \qquad a_3 = \tfrac{18}{24}, \qquad a_4 = -\tfrac{17}{24}$$

which means that the polynomial is

$$\begin{aligned}
p(x) &= 1 - \tfrac{30}{24}x + \tfrac{101}{24}x^2 + \tfrac{18}{24}x^3 - \tfrac{17}{24}x^4 \\
&= \tfrac{1}{24}(24 - 30x + 101x^2 + 18x^3 - 17x^4).
\end{aligned}$$

The graph of p is shown in Figure 1.6.

The system of linear equations in Example 2 is relatively easy to solve because the x-values are small. Given a set of points with large x-values, it is usually best to *translate* the values before attempting the curve-fitting procedure. This approach is demonstrated in the next example.

EXAMPLE 3   *Translating Large x-Values before Curve Fitting*

Find a polynomial that fits the points

$$(x_1, y_1) \qquad (x_2, y_2) \qquad (x_3, y_3) \qquad (x_4, y_4) \qquad (x_5, y_5)$$

$$(2000, 3), \qquad (2001, 5), \qquad (2002, 1), \qquad (2003, 4), \qquad (2004, 10).$$

Solution   Because the given x-values are large, use the translation $z = x - 2002$ to obtain

$$(z_1, y_1) \qquad (z_2, y_2) \qquad (z_3, y_3) \qquad (z_4, y_4) \qquad (z_5, y_5)$$

$$(-2, 3), \qquad (-1, 5), \qquad (0, 1), \qquad (1, 4), \qquad (2, 10).$$

This is the same set of points as in Example 2. Thus the polynomial that fits these points is

$$\begin{aligned}
p(z) &= \tfrac{1}{24}(24 - 30z + 101z^2 + 18z^3 - 17z^4) \\
&= 1 - \tfrac{5}{4}z + \tfrac{101}{24}z^2 + \tfrac{3}{4}z^3 - \tfrac{17}{24}z^4.
\end{aligned}$$

Letting $z = x - 2002$, you have

$$p(x) = 1 - \tfrac{5}{4}(x - 2002) + \tfrac{101}{24}(x - 2002)^2 + \tfrac{3}{4}(x - 2002)^3 - \tfrac{17}{24}(x - 2002)^4.$$

EXAMPLE 4   *An Application of Curve Fitting*

Find a polynomial that relates the periods of the first three planets to their mean distances from the sun, as shown in Table 1.1. Then test the accuracy of the fit by using the polynomial to calculate the period of Mars. (Distance is measured in astronomical units, and period is measured in years.)

TABLE 1.1

Planet	Mercury	Venus	Earth	Mars	Jupiter	Saturn
Mean Distance	0.387	0.723	1.0	1.523	5.203	9.541
Period	0.241	0.615	1.0	1.881	11.861	29.457

Solution   Begin by fitting a quadratic polynomial

$$p(x) = a_0 + a_1x + a_2x^2$$

to the points $(0.387, 0.241)$, $(0.723, 0.615)$, and $(1, 1)$. The system of linear equations obtained by substituting these points into $p(x)$ is

$$a_0 + 0.387a_1 + (0.387)^2a_2 = 0.241$$
$$a_0 + 0.723a_1 + (0.723)^2a_2 = 0.615$$
$$a_0 + \quad a_1 + \quad\quad a_2 = 1.$$

The approximate solution of the system is

$$a_0 \approx -0.0634, \qquad a_1 \approx 0.6119, \qquad a_2 \approx 0.4515$$

which means that the polynomial can be approximated by

$$p(x) = -0.0634 + 0.6119x + 0.4515x^2.$$

Using $p(x)$ to evaluate the period of Mars produces

$$p(1.523) \approx 1.916 \text{ years.}$$

This estimate is compared graphically with the actual period of Mars in Figure 1.7. Note that the actual period (from Table 1.1) is 1.881 years.

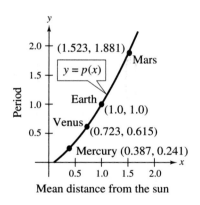

Figure 1.7

An important lesson may be learned from the application shown in Example 4: The polynomial that fits the given data points does not necessarily give an accurate model for the relationship between x and y for x-values other than those corresponding to the given points. Generally, the farther the additional points are from the given points, the worse the fit. For instance, in Example 4 the mean distance of Jupiter is 5.203. The corresponding polynomial approximation for the period is 15.343 years—a poor estimate of Jupiter's actual period of 11.861 years.

The problem of curve fitting can be difficult. Other types of functions often provide better fits than do polynomial functions. To see this, look again at the curve-fitting problem

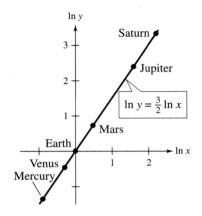

Figure 1.8

in Example 4. Taking the natural logarithm of each of the distances and periods of the first six planets produces the results shown in Table 1.2 and Figure 1.8.

TABLE 1.2

Planet	Mercury	Venus	Earth	Mars	Jupiter	Saturn
Mean Distance (x)	0.387	0.723	1.0	1.523	5.203	9.541
Natural Log of Mean	−0.949	−0.324	0.0	0.421	1.649	2.256
Period (y)	0.241	0.615	1.0	1.881	11.861	29.457
Natural Log of Period	−1.423	−0.486	0.0	0.632	2.473	3.383

Now, fitting a polynomial to the logarithms of the distances and periods produces the following *linear relationship* between $\ln x$ and $\ln y$.

$$\ln y = \tfrac{3}{2} \ln x$$

From this equation it follows that $y = x^{3/2}$, or

$$y^2 = x^3.$$

In other words, the square of the period (in years) of each planet is equal to the cube of its mean distance (in astronomical units) from the sun. This relationship was first discovered by Johannes Kepler in 1619.

Network Analysis

Networks composed of branches and junctions are used as models in fields as diverse as economics, traffic analysis, and electrical engineering.

It is assumed in such models that the total flow into a junction is equal to the total flow out of the junction. For example, because the junction shown in Figure 1.9 has 25 units flowing into it, there must be 25 units flowing out of it. This is represented by the linear equation

$$x_1 + x_2 = 25.$$

Because each junction in a network gives rise to a linear equation, you can analyze the flow through a network composed of several junctions by solving a system of linear equations. This procedure is illustrated in Example 5.

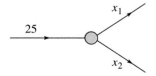

Figure 1.9

EXAMPLE 5   *Analysis of a Network*

Set up a system of linear equations to represent the network shown in Figure 1.10, and solve the system.

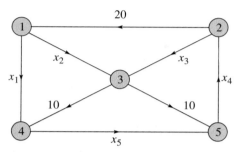

Figure 1.10

Solution   Each of the network's five junctions gives rise to a linear equation, as follows.

$$x_1 + x_2 = 20 \qquad \textbf{Junction 1}$$
$$x_3 + 20 = x_4 \qquad \textbf{Junction 2}$$
$$x_2 + x_3 = 20 \qquad \textbf{Junction 3}$$
$$x_1 + 10 = x_5 \qquad \textbf{Junction 4}$$
$$x_5 + 10 = x_4 \qquad \textbf{Junction 5}$$

The augmented matrix for this system is

$$\begin{bmatrix} 1 & 1 & 0 & 0 & 0 & 20 \\ 0 & 0 & 1 & -1 & 0 & -20 \\ 0 & 1 & 1 & 0 & 0 & 20 \\ 1 & 0 & 0 & 0 & -1 & -10 \\ 0 & 0 & 0 & -1 & 1 & -10 \end{bmatrix}.$$

Gauss-Jordan elimination produces the matrix

$$\begin{bmatrix} 1 & 0 & 0 & 0 & -1 & -10 \\ 0 & 1 & 0 & 0 & 1 & 30 \\ 0 & 0 & 1 & 0 & -1 & -10 \\ 0 & 0 & 0 & 1 & -1 & 10 \\ 0 & 0 & 0 & 0 & 0 & 0 \end{bmatrix}.$$

Letting $t = x_5$, you have

$$x_1 = t - 10, \qquad x_2 = -t + 30, \qquad x_3 = t - 10, \qquad x_4 = t + 10, \qquad x_5 = t$$

where t is a real number. Thus this system has an infinite number of solutions.

In Example 5, suppose that you could control the amount of flow along the branch labeled x_5. Using the solution given in Example 5, you could then control the flow represented by each of the other variables. For instance, letting $t = 10$ would reduce the flow of x_1 and x_3 to zero, as shown in Figure 1.11. Similarly, letting $t = 20$ would produce the network shown in Figure 1.12.

Figure 1.11

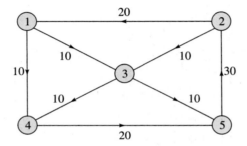

Figure 1.12

You can see how the type of network analysis demonstrated in Example 5 could be used in problems dealing with the flow of traffic through the streets of a city or the flow of water through an irrigation system.

REMARK: A *closed* path is a sequence of branches such that the beginning point of the first branch coincides with the end point of the last branch.

Another type of network to which network analysis is commonly applied is an electrical network. An analysis of such a system uses two properties of electrical networks known as **Kirchhoff's Laws.**

1. All the current flowing into a junction must flow out of it.
2. The sum of the products *IR* (*I* is current and *R* is resistance) around a closed path is equal to the total voltage in the path.

In an electrical network, current is measured in amps, resistance in ohms, and the product of current and resistance in volts. Batteries are represented by the symbol ⊣⊢, where the current flows out of the terminal denoted by the larger vertical bar. Resistance is denoted by the symbol ⌇⌇⌇⌇. The direction of the current is indicated by the arrow in the branch.

EXAMPLE 6   *Analysis of an Electrical Network*

Determine the currents I_1, I_2, and I_3 for the electrical network shown in Figure 1.13.

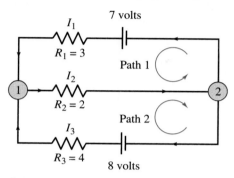

Figure 1.13

Solution   Applying Kirchhoff's first law to either junction produces

$$I_1 + I_3 = I_2 \qquad \text{\textbf{Junction 1 or Junction 2}}$$

and applying Kirchhoff's second law to the two paths produces

$$R_1 I_1 + R_2 I_2 = 3I_1 + 2I_2 = 7 \qquad \text{\textbf{Path 1}}$$
$$R_2 I_2 + R_3 I_3 = 2I_2 + 4I_3 = 8. \qquad \text{\textbf{Path 2}}$$

Thus you have the following system of three linear equations in the variables I_1, I_2, and I_3.

$$\begin{aligned} I_1 - I_2 + I_3 &= 0 \\ 3I_1 + 2I_2 &= 7 \\ 2I_2 + 4I_3 &= 8 \end{aligned}$$

Applying Gauss-Jordan elimination to the augmented matrix

$$\begin{bmatrix} 1 & -1 & 1 & 0 \\ 3 & 2 & 0 & 7 \\ 0 & 2 & 4 & 8 \end{bmatrix}$$

produces the reduced row-echelon form

$$\begin{bmatrix} 1 & 0 & 0 & 1 \\ 0 & 1 & 0 & 2 \\ 0 & 0 & 1 & 1 \end{bmatrix}$$

which means that $I_1 = 1$ amp, $I_2 = 2$ amps, and $I_3 = 1$ amp.

EXAMPLE 7 *Analysis of an Electrical Network*

Determine the currents I_1, I_2, I_3, I_4, I_5, and I_6 for the electrical network shown in Figure 1.14.

Figure 1.14

Solution   Applying Kirchhoff's first law to the four junctions produces

$$I_1 + I_3 = I_2 \qquad \text{Junction 1}$$
$$I_1 + I_4 = I_2 \qquad \text{Junction 2}$$
$$I_3 + I_6 = I_5 \qquad \text{Junction 3}$$
$$I_4 + I_6 = I_5 \qquad \text{Junction 4}$$

and applying Kirchhoff's second law to the three paths produces

$$2I_1 + 4I_2 = 10 \qquad\qquad \text{Path 1}$$
$$4I_2 + I_3 + 2I_4 + 2I_5 = 17 \qquad \text{Path 2}$$
$$2I_5 + 4I_6 = 14. \qquad\qquad \text{Path 3}$$

Thus you have the following system of seven linear equations in the variables I_1, I_2, I_3, I_4, I_5, and I_6.

$$
\begin{aligned}
I_1 - I_2 + I_3 &&&&&= 0 \\
I_1 - I_2 &&+ I_4 &&&= 0 \\
&& I_3 && - I_5 + I_6 &= 0 \\
&&&& I_4 - I_5 + I_6 &= 0 \\
2I_1 + 4I_2 &&&&&= 10 \\
4I_2 &+ I_3 + 2I_4 + 2I_5 &&&&= 17 \\
&&&& 2I_5 + 4I_6 &= 14
\end{aligned}
$$

Using Gauss-Jordan elimination, a graphing calculator, or a computer software program, you can solve this system to obtain

$$I_1 = 1, \qquad I_2 = 2, \qquad I_3 = 1, \qquad I_4 = 1, \qquad I_5 = 3, \quad \text{and} \quad I_6 = 2$$

meaning $I_1 = 1$ amp, $I_2 = 2$ amps, $I_3 = 1$ amp, $I_4 = 1$ amp, $I_5 = 3$ amps, and $I_6 = 2$ amps.

Polynomial Curve Fitting

In Exercises 1–6, (a) determine the polynomial whose graph passes through the given points, and (b) sketch the graph of the polynomial, showing the given points.

1. $(2, 5), (3, 2), (4, 5)$
2. $(2, 4), (3, 4), (4, 4)$
3. $(2, 4), (3, 6), (5, 10)$
4. $(-1, 3), (0, 0), (1, 1), (4, 58)$
5. $(2001, 5), (2002, 7), (2003, 12)$ (Let $z = x - 2002$.)
6. $(2001, 150), (2002, 180), (2003, 240), (2004, 360)$
   (Let $z = x - 2002$.)

7. *Writing* Try to fit the graph of a polynomial function to the values shown in the table. What happens, and why?

x	1	2	3	3	4
y	1	1	2	3	4

8. The graph of a function f passes through the points $(0, 1)$, $\left(2, \frac{1}{3}\right)$, and $\left(4, \frac{1}{5}\right)$. Find a quadratic function whose graph passes through these points.

9. Find a polynomial function p of degree 2 or less that passes through the points $(0, 1), (2, 3),$ and $(4, 5)$. Then sketch the graph of $y = 1/p(x)$ and compare this graph with the graph of the polynomial found in Exercise 8.

10. *Calculus* The graph of a parabola passes through the points $(0, 1)$ and $\left(\frac{1}{2}, \frac{1}{2}\right)$ and has a horizontal tangent at $\left(\frac{1}{2}, \frac{1}{2}\right)$. Find an equation for the parabola and sketch its graph.

11. *Calculus* The graph of a cubic polynomial has horizontal tangents at $(1, -2)$ and $(-1, 2)$. Find an equation for the cubic and sketch its graph.

12. Find an equation of the circle passing through the points $(1, 3), (-2, 6),$ and $(4, 2)$.

13. The U.S. census lists the population of the United States as 227 million in 1980, 249 million in 1990, and 281 million in 2000. Fit a second-degree polynomial to these three points and use it to predict the population in 2005 and in 2010. (Source: U.S. Census Bureau)

14. The U.S. population for the years 1920, 1930, 1940, and 1950 is shown in the table. (Source: U.S. Census Bureau)

Year	1920	1930	1940	1950
Population (in millions)	106	123	132	151

(a) Find a cubic polynomial that fits these data and use it to estimate the population in 1960.

(b) How does your estimate compare with the actual 1960 population of 179 million?

15. The net profit (in millions of dollars) for Microsoft from 1991 to 2002 is shown in the table. (Source: Microsoft Corporation)

Year	1991	1992	1993	1994
Net Profit	462.7	708.1	953.0	1210.0

Year	1995	1996	1997	1998
Net Profit	1453.0	2176.0	3454.0	4786.0

Year	1999	2000	2001	2002
Net Profit	7625.0	9421.0	10,003.0	10,530.0

(a) Set up a system of equations to fit the data for the years 1999, 2000, 2001, and 2002 to a cubic model

(b) Solve the system. Does the solution produce a reasonable model for predicting future net profits? Explain.

16. The sales (in billions of dollars) for Coca-Cola Entertainment from 1991 to 2002 are shown in the table. (Source: Coca-Cola Entertainment)

Year	1991	1992	1993	1994
Sales	4.051	5.127	5.465	6.011

Year	1995	1996	1997	1998
Sales	6.773	7.921	11.278	13.414

Year	1999	2000	2001	2002
Sales	14.406	14.750	15.700	16.800

(a) Set up a system of equations to fit the data for the years 1998, 1999, 2000, 2001, and 2002 to a fourth-degree polynomial model.

(b) Solve the system. Does the solution produce a reasonable model for predicting future sales? Explain.

17. Use $\sin 0 = 0$, $\sin(\pi/2) = 1$, and $\sin \pi = 0$ to estimate $\sin(\pi/3)$.

18. Use $\log_2 1 = 0$, $\log_2 2 = 1$, and $\log_2 4 = 2$ to estimate $\log_2 3$.

19. *Guided Proof* Prove that if a polynomial function $p(x) = a_0 + a_1 x + a_2 x^2$ is zero for $x = -1$, $x = 0$, and $x = 1$, then $a_0 = a_1 = a_2 = 0$.

Getting Started: Write a system of linear equations and solve the system for a_0, a_1, and a_2.

(i) Substitute $x = -1$, 0, and 1 into $p(x)$.

(ii) Set the result equal to 0.

(iii) Solve the resulting system of linear equations in the variables a_0, a_1, and a_2.

20. The statement in Exercise 19 can be generalized: If a polynomial function $p(x) = a_0 + a_1 x + \cdots + a_{n-1}x^{n-1}$ is zero for more than $n - 1$ x-values, then $a_0 = a_1 = \cdots = a_{n-1} = 0$. Use this result to prove that there is at most one polynomial function of degree $n - 1$ (or less) whose graph passes through n points in the plane with distinct x-coordinates.

Network Analysis

21. Water is flowing through a network of pipes (in thousands of cubic meters per hour), as shown in Figure 1.15.

(a) Solve this system for the water flow represented by x_i, $i = 1, 2, \ldots, 7$.

(b) Find the water flow when $x_6 = x_7 = 0$.

(c) Find the water flow when $x_5 = 1000$ and $x_6 = 0$.

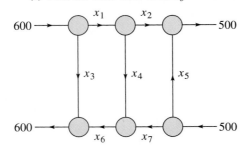

Figure 1.15

22. The flow of traffic (in vehicles per hour) through a network of streets is shown in Figure 1.16.

(a) Solve this system for x_i, $i = 1, 2, \ldots, 5$.

(b) Find the traffic flow when $x_2 = 200$ and $x_3 = 50$.

(c) Find the traffic flow when $x_2 = 150$ and $x_3 = 0$.

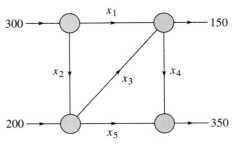

Figure 1.16

23. The flow of traffic (in vehicles per hour) through a network of streets is shown in Figure 1.17.

(a) Solve this system for x_i, $i = 1, 2, 3, 4$.

(b) Find the traffic flow when $x_4 = 0$.

(c) Find the traffic flow when $x_4 = 100$.

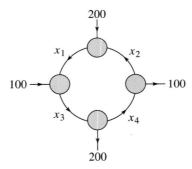

Figure 1.17

24. The flow of traffic (in vehicles per hour) through a network of streets is shown in Figure 1.18.

(a) Solve this system for x_i, $i = 1, 2, \ldots, 5$.

(b) Find the traffic flow when $x_3 = 0$ and $x_5 = 100$.

(c) Find the traffic flow when $x_3 = x_5 = 100$.

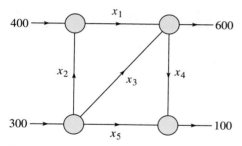

Figure 1.18

25. Determine the currents I_1, I_2, and I_3 for the electrical network shown in Figure 1.19.

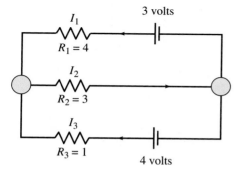

Figure 1.19

26. Determine the currents I_1, I_2, and I_3 for the electrical network shown in Figure 1.20.

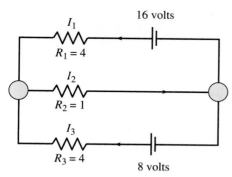

Figure 1.20

27. (a) Determine the currents I_1, I_2, and I_3 for the electrical network shown in Figure 1.21.
   (b) How is the result affected when A is changed to 2 volts and B is changed to 6 volts?

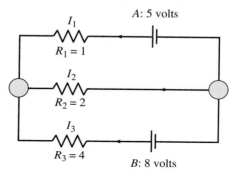

Figure 1.21

28. Determine the currents I_1, I_2, I_3, I_4, I_5, and I_6 for the electrical network shown in Figure 1.22.

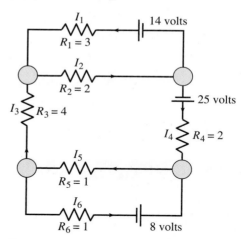

Figure 1.22

CHAPTER 1   Summary

After studying this chapter, you should have acquired the following skills.

- ■ Recognize a linear equation in n variables.

- ■ Find a parametric representation of a solution set of a linear equation.

- ■ Use back-substitution to solve a system of linear equations.

- ■ Graph and solve a system of linear equations in two or three variables.

- ■ Use a computer or graphing calculator to solve a system of linear equations.

- ■ Determine whether a system of linear equations is consistent or inconsistent.

- ■ Use back-substitution to solve a system in row-echelon form.

- ■ Use elimination to rewrite a system in row-echelon form.

- ■ Determine the size of a matrix.

- ■ Write an augmented or coefficient matrix from a system of linear equations.

- ■ Translate a matrix into a system of linear equations.

- ■ Determine whether a matrix is in row-echelon form or reduced row-echelon form.

- ■ Solve a system of linear equations using Gaussian elimination with back-substitution.

- ■ Solve a system of linear equations using Gauss-Jordan elimination.

- ■ Solve a homogeneous system of linear equations.

- ■ Set up and solve a system of equations to fit a polynomial to a set of data points.

- ■ Set up and solve a system of equations to represent a network.

CHAPTER 1   Review Exercises

In Exercises 1 and 2, find a parametric representation of the solution set of the linear equation.

1. $-4x + 2y - 6z = 1$ **2.** $3x_1 + 2x_2 - 4x_3 = 0$

In Exercises 3–6, determine whether the matrix is in row-echelon form. If it is, determine whether it is also in reduced row-echelon form.

3. $\begin{bmatrix} 1 & 0 & 1 & 1 \\ 0 & 1 & 2 & 1 \\ 0 & 0 & 0 & 1 \end{bmatrix}$ **4.** $\begin{bmatrix} 1 & 2 & -3 & 0 \\ 0 & 0 & 0 & 1 \\ 0 & 0 & 0 & 0 \end{bmatrix}$

5. $\begin{bmatrix} -1 & 2 & 1 \\ 0 & 1 & 0 \\ 0 & 0 & 1 \end{bmatrix}$ **6.** $\begin{bmatrix} 0 & 1 & 0 & 0 \\ 0 & 0 & 1 & 2 \\ 0 & 0 & 0 & 0 \end{bmatrix}$

In Exercises 7 and 8, find the solution set of the system of linear equations represented by the augmented matrix.

7. $\begin{bmatrix} 1 & 2 & 0 & 0 \\ 0 & 0 & 1 & 0 \\ 0 & 0 & 0 & 0 \end{bmatrix}$ **8.** $\begin{bmatrix} 1 & 2 & 3 & 0 \\ 0 & 0 & 0 & 1 \\ 0 & 0 & 0 & 0 \end{bmatrix}$

In Exercises 9–30, solve the system of linear equations.

9. $\begin{aligned} x + y &= 2 \\ 3x - y &= 0 \end{aligned}$ **10.** $\begin{aligned} x + y &= -1 \\ 3x + 2y &= 0 \end{aligned}$

11. $\begin{aligned} 3y &= 2x \\ y &= x + 4 \end{aligned}$ **12.** $\begin{aligned} x &= y + 3 \\ 4x &= y + 10 \end{aligned}$

13. $y + x = 0$
$2x + y = 0$

14. $y = -4x$
$y = \quad x$

15. $x - y = 9$
$-x + y = 1$

16. $40x_1 + 30x_2 = \quad 24$
$20x_1 + 15x_2 = -14$

17. $0.2x_1 + 0.3x_2 = 0.14$
$0.4x_1 + 0.5x_2 = 0.20$

18. $0.2x - 0.1y = \quad 0.07$
$0.4x - 0.5y = -0.01$

19. $\frac{1}{2}x - \frac{1}{3}y \quad = 0$
$3x + 2(y + 5) = 10$

20. $\frac{1}{3}x + \frac{4}{7}y = \quad 3$
$2x + 3y = 15$

21. $-x + y + 2z = \quad 1$
$2x + 3y + z = -2$
$5x + 4y + 2z = \quad 4$

22. $2x + 3y + z = \quad 10$
$2x - 3y - 3z = \quad 22$
$4x - 2y + 3z = -2$

23. $2x + 3y + 3z = \quad 3$
$6x + 6y + 12z = 13$
$12x + 9y - \quad z = \quad 2$

24. $2x + \qquad 6z = -9$
$3x - 2y + 11z = -16$
$3x - y + 7z = -11$

25. $x - 2y + z = -6$
$2x - 3y \qquad = -7$
$-x + 3y - 3z = \quad 11$

26. $x + 2y + 6z = \quad 1$
$2x + 5y + 15z = \quad 4$
$3x + y + 3z = -6$

27. $2x + y + 2z = 4$
$2x + 2y \quad = 5$
$2x - y + 6z = 2$

28. $2x_1 + 5x_2 - 19x_3 = 34$
$3x_1 + 8x_2 - 31x_3 = 54$

29. $2x_1 + x_2 + x_3 + 2x_4 = -1$
$5x_1 - 2x_2 + x_3 - 3x_4 = \quad 0$
$-x_1 + 3x_2 + 2x_3 + 2x_4 = \quad 1$
$3x_1 + 2x_2 + 3x_3 - 5x_4 = \quad 12$

30. $x_1 + 5x_2 + 3x_3 \qquad = 14$
$4x_2 + 2x_3 + 5x_4 \qquad = 3$
$3x_3 + 8x_4 + 6x_5 = 16$
$2x_1 + 4x_2 \qquad - 2x_5 = 0$
$2x_1 \qquad - x_3 \qquad = 0$

In Exercises 31–34, solve the homogeneous system of linear equations.

31. $x_1 - 2x_2 - 8x_3 = 0$
$3x_1 + 2x_2 \qquad = 0$
$-x_1 + x_2 + 7x_3 = 0$

32. $2x_1 + 4x_2 - 7x_3 = 0$
$x_1 - 3x_2 + 9x_3 = 0$
$6x_1 \qquad + 9x_3 = 0$

33. $2x_1 - 8x_2 + 4x_3 = 0$
$3x_1 - 10x_2 + 7x_3 = 0$
$10x_2 + 5x_3 = 0$

34. $x_1 + 3x_2 + 5x_3 = 0$
$x_1 + 4x_2 + \frac{1}{2}x_3 = 0$

35. Determine the value of k such that the system of linear equations is inconsistent.
$kx + y = 0$
$x + ky = 1$

36. Determine the value of k such that the system of linear equations has exactly one solution.
$x - y + 2z = 0$
$-x + y - z = 0$
$x + ky + z = 0$

37. Find conditions on a and b such that the system of linear equations has (a) no solution, (b) exactly one solution, and (c) an infinite number of solutions.
$x + 2y = \quad 3$
$ax + by = -9$

38. Find (if possible) conditions on a, b, and c such that the system of linear equations has (a) no solution, (b) exactly one solution, and (c) an infinite number of solutions.
$2x - y + z = a$
$x + y + 2z = b$
$3y + 3z = c$

39. *Writing* Describe a method for showing that two matrices are row-equivalent. Are the two matrices row-equivalent?
$$\begin{bmatrix} 1 & 1 & 2 \\ 0 & -1 & 2 \\ 3 & 1 & 2 \end{bmatrix} \quad \text{and} \quad \begin{bmatrix} 1 & 2 & 3 \\ 4 & 3 & 6 \\ 5 & 5 & 10 \end{bmatrix}$$

40. *Writing* Describe all possible 2×3 reduced row-echelon matrices. Support your answer with examples.

41. Let $n \geq 3$. Find the reduced row-echelon form of the $n \times n$ matrix.
$$\begin{bmatrix} 1 & 2 & 3 & \cdots & n \\ n+1 & n+2 & n+3 & \cdots & 2n \\ 2n+1 & 2n+2 & 2n+3 & \cdots & 3n \\ \vdots & \vdots & \vdots & & \vdots \\ n^2-n+1 & n^2-n+2 & n^2-n+3 & \cdots & n^2 \end{bmatrix}$$

42. Find all values of λ for which the homogeneous system of linear equations has nontrivial solutions.
$(\lambda + 2)x_1 - 2x_2 + 3x_3 = 0$
$-2x_1 + (\lambda - 1)x_2 + 6x_3 = 0$
$x_1 + 2x_2 + \lambda x_3 = 0$

True or False? In Exercises 43 and 44, determine whether each statement is true or false. If a statement is true, give a reason or cite an appropriate statement in the text. If a statement is false, provide an example that shows that the statement is not true in all cases or cite an appropriate statement in the text.

43. (a) The solution set of a linear equation can be represented parametrically in only one way.
(b) A consistent system of linear equations can have an infinite number of solutions.

44. (a) A homogeneous system of linear equations must have at least one solution.
(b) A system of linear equations with fewer equations than variables always has at least one solution.

45. The University of Michigan and the University of Tennessee scored a total of 62 points during the 2002 Florida Citrus Bowl. The points came from a total of 18 different scoring plays, which were a combination of touchdowns, extra-point kicks, and field goals, worth 6 points, 1 point, and 3 points, respectively. The same number of extra points were scored as touchdowns were scored. (Source: National Collegiate Athletic Association)

(a) Set up a system of linear equations to find the number of touchdowns, extra-point kicks, and field goals scored in the 2002 Citrus Bowl.

(b) Solve the system.

46. During the second game of the Western Conference finals in 2002, the Los Angeles Lakers scored a total of 90 points. The scoring resulted from a combination of three-point baskets, two-point baskets, and one-point free-throws. There were 11 times as many two-point baskets as three-point baskets scored and five times as many free-throws as three-point baskets. (Source: National Basketball Association)

(a) Set up a system of linear equations to find the number of three-point baskets, two-point baskets, and one-point free-throws scored by the Los Angeles Lakers in the second game of the 2002 Western Conference finals.

(b) Solve the system.

Polynomial Curve Fitting

In Exercises 47 and 48, (a) determine the polynomial whose graph passes through the given points, and (b) sketch the graph of the polynomial, showing the given points.

47. $(2, 5), (3, 0), (4, 20)$

48. $(-1, -1), (0, 0), (1, 1), (2, 4)$

49. A company has sales (measured in millions) of $50, $60, and $75 during three consecutive years. Find a quadratic function that fits these data, and use it to predict the sales during the fourth year.

50. The polynomial $p(x) = a_0 + a_1x + a_2x^2 + a_3x^3$ is zero when $x = 1, 2, 3,$ and 4. What are the values of $a_0, a_1, a_2,$ and a_3?

51. A wildlife management team studied the population of deer in one small tract of a wildlife preserve. The population and the number of years since the study began are shown in the table.

Year	0	4	80
Population	80	68	30

(a) Set up a system of equations to fit the data to a quadratic polynomial.

(b) Solve your system.

(c) Use a graphing utility to fit a quadratic model to the data.

(d) Compare the quadratic polynomial in part (b) with the model in part (c).

(e) Cite the statement in the text that verifies your results.

52. A research team studied the average monthly temperature of a small lake over a period of about one year. The temperature and the number of months since the study began are shown in the table.

Month	0	6	12
Temperature	40	73	52

(a) Set up a system of equations to fit the data to a quadratic polynomial.

(b) Solve your system.

(c) Use a graphing utility to fit a quadratic model to the data.

(d) Compare the quadratic polynomial in part (b) with the model in part (c).

(e) Cite the statement in the text that verifies your results.

Network Analysis

53. The flow through a network is shown in Figure 1.23.

(a) Solve the system for $x_i, i = 1, 2, \ldots, 6$.

(b) Find the flow when $x_3 = 100$, $x_5 = 50$, and $x_6 = 50$.

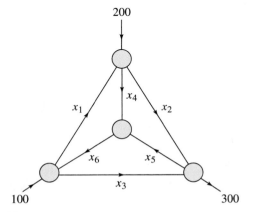

Figure 1.23

54. Determine the currents I_1, I_2, and I_3 for the electrical network shown in Figure 1.24.

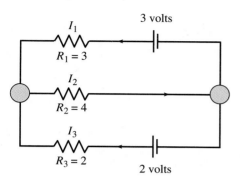

Figure 1.24

---

1   GRAPHING LINEAR EQUATIONS

You saw in Section 1.1 that a system of two linear equations in two unknowns x and y can be represented geometrically as two lines in the plane. These lines can intersect in a point, coincide, or be parallel, as indicated in Figure 1.25.

Figure 1.25

1. Consider the following system, where a and b are constants.

$$2x - y = 3$$
$$ax + by = 6$$

(a) Can you find values of a and b for which the resulting system has a unique solution?

(b) Can you find values of a and b for which the resulting system has an infinite number of solutions?

(a)

(b)

(c)

Figure 1.26

(c) Can you find values of a and b for which the resulting system has no solution?

(d) Graph the resulting lines for each of the systems in parts (a), (b), and (c).

2. Now consider a system of three linear equations in three unknowns x, y, and z. Each equation represents a plane in the three-dimensional coordinate system.

(a) Find an example of a system represented by three planes intersecting in a line, as shown in Figure 1.26(a).

(b) Find an example of a system represented by three planes intersecting in a point, as shown in Figure 1.26(b).

(c) Find an example of a system represented by three planes with no common intersection, as shown in Figure 1.26(c).

(d) Are there other configurations of three planes not covered by the three examples in parts (a), (b), and (c)? Explain.

2   UNDERDETERMINED AND OVERDETERMINED SYSTEMS OF EQUATIONS

The following system of linear equations is said to be **underdetermined** because there are more unknowns than equations.

$$\begin{aligned} x_1 + 2x_2 - 3x_3 &= 4 \\ 2x_1 - x_2 + 4x_3 &= -3 \end{aligned}$$

Similarly, the following system is **overdetermined** because there are more equations than unknowns.

$$\begin{aligned} x_1 + 3x_2 &= 5 \\ 2x_1 - 2x_2 &= -3 \\ -x_1 + 7x_2 &= 0 \end{aligned}$$

You can explore whether the number of unknowns and the number of equations have any bearing on the consistency of a system of linear equations.

1. Can you find a consistent underdetermined linear system? If so, give an example. Otherwise, explain why it is impossible.

2. Can you find a consistent overdetermined linear system? If so, give an example. Otherwise, explain why it is impossible.

3. Can you find an inconsistent underdetermined linear system? If so, give an example. Otherwise, explain why it is impossible.

4. Can you find an inconsistent overdetermined linear system? If so, give an example. Otherwise, explain why it is impossible.

5. Explain why you would expect an overdetermined linear system to be inconsistent. Must this always be the case?

6. Explain why you would expect an underdetermined linear system to have an infinite number of solutions. Must this always be the case?

ARTHUR CAYLEY was the second son of an English merchant. Like Gauss, Cayley showed signs of mathematical genius at an early age. He attended Trinity College, Cambridge, where he soon advanced beyond his fellows. He won a series of honors culminating in his election as a Fellow of Trinity and an assistant tutor, a position to which he could have been re-elected had he been willing to take holy vows as a minister in the Church of England. After graduation from Cambridge, Cayley began a prolific writing career, publishing 25 papers in 3 years. The work he began in these papers directed his mathematical interest for nearly 50 years.

Cayley was unable to find a position as a mathematician, so at 25 he decided to enter law. For 14 years Cayley had a modestly successful legal career, but he continued to spend considerable time on mathematics, publishing more than 200 papers.

In 1863, Cayley was appointed professor of mathematics at Cambridge, a position he held until his death in 1895. While teaching at Cambridge, Cayley was instrumental in persuading the administration to begin accepting women as students.

Cayley's writings and mathematical creations involved many branches of mathematics, including analytic geometry. He is most remembered, however, for his work in linear algebra. He introduced matrix theory in a paper titled *A Memoir on the Theory of Matrices* (1858) and originated the theory of invariants.

ARTHUR CAYLEY

1 8 2 1 – 1 8 9 5

MATRICES CHAPTER 2

2.1 | Operations with Matrices

In Section 1.2 you used matrices to solve systems of linear equations. Matrices, however, can be used to do much more than that. There is a rich mathematical theory of matrices, and its applications are numerous. This section and the next introduce some fundamentals of matrix theory.

It is standard mathematical convention to represent matrices in any one of the following three ways.

1. A matrix can be denoted by an uppercase letter such as

   $A, B, C, \ldots.$

2. A matrix can be denoted by a representative element enclosed in brackets, such as

   $[a_{ij}], [b_{ij}], [c_{ij}], \ldots.$

3. A matrix can be denoted by a rectangular array of numbers

$$A = [a_{ij}] = \begin{bmatrix} a_{11} & a_{12} & a_{13} & \cdots & a_{1n} \\ a_{21} & a_{22} & a_{23} & \cdots & a_{2n} \\ a_{31} & a_{32} & a_{33} & \cdots & a_{3n} \\ \vdots & \vdots & \vdots & & \vdots \\ a_{m1} & a_{m2} & a_{m3} & \cdots & a_{mn} \end{bmatrix}.$$

As mentioned in Chapter 1, the matrices in this text are primarily *real matrices*. That is, their entries are real numbers.

Two matrices are said to be **equal** if their corresponding entries are equal.

Definition of Equality of Matrices

Two matrices $A = [a_{ij}]$ and $B = [b_{ij}]$ are **equal** if they have the same size $(m \times n)$ and

$$a_{ij} = b_{ij}$$

for $1 \le i \le m$ and $1 \le j \le n$.

EXAMPLE 1   *Equality of Matrices*

Consider the four matrices

$$A = \begin{bmatrix} 1 & 2 \\ 3 & 4 \end{bmatrix}, \qquad B = \begin{bmatrix} 1 \\ 3 \end{bmatrix},$$

$$C = \begin{bmatrix} 1 & 3 \end{bmatrix}, \qquad \text{and} \qquad D = \begin{bmatrix} 1 & 2 \\ x & 4 \end{bmatrix}.$$

Matrices A and B are **not** equal because they are of different sizes. Similarly, B and C are not equal. Matrices A and D are equal if and only if $x = 3$.

REMARK: The phrase "if and only if" means that the statement is true in both directions. For example, "p if and only if q" means that p implies q and q implies p.

A matrix that has only one column, such as matrix B in Example 1, is called a **column matrix** or **column vector.** Similarly, a matrix that has only one row, such as matrix C in Example 1, is called a **row matrix** or **row vector.** Boldface lowercase letters are often used to designate column matrices and row matrices. For instance, matrix A in Example 1 can be considered to be partitioned into the two column matrices $\mathbf{a}_1 = \begin{bmatrix} 1 \\ 3 \end{bmatrix}$ and $\mathbf{a}_2 = \begin{bmatrix} 2 \\ 4 \end{bmatrix}$ as follows.

$$A = \begin{bmatrix} 1 & 2 \\ 3 & 4 \end{bmatrix} = \begin{bmatrix} 1 & \vdots & 2 \\ 3 & \vdots & 4 \end{bmatrix} = \begin{bmatrix} \mathbf{a}_1 & \vdots & \mathbf{a}_2 \end{bmatrix}$$

Matrix Addition

You can **add** two matrices (of the same size) by adding their corresponding entries.

Definition of Matrix Addition

If $A = [a_{ij}]$ and $B = [b_{ij}]$ are matrices of size $m \times n$, then their **sum** is the $m \times n$ matrix given by

$$A + B = [a_{ij} + b_{ij}].$$

The sum of two matrices of different sizes is undefined.

EXAMPLE 2   *Addition of Matrices*

(a) $\begin{bmatrix} -1 & 2 \\ 0 & 1 \end{bmatrix} + \begin{bmatrix} 1 & 3 \\ -1 & 2 \end{bmatrix} = \begin{bmatrix} -1+1 & 2+3 \\ 0-1 & 1+2 \end{bmatrix} = \begin{bmatrix} 0 & 5 \\ -1 & 3 \end{bmatrix}$

(b) $\begin{bmatrix} 0 & 1 & -2 \\ 1 & 2 & 3 \end{bmatrix} + \begin{bmatrix} 0 & 0 & 0 \\ 0 & 0 & 0 \end{bmatrix} = \begin{bmatrix} 0 & 1 & -2 \\ 1 & 2 & 3 \end{bmatrix}$

(c) $\begin{bmatrix} 1 \\ -3 \\ -2 \end{bmatrix} + \begin{bmatrix} -1 \\ 3 \\ 2 \end{bmatrix} = \begin{bmatrix} 0 \\ 0 \\ 0 \end{bmatrix}$

(d) The sum of

$$A = \begin{bmatrix} 2 & 1 & 0 \\ 4 & 0 & -1 \\ 3 & -2 & 2 \end{bmatrix} \quad \text{and} \quad B = \begin{bmatrix} 0 & 1 \\ -1 & 3 \\ 2 & 4 \end{bmatrix}$$

is undefined.

Scalar Multiplication

When working with matrices, real numbers are referred to as **scalars.** You can multiply a matrix A by a scalar c by multiplying each entry in A by c.

Definition of Scalar Multiplication

If $A = [a_{ij}]$ is an $m \times n$ matrix and c is a scalar, then the **scalar multiple** of A by c is the $m \times n$ matrix given by

$$cA = [ca_{ij}].$$

You can use $-A$ to represent the scalar product $(-1)A$. If A and B are of the same size, $A - B$ represents the sum of A and $(-1)B$. That is,

$$A - B = A + (-1)B. \qquad \text{Subtraction of matrices}$$

EXAMPLE 3 *Scalar Multiplication and Matrix Subtraction*

For the matrices

$$A = \begin{bmatrix} 1 & 2 & 4 \\ -3 & 0 & -1 \\ 2 & 1 & 2 \end{bmatrix} \quad \text{and} \quad B = \begin{bmatrix} 2 & 0 & 0 \\ 1 & -4 & 3 \\ -1 & 3 & 2 \end{bmatrix}$$

find (a) $3A$, (b) $-B$, and (c) $3A - B$.

Solution

(a) $3A = 3\begin{bmatrix} 1 & 2 & 4 \\ -3 & 0 & -1 \\ 2 & 1 & 2 \end{bmatrix} = \begin{bmatrix} 3(1) & 3(2) & 3(4) \\ 3(-3) & 3(0) & 3(-1) \\ 3(2) & 3(1) & 3(2) \end{bmatrix} = \begin{bmatrix} 3 & 6 & 12 \\ -9 & 0 & -3 \\ 6 & 3 & 6 \end{bmatrix}$

(b) $-B = (-1)\begin{bmatrix} 2 & 0 & 0 \\ 1 & -4 & 3 \\ -1 & 3 & 2 \end{bmatrix} = \begin{bmatrix} -2 & 0 & 0 \\ -1 & 4 & -3 \\ 1 & -3 & -2 \end{bmatrix}$

(c) $3A - B = \begin{bmatrix} 3 & 6 & 12 \\ -9 & 0 & -3 \\ 6 & 3 & 6 \end{bmatrix} - \begin{bmatrix} 2 & 0 & 0 \\ 1 & -4 & 3 \\ -1 & 3 & 2 \end{bmatrix} = \begin{bmatrix} 1 & 6 & 12 \\ -10 & 4 & -6 \\ 7 & 0 & 4 \end{bmatrix}$

Matrix Multiplication

REMARK: It is often convenient to rewrite a matrix B as cA by factoring c out of every entry in matrix B. For instance, in the following example, the scalar $\frac{1}{2}$ has been factored out of the matrix.

$$\begin{bmatrix} \frac{1}{2} & -\frac{3}{2} \\ \frac{5}{2} & \frac{1}{2} \end{bmatrix} = \frac{1}{2}\begin{bmatrix} 1 & -3 \\ 5 & 1 \end{bmatrix}$$

$$B = cA$$

The third basic matrix operation is **matrix multiplication.** To see the usefulness of this operation, consider the following application in which matrices are helpful for organizing information.

A football stadium has three concession areas, located in the south, north, and west stands. The top-selling items are peanuts, hot dogs, and soda. Sales for a certain day are recorded in the first matrix below, and the price (in dollars) of each item is given in the second matrix.

	Number of Items Sold			Selling Price	
	Peanuts	*Hot Dogs*	*Soda*		
South stand	120	250	305	2.00	*Peanuts*
North stand	207	140	419	3.00	*Hot Dogs*
West stand	29	120	190	2.75	*Soda*

To calculate the total sales of the three top-selling items at the south stand, you can multiply each entry in the first row of the matrix on the left by the corresponding entry in the price column matrix on the right and add the results. That is, the south stand sales are

$$(120)(2.00) + (250)(3.00) + (305)(2.75) = \$1828.75.$$ South stand sales

Similarly, you can calculate the sales for the other two stands as follows.

$$(207)(2.00) + (140)(3.00) + (419)(2.75) = \$1986.25$$ North stand sales
$$(29)(2.00) + (120)(3.00) + (190)(2.75) = \$940.50$$ West stand sales

The preceding computation is an example of matrix multiplication. You can write the product of the 3×3 matrix indicating the number of items sold and the 3×1 matrix indicating the selling prices as follows.

$$\begin{bmatrix} 120 & 250 & 305 \\ 207 & 140 & 419 \\ 29 & 120 & 190 \end{bmatrix} \begin{bmatrix} 2.00 \\ 3.00 \\ 2.75 \end{bmatrix} = \begin{bmatrix} 1828.75 \\ 1986.25 \\ 940.50 \end{bmatrix}$$

The product of these matrices is the 3×1 matrix giving the total sales for each of the three stands.

The following general definition of the product of two matrices is based on the ideas just developed. Although at first glance this definition may seem unusual, you will see that it has many practical applications.

Definition of Matrix Multiplication

If $A = [a_{ij}]$ is an $m \times n$ matrix and $B = [b_{ij}]$ is an $n \times p$ matrix, then the **product** AB is an $m \times p$ matrix

$$AB = [c_{ij}]$$

where

$$c_{ij} = \sum_{k=1}^{n} a_{ik}b_{kj} = a_{i1}b_{1j} + a_{i2}b_{2j} + a_{i3}b_{3j} + \cdots + a_{in}b_{nj}.$$

This definition means that the entry in the ith row and the jth column of the product AB is obtained by multiplying the entries in the ith row of A by the corresponding entries in the jth column of B and then adding the results. The following example illustrates this process.

EXAMPLE 4   *Finding the Product of Two Matrices*

Find the product AB, where

$$A = \begin{bmatrix} -1 & 3 \\ 4 & -2 \\ 5 & 0 \end{bmatrix} \quad \text{and} \quad B = \begin{bmatrix} -3 & 2 \\ -4 & 1 \end{bmatrix}.$$

Solution First note that the product AB is defined because A has size 3×2 and B has size 2×2. Moreover, the product AB has size 3×2 and will take the form

$$\begin{bmatrix} -1 & 3 \\ 4 & -2 \\ 5 & 0 \end{bmatrix} \begin{bmatrix} -3 & 2 \\ -4 & 1 \end{bmatrix} = \begin{bmatrix} c_{11} & c_{12} \\ c_{21} & c_{22} \\ c_{31} & c_{32} \end{bmatrix}.$$

To find c_{11} (the entry in the first row and first column of the product), multiply corresponding entries in the first row of A and the first column of B. That is,

$$c_{11} = (-1)(-3) + (3)(-4) = -9$$

$$\begin{bmatrix} -1 & 3 \\ 4 & -2 \\ 5 & 0 \end{bmatrix} \begin{bmatrix} -3 & 2 \\ -4 & 1 \end{bmatrix} = \begin{bmatrix} -9 & c_{12} \\ c_{21} & c_{22} \\ c_{31} & c_{32} \end{bmatrix}.$$

Similarly, to find c_{12}, multiply corresponding entries in the first row of A and the second column of B to obtain

$$c_{12} = (-1)(2) + (3)(1) = 1$$

$$\begin{bmatrix} -1 & 3 \\ 4 & -2 \\ 5 & 0 \end{bmatrix} \begin{bmatrix} -3 & 2 \\ -4 & 1 \end{bmatrix} = \begin{bmatrix} -9 & 1 \\ c_{21} & c_{22} \\ c_{31} & c_{32} \end{bmatrix}.$$

Continuing this pattern produces the following results.

$$\begin{aligned} c_{21} &= (4)(-3) + (-2)(-4) = -4 \\ c_{22} &= (4)(2) + (-2)(1) = 6 \\ c_{31} &= (5)(-3) + (0)(-4) = -15 \\ c_{32} &= (5)(2) + (0)(1) = 10 \end{aligned}$$

Thus the product is

$$AB = \begin{bmatrix} -1 & 3 \\ 4 & -2 \\ 5 & 0 \end{bmatrix} \begin{bmatrix} -3 & 2 \\ -4 & 1 \end{bmatrix} = \begin{bmatrix} -9 & 1 \\ -4 & 6 \\ -15 & 10 \end{bmatrix}.$$

Be sure you understand that for the product of two matrices to be defined, the number of columns of the first matrix must equal the number of rows of the second matrix. That is,

$$\begin{array}{ccc} A & B & = & AB. \\ m \times n & n \times p & & m \times p \end{array}$$

equal

size of AB

Thus, the product BA is not defined for matrices such as A and B in Example 4.

The general pattern for matrix multiplication is as follows. To obtain the element in the ith row and the jth column of the product AB, use the ith row of A and the jth column of B.

$$
\begin{bmatrix}
a_{11} & a_{12} & a_{13} & \cdots & a_{1n} \\
a_{21} & a_{22} & a_{23} & \cdots & a_{2n} \\
 & & \vdots & & \\
a_{i1} & a_{i2} & a_{i3} & \cdots & a_{in} \\
 & & \vdots & & \\
a_{m1} & a_{m2} & a_{m3} & \cdots & a_{mn}
\end{bmatrix}
\begin{bmatrix}
b_{11} & b_{12} & \cdots & b_{1j} & \cdots & b_{1p} \\
b_{21} & b_{22} & \cdots & b_{2j} & \cdots & b_{2p} \\
b_{31} & b_{32} & \cdots & b_{3j} & \cdots & b_{3p} \\
 & & & \vdots & & \\
b_{n1} & b_{n2} & \cdots & b_{nj} & \cdots & b_{np}
\end{bmatrix}
=
\begin{bmatrix}
c_{11} & c_{12} & \cdots & c_{1j} & \cdots & c_{1p} \\
c_{21} & c_{22} & \cdots & c_{2j} & \cdots & c_{2p} \\
 & & & \vdots & & \\
c_{i1} & c_{i2} & \cdots & c_{ij} & \cdots & c_{ip} \\
 & & & \vdots & & \\
c_{m1} & c_{m2} & \cdots & c_{mj} & \cdots & c_{mp}
\end{bmatrix}
$$

$$a_{i1}b_{1j} + a_{i2}b_{2j} + a_{i3}b_{3j} + \cdots + a_{in}b_{nj} = c_{ij}$$

DISCOVERY

Let

$$A = \begin{bmatrix} 1 & 2 \\ 3 & 4 \end{bmatrix} \quad and \quad B = \begin{bmatrix} 0 & 1 \\ 1 & 2 \end{bmatrix}.$$

Calculate $A + B$ and $B + A$.

In general, is the operation of matrix addition commutative? Now calculate AB and BA. Is matrix multiplication commutative?

EXAMPLE 5   *Matrix Multiplication*

(a) $\begin{bmatrix} 1 & 0 & 3 \\ 2 & -1 & -2 \end{bmatrix} \begin{bmatrix} -2 & 4 & 2 \\ 1 & 0 & 0 \\ -1 & 1 & -1 \end{bmatrix} = \begin{bmatrix} -5 & 7 & -1 \\ -3 & 6 & 6 \end{bmatrix}$

$\qquad\qquad 2 \times 3 \qquad\qquad\quad 3 \times 3 \qquad\qquad\quad 2 \times 3$

(b) $\begin{bmatrix} 3 & 4 \\ -2 & 5 \end{bmatrix} \begin{bmatrix} 1 & 0 \\ 0 & 1 \end{bmatrix} = \begin{bmatrix} 3 & 4 \\ -2 & 5 \end{bmatrix}$

$\qquad\quad 2 \times 2 \qquad\quad 2 \times 2 \qquad\quad 2 \times 2$

(c) $\begin{bmatrix} 1 & 2 \\ 1 & 1 \end{bmatrix} \begin{bmatrix} -1 & 2 \\ 1 & -1 \end{bmatrix} = \begin{bmatrix} 1 & 0 \\ 0 & 1 \end{bmatrix}$

$\qquad\quad 2 \times 2 \qquad\quad 2 \times 2 \qquad\quad 2 \times 2$

(d) $\begin{bmatrix} 1 & -2 & -3 \end{bmatrix} \begin{bmatrix} 2 \\ -1 \\ 1 \end{bmatrix} = \begin{bmatrix} 1 \end{bmatrix}$

$\qquad\quad 1 \times 3 \qquad\quad 3 \times 1 \quad 1 \times 1$

(e) $\begin{bmatrix} 2 \\ -1 \\ 1 \end{bmatrix} \begin{bmatrix} 1 & -2 & -3 \end{bmatrix} = \begin{bmatrix} 2 & -4 & -6 \\ -1 & 2 & 3 \\ 1 & -2 & -3 \end{bmatrix}$

$\quad\quad 3 \times 1 \quad\quad 1 \times 3 \quad\quad\quad\quad 3 \times 3$

REMARK: Note the difference between the two products in parts (d) and (e) of Example 5. Matrix multiplication is not, in general, commutative. That is, it is usually not true that the product AB is equal to the product BA. (See Section 2.2 for further discussion of the noncommutativity of matrix multiplication.)

Technology Note

Most graphing calculators and computer software can perform matrix addition, scalar multiplication, and matrix multiplication. If you are using a TI-86, your screens for Example 5(c) will look like the following:

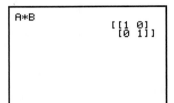

Systems of Linear Equations

One practical application of matrix multiplication is that of representing a system of linear equations. Note how the system

$$a_{11}x_1 + a_{12}x_2 + a_{13}x_3 = b_1$$
$$a_{21}x_1 + a_{22}x_2 + a_{23}x_3 = b_2$$
$$a_{31}x_1 + a_{32}x_2 + a_{33}x_3 = b_3$$

can be written as the matrix equation $A\mathbf{x} = \mathbf{b}$, where A is the coefficient matrix of the system, and $\mathbf{x}$ and $\mathbf{b}$ are column matrices. That is, you can write

$$\underbrace{\begin{bmatrix} a_{11} & a_{12} & a_{13} \\ a_{21} & a_{22} & a_{23} \\ a_{31} & a_{32} & a_{33} \end{bmatrix}}_{A} \underbrace{\begin{bmatrix} x_1 \\ x_2 \\ x_3 \end{bmatrix}}_{\mathbf{x}} = \underbrace{\begin{bmatrix} b_1 \\ b_2 \\ b_3 \end{bmatrix}}_{\mathbf{b}}.$$

EXAMPLE 6   *Solving a System of Linear Equations*

Solve the matrix equation $A\mathbf{x} = \mathbf{0}$, where

$$A = \begin{bmatrix} 1 & -2 & 1 \\ 2 & 3 & -2 \end{bmatrix}, \qquad \mathbf{x} = \begin{bmatrix} x_1 \\ x_2 \\ x_3 \end{bmatrix}, \qquad \text{and} \qquad \mathbf{0} = \begin{bmatrix} 0 \\ 0 \end{bmatrix}.$$

Solution   As a system of linear equations, $A\mathbf{x} = \mathbf{0}$ looks like

$$\begin{aligned} x_1 - 2x_2 + x_3 &= 0 \\ 2x_1 + 3x_2 - 2x_3 &= 0. \end{aligned}$$

Using Gauss-Jordan elimination on the augmented matrix of this system, you obtain

$$\begin{bmatrix} 1 & 0 & -\frac{1}{7} & 0 \\ 0 & 1 & -\frac{4}{7} & 0 \end{bmatrix}.$$

Thus, the system has an infinite number of solutions. Here a convenient choice of parameter is $x_3 = 7t$, and you can write the solution set as

$$x_1 = t, \qquad x_2 = 4t, \qquad x_3 = 7t, \quad t \text{ is any real number.}$$

In matrix terminology, you have found that the matrix equation

$$\begin{bmatrix} 1 & -2 & 1 \\ 2 & 3 & -2 \end{bmatrix} \begin{bmatrix} x_1 \\ x_2 \\ x_3 \end{bmatrix} = \begin{bmatrix} 0 \\ 0 \end{bmatrix}$$

has an infinite number of solutions represented by

$$\mathbf{x} = \begin{bmatrix} x_1 \\ x_2 \\ x_3 \end{bmatrix} = \begin{bmatrix} t \\ 4t \\ 7t \end{bmatrix} = t \begin{bmatrix} 1 \\ 4 \\ 7 \end{bmatrix}, \quad t \text{ is any scalar.}$$

That is, any scalar multiple of the column matrix on the right is a solution.

Partitioned Matrices

The system $A\mathbf{x} = \mathbf{b}$ can be represented in a more convenient way by partitioning the matrices A and $\mathbf{x}$ in the following manner. If

$$A = \begin{bmatrix} a_{11} & a_{12} & \cdots & a_{1n} \\ a_{21} & a_{22} & \cdots & a_{2n} \\ \vdots & \vdots & & \vdots \\ a_{m1} & a_{m2} & \cdots & a_{mn} \end{bmatrix}, \qquad \mathbf{x} = \begin{bmatrix} x_1 \\ x_2 \\ \vdots \\ x_n \end{bmatrix}, \qquad \text{and} \qquad \mathbf{b} = \begin{bmatrix} b_1 \\ b_2 \\ \vdots \\ b_m \end{bmatrix}$$

are the coefficient matrix, the column matrix of unknowns, and the right-hand side, respectively, of the $m \times n$ linear system $A\mathbf{x} = \mathbf{b}$, then you can write

$$A\mathbf{x} = \mathbf{b}$$

$$\begin{bmatrix} a_{11} & a_{12} & \cdots & a_{1n} \\ a_{21} & a_{22} & \cdots & a_{2n} \\ \vdots & \vdots & & \vdots \\ a_{m1} & a_{m2} & \cdots & a_{mn} \end{bmatrix} \begin{bmatrix} x_1 \\ x_2 \\ \vdots \\ x_n \end{bmatrix} = \mathbf{b}$$

$$\begin{bmatrix} a_{11}x_1 + a_{12}x_2 + \cdots + a_{1n}x_n \\ a_{21}x_1 + a_{22}x_2 + \cdots + a_{2n}x_n \\ \vdots \\ a_{m1}x_1 + a_{m2}x_2 + \cdots + a_{mn}x_n \end{bmatrix} = \mathbf{b}$$

$$x_1 \begin{bmatrix} a_{11} \\ a_{21} \\ \vdots \\ a_{m1} \end{bmatrix} + x_2 \begin{bmatrix} a_{12} \\ a_{22} \\ \vdots \\ a_{m2} \end{bmatrix} + \cdots + x_n \begin{bmatrix} a_{1n} \\ a_{2n} \\ \vdots \\ a_{mn} \end{bmatrix} = \mathbf{b}.$$

In other words,

$$A\mathbf{x} = x_1\mathbf{a}_1 + x_2\mathbf{a}_2 + \cdots + x_n\mathbf{a}_n = \mathbf{b},$$

where $\mathbf{a}_1, \mathbf{a}_2, \ldots, \mathbf{a}_n$ are the columns of the matrix A. The expression

$$x_1 \begin{bmatrix} a_{11} \\ a_{21} \\ \vdots \\ a_{m1} \end{bmatrix} + x_2 \begin{bmatrix} a_{12} \\ a_{22} \\ \vdots \\ a_{m2} \end{bmatrix} + \cdots + x_n \begin{bmatrix} a_{1n} \\ a_{2n} \\ \vdots \\ a_{mn} \end{bmatrix}$$

is called a **linear combination** of the column matrices $\mathbf{a}_1, \mathbf{a}_2, \ldots, \mathbf{a}_n$ with **coefficients** $x_1, x_2, \ldots, x_n$.

In general, the matrix product $A\mathbf{x}$ is a linear combination of the column vectors $\mathbf{a}_1, \mathbf{a}_2, \ldots, \mathbf{a}_n$ that form the coefficient matrix A. Furthermore, the system $A\mathbf{x} = \mathbf{b}$ is consistent if and only if $\mathbf{b}$ can be expressed as such a linear combination, where the coefficients of the linear combination are a solution of the system.

EXAMPLE 7   *Solving a System of Linear Equations*

The linear system

$$\begin{aligned} x_1 + 2x_2 + 3x_3 &= 0 \\ 4x_1 + 5x_2 + 6x_3 &= 3 \\ 7x_1 + 8x_2 + 9x_3 &= 6 \end{aligned}$$

Technology Note

Many real-life applications of linear systems involve enormous numbers of equations and unknowns. For example, a flight crew scheduling problem for American Airlines required the manipulation of matrices with 837 rows and more than 12,750,000 columns. This application of *linear programming* required that the problem be partitioned into smaller pieces and then solved on a CRAY Y-MP supercomputer.

(Source: Very Large-Scale Linear Programming: A Case Study in Combining Interior Point and Simplex Methods, Bixby, Robert E., et al., *Operations Research,* 40, no. 5, 1992.)

can be rewritten as a matrix equation $A\mathbf{x} = \mathbf{b}$ as follows.

$$x_1 \begin{bmatrix} 1 \\ 4 \\ 7 \end{bmatrix} + x_2 \begin{bmatrix} 2 \\ 5 \\ 8 \end{bmatrix} + x_3 \begin{bmatrix} 3 \\ 6 \\ 9 \end{bmatrix} = \begin{bmatrix} 0 \\ 3 \\ 6 \end{bmatrix}$$

Using Gaussian elimination, you can show that this system has an infinite number of solutions, one of which is $x_1 = 1$, $x_2 = 1$, $x_3 = -1$.

$$1 \begin{bmatrix} 1 \\ 4 \\ 7 \end{bmatrix} + 1 \begin{bmatrix} 2 \\ 5 \\ 8 \end{bmatrix} + (-1) \begin{bmatrix} 3 \\ 6 \\ 9 \end{bmatrix} = \begin{bmatrix} 0 \\ 3 \\ 6 \end{bmatrix}$$

That is, $\mathbf{b}$ can be expressed as a linear combination of the columns of A. This representation of one column vector in terms of others is a fundamental theme of linear algebra.

Just as you partitioned A into columns and $\mathbf{x}$ into rows, it is often useful to consider an $m \times n$ matrix partitioned into smaller matrices. For example, the matrix below at the left can be partitioned as shown below at the right.

$$\begin{bmatrix} 1 & 2 & 0 & 0 \\ 3 & 4 & 0 & 0 \\ -1 & -2 & 2 & 1 \end{bmatrix} \qquad \left[\begin{array}{cc|cc} 1 & 2 & 0 & 0 \\ 3 & 4 & 0 & 0 \\ \hline -1 & -2 & 2 & 1 \end{array} \right]$$

The matrix could also be partitioned into column matrices

$$\left[\begin{array}{c|c|c|c} 1 & 2 & 0 & 0 \\ 3 & 4 & 0 & 0 \\ -1 & -2 & 2 & 1 \end{array} \right] = \begin{bmatrix} \mathbf{c}_1 & \mathbf{c}_2 & \mathbf{c}_3 & \mathbf{c}_4 \end{bmatrix}$$

or row matrices

$$\left[\begin{array}{cccc} 1 & 2 & 0 & 0 \\ \hline 3 & 4 & 0 & 0 \\ \hline -1 & -2 & 2 & 1 \end{array} \right] = \begin{bmatrix} \mathbf{r}_1 \\ \mathbf{r}_2 \\ \mathbf{r}_3 \end{bmatrix}.$$

SECTION 2.1   Exercises

In Exercises 1–4, find (a) $A + B$, (b) $A - B$, (c) $2A$, and (d) $2A - B$.

1. $A = \begin{bmatrix} 1 & -1 \\ 2 & -1 \end{bmatrix}$, $B = \begin{bmatrix} 2 & -1 \\ -1 & 8 \end{bmatrix}$

2. $A = \begin{bmatrix} 1 & 2 \\ 2 & 1 \end{bmatrix}$, $B = \begin{bmatrix} -3 & -2 \\ 4 & 2 \end{bmatrix}$

3. $A = \begin{bmatrix} 6 & -1 \\ 2 & 4 \\ -3 & 5 \end{bmatrix}$, $B = \begin{bmatrix} 1 & 4 \\ -1 & 5 \\ 1 & 10 \end{bmatrix}$

4. $A = \begin{bmatrix} 2 & 1 & 1 \\ -1 & -1 & 4 \end{bmatrix}$, $B = \begin{bmatrix} 2 & -3 & 4 \\ -3 & 1 & -2 \end{bmatrix}$

5. Find (a) c_{21} and (b) c_{13}, where $C = 2A - 3B$,

$$A = \begin{bmatrix} 5 & 4 & 4 \\ -3 & 1 & 2 \end{bmatrix}, \quad \text{and} \quad B = \begin{bmatrix} 1 & 2 & -7 \\ 0 & -5 & 1 \end{bmatrix}.$$

6. Find (a) c_{23} and (b) c_{32}, where $C = 5A + 2B$,

$$A = \begin{bmatrix} 4 & 11 & -9 \\ 0 & 3 & 2 \\ -3 & 1 & 1 \end{bmatrix}, \quad \text{and} \quad B = \begin{bmatrix} 1 & 0 & 5 \\ -4 & 6 & 11 \\ -6 & 4 & 9 \end{bmatrix}.$$

7. Solve for x, y, and z in the matrix equation

$$4\begin{bmatrix} x & y \\ z & -1 \end{bmatrix} = 2\begin{bmatrix} y & z \\ -x & 1 \end{bmatrix} + 2\begin{bmatrix} 4 & x \\ 5 & -x \end{bmatrix}.$$

8. Solve for x, y, z, and w in the matrix equation

$$\begin{bmatrix} w & x \\ y & x \end{bmatrix} = \begin{bmatrix} -4 & 3 \\ 2 & -1 \end{bmatrix} + 2\begin{bmatrix} y & w \\ z & x \end{bmatrix}.$$

In Exercises 9–18, find (a) AB and (b) BA (if they are defined).

9. $A = \begin{bmatrix} 1 & 2 \\ 4 & 2 \end{bmatrix}$, $B = \begin{bmatrix} 2 & -1 \\ -1 & 8 \end{bmatrix}$

10. $A = \begin{bmatrix} 2 & -1 \\ 1 & 4 \end{bmatrix}$, $B = \begin{bmatrix} 0 & 0 \\ 3 & -3 \end{bmatrix}$

11. $A = \begin{bmatrix} 1 & -1 & 7 \\ 2 & -1 & 8 \\ 3 & 1 & -1 \end{bmatrix}$, $B = \begin{bmatrix} 1 & 1 & 2 \\ 2 & 1 & 1 \\ 1 & -3 & 2 \end{bmatrix}$

12. $A = \begin{bmatrix} 3 & 2 & 1 \end{bmatrix}$, $B = \begin{bmatrix} 2 \\ 3 \\ 0 \end{bmatrix}$

13. $A = \begin{bmatrix} 2 & 1 \\ -3 & 4 \\ 1 & 6 \end{bmatrix}$, $B = \begin{bmatrix} 0 & -1 & 0 \\ 4 & 0 & 2 \\ 8 & -1 & 7 \end{bmatrix}$

14. $A = \begin{bmatrix} 0 & -1 & 0 \\ 4 & 0 & 2 \\ 8 & -1 & 7 \end{bmatrix}$, $B = \begin{bmatrix} 2 \\ -3 \\ 1 \end{bmatrix}$

15. $A = \begin{bmatrix} -1 & 3 \\ 4 & -5 \\ 0 & 2 \end{bmatrix}$, $B = \begin{bmatrix} 1 & 2 \\ 0 & 7 \end{bmatrix}$

16. $A = \begin{bmatrix} 5 & 0 & 0 \\ 0 & -8 & 0 \\ 0 & 0 & 7 \end{bmatrix}$, $B = \begin{bmatrix} \frac{1}{5} & 0 & 0 \\ 0 & -\frac{1}{8} & 0 \\ 0 & 0 & \frac{1}{7} \end{bmatrix}$

17. $A = \begin{bmatrix} 6 \\ -2 \\ 1 \\ 6 \end{bmatrix}$, $B = \begin{bmatrix} 10 & 12 \end{bmatrix}$

18. $A = \begin{bmatrix} 1 & 0 & 3 & -2 & 4 \\ 6 & 13 & 8 & -17 & 20 \end{bmatrix}$, $B = \begin{bmatrix} 1 & 6 \\ 4 & 2 \end{bmatrix}$

In Exercises 19 and 20, find (a) $2A + B$, (b) $3B - A$, (c) AB, and (d) BA (if they are defined).

19. $A = \begin{bmatrix} 2 & -2 & 4 & 1 & 0 & 3 \\ -1 & 4 & 2 & -2 & -1 & 3 \\ 3 & -3 & 1 & 2 & 3 & -4 \\ 2 & -1 & 3 & 0 & 1 & 2 \\ 5 & 1 & -2 & -4 & 1 & 3 \\ 2 & 2 & 3 & -4 & -1 & -2 \end{bmatrix}$

$B = \begin{bmatrix} 1 & 2 & -3 & 4 & 1 & 2 \\ 2 & -3 & 1 & 3 & -1 & 2 \\ 0 & -2 & -3 & 0 & 1 & -1 \\ 1 & 2 & 3 & 2 & 1 & -1 \\ 2 & -1 & -3 & 0 & 4 & 2 \\ 1 & -2 & 4 & -2 & -4 & -1 \end{bmatrix}$

20. $A = \begin{bmatrix} 2 & 1 & 3 & 2 & -1 \\ 3 & -1 & 0 & 1 & 2 \\ 2 & 1 & -3 & 3 & -2 \\ -4 & 0 & 2 & -3 & 1 \\ 1 & 0 & -1 & 2 & 4 \\ 2 & -3 & 2 & 1 & -4 \end{bmatrix}$

$B = \begin{bmatrix} 5 & 2 & 1 & 3 & 2 & 1 \\ -4 & -2 & 2 & -1 & 3 & -1 \\ 4 & 0 & 1 & 3 & -2 & 1 \\ -1 & 2 & -3 & -1 & 2 & 3 \\ -2 & 1 & 4 & 3 & -2 & 2 \\ 1 & -2 & 3 & 4 & -2 & -1 \end{bmatrix}$

In Exercises 21–26, write the system of linear equations in the form $A\mathbf{x} = \mathbf{b}$ and solve this matrix equation for $\mathbf{x}$.

21. $\begin{aligned} x_1 + 3x_2 &= -1 \\ 2x_1 - x_2 &= 3 \end{aligned}$

22. $\begin{aligned} 2x_1 + 2x_2 &= 7 \\ -6x_1 - 6x_2 &= -21 \end{aligned}$

23. $\begin{aligned} 8x_1 - 8x_2 &= 0 \\ -3x_1 + 2x_2 &= 0 \end{aligned}$

24. $\begin{aligned} 6x_2 + 4x_3 &= -12 \\ 3x_1 + 3x_2 &= 9 \\ 2x_1 - 3x_3 &= 10 \end{aligned}$

25. $\begin{aligned} x_1 + 2x_3 &= 5 \\ 3x_1 - 2x_2 + x_3 &= 8 \\ -2x_1 + 2x_2 - x_3 &= -3 \end{aligned}$

26. $\begin{aligned} x_1 + 2x_3 &= 4 \\ 2x_1 - 4x_2 &= -6 \\ 3x_1 + 2x_2 + 5x_3 &= 10 \end{aligned}$

In Exercises 27 and 28, solve the matrix equation for A.

27. $\begin{bmatrix} 1 & 2 \\ 3 & 5 \end{bmatrix} A = \begin{bmatrix} 1 & 0 \\ 0 & 1 \end{bmatrix}$

28. $\begin{bmatrix} 2 & -1 \\ 3 & -2 \end{bmatrix} A = \begin{bmatrix} 1 & 0 \\ 0 & 1 \end{bmatrix}$

In Exercises 29 and 30, solve the matrix equation for a, b, c, and d.

29. $\begin{bmatrix} 1 & 2 \\ 3 & 4 \end{bmatrix} \begin{bmatrix} a & b \\ c & d \end{bmatrix} = \begin{bmatrix} 6 & 3 \\ 19 & 2 \end{bmatrix}$

30. $\begin{bmatrix} a & b \\ c & d \end{bmatrix} \begin{bmatrix} 2 & 1 \\ 3 & 1 \end{bmatrix} = \begin{bmatrix} 3 & 17 \\ 4 & -1 \end{bmatrix}$

31. Find conditions on w, x, y, and z such that $AB = BA$ for the following matrices.

$$A = \begin{bmatrix} w & x \\ y & z \end{bmatrix} \quad \text{and} \quad B = \begin{bmatrix} 1 & 1 \\ -1 & 1 \end{bmatrix}$$

32. Verify that $AB = BA$ for the following matrices.

$$A = \begin{bmatrix} \cos\alpha & -\sin\alpha \\ \sin\alpha & \cos\alpha \end{bmatrix} \quad \text{and} \quad B = \begin{bmatrix} \cos\beta & -\sin\beta \\ \sin\beta & \cos\beta \end{bmatrix}$$

A square matrix

$$A = \begin{bmatrix} a_{11} & 0 & 0 & \cdots & 0 \\ 0 & a_{22} & 0 & \cdots & 0 \\ 0 & 0 & a_{33} & \cdots & 0 \\ \vdots & \vdots & \vdots & & \vdots \\ 0 & 0 & \cdots & 0 & a_{nn} \end{bmatrix}$$

is called a **diagonal matrix** if all entries that are not on the main diagonal are zero. In Exercises 33 and 34, find the product AA for the diagonal matrix.

33. $A = \begin{bmatrix} -1 & 0 & 0 \\ 0 & 2 & 0 \\ 0 & 0 & 3 \end{bmatrix}$ **34.** $A = \begin{bmatrix} 2 & 0 & 0 \\ 0 & -3 & 0 \\ 0 & 0 & 0 \end{bmatrix}$

In Exercises 35 and 36, find the products AB and BA for the diagonal matrices.

35. $A = \begin{bmatrix} 2 & 0 \\ 0 & -3 \end{bmatrix}$, $B = \begin{bmatrix} -5 & 0 \\ 0 & 4 \end{bmatrix}$

36. $A = \begin{bmatrix} 3 & 0 & 0 \\ 0 & -5 & 0 \\ 0 & 0 & 0 \end{bmatrix}$, $B = \begin{bmatrix} -7 & 0 & 0 \\ 0 & 4 & 0 \\ 0 & 0 & 12 \end{bmatrix}$

37. *Guided Proof*   Prove that if A and B are diagonal matrices (of the same size), then $AB = BA$.

Getting Started: To prove that the two matrices AB and BA are equal, you need to show that their corresponding entries are equal.

   (i) Begin your proof by letting $A = [a_{ij}]$ and $B = [b_{ij}]$ be two diagonal $n \times n$ matrices.

   (ii) The ijth entry of the product AB is $c_{ij} = \displaystyle\sum_{k=1}^{n} a_{ik}b_{kj}$.

   (iii) Evaluate the entries c_{ij} for the two cases $i \neq j$ and $i = j$.

   (iv) Repeat this analysis for the product BA.

38. *Writing*   Let A and B be 3×3 matrices, where A is diagonal.

   (a) Describe the product AB. Illustrate your answer with examples.

   (b) Describe the product BA. Illustrate your answer with examples.

   (c) How do the results in parts (a) and (b) change if the diagonal entries of A are all equal?

In Exercises 39 and 40, find the trace of the matrix. The **trace** of an $n \times n$ matrix A is the sum of the main diagonal entries. That is, $\text{Tr}(A) = a_{11} + a_{22} + \cdots + a_{nn}$.

39. $\begin{bmatrix} 1 & 2 & 3 \\ 0 & -2 & 4 \\ 3 & 1 & 3 \end{bmatrix}$ **40.** $\begin{bmatrix} 1 & 0 & 0 \\ 0 & 1 & 0 \\ 0 & 0 & 1 \end{bmatrix}$

41. Prove that each statement is true if A and B are square matrices of order n and c is a scalar.

   (a) $\text{Tr}(A + B) = \text{Tr}(A) + \text{Tr}(B)$

   (b) $\text{Tr}(cA) = c\text{Tr}(A)$

42. Prove that if A and B are square matrices of order n, then $\text{Tr}(AB) = \text{Tr}(BA)$.

43. Show that the matrix equation has no solution.

$$\begin{bmatrix} 1 & 1 \\ 1 & 1 \end{bmatrix} A = \begin{bmatrix} 1 & 0 \\ 0 & 1 \end{bmatrix}$$

44. Show that no 2×2 matrices A and B exist that satisfy the matrix equation

$$AB - BA = \begin{bmatrix} 1 & 0 \\ 0 & 1 \end{bmatrix}.$$

45. Let $i = \sqrt{-1}$ and let $A = \begin{bmatrix} i & 0 \\ 0 & i \end{bmatrix}$ and $B = \begin{bmatrix} 0 & -i \\ i & 0 \end{bmatrix}$.

   (a) Find A^2, A^3, and A^4. Identify any similarities with i^2, i^3, and i^4.

   (b) Find and identify B^2.

46. *Guided Proof*   Prove that if the product AB is a square matrix, then the product BA is defined.

Getting Started: To prove that the product BA is defined, you need to show that the number of columns of B equals the number of rows of A.

   (i) Begin your proof by noting that the number of columns of A equals the number of rows of B.

   (ii) Hence, you can assume that A has size $m \times n$ and B has size $n \times p$.

   (iii) Use the hypothesis that the product AB is a square matrix.

47. Prove that if both products AB and BA are defined, then AB and BA are square matrices.

48. Let A and B be two matrices such that the product AB is defined. Show that if A has two identical rows, then the corresponding two rows of AB are also identical.

49. Let A and B be $n \times n$ matrices. Show that if the ith row of A has all zero entries, then the ith row of AB will have all zero entries. Give an example using 2×2 matrices to show that the converse is not true.

50. The columns of matrix T show the coordinates of the vertices of a triangle. Matrix A is a transformation matrix.

$$A = \begin{bmatrix} 0 & -1 \\ 1 & 0 \end{bmatrix} \qquad T = \begin{bmatrix} 1 & 2 & 3 \\ 1 & 4 & 2 \end{bmatrix}$$

(a) Find AT and AAT. Then sketch the original triangle and the two transformed triangles. What transformation does A represent?

(b) Given the triangle determined by AAT, describe the transformation process that produces the triangle determined by AT and then the triangle determined by T.

True or False? In Exercises 51 and 52, determine whether each statement is true or false. If a statement is true, give a reason or cite an appropriate statement in the text. If a statement is false, provide an example that shows that the statement is not true in all cases or cite an appropriate statement in the text.

51. (a) For the product of two matrices to be defined, the number of columns of the first matrix must equal the number of rows of the second matrix.

(b) The system $A\mathbf{x} = \mathbf{b}$ is consistent if and only if $\mathbf{b}$ can be expressed as a linear combination, where the coefficients of the linear combination are a solution of the system.

52. (a) If A is an $m \times n$ matrix and B is an $n \times r$ matrix, then the product AB is an $m \times r$ matrix.

(b) The matrix equation $A\mathbf{x} = \mathbf{b}$, where A is the coefficient matrix and $\mathbf{x}$ and $\mathbf{b}$ are column matrices, can be used to represent a system of linear equations.

53. A certain corporation has four factories. Each produces two products. The number of units of Product i produced at Factory j in 1 day is represented by a_{ij} in the matrix

$$A = \begin{bmatrix} 100 & 90 & 70 & 30 \\ 40 & 20 & 60 & 60 \end{bmatrix}.$$

Use scalar multiplication (multiply by 1.10) to determine what the production levels would be if production were increased by 10%.

54. A company manufactures tables and chairs at two locations. Matrix C gives the total cost for manufacturing each product at each location.

$$C = \begin{array}{c} \\ Table \\ Chair \end{array} \begin{bmatrix} \overset{Location\ 1}{627} & \overset{Location\ 2}{681} \\ 135 & 150 \end{bmatrix}$$

(a) Given that labor accounts for about $\frac{2}{3}$ of the total cost, determine the matrix L that gives the labor costs for each product at each location. What matrix operation did you use?

(b) Find the matrix M that gives material costs for each product at each location. (Assume there are only labor and material costs.)

55. *Writing* A fruit grower raises two crops, which are shipped to three outlets. The number of units of Product i that are shipped to Outlet j is represented by a_{ij} in the matrix

$$A = \begin{bmatrix} 100 & 75 & 75 \\ 125 & 150 & 100 \end{bmatrix}.$$

The profit on one unit of Product i is represented by b_{1i} in the matrix

$$B = [\$3.75 \quad \$7.00].$$

Find the matrix product BA, and explain what each entry of this product represents.

56. *Writing* The matrix

$$P = \begin{array}{c} From\ R \\ From\ D \\ From\ I \end{array} \begin{bmatrix} \overset{To\ R}{0.75} & \overset{To\ D}{0.15} & \overset{To\ I}{0.10} \\ 0.20 & 0.60 & 0.20 \\ 0.30 & 0.40 & 0.30 \end{bmatrix}$$

represents the proportion of a voting population that changes from Party i to Party j in a given election. That is, $p_{ij}(i \neq j)$ represents the proportion of the voting population that changes from Party i to Party j, and p_{ii} represents the proportion that remains loyal to Party i from one election to the next. Find the product of P with itself. What does this product represent?

57. The matrices on the following page show the number of people (in thousands) who lived in each region of the United States in 2000 and the number of people (in thousands) projected to live in each region in 2015. The regional populations are separated into three age categories. (Source: U.S. Census Bureau)

	2000		
	0–17	*18–64*	*65+*
Northeast	13,049	33,175	7,372
Midwest	16,646	39,386	8,263
South	25,569	62,235	12,437
Mountain	4,935	11,210	2,031
Pacific	12,098	28,036	4,893

	2015		
	0–17	*18–64*	*65+*
Northeast	12,589	34,081	8,165
Midwest	15,886	41,038	10,101
South	25,916	68,998	17,470
Mountain	5,226	12,626	3,270
Pacific	14,906	33,296	6,565

(a) The total population in 2000 was 281,335,000 and the projected total population in 2015 is 310,133,000. Rewrite the matrices to give the information as percents of the total population.

(b) Write a matrix that gives the projected change in the percent of the population in each region and age group from 2000 to 2015.

(c) Based on the result of part (b), which region(s) and age group(s) are projected to show relative growth from 2000 to 2015?

58. **Writing**   Given an $m \times n$ matrix A and an $n \times p$ matrix B, describe step by step how you can form their product column by column.

$$AB = A[\mathbf{b}_1 \ \mathbf{b}_2 \ \ldots \ \mathbf{b}_p] = [A\mathbf{b}_1 \ A\mathbf{b}_2 \ \ldots \ A\mathbf{b}_p]$$

Similarly, describe how you can form their product row by row.

$$AB = \begin{bmatrix} \mathbf{a}_1 \\ \mathbf{a}_2 \\ \vdots \\ \mathbf{a}_m \end{bmatrix} B = \begin{bmatrix} \mathbf{a}_1 B \\ \mathbf{a}_2 B \\ \vdots \\ \mathbf{a}_m B \end{bmatrix}$$

In Exercises 59 and 60, perform the indicated block multiplication of matrices A and B. If matrices A and B have each been partitioned into four submatrices

$$A = \begin{bmatrix} A_{11} & A_{12} \\ A_{21} & A_{22} \end{bmatrix} \quad \text{and} \quad B = \begin{bmatrix} B_{11} & B_{12} \\ B_{21} & B_{22} \end{bmatrix},$$

then you can **block multiply** A and B provided the sizes of the submatrices are such that the matrix multiplications and additions are defined.

$$AB = \begin{bmatrix} A_{11} & A_{12} \\ A_{21} & A_{22} \end{bmatrix} \begin{bmatrix} B_{11} & B_{12} \\ B_{21} & B_{22} \end{bmatrix}$$

$$= \begin{bmatrix} A_{11}B_{11} + A_{12}B_{21} & A_{11}B_{12} + A_{12}B_{22} \\ A_{21}B_{11} + A_{22}B_{21} & A_{21}B_{12} + A_{22}B_{22} \end{bmatrix}$$

59. $A = \begin{bmatrix} 1 & 2 & 0 & 0 \\ 0 & 1 & 0 & 0 \\ 0 & 0 & 2 & 1 \end{bmatrix}$, $B = \begin{bmatrix} 1 & 2 & 0 \\ -1 & 1 & 0 \\ 0 & 0 & 1 \\ 0 & 0 & 3 \end{bmatrix}$

60. $A = \begin{bmatrix} 0 & 0 & 1 & 0 \\ 0 & 0 & 0 & 1 \\ -1 & 0 & 0 & 0 \\ 0 & -1 & 0 & 0 \end{bmatrix}$, $B = \begin{bmatrix} 1 & 2 & 3 & 4 \\ 5 & 6 & 7 & 8 \\ 1 & 2 & 3 & 4 \\ 5 & 6 & 7 & 8 \end{bmatrix}$

In Exercises 61–64, express the column matrix $\mathbf{b}$ as a linear combination of the columns of A.

61. $A = \begin{bmatrix} 1 & -1 & 2 \\ 3 & -3 & 1 \end{bmatrix}$, $\mathbf{b} = \begin{bmatrix} -1 \\ 7 \end{bmatrix}$

62. $A = \begin{bmatrix} 1 & 2 & 4 \\ -1 & 0 & 2 \\ 0 & 1 & 3 \end{bmatrix}$, $\mathbf{b} = \begin{bmatrix} 1 \\ 3 \\ 2 \end{bmatrix}$

63. $A = \begin{bmatrix} 1 & 1 & -5 \\ 1 & 0 & -1 \\ 2 & -1 & -1 \end{bmatrix}$, $\mathbf{b} = \begin{bmatrix} 3 \\ 1 \\ 0 \end{bmatrix}$

64. $A = \begin{bmatrix} -3 & 5 \\ 3 & 4 \\ 4 & -8 \end{bmatrix}$, $\mathbf{b} = \begin{bmatrix} -22 \\ 4 \\ 32 \end{bmatrix}$

2.2 | Properties of Matrix Operations

In Section 2.1 you concentrated on the mechanics of the three basic matrix operations: matrix addition, scalar multiplication, and matrix multiplication. This section begins to develop the **algebra of matrices.** You will see that this algebra shares many (but not all) of the properties of the algebra of real numbers. Several properties of matrix addition and scalar multiplication are listed below.

Theorem 2.1	*Properties of Matrix Addition and Scalar Multiplication*

If A, B, and C are $m \times n$ matrices and c and d are scalars, then the following properties are true.

1. $A + B = B + A$ **Commutative property of addition**
2. $A + (B + C) = (A + B) + C$ **Associative property of addition**
3. $(cd)A = c(dA)$
4. $1A = A$
5. $c(A + B) = cA + cB$ **Distributive property**
6. $(c + d)A = cA + dA$ **Distributive property**

Proof   The proofs of these six properties follow directly from the definitions of matrix addition and scalar multiplication, and the corresponding properties of real numbers. For example, to prove the commutative property of *matrix addition*, let $A = [a_{ij}]$ and $B = [b_{ij}]$. Then, using the commutative property of *addition of real numbers*, write

$$A + B = [a_{ij} + b_{ij}] = [b_{ij} + a_{ij}] = B + A.$$

Similarly, to prove Property 5, use the distributive property (for real numbers) of multiplication over addition to write

$$c(A + B) = [c(a_{ij} + b_{ij})] = [ca_{ij} + cb_{ij}] = cA + cB.$$

The proofs of the remaining four properties are left as exercises. (See Exercises 43–46.)

In the preceding section, matrix addition was defined as the sum of *two* matrices, making it a binary operation. The associative property of matrix addition now allows you to write expressions such as $A + B + C$ as $(A + B) + C$ or as $A + (B + C)$. This same reasoning applies to sums of four or more matrices.

EXAMPLE 1   *Addition of More Than Two Matrices*

By adding corresponding entries, you can obtain the following sum of four matrices.

$$\begin{bmatrix} 1 \\ 2 \\ -3 \end{bmatrix} + \begin{bmatrix} -1 \\ -1 \\ 2 \end{bmatrix} + \begin{bmatrix} 0 \\ 1 \\ 4 \end{bmatrix} + \begin{bmatrix} 2 \\ -3 \\ -2 \end{bmatrix} = \begin{bmatrix} 2 \\ -1 \\ 1 \end{bmatrix}$$

One important property of the addition of real numbers is that the number 0 serves as the additive identity. That is, $c + 0 = c$ for any real number c. For matrices, a similar property holds. Specifically, if A is an $m \times n$ matrix and O_{mn} is the $m \times n$ matrix consisting entirely of zeros, then $A + O_{mn} = A$. The matrix O_{mn} is called a **zero matrix,** and it serves as the **additive identity** for the set of all $m \times n$ matrices. For example, the following matrix serves as the additive identity for the set of all 2×3 matrices.

$$O_{23} = \begin{bmatrix} 0 & 0 & 0 \\ 0 & 0 & 0 \end{bmatrix}$$

When the size of the matrix is understood, you may denote a zero matrix simply by O.

The following properties of zero matrices are easy to prove, and their proofs are left as an exercise. (See Exercise 47.)

Theorem 2.2	*Properties of Zero Matrices*

If A is an $m \times n$ matrix and c is a scalar, then the following properties are true.
1. $A + O_{mn} = A$
2. $A + (-A) = O_{mn}$
3. If $cA = O_{mn}$, then $c = 0$ or $A = O_{mn}$.

REMARK: Property 2 can be described by saying that the matrix $-A$ is the **additive inverse** of A.

The algebra of real numbers and the algebra of matrices have many similarities. For example, compare the following solutions.

Real Numbers	*$m \times n$ Matrices*
(Solve for x.)	*(Solve for X.)*
$x + a = b$	$X + A = B$
$x + a + (-a) = b + (-a)$	$X + A + (-A) = B + (-A)$
$x + 0 = b - a$	$X + O = B - A$
$x = b - a$	$X = B - A$

The process of solving a matrix equation is demonstrated in Example 2.

EXAMPLE 2   *Solving a Matrix Equation*

Solve for X in the equation $3X + A = B$, where

$$A = \begin{bmatrix} 1 & -2 \\ 0 & 3 \end{bmatrix} \quad \text{and} \quad B = \begin{bmatrix} -3 & 4 \\ 2 & 1 \end{bmatrix}.$$

Solution   Begin by solving the given equation for X to obtain

$$3X = B - A \quad \longrightarrow \quad X = \tfrac{1}{3}(B - A).$$

Now, using the given matrices A and B, you have

$$X = \tfrac{1}{3}\left(\begin{bmatrix} -3 & 4 \\ 2 & 1 \end{bmatrix} - \begin{bmatrix} 1 & -2 \\ 0 & 3 \end{bmatrix} \right) = \tfrac{1}{3}\begin{bmatrix} -4 & 6 \\ 2 & -2 \end{bmatrix} = \begin{bmatrix} -\tfrac{4}{3} & 2 \\ \tfrac{2}{3} & -\tfrac{2}{3} \end{bmatrix}.$$

Properties of Matrix Multiplication

In the next theorem, the algebra of matrices is extended to include some useful properties of matrix multiplication. The proof of Property 2 is given below. The proofs of the remaining properties are left as an exercise. (See Exercise 48.)

Theorem 2.3

Properties of Matrix Multiplication

If A, B, and C are matrices (with sizes such that the given matrix products are defined) and c is a scalar, then the following properties are true.
1. $A(BC) = (AB)C$ **Associative property of multiplication**
2. $A(B + C) = AB + AC$ **Distributive property**
3. $(A + B)C = AC + BC$ **Distributive property**
4. $c(AB) = (cA)B = A(cB)$

Proof
To prove Property 2, show that the matrices $A(B + C)$ and $AB + AC$ are equal by showing that their corresponding entries are equal. Assume that A has size $m \times n$, B has size $n \times p$, and C has size $n \times p$. Using the definition of matrix multiplication, the entry in the ith row and jth column of $A(B + C)$ is

$$a_{i1}(b_{1j} + c_{1j}) + \cdots + a_{in}(b_{nj} + c_{nj}).$$

Moreover, the entry in the ith row and jth column of $AB + AC$ is

$$(a_{i1}b_{1j} + \cdots + a_{in}b_{nj}) + (a_{i1}c_{1j} + \cdots + a_{in}c_{nj}).$$

By distributing and regrouping, you can see that these two ijth entries are equal. So,

$$A(B + C) = AB + AC.$$

The associative property of matrix multiplication permits you to write such matrix products as ABC without ambiguity, as demonstrated in Example 3.

EXAMPLE 3   *Matrix Multiplication Is Associative*

Find the matrix product ABC by grouping the factors first as $(AB)C$ and then as $A(BC)$, and show that the same result is obtained.

$$A = \begin{bmatrix} 1 & -2 \\ 2 & -1 \end{bmatrix}, \quad B = \begin{bmatrix} 1 & 0 & 2 \\ 3 & -2 & 1 \end{bmatrix}, \quad C = \begin{bmatrix} -1 & 0 \\ 3 & 1 \\ 2 & 4 \end{bmatrix}$$

Solution   Grouping the factors as $(AB)C$, you have

$$(AB)C = \left(\begin{bmatrix} 1 & -2 \\ 2 & -1 \end{bmatrix} \begin{bmatrix} 1 & 0 & 2 \\ 3 & -2 & 1 \end{bmatrix} \right) \begin{bmatrix} -1 & 0 \\ 3 & 1 \\ 2 & 4 \end{bmatrix}$$

$$= \begin{bmatrix} -5 & 4 & 0 \\ -1 & 2 & 3 \end{bmatrix} \begin{bmatrix} -1 & 0 \\ 3 & 1 \\ 2 & 4 \end{bmatrix}$$

$$= \begin{bmatrix} 17 & 4 \\ 13 & 14 \end{bmatrix}.$$

Grouping the factors as $A(BC)$, you obtain the same result.

$$A(BC) = \begin{bmatrix} 1 & -2 \\ 2 & -1 \end{bmatrix} \left(\begin{bmatrix} 1 & 0 & 2 \\ 3 & -2 & 1 \end{bmatrix} \begin{bmatrix} -1 & 0 \\ 3 & 1 \\ 2 & 4 \end{bmatrix} \right)$$

$$= \begin{bmatrix} 1 & -2 \\ 2 & -1 \end{bmatrix} \begin{bmatrix} 3 & 8 \\ -7 & 2 \end{bmatrix}$$

$$= \begin{bmatrix} 17 & 4 \\ 13 & 14 \end{bmatrix}$$

Note that a commutative property for matrix multiplication was *not* listed in Theorem 2.3. It can easily happen that, although the product AB is defined, A and B are not of the proper sizes to define the product BA. For instance, if A is of size 2×3 and B is of size 3×3, then the product AB is defined but the product BA is not. The next example shows that even if both products AB and BA are defined, they need not be equal.

EXAMPLE 4   *Noncommutativity of Matrix Multiplication*

Show that AB and BA are not equal for the matrices

$$A = \begin{bmatrix} 1 & 3 \\ 2 & -1 \end{bmatrix} \quad \text{and} \quad B = \begin{bmatrix} 2 & -1 \\ 0 & 2 \end{bmatrix}.$$

Solution $AB = \begin{bmatrix} 1 & 3 \\ 2 & -1 \end{bmatrix} \begin{bmatrix} 2 & -1 \\ 0 & 2 \end{bmatrix} = \begin{bmatrix} 2 & 5 \\ 4 & -4 \end{bmatrix}$

$BA = \begin{bmatrix} 2 & -1 \\ 0 & 2 \end{bmatrix} \begin{bmatrix} 1 & 3 \\ 2 & -1 \end{bmatrix} = \begin{bmatrix} 0 & 7 \\ 4 & -2 \end{bmatrix}$

Note that $AB \neq BA$.

Do not conclude from Example 4 that the matrix products AB and BA are *never* the same. Sometimes they are the same. For example, try multiplying the following matrices, first in the order AB and then in the order BA.

$$A = \begin{bmatrix} 1 & 2 \\ 1 & 1 \end{bmatrix} \quad \text{and} \quad B = \begin{bmatrix} -2 & 4 \\ 2 & -2 \end{bmatrix}$$

You will see that the two products are equal. The point is this: Although AB and BA are sometimes equal, usually AB and BA are not equal.

Another important quality of matrix algebra is that it does not have a general cancellation property for matrix multiplication. That is, if $AC = BC$, it is not necessarily true that $A = B$. This is demonstrated in Example 5. (In the next section you will see that, for some special types of matrices, cancellation is valid.)

EXAMPLE 5   *An Example in Which Cancellation Is Not Valid*

Show that $AC = BC$.

$$A = \begin{bmatrix} 1 & 3 \\ 0 & 1 \end{bmatrix}, \quad B = \begin{bmatrix} 2 & 4 \\ 2 & 3 \end{bmatrix}, \quad C = \begin{bmatrix} 1 & -2 \\ -1 & 2 \end{bmatrix}$$

Solution

$$AC = \begin{bmatrix} 1 & 3 \\ 0 & 1 \end{bmatrix}\begin{bmatrix} 1 & -2 \\ -1 & 2 \end{bmatrix} = \begin{bmatrix} -2 & 4 \\ -1 & 2 \end{bmatrix}$$

$$BC = \begin{bmatrix} 2 & 4 \\ 2 & 3 \end{bmatrix}\begin{bmatrix} 1 & -2 \\ -1 & 2 \end{bmatrix} = \begin{bmatrix} -2 & 4 \\ -1 & 2 \end{bmatrix}$$

Thus $AC = BC$, and yet $A \neq B$.

You will now look at a special type of *square* matrix that has 1's on the main diagonal and 0's elsewhere.

$$I_n = \begin{bmatrix} 1 & 0 & 0 & \cdots & 0 \\ 0 & 1 & 0 & \cdots & 0 \\ 0 & 0 & 1 & \cdots & 0 \\ \vdots & \vdots & \vdots & & \vdots \\ 0 & 0 & 0 & \cdots & 1 \end{bmatrix}$$

For instance, if $n = 1, 2,$ or 3, we have

$$I_1 = [1], \quad I_2 = \begin{bmatrix} 1 & 0 \\ 0 & 1 \end{bmatrix}, \quad I_3 = \begin{bmatrix} 1 & 0 & 0 \\ 0 & 1 & 0 \\ 0 & 0 & 1 \end{bmatrix}.$$

$$1 \times 1 \qquad\qquad 2 \times 2 \qquad\qquad 3 \times 3$$

When the order of the matrix is understood to be n, you may denote I_n simply as I.

As stated in Theorem 2.4 on the next page, the matrix I_n serves as the **identity** for matrix multiplication; it is called the **identity matrix of order *n*.** The proof of this theorem is left as an exercise. (See Exercise 49.)

| Theorem 2.4 | *Properties of the Identity Matrix* |

If A is a matrix of size $m \times n$, then the following properties are true.
1. $AI_n = A$
2. $I_mA = A$

As a special case of this theorem, note that if A is a *square* matrix of order n, then

$$AI_n = I_nA = A.$$

EXAMPLE 6   *Multiplication by an Identity Matrix*

(a) $\begin{bmatrix} 3 & -2 \\ 4 & 0 \\ -1 & 1 \end{bmatrix} \begin{bmatrix} 1 & 0 \\ 0 & 1 \end{bmatrix} = \begin{bmatrix} 3 & -2 \\ 4 & 0 \\ -1 & 1 \end{bmatrix}$

(b) $\begin{bmatrix} 1 & 0 & 0 \\ 0 & 1 & 0 \\ 0 & 0 & 1 \end{bmatrix} \begin{bmatrix} -2 \\ 1 \\ 4 \end{bmatrix} = \begin{bmatrix} -2 \\ 1 \\ 4 \end{bmatrix}$

For repeated multiplication of *square* matrices, you can use the same exponential notation used with real numbers. That is, $A^1 = A$, $A^2 = AA$, and for a positive integer k, A^k is

$$A^k = \underbrace{AA \cdots A}_{k \text{ factors}}.$$

It is convenient also to define $A^0 = I_n$ (where A is a square matrix of order n). These definitions allow you to establish the properties

1. $A^jA^k = A^{j+k}$ and   2. $(A^j)^k = A^{jk}$

where j and k are nonnegative integers.

EXAMPLE 7   *Repeated Multiplication of a Square Matrix*

Find A^3 for the matrix $A = \begin{bmatrix} 2 & -1 \\ 3 & 0 \end{bmatrix}$.

Solution $A^3 = \left(\begin{bmatrix} 2 & -1 \\ 3 & 0 \end{bmatrix} \begin{bmatrix} 2 & -1 \\ 3 & 0 \end{bmatrix} \right) \begin{bmatrix} 2 & -1 \\ 3 & 0 \end{bmatrix} = \begin{bmatrix} 1 & -2 \\ 6 & -3 \end{bmatrix} \begin{bmatrix} 2 & -1 \\ 3 & 0 \end{bmatrix} = \begin{bmatrix} -4 & -1 \\ 3 & -6 \end{bmatrix}$

In Section 1.1 you saw that a system of linear equations must have exactly one solution, an infinite number of solutions, or no solution. Using the matrix algebra developed so far, you can now prove this result.

Theorem 2.5

Number of Solutions of a System of Linear Equations

For a system of linear equations in n variables, precisely one of the following is true.
1. The system has exactly one solution.
2. The system has an infinite number of solutions.
3. The system has no solution.

Proof   Represent the system by the matrix equation $A\mathbf{x} = \mathbf{b}$. If the system has exactly one solution or no solution, there is nothing to prove. Therefore, you can assume that the system has at least two distinct solutions $\mathbf{x}_1$ and $\mathbf{x}_2$. The proof will be complete if you can show that this assumption implies that the system has an infinite number of solutions. Because $\mathbf{x}_1$ and $\mathbf{x}_2$ are solutions, you have $A\mathbf{x}_1 = A\mathbf{x}_2 = \mathbf{b}$ and hence $A(\mathbf{x}_1 - \mathbf{x}_2) = O$. This implies that the (nonzero) column matrix $\mathbf{x}_h = \mathbf{x}_1 - \mathbf{x}_2$ is a solution of the homogeneous system of linear equations $A\mathbf{x} = O$. Moreover, for any scalar c, it follows that

$$A(\mathbf{x}_1 + c\mathbf{x}_h) = A\mathbf{x}_1 + A(c\mathbf{x}_h) = \mathbf{b} + c(A\mathbf{x}_h) = \mathbf{b} + cO = \mathbf{b}.$$

Therefore $\mathbf{x}_1 + c\mathbf{x}_h$ is a solution of $A\mathbf{x} = \mathbf{b}$ for any scalar c. Because there are an infinite number of possible values of c and each value produces a different solution, you can conclude that the system has an infinite number of solutions.

The Transpose of a Matrix

The **transpose** of a matrix is formed by writing its columns as rows. For instance, if A is the $m \times n$ matrix given by

$$A = \begin{bmatrix} a_{11} & a_{12} & a_{13} & \cdots & a_{1n} \\ a_{21} & a_{22} & a_{23} & \cdots & a_{2n} \\ a_{31} & a_{32} & a_{33} & \cdots & a_{3n} \\ \vdots & \vdots & \vdots & & \vdots \\ a_{m1} & a_{m2} & a_{m3} & \cdots & a_{mn} \end{bmatrix},$$

Size: $m \times n$

then the transpose, denoted by A^T, is the $n \times m$ matrix given by

$$A^T = \begin{bmatrix} a_{11} & a_{21} & a_{31} & \cdots & a_{m1} \\ a_{12} & a_{22} & a_{32} & \cdots & a_{m2} \\ a_{13} & a_{23} & a_{33} & \cdots & a_{m3} \\ \vdots & \vdots & \vdots & & \vdots \\ a_{1n} & a_{2n} & a_{3n} & \cdots & a_{mn} \end{bmatrix}.$$

Size: $n \times m$

EXAMPLE 8   *The Transpose of a Matrix*

Find the transpose of each matrix.

(a) $A = \begin{bmatrix} 2 \\ 8 \end{bmatrix}$ (b) $B = \begin{bmatrix} 1 & 2 & 3 \\ 4 & 5 & 6 \\ 7 & 8 & 9 \end{bmatrix}$ (c) $C = \begin{bmatrix} 1 & 2 & 0 \\ 2 & 1 & 0 \\ 0 & 0 & 1 \end{bmatrix}$ (d) $D = \begin{bmatrix} 0 & 1 \\ 2 & 4 \\ 1 & -1 \end{bmatrix}$

Solution   (a) $A^T = \begin{bmatrix} 2 & 8 \end{bmatrix}$ (b) $B^T = \begin{bmatrix} 1 & 4 & 7 \\ 2 & 5 & 8 \\ 3 & 6 & 9 \end{bmatrix}$ (c) $C^T = \begin{bmatrix} 1 & 2 & 0 \\ 2 & 1 & 0 \\ 0 & 0 & 1 \end{bmatrix}$

(d) $D^T = \begin{bmatrix} 0 & 2 & 1 \\ 1 & 4 & -1 \end{bmatrix}$

DISCOVERY

Let $A = \begin{bmatrix} 1 & 2 \\ 3 & 4 \end{bmatrix}$ and $B = \begin{bmatrix} 3 & 5 \\ 1 & -1 \end{bmatrix}$.

Calculate $(AB)^T$, $A^T B^T$, and $B^T A^T$. Make a conjecture about the transpose of a product of two square matrices. Select two other square matrices to check your conjecture.

REMARK: Note that the square matrix in part (c) of Example 8 is equal to its transpose. Such a matrix is called **symmetric.** That is, a matrix A is symmetric if $A = A^T$. From this definition it is clear that a symmetric matrix must be square. Also, if $A = [a_{ij}]$ is a symmetric matrix, then $a_{ij} = a_{ji}$ for all $i \neq j$.

Theorem 2.6   *Properties of Transposes*

If A and B are matrices (with sizes such that the given matrix operations are defined) and c is a scalar, then the following properties are true.
1. $(A^T)^T = A$
2. $(A + B)^T = A^T + B^T$ Transpose of a sum
3. $(cA)^T = c(A^T)$ Transpose of a scalar multiple
4. $(AB)^T = B^T A^T$ Transpose of a product

Proof   Because the transpose operation interchanges rows and columns, Property 1 seems to make sense. To prove Property 1, let A be an $m \times n$ matrix. Observe that A^T has size $n \times m$ and hence $(A^T)^T$ has size $m \times n$, the same as A. To show that $(A^T)^T = A$ you must show that the ijth entries are the same. Let a_{ij} be the ijth entry of A. Then a_{ij} is the jith entry of A^T, and hence the ijth entry of $(A^T)^T$. This proves Property 1. The proofs of the remaining properties are left as an exercise. (See Exercise 50.)

REMARK: Remember that you *reverse the order* of multiplication when forming the transpose of a product. That is, the transpose of AB is given by $(AB)^T = B^TA^T$ and is *not* usually equal to A^TB^T.

Properties 2 and 4 can be generalized to cover sums or products of any finite number of matrices. For instance, the transpose of the sum of three matrices is given by

$$(A + B + C)^T = A^T + B^T + C^T,$$

and the transpose of the product of three matrices is given by

$$(ABC)^T = C^TB^TA^T.$$

EXAMPLE 9   *Finding the Transpose of a Product*

Show that $(AB)^T$ and B^TA^T are equal.

$$A = \begin{bmatrix} 2 & 1 & -2 \\ -1 & 0 & 3 \\ 0 & -2 & 1 \end{bmatrix} \quad \text{and} \quad B = \begin{bmatrix} 3 & 1 \\ 2 & -1 \\ 3 & 0 \end{bmatrix}$$

Solution $AB = \begin{bmatrix} 2 & 1 & -2 \\ -1 & 0 & 3 \\ 0 & -2 & 1 \end{bmatrix} \begin{bmatrix} 3 & 1 \\ 2 & -1 \\ 3 & 0 \end{bmatrix} = \begin{bmatrix} 2 & 1 \\ 6 & -1 \\ -1 & 2 \end{bmatrix}$

$(AB)^T = \begin{bmatrix} 2 & 6 & -1 \\ 1 & -1 & 2 \end{bmatrix}$

$B^TA^T = \begin{bmatrix} 3 & 2 & 3 \\ 1 & -1 & 0 \end{bmatrix} \begin{bmatrix} 2 & -1 & 0 \\ 1 & 0 & -2 \\ -2 & 3 & 1 \end{bmatrix} = \begin{bmatrix} 2 & 6 & -1 \\ 1 & -1 & 2 \end{bmatrix}$

Note that $(AB)^T = B^TA^T$.

EXAMPLE 10   *The Product of a Matrix and Its Transpose*

For the matrix

$$A = \begin{bmatrix} 1 & 3 \\ 0 & -2 \\ -2 & -1 \end{bmatrix}$$

find the product AA^T and show that it is symmetric.

Solution   Because

$$AA^T = \begin{bmatrix} 1 & 3 \\ 0 & -2 \\ -2 & -1 \end{bmatrix} \begin{bmatrix} 1 & 0 & -2 \\ 3 & -2 & -1 \end{bmatrix} = \begin{bmatrix} 10 & -6 & -5 \\ -6 & 4 & 2 \\ -5 & 2 & 5 \end{bmatrix}$$

it follows that $AA^T = (AA^T)^T$, so AA^T is symmetric.

REMARK: The property demonstrated in Example 10 is true in general. That is, for any matrix A, the matrix given by $B = AA^T$ is symmetric. You are asked to prove this result in Exercise 51.

SECTION 2.2 Exercises

In Exercises 1–6, perform the indicated operations when $a = 3$, $b = -4$, and

$$A = \begin{bmatrix} 1 & 2 \\ 3 & 4 \end{bmatrix}, \quad B = \begin{bmatrix} 0 & 1 \\ -1 & 2 \end{bmatrix}, \quad O = \begin{bmatrix} 0 & 0 \\ 0 & 0 \end{bmatrix}.$$

1. $aA + bB$ **2.** $A + B$ **3.** $ab(B)$
4. $(a + b)B$ **5.** $(a - b)(A - B)$ **6.** $(ab)O$

In Exercises 7–10, solve for X when

$$A = \begin{bmatrix} -4 & 0 \\ 1 & -5 \\ -3 & 2 \end{bmatrix} \quad \text{and} \quad B = \begin{bmatrix} 1 & 2 \\ -2 & 1 \\ 4 & 4 \end{bmatrix}.$$

7. $3X + 2A = B$ **8.** $2A - 5B = 3X$
9. $X - 3A + 2B = O$ **10.** $6X - 4A - 3B = O$

In Exercises 11–16, perform the indicated operations, given that $c = -2$ and

$$A = \begin{bmatrix} 1 & 2 & 3 \\ 0 & 1 & -1 \end{bmatrix}, \quad B = \begin{bmatrix} 1 & 3 \\ -1 & 2 \end{bmatrix}, \quad C = \begin{bmatrix} 0 & 1 \\ -1 & 0 \end{bmatrix},$$
$$O = \begin{bmatrix} 0 & 0 \\ 0 & 0 \end{bmatrix}.$$

11. $B(CA)$ **12.** $C(BC)$ **13.** $(B + C)A$
14. $B(C + O)$ **15.** $(cB)(C + C)$ **16.** $B(cA)$

17. If $AC = BC$, then A is *not* necessarily equal to B. Demonstrate this using the matrices

$$A = \begin{bmatrix} 1 & 2 & 3 \\ 0 & 5 & 4 \\ 3 & -2 & 1 \end{bmatrix}, \quad B = \begin{bmatrix} 4 & -6 & 3 \\ 5 & 4 & 4 \\ -1 & 0 & 1 \end{bmatrix},$$
$$C = \begin{bmatrix} 0 & 0 & 0 \\ 0 & 0 & 0 \\ 4 & -2 & 3 \end{bmatrix}.$$

18. If $AB = O$, then it is *not* necessarily true that $A = O$ or $B = O$. Demonstrate this using the matrices

$$A = \begin{bmatrix} 3 & 3 \\ 4 & 4 \end{bmatrix} \quad \text{and} \quad B = \begin{bmatrix} 1 & -1 \\ -1 & 1 \end{bmatrix}.$$

In Exercises 19–22, perform the indicated operations when

$$A = \begin{bmatrix} 1 & 2 \\ 0 & -1 \end{bmatrix} \quad \text{and} \quad I = \begin{bmatrix} 1 & 0 \\ 0 & 1 \end{bmatrix}.$$

19. A^2 **20.** A^4
21. $A(I + A)$ **22.** $A + IA$

In Exercises 23–28, find (a) A^T, (b) $A^T A$, and (c) AA^T.

23. $A = \begin{bmatrix} 2 & -1 & 3 \\ 4 & 3 & -5 \end{bmatrix}$ **24.** $A = \begin{bmatrix} -7 & 11 & 12 \\ 4 & -3 & 1 \\ 6 & -1 & 3 \end{bmatrix}$

25. $A = \begin{bmatrix} 6 & 0 \\ 0 & -4 \\ 7 & 5 \end{bmatrix}$ **26.** $A = \begin{bmatrix} 2 \\ -1 \\ -3 \end{bmatrix}$

27. $A = \begin{bmatrix} 0 & -4 & 3 & 2 \\ 8 & 4 & 0 & 1 \\ -2 & 3 & 5 & 1 \\ 0 & 0 & -3 & 2 \end{bmatrix}$

28. $A = \begin{bmatrix} 4 & -3 & 2 & 0 \\ 2 & 0 & 11 & -1 \\ -1 & -2 & 0 & 3 \\ 14 & -2 & 12 & -9 \\ 6 & 8 & -5 & 4 \end{bmatrix}$

Writing   In Exercises 29 and 30, explain why the given formula is *not* valid for matrices. Illustrate your argument with examples.

29. $(A + B)(A - B) = A^2 - B^2$
30. $(A + B)(A + B) = A^2 + 2AB + B^2$

31. Verify that $(AB)^T = B^T A^T$ for the following matrices.

$$A = \begin{bmatrix} -1 & 1 & -2 \\ 2 & 0 & 1 \end{bmatrix} \quad \text{and} \quad B = \begin{bmatrix} -3 & 0 \\ 1 & 2 \\ 1 & -1 \end{bmatrix}$$

32. Verify that $(AB)^T = B^T A^T$ for the following matrices.

$$A = \begin{bmatrix} 1 & 2 \\ 0 & -2 \end{bmatrix} \quad \text{and} \quad B = \begin{bmatrix} -3 & -1 \\ 2 & 1 \end{bmatrix}$$

True or False?   In Exercises 33 and 34, determine whether each statement is true or false. If a statement is true, give a reason or cite an appropriate statement in the text. If a statement is false, provide an example that shows that the statement is not true in all cases or cite an appropriate statement in the text.

33. (a) Matrix addition is commutative.

(b) Matrix multiplication is associative.

(c) The transpose of the product of two matrices equals the product of their transposes; that is, $(AB)^T = A^T B^T$.

(d) For any matrix C, the matrix CC^T is symmetric.

34. (a) Matrix multiplication is commutative.

(b) Every matrix A has an additive inverse.

(c) If the matrices A, B, and C satisfy $AB = AC$, then $B = C$.

(d) The transpose of the sum of two matrices equals the sum of their transposes.

35. Consider the following matrices.

$$X = \begin{bmatrix} 1 \\ 0 \\ 1 \end{bmatrix}, \quad Y = \begin{bmatrix} 1 \\ 1 \\ 0 \end{bmatrix}, \quad Z = \begin{bmatrix} 2 \\ -1 \\ 3 \end{bmatrix}, \quad W = \begin{bmatrix} 1 \\ 1 \\ 1 \end{bmatrix}, \quad O = \begin{bmatrix} 0 \\ 0 \\ 0 \end{bmatrix}$$

(a) Find scalars a and b such that $Z = aX + bY$.

(b) Show that there do not exist scalars a and b such that $W = aX + bY$.

(c) Show that if $aX + bY + cW = O$, then $a = b = c = 0$.

(d) Find scalars a, b, and c, not all equal to zero, such that $aX + bY + cZ = O$.

36. Consider the following matrices.

$$X = \begin{bmatrix} 1 \\ 2 \\ 3 \end{bmatrix}, \quad Y = \begin{bmatrix} 1 \\ 0 \\ 2 \end{bmatrix}, \quad Z = \begin{bmatrix} 1 \\ 4 \\ 4 \end{bmatrix}, \quad W = \begin{bmatrix} 0 \\ 0 \\ 1 \end{bmatrix}, \quad O = \begin{bmatrix} 0 \\ 0 \\ 0 \end{bmatrix}$$

(a) Find scalars a and b such that $Z = aX + bY$.

(b) Show that there do not exist scalars a and b such that $W = aX + bY$.

(c) Show that if $aX + bY + cW = O$, then $a = b = c = 0$.

(d) Find scalars a, b, and c, not all equal to zero, such that $aX + bY + cZ = O$.

In Exercises 37 and 38, compute the power of A for the matrix

$$A = \begin{bmatrix} 1 & 0 & 0 \\ 0 & -1 & 0 \\ 0 & 0 & 1 \end{bmatrix}.$$

37. A^{19}

38. A^{20}

An nth root of a matrix B is a matrix A such that $A^n = B$. In Exercises 39 and 40, find the nth root of the matrix B.

39. $B = \begin{bmatrix} 9 & 0 \\ 0 & 4 \end{bmatrix}$, $n = 2$ **40.** $B = \begin{bmatrix} 8 & 0 & 0 \\ 0 & -1 & 0 \\ 0 & 0 & 27 \end{bmatrix}$, $n = 3$

In Exercises 41–44, use the following definition to find $f(A)$: If f is the polynomial function given by

$$f(x) = a_0 + a_1 x + a_2 x^2 + \cdots + a_n x^n$$

then for an $n \times n$ matrix A, $f(A)$ is defined to be

$$f(A) = a_0 I_n + a_1 A + a_2 A^2 + \cdots + a_n A^n.$$

41. $f(x) = x^2 - 5x + 2$, $A = \begin{bmatrix} 2 & 0 \\ 4 & 5 \end{bmatrix}$

42. $f(x) = x^2 - 7x + 6$, $A = \begin{bmatrix} 5 & 4 \\ 1 & 2 \end{bmatrix}$

43. $f(x) = x^3 - 10x^2 + 31x - 30$, $A = \begin{bmatrix} 3 & 1 & 4 \\ 0 & 2 & 6 \\ 0 & 0 & 5 \end{bmatrix}$

44. $f(x) = x^2 - 10x + 24$, $A = \begin{bmatrix} 8 & -4 \\ 2 & 2 \end{bmatrix}$

45. *Guided Proof*   Prove the associative property of matrix addition: $A + (B + C) = (A + B) + C$.

Getting Started: To prove that $A + (B + C)$ and $(A + B) + C$ are equal, show that their corresponding entries are the same.

(i) Begin your proof by letting A, B, and C be $m \times n$ matrices.

(ii) Observe that the ijth entry of $B + C$ is $b_{ij} + c_{ij}$.

(iii) Furthermore, the ijth entry of $A + (B + C)$ is $a_{ij} + (b_{ij} + c_{ij})$.

(iv) Determine the ijth entry of $(A + B) + C$.

46. Prove the associative property of scalar multiplication: $(cd)A = c(dA)$.

47. Prove that the scalar 1 is the identity for scalar multiplication: $1A = A$.

48. Prove the following distributive property: $(c + d)A = cA + dA$.

49. Prove Theorem 2.2.

50. Complete the proof of Theorem 2.3.

(a) Prove the following associative property of multiplication: $A(BC) = (AB)C$.

(b) Prove the following distributive property: $(A + B)C = AC + BC$.

(c) Prove the following property: $c(AB) = (cA)B = A(cB)$.

51. Prove Theorem 2.4.

52. Prove Properties 2, 3, and 4 of Theorem 2.6.

53. *Guided Proof* Prove that if A is an $m \times n$ matrix, then AA^T and A^TA are symmetric matrices.

Getting Started: To prove that AA^T is symmetric, you need to show that it is equal to its transpose, $(AA^T)^T = AA^T$.

   (i) Begin your proof with the left-hand matrix expression $(AA^T)^T$.

   (ii) Use the properties of the transpose operation to show that it can be simplified to equal the right-hand expression, AA^T.

   (iii) Repeat this analysis for the product A^TA.

54. Give an example of two 2×2 matrices A and B such that $(AB)^T \neq A^TB^T$.

In Exercises 55–58, determine whether the matrix is symmetric, skew-symmetric, or neither. A square matrix is called **skew-symmetric** if $A^T = -A$.

55. $A = \begin{bmatrix} 0 & 2 \\ -2 & 0 \end{bmatrix}$ **56.** $A = \begin{bmatrix} 2 & 1 \\ 1 & 3 \end{bmatrix}$

57. $A = \begin{bmatrix} 0 & 2 & 1 \\ 2 & 0 & 3 \\ 1 & 3 & 0 \end{bmatrix}$ **58.** $A = \begin{bmatrix} 0 & 2 & -1 \\ -2 & 0 & -3 \\ 1 & 3 & 0 \end{bmatrix}$

59. Prove that the main diagonal of a skew-symmetric matrix consists entirely of zeros.

60. Prove that if A and B are $n \times n$ skew-symmetric matrices, then $A + B$ is skew-symmetric.

61. Let A be a square matrix of order n.

   (a) Show that $\frac{1}{2}(A + A^T)$ is symmetric.

   (b) Show that $\frac{1}{2}(A - A^T)$ is skew-symmetric.

   (c) Prove that A can be written as the sum of a symmetric matrix B and a skew-symmetric matrix C, $A = B + C$.

   (d) Write the matrix

$$A = \begin{bmatrix} 2 & 5 & 3 \\ -3 & 6 & 0 \\ 4 & 1 & 1 \end{bmatrix}$$

as the sum of a skew-symmetric matrix and a symmetric matrix.

62. Prove that if A is an $n \times n$ matrix, then $A - A^T$ is skew-symmetric.

63. Let A and B be two $n \times n$ symmetric matrices.

   (a) Give an example to show that the product AB is not necessarily symmetric.

   (b) Prove that AB is symmetric if and only if $AB = BA$.

64. Consider matrices of the form

$$A = \begin{bmatrix} 0 & a_{12} & a_{13} & a_{14} & \cdots & a_{1n} \\ 0 & 0 & a_{23} & a_{24} & \cdots & a_{2n} \\ 0 & 0 & 0 & a_{34} & \cdots & a_{3n} \\ \vdots & \vdots & \vdots & \vdots & \cdots & \vdots \\ 0 & 0 & 0 & 0 & \cdots & a_{(n-1)n} \\ 0 & 0 & 0 & 0 & \cdots & 0 \end{bmatrix}$$

   (a) Write a 2×2 matrix and a 3×3 matrix in the form of A.

   (b) Use a graphing utility to raise each of the matrices to higher powers. Describe the result.

   (c) Use the result of part (b) to make a conjecture about powers of A if A is a 4×4 matrix. Use a graphing utility to test your conjecture.

   (d) Use the results of parts (b) and (c) to make a conjecture about powers of A if A is an $n \times n$ matrix.

2.3 | The Inverse of a Matrix

Section 2.2 discussed some of the similarities between the algebra of real numbers and the algebra of matrices. This section further develops the algebra of matrices to include the solutions of matrix equations involving matrix multiplication. To begin, consider the real number equation $ax = b$. To solve this equation for x, multiply both sides of the equation by a^{-1} (provided $a \neq 0$).

$$ax = b$$
$$(a^{-1}a)x = a^{-1}b$$
$$(1)x = a^{-1}b$$
$$x = a^{-1}b$$

The number a^{-1} is called the *multiplicative inverse* of a because $a^{-1}a$ yields 1 (the identity element for multiplication). The definition of a multiplicative inverse of a matrix is similar.

Definition of an Inverse of a Matrix

An $n \times n$ matrix A is **invertible** (or **nonsingular**) if there exists an $n \times n$ matrix B such that

$$AB = BA = I_n$$

where I_n is the identity matrix of order n. The matrix B is called the (multiplicative) **inverse** of A. A matrix that does not have an inverse is called **noninvertible** (or **singular**).

Nonsquare matrices do not have inverses. To see this, note that if A is of size $m \times n$ and B is of size $n \times m$ (where $m \neq n$), then the products AB and BA are of different sizes and therefore cannot be equal to each other. Indeed, not all square matrices possess inverses. (See Example 4.) However, the following theorem tells you that if a matrix *does* possess an inverse, then that inverse is unique.

Theorem 2.7

Uniqueness of an Inverse Matrix

If A is an invertible matrix, then its inverse is unique. The inverse of A is denoted by A^{-1}.

Proof   Because A is invertible, you know that it has at least one inverse B such that

$$AB = I = BA.$$

Suppose that A has another inverse C such that

$$AC = I = CA.$$

Then you can show that B and C are equal as follows.

$$
\begin{aligned}
AB &= I \\
C(AB) &= CI \\
(CA)B &= C \\
IB &= C \\
B &= C
\end{aligned}
$$

Consequently $B = C$, and it follows that the inverse of a matrix is unique.

Because the inverse A^{-1} of an invertible matrix A is unique, you can call it ***the*** inverse of A and write

$$AA^{-1} = A^{-1}A = I.$$

EXAMPLE 1   *The Inverse of a Matrix*

Show that B is the inverse of A, where

$$A = \begin{bmatrix} -1 & 2 \\ -1 & 1 \end{bmatrix} \quad \text{and} \quad B = \begin{bmatrix} 1 & -2 \\ 1 & -1 \end{bmatrix}.$$

Solution   Using the definition of an inverse matrix, you can show that B is the inverse of A by showing that $AB = I = BA$ as follows.

$$AB = \begin{bmatrix} -1 & 2 \\ -1 & 1 \end{bmatrix} \begin{bmatrix} 1 & -2 \\ 1 & -1 \end{bmatrix} = \begin{bmatrix} -1+2 & 2-2 \\ -1+1 & 2-1 \end{bmatrix} = \begin{bmatrix} 1 & 0 \\ 0 & 1 \end{bmatrix}$$

$$BA = \begin{bmatrix} 1 & -2 \\ 1 & -1 \end{bmatrix} \begin{bmatrix} -1 & 2 \\ -1 & 1 \end{bmatrix} = \begin{bmatrix} -1+2 & 2-2 \\ -1+1 & 2-1 \end{bmatrix} = \begin{bmatrix} 1 & 0 \\ 0 & 1 \end{bmatrix}$$

REMARK: Recall that it is not always true that $AB = BA$, even if both products are defined. However, if A and B are both square matrices and $AB = I_n$, then it can be shown that $BA = I_n$. Although the proof of this fact is omitted, it implies that in Example 1 you needed only to check that $AB = I_2$.

The following example shows how to use a system of equations to find the inverse of a matrix.

EXAMPLE 2   *Finding the Inverse of a Matrix*

Find the inverse of the matrix

$$A = \begin{bmatrix} 1 & 4 \\ -1 & -3 \end{bmatrix}.$$

Solution   To find the inverse of A, try to solve the matrix equation $AX = I$ for X.

$$\begin{bmatrix} 1 & 4 \\ -1 & -3 \end{bmatrix} \begin{bmatrix} x_{11} & x_{12} \\ x_{21} & x_{22} \end{bmatrix} = \begin{bmatrix} 1 & 0 \\ 0 & 1 \end{bmatrix}$$

$$\begin{bmatrix} x_{11} + 4x_{21} & x_{12} + 4x_{22} \\ -x_{11} - 3x_{21} & -x_{12} - 3x_{22} \end{bmatrix} = \begin{bmatrix} 1 & 0 \\ 0 & 1 \end{bmatrix}$$

Now, by equating corresponding entries, you obtain the following two systems of linear equations.

$$\begin{aligned} x_{11} + 4x_{21} &= 1 \\ -x_{11} - 3x_{21} &= 0 \end{aligned} \qquad \begin{aligned} x_{12} + 4x_{22} &= 0 \\ -x_{12} - 3x_{22} &= 1 \end{aligned}$$

Solving the first system, you find that the first column of X is $x_{11} = -3$ and $x_{21} = 1$. Similarly, solving the second system, you find that the second column of X is $x_{12} = -4$ and $x_{22} = 1$. Therefore the inverse of A is

$$X = A^{-1} = \begin{bmatrix} -3 & -4 \\ 1 & 1 \end{bmatrix}.$$

Try using matrix multiplication to check this result.

Generalizing the method used to solve Example 2 provides a convenient method for finding an inverse. Notice first that the two systems of linear equations

$$x_{11} + 4x_{21} = 1 \qquad x_{12} + 4x_{22} = 0$$
$$-x_{11} - 3x_{21} = 0 \qquad -x_{12} - 3x_{22} = 1$$

have the *same coefficient matrix.* Rather than solve the two systems represented by

$$\begin{bmatrix} 1 & 4 & \vdots & 1 \\ -1 & -3 & \vdots & 0 \end{bmatrix} \quad \text{and} \quad \begin{bmatrix} 1 & 4 & \vdots & 0 \\ -1 & -3 & \vdots & 1 \end{bmatrix}$$

separately, you can solve them simultaneously. You can do this by **adjoining** the identity matrix to the coefficient matrix to obtain

$$\begin{bmatrix} 1 & 4 & \vdots & 1 & 0 \\ -1 & -3 & \vdots & 0 & 1 \end{bmatrix}.$$

By applying Gauss-Jordan elimination to this matrix, you can solve *both* systems with a single elimination process as follows.

$$\begin{bmatrix} 1 & 4 & \vdots & 1 & 0 \\ 0 & 1 & \vdots & 1 & 1 \end{bmatrix} \qquad R_2 + R_1 \rightarrow R_2$$

$$\begin{bmatrix} 1 & 0 & \vdots & -3 & -4 \\ 0 & 1 & \vdots & 1 & 1 \end{bmatrix} \qquad R_1 + (-4)R_2 \rightarrow R_1$$

Thus, by applying Gauss-Jordan elimination to the "doubly augmented" matrix $[A \; \vdots \; I]$, you obtain the matrix $[I \; \vdots \; A^{-1}]$.

$$\underset{A \qquad\qquad I}{\begin{bmatrix} 1 & 4 & \vdots & 1 & 0 \\ -1 & -3 & \vdots & 0 & 1 \end{bmatrix}} \longrightarrow \underset{I \qquad\qquad A^{-1}}{\begin{bmatrix} 1 & 0 & \vdots & -3 & -4 \\ 0 & 1 & \vdots & 1 & 1 \end{bmatrix}}$$

This procedure (or algorithm) works for an arbitrary $n \times n$ matrix. If A cannot be row reduced to I_n, then A is singular. This procedure will be formally justified in the next section, after introducing the concept of an elementary matrix. For now the algorithm is summarized as follows.

Finding the Inverse of a Matrix by Gauss-Jordan Elimination	Let A be a square matrix of order n. 1. Write the $n \times 2n$ matrix that consists of the given matrix A on the left and the $n \times n$ identity matrix I on the right to obtain $[A \; \vdots \; I]$. Note that you separate the matrices A and I by a dotted line. This process is called **adjoining** the matrices A and I. 2. If possible, row reduce A to I using elementary row operations on the *entire* matrix $[A \; \vdots \; I]$. The result will be the matrix $[I \; \vdots \; A^{-1}]$. If this is not possible, then A is not invertible. 3. Check your work by multiplying to see that $AA^{-1} = I = A^{-1}A$.

EXAMPLE 3   *Finding the Inverse of a Matrix*

Find the inverse of the matrix.

$$A = \begin{bmatrix} 1 & -1 & 0 \\ 1 & 0 & -1 \\ -6 & 2 & 3 \end{bmatrix}$$

Solution   Begin by adjoining the identity matrix to A to form the matrix

$$[A \vdots I] = \begin{bmatrix} 1 & -1 & 0 & \vdots & 1 & 0 & 0 \\ 1 & 0 & -1 & \vdots & 0 & 1 & 0 \\ -6 & 2 & 3 & \vdots & 0 & 0 & 1 \end{bmatrix}.$$

Now, using elementary row operations, rewrite this matrix in the form $[I \vdots A^{-1}]$ as follows.

$$\begin{bmatrix} 1 & -1 & 0 & \vdots & 1 & 0 & 0 \\ 0 & 1 & -1 & \vdots & -1 & 1 & 0 \\ -6 & 2 & 3 & \vdots & 0 & 0 & 1 \end{bmatrix} \qquad R_2 + (-1)R_1 \rightarrow R_2$$

$$\begin{bmatrix} 1 & -1 & 0 & \vdots & 1 & 0 & 0 \\ 0 & 1 & -1 & \vdots & -1 & 1 & 0 \\ 0 & -4 & 3 & \vdots & 6 & 0 & 1 \end{bmatrix} \qquad R_3 + (6)R_1 \rightarrow R_3$$

$$\begin{bmatrix} 1 & -1 & 0 & \vdots & 1 & 0 & 0 \\ 0 & 1 & -1 & \vdots & -1 & 1 & 0 \\ 0 & 0 & -1 & \vdots & 2 & 4 & 1 \end{bmatrix} \qquad R_3 + (4)R_2 \rightarrow R_3$$

$$\begin{bmatrix} 1 & -1 & 0 & \vdots & 1 & 0 & 0 \\ 0 & 1 & -1 & \vdots & -1 & 1 & 0 \\ 0 & 0 & 1 & \vdots & -2 & -4 & -1 \end{bmatrix} \qquad (-1)R_3 \rightarrow R_3$$

$$\begin{bmatrix} 1 & -1 & 0 & \vdots & 1 & 0 & 0 \\ 0 & 1 & 0 & \vdots & -3 & -3 & -1 \\ 0 & 0 & 1 & \vdots & -2 & -4 & -1 \end{bmatrix} \qquad R_2 + R_3 \rightarrow R_2$$

$$\begin{bmatrix} 1 & 0 & 0 & \vdots & -2 & -3 & -1 \\ 0 & 1 & 0 & \vdots & -3 & -3 & -1 \\ 0 & 0 & 1 & \vdots & -2 & -4 & -1 \end{bmatrix} \qquad R_1 + R_2 \rightarrow R_1$$

Therefore the matrix A is invertible, and its inverse is

$$A^{-1} = \begin{bmatrix} -2 & -3 & -1 \\ -3 & -3 & -1 \\ -2 & -4 & -1 \end{bmatrix}.$$

Try confirming this by showing that $AA^{-1} = I = A^{-1}A$.

Technology   Note

Most graphing calculators and computer software programs can calculate the inverse of a square matrix. If you are using a TI-86, your screens for Example 3 will look like the following.

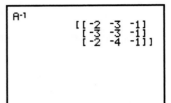

The process shown in Example 3 applies to any $n \times n$ matrix. If the matrix A has an inverse, this process will find it. On the other hand, if the matrix A has no inverse, the process will tell you that. The next example applies the process to a singular matrix (one that has no inverse).

EXAMPLE 4   *A Singular Matrix*

Show that the matrix has no inverse.

$$A = \begin{bmatrix} 1 & 2 & 0 \\ 3 & -1 & 2 \\ -2 & 3 & -2 \end{bmatrix}$$

Solution   Adjoin the identity matrix to A to form

$$[A \vdots I] = \begin{bmatrix} 1 & 2 & 0 & \vdots & 1 & 0 & 0 \\ 3 & -1 & 2 & \vdots & 0 & 1 & 0 \\ -2 & 3 & -2 & \vdots & 0 & 0 & 1 \end{bmatrix}$$

and apply Gauss-Jordan elimination as follows.

$$\begin{bmatrix} 1 & 2 & 0 & \vdots & 1 & 0 & 0 \\ 0 & -7 & 2 & \vdots & -3 & 1 & 0 \\ -2 & 3 & -2 & \vdots & 0 & 0 & 1 \end{bmatrix} \qquad R_2 + (-3)R_1 \rightarrow R_2$$

$$\begin{bmatrix} 1 & 2 & 0 & \vdots & 1 & 0 & 0 \\ 0 & -7 & 2 & \vdots & -3 & 1 & 0 \\ 0 & 7 & -2 & \vdots & 2 & 0 & 1 \end{bmatrix} \qquad R_3 + (2)R_1 \rightarrow R_3$$

Now, notice that adding the second row to the third row produces a row of zeros on the left side of the matrix.

$$\begin{bmatrix} 1 & 2 & 0 & \vdots & 1 & 0 & 0 \\ 0 & -7 & 2 & \vdots & -3 & 1 & 0 \\ 0 & 0 & 0 & \vdots & -1 & 1 & 1 \end{bmatrix} \qquad R_3 + R_2 \to R_3$$

Because the "*A* portion" of the matrix has a row of zeros, you can conclude that it is not possible to rewrite the matrix $[A \vdots I]$ in the form $[I \vdots A^{-1}]$. This means that *A* has no inverse.

Using Gauss-Jordan elimination to find the inverse of a matrix works well (even as a computer technique) for matrices of size 3×3 or greater. For 2×2 matrices, however, you can use a formula for the inverse rather than find it by Gauss-Jordan elimination. This simple formula is explained as follows. If *A* is a 2×2 matrix given by

$$A = \begin{bmatrix} a & b \\ c & d \end{bmatrix}$$

then *A* is invertible if and only if $ad - bc \neq 0$. Moreover, if $ad - bc \neq 0$, then the inverse is given by

$$A^{-1} = \frac{1}{ad - bc} \begin{bmatrix} d & -b \\ -c & a \end{bmatrix}.$$

Try verifying this inverse by finding the product AA^{-1}.

REMARK: The denominator $ad - bc$ is called the **determinant** of *A*. You will study determinants in detail in Chapter 3.

EXAMPLE 5   *Finding the Inverse of a 2 × 2 Matrix*

If possible, find the inverse of each matrix.

(a) $A = \begin{bmatrix} 3 & -1 \\ -2 & 2 \end{bmatrix}$ (b) $B = \begin{bmatrix} 3 & -1 \\ -6 & 2 \end{bmatrix}$

Solution   (a) For the matrix *A*, apply the formula for the inverse of a 2×2 matrix to obtain $ad - bc = (3)(2) - (-1)(-2) = 4$. Because this quantity is not zero, the inverse is formed by interchanging the entries on the main diagonal and changing the signs of the other two entries as follows.

$$A^{-1} = \tfrac{1}{4} \begin{bmatrix} 2 & 1 \\ 2 & 3 \end{bmatrix} = \begin{bmatrix} \frac{1}{2} & \frac{1}{4} \\ \frac{1}{2} & \frac{3}{4} \end{bmatrix}$$

(b) For the matrix *B*, you have $ad - bc = (3)(2) - (-1)(-6) = 0$, which means that *B* is not invertible.

Properties of Inverses

Some important properties of inverse matrices are listed below.

Theorem 2.8	*Properties of Inverse Matrices*

If A is an invertible matrix, k is a positive integer, and c is a scalar, then A^{-1}, A^k, cA, and A^T are invertible and the following are true.

1. $(A^{-1})^{-1} = A$

2. $(A^k)^{-1} = \underbrace{A^{-1}A^{-1} \cdots A^{-1}}_{k \text{ factors}} = (A^{-1})^k$

3. $(cA)^{-1} = \dfrac{1}{c}A^{-1}, \ c \neq 0$

4. $(A^T)^{-1} = (A^{-1})^T$

Proof   The key to the proofs of Properties 1, 3, and 4 is the fact that the inverse of a matrix is unique (Theorem 2.7). That is, if $BC = CB = I$, then C is the inverse of B.

Property 1 states that the inverse of A^{-1} is A itself. To prove this, observe that $A^{-1}A = AA^{-1} = I$, which means that A is the inverse of A^{-1}. Thus, $A = (A^{-1})^{-1}$.

Similarly, Property 3 states that $\dfrac{1}{c}A^{-1}$ is the inverse of (cA), $c \neq 0$. To prove this, use the properties of scalar multiplication given in Theorems 2.1 and 2.3 as follows.

$$(cA)\left(\frac{1}{c}A^{-1}\right) = \left(c\frac{1}{c}\right)AA^{-1} = (1)I = I$$

and

$$\left(\frac{1}{c}A^{-1}\right)(cA) = \left(\frac{1}{c}c\right)A^{-1}A = (1)I = I$$

Hence $\dfrac{1}{c}A^{-1}$ is the inverse of (cA), which implies that

$$\frac{1}{c}A^{-1} = (cA)^{-1}.$$

Properties 2 and 4 are left for you to prove. (See Exercises 41 and 42.)

For nonsingular matrices, the exponential notation used for repeated multiplication of *square* matrices can be extended to include exponents that are negative integers. This may be done by defining A^{-k} to be

$$A^{-k} = \underbrace{A^{-1}A^{-1} \cdots A^{-1}}_{k \text{ factors}} = (A^{-1})^k.$$

Using this convention you can show that the properties $A^j A^k = A^{j+k}$ and $(A^j)^k = A^{jk}$ hold true for any integers j and k.

EXAMPLE 6   *The Inverse of the Square of a Matrix*

Compute A^{-2} in two different ways and show that the results are equal.

$$A = \begin{bmatrix} 1 & 1 \\ 2 & 4 \end{bmatrix}$$

Solution   One way to find A^{-2} is to find $(A^2)^{-1}$ by squaring the matrix A to obtain

$$A^2 = \begin{bmatrix} 3 & 5 \\ 10 & 18 \end{bmatrix}$$

and using the formula for the inverse of a 2×2 matrix to obtain

$$(A^2)^{-1} = \tfrac{1}{4} \begin{bmatrix} 18 & -5 \\ -10 & 3 \end{bmatrix}$$

$$= \begin{bmatrix} \frac{9}{2} & -\frac{5}{4} \\ -\frac{5}{2} & \frac{3}{4} \end{bmatrix}.$$

Another way to find A^{-2} is to find $(A^{-1})^2$ by finding A^{-1}

$$A^{-1} = \tfrac{1}{2} \begin{bmatrix} 4 & -1 \\ -2 & 1 \end{bmatrix}$$

$$= \begin{bmatrix} 2 & -\frac{1}{2} \\ -1 & \frac{1}{2} \end{bmatrix}$$

and then squaring this matrix to obtain

$$(A^{-1})^2 = \begin{bmatrix} \frac{9}{2} & -\frac{5}{4} \\ -\frac{5}{2} & \frac{3}{4} \end{bmatrix}.$$

Note that each method produces the same result.

DISCOVERY

Let $A = \begin{bmatrix} 1 & 2 \\ 1 & 3 \end{bmatrix}$ and $B = \begin{bmatrix} 2 & -1 \\ 1 & -1 \end{bmatrix}.$

Calculate $(AB)^{-1}$, $A^{-1}B^{-1}$, and $B^{-1}A^{-1}$. Make a conjecture about the inverse of a product of two nonsingular matrices. Select two other nonsingular matrices and see whether your conjecture holds.

The next theorem gives a formula for computing the inverse of a product of two matrices.

Theorem 2.9

The Inverse of a Product

If A and B are invertible matrices of size n, then AB is invertible and

$$(AB)^{-1} = B^{-1}A^{-1}.$$

Proof To show that $B^{-1}A^{-1}$ is the inverse of AB, you need only show that it conforms to the definition of an inverse matrix. That is,

$$(AB)(B^{-1}A^{-1}) = A(BB^{-1})A^{-1} = A(I)A^{-1} = (AI)A^{-1} = AA^{-1} = I.$$

In a similar way you can show that $(B^{-1}A^{-1})(AB) = I$ and conclude that AB is invertible and has the indicated inverse.

Theorem 2.9 states that the inverse of a product of two invertible matrices is the product of their inverses taken in the *reverse* order. This can be generalized to include the product of several invertible matrices:

$$(A_1A_2A_3 \cdots A_n)^{-1} = A_n^{-1} \cdots A_3^{-1}A_2^{-1}A_1^{-1}.$$

(See Example 4 in Appendix A.)

EXAMPLE 7 *Finding the Inverse of a Matrix Product*

Find $(AB)^{-1}$ for the matrices

$$A = \begin{bmatrix} 1 & 3 & 3 \\ 1 & 4 & 3 \\ 1 & 3 & 4 \end{bmatrix} \quad \text{and} \quad B = \begin{bmatrix} 1 & 2 & 3 \\ 1 & 3 & 3 \\ 2 & 4 & 3 \end{bmatrix}$$

using the fact that A^{-1} and B^{-1} are given by

$$A^{-1} = \begin{bmatrix} 7 & -3 & -3 \\ -1 & 1 & 0 \\ -1 & 0 & 1 \end{bmatrix} \quad \text{and} \quad B^{-1} = \begin{bmatrix} 1 & -2 & 1 \\ -1 & 1 & 0 \\ \frac{2}{3} & 0 & -\frac{1}{3} \end{bmatrix}.$$

Solution Using Theorem 2.9 produces the following.

$$(AB)^{-1} = B^{-1}A^{-1} = \overset{B^{-1}}{\begin{bmatrix} 1 & -2 & 1 \\ -1 & 1 & 0 \\ \frac{2}{3} & 0 & -\frac{1}{3} \end{bmatrix}} \overset{A^{-1}}{\begin{bmatrix} 7 & -3 & -3 \\ -1 & 1 & 0 \\ -1 & 0 & 1 \end{bmatrix}} = \begin{bmatrix} 8 & -5 & -2 \\ -8 & 4 & 3 \\ 5 & -2 & -\frac{7}{3} \end{bmatrix}$$

REMARK: Note that you *reverse the order* of multiplication to find the inverse of AB. That is, $(AB)^{-1} = B^{-1}A^{-1}$, and the inverse of AB is usually *not* equal to $A^{-1}B^{-1}$.

One important property in the algebra of real numbers is the cancellation property. That is, if $ac = bc$ $(c \neq 0)$, then $a = b$. *Invertible* matrices have similar cancellation properties.

Theorem 2.10	*Cancellation Properties*

If C is an invertible matrix, then the following properties hold.
1. If $AC = BC$, then $A = B$.   **Right cancellation property**
2. If $CA = CB$, then $A = B$.   **Left cancellation property**

Proof   To prove Property 1, use the fact that C is invertible and write

$$AC = BC$$
$$(AC)C^{-1} = (BC)C^{-1}$$
$$A(CC^{-1}) = B(CC^{-1})$$
$$AI = BI$$
$$A = B.$$

The second property can be proved in a similar way; this is left to you. (See Exercise 44.)

Be sure to remember that Theorem 2.10 can be applied only if C is an *invertible* matrix. If C is not invertible, then cancellation is not usually valid. For instance, Example 5 in Section 2.2 gives an example of a matrix equation $AC = BC$ in which $A \neq B$. Note that C is not invertible in this example.

Systems of Equations

In Theorem 2.5 you were able to prove that a system of linear equations can have exactly one solution, an infinite number of solutions, or no solution. For *square* systems (those having the same number of equations as variables), you can use the following theorem to determine whether the system has a unique solution.

Theorem 2.11	*Systems of Equations with Unique Solutions*

If A is an invertible matrix, then the system of linear equations $A\mathbf{x} = \mathbf{b}$ has a unique solution given by

$$\mathbf{x} = A^{-1}\mathbf{b}.$$

Proof   Because A is nonsingular, the following steps are valid.

$$A\mathbf{x} = \mathbf{b}$$
$$A^{-1}A\mathbf{x} = A^{-1}\mathbf{b}$$
$$I\mathbf{x} = A^{-1}\mathbf{b}$$
$$\mathbf{x} = A^{-1}\mathbf{b}$$

This solution is unique because if $\mathbf{x}_1$ and $\mathbf{x}_2$ were two solutions, you could apply the cancellation property to the equation $A\mathbf{x}_1 = \mathbf{b} = A\mathbf{x}_2$ to conclude that $\mathbf{x}_1 = \mathbf{x}_2$.

Theorem 2.11 is important theoretically, but it is not very practical for solving a system of linear equations. That is, it would be more work to find A^{-1} and then multiply by $\mathbf{b}$ than simply to solve the system using Gaussian elimination with back-substitution. A situation in which you might consider using Theorem 2.11 as a computational technique would be one in which you had *several* systems of linear equations, all of which had the same coefficient matrix A. In such a case, you could find the inverse matrix once and then solve each system by computing the product $A^{-1}\mathbf{b}$. This is demonstrated in Example 8.

EXAMPLE 8   *Solving a System of Equations Using an Inverse Matrix*

Use an inverse matrix to solve each system.

(a) $2x + 3y + z = -1$
$\ 3x + 3y + z = 1$
$\ 2x + 4y + z = -2$

(b) $2x + 3y + z = 4$
$\ 3x + 3y + z = 8$
$\ 2x + 4y + z = 5$

(c) $2x + 3y + z = 0$
$\ 3x + 3y + z = 0$
$\ 2x + 4y + z = 0$

Solution   First note that the coefficient matrix for each system is

$$A = \begin{bmatrix} 2 & 3 & 1 \\ 3 & 3 & 1 \\ 2 & 4 & 1 \end{bmatrix}.$$

Using Gauss-Jordan elimination, you can find A^{-1} to be

$$A^{-1} = \begin{bmatrix} -1 & 1 & 0 \\ -1 & 0 & 1 \\ 6 & -2 & -3 \end{bmatrix}.$$

To solve each system, use matrix multiplication as follows.

(a) $\mathbf{x} = A^{-1}\mathbf{b} = \begin{bmatrix} -1 & 1 & 0 \\ -1 & 0 & 1 \\ 6 & -2 & -3 \end{bmatrix} \begin{bmatrix} -1 \\ 1 \\ -2 \end{bmatrix}$

$\phantom{(a)\ \mathbf{x}} = \begin{bmatrix} 2 \\ -1 \\ -2 \end{bmatrix}$

The solution is $x = 2$, $y = -1$, and $z = -2$.

(b) $\mathbf{x} = A^{-1}\mathbf{b} = \begin{bmatrix} -1 & 1 & 0 \\ -1 & 0 & 1 \\ 6 & -2 & -3 \end{bmatrix} \begin{bmatrix} 4 \\ 8 \\ 5 \end{bmatrix} = \begin{bmatrix} 4 \\ 1 \\ -7 \end{bmatrix}$

The solution is $x = 4$, $y = 1$, and $z = -7$.

(c) $\mathbf{x} = A^{-1}\mathbf{b} = \begin{bmatrix} -1 & 1 & 0 \\ -1 & 0 & 1 \\ 6 & -2 & -3 \end{bmatrix} \begin{bmatrix} 0 \\ 0 \\ 0 \end{bmatrix} = \begin{bmatrix} 0 \\ 0 \\ 0 \end{bmatrix}$

The solution is trivial: $x = 0$, $y = 0$, and $z = 0$.

SECTION 2.3   Exercises

In Exercises 1–4, show that B is the inverse of A.

1. $A = \begin{bmatrix} 1 & 2 \\ 3 & 4 \end{bmatrix}$, $B = \begin{bmatrix} -2 & 1 \\ \frac{3}{2} & -\frac{1}{2} \end{bmatrix}$

2. $A = \begin{bmatrix} 1 & -1 \\ 2 & 3 \end{bmatrix}$, $B = \begin{bmatrix} \frac{3}{5} & \frac{1}{5} \\ -\frac{2}{5} & \frac{1}{5} \end{bmatrix}$

3. $A = \begin{bmatrix} -2 & 2 & 3 \\ 1 & -1 & 0 \\ 0 & 1 & 4 \end{bmatrix}$, $B = \frac{1}{3}\begin{bmatrix} -4 & -5 & 3 \\ -4 & -8 & 3 \\ 1 & 2 & 0 \end{bmatrix}$

4. $A = \begin{bmatrix} 2 & -17 & 11 \\ -1 & 11 & -7 \\ 0 & 3 & -2 \end{bmatrix}$, $B = \begin{bmatrix} 1 & 1 & 2 \\ 2 & 4 & -3 \\ 3 & 6 & -5 \end{bmatrix}$

In Exercises 5–24, find the inverse of the matrix (if it exists).

5. $\begin{bmatrix} 1 & 2 \\ 3 & 7 \end{bmatrix}$

6. $\begin{bmatrix} 1 & -2 \\ 2 & -3 \end{bmatrix}$

7. $\begin{bmatrix} -7 & 33 \\ 4 & -19 \end{bmatrix}$

8. $\begin{bmatrix} -1 & 1 \\ 3 & -3 \end{bmatrix}$

9. $\begin{bmatrix} 1 & 1 & 1 \\ 3 & 5 & 4 \\ 3 & 6 & 5 \end{bmatrix}$

10. $\begin{bmatrix} 1 & 2 & 2 \\ 3 & 7 & 9 \\ -1 & -4 & -7 \end{bmatrix}$

11. $\begin{bmatrix} 1 & 2 & -1 \\ 3 & 7 & -10 \\ 7 & 16 & -21 \end{bmatrix}$

12. $\begin{bmatrix} 10 & 5 & -7 \\ -5 & 1 & 4 \\ 3 & 2 & -2 \end{bmatrix}$

13. $\begin{bmatrix} 1 & 1 & 2 \\ 3 & 1 & 0 \\ -2 & 0 & 3 \end{bmatrix}$

14. $\begin{bmatrix} 3 & 2 & 5 \\ 2 & 2 & 4 \\ -4 & 4 & 0 \end{bmatrix}$

15. $\begin{bmatrix} 0.1 & 0.2 & 0.3 \\ -0.3 & 0.2 & 0.2 \\ 0.5 & 0.5 & 0.5 \end{bmatrix}$

16. $\begin{bmatrix} 2 & 0 & 0 \\ 0 & 3 & 0 \\ 0 & 0 & 5 \end{bmatrix}$

17. $\begin{bmatrix} 1 & 0 & 0 \\ 3 & 4 & 0 \\ 2 & 5 & 5 \end{bmatrix}$

18. $\begin{bmatrix} 1 & 0 & 0 \\ 3 & 0 & 0 \\ 2 & 5 & 5 \end{bmatrix}$

19. $\begin{bmatrix} -8 & 0 & 0 & 0 \\ 0 & 1 & 0 & 0 \\ 0 & 0 & 0 & 0 \\ 0 & 0 & 0 & -5 \end{bmatrix}$

20. $\begin{bmatrix} \frac{1}{a} & 0 \\ a & a \end{bmatrix}$, $a \neq 0$

21. $\begin{bmatrix} 1 & -2 & -1 & -2 \\ 3 & -5 & -2 & -3 \\ 2 & -5 & -2 & -5 \\ -1 & 4 & 4 & 11 \end{bmatrix}$

22. $\begin{bmatrix} 4 & 8 & -7 & 14 \\ 2 & 5 & -4 & 6 \\ 0 & 2 & 1 & -7 \\ 3 & 6 & -5 & 10 \end{bmatrix}$

23. $\begin{bmatrix} 1 & 0 & 3 & 0 \\ 0 & 2 & 0 & 4 \\ 1 & 0 & 3 & 0 \\ 0 & 2 & 0 & 4 \end{bmatrix}$

24. $\begin{bmatrix} 1 & 3 & -2 & 0 \\ 0 & 2 & 4 & 6 \\ 0 & 0 & -2 & 1 \\ 0 & 0 & 0 & 5 \end{bmatrix}$

In Exercises 25–28, use an inverse matrix to solve each system of linear equations.

25. (a) $\begin{aligned} -x + y &= 4 \\ -2x + y &= 0 \end{aligned}$

(b) $\begin{aligned} -x + y &= -3 \\ -2x + y &= 5 \end{aligned}$

(c) $\begin{aligned} -x + y &= 0 \\ -2x + y &= 0 \end{aligned}$

26. (a) $\begin{aligned} 2x + 3y &= 5 \\ x + 4y &= 10 \end{aligned}$

(b) $\begin{aligned} 2x + 3y &= 0 \\ x + 4y &= 0 \end{aligned}$

(c) $\begin{aligned} 2x + 3y &= 1 \\ x + 4y &= -2 \end{aligned}$

27. (a) $\begin{aligned} 3x_1 + 2x_2 + 2x_3 &= 0 \\ 2x_1 + 2x_2 + 2x_3 &= 5 \\ -4x_1 + 4x_2 + 3x_3 &= 2 \end{aligned}$ (b) $\begin{aligned} 3x_1 + 2x_2 + 2x_3 &= 0 \\ 2x_1 + 2x_2 + 2x_3 &= 0 \\ -4x_1 + 4x_2 + 3x_3 &= 0 \end{aligned}$

28. (a) $\begin{aligned} x_1 - 2x_2 - x_3 &= 0 \\ 3x_1 - 5x_2 - 2x_3 &= 1 \\ 2x_1 - 5x_2 - 2x_3 &= -1 \end{aligned}$

(b) $\begin{aligned} x_1 - 2x_2 - x_3 &= 1 \\ 3x_1 - 5x_2 - 2x_3 &= -2 \\ 2x_1 - 5x_2 - 2x_3 &= 0 \end{aligned}$

In Exercises 29 and 30, use a graphing utility with matrix capabilities to solve the system of linear equations using an inverse matrix.

29. $\begin{aligned} 2x_1 - 3x_2 + x_3 - 2x_4 + x_5 - 4x_6 &= 20 \\ 3x_1 + x_2 - 4x_3 + x_4 - x_5 + 2x_6 &= -16 \\ 4x_1 + x_2 - 3x_3 + 4x_4 - x_5 + 2x_6 &= -12 \\ -5x_1 - x_2 + 4x_3 + 2x_4 - 5x_5 + 3x_6 &= -2 \\ x_1 + x_2 - 3x_3 + 4x_4 - 3x_5 + x_6 &= -15 \\ 3x_1 - x_2 + 2x_3 - 3x_4 + 2x_5 - 6x_6 &= 25 \end{aligned}$

30. $\begin{aligned} 4x_1 - 2x_2 + 4x_3 + 2x_4 - 5x_5 - x_6 &= 1 \\ 3x_1 + 6x_2 - 5x_3 - 6x_4 + 3x_5 + 3x_6 &= -11 \\ 2x_1 - 3x_2 + x_3 + 3x_4 - x_5 - 2x_6 &= 0 \\ -x_1 + 4x_2 - 4x_3 - 6x_4 + 2x_5 + 4x_6 &= -9 \\ 3x_1 - x_2 + 5x_3 + 2x_4 - 3x_5 - 5x_6 &= 1 \\ -2x_1 + 3x_2 - 4x_3 - 6x_4 + x_5 + 2x_6 &= -12 \end{aligned}$

In Exercises 31–34, use the given inverse matrices to find (a) $(AB)^{-1}$, (b) $(A^T)^{-1}$, (c) A^{-2}, and (d) $(2A)^{-1}$.

31. $A^{-1} = \begin{bmatrix} 2 & 5 \\ -7 & 6 \end{bmatrix}$, $B^{-1} = \begin{bmatrix} 7 & -3 \\ 2 & 0 \end{bmatrix}$

32. $A^{-1} = \begin{bmatrix} -\frac{2}{7} & \frac{1}{7} \\ \frac{3}{7} & \frac{2}{7} \end{bmatrix}$, $B^{-1} = \begin{bmatrix} \frac{5}{11} & \frac{2}{11} \\ \frac{3}{11} & -\frac{1}{11} \end{bmatrix}$

33. $A^{-1} = \begin{bmatrix} 1 & -\frac{1}{2} & \frac{3}{4} \\ \frac{3}{2} & \frac{1}{2} & -2 \\ \frac{1}{4} & 1 & \frac{1}{2} \end{bmatrix}$, $B^{-1} = \begin{bmatrix} 2 & 4 & \frac{5}{2} \\ -\frac{3}{4} & 2 & \frac{1}{4} \\ \frac{1}{4} & \frac{1}{2} & 2 \end{bmatrix}$

34. $A^{-1} = \begin{bmatrix} 1 & -4 & 2 \\ 0 & 1 & 3 \\ 4 & 2 & 1 \end{bmatrix}$, $B^{-1} = \begin{bmatrix} 6 & 5 & -3 \\ -2 & 4 & -1 \\ 1 & 3 & 4 \end{bmatrix}$

35. Find x such that the matrix is equal to its own inverse.
$$A = \begin{bmatrix} 3 & x \\ -2 & -3 \end{bmatrix}$$

36. Find x such that the matrix is singular.
$$A = \begin{bmatrix} 4 & x \\ -2 & -3 \end{bmatrix}$$

37. Find A given that
$$(2A)^{-1} = \begin{bmatrix} 1 & 2 \\ 3 & 4 \end{bmatrix}.$$

38. Show that the matrix is invertible and find its inverse.
$$A = \begin{bmatrix} \sin\theta & \cos\theta \\ -\cos\theta & \sin\theta \end{bmatrix}$$

True or False? In Exercises 39 and 40, determine whether each statement is true or false. If a statement is true, give a reason or cite an appropriate statement in the text. If a statement is false, provide an example that shows that the statement is not true in all cases or cite an appropriate statement in the text.

39. (a) The inverse of a nonsingular matrix is unique.
 (b) If the matrices A, B, and C satisfy $BA = CA$ and A is invertible, then $B = C$.
 (c) The inverse of the product of two matrices is the product of their inverses; that is, $(AB)^{-1} = A^{-1}B^{-1}$.
 (d) If A can be row reduced to the identity matrix, then A is nonsingular.

40. (a) The product of four invertible 7×7 matrices is invertible.
 (b) The transpose of the inverse of a nonsingular matrix is equal to the inverse of the transpose.
 (c) The matrix $\begin{bmatrix} a & b \\ c & d \end{bmatrix}$ is invertible if $ab - dc \neq 0$.
 (d) If A is a square matrix, then the system of linear equations $A\mathbf{x} = \mathbf{b}$ has a unique solution.

41. Prove Property 2 of Theorem 2.8: If A is an invertible matrix and k is a positive integer, then
$$(A^k)^{-1} = \underbrace{A^{-1}A^{-1} \cdots A^{-1}}_{k \text{ factors}}.$$

42. Prove Property 4 of Theorem 2.8: If A is an invertible matrix, then $(A^T)^{-1} = (A^{-1})^T$.

43. *Guided Proof* Prove that the inverse of a symmetric nonsingular matrix is symmetric.
 Getting Started: To prove that the inverse of A is symmetric, you need to show that $(A^{-1})^T = A^{-1}$.
 (i) Let A be a symmetric, nonsingular matrix.
 (ii) This means that $A^T = A$ and A^{-1} exists.
 (iii) Use the properties of the transpose to show that $(A^{-1})^T$ is equal to A^{-1}.

44. Prove Property 2 of Theorem 2.10: If C is an invertible matrix such that $CA = CB$, then $A = B$.

45. Prove that if $A^2 = A$, then $I - 2A = (I - 2A)^{-1}$.

46. Prove that if A, B, and C are square matrices and $ABC = I$, then B is invertible and $B^{-1} = CA$.

47. Prove that if A is invertible and $AB = O$, then $B = O$.

48. *Guided Proof*   Prove that if $A^2 = A$, then either $A = I$ or A is singular.

Getting Started: You must show that either A is singular or that A equals the identity matrix.

(i) Begin your proof by observing that A is either singular or nonsingular.

(ii) If A is singular, then you are done.

(iii) If A is nonsingular, then use the inverse matrix A^{-1} and the hypothesis $A^2 = A$ to show that $A = I$.

49. *Writing*   Is the sum of two invertible matrices invertible? Explain why or why not. Illustrate your conclusion with appropriate examples.

50. *Writing*   Under what conditions will the diagonal matrix

$$A = \begin{bmatrix} a_{11} & 0 & 0 & \cdots & 0 \\ 0 & a_{22} & 0 & \cdots & 0 \\ \vdots & \vdots & \vdots & & \vdots \\ 0 & 0 & 0 & \cdots & a_{nn} \end{bmatrix}$$

be invertible? If A is invertible, find its inverse.

51. Use the result of Exercise 50 to find A^{-1} for each matrix.

(a) $A = \begin{bmatrix} -1 & 0 & 0 \\ 0 & 3 & 0 \\ 0 & 0 & 2 \end{bmatrix}$ (b) $A = \begin{bmatrix} \frac{1}{2} & 0 & 0 \\ 0 & \frac{1}{3} & 0 \\ 0 & 0 & \frac{1}{4} \end{bmatrix}$

52. Let $A = \begin{bmatrix} 1 & 2 \\ -2 & 1 \end{bmatrix}$.

(a) Show that $A^2 - 2A + 5I = O$, where I is the identity matrix of order 2.

(b) Show that $A^{-1} = \frac{1}{5}(2I - A)$.

(c) Show that in general, for any square matrix satisfying $A^2 - 2A + 5I = 0$, the inverse of A is given by $A^{-1} = \frac{1}{5}(2I - A)$.

53. Let $\mathbf{u}$ be an $n \times 1$ column matrix satisfying $\mathbf{u}^T\mathbf{u} = 1$. The $n \times n$ matrix $H = I_n - 2\mathbf{u}\mathbf{u}^T$ is called a **Householder matrix.**

(a) Prove that H is symmetric and nonsingular.

(b) Let $\mathbf{u} = \begin{bmatrix} \sqrt{2}/2 \\ \sqrt{2}/2 \\ 0 \end{bmatrix}$. Show that $\mathbf{u}^T\mathbf{u} = 1$ and calculate the Householder matrix H.

54. Prove that if the matrix $I - AB$ is nonsingular, then so is $I - BA$.

55. Let A, D, and P be $n \times n$ matrices satisfying $AP = PD$. If P is nonsingular, solve this equation for A. Must it be true that $A = D$?

56. Let A, D, and P be $n \times n$ matrices satisfying $P^{-1}AP = D$. Solve this equation for A. Must it be true that $A = D$?

57. Find an example of a singular 2×2 matrix satisfying $A^2 = A$.

2.4 | Elementary Matrices

In Section 1.2, the following three elementary row operations for matrices were introduced.

1. Interchange two rows.
2. Multiply a row by a nonzero constant.
3. Add a multiple of a row to another row.

In this section, you will see how matrix multiplication can be used to perform these operations.

Definition of an Elementary Matrix

An $n \times n$ matrix is called an **elementary matrix** if it can be obtained from the identity matrix I_n by a single elementary row operation.

REMARK: The identity matrix I_n is elementary by this definition because it can be obtained from itself by multiplying any one of its rows by 1.

EXAMPLE 1 *Elementary Matrices and Nonelementary Matrices*

Which matrices are elementary? For those that are, describe the corresponding elementary row operation.

(a) $\begin{bmatrix} 1 & 0 & 0 \\ 0 & 3 & 0 \\ 0 & 0 & 1 \end{bmatrix}$ (b) $\begin{bmatrix} 1 & 0 & 0 \\ 0 & 1 & 0 \end{bmatrix}$ (c) $\begin{bmatrix} 1 & 0 & 0 \\ 0 & 1 & 0 \\ 0 & 0 & 0 \end{bmatrix}$

(d) $\begin{bmatrix} 1 & 0 & 0 \\ 0 & 0 & 1 \\ 0 & 1 & 0 \end{bmatrix}$ (e) $\begin{bmatrix} 1 & 0 \\ 2 & 1 \end{bmatrix}$ (f) $\begin{bmatrix} 1 & 0 & 0 \\ 0 & 2 & 0 \\ 0 & 0 & -1 \end{bmatrix}$

Solution

(a) This matrix *is* elementary. It can be obtained by multiplying the second row of I_3 by 3.

(b) This matrix is *not* elementary because it is not square.

(c) This matrix is *not* elementary because it was obtained by multiplying the third row of I_3 by 0 (row multiplication must be by a *nonzero* constant).

(d) This matrix *is* elementary. It can be obtained by interchanging the second and third rows of I_3.

(e) This matrix *is* elementary. It can be obtained by multiplying the first row of I_2 by 2 and adding the result to the second row.

(f) This matrix is *not* elementary because two elementary row operations are required to obtain it from I_3.

Elementary matrices are useful because they enable you to use matrix multiplication to perform elementary row operations, as demonstrated in Example 2.

EXAMPLE 2 *Elementary Matrices and Elementary Row Operations*

(a) In the following matrix product, E is the elementary matrix in which the first two rows of I_3 have been interchanged.

$$\overset{E}{\begin{bmatrix} 0 & 1 & 0 \\ 1 & 0 & 0 \\ 0 & 0 & 1 \end{bmatrix}} \overset{A}{\begin{bmatrix} 0 & 2 & 1 \\ 1 & -3 & 6 \\ 3 & 2 & -1 \end{bmatrix}} = \begin{bmatrix} 1 & -3 & 6 \\ 0 & 2 & 1 \\ 3 & 2 & -1 \end{bmatrix}$$

Note that the first two rows of A have been interchanged by multiplying *on the left* by E.

(b) In the following matrix product, E is the elementary matrix in which the second row of I_3 has been multiplied by $\frac{1}{2}$.

$$\begin{matrix} E \end{matrix} \qquad \begin{matrix} A \end{matrix}$$

$$\begin{bmatrix} 1 & 0 & 0 \\ 0 & \frac{1}{2} & 0 \\ 0 & 0 & 1 \end{bmatrix} \begin{bmatrix} 1 & 0 & -4 & 1 \\ 0 & 2 & 6 & -4 \\ 0 & 1 & 3 & 1 \end{bmatrix} = \begin{bmatrix} 1 & 0 & -4 & 1 \\ 0 & 1 & 3 & -2 \\ 0 & 1 & 3 & 1 \end{bmatrix}$$

Here the size of A is 3×4. However, A could be *any* $3 \times n$ matrix and multiplication on the left by E would still result in multiplying the second row of A by $\frac{1}{2}$.

(c) In the following product, E is the elementary matrix in which 2 times the first row of I_3 has been added to the second row.

$$\begin{matrix} E \end{matrix} \qquad \begin{matrix} A \end{matrix}$$

$$\begin{bmatrix} 1 & 0 & 0 \\ 2 & 1 & 0 \\ 0 & 0 & 1 \end{bmatrix} \begin{bmatrix} 1 & 0 & -1 \\ -2 & -2 & 3 \\ 0 & 4 & 5 \end{bmatrix} = \begin{bmatrix} 1 & 0 & -1 \\ 0 & -2 & 1 \\ 0 & 4 & 5 \end{bmatrix}$$

Note that in the product EA, 2 times the first row of A has been added to the second row.

In each of the three products in Example 2, you were able to perform elementary row operations by multiplying *on the left* by an elementary matrix. This property of elementary matrices is generalized in the following theorem, which is stated without proof.

Theorem 2.12 *Representing Elementary Row Operations*

Let E be the elementary matrix obtained by performing an elementary row operation on I_m. If that same elementary row operation is performed on an $m \times n$ matrix A, then the resulting matrix is given by the product EA.

R E M A R K : Be sure to remember that in Theorem 2.12, A is multiplied *on the left* by the elementary matrix E. Right multiplication by elementary matrices, which involves column operations, will not be considered in this text.

Most applications of elementary row operations require a sequence of operations. For instance, Gaussian elimination usually requires several elementary row operations to row reduce a matrix A. For elementary matrices, this sequence translates into multiplication (on the left) by several elementary matrices. The order of multiplication is important; the elementary matrix immediately to the left of A corresponds to the row operation performed first. This process is demonstrated in Example 3.

EXAMPLE 3 *Using Elementary Matrices*

Find a sequence of elementary matrices that can be used to write the matrix A in row-echelon form.

$$A = \begin{bmatrix} 0 & 1 & 3 & 5 \\ 1 & -3 & 0 & 2 \\ 2 & -6 & 2 & 0 \end{bmatrix}$$

Solution

Matrix	*Elementary Row Operation*	*Elementary Matrix*
$\begin{bmatrix} 1 & -3 & 0 & 2 \\ 0 & 1 & 3 & 5 \\ 2 & -6 & 2 & 0 \end{bmatrix}$	$R_1 \leftrightarrow R_2$	$E_1 = \begin{bmatrix} 0 & 1 & 0 \\ 1 & 0 & 0 \\ 0 & 0 & 1 \end{bmatrix}$
$\begin{bmatrix} 1 & -3 & 0 & 2 \\ 0 & 1 & 3 & 5 \\ 0 & 0 & 2 & -4 \end{bmatrix}$	$R_3 + (-2)R_1 \rightarrow R_3$	$E_2 = \begin{bmatrix} 1 & 0 & 0 \\ 0 & 1 & 0 \\ -2 & 0 & 1 \end{bmatrix}$
$\begin{bmatrix} 1 & -3 & 0 & 2 \\ 0 & 1 & 3 & 5 \\ 0 & 0 & 1 & -2 \end{bmatrix}$	$\left(\frac{1}{2}\right)R_3 \rightarrow R_3$	$E_3 = \begin{bmatrix} 1 & 0 & 0 \\ 0 & 1 & 0 \\ 0 & 0 & \frac{1}{2} \end{bmatrix}$

The three elementary matrices E_1, E_2, and E_3 can be used to perform the same elimination.

$$B = E_3 E_2 E_1 A = \begin{bmatrix} 1 & 0 & 0 \\ 0 & 1 & 0 \\ 0 & 0 & \frac{1}{2} \end{bmatrix} \begin{bmatrix} 1 & 0 & 0 \\ 0 & 1 & 0 \\ -2 & 0 & 1 \end{bmatrix} \begin{bmatrix} 0 & 1 & 0 \\ 1 & 0 & 0 \\ 0 & 0 & 1 \end{bmatrix} \begin{bmatrix} 0 & 1 & 3 & 5 \\ 1 & -3 & 0 & 2 \\ 2 & -6 & 2 & 0 \end{bmatrix}$$

$$= \begin{bmatrix} 1 & 0 & 0 \\ 0 & 1 & 0 \\ 0 & 0 & \frac{1}{2} \end{bmatrix} \begin{bmatrix} 1 & 0 & 0 \\ 0 & 1 & 0 \\ -2 & 0 & 1 \end{bmatrix} \begin{bmatrix} 1 & -3 & 0 & 2 \\ 0 & 1 & 3 & 5 \\ 2 & -6 & 2 & 0 \end{bmatrix}$$

$$= \begin{bmatrix} 1 & 0 & 0 \\ 0 & 1 & 0 \\ 0 & 0 & \frac{1}{2} \end{bmatrix} \begin{bmatrix} 1 & -3 & 0 & 2 \\ 0 & 1 & 3 & 5 \\ 0 & 0 & 2 & -4 \end{bmatrix}$$

$$= \begin{bmatrix} 1 & -3 & 0 & 2 \\ 0 & 1 & 3 & 5 \\ 0 & 0 & 1 & -2 \end{bmatrix}$$

R E M A R K : The procedure demonstrated in Example 3 is primarily of theoretical interest. In other words, this procedure is not suggested as a practical method for performing Gaussian elimination.

The two matrices in Example 3

$$A = \begin{bmatrix} 0 & 1 & 3 & 5 \\ 1 & -3 & 0 & 2 \\ 2 & -6 & 2 & 0 \end{bmatrix} \quad \text{and} \quad B = \begin{bmatrix} 1 & -3 & 0 & 2 \\ 0 & 1 & 3 & 5 \\ 0 & 0 & 1 & -2 \end{bmatrix}$$

are row-equivalent because you can obtain B by performing a sequence of row operations on A. That is,

$$B = E_3 E_2 E_1 A.$$

The definition of row-equivalent matrices can be restated using elementary matrices, as follows.

Definition of Row Equivalence

Let A and B be $m \times n$ matrices. Matrix B is **row-equivalent** to A if there exists a finite number of elementary matrices $E_1, E_2, \ldots, E_k$ such that

$$B = E_k E_{k-1} \cdots E_2 E_1 A.$$

You know from Section 2.3 that not all square matrices are invertible. Every elementary matrix, however, is invertible. Moreover, the inverse of an elementary matrix is itself an elementary matrix.

Theorem 2.13

Elementary Matrices Are Invertible

If E is an elementary matrix, then E^{-1} exists and is an elementary matrix.

To find the inverse of an elementary matrix E, simply reverse the elementary row operation used to obtain E. For instance, you can find the inverse of each of the three elementary matrices shown in Example 3 as follows.

Elementary Matrix

$$E_1 = \begin{bmatrix} 0 & 1 & 0 \\ 1 & 0 & 0 \\ 0 & 0 & 1 \end{bmatrix} \quad R_1 \leftrightarrow R_2$$

$$E_2 = \begin{bmatrix} 1 & 0 & 0 \\ 0 & 1 & 0 \\ -2 & 0 & 1 \end{bmatrix} \quad R_3 + (-2)R_1 \to R_3$$

$$E_3 = \begin{bmatrix} 1 & 0 & 0 \\ 0 & 1 & 0 \\ 0 & 0 & \frac{1}{2} \end{bmatrix} \quad (\frac{1}{2})R_3 \to R_3$$

Inverse Matrix

$$E_1^{-1} = \begin{bmatrix} 0 & 1 & 0 \\ 1 & 0 & 0 \\ 0 & 0 & 1 \end{bmatrix} \quad R_1 \leftrightarrow R_2$$

$$E_2^{-1} = \begin{bmatrix} 1 & 0 & 0 \\ 0 & 1 & 0 \\ 2 & 0 & 1 \end{bmatrix} \quad R_3 + (2)R_1 \to R_3$$

$$E_3^{-1} = \begin{bmatrix} 1 & 0 & 0 \\ 0 & 1 & 0 \\ 0 & 0 & 2 \end{bmatrix} \quad (2)R_3 \to R_3$$

The following theorem tells you that every invertible matrix can be written as the product of elementary matrices.

Theorem 2.14   *A Property of Invertible Matrices*

A square matrix A is invertible if and only if it can be written as the product of elementary matrices.

Proof   The phrase "if and only if" means that there are actually two parts to the theorem. On the one hand, you have to show that *if A is invertible,* *then* it can be written as the product of elementary matrices. Then you have to show that *if A can be written as the product of elementary matrices, then A is invertible.*

To prove the theorem in one direction, assume that A is the product of elementary matrices. Then, because every elementary matrix is invertible and the product of invertible matrices is invertible, it follows that A is invertible.

To prove the theorem in the other direction, assume that A is invertible. From Theorem 2.11 you know that the system of linear equations represented by $A\mathbf{x} = O$ has only the trivial solution. But this implies that the augmented matrix $[A \vdots O]$ can be rewritten in the form $[I \vdots O]$ (using elementary row operations corresponding to $E_1, E_2, \ldots,$ and E_k). Therefore we have

$$E_k \cdots E_3 E_2 E_1 A = I$$

and it follows that

$$A = E_1^{-1} E_2^{-1} E_3^{-1} \cdots E_k^{-1}.$$

Thus, A can be written as the product of elementary matrices, and the proof is complete.

The first part of this proof is illustrated in Example 4.

EXAMPLE 4   *Writing a Matrix As the Product of Elementary Matrices*

Find a sequence of elementary matrices whose product is

$$A = \begin{bmatrix} -1 & -2 \\ 3 & 8 \end{bmatrix}.$$

Solution   Begin by finding a sequence of elementary row operations that can be used to rewrite A in reduced row-echelon form.

Matrix	*Elementary Row Operation*	*Elementary Matrix*
$\begin{bmatrix} 1 & 2 \\ 3 & 8 \end{bmatrix}$	$(-1)R_1 \rightarrow R_1$	$E_1 = \begin{bmatrix} -1 & 0 \\ 0 & 1 \end{bmatrix}$

Matrix	Elementary Row Operation	Elementary Matrix
$\begin{bmatrix} 1 & 2 \\ 0 & 2 \end{bmatrix}$	$R_2 + (-3)R_1 \rightarrow R_2$	$E_2 = \begin{bmatrix} 1 & 0 \\ -3 & 1 \end{bmatrix}$
$\begin{bmatrix} 1 & 2 \\ 0 & 1 \end{bmatrix}$	$(\frac{1}{2})R_2 \rightarrow R_2$	$E_3 = \begin{bmatrix} 1 & 0 \\ 0 & \frac{1}{2} \end{bmatrix}$
$\begin{bmatrix} 1 & 0 \\ 0 & 1 \end{bmatrix}$	$R_1 + (-2)R_2 \rightarrow R_1$	$E_4 = \begin{bmatrix} 1 & -2 \\ 0 & 1 \end{bmatrix}$

Now, from the matrix product

$$E_4 E_3 E_2 E_1 A = I$$

solve for A to obtain

$$A = E_1^{-1} E_2^{-1} E_3^{-1} E_4^{-1}.$$

This implies that A is a product of elementary matrices.

$$A = \overset{E_1^{-1}}{\begin{bmatrix} -1 & 0 \\ 0 & 1 \end{bmatrix}} \overset{E_2^{-1}}{\begin{bmatrix} 1 & 0 \\ 3 & 1 \end{bmatrix}} \overset{E_3^{-1}}{\begin{bmatrix} 1 & 0 \\ 0 & 2 \end{bmatrix}} \overset{E_4^{-1}}{\begin{bmatrix} 1 & 2 \\ 0 & 1 \end{bmatrix}}$$

$$= \begin{bmatrix} -1 & -2 \\ 3 & 8 \end{bmatrix}$$

In Section 2.3 you learned a process for finding the inverse of a nonsingular matrix A. There, you used Gauss-Jordan elimination to reduce the augmented matrix $[A \vdots I]$ to $[I \vdots A^{-1}]$. You can now use Theorem 2.14 to justify this procedure. Specifically, the proof of Theorem 2.14 allows you to write the product

$$I = E_k \cdots E_3 E_2 E_1 A.$$

Multiplying both sides of this equation (on the right) by A^{-1}, we can write

$$A^{-1} = E_k \cdots E_3 E_2 E_1 I.$$

In other words, a sequence of elementary matrices that reduces A to the identity also can be used to reduce the identity I to A^{-1}. Applying the corresponding sequence of elementary row operations to the matrices A and I simultaneously, you have

$$E_k \cdots E_3 E_2 E_1 [A \vdots I] = [I \vdots A^{-1}].$$

Of course, if A is singular, then no such sequence can be found.

The next theorem ties together some important relationships between $n \times n$ matrices and systems of linear equations. The essential parts of this theorem have already been proved (see Theorems 2.11 and 2.14); it is left to you to fill in the other parts of the proof.

Theorem 2.15

Equivalent Conditions

If A is an $n \times n$ matrix, then the following statements are equivalent.
1. A is invertible.
2. $A\mathbf{x} = \mathbf{b}$ has a unique solution for every $n \times 1$ column matrix $\mathbf{b}$.
3. $A\mathbf{x} = O$ has only the trivial solution.
4. A is row-equivalent to I_n.
5. A can be written as the product of elementary matrices.

The LU–Factorization

Solving systems of linear equations is the most important application of linear algebra. At the heart of the most efficient and modern algorithms for solving linear systems $A\mathbf{x} = \mathbf{b}$ is the so-called *LU*–factorization, in which the square matrix A is expressed as a product, $A = LU$. In this product, the square matrix L is **lower triangular,** which means that all the entries above the main diagonal are zero. The square matrix U is **upper triangular,** which means that all the entries below the main diagonal are zero.

$$\begin{bmatrix} a_{11} & 0 & 0 \\ a_{21} & a_{22} & 0 \\ a_{31} & a_{32} & a_{33} \end{bmatrix} \qquad \begin{bmatrix} a_{11} & a_{12} & a_{13} \\ 0 & a_{22} & a_{23} \\ 0 & 0 & a_{33} \end{bmatrix}$$

3×3 lower triangular matrix $\qquad$ 3×3 upper triangular matrix

By writing $A\mathbf{x} = LU\mathbf{x}$ and letting $U\mathbf{x} = \mathbf{y}$, you can solve for $\mathbf{x}$ in two stages. First solve $L\mathbf{y} = \mathbf{b}$ for $\mathbf{y}$; then solve $U\mathbf{x} = \mathbf{y}$ for $\mathbf{x}$. Each system is easy to solve because the coefficient matrices are triangular. In particular, neither system requires any row operations.

**Definition of
LU–Factorization**

If the $n \times n$ matrix A can be written as the product of a lower triangular matrix L and an upper triangular matrix U, then $A = LU$ is an **LU–factorization** of A.

EXAMPLE 5 *LU-Factorizations*

(a) $\begin{bmatrix} 1 & 2 \\ 1 & 0 \end{bmatrix} = \begin{bmatrix} 1 & 0 \\ 1 & 1 \end{bmatrix} \begin{bmatrix} 1 & 2 \\ 0 & -2 \end{bmatrix}$

is an *LU*–factorization of the matrix $A = \begin{bmatrix} 1 & 2 \\ 1 & 0 \end{bmatrix}$ as the product of the lower triangular matrix $L = \begin{bmatrix} 1 & 0 \\ 1 & 1 \end{bmatrix}$ and the upper triangular matrix $U = \begin{bmatrix} 1 & 2 \\ 0 & -2 \end{bmatrix}$.

(b) $A = \begin{bmatrix} 1 & -3 & 0 \\ 0 & 1 & 3 \\ 2 & -10 & 2 \end{bmatrix} = \begin{bmatrix} 1 & 0 & 0 \\ 0 & 1 & 0 \\ 2 & -4 & 1 \end{bmatrix} \begin{bmatrix} 1 & -3 & 0 \\ 0 & 1 & 3 \\ 0 & 0 & 14 \end{bmatrix} = LU$

is an *LU*–factorization of the matrix *A*.

If a square matrix *A* can be row reduced to an upper triangular matrix *U* using only the row operation of adding a multiple of one row to another row, then it is easy to find an *LU*–factorization of the matrix *A*. All you need to do is keep track of the individual row operations, as indicated in the following example.

EXAMPLE 6   *Finding the LU-Factorization of a Matrix*

Find the *LU*–factorization of the matrix $A = \begin{bmatrix} 1 & -3 & 0 \\ 0 & 1 & 3 \\ 2 & -10 & 2 \end{bmatrix}$.

Solution   Begin by row reducing *A* to upper triangular form while keeping track of the elementary matrices used for each row operation.

Matrix	*Elementary Matrix*
$\begin{bmatrix} 1 & -3 & 0 \\ 0 & 1 & 3 \\ 0 & -4 & 2 \end{bmatrix}$	$E_1 = \begin{bmatrix} 1 & 0 & 0 \\ 0 & 1 & 0 \\ -2 & 0 & 1 \end{bmatrix}$
$\begin{bmatrix} 1 & -3 & 0 \\ 0 & 1 & 3 \\ 0 & 0 & 14 \end{bmatrix}$	$E_2 = \begin{bmatrix} 1 & 0 & 0 \\ 0 & 1 & 0 \\ 0 & 4 & 1 \end{bmatrix}$

The matrix *U* on the left is upper triangular, and it follows that $E_2 E_1 A = U$, or $A = E_1^{-1} E_2^{-1} U$. Because the product of the lower triangular matrices

$$E_1^{-1} E_2^{-1} = \begin{bmatrix} 1 & 0 & 0 \\ 0 & 1 & 0 \\ 2 & 0 & 1 \end{bmatrix} \begin{bmatrix} 1 & 0 & 0 \\ 0 & 1 & 0 \\ 0 & -4 & 1 \end{bmatrix} = \begin{bmatrix} 1 & 0 & 0 \\ 0 & 1 & 0 \\ 2 & -4 & 1 \end{bmatrix}$$

is again a lower triangular matrix *L*, the factorization $A = LU$ is complete. Note that this is the same *LU*–factorization that was shown in Example 5(b) at the top of the page.

In general, if *A* can be row reduced to an upper triangular matrix *U* using only the row operation of adding a multiple of one row to another row, then *A* has an *LU*–factorization.

$$E_k \cdots E_2 E_1 A = U$$
$$A = E_1^{-1} E_2^{-1} \cdots E_k^{-1} U \quad .$$
$$A = LU$$

Here *L* is the product of the inverses of the elementary matrices used in the row reduction.

Note that the multipliers in Example 6 were -2 and 4, which are the negatives of the corresponding entries in L. This is true in general. If U can be obtained from A using only the row operation of adding a multiple of one row to another row, then the matrix L is lower triangular with 1's along the diagonal. Furthermore, the negative of each multiplier is in the same position used to introduce the corresponding zero in U.

Once you have obtained an LU–factorization of a matrix A, then you can solve the system of n linear equations in n unknowns $A\mathbf{x} = \mathbf{b}$ very efficiently in two steps.

1. Write $\mathbf{y} = U\mathbf{x}$ and solve $L\mathbf{y} = \mathbf{b}$ for $\mathbf{y}$.
2. Solve $U\mathbf{x} = \mathbf{y}$ for $\mathbf{x}$.

The column matrix $\mathbf{x}$ is the solution of the original system because

$$A\mathbf{x} = LU\mathbf{x} = L\mathbf{y} = \mathbf{b}.$$

The second step in this algorithm is just back-substitution, because the matrix U is upper triangular. The first step is similar, except that it starts at the top of the matrix, because L is lower triangular. For this reason, the first step is often called **forward substitution.**

EXAMPLE 7   *Solving a Linear System Using LU-Factorization*

Solve the linear system.

$$\begin{aligned} x_1 - 3x_2 \qquad\quad &= -5 \\ x_2 + 3x_3 &= -1 \\ 2x_1 - 10x_2 + 2x_3 &= -20 \end{aligned}$$

Solution   You obtained the LU–factorization of the coefficient matrix A in Example 6.

$$A = \begin{bmatrix} 1 & -3 & 0 \\ 0 & 1 & 3 \\ 2 & -10 & 2 \end{bmatrix} = \begin{bmatrix} 1 & 0 & 0 \\ 0 & 1 & 0 \\ 2 & -4 & 1 \end{bmatrix}\begin{bmatrix} 1 & -3 & 0 \\ 0 & 1 & 3 \\ 0 & 0 & 14 \end{bmatrix}$$

First, let $\mathbf{y} = U\mathbf{x}$ and solve the system $L\mathbf{y} = \mathbf{b}$ for $\mathbf{y}$.

$$\begin{bmatrix} 1 & 0 & 0 \\ 0 & 1 & 0 \\ 2 & -4 & 1 \end{bmatrix}\begin{bmatrix} y_1 \\ y_2 \\ y_3 \end{bmatrix} = \begin{bmatrix} -5 \\ -1 \\ -20 \end{bmatrix}$$

This system is easy to solve using forward substitution. Starting with the first equation, you have $y_1 = -5$. The second equation gives $y_2 = -1$. Finally, from the third equation,

$$\begin{aligned} 2y_1 - 4y_2 + y_3 &= -20 \\ y_3 &= -20 - 2y_1 + 4y_2 \\ y_3 &= -20 - 2(-5) + 4(-1) \\ y_3 &= -14. \end{aligned}$$

Thus, the solution of $L\mathbf{y} = \mathbf{b}$ is

$$\mathbf{y} = \begin{bmatrix} -5 \\ -1 \\ -14 \end{bmatrix}.$$

Now solve the system $U\mathbf{x} = \mathbf{y}$ for $\mathbf{x}$ using back-substitution.

$$\begin{bmatrix} 1 & -3 & 0 \\ 0 & 1 & 3 \\ 0 & 0 & 14 \end{bmatrix} \begin{bmatrix} x_1 \\ x_2 \\ x_3 \end{bmatrix} = \begin{bmatrix} -5 \\ -1 \\ -14 \end{bmatrix}$$

From the bottom equation, $x_3 = -1$. Then, the second equation gives $x_2 + 3(-1) = -1$, or $x_2 = 2$. Finally, the first equation is $x_1 - 3(2) = -5$, or $x_1 = 1$. Thus, the solution of the original system of equations is

$$\mathbf{x} = \begin{bmatrix} 1 \\ 2 \\ -1 \end{bmatrix}.$$

SECTION 2.4   Exercises

In Exercises 1–8, determine whether the matrix is elementary. If it is, state the elementary row operation that was used to produce it.

1. $\begin{bmatrix} 1 & 0 \\ 0 & 2 \end{bmatrix}$

2. $\begin{bmatrix} 1 & 0 & 0 \\ 0 & 0 & 1 \end{bmatrix}$

3. $\begin{bmatrix} 1 & 0 \\ 2 & 1 \end{bmatrix}$

4. $\begin{bmatrix} 0 & 1 \\ 1 & 0 \end{bmatrix}$

5. $\begin{bmatrix} 2 & 0 & 0 \\ 0 & 0 & 1 \\ 0 & 1 & 0 \end{bmatrix}$

6. $\begin{bmatrix} 1 & 0 & 0 \\ 0 & 1 & 0 \\ 2 & 0 & 1 \end{bmatrix}$

7. $\begin{bmatrix} 1 & 0 & 0 & 0 \\ 0 & 1 & 0 & 0 \\ 0 & -5 & 1 & 0 \\ 0 & 0 & 0 & 1 \end{bmatrix}$

8. $\begin{bmatrix} 1 & 0 & 0 & 0 \\ 2 & 1 & 0 & 0 \\ 0 & 0 & 1 & 0 \\ 0 & 0 & -3 & 1 \end{bmatrix}$

In Exercises 9–12, let A, B, and C be given by

$$A = \begin{bmatrix} 1 & 2 & -3 \\ 0 & 1 & 2 \\ -1 & 2 & 0 \end{bmatrix}, \quad B = \begin{bmatrix} -1 & 2 & 0 \\ 0 & 1 & 2 \\ 1 & 2 & -3 \end{bmatrix},$$

$$C = \begin{bmatrix} 0 & 4 & -3 \\ 0 & 1 & 2 \\ -1 & 2 & 0 \end{bmatrix}.$$

9. Find an elementary matrix E such that $EA = B$.

10. Find an elementary matrix E such that $EA = C$.

11. Find an elementary matrix E such that $EB = A$.

12. Find an elementary matrix E such that $EC = A$.

In Exercises 13–20, find the inverse of the elementary matrix.

13. $\begin{bmatrix} 0 & 1 \\ 1 & 0 \end{bmatrix}$

14. $\begin{bmatrix} 5 & 0 \\ 0 & 1 \end{bmatrix}$

15. $\begin{bmatrix} 0 & 0 & 1 \\ 0 & 1 & 0 \\ 1 & 0 & 0 \end{bmatrix}$

16. $\begin{bmatrix} 1 & 0 & 0 \\ 0 & 1 & 0 \\ 0 & -3 & 1 \end{bmatrix}$

17. $\begin{bmatrix} k & 0 & 0 \\ 0 & 1 & 0 \\ 0 & 0 & 1 \end{bmatrix}, \quad k \neq 0$

18. $\begin{bmatrix} 1 & 0 & 0 \\ 0 & 0 & 1 \\ 0 & 1 & 0 \end{bmatrix}$

19. $\begin{bmatrix} 1 & 0 & 0 & 0 \\ 0 & 1 & k & 0 \\ 0 & 0 & 1 & 0 \\ 0 & 0 & 0 & 1 \end{bmatrix}$

20. $\begin{bmatrix} 1 & 0 & 0 & 0 \\ 0 & 1 & 0 & 0 \\ 0 & 0 & 1/k & 0 \\ 0 & 0 & 0 & 1 \end{bmatrix}, \quad k \neq 0$

In Exercises 21–28, factor the matrix A into a product of elementary matrices.

21. $A = \begin{bmatrix} 1 & 2 \\ 1 & 0 \end{bmatrix}$

22. $A = \begin{bmatrix} 0 & 1 \\ 1 & 0 \end{bmatrix}$

23. $A = \begin{bmatrix} 4 & -1 \\ 3 & -1 \end{bmatrix}$

24. $A = \begin{bmatrix} 1 & 1 \\ 2 & 1 \end{bmatrix}$

25. $A = \begin{bmatrix} 1 & -2 & 0 \\ -1 & 3 & 0 \\ 0 & 0 & 1 \end{bmatrix}$

26. $A = \begin{bmatrix} 1 & 2 & 3 \\ 2 & 5 & 6 \\ 1 & 3 & 4 \end{bmatrix}$

27. $A = \begin{bmatrix} 2 & 0 & -8 \\ 2 & 2 & 2 \\ 0 & 2 & 7 \end{bmatrix}$ **28.** $A = \begin{bmatrix} -1 & 2 & 0 \\ 0 & 2 & 0 \\ 1 & 0 & 1 \end{bmatrix}$

True or False? In Exercises 29 and 30, determine whether each statement is true or false. If a statement is true, give a reason or cite an appropriate statement in the text. If a statement is false, provide an example that shows that the statement is not true in all cases or cite an appropriate statement in the text.

29. (a) The identity matrix is an elementary matrix.
(b) If E is an elementary matrix, then $2E$ is an elementary matrix.
(c) The matrix A is row-equivalent to the matrix B if there exists a finite number of elementary matrices $E_1, E_2, \ldots, E_k$ such that $A = E_k E_{k-1} \cdots E_2 E_1 B$.
(d) The inverse of an elementary matrix is an elementary matrix.

30. (a) The zero matrix is an elementary matrix.
(b) A square matrix is nonsingular if it can be written as the product of elementary matrices.
(c) $A\mathbf{x} = O$ has only the trivial solution if and only if $A\mathbf{x} = \mathbf{b}$ has a unique solution for every $n \times 1$ column matrix $\mathbf{b}$.

31. ***Writing*** E is the elementary matrix obtained by interchanging two rows in I_n. A is an $n \times n$ matrix.
(a) How will EA compare with A?
(b) Find E^2.

32. ***Writing*** E is the elementary matrix obtained by multiplying a row in I_n by a nonzero constant c. A is an $n \times n$ matrix.
(a) How will EA compare with A?
(b) Find E^2.

33. Use elementary matrices to find the inverse of
$$A = \begin{bmatrix} 1 & a & 0 \\ 0 & 1 & 0 \\ 0 & 0 & 1 \end{bmatrix} \begin{bmatrix} 1 & 0 & 0 \\ b & 1 & 0 \\ 0 & 0 & 1 \end{bmatrix} \begin{bmatrix} 1 & 0 & 0 \\ 0 & 1 & 0 \\ 0 & 0 & c \end{bmatrix}, \; c \neq 0.$$

34. Use elementary matrices to find the inverse of
$$A = \begin{bmatrix} 1 & 0 & 0 \\ 0 & 1 & 0 \\ a & b & c \end{bmatrix}, \; c \neq 0.$$

35. ***Writing*** Is the product of two elementary matrices always elementary? Explain why or why not and provide appropriate examples to illustrate your conclusion.

36. ***Writing*** Is the sum of two elementary matrices always elementary? Explain why or why not and provide appropriate examples to illustrate your conclusion.

In Exercises 37–40, find an *LU*–factorization of the matrix.

37. $\begin{bmatrix} 1 & 0 \\ -2 & 1 \end{bmatrix}$ **38.** $\begin{bmatrix} -2 & 1 \\ -6 & 4 \end{bmatrix}$

39. $\begin{bmatrix} 3 & 0 & 1 \\ 6 & 1 & 1 \\ -3 & 1 & 0 \end{bmatrix}$ **40.** $\begin{bmatrix} 2 & 0 & 0 \\ 0 & -3 & 1 \\ 10 & 12 & 3 \end{bmatrix}$

In Exercises 41 and 42, solve the linear system $A\mathbf{x} = \mathbf{b}$ by
(a) finding the *LU*–factorization of the coefficient matrix A,
(b) solving the lower triangular system $L\mathbf{y} = \mathbf{b}$, and
(c) solving the upper triangular system $U\mathbf{x} = \mathbf{y}$.

41.
$$\begin{aligned} 2x + y & = 1 \\ y - z & = 2 \\ -2x + y + z & = -2 \end{aligned}$$

42.
$$\begin{aligned} 2x_1 & = 4 \\ -2x_1 + x_2 - x_3 & = -4 \\ 6x_1 + 2x_2 + x_3 & = 15 \\ -x_4 & = -1 \end{aligned}$$

43. ***Writing*** Suppose you needed to solve many systems of linear equations $A\mathbf{x} = \mathbf{b}_i$, each having the same coefficient matrix A. Explain how you could use the *LU*–factorization technique to make the task easier, rather than solving each system individually using Gaussian elimination.

44. (a) Show that the matrix
$$A = \begin{bmatrix} 0 & 1 \\ 1 & 0 \end{bmatrix}$$
does not have an *LU*–factorization.
(b) Find the *LU*–factorization of the matrix
$$A = \begin{bmatrix} a & b \\ c & d \end{bmatrix}$$
that has 1's along the main diagonal of L. Are there any restrictions on the matrix A?

In Exercises 45–50, determine whether the matrix is idempotent. A square matrix A is **idempotent** if $A^2 = A$.

45. $\begin{bmatrix} 1 & 0 \\ 0 & 0 \end{bmatrix}$ **46.** $\begin{bmatrix} 0 & 1 \\ 1 & 0 \end{bmatrix}$

47. $\begin{bmatrix} 2 & 3 \\ -1 & -2 \end{bmatrix}$ **48.** $\begin{bmatrix} 2 & 3 \\ 1 & 2 \end{bmatrix}$

49. $\begin{bmatrix} 0 & 0 & 1 \\ 0 & 1 & 0 \\ 1 & 0 & 0 \end{bmatrix}$ **50.** $\begin{bmatrix} 0 & 1 & 0 \\ 1 & 0 & 0 \\ 0 & 0 & 1 \end{bmatrix}$

51. Determine a and b such that A is idempotent.

$$A = \begin{bmatrix} 1 & 0 \\ a & b \end{bmatrix}$$

52. Determine conditions on a, b, and c such that A is idempotent.

$$A = \begin{bmatrix} a & 0 \\ b & c \end{bmatrix}$$

53. Prove that if A is an $n \times n$ matrix that is idempotent and invertible, then $A = I_n$.

54. *Guided Proof*   Prove that A is idempotent if and only if A^T is idempotent.

Getting Started: The phrase "if and only if" means that you have to prove two statements:

1. If A is idempotent, then A^T is idempotent.
2. If A^T is idempotent, then A is idempotent.

 (i) Begin your proof of the first statement by assuming that A is idempotent.
 (ii) This means that $A^2 = A$.
 (iii) Use the properties of the transpose to show that A^T is idempotent.
 (iv) Begin your proof of the second statement by assuming that A^T is idempotent.

55. Prove that if A and B are idempotent and $AB = BA$, then AB is idempotent.

56. Prove that if A is row-equivalent to B, then B is row-equivalent to A.

57. *Guided Proof*   Prove that if A is row-equivalent to B and B is row-equivalent to C, then A is row-equivalent to C.

Getting Started: To prove that A is row-equivalent to C, you have to find elementary matrices $E_1, \ldots, E_k$ such that $A = E_k \cdots E_1 C$.

 (i) Begin your proof by observing that A is row-equivalent to B.
 (ii) Hence, there exist elementary matrices $F_1, \ldots, F_n$ such that $A = F_n \cdots F_1 B$.
 (iii) There exist elementary matrices $G_1, \ldots, G_m$ such that $B = G_1 \cdots G_m C$.
 (iv) Combine the matrix equations from steps (ii) and (iii).

58. Let A be a nonsingular matrix. Prove that if B is row-equivalent to A, then B is also nonsingular.

2.5 | Applications of Matrix Operations

Stochastic Matrices

Many types of applications involve a finite set of *states* $\{S_1, S_2, \ldots, S_n\}$ of a given population. For instance, residents of a city may live downtown or in the suburbs. Voters may vote Democrat, Republican, or for a third party. Soft drink consumers may buy Coca-Cola, Pepsi Cola, or another brand.

The probability that a member of a population will change from the jth state to the ith state is represented by a number p_{ij}, where $0 \le p_{ij} \le 1$. A probability of $p_{ij} = 0$ means that the member is certain *not* to change from the jth state to the ith state, whereas a probability of $p_{ij} = 1$ means that the member is certain to change from the jth state to the ith state.

$$P = \begin{bmatrix} p_{11} & p_{12} & \cdots & p_{1n} \\ p_{21} & p_{22} & \cdots & p_{2n} \\ \vdots & \vdots & & \vdots \\ p_{n1} & p_{n2} & \cdots & p_{nn} \end{bmatrix} \begin{matrix} S_1 \\ S_2 \\ \vdots \\ S_n \end{matrix}$$

From $S_1 \ S_2 \ \cdots \ S_n$ *To*

P is called the **matrix of transition probabilities** because it gives the probabilities of each possible type of transition (or change) within the population.

At each transition, each member in a given state must either stay in that state or change to another state. For probabilities, this means that the sum of the entries in any column of P is 1. For instance, in the first column we have

$$p_{11} + p_{21} + \cdots + p_{n1} = 1.$$

In general, such a matrix is called **stochastic** (the term "stochastic" means "regarding conjecture"). That is, an $n \times n$ matrix P is called a **stochastic matrix** if each entry is a number between 0 and 1 and each column of P adds up to 1.

EXAMPLE 1   *Examples of Stochastic Matrices and Nonstochastic Matrices*

The matrices in parts (a) and (b) are stochastic, but the matrix in part (c) is not.

$$\text{(a)} \begin{bmatrix} 1 & 0 & 0 \\ 0 & 1 & 0 \\ 0 & 0 & 1 \end{bmatrix} \quad \text{(b)} \begin{bmatrix} \frac{1}{2} & \frac{1}{3} & \frac{1}{4} \\ \frac{1}{4} & 0 & \frac{3}{4} \\ \frac{1}{4} & \frac{2}{3} & 0 \end{bmatrix} \quad \text{(c)} \begin{bmatrix} 0.1 & 0.2 & 0.3 \\ 0.2 & 0.3 & 0.4 \\ 0.3 & 0.4 & 0.5 \end{bmatrix}$$

Example 2 describes the use of a stochastic matrix to measure consumer preference.

EXAMPLE 2   *A Consumer Preference Model*

Two competing companies offer cable television service to a city of 100,000 households. The change in cable subscriptions each year is shown in Figure 2.1. Company A now has 15,000 subscribers and Company B has 20,000 subscribers. How many subscribers will each company have 1 year from now?

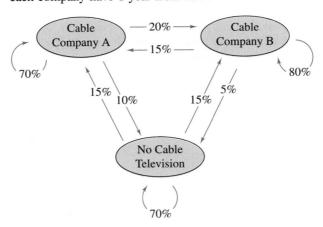

Figure 2.1

Solution   The matrix representing the given transition probabilities is

$$P = \begin{array}{c} \overset{\textit{From}}{\overbrace{\begin{array}{ccc} A & B & \textit{None} \end{array}}} \\ \begin{bmatrix} 0.70 & 0.15 & 0.15 \\ 0.20 & 0.80 & 0.15 \\ 0.10 & 0.05 & 0.70 \end{bmatrix} \begin{array}{l} A \\ B \\ \textit{None} \end{array} \left. \vphantom{\begin{bmatrix} 0.70 \\ 0.20 \\ 0.10 \end{bmatrix}} \right\} \textit{To} \end{array}$$

and the **state matrix** representing the current population in each of the three states is

$$X = \begin{bmatrix} 15,000 \\ 20,000 \\ 65,000 \end{bmatrix} \begin{array}{l} \text{A} \\ \text{B} \\ \text{None} \end{array}.$$

To find the state matrix representing the population in each of the three states after 1 year, multiply P by X to obtain

$$PX = \begin{bmatrix} 0.70 & 0.15 & 0.15 \\ 0.20 & 0.80 & 0.15 \\ 0.10 & 0.05 & 0.70 \end{bmatrix} \begin{bmatrix} 15,000 \\ 20,000 \\ 65,000 \end{bmatrix} = \begin{bmatrix} 23,250 \\ 28,750 \\ 48,000 \end{bmatrix}.$$

After 1 year Company A will have 23,250 subscribers and Company B will have 28,750 subscribers.

One of the appeals of the matrix solution given in Example 2 is that once the model has been created, it becomes easy to find the state matrices representing future years by repeatedly multiplying by the matrix P. This process is demonstrated in Example 3.

EXAMPLE 3   *A Consumer Preference Model*

Assuming that the matrix of transition probabilities given in Example 2 remains the same year after year, find the number of subscribers each cable television company will have after (a) 3 years, (b) 5 years, and (c) 10 years. (The answers in this example have been rounded to the nearest person.)

Solution   (a) From Example 2 you know that the number of subscribers after 1 year is

$$PX = \begin{bmatrix} 23,250 \\ 28,750 \\ 48,000 \end{bmatrix} \begin{array}{l} \text{A} \\ \text{B} \\ \text{None} \end{array}. \qquad \text{After 1 year}$$

Because the matrix of transition probabilities is the same from the first to the third year, the number of subscribers after 3 years is

$$P^3X = \begin{bmatrix} 30,283 \\ 39,042 \\ 30,675 \end{bmatrix} \begin{array}{l} \text{A} \\ \text{B} \\ \text{None} \end{array}. \qquad \text{After 3 years}$$

(b) The number of subscribers after 5 years is

$$P^5 X = \begin{bmatrix} 32,411 \\ 43,812 \\ 23,777 \end{bmatrix}. \quad \begin{matrix} \text{A} \\ \text{B} \\ \text{None} \end{matrix} \qquad \text{After 5 years}$$

(c) The number of subscribers after 10 years is

$$P^{10} X = \begin{bmatrix} 33,287 \\ 47,147 \\ 19,566 \end{bmatrix}. \quad \begin{matrix} \text{A} \\ \text{B} \\ \text{None} \end{matrix} \qquad \text{After 10 years}$$

In Example 3, notice that there is little difference between the number of subscribers after 5 years and after 10 years. If the process shown in this example is continued, the number of subscribers eventually reaches a **steady state.** That is, as long as the matrix P doesn't change, the matrix product $P^n X$ approaches a limit $\overline{X}$. In this particular example, the limit is given by the state matrix

$$\overline{X} = \begin{bmatrix} 33,333 \\ 47,619 \\ 19,048 \end{bmatrix}. \quad \begin{matrix} \text{A} \\ \text{B} \\ \text{None} \end{matrix} \qquad \text{Steady state}$$

You can check to see that $P\overline{X} = \overline{X}$.

Cryptography

A **cryptogram** is a message written according to a secret code (the Greek word *kryptos* means "hidden"). This section describes a method of using matrix multiplication to **encode** and **decode** messages.

Begin by assigning a number to each letter in the alphabet (with 0 assigned to a blank space) as follows.

0 = __	14 = N
1 = A	15 = O
2 = B	16 = P
3 = C	17 = Q
4 = D	18 = R
5 = E	19 = S
6 = F	20 = T
7 = G	21 = U
8 = H	22 = V
9 = I	23 = W
10 = J	24 = X
11 = K	25 = Y
12 = L	26 = Z
13 = M	

Then the message is converted to numbers and partitioned into **uncoded row matrices,** each having n entries, as demonstrated in Example 4.

EXAMPLE 4   *Forming Uncoded Row Matrices*

Write the uncoded row matrices of size 1×3 for the message MEET ME MONDAY.

Solution   Partitioning the message (including blank spaces, but ignoring other punctuation) into groups of three produces the following uncoded row matrices.

$$[13 \quad 5 \quad 5] \quad [20 \quad 0 \quad 13] \quad [5 \quad 0 \quad 13] \quad [15 \quad 14 \quad 4] \quad [1 \quad 25 \quad 0]$$
$$\ \ M \quad E \quad E \qquad T \quad \_ \quad M \qquad E \quad \_ \quad M \qquad O \quad N \quad D \qquad A \quad Y \quad \_$$

Note that a blank space is used to fill out the last uncoded row matrix.

To **encode** a message, choose an $n \times n$ invertible matrix A and multiply the uncoded row matrices (on the right) by A to obtain **coded row matrices.** This process is demonstrated in Example 5.

EXAMPLE 5   *Encoding a Message*

Use the matrix

$$A = \begin{bmatrix} 1 & -2 & 2 \\ -1 & 1 & 3 \\ 1 & -1 & -4 \end{bmatrix}$$

to encode the message MEET ME MONDAY.

Solution   The coded row matrices are obtained by multiplying each of the uncoded row matrices found in Example 4 by the matrix A as follows.

Uncoded Row Matrix	*Encoding Matrix A*	*Coded Row Matrix*

$$[13 \quad 5 \quad 5] \begin{bmatrix} 1 & -2 & 2 \\ -1 & 1 & 3 \\ 1 & -1 & -4 \end{bmatrix} = [13 \quad -26 \quad 21]$$

$$[20 \quad 0 \quad 13] \begin{bmatrix} 1 & -2 & 2 \\ -1 & 1 & 3 \\ 1 & -1 & -4 \end{bmatrix} = [33 \quad -53 \quad -12]$$

$$[5 \quad 0 \quad 13] \begin{bmatrix} 1 & -2 & 2 \\ -1 & 1 & 3 \\ 1 & -1 & -4 \end{bmatrix} = [18 \quad -23 \quad -42]$$

	Uncoded *Row Matrix*	*Encoding* *Matrix A*	*Coded Row* *Matrix*

$$[15 \quad 14 \quad 4] \begin{bmatrix} 1 & -2 & 2 \\ -1 & 1 & 3 \\ 1 & -1 & -4 \end{bmatrix} = [5 \ -20 \quad 56]$$

$$[1 \quad 25 \quad 0] \begin{bmatrix} 1 & -2 & 2 \\ -1 & 1 & 3 \\ 1 & -1 & -4 \end{bmatrix} = [-24 \quad 23 \quad 77]$$

Thus the sequence of coded row matrices is

$$[13 \ -26 \quad 21][33 \ -53 \ -12][18 \ -23 \ -42][5 \ -20 \quad 56][-24 \quad 23 \quad 77].$$

Finally, removing the matrix notation produces the following cryptogram.

$$13 \ -26 \ 21 \ 33 \ -53 \ -12 \ 18 \ -23 \ -42 \ 5 \ -20 \ 56 \ -24 \ 23 \ 77$$

For those who do not know the matrix A, decoding the cryptogram found in Example 5 is difficult. But for an authorized receiver who knows the matrix A, decoding is simple. The receiver need only multiply the coded row matrices by A^{-1} to retrieve the uncoded row matrices. In other words, if

$$X = [x_1 \ x_2 \ \cdots \ x_n]$$

is an uncoded $1 \times n$ matrix, then $Y = XA$ is the corresponding encoded matrix. The receiver of the encoded matrix can decode Y by multiplying on the right by A^{-1} to obtain

$$YA^{-1} = (XA)A^{-1} = X.$$

This procedure is demonstrated in Example 6.

EXAMPLE 6   *Decoding a Message*

Use the inverse of the matrix

$$A = \begin{bmatrix} 1 & -2 & 2 \\ -1 & 1 & 3 \\ 1 & -1 & -4 \end{bmatrix}$$

to decode the cryptogram

$$13 \ -26 \ 21 \ 33 \ -53 \ -12 \ 18 \ -23 \ -42 \ 5 \ -20 \ 56 \ -24 \ 23 \ 77.$$

Solution   Begin by using Gauss-Jordan elimination to find A^{-1}.

Simulation

Explore this concept further with an electronic simulation available on the *Elementary Linear Algebra* Learning Tools Student CD-ROM.

$$[A \ \vdots \ I]$$

$$\begin{bmatrix} 1 & -2 & 2 & \vdots & 1 & 0 & 0 \\ -1 & 1 & 3 & \vdots & 0 & 1 & 0 \\ 1 & -1 & -4 & \vdots & 0 & 0 & 1 \end{bmatrix} \longrightarrow$$

$$[I \ \vdots \ A^{-1}]$$

$$\begin{bmatrix} 1 & 0 & 0 & \vdots & -1 & -10 & -8 \\ 0 & 1 & 0 & \vdots & -1 & -6 & -5 \\ 0 & 0 & 1 & \vdots & 0 & -1 & -1 \end{bmatrix}$$

Now, to decode the message, partition the message into groups of three to form the coded row matrices

$$[13 \; -26 \; 21][33 \; -53 \; -12][18 \; -23 \; -42][5 \; -20 \; 56][-24 \; 23 \; 77]$$

To obtain the decoded row matrices, multiply each coded row matrix by A^{-1} (on the right).

Coded Row Matrix	Decoding Matrix A^{-1}	Decoded Row Matrix

$$[13 \; -26 \; 21]\begin{bmatrix} -1 & -10 & -8 \\ -1 & -6 & -5 \\ 0 & -1 & -1 \end{bmatrix} = [13 \; 5 \; 5]$$

$$[33 \; -53 \; -12]\begin{bmatrix} -1 & -10 & -8 \\ -1 & -6 & -5 \\ 0 & -1 & -1 \end{bmatrix} = [20 \; 0 \; 13]$$

$$[18 \; -23 \; -42]\begin{bmatrix} -1 & -10 & -8 \\ -1 & -6 & -5 \\ 0 & -1 & -1 \end{bmatrix} = [5 \; 0 \; 13]$$

$$[5 \; -20 \; 56]\begin{bmatrix} -1 & -10 & -8 \\ -1 & -6 & -5 \\ 0 & -1 & -1 \end{bmatrix} = [15 \; 14 \; 4]$$

$$[-24 \; 23 \; 77]\begin{bmatrix} -1 & -10 & -8 \\ -1 & -6 & -5 \\ 0 & -1 & -1 \end{bmatrix} = [1 \; 25 \; 0]$$

Thus the sequence of decoded row matrices is

$$[13 \; 5 \; 5][20 \; 0 \; 13][5 \; 0 \; 13][15 \; 14 \; 4][1 \; 25 \; 0]$$

and the message is

13	5	5	20	0	13	5	0	13	15	14	4	1	25	0.
M	*E*	*E*	*T*	_	*M*	*E*	_	*M*	*O*	*N*	*D*	*A*	*Y*	_

Leontief Input-Output Models

Matrix algebra has proved effective in analyzing problems concerning the input and output of an economic system. The model discussed here, developed by the American economist Wassily W. Leontief (1906–1999), was first published in 1936. In 1973, Leontief was awarded a Nobel prize for his work in economics.

Suppose that an economic system has n different industries $I_1, I_2, \ldots, I_n$, each of which has **input** needs (raw materials, utilities, etc.) and an **output** (finished product). In producing each unit of output, an industry may use the outputs of other industries, including itself.

For example, an electric utility uses outputs from other industries, such as coal and water, and even uses its own electricity.

Let d_{ij} be the amount of output the jth industry needs from the ith industry to produce one unit of output per year. The matrix of these coefficients is called the **input-output matrix.**

$$D = \begin{array}{c} \overbrace{\begin{array}{cccc} I_1 & I_2 & \cdots & I_n \end{array}}^{User\ (Output)} \\ \begin{bmatrix} d_{11} & d_{12} & \cdots & d_{1n} \\ d_{21} & d_{22} & \cdots & d_{2n} \\ \vdots & \vdots & & \vdots \\ d_{n1} & d_{n2} & \cdots & d_{nn} \end{bmatrix} \begin{array}{c} I_1 \\ I_2 \\ \vdots \\ I_n \end{array} \end{array} \quad Supplier\ (Input)$$

To understand how to use this matrix, imagine that $d_{12} = 0.4$. This means that 0.4 unit of Industry 1's product must be used to produce one unit of Industry 2's product. If $d_{33} = 0.2$, then 0.2 unit of Industry 3's product is needed to produce one unit of its own product. For this model to work, the values of d_{ij} must satisfy $0 \le d_{ij} \le 1$ and the sum of the entries in any column must be less than or equal to 1.

EXAMPLE 7   *Forming an Input-Output Matrix*

Consider a simple economic system consisting of three industries: electricity, water, and coal. Production, or output, of one unit of electricity requires 0.5 unit of itself, 0.25 unit of water, and 0.25 unit of coal. Production, or output, of one unit of water requires 0.1 unit of electricity, 0.6 unit of itself, and 0 units of coal. Production, or output, of one unit of coal requires 0.2 unit of electricity, 0.15 unit of water, and 0.5 unit of itself. Find the input-output matrix for this system.

Solution   The column entries show the amount each industry requires from the others, as well as itself, in order to produce one unit of output.

$$\begin{array}{c} \overbrace{\begin{array}{ccc} E & W & C \end{array}}^{User\ (Output)} \\ \begin{bmatrix} 0.5 & 0.1 & 0.2 \\ 0.25 & 0.6 & 0.15 \\ 0.25 & 0 & 0.5 \end{bmatrix} \begin{array}{c} E \\ W \\ C \end{array} \end{array} \quad Supplier\ (Input)$$

The row entries show the amount each industry supplies to the other industries, as well as itself, in order for that particular industry to produce output of one unit. For instance, the electricity industry supplies 0.5 unit to itself, 0.1 unit to water, and 0.2 unit to coal.

To develop the Leontief input-output model further, let the total output of the ith industry be denoted by x_i. If the economic system is **closed** (meaning that it sells its products only to industries within the system, as in the example above), then the total output of the ith industry is given by the linear equation

$$x_i = d_{i1}x_1 + d_{i2}x_2 + \cdots + d_{in}x_n. \qquad \textbf{Closed system}$$

On the other hand, if the industries within the system sell products to nonproducing groups (such as governments or charitable organizations) outside the system, then the system is called **open** and the total output of the ith industry is given by

$$x_i = d_{i1}x_1 + d_{i2}x_2 + \cdots + d_{in}x_n + e_i, \qquad \text{Open system}$$

where e_i represents the external demand for the ith industry's product. The collection of total outputs for an open system is therefore represented by the following system of n linear equations.

$$x_1 = d_{11}x_1 + d_{12}x_2 + \cdots + d_{1n}x_n + e_1$$
$$x_2 = d_{21}x_1 + d_{22}x_2 + \cdots + d_{2n}x_n + e_2$$
$$\vdots$$
$$x_n = d_{n1}x_1 + d_{n2}x_2 + \cdots + d_{nn}x_n + e_n$$

The matrix form of this system is

$$X = DX + E,$$

where X is called the **output matrix** and E is called the **external demand matrix.**

EXAMPLE 8   *Solving for the Output Matrix of an Open System*

An economic system composed of three industries has the following input-output matrix.

$$D = \begin{array}{c} \\ \overbrace{}^{User\ (Output)} \\ \begin{array}{ccc} A & B & C \end{array} \\ \begin{bmatrix} 0.1 & 0.43 & 0 \\ 0.15 & 0 & 0.37 \\ 0.23 & 0.03 & 0.02 \end{bmatrix} \begin{array}{l} A \\ B \\ C \end{array} \end{array} \quad Supplier\ (Input)$$

Find the output matrix X if the external demand is given by

$$E = \begin{bmatrix} 20,000 \\ 30,000 \\ 25,000 \end{bmatrix}.$$

(The answers in this example have been rounded to the nearest unit.)

Solution   Letting I be the identity matrix, write the equation $X = DX + E$ as $IX - DX = E$, which means that

$$(I - D)X = E.$$

Using the matrix D above produces

$$I - D = \begin{bmatrix} 0.9 & -0.43 & 0 \\ -0.15 & 1 & -0.37 \\ -0.23 & -0.03 & 0.98 \end{bmatrix}.$$

Finally, applying Gauss-Jordan elimination to the system of linear equations represented by $(I - D)X = E$ produces

$$\begin{bmatrix} 0.9 & -0.43 & 0 & 20{,}000 \\ -0.15 & 1 & -0.37 & 30{,}000 \\ -0.23 & -0.03 & 0.98 & 25{,}000 \end{bmatrix} \longrightarrow \begin{bmatrix} 1 & 0 & 0 & 46{,}616 \\ 0 & 1 & 0 & 51{,}058 \\ 0 & 0 & 1 & 38{,}014 \end{bmatrix}.$$

Therefore, the output matrix is

$$X = \begin{bmatrix} 46{,}616 \\ 51{,}058 \\ 38{,}014 \end{bmatrix}. \quad \begin{matrix} \text{A} \\ \text{B} \\ \text{C} \end{matrix}$$

To produce the given external demand, the output of the three industries must be as follows.

> Output for Industry A: 46,616 units
>
> Output for Industry B: 51,058 units
>
> Output for Industry C: 38,014 units

The economic systems described in Examples 7 and 8 are, of course, simple ones. In the real world, an economic system would have many industries or industrial groups. For instance, an economic analysis of some of the producing groups in the United States would include the following list (taken from the *Statistical Abstract of the United States*).

1. Farm products (grain, livestock, poultry, bulk milk)
2. Processed foods and feeds (beverages, dairy products)
3. Textile products and apparel (yarns, threads, clothing)
4. Hides, skins, and leather (shoes, upholstery)
5. Fuels and power (coal, gasoline, electricity)
6. Chemicals and allied products (drugs, plastic resins)
7. Rubber and plastic products (tires, plastic containers)
8. Lumber and wood products (plywood, pencils)
9. Pulp, paper, and allied products (cardboard, newsprint)
10. Metals and metal products (plumbing fixtures, cans)
11. Machinery and equipment (tractors, drills, computers)
12. Furniture and household durables (carpets, appliances)
13. Nonmetallic mineral products (glass, concrete, bricks)
14. Transportation equipment (automobiles, trucks, planes)
15. Miscellaneous products (toys, cameras, linear algebra texts)

A matrix of order 15×15 would be required to represent even these broad industrial groupings using the Leontief input-output model. A more detailed analysis could easily require an input-output matrix of order greater than 100×100. Clearly, this type of analysis could be done only with the aid of a computer.

Least Squares Regression Analysis

You will now look at a procedure that is used in statistics to develop linear models. The following example demonstrates a visual method for approximating a line of best fit for a given set of data points.

EXAMPLE 9 *A Visual Straight-Line Approximation*

Determine the straight line that best fits the points.

$$(1, 1), (2, 2), (3, 4), (4, 4), \text{ and } (5, 6)$$

Solution Plot the points as shown in Figure 2.2. It appears that a good choice would be the line whose slope is 1 and whose y-intercept is 0.5. The equation of this line is

$$y = 0.5 + x.$$

An examination of the line shown in Figure 2.2 reveals that you can improve the fit by rotating the line counterclockwise slightly, as shown in Figure 2.3. It seems clear that this new line, the equation of which is $y = 1.2x$, fits the given points better than the original line.

Figure 2.2

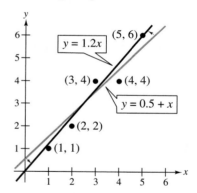

Figure 2.3

One way of measuring how well a function $y = f(x)$ fits a set of points

$$(x_1, y_1), (x_2, y_2), \ldots, (x_n, y_n)$$

is to compute the differences between the values given by the function $f(x_i)$ and the actual values y_i, as shown in Figure 2.4 on the next page. By squaring these differences and summing the results, you obtain a measure of error that is called the **sum of squared error.** The sums of squared errors for our two linear models are shown in Table 2.1 on the next page.

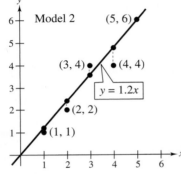

Figure 2.4

TABLE 2.1

Model 1: $f(x) = 0.5 + x$				Model 2: $f(x) = 1.2x$			
x_i	y_i	$f(x_i)$	$[y_i - f(x_i)]^2$	x_i	y_i	$f(x_i)$	$[y_i - f(x_i)]^2$
1	1	1.5	$(-0.5)^2$	1	1	1.2	$(-0.2)^2$
2	2	2.5	$(-0.5)^2$	2	2	2.4	$(-0.4)^2$
3	4	3.5	$(+0.5)^2$	3	4	3.6	$(+0.4)^2$
4	4	4.5	$(-0.5)^2$	4	4	4.8	$(-0.8)^2$
5	6	5.5	$(+0.5)^2$	5	6	6.0	$(+0.0)^2$
Total			1.25	Total			1.00

The sums of squared errors confirm that the second model fits the given points better than the first.

Of all possible linear models for a given set of points, the model that has the best fit is defined to be the one that minimizes the sum of squared error. This model is called the **least squares regression line,** and the procedure for finding it is called the **method of least squares.**

Definition of Least Squares Regression Line

For a set of points (x_1, y_1), (x_2, y_2), . . . , (x_n, y_n), the **least squares regression line** is given by the linear function

$$f(x) = a_0 + a_1x$$

that minimizes the sum of squared error

$$[y_1 - f(x_1)]^2 + [y_2 - f(x_2)]^2 + \cdots + [y_n - f(x_n)]^2.$$

To find the least squares regression line for a set of points, begin by forming the system of linear equations

$$y_1 = f(x_1) + [y_1 - f(x_1)]$$
$$y_2 = f(x_2) + [y_2 - f(x_2)]$$
$$\vdots$$
$$y_n = f(x_n) + [y_n - f(x_n)]$$

where the right-hand term, $[y_i - f(x_i)]$, of each equation is thought of as the error in the approximation of y_i by $f(x_i)$. Then write this error as

$$e_i = y_i - f(x_i)$$

so that the system of equations takes the form

$$y_1 = (a_0 + a_1x_1) + e_1$$
$$y_2 = (a_0 + a_1x_2) + e_2$$
$$\vdots$$
$$y_n = (a_0 + a_1x_n) + e_n.$$

Now, if you define Y, X, A, and E as

$$Y = \begin{bmatrix} y_1 \\ y_2 \\ \vdots \\ y_n \end{bmatrix}, \qquad X = \begin{bmatrix} 1 & x_1 \\ 1 & x_2 \\ \vdots & \vdots \\ 1 & x_n \end{bmatrix}, \qquad A = \begin{bmatrix} a_0 \\ a_1 \end{bmatrix}, \qquad E = \begin{bmatrix} e_1 \\ e_2 \\ \vdots \\ e_n \end{bmatrix}$$

the n linear equations may be replaced by the matrix equation

$$Y = XA + E.$$

Note that the matrix X has two columns, a column of 1's (corresponding to a_0) and a column containing the x_i's. This matrix equation can be used to determine the coefficients of the least squares regression line as follows.

Matrix Form for
Linear Regression

For the regression model $Y = XA + E$, the coefficients of the least squares regression line are given by the matrix equation

$$A = (X^TX)^{-1}X^TY$$

and the sum of squared error is

$$E^TE.$$

REMARK: You will learn more about this result in Section 5.4.

Example 10 demonstrates the use of this procedure to find the least squares regression line for the set of points given in Example 9.

EXAMPLE 10   *Finding the Least Squares Regression Line*

Find the least squares regression line for the points $(1, 1)$, $(2, 2)$, $(3, 4)$, $(4, 4)$, and $(5, 6)$. Then find the sum of squared error for this regression line.

Solution　Using the five given points, the matrices X and Y are

$$X = \begin{bmatrix} 1 & 1 \\ 1 & 2 \\ 1 & 3 \\ 1 & 4 \\ 1 & 5 \end{bmatrix} \quad \text{and} \quad Y = \begin{bmatrix} 1 \\ 2 \\ 4 \\ 4 \\ 6 \end{bmatrix}.$$

This means that

$$X^T X = \begin{bmatrix} 1 & 1 & 1 & 1 & 1 \\ 1 & 2 & 3 & 4 & 5 \end{bmatrix} \begin{bmatrix} 1 & 1 \\ 1 & 2 \\ 1 & 3 \\ 1 & 4 \\ 1 & 5 \end{bmatrix} = \begin{bmatrix} 5 & 15 \\ 15 & 55 \end{bmatrix}$$

and

$$X^T Y = \begin{bmatrix} 1 & 1 & 1 & 1 & 1 \\ 1 & 2 & 3 & 4 & 5 \end{bmatrix} \begin{bmatrix} 1 \\ 2 \\ 4 \\ 4 \\ 6 \end{bmatrix} = \begin{bmatrix} 17 \\ 63 \end{bmatrix}.$$

Now, using $(X^T X)^{-1}$ to find the coefficient matrix A, you have

$$A = (X^T X)^{-1} X^T Y = \tfrac{1}{50} \begin{bmatrix} 55 & -15 \\ -15 & 5 \end{bmatrix} \begin{bmatrix} 17 \\ 63 \end{bmatrix} = \begin{bmatrix} -0.2 \\ 1.2 \end{bmatrix}.$$

Thus the least squares regression line is

$$y = -0.2 + 1.2x.$$

See Figure 2.5. The sum of squared error for this line can be shown to be 0.8, which means that this line fits the data better than either of the two experimental linear models determined earlier.

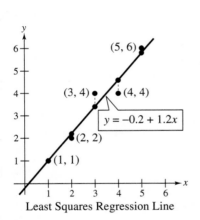

Least Squares Regression Line

Figure 2.5

Exercises

Stochastic Matrices

In Exercises 1–6, determine whether the matrix is stochastic.

1. $\begin{bmatrix} \dfrac{2}{5} & -\dfrac{2}{5} \\ \dfrac{3}{5} & \dfrac{7}{5} \end{bmatrix}$

2. $\begin{bmatrix} \dfrac{\sqrt{2}}{2} & \dfrac{\sqrt{2}}{2} \\ -\dfrac{\sqrt{2}}{2} & \dfrac{\sqrt{2}}{2} \end{bmatrix}$

3. $\begin{bmatrix} 0 & 1 & 0 \\ 0 & 0 & 1 \\ 1 & 0 & 0 \end{bmatrix}$

4. $\begin{bmatrix} 0.3 & 0.1 & 0.8 \\ 0.5 & 0.2 & 0.1 \\ 0.2 & 0.7 & 0.1 \end{bmatrix}$

5. $\begin{bmatrix} \dfrac{1}{3} & \dfrac{1}{6} & \dfrac{1}{4} \\ \dfrac{1}{3} & \dfrac{2}{3} & \dfrac{1}{4} \\ \dfrac{1}{3} & \dfrac{1}{6} & \dfrac{1}{2} \end{bmatrix}$

6. $\begin{bmatrix} 1 & 0 & 0 & 0 \\ 0 & 1 & 0 & 0 \\ 0 & 0 & 1 & 0 \\ 0 & 0 & 0 & 1 \end{bmatrix}$

7. The market research department at a manufacturing plant determines that 20% of the people who purchase the plant's product during any given month will not purchase it the next month. On the other hand, 30% of the people who do not purchase the product during any given month will purchase it the next month. In a population of 1000 people, 100 people purchased the product this month. How many will purchase the product next month? In 2 months?

8. A medical researcher is studying the spread of a virus in a population of 1000 laboratory mice. During any given week there is an 80% probability that an infected mouse will overcome the virus, and during the same week there is a 10% probability that a noninfected mouse will become infected. One hundred mice are currently infected with the virus. How many will be infected next week? In 2 weeks?

9. A population of 10,000 is grouped as follows: 5000 nonsmokers, 2500 smokers of one pack or less per day, and 2500 smokers of more than one pack per day. During any given month there is a 5% probability that a nonsmoker will begin smoking a pack or less per day, and a 2% probability that a nonsmoker will begin smoking more than a pack per day. For smokers who smoke a pack or less per day, there is a 10% probability of quitting and a 10% probability of increasing to more than a pack per day. For smokers who smoke more than a pack per day, there is a 5% probability of quitting and a 10% probability of dropping to a pack or less per day. How many people will be in each of the three groups in 1 month? In 2 months?

10. A population of 100,000 consumers is grouped as follows: 20,000 users of Brand A, 30,000 users of Brand B, and 50,000 who use neither brand. During any given month a Brand A user has a 20% probability of switching to Brand B and a 5% probability of not using either brand. A Brand B user has a 15% probability of switching to Brand A and a 10% probability of not using either brand. A nonuser has a 10% probability of purchasing Brand A and a 15% probability of purchasing Brand B. How many people will be in each group in 1 month? In 2 months? In 3 months?

11. A college dormitory has 200 students. Those who watch an hour or more of television on any given day always watch for less than an hour the next day. One-fourth of those who watch television for less than an hour one day will watch an hour or more the next day. Half of the students watched television for an hour or more today. How many will watch television for an hour or more tomorrow? In 2 days? In 30 days?

12. A matrix of transition probabilities is given by

$$P = \begin{bmatrix} 0.6 & 0.1 & 0.1 \\ 0.2 & 0.7 & 0.1 \\ 0.2 & 0.2 & 0.8 \end{bmatrix}.$$

Find P^2X and P^3X for the state matrix

$$X = \begin{bmatrix} 100 \\ 100 \\ 800 \end{bmatrix}.$$

Then find the steady state matrix for P.

13. Prove that the product of two 2×2 stochastic matrices is stochastic.

14. Let P be a 2×2 stochastic matrix. Prove that there exists a 2×1 state matrix X with nonnegative entries such that $PX = X$.

Cryptography

In Exercises 15–18, find the uncoded row matrices of indicated size for the given messages. Then encode the message using the matrix A.

15. *Message:* **SELL CONSOLIDATED**
Row Matrix Size: 1×3

Encoding Matrix: $A = \begin{bmatrix} 1 & -1 & 0 \\ 1 & 0 & -1 \\ -6 & 2 & 3 \end{bmatrix}$

16. *Message:* **PLEASE SEND MONEY**
Row Matrix Size: 1×3

Encoding Matrix: $A = \begin{bmatrix} 4 & 2 & 1 \\ -3 & -3 & -1 \\ 3 & 2 & 1 \end{bmatrix}$

17. *Message:* **COME HOME SOON**
Row Matrix Size: 1×2

Encoding Matrix: $A = \begin{bmatrix} 1 & 2 \\ 3 & 5 \end{bmatrix}$

18. *Message:* **HELP IS COMING**
Row Matrix Size: 1×4

Encoding Matrix: $A = \begin{bmatrix} -2 & 3 & -1 & -1 \\ -1 & 1 & 1 & 1 \\ -1 & -1 & 1 & 2 \\ 3 & 1 & -2 & -4 \end{bmatrix}$

In Exercises 19–22, find A^{-1} to decode the cryptogram.

19. $A = \begin{bmatrix} 1 & 2 \\ 3 & 5 \end{bmatrix}$,

11, 21, 64, 112, 25, 50, 29, 53, 23, 46, 40, 75, 55, 92

20. $A = \begin{bmatrix} 2 & 3 \\ 3 & 4 \end{bmatrix}$,

85, 120, 6, 8, 10, 15, 84, 117, 42, 56, 90, 125, 60, 80, 30, 45, 19, 26

21. $A = \begin{bmatrix} 4 & 2 & 1 \\ -3 & -3 & -1 \\ 3 & 2 & 1 \end{bmatrix}$,

33, 9, 9, 55, 28, 14, 95, 50, 25, 99, 53, 29, −22, −32, −9

22. $A = \begin{bmatrix} 1 & -1 & 0 \\ 1 & 0 & -1 \\ -6 & 2 & 3 \end{bmatrix}$,

9, −1, −9, 38, −19, −19, 28, −9, −19, −80, 25, 41, −64, 21, 31, −7, −4, 7

23. The following cryptogram was encoded with a 2 × 2 matrix.

8, 21, −15, −10, −13, −13, 5, 10, 5, 25, 5, 19, −1, 6, 20, 40, −18, −18, 1, 16

The last word of the message is __RON. What is the message?

24. The following cryptogram was encoded with a 2 × 2 matrix.

5, 2, 25, 11, −2, −7, −15, −15, 32, 14, −8, −13, 38, 19, −19, −19, 37, 16

The last word of the message is __SUE. What is the message?

25. Use a graphing utility with matrix capabilities to find A^{-1}. Then decode the cryptogram.

$A = \begin{bmatrix} 1 & 0 & 2 \\ 2 & -1 & 1 \\ 0 & 1 & 2 \end{bmatrix}$

38, −14, 29, 56, −15, 62, 17, 3, 38, 18, 20, 76, 18, −5, 21, 29, −7, 32, 32, 9, 77, 36, −8, 48, 33, −5, 51, 41, 3, 79, 12, 1, 26, 58, −22, 49, 63, −19, 69, 28, 8, 67, 31, −11, 27, 41, −18, 28

26. A code breaker intercepted the encoded message below.

45 −35 38 −30 18 −18 35 −30 81 −60
42 −28 75 −55 2   −2 22 −21 15 −10

Let $A^{-1} = \begin{bmatrix} w & x \\ y & z \end{bmatrix}$.

(a) You know that $[45 \ -35]A^{-1} = [10 \ \ 15]$ and $[38 \ -30]A^{-1} = [8 \ \ 14]$, where A^{-1} is the inverse of the encoding matrix A. Write and solve two systems of equations to find w, x, y, and z.

(b) Decode the message.

Leontief Input-Output Models

27. A system composed of two industries, coal and steel, has the following inputs.

(a) To produce 1 dollar's worth of output, the coal industry requires $0.10 of its own product and $0.80 of steel.

(b) To produce 1 dollar's worth of output, the steel industry requires $0.10 of its own product and $0.20 of coal.

Find D, the input-output matrix for this system. Then solve for the output matrix X in the equation $X = DX + E$, where the external demand is given by

$$E = \begin{bmatrix} 10{,}000 \\ 20{,}000 \end{bmatrix}.$$

28. An industrial system has two industries with the given inputs.

(a) To produce 1 dollar's worth of output, Industry A requires $0.30 of its own product and $0.40 of Industry B's product.

(b) To produce 1 dollar's worth of output, Industry B requires $0.20 of its own product and $0.40 of Industry A's product.

Find D, the input-output matrix for this system. Then solve for the output matrix X in the equation $X = DX + E$, where the external demand is given by

$$E = \begin{bmatrix} 50{,}000 \\ 30{,}000 \end{bmatrix}.$$

29. A small community includes a farmer, a baker, and a grocer and has the following input-output matrix D and external demand matrix E.

$$D = \begin{bmatrix} 0.40 & 0.50 & 0.50 \\ 0.30 & 0.00 & 0.30 \\ 0.20 & 0.20 & 0.00 \end{bmatrix} \begin{matrix} Farmer \\ Baker \\ Grocer \end{matrix} \quad \text{and} \quad E = \begin{bmatrix} 1000 \\ 1000 \\ 1000 \end{bmatrix}$$

with column headers *Farmer   Baker   Grocer*.

Solve for the output matrix X in the equation $X = DX + E$.

30. An industrial system has three industries and the following input-output matrix D and external demand matrix E.

$$D = \begin{bmatrix} 0.2 & 0.4 & 0.4 \\ 0.4 & 0.2 & 0.2 \\ 0.0 & 0.2 & 0.2 \end{bmatrix} \quad \text{and} \quad E = \begin{bmatrix} 5000 \\ 2000 \\ 8000 \end{bmatrix}$$

Solve for the output matrix X in the equation $X = DX + E$.

Least Squares Regression Analysis

In Exercises 31–34, (a) sketch the line that appears to be the best fit to the given points, (b) use the method of least squares to find the least squares regression line, and (c) calculate the sum of the squared error.

31.

32.

33.

34.

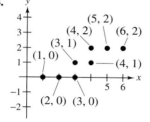

In Exercises 35–42, find the least squares regression line.

35. $(0, 0), (1, 1), (2, 4)$

36. $(1, 0), (3, 3), (5, 6)$

37. $(-2, 0), (-1, 1), (0, 1), (1, 2)$

38. $(-4, -1), (-2, 0), (2, 4), (4, 5)$

39. $(-5, 1), (1, 3), (2, 3), (2, 5)$

40. $(-3, 4), (-1, 2), (1, 1), (3, 0)$

41. $(-5, 10), (-1, 8), (3, 6), (7, 4), (5, 5)$

42. $(0, 6), (4, 3), (5, 0), (8, -4), (10, -5)$

43. A store manager wants to know the demand for a certain product as a function of the price. The daily sales for three different prices of the product are shown in the table.

Price (x)	$1.00	$1.25	$1.50
Demand (y)	450	375	330

(a) Find the least squares regression line for these data.

(b) Estimate the demand when the price is $1.40.

44. A hardware retailer wants to know the demand for a certain tool as a function of the price. The monthly sales for four different prices of the tool are shown in the table.

Price (x)	$25	$30	$35	$40
Demand (y)	82	75	67	55

(a) Find the least squares regression line for these data.

(b) Estimate the demand when the price is $32.95.

45. The table shows the numbers y of electric-powered vehicles in use in the United States for the years 1997 through 2001. (Source: U.S. Energy Information Administration)

Year	1997	1998	1999	2000	2001
Number (y)	4453	5243	6964	11,834	17,848

(a) Use the method of least squares to find the least squares regression line for the data. Let t represent the year, with $t = 7$ corresponding to 1997.

(b) Use the linear regression capabilities of a graphing utility to find a linear model for the data. Let t represent the year, with $t = 7$ corresponding to 1997.

46. A wildlife management team studied the reproduction rates of deer in three tracts of a wildlife preserve. Each tract contained 5 acres. In each tract the number of females x and the percent of females y that had offspring the following year were recorded. The results are shown in the table.

Number (x)	100	120	140
Percent (y)	75	68	55

(a) Use the method of least squares to find the least squares regression line that models the data.

(b) Use a graphing utility to graph the model and the data in the same viewing window.

(c) Use the model to create a table of estimated values of y. Compare the estimated values with the actual data.

(d) Use the model to estimate the percent of females that had offspring when there were 170 females.

(e) Use the model to estimate the number of females when 40% of the females had offspring.

CHAPTER 2 Summary

After studying this chapter, you should have acquired the following skills.

- Perform matrix operations.
- Write the system of linear equations represented by a given matrix.
- Write a given system of linear equations in matrix form.
- Write a given system of linear equations in the form $A\mathbf{x} = \mathbf{b}$ and use it to solve for $\mathbf{x}$.
- Use properties of matrix operations to solve matrix equations.
- Find the transpose of a given matrix.
- Find the inverse of a given matrix (if it exists).
- Find the inverse of a matrix product (if it exists).
- Factor a given matrix into a product of elementary matrices.
- Find the LU-factorization of a matrix.
- Use LU-factorization to solve a system of linear equations.
- Use a stochastic matrix to measure consumer preference.
- Use matrix multiplication to encode and decode messages.
- Use matrix algebra to analyze economic systems (Leontief input-output models).
- Use the method of least squares to find the least squares regression line for a set of data.

CHAPTER 2 Review Exercises

In Exercises 1–6, perform the indicated matrix operations.

1. $\begin{bmatrix} 2 & 1 & 0 \\ 0 & 5 & -4 \end{bmatrix} - 3\begin{bmatrix} 5 & 3 & -6 \\ 0 & -2 & 5 \end{bmatrix}$

2. $-2\begin{bmatrix} 1 & 2 \\ 5 & -4 \\ 6 & 0 \end{bmatrix} + 8\begin{bmatrix} 7 & 1 \\ 1 & 2 \\ 1 & 4 \end{bmatrix}$

3. $\begin{bmatrix} 1 & 2 \\ 5 & -4 \\ 6 & 0 \end{bmatrix}\begin{bmatrix} 6 & -2 & 8 \\ 4 & 0 & 0 \end{bmatrix}$

4. $\begin{bmatrix} 1 & 5 \\ 2 & -4 \end{bmatrix}\begin{bmatrix} 6 & -2 & 8 \\ 4 & 0 & 0 \end{bmatrix}$

5. $\begin{bmatrix} 1 & 3 & 2 \\ 0 & 2 & -4 \\ 0 & 0 & 3 \end{bmatrix}\begin{bmatrix} 4 & -3 & 2 \\ 0 & 3 & -1 \\ 0 & 0 & 2 \end{bmatrix}$

6. $\begin{bmatrix} 2 & 1 \\ 6 & 0 \end{bmatrix}\begin{bmatrix} 4 & 2 \\ -3 & 1 \end{bmatrix} + \begin{bmatrix} -2 & 4 \\ 0 & 4 \end{bmatrix}$

In Exercises 7 and 8, write out the system of linear equations represented by the matrix equation.

7. $\begin{bmatrix} 5 & 4 \\ -1 & 1 \end{bmatrix}\begin{bmatrix} x \\ y \end{bmatrix} = \begin{bmatrix} 2 \\ -22 \end{bmatrix}$

8. $\begin{bmatrix} 0 & 1 & -2 \\ -1 & 3 & 1 \\ 2 & -2 & 4 \end{bmatrix}\begin{bmatrix} x_1 \\ x_2 \\ x_3 \end{bmatrix} = \begin{bmatrix} -1 \\ 0 \\ 2 \end{bmatrix}$

In Exercises 9 and 10, write the system of linear equations in matrix form.

9. $\begin{aligned} 2x_1 + 3x_2 + x_3 &= 10 \\ 2x_1 - 3x_2 - 3x_3 &= 22 \\ 4x_1 - 2x_2 + 3x_3 &= -2 \end{aligned}$

10. $\begin{aligned} -3x_1 - x_2 + x_3 &= 0 \\ 2x_1 + 4x_2 - 5x_3 &= -3 \\ x_1 - 2x_2 + 3x_3 &= 1 \end{aligned}$

In Exercises 11 and 12, find A^T, A^TA, and AA^T.

11. $\begin{bmatrix} 1 & 2 & -3 \\ 0 & 1 & 2 \end{bmatrix}$

12. $\begin{bmatrix} 1 \\ 3 \\ -1 \end{bmatrix}$

In Exercises 13–16, find the inverse of the matrix (if it exists).

13. $\begin{bmatrix} 3 & -1 \\ 2 & -1 \end{bmatrix}$

14. $\begin{bmatrix} 4 & -1 \\ -8 & 2 \end{bmatrix}$

15. $\begin{bmatrix} 2 & 3 & 1 \\ 2 & -3 & -3 \\ 4 & 0 & 3 \end{bmatrix}$

16. $\begin{bmatrix} 1 & 1 & 1 & 1 \\ 0 & 1 & 1 & 1 \\ 0 & 0 & 1 & 1 \\ 0 & 0 & 0 & 1 \end{bmatrix}$

In Exercises 17 and 18, write the system of linear equations in the form $A\mathbf{x} = \mathbf{b}$. Then find A^{-1} and use it to solve for $\mathbf{x}$.

17. $\begin{aligned} 5x_1 + 4x_2 &= 2 \\ -x_1 + x_2 &= -22 \end{aligned}$

18. $\begin{aligned} -x_1 + x_2 + 2x_3 &= 1 \\ 2x_1 + 3x_2 + x_3 &= -2 \\ 5x_1 + 4x_2 + 2x_3 &= 4 \end{aligned}$

19. Find A when
$$(3A)^{-1} = \begin{bmatrix} 4 & -1 \\ 2 & 3 \end{bmatrix}.$$

20. Find x such that the matrix A is nonsingular.
$$A = \begin{bmatrix} 3 & 1 \\ x & -1 \end{bmatrix}$$

In Exercises 21 and 22, find the inverse of the elementary matrix.

21. $\begin{bmatrix} 1 & 0 & 4 \\ 0 & 1 & 0 \\ 0 & 0 & 1 \end{bmatrix}$

22. $\begin{bmatrix} 1 & 0 & 0 \\ 0 & 6 & 0 \\ 0 & 0 & 1 \end{bmatrix}$

In Exercises 23–26, factor A into a product of elementary matrices.

23. $A = \begin{bmatrix} 2 & 3 \\ 0 & 1 \end{bmatrix}$

24. $A = \begin{bmatrix} -3 & 13 \\ 1 & -4 \end{bmatrix}$

25. $A = \begin{bmatrix} 1 & 0 & 1 \\ 0 & 1 & -2 \\ 0 & 0 & 4 \end{bmatrix}$

26. $A = \begin{bmatrix} 3 & 0 & 6 \\ 0 & 2 & 0 \\ 1 & 0 & 3 \end{bmatrix}$

27. Find two 2×2 matrices A such that $A^2 = I$.

28. Find two 2×2 matrices A such that $A^2 = O$.

29. Find three 2×2 matrices A such that $A^2 = A$.

30. Find 2×2 matrices A and B such that $AB = O$ but $BA \neq O$.

In Exercises 31 and 32, let the matrices X, Y, Z, and W be given by

$$X = \begin{bmatrix} 1 \\ 2 \\ 0 \\ 1 \end{bmatrix}, \quad Y = \begin{bmatrix} -1 \\ 0 \\ 3 \\ 2 \end{bmatrix}, \quad Z = \begin{bmatrix} 3 \\ 4 \\ -1 \\ 2 \end{bmatrix}, \quad W = \begin{bmatrix} 3 \\ 2 \\ -4 \\ -1 \end{bmatrix}.$$

31. (a) Find scalars a, b, and c such that $W = aX + bY + cZ$.
(b) Show that there do not exist scalars a and b such that $Z = aX + bY$.

32. Show that if $aX + bY + cZ = O$, then $a = b = c = 0$.

33. Let A, B, and $A + B$ be nonsingular matrices. Prove that $A^{-1} + B^{-1}$ is nonsingular by showing that
$$(A^{-1} + B^{-1})^{-1} = A(A + B)^{-1}B.$$

34. *Writing*   Let A, B, and C be $n \times n$ matrices and C be nonsingular. If $AC = CB$, does it follow that $A = B$? If so, prove it. If not, explain why and find an example for which the hypothesis is false.

In Exercises 35 and 36, find the LU–factorization of the matrix.

35. $\begin{bmatrix} 2 & 5 \\ 6 & 14 \end{bmatrix}$

36. $\begin{bmatrix} 1 & 1 & 1 \\ 1 & 2 & 2 \\ 1 & 2 & 3 \end{bmatrix}$

In Exercises 37 and 38, use the LU–factorization of the coefficient matrix to solve the linear system.

37. $\begin{aligned} x \quad + z &= 3 \\ 2x + y + 2z &= 7 \\ 3x + 2y + 6z &= 8 \end{aligned}$

38. $\begin{aligned} 2x_1 + x_2 + x_3 - x_4 &= 7 \\ 3x_2 + x_3 - x_4 &= -3 \\ - 2x_3 \quad &= 2 \\ 2x_1 + x_2 + x_3 - 2x_4 &= 8 \end{aligned}$

True or False?   In Exercises 39–42, determine whether each statement is true or false. If a statement is true, give a reason or cite an appropriate statement in the text. If a statement is false, provide an example that shows that the statement is not true in all cases or cite an appropriate statement in the text.

39. (a) Addition of matrices is not commutative.
(b) The transpose of the sum of matrices is equal to the sum of the transposes of the matrices.

40. (a) The product of a 2×3 matrix and a 3×5 matrix is a matrix that is 5×2.
(b) The transpose of a product is equal to the product of transposes in reverse order.

41. (a) All $n \times n$ matrices are invertible.
(b) If an $n \times n$ matrix A is not symmetric, then A^TA is not symmetric.

42. (a) If A and B are $n \times n$ matrices and A is invertible, then $(ABA^{-1})^2 = AB^2A^{-1}$.
(b) If A and B are nonsingular $n \times n$ matrices, then $A + B$ is a nonsingular matrix.

Stochastic Matrices

In Exercises 43 and 44, determine whether the matrix is stochastic.

43. $\begin{bmatrix} 1 & 0 & 0 \\ 0 & 0.5 & 0.1 \\ 0 & 0.1 & 0.5 \end{bmatrix}$

44. $\begin{bmatrix} 0.3 & 0.4 & 0.1 \\ 0.2 & 0.4 & 0.5 \\ 0.5 & 0.2 & 0.4 \end{bmatrix}$

In Exercises 45 and 46, use the given matrix of transition probabilities P and state matrix X to find the state matrices PX, P^2X, and P^3X.

45. $P = \begin{bmatrix} \frac{1}{2} & \frac{1}{4} \\ \frac{1}{2} & \frac{3}{4} \end{bmatrix}$, $X = \begin{bmatrix} 128 \\ 64 \end{bmatrix}$

46. $P = \begin{bmatrix} 0.6 & 0.2 & 0.0 \\ 0.2 & 0.7 & 0.1 \\ 0.2 & 0.1 & 0.9 \end{bmatrix}$, $X = \begin{bmatrix} 1000 \\ 1000 \\ 1000 \end{bmatrix}$

47. A country is divided into three regions. Each year, 10% of the residents of Region 1 move to Region 2 and 5% move to Region 3; 15% of the residents of Region 2 move to Region 1 and 5% move to Region 3; and 10% of the residents of Region 3 move to Region 1 and 10% move to Region 2. This year each region has a population of 100,000. Find the population of each region (a) in 1 year and (b) in 3 years.

48. Find the steady state matrix for the populations described in Exercise 47.

Cryptography

In Exercises 49 and 50, find the uncoded row matrices of indicated size for the given message. Then encode the message using the matrix A.

49. *Message:* ONE IF BY LAND

Row Matrix Size: 1×2

Encoding Matrix: $A = \begin{bmatrix} 5 & 2 \\ 2 & 1 \end{bmatrix}$

50. *Message:* BEAM ME UP SCOTTY

Row Matrix Size: 1×3

Encoding Matrix: $A = \begin{bmatrix} 2 & 1 & 4 \\ 3 & 1 & 3 \\ -2 & -1 & -3 \end{bmatrix}$

In Exercises 51–54, find A^{-1} to decode the cryptogram.

51. $A = \begin{bmatrix} 3 & -2 \\ -4 & 3 \end{bmatrix}$,

$-45, 34, 36, -24, -43, 37, -23, 22, -37, 29, 57, -38, -39, 31$

52. $A = \begin{bmatrix} 1 & 4 \\ -1 & -3 \end{bmatrix}$,

$11, 52, -8, -9, -13, -39, 5, 20, 12, 56, 5, 20, -2, 7, 9, 41, 25, 100$

53. $A = \begin{bmatrix} 2 & -1 & -1 \\ -5 & 2 & 2 \\ 5 & -1 & -2 \end{bmatrix}$,

$58, -3, -25, -48, 28, 19, -40, 13, 13, -98, 39, 39, 118, -25, -48, 28, -14, -14$

54. $A = \begin{bmatrix} 1 & 2 & 3 \\ 2 & 5 & 3 \\ 1 & 0 & 8 \end{bmatrix}$,

$23, 20, 132, 54, 128, 102, 32, 21, 203, 6, 10, 23, 21, 15, 129, 36, 46, 173, 29, 72, 45$

Leontief Input-Output Models

55. An industrial system has two industries with the following inputs.

(a) To produce 1 dollar's worth of output, Industry A requires $0.20 of its own product and $0.30 of Industry B's product.

(b) To produce 1 dollar's worth of output, Industry B requires $0.10 of its own product and $0.50 of Industry A's product.

Find D, the input-output matrix for this system. Then solve for the output matrix X in the equation $X = DX + E$, where the external demand is given by

$$E = \begin{bmatrix} 40,000 \\ 80,000 \end{bmatrix}.$$

56. An industrial system has three industries and the following input-output matrix D and external demand matrix E.

$$D = \begin{bmatrix} 0.1 & 0.3 & 0.2 \\ 0.0 & 0.2 & 0.3 \\ 0.4 & 0.1 & 0.1 \end{bmatrix} \quad \text{and} \quad E = \begin{bmatrix} 3000 \\ 3500 \\ 8500 \end{bmatrix}$$

Solve for the output matrix X in the equation $X = DX + E$.

Least Squares Regression Analysis

In Exercises 57–60, find the least squares regression line for the points.

57. $(1, 5), (2, 4), (3, 2)$

58. $(2, 1), (3, 3), (4, 2), (5, 4), (6, 4)$

59. $(-2, 4), (-1, 2), (0, 1), (1, -2), (2, -3)$

60. $(1, 1), (1, 3), (1, 2), (1, 4), (2, 5)$

61. A farmer used four test plots to determine the relationship between wheat yield in kilograms per square kilometer and the amount of fertilizer in hundreds of kilograms per square kilometer. The results are shown in the table.

Fertilizer (x)	1.0	1.5	2.0	2.5
Yield (y)	32	41	48	53

(a) Find the least squares regression line for these data.
(b) Estimate the yield for a fertilizer application of 160 kilograms per square kilometer.

62. The Consumer Price Index (CPI) for all items for 5 years is shown in the table. (Source: Bureau of Labor Statistics)

Year	1996	1997	1998	1999	2000
CPI (y)	156.9	160.5	163.0	166.6	172.2

(a) Find the least squares regression line for these data. (Let $x = 6$ represent 1996.)
(b) Estimate the CPI for the years 2005 and 2010.

63. The table shows the average room rates y for a hotel room in the United States for the years 1996 through 2000. (Source: American Hotel & Lodging Association)

Year	1996	1997	1998	1999	2000
Average Room Rate (y)	$70.93	$75.31	$78.62	$81.33	$85.89

(a) Use the method of least squares to find the least squares regression line for the data. Let x represent the year, with $x = 6$ corresponding to 1996.
(b) Use the linear regression capabilities of a graphing utility to find a linear model for the data. How does this model compare to the model obtained in part (a)?
(c) Use the linear model to create a table of estimated values for y. Compare the estimated values with the actual data.
(d) Use the linear model to predict the average room rate in 2005.
(e) Use the linear model to predict when the average room rate will be $100.00.

CHAPTER 2 Projects

1 EXPLORING MATRIX MULTIPLICATION

The first two test scores for Anna, Bruce, Chris, and David are shown in the table. Use the table to create a matrix M to represent the data. Input matrix M into a graphing utility and use it to answer the following.

	Test 1	Test 2
Anna	84	96
Bruce	56	72
Chris	78	83
David	82	91

1. Which test was more difficult? Which was easier? Explain.
2. How would you rank the performances of the four students?

3. Describe the meaning of the matrix products $M\begin{bmatrix} 1 \\ 0 \end{bmatrix}$ and $M\begin{bmatrix} 0 \\ 1 \end{bmatrix}$.

4. Describe the meaning of the matrix products $\begin{bmatrix} 1 & 0 & 0 & 0 \end{bmatrix}M$ and $\begin{bmatrix} 0 & 0 & 1 & 0 \end{bmatrix}M$.

5. Describe the meaning of the matrix products $M\begin{bmatrix} 1 \\ 1 \end{bmatrix}$ and $\frac{1}{2}M\begin{bmatrix} 1 \\ 1 \end{bmatrix}$.

6. Describe the meaning of the matrix products $\begin{bmatrix} 1 & 1 & 1 & 1 \end{bmatrix}M$ and $\frac{1}{4}\begin{bmatrix} 1 & 1 & 1 & 1 \end{bmatrix}M$.

7. Describe the meaning of the matrix product $\begin{bmatrix} 1 & 1 & 1 & 1 \end{bmatrix}M\begin{bmatrix} 1 \\ 1 \end{bmatrix}$.

8. Use matrix multiplication to express the combined overall average score on both tests.

9. As the instructor, you would like to raise the scores on test 1 for all the students. How could you use matrix multiplication to scale the scores by a factor of 1.1?

2 NILPOTENT MATRICES

Let A be a nonzero square $n \times n$ matrix. Is it possible that a positive integer k exists such that $A^k = 0$? For example, find A^3 for the matrix

$$A = \begin{bmatrix} 0 & 1 & 2 \\ 0 & 0 & 1 \\ 0 & 0 & 0 \end{bmatrix}.$$

A square matrix A is said to be **nilpotent of index k** if $A \neq 0$, $A^2 \neq 0, \ldots,$ $A^{k-1} \neq 0$, but $A^k = 0$. In this project you will explore the world of nilpotent matrices.

1. What is the index of the nilpotent matrix A?

2. Use a graphing utility to determine which of the following matrices are nilpotent and to find their indices.

(a) $\begin{bmatrix} 0 & 1 \\ 0 & 0 \end{bmatrix}$ (b) $\begin{bmatrix} 0 & 1 \\ 1 & 0 \end{bmatrix}$ (c) $\begin{bmatrix} 0 & 0 \\ 1 & 0 \end{bmatrix}$

(d) $\begin{bmatrix} 1 & 0 \\ 1 & 0 \end{bmatrix}$ (e) $\begin{bmatrix} 0 & 0 & 1 \\ 0 & 0 & 0 \\ 0 & 0 & 0 \end{bmatrix}$ (f) $\begin{bmatrix} 0 & 0 & 0 \\ 1 & 0 & 0 \\ 1 & 1 & 0 \end{bmatrix}$

3. Find 3×3 nilpotent matrices of indices 2 and 3.

4. Find 4×4 nilpotent matrices of indices 2, 3, and 4.

5. Find a nilpotent matrix of index 5.

6. Are nilpotent matrices invertible? Prove your answer.

7. If A is nilpotent, what can you say about A^T? Prove your answer.

8. If A is nilpotent, show that $I - A$ is invertible.

AUGUSTIN-LOUIS CAUCHY was the eldest of six children of a French parliamentary lawyer. He was born in Paris about 6 weeks after the fall of the Bastille during the French Revolution. To escape the danger of the guillotine, Cauchy's father moved the family from Paris to their country home in Arcueil, which bordered the home of one of France's leading mathematicians, Pierre Simon de Laplace. Laplace was the first to recognize young Cauchy's gift for mathematics and encouraged him to develop his talent.

In 1800 the Cauchy family moved back to Paris, where Cauchy began his formal schooling. He won several prizes, and his mathematical prowess began attracting attention. In 1810 Cauchy graduated as a civil engineer and was given a military commission that he reluctantly held for a few years. Political turmoil then led

Cauchy through a series of teaching positions that culminated with his appointment as professor of mathematical astronomy at the Sorbonne. By the age of 27 he was recognized as one of the greatest mathematicians of his day.

Cauchy was a prolific writer, publishing nearly 800 papers. His contributions to the study of mathematics were revolutionary, and he is often credited with bringing rigor to modern mathematics. For example, Cauchy was the first to rigorously define limits, continuity, and the convergence of an infinite series. In addition to founding complex number theory, he contributed to the theories of determinants and differential equations. It is interesting to note that Cauchy's work on determinants preceded Cayley's development of matrices.

AUGUSTIN-LOUIS CAUCHY

1 7 8 9 – 1 8 5 7

DETERMINANTS CHAPTER 3

3.1 | The Determinant of a Matrix

Every *square* matrix can be associated with a real number called its *determinant*. Determinants have many uses, several of which will be explored in this chapter. The first two sections of this chapter concentrate on procedures for evaluating the determinant of a matrix.

Historically, the use of determinants arose from the recognition of special patterns that occur in the solutions of systems of linear equations. For instance, the general solution of the system

$$a_{11}x_1 + a_{12}x_2 = b_1$$
$$a_{21}x_1 + a_{22}x_2 = b_2$$

can be shown to be

$$x_1 = \frac{b_1 a_{22} - b_2 a_{12}}{a_{11} a_{22} - a_{21} a_{12}} \quad \text{and} \quad x_2 = \frac{b_2 a_{11} - b_1 a_{21}}{a_{11} a_{22} - a_{21} a_{12}},$$

provided that $a_{11} a_{22} - a_{21} a_{12} \neq 0$. Note that both fractions have the same denominator, $a_{11} a_{22} - a_{21} a_{12}$. This quantity is called the determinant of the coefficient matrix A.

Definition of the Determinant of a 2×2 Matrix	The **determinant** of the matrix $$A = \begin{bmatrix} a_{11} & a_{12} \\ a_{21} & a_{22} \end{bmatrix}$$ is given by $$\det(A) = |A| = a_{11} a_{22} - a_{21} a_{12}.$$

REMARK: In this text, $\det(A)$ and $|A|$ are used interchangeably to represent the determinant of a matrix. Vertical bars are also used to denote the absolute value of a real number; the context will show which use is intended. Furthermore, it is common practice to delete the matrix brackets and write

$$\begin{vmatrix} a_{11} & a_{12} \\ a_{21} & a_{22} \end{vmatrix} \quad \text{instead of} \quad \left| \begin{bmatrix} a_{11} & a_{12} \\ a_{21} & a_{22} \end{bmatrix} \right|.$$

A convenient method for remembering the formula for the determinant of a 2×2 matrix is shown in the following diagram.

$$|A| = \begin{vmatrix} a_{11} & a_{12} \\ a_{21} & a_{22} \end{vmatrix} = a_{11} a_{22} - a_{21} a_{12}$$

Note that the determinant is given by the difference of the products of the two diagonals of the matrix.

EXAMPLE 1 *The Determinant of a Matrix of Order 2*

Find the determinant of each matrix.

(a) $A = \begin{bmatrix} 2 & -3 \\ 1 & 2 \end{bmatrix}$ (b) $B = \begin{bmatrix} 2 & 1 \\ 4 & 2 \end{bmatrix}$ (c) $C = \begin{bmatrix} 0 & 3 \\ 2 & 4 \end{bmatrix}$

Solution

(a) $|A| = \begin{vmatrix} 2 & -3 \\ 1 & 2 \end{vmatrix} = 2(2) - 1(-3) = 4 + 3 = 7$

(b) $|B| = \begin{vmatrix} 2 & 1 \\ 4 & 2 \end{vmatrix} = 2(2) - 4(1) = 4 - 4 = 0$

(c) $|C| = \begin{vmatrix} 0 & 3 \\ 2 & 4 \end{vmatrix} = 0(4) - 2(3) = 0 - 6 = -6$

REMARK: Note that the determinant of a matrix can be positive, zero, or negative.

The determinant of a matrix of order 1 is defined simply as the entry of the matrix. For instance, if $A = [-2]$, then

$$\det(A) = -2.$$

To define the determinant of a matrix of order higher than 2, it is convenient to use the notions of *minors* and *cofactors*.

Definition of Minors and Cofactors of a Matrix	If A is a square matrix, then the **minor** M_{ij} of the element a_{ij} is the determinant of the matrix obtained by deleting the ith row and jth column of A. The **cofactor** C_{ij} is given by $$C_{ij} = (-1)^{i+j}M_{ij}.$$

For example, if A is a 3×3 matrix, then the minors and cofactors of a_{21} and a_{22} are as shown in the following diagram.

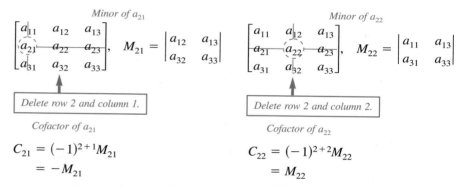

As you can see, the minors and cofactors of a matrix differ at most in sign. To obtain the cofactors of a matrix, first find the minors and then apply the following checkerboard pattern of $+$'s and $-$'s.

Sign Pattern for Cofactors

$$\begin{bmatrix} + & - & + \\ - & + & - \\ + & - & + \end{bmatrix}$$
3×3 *matrix*

$$\begin{bmatrix} + & - & + & - \\ - & + & - & + \\ + & - & + & - \\ - & + & - & + \end{bmatrix}$$
4×4 *matrix*

$$\begin{bmatrix} + & - & + & - & + & \cdots \\ - & + & - & + & - & \cdots \\ + & - & + & - & + & \cdots \\ - & + & - & + & - & \cdots \\ + & - & + & - & + & \cdots \\ \vdots & \vdots & \vdots & \vdots & \vdots & \end{bmatrix}$$
$n \times n$ *matrix*

Note that *odd* positions (where $i + j$ is odd) have negative signs, and *even* positions (where $i + j$ is even) have positive signs.

EXAMPLE 2   *Finding the Minors and Cofactors of a Matrix*

Find all the minors and cofactors of

$$A = \begin{bmatrix} 0 & 2 & 1 \\ 3 & -1 & 2 \\ 4 & 0 & 1 \end{bmatrix}.$$

Solution   To find the minor M_{11}, delete the first row and first column of A and evaluate the determinant of the resulting matrix.

$$\begin{bmatrix} 0 & 2 & 1 \\ 3 & -1 & 2 \\ 4 & 0 & 1 \end{bmatrix}, \quad M_{11} = \begin{vmatrix} -1 & 2 \\ 0 & 1 \end{vmatrix} = -1(1) - 0(2) = -1$$

Similarly, to find M_{12}, delete the first row and second column.

$$\begin{bmatrix} 0 & 2 & 1 \\ 3 & -1 & 2 \\ 4 & 0 & 1 \end{bmatrix}, \quad M_{12} = \begin{vmatrix} 3 & 2 \\ 4 & 1 \end{vmatrix} = 3(1) - 4(2) = -5$$

Continuing this pattern, you obtain the following minors.

$$\begin{array}{lll} M_{11} = -1 & M_{12} = -5 & M_{13} = 4 \\ M_{21} = 2 & M_{22} = -4 & M_{23} = -8 \\ M_{31} = 5 & M_{32} = -3 & M_{33} = -6 \end{array}$$

Now, to find the cofactors, combine the checkerboard pattern of signs with these minors to obtain

$$\begin{array}{lll} C_{11} = -1 & C_{12} = 5 & C_{13} = 4 \\ C_{21} = -2 & C_{22} = -4 & C_{23} = 8 \\ C_{31} = 5 & C_{32} = 3 & C_{33} = -6. \end{array}$$

Now that the minors and cofactors of a matrix have been defined, you are ready for a general definition of the determinant of a matrix. The following definition is called **inductive** because it uses determinants of matrices of order $n - 1$ to define the determinant of a matrix of order n.

Definition of the Determinant of a Matrix

If A is a square matrix (of order 2 or greater), then the determinant of A is the sum of the entries in the first row of A multiplied by their cofactors. That is,

$$\det(A) = |A| = \sum_{j=1}^{n} a_{1j}C_{1j} = a_{11}C_{11} + a_{12}C_{12} + \cdots + a_{1n}C_{1n}.$$

REMARK: Try checking that for 2×2 matrices this definition yields $|A| = a_{11}a_{22} - a_{21}a_{12}$, as previously defined.

When you use this definition to evaluate a determinant, you are **expanding by cofactors in the first row.** This procedure is demonstrated in Example 3.

EXAMPLE 3   *The Determinant of a Matrix of Order 3*

Find the determinant of

$$A = \begin{bmatrix} 0 & 2 & 1 \\ 3 & -1 & 2 \\ 4 & 0 & 1 \end{bmatrix}.$$

Solution   Note that this matrix is the same as that given in Example 2. There you found the cofactors of the entries in the first row to be

$$C_{11} = -1, \qquad C_{12} = 5, \qquad C_{13} = 4.$$

Therefore, by the definition of a determinant, you have the following.

$$|A| = a_{11}C_{11} + a_{12}C_{12} + a_{13}C_{13} \qquad \text{First row expansion}$$
$$= 0(-1) + 2(5) + 1(4) = 14$$

Although the determinant is defined as an expansion by the cofactors in the first row, it can be shown that the determinant can be evaluated by *any* row or column. For instance, you could expand the 3×3 matrix in Example 3 by the second row to obtain

$$|A| = a_{21}C_{21} + a_{22}C_{22} + a_{23}C_{23} \qquad \text{Second row expansion}$$
$$= 3(-2) + (-1)(-4) + 2(8) = 14$$

or by the first column to obtain

$$|A| = a_{11}C_{11} + a_{21}C_{21} + a_{31}C_{31} \qquad \text{First column expansion}$$
$$= 0(-1) + 3(-2) + 4(5) = 14.$$

Try other possibilities to see that the determinant of A can be evaluated by expanding by *any* row or column. This result is stated in the following theorem, Laplace's Expansion of a Determinant, named after the French mathematician Pierre Simon de Laplace (1749–1827).

Theorem 3.1   *Expansion by Cofactors*

Let A be a square matrix of order n. Then the determinant of A is given by

$$\det(A) = |A| = \sum_{j=1}^{n} a_{ij}C_{ij} = a_{i1}C_{i1} + a_{i2}C_{i2} + \cdots + a_{in}C_{in} \qquad \textit{i}\text{th row expansion}$$

or

$$\det(A) = |A| = \sum_{i=1}^{n} a_{ij}C_{ij} = a_{1j}C_{1j} + a_{2j}C_{2j} + \cdots + a_{nj}C_{nj}. \qquad \textit{j}\text{th column expansion}$$

When expanding by cofactors you do not need to evaluate the cofactors of zero entries, because a zero entry times its cofactor is zero.

$$a_{ij}C_{ij} = (0)C_{ij} = 0$$

Thus, the row (or column) containing the most zeros is usually the best choice for expansion by cofactors. This is demonstrated in the next example.

EXAMPLE 4   *The Determinant of a Matrix of Order 4*

Find the determinant of

$$A = \begin{bmatrix} 1 & -2 & 3 & 0 \\ -1 & 1 & 0 & 2 \\ 0 & 2 & 0 & 3 \\ 3 & 4 & 0 & -2 \end{bmatrix}.$$

Solution   Inspecting this matrix, you can see that three of the entries in the third column are zeros. Thus, you can eliminate some of the work in the expansion by using the third column.

$$|A| = 3(C_{13}) + 0(C_{23}) + 0(C_{33}) + 0(C_{43})$$

Because C_{23}, C_{33}, and C_{43} have zero coefficients, you need only find the cofactor C_{13}. To do this, delete the first row and third column of A and evaluate the determinant of the resulting matrix.

$$C_{13} = (-1)^{1+3} \begin{vmatrix} -1 & 1 & 2 \\ 0 & 2 & 3 \\ 3 & 4 & -2 \end{vmatrix}$$

$$= \begin{vmatrix} -1 & 1 & 2 \\ 0 & 2 & 3 \\ 3 & 4 & -2 \end{vmatrix}$$

Expanding by cofactors in the second row yields the following.

$$C_{13} = (0)(-1)^{2+1} \begin{vmatrix} 1 & 2 \\ 4 & -2 \end{vmatrix} + (2)(-1)^{2+2} \begin{vmatrix} -1 & 2 \\ 3 & -2 \end{vmatrix} + (3)(-1)^{2+3} \begin{vmatrix} -1 & 1 \\ 3 & 4 \end{vmatrix}$$

$$= 0 + 2(1)(-4) + 3(-1)(-7)$$

$$= 13$$

Thus, you obtain

$$|A| = 3(13) = 39.$$

Technology   Note

Most graphing calculators and computer software programs have the capability to calculate the determinant of a square matrix. If you use the determinant command of a TI-86 to verify that the determinant of the matrix A from Example 4 is 39, your screen will look like this.

```
A
        [[1  -2 3  0 ]
         [-1 1  0  2 ]
         [0  2  0  3 ]
         [3  4  0  -2]]
det A
                    39
```

There is an alternative method that is commonly used for evaluating the determinant of a 3×3 matrix A. To apply this method, copy the first and second columns of A to form fourth and fifth columns. The determinant of A is then obtained by adding (or subtracting) the products of the six diagonals, as shown in the diagram on the next page.

Subtract these three products.

Add these three products.

Try confirming that the determinant of A is given by

$$|A| = a_{11}a_{22}a_{33} + a_{12}a_{23}a_{31} + a_{13}a_{21}a_{32} - a_{31}a_{22}a_{13} - a_{32}a_{23}a_{11} - a_{33}a_{21}a_{12}.$$

Simulation

Explore this concept further with an electronic simulation available on the *Elementary Linear Algebra* Learning Tools Student CD-ROM.

EXAMPLE 5   *The Determinant of a Matrix of Order 3*

Find the determinant of

$$A = \begin{bmatrix} 0 & 2 & 1 \\ 3 & -1 & 2 \\ 4 & -4 & 1 \end{bmatrix}.$$

Solution   Begin by recopying the first two columns and then computing the six diagonal products as follows.

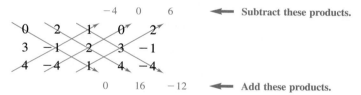

−4   0   6   ⬅ **Subtract these products.**

0   16   −12   ⬅ **Add these products.**

R E M A R K : The diagonal process illustrated in Example 5 is valid *only* for matrices of order 3. For matrices of higher orders, another method must be used.

Now, by adding the lower three products and subtracting the upper three products, you can find the determinant of A to be

$$|A| = 0 + 16 + (-12) - (-4) - 0 - 6 = 2.$$

Triangular Matrices

Evaluating determinants of matrices of order 4 or higher can be tedious. There is, however, an important exception: the determinant of a *triangular* matrix. Recall from Section 2.4 that a square matrix is called *upper triangular* if it has all zero entries below its main diagonal, and *lower triangular* if it has all zero entries above its main diagonal. A matrix that is both upper and lower triangular is called **diagonal.** That is, a diagonal matrix is one in which all entries above and below the main diagonal are zero.

Upper Triangular Matrix

$$\begin{bmatrix} a_{11} & a_{12} & a_{13} & \cdots & a_{1n} \\ 0 & a_{22} & a_{23} & \cdots & a_{2n} \\ 0 & 0 & a_{33} & \cdots & a_{3n} \\ \vdots & \vdots & \vdots & & \vdots \\ 0 & 0 & 0 & \cdots & a_{nn} \end{bmatrix}$$

Lower Triangular Matrix

$$\begin{bmatrix} a_{11} & 0 & 0 & \cdots & 0 \\ a_{21} & a_{22} & 0 & \cdots & 0 \\ a_{31} & a_{32} & a_{33} & \cdots & 0 \\ \vdots & \vdots & \vdots & & \vdots \\ a_{n1} & a_{n2} & a_{n3} & \cdots & a_{nn} \end{bmatrix}$$

To find the determinant of a triangular matrix, simply form the product of the entries on the main diagonal. It is easy to see that this procedure is valid for triangular matrices of order 2 or 3. For instance, the determinant of

$$A = \begin{bmatrix} 2 & 3 & -1 \\ 0 & -1 & 2 \\ 0 & 0 & 3 \end{bmatrix}$$

can be found by expanding by the third row to obtain

$$|A| = 0 \begin{vmatrix} 3 & -1 \\ -1 & 2 \end{vmatrix} - 0 \begin{vmatrix} 2 & -1 \\ 0 & 2 \end{vmatrix} + 3 \begin{vmatrix} 2 & 3 \\ 0 & -1 \end{vmatrix} = 3(2)(-1) = -6,$$

which is the product of the entries on the main diagonal.

| Theorem 3.2 | *Determinant of a Triangular Matrix* |

If A is a triangular matrix of order n, then its determinant is the product of the entries on the main diagonal. That is,

$$\det(A) = |A| = a_{11}a_{22}a_{33} \cdots a_{nn}.$$

Proof   You can use *mathematical induction** to prove this theorem for the case in which A is an upper triangular matrix. The case in which A is lower triangular can be proven similarly. If A has order 1, then $A = [a_{11}]$ and the determinant is given by

$$|A| = a_{11}.$$

Assuming that the theorem is true for any upper triangular matrix of order $k - 1$, consider an upper triangular matrix A of order k. Expanding by the kth row, you obtain

$$|A| = 0C_{k1} + 0C_{k2} + \cdots + 0C_{k\,k-1} + a_{kk}C_{kk} = a_{kk}C_{kk}.$$

Now, note that $C_{kk} = (-1)^{2k}M_{kk} = M_{kk}$, where M_{kk} is the determinant of the upper triangular matrix formed by deleting the kth row and kth column of A. Because this matrix is of order $k - 1$, you can apply the induction assumption to write the following.

$$|A| = a_{kk}M_{kk} = a_{kk}(a_{11}a_{22}a_{33} \cdots a_{k-1\,k-1})$$
$$= a_{11}a_{22}a_{33} \cdots a_{kk}$$

*A discussion of mathematical induction can be found in Appendix A.

EXAMPLE 6   *The Determinant of a Triangular Matrix*

Find the determinant of each matrix.

(a) $A = \begin{bmatrix} 2 & 0 & 0 & 0 \\ 4 & -2 & 0 & 0 \\ -5 & 6 & 1 & 0 \\ 1 & 5 & 3 & 3 \end{bmatrix}$
(b) $B = \begin{bmatrix} -1 & 0 & 0 & 0 & 0 \\ 0 & 3 & 0 & 0 & 0 \\ 0 & 0 & 2 & 0 & 0 \\ 0 & 0 & 0 & 4 & 0 \\ 0 & 0 & 0 & 0 & -2 \end{bmatrix}$

Solution   (a) The determinant of this triangular matrix is given by $|A| = (2)(-2)(1)(3) = -12$.

(b) This matrix is *diagonal* and its determinant is given by $|B| = (-1)(3)(2)(4)(-2) = 48$.

SECTION 3.1   Exercises

In Exercises 1–12, find the determinant of the matrix.

1. $[1]$

2. $[-3]$

3. $\begin{bmatrix} 2 & 1 \\ 3 & 4 \end{bmatrix}$

4. $\begin{bmatrix} -3 & 1 \\ 5 & 2 \end{bmatrix}$

5. $\begin{bmatrix} 5 & 2 \\ -6 & 3 \end{bmatrix}$

6. $\begin{bmatrix} 2 & -2 \\ 4 & 3 \end{bmatrix}$

7. $\begin{bmatrix} -7 & 6 \\ \frac{1}{2} & 3 \end{bmatrix}$

8. $\begin{bmatrix} \frac{1}{3} & 5 \\ 4 & -9 \end{bmatrix}$

9. $\begin{bmatrix} 2 & 6 \\ 0 & 3 \end{bmatrix}$

10. $\begin{bmatrix} 2 & -3 \\ -6 & 9 \end{bmatrix}$

11. $\begin{bmatrix} \lambda - 3 & 2 \\ 4 & \lambda - 1 \end{bmatrix}$

12. $\begin{bmatrix} \lambda - 2 & 0 \\ 4 & \lambda - 4 \end{bmatrix}$

In Exercises 13–16, find (a) the minors and (b) the cofactors of the matrix.

13. $\begin{bmatrix} 1 & 2 \\ 3 & 4 \end{bmatrix}$

14. $\begin{bmatrix} -1 & 0 \\ 2 & 1 \end{bmatrix}$

15. $\begin{bmatrix} -3 & 2 & 1 \\ 4 & 5 & 6 \\ 2 & -3 & 1 \end{bmatrix}$

16. $\begin{bmatrix} -3 & 4 & 2 \\ 6 & 3 & 1 \\ 4 & -7 & -8 \end{bmatrix}$

17. Find the determinant of the matrix in Exercise 15 using the method of expansion by cofactors. Use (a) the second row and (b) the second column.

18. Find the determinant of the matrix in Exercise 16 using the method of expansion by cofactors. Use (a) the third row and (b) the first column.

In Exercises 19–34, use expansion by cofactors to find the determinant of the matrix.

19. $\begin{bmatrix} 1 & 4 & -2 \\ 3 & 2 & 0 \\ -1 & 4 & 3 \end{bmatrix}$

20. $\begin{bmatrix} 2 & -1 & 3 \\ 1 & 4 & 4 \\ 1 & 0 & 2 \end{bmatrix}$

21. $\begin{bmatrix} 2 & 4 & 6 \\ 0 & 3 & 1 \\ 0 & 0 & -5 \end{bmatrix}$

22. $\begin{bmatrix} -3 & 0 & 0 \\ 7 & 11 & 0 \\ 1 & 2 & 2 \end{bmatrix}$

23. $\begin{bmatrix} 0.1 & 0.2 & 0.3 \\ -0.3 & 0.2 & 0.2 \\ 0.5 & 0.4 & 0.4 \end{bmatrix}$

24. $\begin{bmatrix} -0.4 & 0.4 & 0.3 \\ 0.2 & 0.2 & 0.2 \\ 0.3 & 0.2 & 0.2 \end{bmatrix}$

25. $\begin{bmatrix} x & y & 1 \\ 2 & 3 & 1 \\ 0 & -1 & 1 \end{bmatrix}$

26. $\begin{bmatrix} x & y & 1 \\ -2 & -2 & 1 \\ 1 & 5 & 1 \end{bmatrix}$

27. $\begin{bmatrix} 2 & 6 & 6 & 2 \\ 2 & 7 & 3 & 6 \\ 1 & 5 & 0 & 1 \\ 3 & 7 & 0 & 7 \end{bmatrix}$

28. $\begin{bmatrix} 1 & 4 & 3 & 2 \\ -5 & 6 & 2 & 1 \\ 0 & 0 & 0 & 0 \\ 3 & -2 & 1 & 5 \end{bmatrix}$

29. $\begin{bmatrix} 5 & 3 & 0 & 6 \\ 4 & 6 & 4 & 12 \\ 0 & 2 & -3 & 4 \\ 0 & 1 & -2 & 2 \end{bmatrix}$

30. $\begin{bmatrix} 3 & 0 & 7 & 0 \\ 2 & 6 & 11 & 12 \\ 4 & 1 & -1 & 2 \\ 1 & 5 & 2 & 10 \end{bmatrix}$

31. $\begin{bmatrix} & w & x & y & z \\ & 21 & -15 & 24 & 30 \\ & -10 & 24 & -32 & 18 \\ & -40 & 22 & 32 & -35 \end{bmatrix}$

32. $\begin{bmatrix} & w & x & y & z \\ & 10 & 15 & -25 & 30 \\ & -30 & 20 & -15 & -10 \\ & 30 & 35 & -25 & -40 \end{bmatrix}$

33. $\begin{bmatrix} 5 & 2 & 0 & 0 & -2 \\ 0 & 1 & 4 & 3 & 2 \\ 0 & 0 & 2 & 6 & 3 \\ 0 & 0 & 3 & 4 & 1 \\ 0 & 0 & 0 & 0 & 2 \end{bmatrix}$

34. $\begin{bmatrix} 4 & 3 & -2 & 1 & 2 \\ 0 & 0 & 0 & 0 & 0 \\ 1 & 2 & -7 & 13 & 12 \\ 6 & -2 & 5 & 6 & 7 \\ 1 & 4 & 2 & 0 & 9 \end{bmatrix}$

In Exercises 35–40, use a graphing utility with matrix capabilities to find the determinant of the matrix.

35. $\begin{bmatrix} \frac{1}{2} & 1 & -5 \\ 4 & -\frac{1}{4} & -4 \\ 3 & 2 & -2 \end{bmatrix}$

36. $\begin{bmatrix} 0.25 & -1 & 0.6 \\ 0.50 & 0.8 & -0.2 \\ 0.75 & 0.9 & -0.4 \end{bmatrix}$

37. $\begin{bmatrix} 4 & 3 & 2 & 5 \\ 1 & 6 & -1 & 2 \\ -3 & 2 & 4 & 5 \\ 6 & 1 & 3 & -2 \end{bmatrix}$

38. $\begin{bmatrix} 1 & 2 & 4 & -1 \\ 6 & 2 & 3 & -2 \\ 4 & -1 & 4 & 3 \\ 5 & -2 & 2 & 4 \end{bmatrix}$

Data 39. $\begin{bmatrix} 1 & 2 & -1 & 4 & 2 & -1 \\ 0 & 1 & 2 & -2 & -3 & 1 \\ 0 & 3 & 2 & -1 & 3 & -2 \\ 1 & 2 & 0 & -2 & 3 & 2 \\ 1 & -2 & 3 & 1 & 2 & -1 \\ 2 & 0 & 2 & 3 & 1 & 1 \end{bmatrix}$

Data 40. $\begin{bmatrix} 8 & 5 & 1 & -2 & 0 & 3 \\ -1 & 0 & 7 & 1 & 6 & -5 \\ 0 & 8 & 6 & 5 & -3 & 1 \\ 1 & 2 & 5 & -8 & 4 & 3 \\ 2 & 6 & -2 & 0 & 6 & 7 \\ 8 & -3 & 1 & 2 & -5 & 1 \end{bmatrix}$

In Exercises 41–46, find the determinant of the triangular matrix.

41. $\begin{bmatrix} -2 & 0 & 0 \\ 4 & 6 & 0 \\ -3 & 7 & 2 \end{bmatrix}$

42. $\begin{bmatrix} 5 & 0 & 0 \\ 0 & 6 & 0 \\ 0 & 0 & -3 \end{bmatrix}$

43. $\begin{bmatrix} 5 & 8 & -4 & 2 \\ 0 & 0 & 6 & 0 \\ 0 & 0 & 2 & 2 \\ 0 & 0 & 0 & -1 \end{bmatrix}$

44. $\begin{bmatrix} 4 & 0 & 0 & 0 \\ -1 & \frac{1}{2} & 0 & 0 \\ 3 & 5 & 3 & 0 \\ -8 & 7 & 0 & -2 \end{bmatrix}$

45. $\begin{bmatrix} -1 & 4 & 2 & 1 & -3 \\ 0 & 3 & -4 & 5 & 2 \\ 0 & 0 & 2 & 7 & 0 \\ 0 & 0 & 0 & 5 & -1 \\ 0 & 0 & 0 & 0 & 1 \end{bmatrix}$

46. $\begin{bmatrix} 7 & 0 & 0 & 0 & 0 \\ -8 & \frac{1}{4} & 0 & 0 & 0 \\ 4 & 5 & 2 & 0 & 0 \\ 3 & -3 & 5 & -1 & 0 \\ 1 & 13 & 4 & 1 & -2 \end{bmatrix}$

True or False? In Exercises 47 and 48, determine whether each statement is true or false. If a statement is true, give a reason or cite an appropriate statement in the text. If a statement is false, provide an example that shows that the statement is not true in all cases or cite an appropriate statement in the text.

47. (a) The determinant of the 2×2 matrix A is $a_{21}a_{12} - a_{11}a_{22}$.
(b) The determinant of a matrix of order 1 is the entry of the matrix.
(c) The ij-cofactor of a square matrix A is the matrix defined by deleting the ith row and the jth column of A.

48. (a) To find the determinant of a triangular matrix, add the entries on the main diagonal.
(b) The determinant of a matrix can be evaluated using expansion by cofactors in any row or column.
(c) When expanding by cofactors, you need not evaluate the cofactors of zero entries.

In Exercises 49–52, find the values of λ for which the determinant is zero.

49. $\begin{vmatrix} \lambda - 1 & -4 \\ -2 & \lambda + 1 \end{vmatrix}$

50. $\begin{vmatrix} \lambda - 2 & -2 \\ -6 & \lambda + 2 \end{vmatrix}$

51. $\begin{vmatrix} \lambda + 1 & 0 & 0 \\ 4 & \lambda & 3 \\ 2 & 8 & \lambda + 5 \end{vmatrix}$

52. $\begin{vmatrix} \lambda + 2 & -2 & 4 \\ 0 & \lambda & -2 \\ 0 & -1 & \lambda - 1 \end{vmatrix}$

53. The determinant of a 2×2 matrix involves two products. The determinant of a 3×3 matrix involves six triple products. Show that the determinant of a 4×4 matrix involves 24 quadruple products. (In general, the determinant of an $n \times n$ matrix involves $n!$ n-fold products.)

54. Show that the system of linear equations

$$ax + by = e$$
$$cx + dy = f$$

has a unique solution if and only if the determinant of the coefficient matrix is nonzero.

55. Verify the equation.

$$\begin{vmatrix} 1 & 1 & 1 \\ a & b & c \\ a^2 & b^2 & c^2 \end{vmatrix} = (a - b)(b - c)(c - a)$$

56. Verify the equation.

$$\begin{vmatrix} 1 & 1 & 1 \\ a & b & c \\ a^3 & b^3 & c^3 \end{vmatrix} = (a - b)(b - c)(c - a)(a + b + c)$$

57. Verify the equation.

$$\begin{vmatrix} x & 0 & c \\ -1 & x & b \\ 0 & -1 & a \end{vmatrix} = ax^2 + bx + c$$

58. Use the equation given in Exercise 57 as a model to find a determinant that is equal to $ax^3 + bx^2 + cx + d$.

59. *Writing*   Explain why it is easy to calculate the determinant of a matrix that has an entire row of zeros.

3.2 | Evaluation of a Determinant Using Elementary Operations

Which of the following two determinants is easier to evaluate?

$$|A| = \begin{vmatrix} 1 & -2 & 3 & 1 \\ 4 & -6 & 3 & 2 \\ -2 & 4 & -9 & -3 \\ 3 & -6 & 9 & 2 \end{vmatrix} \quad \text{or} \quad |B| = \begin{vmatrix} 1 & -2 & 3 & 1 \\ 0 & 2 & -9 & -2 \\ 0 & 0 & -3 & -1 \\ 0 & 0 & 0 & -1 \end{vmatrix}$$

Given what you now know about the determinant of a triangular matrix, it is clear that the second determinant is *much* easier to evaluate. This determinant is simply the product of the entries on the main diagonal. That is,

$$|B| = (1)(2)(-3)(-1) = 6.$$

On the other hand, using expansion by cofactors (the only technique discussed so far) to evaluate the first determinant is messy. For instance, if you expand by cofactors across the first row, you have the following.

$$|A| = 1\begin{vmatrix} -6 & 3 & 2 \\ 4 & -9 & -3 \\ -6 & 9 & 2 \end{vmatrix} + 2\begin{vmatrix} 4 & 3 & 2 \\ -2 & -9 & -3 \\ 3 & 9 & 2 \end{vmatrix} + 3\begin{vmatrix} 4 & -6 & 2 \\ -2 & 4 & -3 \\ 3 & -6 & 2 \end{vmatrix} - 1\begin{vmatrix} 4 & -6 & 3 \\ -2 & 4 & -9 \\ 3 & -6 & 9 \end{vmatrix}$$

Evaluating the determinants of these four 3×3 matrices produces

$$|A| = (1)(-60) + (2)(39) + (3)(-10) - (1)(-18) = 6.$$

It is not coincidental that these two determinants have the same value. In fact, you can obtain the matrix B by performing elementary row operations on matrix A. (Try verifying this.) In this section, you will see the effect of elementary row (and column) operations on the value of a determinant.

EXAMPLE 1 *The Effect of Elementary Row Operations on a Determinant*

(a) The matrix B was obtained from A by interchanging the rows of A.

$$|A| = \begin{vmatrix} 2 & -3 \\ 1 & 4 \end{vmatrix} = 11 \quad \text{and} \quad |B| = \begin{vmatrix} 1 & 4 \\ 2 & -3 \end{vmatrix} = -11$$

(b) The matrix B was obtained from A by adding -2 times the first row of A to the second row of A.

$$|A| = \begin{vmatrix} 1 & -3 \\ 2 & -4 \end{vmatrix} = 2 \quad \text{and} \quad |B| = \begin{vmatrix} 1 & -3 \\ 0 & 2 \end{vmatrix} = 2$$

(c) The matrix B was obtained from A by multiplying the first row of A by $\frac{1}{2}$.

$$|A| = \begin{vmatrix} 2 & -8 \\ -2 & 9 \end{vmatrix} = 2 \quad \text{and} \quad |B| = \begin{vmatrix} 1 & -4 \\ -2 & 9 \end{vmatrix} = 1$$

In Example 1, you can see that interchanging two rows of a matrix changed the sign of its determinant. Adding a multiple of one row to another did not change the determinant. Finally, multiplying a row by a nonzero constant multiplied the determinant by that same constant. The following theorem generalizes these observations. The proof of Property 1 follows the theorem and the proofs of the other two properties are left as exercises. (See Exercises 44 and 45.)

Theorem 3.3

Elementary Row Operations and Determinants

Let A and B be square matrices.

1. If B is obtained from A by interchanging two rows of A, then

$$\det(B) = -\det(A).$$

2. If B is obtained from A by adding a multiple of a row of A to another row of A, then

$$\det(B) = \det(A).$$

3. If B is obtained from A by multiplying a row of A by a nonzero constant c, then

$$\det(B) = c \det(A).$$

Proof   To prove Property 1 of Theorem 3.3, use mathematical induction as follows. Assume that A and B are 2×2 matrices such that

$$A = \begin{bmatrix} a_{11} & a_{12} \\ a_{21} & a_{22} \end{bmatrix} \quad \text{and} \quad B = \begin{bmatrix} a_{21} & a_{22} \\ a_{11} & a_{12} \end{bmatrix}.$$

Then, you have

$$|A| = a_{11}a_{22} - a_{21}a_{12} \quad \text{and} \quad |B| = a_{21}a_{12} - a_{11}a_{22}.$$

Thus, $|B| = -|A|$. Now assume that the property is true for matrices of order $(n - 1)$. Let A be an $n \times n$ matrix such that B is obtained from A by interchanging two rows of A. Then, to find $|A|$ and $|B|$, expand along a row other than the two interchanged rows. By the induction assumption, the cofactors of B will be the negatives of the cofactors of A because the corresponding $(n - 1) \times (n - 1)$ matrices have two rows interchanged. Thus,

$$|B| = -|A|$$

and the proof is complete.

REMARK: Note that the third property enables you to take a common factor out of a row. For instance,

$$\begin{vmatrix} 2 & 4 \\ 1 & 3 \end{vmatrix} = 2 \begin{vmatrix} 1 & 2 \\ 1 & 3 \end{vmatrix}. \qquad \textbf{Factor 2 out of first row.}$$

Theorem 3.3 provides a practical way to evaluate determinants. (This method works particularly well with computers.) To find the determinant of a matrix A, use elementary row operations to obtain a triangular matrix B that is row-equivalent to A. For each step in the elimination process, use Theorem 3.3 to determine the effect of the elementary row operation on the determinant. Finally, find the determinant of B by forming the product of the entries on its main diagonal. This process is demonstrated in the next example.

EXAMPLE 2   *Evaluating a Determinant Using Elementary Row Operations*

Find the determinant of

$$A = \begin{bmatrix} 2 & -3 & 10 \\ 1 & 2 & -2 \\ 0 & 1 & -3 \end{bmatrix}.$$

Solution   Using elementary row operations, rewrite A in triangular form as follows.

$$\begin{vmatrix} 2 & -3 & 10 \\ 1 & 2 & -2 \\ 0 & 1 & -3 \end{vmatrix} = - \begin{vmatrix} 1 & 2 & -2 \\ 2 & -3 & 10 \\ 0 & 1 & -3 \end{vmatrix} \qquad \longleftarrow \textbf{Interchange the first two rows.}$$

$$= - \begin{vmatrix} 1 & 2 & -2 \\ 0 & -7 & 14 \\ 0 & 1 & -3 \end{vmatrix} \qquad \longleftarrow \textbf{Adding } -2 \textbf{ times the first row to the second row produces a new second row.}$$

$$= 7 \begin{vmatrix} 1 & 2 & -2 \\ 0 & 1 & -2 \\ 0 & 1 & -3 \end{vmatrix} \quad \longleftarrow \text{Factor } -7 \text{ out of the second row.}$$

$$= 7 \begin{vmatrix} 1 & 2 & -2 \\ 0 & 1 & -2 \\ 0 & 0 & -1 \end{vmatrix} \quad \longleftarrow \begin{array}{l} \text{Adding } -1 \text{ times the second row to the third} \\ \text{row produces a new third row.} \end{array}$$

Now, because the final matrix is triangular, you can conclude that the determinant is

$$|A| = 7(1)(1)(-1) = -7.$$

Determinants and Elementary Column Operations

Although Theorem 3.3 was stated in terms of elementary *row* operations, the theorem remains valid if the word "row" is replaced by the word "column." Operations performed on the columns of a matrix (rather than the rows) are called **elementary column operations,** and two matrices are called **column-equivalent** if one can be obtained from the other by elementary column operations. The column version of Theorem 3.3 is illustrated as follows.

$$\begin{vmatrix} 2 & 1 & -3 \\ 4 & 0 & 1 \\ 0 & 0 & 2 \end{vmatrix} = - \begin{vmatrix} 1 & 2 & -3 \\ 0 & 4 & 1 \\ 0 & 0 & 2 \end{vmatrix} \qquad \begin{vmatrix} 2 & 3 & -5 \\ 4 & 1 & 0 \\ -2 & 4 & -3 \end{vmatrix} = 2 \begin{vmatrix} 1 & 3 & -5 \\ 2 & 1 & 0 \\ -1 & 4 & -3 \end{vmatrix}$$

Interchange the first two columns.   *Factor 2 out of the first column.*

In evaluating a determinant by hand, it is occasionally convenient to use elementary column operations, as shown in Example 3.

EXAMPLE 3 *Evaluating a Determinant Using Elementary Column Operations*

Find the determinant of

$$A = \begin{bmatrix} -1 & 2 & 2 \\ 3 & -6 & 4 \\ 5 & -10 & -3 \end{bmatrix}.$$

Solution Because the first two columns of A are multiples of each other, you can obtain a column of zeros by adding 2 times the first column to the second column, as follows.

$$\begin{vmatrix} -1 & 2 & 2 \\ 3 & -6 & 4 \\ 5 & -10 & -3 \end{vmatrix} = \begin{vmatrix} -1 & 0 & 2 \\ 3 & 0 & 4 \\ 5 & 0 & -3 \end{vmatrix}$$

At this point you need not continue to rewrite the matrix in triangular form. Because there is an entire column of zeros, simply conclude that the determinant is zero. The validity of this conclusion follows from Theorem 3.1. Specifically, by expanding by cofactors along the second column, you have

$$|A| = (0)C_{12} + (0)C_{22} + (0)C_{32} = 0.$$

Example 3 shows that if one column of a matrix is a scalar multiple of another column, you can immediately conclude that the determinant of the matrix is zero. This is one of three conditions, listed next, that yield a determinant of zero.

Theorem 3.4

Conditions That Yield a Zero Determinant

If A is a square matrix and any of the following conditions is true, then $\det(A) = 0$.
1. An entire row (or an entire column) consists of zeros.
2. Two rows (or columns) are equal.
3. One row (or column) is a multiple of another row (or column).

Proof   Each part of this theorem is easily verified by using elementary row operations and expansion by cofactors. For example, if an entire row or column is zero, then each cofactor in the expansion is multiplied by zero. And if condition 2 or 3 is true, then you can use elementary row or column operations to create a row or column with all zeros.

Recognizing the conditions listed in Theorem 3.4 can make evaluating a determinant much easier. For instance,

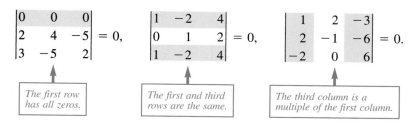

$$\begin{vmatrix} 0 & 0 & 0 \\ 2 & 4 & -5 \\ 3 & -5 & 2 \end{vmatrix} = 0, \qquad \begin{vmatrix} 1 & -2 & 4 \\ 0 & 1 & 2 \\ 1 & -2 & 4 \end{vmatrix} = 0, \qquad \begin{vmatrix} 1 & 2 & -3 \\ 2 & -1 & -6 \\ -2 & 0 & 6 \end{vmatrix} = 0.$$

The first row has all zeros.   *The first and third rows are the same.*   *The third column is a multiple of the first column.*

Do not conclude, however, that Theorem 3.4 gives the *only* conditions that produce a determinant of zero. The theorem is often used indirectly. That is, you can begin with a matrix that does not satisfy any of the conditions of Theorem 3.4 and, through elementary row or column operations, obtain a matrix that does satisfy one of the conditions. Then you can conclude that the original matrix has a determinant of zero. This process is demonstrated in Example 4.

EXAMPLE 4 *A Matrix with a Zero Determinant*

Find the determinant of

$$A = \begin{bmatrix} 1 & 4 & 1 \\ 2 & -1 & 0 \\ 0 & 18 & 4 \end{bmatrix}.$$

Solution Adding -2 times the first row to the second row produces

$$|A| = \begin{vmatrix} 1 & 4 & 1 \\ 2 & -1 & 0 \\ 0 & 18 & 4 \end{vmatrix} = \begin{vmatrix} 1 & 4 & 1 \\ 0 & -9 & -2 \\ 0 & 18 & 4 \end{vmatrix}.$$

Now, because the second and third rows are multiples of each other, you can conclude that the determinant is zero.

In Example 4 you could have obtained a matrix with a row of all zeros by performing an additional elementary row operation (adding 2 times the second row to the third row). This is true in general. That is, a square matrix has a determinant of zero if and only if it is row- (or column-) equivalent to a matrix that has at least one row (or column) consisting entirely of zeros. This result will be proved in the next section.

You have now surveyed two general methods for evaluating determinants. Of these, the method of using elementary row operations to reduce the matrix to triangular form is usually faster than cofactor expansion along a row or column. If the matrix is large, then the number of arithmetic operations needed for cofactor expansion can become extremely large. For this reason, most computer and calculator algorithms use the method involving elementary row operations. The following table shows the number of additions (plus subtractions) and multiplications (plus divisions) needed for each of these two methods for matrices of order 3, 5, and 10.

TABLE 3.1

Order n	Cofactor Expansion		Row Reduction	
	Additions	Multiplications	Additions	Multiplications
3	5	9	5	10
5	119	205	30	45
10	3,628,799	6,235,300	285	339

In fact, the number of operations for the cofactor expansion of an $n \times n$ matrix grows like $n!$. Because $30! \approx 2.65 \times 10^{32}$, even a relatively small 30×30 matrix would require more than 10^{32} operations. If a computer could do one trillion operations per second, it would still take more than one trillion years to compute the determinant of this matrix using cofactor expansion. Yet, row reduction would take only a few seconds.

When evaluating a determinant *by hand*, you can sometimes save steps by using elementary row (or column) operations to create a row (or column) having zeros in all but one position and then using cofactor expansion to reduce the order of the matrix by 1. This approach is illustrated in the next two examples.

EXAMPLE 5   *Evaluating a Determinant*

Find the determinant of

$$A = \begin{bmatrix} -3 & 5 & 2 \\ 2 & -4 & -1 \\ -3 & 0 & 6 \end{bmatrix}.$$

Solution   Notice that the matrix A already has one zero in the third row. You can create another zero in the third row by adding 2 times the first column to the third column as follows.

$$|A| = \begin{vmatrix} -3 & 5 & 2 \\ 2 & -4 & -1 \\ -3 & 0 & 6 \end{vmatrix} = \begin{vmatrix} -3 & 5 & -4 \\ 2 & -4 & 3 \\ -3 & 0 & 0 \end{vmatrix}$$

Expanding by cofactors along the third row produces

$$|A| = \begin{vmatrix} -3 & 5 & -4 \\ 2 & -4 & 3 \\ -3 & 0 & 0 \end{vmatrix}$$

$$= -3(-1)^4 \begin{vmatrix} 5 & -4 \\ -4 & 3 \end{vmatrix}$$

$$= -3(-1)$$

$$= 3.$$

EXAMPLE 6   *Evaluating a Determinant*

Evaluate the determinant of

$$A = \begin{bmatrix} 2 & 0 & 1 & 3 & -2 \\ -2 & 1 & 3 & 2 & -1 \\ 1 & 0 & -1 & 2 & 3 \\ 3 & -1 & 2 & 4 & -3 \\ 1 & 1 & 3 & 2 & 0 \end{bmatrix}.$$

Solution Because the second column of this matrix already has two zeros, choose it for cofactor expansion. Two additional zeros can be created in the second column by adding the second row to the fourth row, and then adding -1 times the second row to the fifth row.

$$|A| = \begin{vmatrix} 2 & 0 & 1 & 3 & -2 \\ -2 & 1 & 3 & 2 & -1 \\ 1 & 0 & -1 & 2 & 3 \\ 3 & -1 & 2 & 4 & -3 \\ 1 & 1 & 3 & 2 & 0 \end{vmatrix}$$

$$= \begin{vmatrix} 2 & 0 & 1 & 3 & -2 \\ -2 & 1 & 3 & 2 & -1 \\ 1 & 0 & -1 & 2 & 3 \\ 1 & 0 & 5 & 6 & -4 \\ 3 & 0 & 0 & 0 & 1 \end{vmatrix}$$

$$= (1)(-1)^4 \begin{vmatrix} 2 & 1 & 3 & -2 \\ 1 & -1 & 2 & 3 \\ 1 & 5 & 6 & -4 \\ 3 & 0 & 0 & 1 \end{vmatrix}$$

(Note that you have now reduced the problem of finding the determinant of a 5×5 matrix to finding the determinant of a 4×4 matrix.) Because you already have two zeros in the fourth row, it is chosen for the next cofactor expansion. Add -3 times the fourth column to the first column to produce the following.

$$|A| = \begin{vmatrix} 2 & 1 & 3 & -2 \\ 1 & -1 & 2 & 3 \\ 1 & 5 & 6 & -4 \\ 3 & 0 & 0 & 1 \end{vmatrix} = \begin{vmatrix} 8 & 1 & 3 & -2 \\ -8 & -1 & 2 & 3 \\ 13 & 5 & 6 & -4 \\ 0 & 0 & 0 & 1 \end{vmatrix}$$

$$= (1)(-1)^8 \begin{vmatrix} 8 & 1 & 3 \\ -8 & -1 & 2 \\ 13 & 5 & 6 \end{vmatrix}$$

Add the second row to the first row and then expand by cofactors along the first row.

$$|A| = \begin{vmatrix} 8 & 1 & 3 \\ -8 & -1 & 2 \\ 13 & 5 & 6 \end{vmatrix} = \begin{vmatrix} 0 & 0 & 5 \\ -8 & -1 & 2 \\ 13 & 5 & 6 \end{vmatrix}$$

$$= 5(-1)^4 \begin{vmatrix} -8 & -1 \\ 13 & 5 \end{vmatrix}$$

$$= 5(-40 + 13) = -135.$$

SECTION 3.2 **Exercises**

In Exercises 1–16, which property of determinants is illustrated by the equation?

1. $\begin{vmatrix} 2 & -6 \\ 1 & -3 \end{vmatrix} = 0$

2. $\begin{vmatrix} -4 & 5 \\ 12 & -15 \end{vmatrix} = 0$

3. $\begin{vmatrix} 1 & 4 & 2 \\ 0 & 0 & 0 \\ 5 & 6 & -7 \end{vmatrix} = 0$

4. $\begin{vmatrix} -4 & 3 & 2 \\ 8 & 0 & 0 \\ -4 & 3 & 2 \end{vmatrix} = 0$

5. $\begin{vmatrix} 1 & 3 & 4 \\ -7 & 2 & -5 \\ 6 & 1 & 2 \end{vmatrix} = - \begin{vmatrix} 1 & 4 & 3 \\ -7 & -5 & 2 \\ 6 & 2 & 1 \end{vmatrix}$

6. $\begin{vmatrix} 1 & 3 & 4 \\ -2 & 2 & 0 \\ 1 & 6 & 2 \end{vmatrix} = - \begin{vmatrix} 1 & 6 & 2 \\ -2 & 2 & 0 \\ 1 & 3 & 4 \end{vmatrix}$

7. $\begin{vmatrix} 5 & 10 \\ 2 & -7 \end{vmatrix} = 5 \begin{vmatrix} 1 & 2 \\ 2 & -7 \end{vmatrix}$

8. $\begin{vmatrix} 1 & 8 & -3 \\ 3 & -12 & 6 \\ 7 & 4 & 9 \end{vmatrix} = 12 \begin{vmatrix} 1 & 2 & -1 \\ 3 & -3 & 2 \\ 7 & 1 & 3 \end{vmatrix}$

9. $\begin{vmatrix} 5 & 0 & 10 \\ 25 & -30 & 40 \\ -15 & 5 & 20 \end{vmatrix} = 5^3 \begin{vmatrix} 1 & 0 & 2 \\ 5 & -6 & 8 \\ -3 & 1 & 4 \end{vmatrix}$

10. $\begin{vmatrix} 6 & 0 & 0 & 0 \\ 0 & 6 & 0 & 0 \\ 0 & 0 & 6 & 0 \\ 0 & 0 & 0 & 6 \end{vmatrix} = 6^4 \begin{vmatrix} 1 & 0 & 0 & 0 \\ 0 & 1 & 0 & 0 \\ 0 & 0 & 1 & 0 \\ 0 & 0 & 0 & 1 \end{vmatrix}$

11. $\begin{vmatrix} 2 & -3 \\ 8 & 7 \end{vmatrix} = \begin{vmatrix} 2 & -3 \\ 0 & 19 \end{vmatrix}$

12. $\begin{vmatrix} 1 & -3 & 2 \\ 5 & 2 & -1 \\ -1 & 0 & 6 \end{vmatrix} = \begin{vmatrix} 1 & -3 & 2 \\ 0 & 17 & -11 \\ -1 & 0 & 6 \end{vmatrix}$

13. $\begin{vmatrix} 3 & 2 & 4 \\ -2 & 1 & 5 \\ 5 & -7 & -20 \end{vmatrix} = \begin{vmatrix} 3 & 2 & -6 \\ -2 & 1 & 0 \\ 5 & -7 & 15 \end{vmatrix}$

14. $\begin{vmatrix} 5 & 4 & 2 \\ 4 & -3 & 4 \\ 7 & 6 & 3 \end{vmatrix} = - \begin{vmatrix} 5 & 4 & 2 \\ -4 & 3 & -4 \\ 7 & 6 & 3 \end{vmatrix}$

15. $\begin{vmatrix} 2 & 1 & -1 & 0 & 3 & 4 \\ 1 & 0 & 1 & 3 & 5 & 2 \\ 3 & 6 & 1 & -3 & 2 & 6 \\ 0 & 4 & 0 & 2 & 4 & 0 \\ -1 & 8 & 5 & 3 & 4 & -2 \\ -2 & 0 & 2 & 6 & 4 & -4 \end{vmatrix} = 0$

16. $\begin{vmatrix} 4 & 3 & 1 & 9 & 6 & 9 \\ 9 & -1 & 2 & 3 & 0 & -3 \\ 3 & 4 & 6 & 9 & -6 & 12 \\ 5 & 2 & 0 & 6 & 12 & 6 \\ 6 & 0 & 3 & 0 & -3 & 0 \\ 1 & -2 & 1 & -3 & 9 & -6 \end{vmatrix} = 0$

In Exercises 17–30, use elementary row or column operations to evaluate the determinant.

17. $\begin{vmatrix} 1 & 7 & -3 \\ 1 & 3 & 1 \\ 4 & 8 & 1 \end{vmatrix}$

18. $\begin{vmatrix} 1 & 1 & 1 \\ 2 & -1 & -2 \\ 1 & -2 & -1 \end{vmatrix}$

19. $\begin{vmatrix} 2 & -1 & -1 \\ 1 & 3 & 2 \\ 1 & 1 & 3 \end{vmatrix}$

20. $\begin{vmatrix} 3 & -1 & -3 \\ -1 & -4 & -2 \\ 3 & -1 & -1 \end{vmatrix}$

21. $\begin{vmatrix} 4 & 3 & -2 \\ 5 & 4 & 1 \\ -2 & 3 & 4 \end{vmatrix}$

22. $\begin{vmatrix} 3 & 8 & -7 \\ 0 & -5 & 4 \\ 6 & 1 & 6 \end{vmatrix}$

23. $\begin{vmatrix} 5 & -8 & 0 \\ 9 & 7 & 4 \\ -8 & 7 & 1 \end{vmatrix}$

24. $\begin{vmatrix} 4 & -8 & 5 \\ 8 & -5 & 3 \\ 8 & 5 & 2 \end{vmatrix}$

25. $\begin{vmatrix} 4 & -7 & 9 & 1 \\ 6 & 2 & 7 & 0 \\ 3 & 6 & -3 & 3 \\ 0 & 7 & 4 & -1 \end{vmatrix}$

26. $\begin{vmatrix} 9 & -4 & 2 & 5 \\ 2 & 7 & 6 & -5 \\ 4 & 1 & -2 & 0 \\ 7 & 3 & 4 & 10 \end{vmatrix}$

27. $\begin{vmatrix} 1 & -2 & 7 & 9 \\ 3 & -4 & 5 & 5 \\ 3 & 6 & 1 & -1 \\ 4 & 5 & 3 & 2 \end{vmatrix}$

28. $\begin{vmatrix} 0 & -3 & 8 & 2 \\ 8 & 1 & -1 & 6 \\ -4 & 6 & 0 & 9 \\ -7 & 0 & 0 & 14 \end{vmatrix}$

29. $\begin{vmatrix} 1 & -1 & 8 & 4 & 2 \\ 2 & 6 & 0 & -4 & 3 \\ 2 & 0 & 2 & 6 & 2 \\ 0 & 2 & 8 & 0 & 0 \\ 0 & 1 & 1 & 2 & 2 \end{vmatrix}$

30. $\begin{vmatrix} 3 & -2 & 4 & 3 & 1 \\ -1 & 0 & 2 & 1 & 0 \\ 5 & -1 & 0 & 3 & 2 \\ 4 & 7 & -8 & 0 & 0 \\ 1 & 2 & 3 & 0 & 2 \end{vmatrix}$

True or False? In Exercises 31 and 32, determine whether each statement is true or false. If a statement is true, give a reason or cite an appropriate statement in the text. If a statement is false, provide an example that shows that the statement is not true in all cases or cite an appropriate statement in the text.

31. (a) Interchanging two rows of a given matrix changes the sign of its determinant.
    (b) Multiplying a row of a matrix by a nonzero constant results in the determinant being multiplied by the same nonzero constant.
    (c) If two rows of a square matrix are equal, then its determinant is equal to 0.

32. (a) Adding a multiple of one row of a matrix to another row changes only the sign of the determinant.
    (b) Two matrices are column-equivalent if one matrix can be obtained from the other by elementary column operations.
    (c) If one row of a square matrix is a multiple of another row, then the determinant is equal to 0.

In Exercises 33–38, find the determinant of the elementary matrix. (Assume $k \neq 0$.)

33. $\begin{bmatrix} 1 & 0 & 0 \\ 0 & k & 0 \\ 0 & 0 & 1 \end{bmatrix}$
34. $\begin{bmatrix} 1 & 0 & 0 \\ 0 & 1 & 0 \\ 0 & 0 & k \end{bmatrix}$
35. $\begin{bmatrix} 0 & 1 & 0 \\ 1 & 0 & 0 \\ 0 & 0 & 1 \end{bmatrix}$

36. $\begin{bmatrix} 0 & 0 & 1 \\ 0 & 1 & 0 \\ 1 & 0 & 0 \end{bmatrix}$
37. $\begin{bmatrix} 1 & 0 & 0 \\ k & 1 & 0 \\ 0 & 0 & 1 \end{bmatrix}$
38. $\begin{bmatrix} 1 & 0 & 0 \\ 0 & 1 & 0 \\ 0 & k & 1 \end{bmatrix}$

39. Prove the property.
$$\begin{vmatrix} a_{11} & a_{12} & a_{13} \\ a_{21} & a_{22} & a_{23} \\ a_{31} & a_{32} & a_{33} \end{vmatrix} + \begin{vmatrix} b_{11} & a_{12} & a_{13} \\ b_{21} & a_{22} & a_{23} \\ b_{31} & a_{32} & a_{33} \end{vmatrix} = \begin{vmatrix} (a_{11} + b_{11}) & a_{12} & a_{13} \\ (a_{21} + b_{21}) & a_{22} & a_{23} \\ (a_{31} + b_{31}) & a_{32} & a_{33} \end{vmatrix}$$

40. Prove the property.
$$\begin{vmatrix} 1+a & 1 & 1 \\ 1 & 1+b & 1 \\ 1 & 1 & 1+c \end{vmatrix} = abc\left(1 + \frac{1}{a} + \frac{1}{b} + \frac{1}{c}\right),$$
$$a \neq 0, \quad b \neq 0, \quad c \neq 0$$

In Exercises 41 and 42, evaluate the determinant.

41. $\begin{vmatrix} \cos\theta & \sin\theta \\ -\sin\theta & \cos\theta \end{vmatrix}$
42. $\begin{vmatrix} \sec\theta & \tan\theta \\ \tan\theta & \sec\theta \end{vmatrix}$

43. ***Writing*** Solve the equation for x, if possible. Explain your result.
$$\begin{vmatrix} \cos x & 0 & \sin x \\ \sin x & 0 & -\cos x \\ \sin x - \cos x & 1 & \sin x + \cos x \end{vmatrix} = 0$$

44. ***Guided Proof*** Prove Property 2 of Theorem 3.3: If B is obtained from A by adding a multiple of a row of A to another row of A, then the determinant of B is equal to the determinant of A.

    Getting Started: To prove that the determinant of B is equal to the determinant of A, you need to show that their respective cofactor expansions are equal.

    (i) Begin your proof by letting B be the matrix obtained by adding c times the jth row of A to the ith row of A.
    (ii) Find the determinant of B by expanding along this ith row.
    (iii) Distribute and then group the terms containing a coefficient of c and those not containing a coefficient of c.
    (iv) Show that the sum of the terms not containing a coefficient of c is the determinant of A and that the sum of the terms containing a coefficient of c is equal to 0.

45. ***Guided Proof*** Prove Property 3 of Theorem 3.3: If B is obtained from A by multiplying a row of A by a nonzero constant c, then $\det(B) = c\det(A)$.

    Getting Started: To prove that the determinant of B is equal to c times the determinant of A, you need to show that the determinant of B is equal to c times the cofactor expansion of the determinant of A.

    (i) Begin your proof by letting B be the matrix obtained by multiplying c times the ith row of A.
    (ii) Find the determinant of B by expanding along this ith row.
    (iii) Factor out the common factor c.
    (iv) Show that the result is c times the determinant of A.

46. ***Writing*** Suppose a computer operator charges $0.001 (one tenth of a cent) for each addition and subtraction, and $0.003 for each multiplication and division. Contrast the costs of calculating the determinant of a 10×10 matrix by cofactor expansion and by row reduction. Which method would you prefer to use for calculating determinants?

3.3 | Properties of Determinants

In this section you will learn several important properties of determinants. You will begin by considering the determinant of the product of two matrices.

EXAMPLE 1   *The Determinant of a Matrix Product*

Find $|A|$, $|B|$, and $|AB|$ for the matrices

$$A = \begin{bmatrix} 1 & -2 & 2 \\ 0 & 3 & 2 \\ 1 & 0 & 1 \end{bmatrix} \quad \text{and} \quad B = \begin{bmatrix} 2 & 0 & 1 \\ 0 & -1 & -2 \\ 3 & 1 & -2 \end{bmatrix}.$$

Solution   Using the techniques described in the preceding section, you can show that $|A|$ and $|B|$ have the following values.

$$|A| = \begin{vmatrix} 1 & -2 & 2 \\ 0 & 3 & 2 \\ 1 & 0 & 1 \end{vmatrix} = -7 \quad \text{and} \quad |B| = \begin{vmatrix} 2 & 0 & 1 \\ 0 & -1 & -2 \\ 3 & 1 & -2 \end{vmatrix} = 11$$

The matrix product AB is given by

$$AB = \begin{bmatrix} 1 & -2 & 2 \\ 0 & 3 & 2 \\ 1 & 0 & 1 \end{bmatrix} \begin{bmatrix} 2 & 0 & 1 \\ 0 & -1 & -2 \\ 3 & 1 & -2 \end{bmatrix} = \begin{bmatrix} 8 & 4 & 1 \\ 6 & -1 & -10 \\ 5 & 1 & -1 \end{bmatrix}.$$

Therefore the determinant of AB is

$$|AB| = \begin{vmatrix} 8 & 4 & 1 \\ 6 & -1 & -10 \\ 5 & 1 & -1 \end{vmatrix} = -77.$$

In Example 1, note that the determinant of the matrix product is given by the product of the determinants. That is,

$$|AB| = |A||B|$$

$$-77 = (-7)(11).$$

This is true in general, as indicated in the following theorem.

Theorem 3.5   *Determinant of a Matrix Product*

If A and B are square matrices of order n, then

$$\det(AB) = \det(A)\det(B).$$

Proof   To begin, observe that if E is an elementary matrix, then, by Theorem 3.3, the following statements are true. If E is obtained from I by interchanging two rows, then $|E| = -1$. If E is obtained by multiplying a row of I by a nonzero constant c, then $|E| = c$. If E is obtained by adding a multiple of one row of I to another row of I, then $|E| = 1$. Additionally, by Theorem 2.12, if E results from performing an elementary row operation on I and the same elementary row operation is performed on B, then the matrix EB results. It follows that

$$|EB| = |E|\,|B|.$$

This can be generalized to conclude that

$$|E_k \cdots E_2 E_1 B| = |E_k| \cdots |E_2|\,|E_1|\,|B|,$$

where E_i is an elementary matrix. Now consider the matrix AB. If A is *nonsingular*, then, by Theorem 2.14, it can be written as the product of elementary matrices $A = E_k \cdots E_2 E_1$ and you can write

$$|AB| = |E_k \cdots E_2 E_1 B|$$
$$= |E_k| \cdots |E_2|\,|E_1|\,|B| = |E_k \cdots E_2 E_1|\,|B| = |A|\,|B|.$$

If A is *singular*, then A is row-equivalent to a matrix with an entire row of zeros. Thus, from Theorem 3.4, you can conclude that $|A| = 0$. Moreover, because A is singular, it follows that AB is also singular. (If AB were nonsingular, then $A[B(AB)^{-1}] = I$ would imply that A is nonsingular.) Thus, $|AB| = 0$, and you can conclude that $|AB| = |A|\,|B|$.

REMARK: Theorem 3.5 can be extended to include the product of any finite number of matrices. That is,

$$|A_1 A_2 A_3 \ldots A_k| = |A_1||A_2||A_3| \ldots |A_k|.$$

The relationship between $|A|$ and $|cA|$ is given in the next theorem.

Theorem 3.6

Determinant of a Scalar Multiple of a Matrix

If A is an $n \times n$ matrix and c is a scalar, then the determinant of cA is given by

$$\det(cA) = c^n \det(A).$$

Proof   This formula can be proven by repeated applications of Property 3 of Theorem 3.3. That is, factor the scalar c out of each of the n rows of $|cA|$ to obtain

$$|cA| = c^n |A|.$$

EXAMPLE 2   *The Determinant of a Scalar Multiple of a Matrix*

Find the determinant of the matrix.

$$A = \begin{bmatrix} 10 & -20 & 40 \\ 30 & 0 & 50 \\ -20 & -30 & 10 \end{bmatrix}$$

Solution   Because

$$A = 10 \begin{bmatrix} 1 & -2 & 4 \\ 3 & 0 & 5 \\ -2 & -3 & 1 \end{bmatrix} \quad \text{and} \quad \begin{vmatrix} 1 & -2 & 4 \\ 3 & 0 & 5 \\ -2 & -3 & 1 \end{vmatrix} = 5,$$

you can apply Theorem 3.6 to conclude that

$$|A| = 10^3 \begin{vmatrix} 1 & -2 & 4 \\ 3 & 0 & 5 \\ -2 & -3 & 1 \end{vmatrix} = 1000(5) = 5000.$$

Theorems 3.5 and 3.6 provide formulas for evaluating the determinants of the product of two matrices and a scalar multiple of a matrix. These theorems do not, however, list a formula for the determinant of the *sum* of two matrices. It is important to note that the sum of the determinants of two matrices usually does not equal the determinant of their sum. That is, in general,

$$|A| + |B| \neq |A + B|.$$

For instance, if

$$A = \begin{bmatrix} 6 & 2 \\ 2 & 1 \end{bmatrix} \quad \text{and} \quad B = \begin{bmatrix} 3 & 7 \\ 0 & -1 \end{bmatrix},$$

then $|A| = 2$ and $|B| = -3$, but $|A + B| = -18$.

Determinants and the Inverse of a Matrix

You saw in Chapter 2 that some square matrices are not invertible. However, it can be difficult to tell simply by inspection whether or not a matrix has an inverse. For instance, can you tell which of the following two matrices is invertible?

$$A = \begin{bmatrix} 0 & 2 & -1 \\ 3 & -2 & 1 \\ 3 & 2 & -1 \end{bmatrix} \quad \text{or} \quad B = \begin{bmatrix} 0 & 2 & -1 \\ 3 & -2 & 1 \\ 3 & 2 & 1 \end{bmatrix}$$

The next theorem shows that determinants are useful for classifying square matrices as invertible or noninvertible.

Theorem 3.7

Determinant of an Invertible Matrix

A square matrix A is invertible (nonsingular) if and only if

$$\det(A) \neq 0.$$

Proof

DISCOVERY

Let

$$A = \begin{bmatrix} 6 & 4 & 1 \\ 0 & 2 & 3 \\ 1 & 1 & 2 \end{bmatrix}.$$

Use a graphing utility to find A^{-1}. Compare $\det(A^{-1})$ with $\det(A)$. Make a conjecture about the determinant of the inverse of a matrix.

To prove the theorem in one direction, assume that A is invertible. Then $AA^{-1} = I$, and by Theorem 3.5 you can write

$$|A||A^{-1}| = |I|.$$

Now, because $|I| = 1$, you know that neither determinant on the left is zero. Specifically, $|A| \neq 0$.

To prove the theorem in the other direction, assume that the determinant of A is nonzero. Then, using Gauss-Jordan elimination, find a matrix B, in reduced row-echelon form, that is row-equivalent to A. Because B is in reduced row-echelon form, it must be the identity matrix I or it must have at least one row that consists entirely of zeros. But if B has a row of all zeros, then by Theorem 3.4 you know that $|B| = 0$, which would imply that $|A| = 0$. Because you assumed $|A|$ is nonzero, you can conclude that $B = I$. Therefore A is row-equivalent to the identity matrix, and by Theorem 2.15 you know that A is invertible.

EXAMPLE 3   *Classifying Square Matrices as Singular or Nonsingular*

Which of the matrices has an inverse?

(a) $\begin{bmatrix} 0 & 2 & -1 \\ 3 & -2 & 1 \\ 3 & 2 & -1 \end{bmatrix}$ (b) $\begin{bmatrix} 0 & 2 & -1 \\ 3 & -2 & 1 \\ 3 & 2 & 1 \end{bmatrix}$

Solution   (a) Because

$$\begin{vmatrix} 0 & 2 & -1 \\ 3 & -2 & 1 \\ 3 & 2 & -1 \end{vmatrix} = 0,$$

you can conclude that this matrix has no inverse (it is singular).

(b) Because

$$\begin{vmatrix} 0 & 2 & -1 \\ 3 & -2 & 1 \\ 3 & 2 & 1 \end{vmatrix} = -12 \neq 0,$$

you can conclude that this matrix has an inverse (it is nonsingular).

The following theorem gives a convenient way to find the determinant of the inverse of a matrix.

Theorem 3.8 *Determinant of an Inverse Matrix*

If A is invertible, then

$$\det(A^{-1}) = \frac{1}{\det(A)}.$$

Proof Because A is invertible, $AA^{-1} = I$, and you can apply Theorem 3.5 to conclude that $|A||A^{-1}| = |I| = 1$. Moreover, because A is invertible, you also know that $|A| \neq 0$, and you can divide by $|A|$ to obtain

$$|A^{-1}| = \frac{1}{|A|}.$$

EXAMPLE 4 *Determinant of the Inverse of a Matrix*

Find $|A^{-1}|$ for the matrix

$$A = \begin{bmatrix} 1 & 0 & 3 \\ 0 & -1 & 2 \\ 2 & 1 & 0 \end{bmatrix}.$$

Solution One way to solve this problem is to find A^{-1} and then evaluate its determinant. It is easier, however, to apply Theorem 3.8 as follows. Find the determinant of A,

$$|A| = \begin{vmatrix} 1 & 0 & 3 \\ 0 & -1 & 2 \\ 2 & 1 & 0 \end{vmatrix} = 4,$$

and then use the formula $|A^{-1}| = 1/|A|$ to conclude that $|A^{-1}| = \frac{1}{4}$.

REMARK: In Example 4, the inverse of A is

$$A^{-1} = \begin{bmatrix} -\frac{1}{2} & \frac{3}{4} & \frac{3}{4} \\ 1 & -\frac{3}{2} & -\frac{1}{2} \\ \frac{1}{2} & -\frac{1}{4} & -\frac{1}{4} \end{bmatrix}.$$

Try evaluating the determinant of this matrix directly. Then compare your answer with that obtained in Example 4.

Note that Theorem 3.7 (a matrix is invertible if and only if its determinant is nonzero) provides another equivalent condition that can be added to the list given in Theorem 2.15. All six conditions are summarized below.

Equivalent Conditions for a Nonsingular Matrix

If A is an $n \times n$ matrix, then the following statements are equivalent.
1. A is invertible.
2. $A\mathbf{x} = \mathbf{b}$ has a unique solution for every $n \times 1$ matrix $\mathbf{b}$.
3. $A\mathbf{x} = O$ has only the trivial solution.
4. A is row-equivalent to I_n.
5. A can be written as the product of elementary matrices.
6. $\det(A) \neq 0$

REMARK: In Section 3.2 you saw that a square matrix A has a determinant of zero if and only if A is row-equivalent to a matrix that has at least one row consisting entirely of zeros. The validity of this statement follows from the equivalence of Properties 4 and 6.

EXAMPLE 5 *Systems of Linear Equations*

Which of the systems has a unique solution?

(a)
$$\begin{aligned} 2x_2 - x_3 &= -1 \\ 3x_1 - 2x_2 + x_3 &= 4 \\ 3x_1 + 2x_2 - x_3 &= -4 \end{aligned}$$

(b)
$$\begin{aligned} 2x_2 - x_3 &= -1 \\ 3x_1 - 2x_2 + x_3 &= 4 \\ 3x_1 + 2x_2 + x_3 &= -4 \end{aligned}$$

Solution From Example 3 you know that the coefficient matrices for these two systems have the following determinants.

(a) $\begin{vmatrix} 0 & 2 & -1 \\ 3 & -2 & 1 \\ 3 & 2 & -1 \end{vmatrix} = 0$ (b) $\begin{vmatrix} 0 & 2 & -1 \\ 3 & -2 & 1 \\ 3 & 2 & 1 \end{vmatrix} = -12$

Therefore, using the preceding list of equivalent conditions, you can conclude that only the second system has a unique solution.

Determinants and the Transpose of a Matrix

The following theorem tells you that the determinant of the transpose of a square matrix is equal to the determinant of the matrix. This theorem can be proven using mathematical induction and Theorem 3.1, which states that a determinant can be evaluated using cofactor expansion along a row *or* a column. The details of the proof are left to you. (See Exercise 46.)

Theorem 3.9   *Determinant of a Transpose*

If A is a square matrix, then

$$\det(A) = \det(A^T).$$

EXAMPLE 6   *The Determinant of a Transpose*

Show that $|A| = |A^T|$ for the following matrix.

$$A = \begin{bmatrix} 3 & 1 & -2 \\ 2 & 0 & 0 \\ -4 & -1 & 5 \end{bmatrix}$$

Solution   To find the determinant of A, expand by cofactors along the second *row* to obtain

$$|A| = 2(-1)^3 \begin{vmatrix} 1 & -2 \\ -1 & 5 \end{vmatrix} = -2(3) = -6.$$

To find the determinant of

$$A^T = \begin{vmatrix} 3 & 2 & -4 \\ 1 & 0 & -1 \\ -2 & 0 & 5 \end{vmatrix},$$

expand by cofactors down the second *column* to obtain

$$|A^T| = 2(-1)^3 \begin{vmatrix} 1 & -1 \\ -2 & 5 \end{vmatrix} = -2(3) = -6.$$

Thus, $|A| = |A^T|$.

SECTION 3.3   Exercises

In Exercises 1–4, find (a) $|A|$, (b) $|B|$, (c) AB, and (d) $|AB|$. Then verify that $|A||B| = |AB|$.

1. $A = \begin{bmatrix} -2 & 1 \\ 4 & -2 \end{bmatrix}$, $B = \begin{bmatrix} 1 & 1 \\ 0 & -1 \end{bmatrix}$

2. $A = \begin{bmatrix} 1 & 2 \\ 2 & 4 \end{bmatrix}$, $B = \begin{bmatrix} -1 & 2 \\ 3 & 0 \end{bmatrix}$

3. $A = \begin{bmatrix} -1 & 2 & 1 \\ 1 & 0 & 1 \\ 0 & 1 & 0 \end{bmatrix}$, $B = \begin{bmatrix} -1 & 0 & 0 \\ 0 & 2 & 0 \\ 0 & 0 & 3 \end{bmatrix}$

4. $A = \begin{bmatrix} 2 & 0 & 1 \\ 1 & -1 & 2 \\ 3 & 1 & 0 \end{bmatrix}$, $B = \begin{bmatrix} 2 & -1 & 4 \\ 0 & 1 & 3 \\ 3 & -2 & 1 \end{bmatrix}$

In Exercises 5–8, use the fact that $|cA| = c^n|A|$ to evaluate the determinant of the $n \times n$ matrix.

5. $A = \begin{bmatrix} 4 & 2 \\ 6 & -8 \end{bmatrix}$

6. $A = \begin{bmatrix} 5 & 15 \\ 10 & -20 \end{bmatrix}$

7. $A = \begin{bmatrix} -3 & 6 & 9 \\ 6 & 9 & 12 \\ 9 & 12 & 15 \end{bmatrix}$

8. $A = \begin{bmatrix} 4 & 16 & 0 \\ 12 & -8 & 8 \\ 16 & 20 & -4 \end{bmatrix}$

In Exercises 9 and 10, find (a) $|A|$, (b) $|B|$, and (c) $|A + B|$. Then verify that $|A| + |B| \neq |A + B|$.

9. $A = \begin{bmatrix} -1 & 1 \\ 2 & 0 \end{bmatrix}$, $B = \begin{bmatrix} 1 & -1 \\ -2 & 0 \end{bmatrix}$

10. $A = \begin{bmatrix} 1 & -2 \\ 1 & 0 \end{bmatrix}$, $B = \begin{bmatrix} 3 & -2 \\ 0 & 0 \end{bmatrix}$

In Exercises 11–14, find (a) $|A^T|$, (b) $|A^2|$, (c) $|AA^T|$, (d) $|2A|$, and (e) $|A^{-1}|$.

11. $A = \begin{bmatrix} 6 & -11 \\ 4 & -5 \end{bmatrix}$

12. $A = \begin{bmatrix} -4 & 10 \\ 5 & 6 \end{bmatrix}$

13. $A = \begin{bmatrix} 2 & 0 & 5 \\ 4 & -1 & 6 \\ 3 & 2 & 1 \end{bmatrix}$

14. $A = \begin{bmatrix} 1 & 5 & 4 \\ 0 & -6 & 2 \\ 0 & 0 & -3 \end{bmatrix}$

In Exercises 15–18, use a graphing utility with matrix capabilities to find (a) $|A|$, (b) $|A^T|$, (c) $|A^2|$, (d) $|2A|$, and (e) $|A^{-1}|$.

15. $A = \begin{bmatrix} 4 & 2 \\ -1 & 5 \end{bmatrix}$

16. $A = \begin{bmatrix} -2 & 4 \\ 6 & 8 \end{bmatrix}$

17. $A = \begin{bmatrix} 4 & -2 & 1 & 5 \\ 3 & 8 & 2 & -1 \\ 6 & 8 & 9 & 2 \\ 2 & 3 & -1 & 0 \end{bmatrix}$

18. $A = \begin{bmatrix} 6 & 5 & 1 & -1 \\ -2 & 4 & 3 & 5 \\ 6 & 1 & -4 & -2 \\ 2 & 2 & 1 & 3 \end{bmatrix}$

19. Let A and B be square matrices of order 4 such that $|A| = -5$ and $|B| = 3$. Find (a) $|AB|$, (b) $|A^3|$, (c) $|3B|$, (d) $|(AB)^T|$, and (e) $|A^{-1}|$.

20. Let A and B be square matrices of order 3 such that $|A| = 10$ and $|B| = 12$. Find (a) $|AB|$, (b) $|A^4|$, (c) $|2B|$, (d) $|(AB)^T|$, and (e) $|A^{-1}|$.

21. Let A and B be square matrices of order 4 such that $|A| = 4$ and $|B| = 2$. Find (a) $|BA|$, (b) $|B^2|$, (c) $|2A|$, (d) $|(AB)^T|$, and (e) $|B^{-1}|$.

22. Let A and B be square matrices of order 3 such that $|A| = -2$ and $|B| = 5$. Find (a) $|BA|$, (b) $|B^4|$, (c) $|2A|$, (d) $|(AB)^T|$, and (e) $|B^{-1}|$.

In Exercises 23–30, use a determinant to decide whether the matrix is singular or nonsingular.

23. $\begin{bmatrix} 5 & 4 \\ 10 & 8 \end{bmatrix}$

24. $\begin{bmatrix} 3 & -6 \\ 4 & 2 \end{bmatrix}$

25. $\begin{bmatrix} 14 & 5 & 7 \\ -2 & 0 & 3 \\ 1 & -5 & -10 \end{bmatrix}$

26. $\begin{bmatrix} 1 & 0 & 4 \\ 0 & 6 & 3 \\ 2 & -1 & 4 \end{bmatrix}$

27. $\begin{bmatrix} \frac{1}{2} & \frac{3}{2} & 2 \\ \frac{2}{3} & -\frac{1}{3} & 0 \\ 1 & 1 & 1 \end{bmatrix}$

28. $\begin{bmatrix} 2 & -\frac{1}{2} & 8 \\ 1 & -\frac{1}{4} & 4 \\ -\frac{5}{2} & \frac{3}{2} & 8 \end{bmatrix}$

29. $\begin{bmatrix} 1 & 0 & -8 & 2 \\ 0 & 8 & -1 & 10 \\ 0 & 0 & 0 & 1 \\ 0 & 0 & 0 & 2 \end{bmatrix}$

30. $\begin{bmatrix} 0.8 & 0.2 & -0.6 & 0.1 \\ -1.2 & 0.6 & 0.6 & 0 \\ 0.7 & -0.3 & 0.1 & 0 \\ 0.2 & -0.3 & 0.6 & 0 \end{bmatrix}$

In Exercises 31–34, use the determinant of the coefficient matrix to determine whether the system of linear equations has a unique solution.

31.
$$\begin{aligned} x_1 - x_2 + x_3 &= 4 \\ 2x_1 - x_2 + x_3 &= 6 \\ 3x_1 - 2x_2 + 2x_3 &= 0 \end{aligned}$$

32.
$$\begin{aligned} x_1 + x_2 - x_3 &= 4 \\ 2x_1 - x_2 + x_3 &= 6 \\ 3x_1 - 2x_2 + 2x_3 &= 0 \end{aligned}$$

33.
$$\begin{aligned} 2x_1 + x_2 + 5x_3 + x_4 &= 5 \\ x_1 + x_2 - 3x_3 - 4x_4 &= -1 \\ 2x_1 + 2x_2 + 2x_3 - 3x_4 &= 2 \\ x_1 + 5x_2 - 6x_3 &= 3 \end{aligned}$$

34.
$$\begin{aligned} x_1 - x_2 - x_3 - x_4 &= 0 \\ x_1 + x_2 - x_3 - x_4 &= 0 \\ x_1 + x_2 + x_3 - x_4 &= 0 \\ x_1 + x_2 + x_3 + x_4 &= 6 \end{aligned}$$

In Exercises 35–38, find the value(s) of k such that A is singular.

35. $A = \begin{bmatrix} k - 1 & 3 \\ 2 & k - 2 \end{bmatrix}$

36. $A = \begin{bmatrix} k - 1 & 2 \\ 2 & k + 2 \end{bmatrix}$

37. $A = \begin{bmatrix} 1 & 0 & 3 \\ 2 & -1 & 0 \\ 4 & 2 & k \end{bmatrix}$

38. $A = \begin{bmatrix} 1 & k & 2 \\ -2 & 0 & -k \\ 3 & 1 & -4 \end{bmatrix}$

39. Let A and B be $n \times n$ matrices such that $AB = I$. Prove that $|A| \neq 0$ and $|B| \neq 0$.

40. Let A and B be $n \times n$ matrices such that AB is singular. Prove that either A or B is singular.

41. Find two 2×2 matrices such that $|A| + |B| = |A + B|$.

42. Verify the equation.
$$\begin{vmatrix} a + b & a & a \\ a & a + b & a \\ a & a & a + b \end{vmatrix} = b^2(3a + b)$$

43. Let A be an $n \times n$ matrix each of whose rows adds up to zero. Find $|A|$.

44. Illustrate the result of Exercise 43 with the matrix
$$A = \begin{bmatrix} 2 & -1 & -1 \\ -3 & 1 & 2 \\ 0 & -2 & 2 \end{bmatrix}.$$

45. *Guided Proof* Prove that the determinant of an invertible matrix A is equal to ± 1 if all of the entries of A and A^{-1} are integers.

Getting Started: Denote $\det(A)$ as x and $\det(A^{-1})$ as y. Note that x and y are real numbers. To prove that $\det(A)$ is equal to ± 1, you must show that both x and y are integers such that their product xy is equal to 1.

   (i) Use the property for the determinant of a matrix product to show that $xy = 1$.
   (ii) Use the definition of a determinant and the fact that the entries of A and A^{-1} are integers to show that both $x = \det(A)$ and $y = \det(A^{-1})$ are integers.
   (iii) Conclude that $x = \det(A)$ must be either 1 or -1 because these are the only integer solutions to the equation $xy = 1$.

46. *Guided Proof* Prove Theorem 3.9: If A is a square matrix, then $\det(A) = \det(A^T)$.

Getting Started: To prove that the determinants of A and A^T are equal, you need to show that their cofactor expansions are equal. Because the cofactors are $\pm$ determinants of smaller matrices, you need to use mathematical induction.

   (i) Initial step for induction: If A is of order 1, then $A = [a_{11}] = A^T$, so $\det(A) = \det(A^T) = a_{11}$.
   (ii) Assume that the inductive hypothesis holds for all matrices of order $n - 1$. Let A be a square matrix of order n. Write an expression for the determinant of A by expanding by the first row.
   (iii) Write an expression for the determinant of A^T by expanding by the first column.
   (iv) Compare these two expansions. The entries of the first row of A are the same as the entries of the first column of A^T. Compare cofactors (these are the $\pm$ determinants of smaller matrices that are transposes of one another) and use the inductive hypothesis to conclude that they are equal as well.

47. *Writing* Let A and P be $n \times n$ matrices, with P invertible. Does $P^{-1}AP = A$? Illustrate your conclusion with appropriate examples. What can you say about the two determinants $|P^{-1}AP|$ and $|A|$?

48. *Writing* Let A be an $n \times n$ nonzero matrix satisfying $A^{10} = O$. Explain why A must be singular. What properties of determinants are you using in your argument?

True or False? In Exercises 49 and 50, determine whether each statement is true or false. If a statement is true, give a reason or cite an appropriate statement in the text. If a statement is false, provide an example that shows that the statement is not true in all cases or cite an appropriate statement in the text.

49. (a) If A is an $n \times n$ matrix and c is a nonzero scalar, then the determinant of the matrix cA is given by $nc \cdot \det(A)$.
   (b) If A is an invertible matrix, then the determinant of A^{-1} is equal to the reciprocal of the determinant of A.
   (c) If A is an invertible $n \times n$ matrix, then $A\mathbf{x} = \mathbf{b}$ has a unique solution for every $\mathbf{b}$.

50. (a) In general, the determinant of the sum of two matrices equals the sum of the determinants of the matrices.
   (b) If A is a square matrix, then the determinant of A is equal to the determinant of the transpose of A.
   (c) If the determinant of an $n \times n$ matrix A is nonzero, then $A\mathbf{x} = \mathbf{0}$ has only the trivial solution.

51. A square matrix is called **skew-symmetric** if $A^T = -A$. Prove that if A is an $n \times n$ skew-symmetric matrix, then $|A| = (-1)^n|A|$.

52. Let A be a skew-symmetric matrix of odd order. Use the result of Exercise 51 to prove that $|A| = 0$.

In Exercises 53–58, determine whether the matrix is orthogonal. An invertible square matrix A is called **orthogonal** if $A^{-1} = A^T$.

53. $\begin{bmatrix} 0 & 1 \\ 1 & 0 \end{bmatrix}$

54. $\begin{bmatrix} 1 & -1 \\ -1 & -1 \end{bmatrix}$

55. $\begin{bmatrix} 0 & 0 \\ 1 & 0 \end{bmatrix}$

56. $\begin{bmatrix} 1/\sqrt{2} & -1/\sqrt{2} \\ -1/\sqrt{2} & -1/\sqrt{2} \end{bmatrix}$

57. $\begin{bmatrix} 1 & 0 & 0 \\ 0 & 0 & 1 \\ 0 & 1 & 0 \end{bmatrix}$

58. $\begin{bmatrix} 1/\sqrt{2} & 0 & -1/\sqrt{2} \\ 0 & 1 & 0 \\ 1/\sqrt{2} & 0 & 1/\sqrt{2} \end{bmatrix}$

59. Prove that if A is an orthogonal matrix, then $|A| = \pm 1$.

In Exercises 60 and 61, use a graphing utility with matrix capabilities to determine whether A is orthogonal. To test for orthogonality, find (a) A^{-1}, (b) A^T, and (c) $|A|$, and verify that $A^{-1} = A^T$ and $|A| = \pm 1$.

60. $A = \begin{bmatrix} \frac{3}{5} & 0 & -\frac{4}{5} \\ 0 & 1 & 0 \\ \frac{4}{5} & 0 & \frac{3}{5} \end{bmatrix}$

61. $A = \begin{bmatrix} \frac{2}{3} & -\frac{2}{3} & \frac{1}{3} \\ \frac{2}{3} & \frac{1}{3} & -\frac{2}{3} \\ \frac{1}{3} & \frac{2}{3} & \frac{2}{3} \end{bmatrix}$

62. If A is an idempotent matrix ($A^2 = A$), prove that the determinant of A is either 0 or 1.

63. Let S be an $n \times n$ singular matrix. Prove that for any $n \times n$ matrix B, the matrix SB is also singular.

64. Let A_{11}, A_{12}, and A_{22} be $n \times n$ matrices. Find the determinant of the partitioned matrix

$$\begin{bmatrix} A_{11} & A_{12} \\ 0 & A_{22} \end{bmatrix}$$

in terms of the determinants of $A_{11}, A_{12},$ and A_{22}.

3.4 | Introduction to Eigenvalues

This chapter continues with a look ahead to one of the most important topics of linear algebra—**eigenvalues.** One application of eigenvalues involves the study of population growth. For example, suppose that, of a population of rabbits raised in a laboratory, half survive their first year, and of those, half survive their second year. Their maximum life span is 3 years. Furthermore, during the first year the rabbits produce no offspring, whereas the average number of offspring is 6 during the second year and 8 during the third year. If there are 24 rabbits in each age class now, what will the distribution be in 1 year? In 20 years?

As you will find in Chapter 7, the solution of this application depends on the concept of eigenvalues. You will see later that eigenvalues are used in a wide variety of real-life applications of linear algebra. Aside from population growth, eigenvalues are used in solving systems of differential equations, in quadratic forms, and in engineering and science.

The central question of the **eigenvalue problem** can be stated as follows. If A is an $n \times n$ matrix, do there exist $n \times 1$ nonzero matrices $\mathbf{x}$ such that $A\mathbf{x}$ is a scalar multiple of $\mathbf{x}$? The scalar is usually denoted by λ (the Greek letter lambda) and is called an **eigenvalue** of A, and the nonzero column matrix $\mathbf{x}$ is called an **eigenvector** of A corresponding to λ. Thus the fundamental equation for the eigenvalue problem is

$$A\mathbf{x} = \lambda\mathbf{x}.$$

EXAMPLE 1 *Verifying Eigenvalues and Eigenvectors*

Let $A = \begin{bmatrix} 1 & 4 \\ 2 & 3 \end{bmatrix}$, $\mathbf{x}_1 = \begin{bmatrix} 1 \\ 1 \end{bmatrix}$, and $\mathbf{x}_2 = \begin{bmatrix} 2 \\ -1 \end{bmatrix}$.

Verify that $\lambda_1 = 5$ is an eigenvalue of A corresponding to $\mathbf{x}_1$, and that $\lambda_2 = -1$ is an eigenvalue of A corresponding to $\mathbf{x}_2$.

Solution   To verify that $\lambda_1 = 5$ is an eigenvalue of A corresponding to $\mathbf{x}_1$, multiply the matrices A and $\mathbf{x}_1$ as follows.

$$A\mathbf{x}_1 = \begin{bmatrix} 1 & 4 \\ 2 & 3 \end{bmatrix}\begin{bmatrix} 1 \\ 1 \end{bmatrix} = \begin{bmatrix} 5 \\ 5 \end{bmatrix} = 5\begin{bmatrix} 1 \\ 1 \end{bmatrix} = \lambda_1\mathbf{x}_1$$

Similarly,

$$A\mathbf{x}_2 = \begin{bmatrix} 1 & 4 \\ 2 & 3 \end{bmatrix}\begin{bmatrix} 2 \\ -1 \end{bmatrix} = \begin{bmatrix} -2 \\ 1 \end{bmatrix} = -1\begin{bmatrix} 2 \\ -1 \end{bmatrix} = \lambda_2\mathbf{x}_2.$$

From this example, you can see that it is easy to verify whether a given scalar λ and an $n \times 1$ matrix $\mathbf{x}$ satisfy the equation $A\mathbf{x} = \lambda\mathbf{x}$. Notice also that if $\mathbf{x}$ is an eigenvector corresponding to λ, then so is any nonzero multiple of $\mathbf{x}$. For instance, the column matrices

$$\begin{bmatrix} 2 \\ 2 \end{bmatrix} \quad \text{and} \quad \begin{bmatrix} -7 \\ -7 \end{bmatrix}$$

are also eigenvectors of A corresponding to $\lambda_1 = 5$.

Given an $n \times n$ matrix A, how can you find the eigenvalues and corresponding eigenvectors? The key is to write the equation $A\mathbf{x} = \lambda\mathbf{x}$ in the equivalent form

$$(\lambda I - A)\mathbf{x} = \mathbf{0},$$

where I is the $n \times n$ identity matrix. This homogeneous system of equations has nonzero solutions if and only if the coefficient matrix $(\lambda I - A)$ is singular; that is, if and only if the determinant of $(\lambda I - A)$ is zero. The equation $\det(\lambda I - A) = 0$ is called the **characteristic equation** of A, and is a polynomial equation of degree n in the variable λ. Once you have found the eigenvalues of A, you can use Gaussian elimination to find the corresponding eigenvectors, as shown in the next two examples.

EXAMPLE 2   *Finding Eigenvalues and Eigenvectors*

Find the eigenvalues and corresponding eigenvectors of the matrix $A = \begin{bmatrix} 1 & 4 \\ 2 & 3 \end{bmatrix}$.

Solution   The characteristic equation of A is

$$\begin{aligned} |\lambda I - A| &= \left| \begin{bmatrix} \lambda & 0 \\ 0 & \lambda \end{bmatrix} - \begin{bmatrix} 1 & 4 \\ 2 & 3 \end{bmatrix} \right| \\ &= \begin{vmatrix} \lambda - 1 & -4 \\ -2 & \lambda - 3 \end{vmatrix} \\ &= \lambda^2 - 4\lambda + 3 - 8 \\ &= \lambda^2 - 4\lambda - 5 \\ &= (\lambda - 5)(\lambda + 1) = 0 \end{aligned}$$

This gives two eigenvalues, $\lambda_1 = 5$ and $\lambda_2 = -1$.

To find the corresponding eigenvectors, solve the homogeneous linear system $(\lambda I - A)\mathbf{x} = \mathbf{0}$. For $\lambda_1 = 5$, the coefficient matrix is

$$5I - A = \begin{bmatrix} 5 - 1 & -4 \\ -2 & 5 - 3 \end{bmatrix} = \begin{bmatrix} 4 & -4 \\ -2 & 2 \end{bmatrix},$$

which row reduces to

$$\begin{bmatrix} 1 & -1 \\ 0 & 0 \end{bmatrix}.$$

Technology   Note

Most graphing calculators and computer software programs have the capability to calculate the eigenvalues of a square matrix. If you use the eigenvalue command of a TI-86 to verify that the eigenvalues of the matrix A from Example 2 are -1 and 5, your screen will look like this.

```
A
               [[1 4]
                [2 3]]
eigVl A
               {-1 5}
```

The solutions of the homogeneous system having this coefficient matrix are all of the form

$$\begin{bmatrix} t \\ t \end{bmatrix},$$

where t is a real number. Thus, the eigenvectors corresponding to $\lambda_1 = 5$ are the nonzero scalar multiples of

$$\begin{bmatrix} 1 \\ 1 \end{bmatrix}.$$

Similarly, for $\lambda_2 = -1$, the corresponding coefficient matrix is

$$-1I - A = \begin{bmatrix} -1 - 1 & -4 \\ -2 & -1 - 3 \end{bmatrix} = \begin{bmatrix} -2 & -4 \\ -2 & -4 \end{bmatrix},$$

which row reduces to

$$\begin{bmatrix} 1 & 2 \\ 0 & 0 \end{bmatrix}.$$

The solutions of the homogeneous system having this coefficient matrix are all of the form

$$\begin{bmatrix} 2t \\ -t \end{bmatrix},$$

where t is a real number. Thus, the eigenvectors corresponding to the eigenvalue $\lambda_2 = -1$ are the nonzero multiples of

$$\begin{bmatrix} 2 \\ -1 \end{bmatrix}.$$

EXAMPLE 3   *Finding Eigenvalues and Eigenvectors*

Find the eigenvalues and corresponding eigenvectors of the matrix $A = \begin{bmatrix} 1 & 2 & -2 \\ 1 & 2 & 1 \\ -1 & -1 & 0 \end{bmatrix}.$

Solution   The characteristic equation of A is

$$|\lambda I - A| = \left| \begin{bmatrix} \lambda & 0 & 0 \\ 0 & \lambda & 0 \\ 0 & 0 & \lambda \end{bmatrix} - \begin{bmatrix} 1 & 2 & -2 \\ 1 & 2 & 1 \\ -1 & -1 & 0 \end{bmatrix} \right| = \begin{vmatrix} \lambda - 1 & -2 & 2 \\ -1 & \lambda - 2 & -1 \\ 1 & 1 & \lambda \end{vmatrix}$$

$$= (\lambda - 1) \begin{vmatrix} \lambda - 2 & -1 \\ 1 & \lambda \end{vmatrix} - (-2) \begin{vmatrix} -1 & -1 \\ 1 & \lambda \end{vmatrix} + 2 \begin{vmatrix} -1 & \lambda - 2 \\ 1 & 1 \end{vmatrix}$$

$$= (\lambda - 1)(\lambda^2 - 2\lambda + 1) + 2(-\lambda + 1) + 2(-1 - \lambda + 2)$$

$$= \lambda^3 - 3\lambda^2 - \lambda + 3$$

$$= (\lambda^2 - 1)(\lambda - 3) = 0$$

This gives three eigenvalues, $\lambda_1 = 1$, $\lambda_2 = -1$, and $\lambda_3 = 3$.

To find the corresponding eigenvectors, solve the homogeneous linear system $(\lambda I - A)\mathbf{x} = \mathbf{0}$. For $\lambda_1 = 1$, the coefficient matrix is

$$(1)I - A = \begin{bmatrix} 1-1 & -2 & 2 \\ -1 & 1-2 & -1 \\ 1 & 1 & 1-0 \end{bmatrix} = \begin{bmatrix} 0 & -2 & 2 \\ -1 & -1 & -1 \\ 1 & 1 & 1 \end{bmatrix},$$

which row reduces to

$$\begin{bmatrix} 1 & 0 & 2 \\ 0 & 1 & -1 \\ 0 & 0 & 0 \end{bmatrix}.$$

The solutions of the homogenous system having this coefficient matrix are all of the form

$$\begin{bmatrix} -2t \\ t \\ t \end{bmatrix},$$

where t is a real number. So, the eigenvectors corresponding to $\lambda_1 = 1$ are nonzero scalar multiples of

$$\begin{bmatrix} -2 \\ 1 \\ 1 \end{bmatrix}.$$

For $\lambda_2 = -1$, the coefficient matrix is

$$(-1)I - A = \begin{bmatrix} -1-1 & -2 & 2 \\ -1 & -1-2 & -1 \\ 1 & 1 & -1-0 \end{bmatrix} = \begin{bmatrix} -2 & -2 & 2 \\ -1 & -3 & -1 \\ 1 & 1 & -1 \end{bmatrix},$$

which row reduces to

$$\begin{bmatrix} 1 & 0 & -2 \\ 0 & 1 & 1 \\ 0 & 0 & 0 \end{bmatrix}.$$

The solutions of the homogeneous system having this coefficient matrix are all of the form

$$\begin{bmatrix} 2t \\ -t \\ t \end{bmatrix},$$

where t is a real number. So, the eigenvectors corresponding to the eigenvalue $\lambda_2 = -1$ are the nonzero multiples of

$$\begin{bmatrix} 2 \\ -1 \\ 1 \end{bmatrix}.$$

For $\lambda_3 = 3$, the coefficient matrix is

$$3I - A = \begin{bmatrix} 3-1 & -2 & 2 \\ -1 & 3-2 & -1 \\ 1 & 1 & 3-0 \end{bmatrix} = \begin{bmatrix} 2 & -2 & 2 \\ -1 & 1 & -1 \\ 1 & 1 & 3 \end{bmatrix},$$

which row reduces to

$$\begin{bmatrix} 1 & 0 & 2 \\ 0 & 1 & 1 \\ 0 & 0 & 0 \end{bmatrix}.$$

The solutions of the homogeneous system having this coefficient matrix are all of the form

$$\begin{bmatrix} -2t \\ -t \\ t \end{bmatrix},$$

where t is a real number. So, the eigenvectors corresponding to the eigenvalue $\lambda_3 = 3$ are the nonzero multiples of

$$\begin{bmatrix} -2 \\ -1 \\ 1 \end{bmatrix}.$$

Eigenvalues and eigenvectors have ample applications in mathematics, engineering, biology, and other sciences. You will see one such application to stochastic processes at the end of this chapter. In Chapter 7 you will study eigenvalues and their applications in more detail and learn how to solve the rabbit population problem presented earlier.

SECTION 3.4   Exercises

Eigenvalues and Eigenvectors

In Exercises 1–4, verify that λ_i is an eigenvalue of A and that $\mathbf{x}_i$ is a corresponding eigenvector.

1. $A = \begin{bmatrix} 1 & 2 \\ 0 & -3 \end{bmatrix}$; $\lambda_1 = 1$, $\mathbf{x}_1 = \begin{bmatrix} 1 \\ 0 \end{bmatrix}$;

$\lambda_2 = -3$, $\mathbf{x}_2 = \begin{bmatrix} -1 \\ 2 \end{bmatrix}$

2. $A = \begin{bmatrix} 4 & 3 \\ 1 & 2 \end{bmatrix}$; $\lambda_1 = 5$, $\mathbf{x}_1 = \begin{bmatrix} 3 \\ 1 \end{bmatrix}$;

$\lambda_2 = 1$, $\mathbf{x}_2 = \begin{bmatrix} -1 \\ 1 \end{bmatrix}$

3. $A = \begin{bmatrix} 1 & 1 & 1 \\ 0 & 1 & 0 \\ 1 & 1 & 1 \end{bmatrix}$; $\lambda_1 = 2$, $\mathbf{x}_1 = \begin{bmatrix} 1 \\ 0 \\ 1 \end{bmatrix}$;

$\lambda_2 = 0$, $\mathbf{x}_2 = \begin{bmatrix} -1 \\ 0 \\ 1 \end{bmatrix}$; $\lambda_3 = 1$, $\mathbf{x}_3 = \begin{bmatrix} -1 \\ 1 \\ -1 \end{bmatrix}$

4. $A = \begin{bmatrix} 1 & -2 & 1 \\ 0 & 1 & 4 \\ 0 & 0 & 2 \end{bmatrix}$; $\lambda_1 = 1$, $\mathbf{x}_1 = \begin{bmatrix} 1 \\ 0 \\ 0 \end{bmatrix}$;

$\lambda_2 = 2$, $\mathbf{x}_2 = \begin{bmatrix} -7 \\ 4 \\ 1 \end{bmatrix}$

In Exercises 5–12, find (a) the characteristic equation, (b) the eigenvalues, and (c) the corresponding eigenvectors of the matrix.

5. $\begin{bmatrix} 4 & -5 \\ 2 & -3 \end{bmatrix}$

6. $\begin{bmatrix} 2 & 2 \\ 2 & 2 \end{bmatrix}$

7. $\begin{bmatrix} 2 & 1 \\ 3 & 0 \end{bmatrix}$

8. $\begin{bmatrix} 2 & 5 \\ 4 & 3 \end{bmatrix}$

9. $\begin{bmatrix} -2 & 4 \\ 2 & 5 \end{bmatrix}$

10. $\begin{bmatrix} 3 & -1 \\ 5 & -3 \end{bmatrix}$

11. $\begin{bmatrix} 1 & -1 & -1 \\ 1 & 3 & 1 \\ -3 & 1 & -1 \end{bmatrix}$

12. $\begin{bmatrix} 2 & 0 & 1 \\ 0 & 3 & 4 \\ 0 & 0 & 1 \end{bmatrix}$

In Exercises 13–18, use a graphing utility with matrix capabilities to find the eigenvalues of the matrix. Then find the corresponding eigenvectors.

13. $\begin{bmatrix} 2 & 5 \\ -1 & -4 \end{bmatrix}$

14. $\begin{bmatrix} 4 & 3 \\ -3 & -2 \end{bmatrix}$

15. $\begin{bmatrix} 4 & -2 & -2 \\ 0 & 1 & 0 \\ 1 & 0 & 1 \end{bmatrix}$

16. $\begin{bmatrix} 4 & 0 & 0 \\ 0 & 0 & -3 \\ 0 & -2 & 1 \end{bmatrix}$

17. $\begin{bmatrix} 3 & 0 & 0 & 0 \\ 0 & -1 & 0 & 0 \\ 0 & 0 & 2 & 5 \\ 0 & 0 & 3 & 0 \end{bmatrix}$

18. $\begin{bmatrix} 1 & 0 & 2 & 3 \\ 0 & 2 & 0 & 0 \\ 0 & 0 & 1 & 3 \\ 0 & -1 & 3 & 1 \end{bmatrix}$

True or False?   In Exercises 19 and 20, determine whether each statement is true or false. If a statement is true, give a reason or cite an appropriate statement in the text. If a statement is false, provide an example that shows that the statement is not true in all cases or cite an appropriate statement in the text.

19. (a) If $\mathbf{x}$ is an eigenvector corresponding to a given eigenvalue λ, then any multiple of $\mathbf{x}$ is also an eigenvector corresponding to that same λ.

(b) If $\lambda = a$ is an eigenvalue of the matrix A, then $\lambda = a$ is a solution of the characteristic equation $\lambda I - A = 0$.

20. (a) The characteristic equation of the matrix $A = \begin{bmatrix} 2 & -1 \\ 1 & 0 \end{bmatrix}$ yields eigenvalues $\lambda_1 = \lambda_2 = 1$.

(b) The matrix $A = \begin{bmatrix} 4 & -2 \\ -1 & 0 \end{bmatrix}$ has irrational eigenvalues $\lambda_1 = 2 + \sqrt{6}$ and $\lambda_2 = 2 - \sqrt{6}$.

3.5 | Applications of Determinants

So far in this chapter, you have examined procedures for evaluating determinants, studied properties of determinants, and learned how determinants are used to find eigenvalues. In this section, you will study an explicit formula for the inverse of a nonsingular matrix and then use this formula to derive a theorem known as Cramer's Rule. You will then solve several applications of determinants using Cramer's Rule.

The Adjoint of a Matrix

Recall from Section 3.1 that the cofactor C_{ij} of a matrix A is defined to be $(-1)^{i+j}$ times the determinant of the matrix obtained by deleting the ith row and the jth column of A. If A is a square matrix, then the **matrix of cofactors** of A has the form

$$\begin{bmatrix} C_{11} & C_{12} & \cdots & C_{1n} \\ C_{21} & C_{22} & \cdots & C_{2n} \\ \vdots & \vdots & & \vdots \\ C_{n1} & C_{n2} & \cdots & C_{nn} \end{bmatrix}.$$

The transpose of this matrix is called the **adjoint** of A and is denoted by adj(A). That is,

$$\text{adj}(A) = \begin{bmatrix} C_{11} & C_{21} & \cdots & C_{n1} \\ C_{12} & C_{22} & \cdots & C_{n2} \\ \vdots & \vdots & & \vdots \\ C_{1n} & C_{2n} & \cdots & C_{nn} \end{bmatrix}.$$

EXAMPLE 1   *Finding the Adjoint of a Square Matrix*

Find the adjoint of

$$A = \begin{bmatrix} -1 & 3 & 2 \\ 0 & -2 & 1 \\ 1 & 0 & -2 \end{bmatrix}.$$

Solution   The cofactor C_{11} is given by

$$\begin{bmatrix} -1 & 3 & 2 \\ 0 & -2 & 1 \\ 1 & 0 & -2 \end{bmatrix} \longrightarrow C_{11} = (-1)^2 \begin{vmatrix} -2 & 1 \\ 0 & -2 \end{vmatrix} = 4.$$

Continuing this process produces the following matrix of cofactors of A.

$$\begin{bmatrix} \begin{vmatrix} -2 & 1 \\ 0 & -2 \end{vmatrix} & -\begin{vmatrix} 0 & 1 \\ 1 & -2 \end{vmatrix} & \begin{vmatrix} 0 & -2 \\ 1 & 0 \end{vmatrix} \\[6pt] -\begin{vmatrix} 3 & 2 \\ 0 & -2 \end{vmatrix} & \begin{vmatrix} -1 & 2 \\ 1 & -2 \end{vmatrix} & -\begin{vmatrix} -1 & 3 \\ 1 & 0 \end{vmatrix} \\[6pt] \begin{vmatrix} 3 & 2 \\ -2 & 1 \end{vmatrix} & -\begin{vmatrix} -1 & 2 \\ 0 & 1 \end{vmatrix} & \begin{vmatrix} -1 & 3 \\ 0 & -2 \end{vmatrix} \end{bmatrix} = \begin{bmatrix} 4 & 1 & 2 \\ 6 & 0 & 3 \\ 7 & 1 & 2 \end{bmatrix}$$

The transpose of this matrix is the adjoint of A. That is,

$$\text{adj}(A) = \begin{bmatrix} 4 & 6 & 7 \\ 1 & 0 & 1 \\ 2 & 3 & 2 \end{bmatrix}.$$

The adjoint of a matrix A can be used to find the inverse of A, as indicated in the following theorem.

Theorem 3.10   *The Inverse of a Matrix Given by Its Adjoint*

If A is an $n \times n$ invertible matrix, then

$$A^{-1} = \frac{1}{\det(A)} \text{adj}(A).$$

Proof   Begin by proving that the product of A and its adjoint is equal to the product of the determinant of A and I_n.

Consider the product

$$
A\,[\mathrm{adj}(A)] =
\begin{bmatrix}
a_{11} & a_{12} & \cdots & a_{1n} \\
a_{21} & a_{22} & \cdots & a_{2n} \\
\vdots & \vdots & & \vdots \\
a_{i1} & a_{i2} & \cdots & a_{in} \\
\vdots & \vdots & & \vdots \\
a_{n1} & a_{n2} & \cdots & a_{nn}
\end{bmatrix}
\begin{bmatrix}
C_{11} & C_{21} & \cdots & C_{j1} & \cdots & C_{n1} \\
C_{12} & C_{22} & \cdots & C_{j2} & \cdots & C_{n2} \\
\vdots & \vdots & & \vdots & & \vdots \\
C_{1n} & C_{2n} & \cdots & C_{jn} & \cdots & C_{nn}
\end{bmatrix}.
$$

The entry in the ith row and jth column of this product is given by

$$
a_{i1}C_{j1} + a_{i2}C_{j2} + \cdots + a_{in}C_{jn}.
$$

If $i = j$, then this sum is simply the cofactor expansion of A along its ith row, which means that the sum is the determinant of A. On the other hand, if $i \neq j$, then the sum is zero.

Therefore

$$
A[\mathrm{adj}(A)] =
\begin{bmatrix}
\det(A) & 0 & \cdots & 0 \\
0 & \det(A) & \cdots & 0 \\
\vdots & \vdots & & \vdots \\
0 & 0 & \cdots & \det(A)
\end{bmatrix}
= \det(A)I.
$$

Because A is invertible, $\det(A) \neq 0$ and you can write

$$
\frac{1}{\det(A)} A[\mathrm{adj}(A)] = I
$$

or

$$
A\left[\frac{1}{\det(A)}\,\mathrm{adj}(A)\right] = I.
$$

Multiplying both sides of the equation by A^{-1} results in the equation

$$
A^{-1}A\left[\frac{1}{\det(A)}\,\mathrm{adj}(A)\right] = A^{-1}I,
$$

which yields

$$
\frac{1}{\det(A)}\,\mathrm{adj}(A) = A^{-1}.
$$

If A is a 2×2 matrix

$$
A = \begin{bmatrix} a & b \\ c & d \end{bmatrix},
$$

then the adjoint of A is simply

$$\text{adj}(A) = \begin{bmatrix} d & -b \\ -c & a \end{bmatrix}.$$

Moreover, if A is invertible, then from Theorem 3.10 you have

$$A^{-1} = \frac{1}{|A|}\text{adj}(A) = \frac{1}{ad - bc}\begin{bmatrix} d & -b \\ -c & a \end{bmatrix},$$

which agrees with the result given in Section 2.3.

EXAMPLE 2   *Using the Adjoint of a Matrix to Find Its Inverse*

Use the adjoint of

$$A = \begin{bmatrix} -1 & 3 & 2 \\ 0 & -2 & 1 \\ 1 & 0 & -2 \end{bmatrix}$$

to find A^{-1}.

Solution   The determinant of this matrix is $|A| = 3$. Using the adjoint of A (found in Example 1), you can find the inverse of A to be

$$A^{-1} = \frac{1}{|A|}\text{adj}(A) = \frac{1}{3}\begin{bmatrix} 4 & 6 & 7 \\ 1 & 0 & 1 \\ 2 & 3 & 2 \end{bmatrix} = \begin{bmatrix} \frac{4}{3} & 2 & \frac{7}{3} \\ \frac{1}{3} & 0 & \frac{1}{3} \\ \frac{2}{3} & 1 & \frac{2}{3} \end{bmatrix}.$$

You can check to see that this matrix is the inverse of A by multiplying to obtain

$$AA^{-1} = \begin{bmatrix} -1 & 3 & 2 \\ 0 & -2 & 1 \\ 1 & 0 & -2 \end{bmatrix}\begin{bmatrix} \frac{4}{3} & 2 & \frac{7}{3} \\ \frac{1}{3} & 0 & \frac{1}{3} \\ \frac{2}{3} & 1 & \frac{2}{3} \end{bmatrix} = \begin{bmatrix} 1 & 0 & 0 \\ 0 & 1 & 0 \\ 0 & 0 & 1 \end{bmatrix}.$$

REMARK: Theorem 3.10 is not particularly efficient for calculating inverses. The Gauss-Jordan elimination method discussed in Section 2.3 is much better. This theorem is theoretically useful, however, because it provides a concise formula for the inverse of a matrix.

Cramer's Rule

Cramer's Rule, named after Gabriel Cramer (1704–1752), is a formula that uses determinants to solve a system of n linear equations in n variables. This rule can be applied only to systems of linear equations that have unique solutions.

To see how Cramer's Rule arises, look at the solution of a general system involving two linear equations in two unknowns.

$$a_{11}x_1 + a_{12}x_2 = b_1$$
$$a_{21}x_1 + a_{22}x_2 = b_2$$

Multiplying the first equation by $-a_{21}$ and the second by a_{11} and adding the results produces

$$-a_{21}a_{11}x_1 - a_{21}a_{12}x_2 = -a_{21}b_1$$
$$\underline{a_{11}a_{21}x_1 + a_{11}a_{22}x_2 = a_{11}b_2}$$
$$(a_{11}a_{22} - a_{21}a_{12})x_2 = a_{11}b_2 - a_{21}b_1.$$

Solving for x_2 (provided that $a_{11}a_{22} - a_{21}a_{12} \neq 0$) produces

$$x_2 = \frac{a_{11}b_2 - a_{21}b_1}{a_{11}a_{22} - a_{21}a_{12}}.$$

In a similar way, you can solve for x_1 to obtain

$$x_1 = \frac{a_{22}b_1 - a_{12}b_2}{a_{11}a_{22} - a_{21}a_{12}}.$$

Finally, recognizing that the numerator and denominator for both x_1 and x_2 can be represented as determinants, you have

$$x_1 = \frac{\begin{vmatrix} b_1 & a_{12} \\ b_2 & a_{22} \end{vmatrix}}{\begin{vmatrix} a_{11} & a_{12} \\ a_{21} & a_{22} \end{vmatrix}}, \qquad x_2 = \frac{\begin{vmatrix} a_{11} & b_1 \\ a_{21} & b_2 \end{vmatrix}}{\begin{vmatrix} a_{11} & a_{12} \\ a_{21} & a_{22} \end{vmatrix}}, \qquad a_{11}a_{22} - a_{21}a_{12} \neq 0.$$

The denominator for both x_1 and x_2 is simply the determinant of the coefficient matrix A. The determinant forming the numerator for x_1 can be obtained from A by replacing its first column by the column representing the constants of the system. The determinant forming the numerator for x_2 can be obtained in a similar way. These two determinants are denoted by $|A_1|$ and $|A_2|$ as follows.

$$|A_1| = \begin{vmatrix} b_1 & a_{12} \\ b_2 & a_{22} \end{vmatrix}$$

and

$$|A_2| = \begin{vmatrix} a_{11} & b_1 \\ a_{21} & b_2 \end{vmatrix}$$

Thus, you have $x_1 = \dfrac{|A_1|}{|A|}$ and $x_2 = \dfrac{|A_2|}{|A|}$. This determinant form of the solution is called **Cramer's Rule.**

EXAMPLE 3 *Using Cramer's Rule*

Use Cramer's Rule to solve the system of linear equations.

$$4x_1 - 2x_2 = 10$$
$$3x_1 - 5x_2 = 11$$

Solution First find the determinant of the coefficient matrix.

$$|A| = \begin{vmatrix} 4 & -2 \\ 3 & -5 \end{vmatrix}$$
$$= -14$$

Because $|A| \neq 0$, you know that the system has a unique solution, and applying Cramer's Rule produces

$$x_1 = \frac{|A_1|}{|A|} = \frac{\begin{vmatrix} 10 & -2 \\ 11 & -5 \end{vmatrix}}{-14} = \frac{-28}{-14} = 2$$

$$x_2 = \frac{|A_2|}{|A|} = \frac{\begin{vmatrix} 4 & 10 \\ 3 & 11 \end{vmatrix}}{-14} = \frac{14}{-14} = -1.$$

Thus, the solution is $x_1 = 2$ and $x_2 = -1$.

Cramer's Rule generalizes easily to systems of n linear equations in n variables. The value of each variable is given as the quotient of two determinants. The denominator is the determinant of the coefficient matrix, and the numerator is the determinant of the matrix formed by replacing the column corresponding to the variable being solved for with the column representing the constants. For example, the solution for x_3 in the system

$$a_{11}x_1 + a_{12}x_2 + a_{13}x_3 = b_1$$
$$a_{21}x_1 + a_{22}x_2 + a_{23}x_3 = b_2$$
$$a_{31}x_1 + a_{32}x_2 + a_{33}x_3 = b_3$$

is given by

$$x_3 = \frac{|A_3|}{|A|} = \frac{\begin{vmatrix} a_{11} & a_{12} & b_1 \\ a_{21} & a_{22} & b_2 \\ a_{31} & a_{32} & b_3 \end{vmatrix}}{\begin{vmatrix} a_{11} & a_{12} & a_{13} \\ a_{21} & a_{22} & a_{23} \\ a_{31} & a_{32} & a_{33} \end{vmatrix}}.$$

Theorem 3.11	*Cramer's Rule*

If a system of n linear equations in n variables has a coefficient matrix with a nonzero determinant $|A|$, then the solution of the system is given by

$$x_1 = \frac{\det(A_1)}{\det(A)}, \qquad x_2 = \frac{\det(A_2)}{\det(A)}, \qquad \cdots, \qquad x_n = \frac{\det(A_n)}{\det(A)},$$

where the ith column of A_i is the column of constants in the system of equations.

Proof   Let the system be represented by $AX = B$. Because $|A|$ is nonzero, you can write

$$X = A^{-1}B = \frac{1}{|A|} \operatorname{adj}(A)B = \begin{bmatrix} x_1 \\ x_2 \\ \vdots \\ x_n \end{bmatrix}.$$

If the entries of B are $b_1, b_2, \ldots, b_n$, then x_i is given by

$$x_i = \frac{1}{|A|}(b_1 C_{1i} + b_2 C_{2i} + \cdots + b_n C_{ni}).$$

But the sum (in parentheses) is precisely the cofactor expansion of A_i, which means that $x_i = |A_i|/|A|$, and the proof is complete.

EXAMPLE 4   *Using Cramer's Rule*

Use Cramer's Rule to solve the system of linear equations for x.

$$\begin{aligned} -x + 2y - 3z &= 1 \\ 2x \qquad + z &= 0 \\ 3x - 4y + 4z &= 2 \end{aligned}$$

Solution   The determinant of the coefficient matrix is

$$|A| = \begin{vmatrix} -1 & 2 & -3 \\ 2 & 0 & 1 \\ 3 & -4 & 4 \end{vmatrix} = 10.$$

Because $|A| \neq 0$, you know that the solution is unique, and Cramer's Rule can be applied to solve for x as follows.

$$x = \frac{\begin{vmatrix} 1 & 2 & -3 \\ 0 & 0 & 1 \\ 2 & -4 & 4 \end{vmatrix}}{10} = \frac{(-1)^5 \begin{vmatrix} 1 & 2 \\ 2 & -4 \end{vmatrix}}{10} = \frac{(-1)(-8)}{10} = \frac{4}{5}$$

REMARK: Try applying Cramer's Rule in Example 4 to solve for y and z. You will see that the solution is $y = -\frac{3}{2}$ and $z = -\frac{8}{5}$.

Area, Volume, and Equations of Lines and Planes

Determinants have many applications in analytic geometry. Several are presented here. The first application is finding the area of a triangle in the xy-plane.

Area of a Triangle in the xy-Plane

The area of the triangle whose vertices are (x_1, y_1), (x_2, y_2), and (x_3, y_3) is given by

$$\text{Area} = \pm\frac{1}{2}\det\begin{bmatrix} x_1 & y_1 & 1 \\ x_2 & y_2 & 1 \\ x_3 & y_3 & 1 \end{bmatrix},$$

where the sign ($\pm$) is chosen to give a positive area.

Proof

Prove the case for $y_i > 0$. Assume that $x_1 \leq x_3 \leq x_2$ and that (x_3, y_3) lies above the line segment connecting (x_1, y_1) and (x_2, y_2), as shown in Figure 3.1. Consider the three trapezoids whose vertices are as follows.

Trapezoid 1: $(x_1, 0)$, (x_1, y_1), (x_3, y_3), $(x_3, 0)$
Trapezoid 2: $(x_3, 0)$, (x_3, y_3), (x_2, y_2), $(x_2, 0)$
Trapezoid 3: $(x_1, 0)$, (x_1, y_1), (x_2, y_2), $(x_2, 0)$

The area of the given triangle is equal to the sum of the areas of the first two trapezoids less the area of the third. Therefore

$$\begin{aligned}\text{Area} &= \tfrac{1}{2}(y_1 + y_3)(x_3 - x_1) + \tfrac{1}{2}(y_3 + y_2)(x_2 - x_3) - \tfrac{1}{2}(y_1 + y_2)(x_2 - x_1) \\ &= \tfrac{1}{2}(x_1 y_2 + x_2 y_3 + x_3 y_1 - x_1 y_3 - x_2 y_1 - x_3 y_2) \\ &= \tfrac{1}{2}\begin{vmatrix} x_1 & y_1 & 1 \\ x_2 & y_2 & 1 \\ x_3 & y_3 & 1 \end{vmatrix}.\end{aligned}$$

y
(x_3, y_3)
(x_2, y_2)
(x_1, y_1)
x
$(x_1, 0)$ $(x_3, 0)$ $(x_2, 0)$

Figure 3.1

If the vertices do not occur in the order $x_1 \leq x_3 \leq x_2$ or if the vertex (x_3, y_3) is not above the line segment connecting the other two vertices, then the formula may give the negative of the area.

EXAMPLE 5 *Finding the Area of a Triangle*

Find the area of the triangle whose vertices are $(1, 0)$, $(2, 2)$, and $(4, 3)$.

Solution   It is not necessary to know the relative positions of the three vertices. Simply evaluate the determinant

$$\frac{1}{2}\begin{vmatrix} 1 & 0 & 1 \\ 2 & 2 & 1 \\ 4 & 3 & 1 \end{vmatrix} = -\frac{3}{2}$$

and conclude that the area of the triangle is $\frac{3}{2}$.

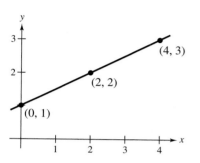

Figure 3.2

Suppose the three points in Example 5 had been on the same line. What would have happened had you applied the area formula to three such points? The answer is that the determinant would have been zero. Consider, for instance, the collinear points $(0, 1)$, $(2, 2)$, and $(4, 3)$, shown in Figure 3.2. The determinant giving the area of the "triangle" having these three points as vertices is

$$\frac{1}{2}\begin{vmatrix} 0 & 1 & 1 \\ 2 & 2 & 1 \\ 4 & 3 & 1 \end{vmatrix} = 0.$$

Thus, if three points in the xy-plane lie on the same line, then the determinant in the formula for the area of a triangle turns out to be zero. This result is generalized in the following test.

Test for Collinear Points in the xy-Plane

Three points (x_1, y_1), (x_2, y_2), and (x_3, y_3) are collinear if and only if

$$\det\begin{bmatrix} x_1 & y_1 & 1 \\ x_2 & y_2 & 1 \\ x_3 & y_3 & 1 \end{bmatrix} = 0.$$

The following determinant form for an equation of the line passing through two points in the xy-plane is derived from the test for collinear points.

Two-Point Form of the Equation of a Line

An equation of the line passing through the distinct points (x_1, y_1) and (x_2, y_2) is given by

$$\det\begin{bmatrix} x & y & 1 \\ x_1 & y_1 & 1 \\ x_2 & y_2 & 1 \end{bmatrix} = 0.$$

EXAMPLE 6   *Finding an Equation of the Line Passing Through Two Points*

Find an equation of the line passing through the points $(2, 4)$ and $(-1, 3)$.

Solution Applying the determinant formula for the equation of a line passing through these two points produces

$$\begin{vmatrix} x & y & 1 \\ 2 & 4 & 1 \\ -1 & 3 & 1 \end{vmatrix} = 0.$$

To evaluate this determinant, expand by cofactors along the top row to obtain

$$x \begin{vmatrix} 4 & 1 \\ 3 & 1 \end{vmatrix} - y \begin{vmatrix} 2 & 1 \\ -1 & 1 \end{vmatrix} + 1 \begin{vmatrix} 2 & 4 \\ -1 & 3 \end{vmatrix} = 0$$

$$x - 3y + 10 = 0.$$

Therefore, an equation of the line is $x - 3y = -10$.

The formula for the area of a triangle in the plane has a straightforward generalization to three-dimensional space, which is presented without proof as follows.

Volume of a Tetrahedron The volume of the tetrahedron whose vertices are (x_1, y_1, z_1), (x_2, y_2, z_2), (x_3, y_3, z_3), and (x_4, y_4, z_4) is given by

$$\text{Volume} = \pm \tfrac{1}{6} \det \begin{bmatrix} x_1 & y_1 & z_1 & 1 \\ x_2 & y_2 & z_2 & 1 \\ x_3 & y_3 & z_3 & 1 \\ x_4 & y_4 & z_4 & 1 \end{bmatrix},$$

where the sign ($\pm$) is chosen to give a positive volume.

EXAMPLE 7 *Finding the Volume of a Tetrahedron*

Find the volume of the tetrahedron whose vertices are $(0, 4, 1)$, $(4, 0, 0)$, $(3, 5, 2)$, and $(2, 2, 5)$, as shown in Figure 3.3.

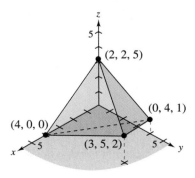

Figure 3.3

Solution   Using the determinant formula for volume produces

$$\frac{1}{6}\begin{vmatrix} 0 & 4 & 1 & 1 \\ 4 & 0 & 0 & 1 \\ 3 & 5 & 2 & 1 \\ 2 & 2 & 5 & 1 \end{vmatrix} = \frac{1}{6}(-72) = -12.$$

Therefore, the volume of the tetrahedron is 12.

If four points in three-dimensional space happen to lie in the same plane, then the determinant in the formula for volume turns out to be zero. Thus, you have the following test.

Test for Coplanar Points in Space

Four points (x_1, y_1, z_1), (x_2, y_2, z_2), (x_3, y_3, z_3), and (x_4, y_4, z_4) are coplanar if and only if

$$\det\begin{bmatrix} x_1 & y_1 & z_1 & 1 \\ x_2 & y_2 & z_2 & 1 \\ x_3 & y_3 & z_3 & 1 \\ x_4 & y_4 & z_4 & 1 \end{bmatrix} = 0.$$

This test now provides the following determinant form for an equation of a plane passing through three points in space.

Three-Point Form of the Equation of a Plane

An equation of the plane passing through the distinct points (x_1, y_1, z_1), (x_2, y_2, z_2), and (x_3, y_3, z_3) is given by

$$\det\begin{bmatrix} x & y & z & 1 \\ x_1 & y_1 & z_1 & 1 \\ x_2 & y_2 & z_2 & 1 \\ x_3 & y_3 & z_3 & 1 \end{bmatrix} = 0.$$

EXAMPLE 8   *Finding an Equation of the Plane Passing Through Three Points*

Find an equation of the plane passing through the points $(0, 1, 0)$, $(-1, 3, 2)$, and $(-2, 0, 1)$.

Solution   Using the determinant form of the equation of a plane passing through three points produces

$$\begin{vmatrix} x & y & z & 1 \\ 0 & 1 & 0 & 1 \\ -1 & 3 & 2 & 1 \\ -2 & 0 & 1 & 1 \end{vmatrix} = 0.$$

To evaluate this determinant, subtract the fourth column from the second column to obtain

$$\begin{vmatrix} x & y-1 & z & 1 \\ 0 & 0 & 0 & 1 \\ -1 & 2 & 2 & 1 \\ -2 & -1 & 1 & 1 \end{vmatrix} = 0.$$

Now, expanding by cofactors along the second row yields

$$4x - 3y + 5z = -3.$$

SECTION 3.5 Exercises

The Adjoint of a Matrix

In Exercises 1–8, find the adjoint of the matrix A. Then use the adjoint to find the inverse of A, if possible.

1. $A = \begin{bmatrix} 1 & 2 \\ 3 & 4 \end{bmatrix}$

2. $A = \begin{bmatrix} -1 & 0 \\ 0 & 4 \end{bmatrix}$

3. $A = \begin{bmatrix} 1 & 0 & 0 \\ 0 & 2 & 6 \\ 0 & -4 & -12 \end{bmatrix}$

4. $A = \begin{bmatrix} 1 & 2 & 3 \\ 0 & 1 & -1 \\ 2 & 2 & 2 \end{bmatrix}$

5. $A = \begin{bmatrix} -3 & -5 & -7 \\ 2 & 4 & 3 \\ 0 & 1 & -1 \end{bmatrix}$

6. $A = \begin{bmatrix} 0 & 1 & 1 \\ 1 & 2 & 3 \\ -1 & -1 & -2 \end{bmatrix}$

7. $A = \begin{bmatrix} -1 & 2 & 0 & 1 \\ 3 & -1 & 4 & 1 \\ 0 & 0 & 1 & 2 \\ -1 & 1 & 1 & 2 \end{bmatrix}$

8. $A = \begin{bmatrix} 1 & 1 & 1 & 0 \\ 1 & 1 & 0 & 1 \\ 1 & 0 & 1 & 1 \\ 0 & 1 & 1 & 1 \end{bmatrix}$

9. Prove that if $|A| = 1$ and all entries of A are integers, then all entries of $|A^{-1}|$ must also be integers.

10. Prove that if an $n \times n$ matrix A is not invertible, then $A[\text{adj}(A)]$ is the zero matrix.

In Exercises 11 and 12, prove the formula for a nonsingular $n \times n$ matrix A. Assume $n \geq 3$.

11. $|\text{adj}(A)| = |A|^{n-1}$

12. $\text{adj}[\text{adj}(A)] = |A|^{n-2}A$

13. Illustrate the formula given in Exercise 11 for the matrix
$$A = \begin{bmatrix} 1 & 0 \\ 1 & -2 \end{bmatrix}.$$

14. Illustrate the formula given in Exercise 12 for the matrix
$$A = \begin{bmatrix} -1 & 3 \\ 1 & 2 \end{bmatrix}.$$

15. Prove that if A is an $n \times n$ invertible matrix, then $\text{adj}(A^{-1}) = [\text{adj}(A)]^{-1}$.

16. Illustrate the formula given in Exercise 15 for the matrix
$$A = \begin{bmatrix} 1 & 3 \\ 1 & 2 \end{bmatrix}.$$

Cramer's Rule

In Exercises 17–32, use Cramer's Rule to solve the system of linear equations, if possible.

17. $\begin{aligned} x_1 + 2x_2 &= 5 \\ -x_1 + x_2 &= 1 \end{aligned}$

18. $\begin{aligned} 2x_1 - x_2 &= -10 \\ 3x_1 + 2x_2 &= -1 \end{aligned}$

19. $\begin{aligned} 3x_1 + 4x_2 &= -2 \\ 5x_1 + 3x_2 &= 4 \end{aligned}$

20. $\begin{aligned} 18x_1 + 12x_2 &= 13 \\ 30x_1 + 24x_2 &= 23 \end{aligned}$

21. $\begin{aligned} 20x_1 + 8x_2 &= 11 \\ 12x_1 - 24x_2 &= 21 \end{aligned}$

22. $\begin{aligned} 13x_1 - 6x_2 &= 17 \\ 26x_1 - 12x_2 &= 8 \end{aligned}$

23. $\begin{aligned} -0.4x_1 + 0.8x_2 &= 1.6 \\ 2x_1 - 4x_2 &= 5.0 \end{aligned}$

24. $\begin{aligned} -0.4x_1 + 0.8x_2 &= 1.6 \\ 0.2x_1 + 0.3x_2 &= 0.6 \end{aligned}$

25. $\begin{aligned} 3x_1 + 6x_2 &= 5 \\ 6x_1 + 12x_2 &= 10 \end{aligned}$

26. $\begin{aligned} 3x_1 + 2x_2 &= 1 \\ 2x_1 + 10x_2 &= 6 \end{aligned}$

27. $\begin{aligned} 4x_1 - x_2 - x_3 &= 1 \\ 2x_1 + 2x_2 + 3x_3 &= 10 \\ 5x_1 - 2x_2 - 2x_3 &= -1 \end{aligned}$

28. $\begin{aligned} 4x_1 - 2x_2 + 3x_3 &= -2 \\ 2x_1 + 2x_2 + 5x_3 &= 16 \\ 8x_1 - 5x_2 - 2x_3 &= 4 \end{aligned}$

29. $\begin{aligned} 3x_1 + 4x_2 + 4x_3 &= 11 \\ 4x_1 - 4x_2 + 6x_3 &= 11 \\ 6x_1 - 6x_2 &= 3 \end{aligned}$

30. $\begin{aligned} 14x_1 - 21x_2 - 7x_3 &= -21 \\ -4x_1 + 2x_2 - 2x_3 &= 2 \\ 56x_1 - 21x_2 + 7x_3 &= 7 \end{aligned}$

31. $\begin{aligned} 3x_1 + 3x_2 + 5x_3 &= 1 \\ 3x_1 + 5x_2 + 9x_3 &= 2 \\ 5x_1 + 9x_2 + 17x_3 &= 4 \end{aligned}$

32. $\begin{aligned} 2x_1 + 3x_2 + 5x_3 &= 4 \\ 3x_1 + 5x_2 + 9x_3 &= 7 \\ 5x_1 + 9x_2 + 17x_3 &= 13 \end{aligned}$

In Exercises 33–36, use a graphing utility with matrix capabilities and Cramer's Rule to solve for x_1.

33.
$$\begin{aligned} 7x_1 - 3x_2 \quad\quad + 2x_4 &= 41 \\ -2x_1 + x_2 \quad\quad - x_4 &= -13 \\ 4x_1 \quad\quad + x_3 - 2x_4 &= 12 \\ -x_1 + x_2 \quad\quad - x_4 &= -8 \end{aligned}$$

34.
$$\begin{aligned} 2x_1 + 5x_2 \quad\quad + x_4 &= 11 \\ x_1 + 4x_2 + 2x_3 - 2x_4 &= -7 \\ 2x_1 - 2x_2 + 5x_3 + x_4 &= 3 \\ x_1 \quad\quad\quad - 3x_4 &= 1 \end{aligned}$$

35.
$$\begin{aligned} 5x_1 - 3x_2 + x_3 &= 2 \\ 2x_1 + 2x_2 - 3x_3 &= 4 \\ x_1 - 7x_2 + 8x_3 &= -6 \end{aligned}$$

36.
$$\begin{aligned} 3x_1 + 2x_2 + 5x_3 &= 4 \\ 4x_1 - 3x_2 - 4x_3 &= 1 \\ -8x_1 + 2x_2 + 3x_3 &= 0 \end{aligned}$$

37. Use Cramer's Rule to solve the system of linear equations for x and y.

$$\begin{aligned} kx + (1 - k)y &= 1 \\ (1 - k)x + ky &= 3 \end{aligned}$$

For what value(s) of k will the system be inconsistent?

38. Verify the following system of linear equations in $\cos A$, $\cos B$, and $\cos C$ for the triangle shown in Figure 3.4.

$$\begin{aligned} c \cos B + b \cos C &= a \\ c \cos A \quad\quad + a \cos C &= b \\ b \cos A + a \cos B \quad\quad &= c \end{aligned}$$

Then use Cramer's Rule to solve for $\cos C$, and use the result to verify the Law of Cosines, $c^2 = a^2 + b^2 - 2ab \cos C$.

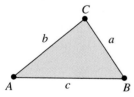

Figure 3.4

Area, Volume, and Equations of Lines and Planes

In Exercises 39–42, find the area of the triangle having the given vertices.

39. $(0, 0), (2, 0), (0, 3)$

40. $(1, 1), (2, 4), (4, 2)$

41. $(-1, 2), (2, 2), (-2, 4)$

42. $(1, 1), (-1, 1), (0, -2)$

In Exercises 43–46, determine whether the points are collinear.

43. $(1, 2), (3, 4), (5, 6)$

44. $(-1, 0), (1, 1), (3, 3)$

45. $(-2, 5), (0, -1), (3, -9)$

46. $(-1, -3), (-4, 7), (2, -13)$

In Exercises 47–50, find an equation of the line passing through the given points.

47. $(0, 0), (3, 4)$

48. $(-4, 7), (2, 4)$

49. $(-2, 3), (-2, -4)$

50. $(1, 4), (3, 4)$

In Exercises 51–54, find the volume of the tetrahedron having the given vertices.

51. $(1, 0, 0), (0, 1, 0), (0, 0, 1), (1, 1, 1)$

52. $(1, 1, 1), (0, 0, 0), (2, 1, -1), (-1, 1, 2)$

53. $(3, -1, 1), (4, -4, 4), (1, 1, 1), (0, 0, 1)$

54. $(0, 0, 0), (0, 2, 0), (3, 0, 0), (1, 1, 4)$

In Exercises 55–58, determine whether the points are coplanar.

55. $(-4, 1, 0), (0, 1, 2), (4, 3, -1), (0, 0, 1)$

56. $(1, 2, 3), (-1, 0, 1), (0, -2, -5), (2, 6, 11)$

57. $(0, 0, -1), (0, -1, 0), (1, 1, 0), (2, 1, 2)$

58. $(1, 2, 7), (-3, 6, 6), (4, 4, 2), (3, 3, 4)$

In Exercises 59–62, find an equation of the plane passing through the given points.

59. $(1, -2, 1), (-1, -1, 7), (2, -1, 3)$

60. $(0, -1, 0), (1, 1, 0), (2, 1, 2)$

61. $(0, 0, 0), (1, -1, 0), (0, 1, -1)$

62. $(1, 2, 7), (4, 4, 2), (3, 3, 4)$

In Exercises 63–65, Cramer's Rule has been used to solve for one of the variables in a system of equations. Determine whether Cramer's Rule was used correctly to solve for the variable. If not, identify the mistake.

63. *System of Equations*

$$\begin{aligned} x + 2y + z &= 2 \\ -x + 3y - 2z &= 4 \\ 4x + y - z &= 6 \end{aligned}$$

Solve for y

$$y = \frac{\begin{vmatrix} 1 & 2 & 1 \\ -1 & 3 & -2 \\ 4 & 1 & -1 \end{vmatrix}}{\begin{vmatrix} 1 & 2 & 1 \\ -1 & 4 & -2 \\ 4 & 6 & -1 \end{vmatrix}}$$

64. *System of Equations*   *Solve for z*

$$
\begin{aligned}
x - 4y - z &= -1 \\
2x - 3y + z &= 6 \\
x + y - 4z &= 1
\end{aligned}
$$

$$
z = \dfrac{\begin{vmatrix} -1 & -4 & -1 \\ 6 & -3 & 1 \\ 1 & 1 & -4 \end{vmatrix}}{\begin{vmatrix} 1 & -4 & -1 \\ 2 & -3 & 1 \\ 1 & 1 & -4 \end{vmatrix}}
$$

65. *System of Equations*   *Solve for x*

$$
\begin{aligned}
5x - 2y + z &= 15 \\
3x - 3y - z &= -7 \\
2x - y - 7z &= -3
\end{aligned}
$$

$$
x = \dfrac{\begin{vmatrix} 15 & -2 & 1 \\ -7 & -3 & -1 \\ -3 & -1 & -7 \end{vmatrix}}{\begin{vmatrix} 5 & -2 & 1 \\ 3 & -3 & -1 \\ 2 & -1 & -7 \end{vmatrix}}
$$

CHAPTER 3 Summary

After studying this chapter, you should have acquired the following skills.

- Find the determinant of a given matrix.
- Find the minors and cofactors of a given matrix.
- Use expansion by cofactors to find the determinant of a given matrix.
- Find the determinant of a given triangular matrix.
- Use elementary row or column operations to evaluate the determinant of a given matrix.
- Recognize conditions that yield zero determinants.
- Use the properties of determinants.
- Find the determinant of a given elementary matrix.
- Use the determinant to decide whether a given matrix is singular or nonsingular.
- Verify an eigenvalue and an eigenvector of a given matrix.
- Find and use the adjoint of a matrix to find its inverse.
- Use Cramer's Rule to solve a given system of linear equations.
- Use determinants to find the area of a triangle defined by three distinct points.
- Use determinants to find an equation of a line passing through two distinct points.
- Use determinants to find the volume of a tetrahedron defined by four distinct points.
- Use determinants to find an equation of a plane passing through three distinct points.

CHAPTER 3 Review Exercises

In Exercises 1–18, find the determinant of the matrix.

1. $\begin{bmatrix} 4 & -1 \\ 2 & 2 \end{bmatrix}$

2. $\begin{bmatrix} 0 & -3 \\ 1 & 2 \end{bmatrix}$

3. $\begin{bmatrix} -3 & 1 \\ 6 & -2 \end{bmatrix}$

4. $\begin{bmatrix} -2 & 0 \\ 0 & 3 \end{bmatrix}$

5. $\begin{bmatrix} 1 & 4 & -2 \\ 0 & -3 & 1 \\ 1 & 1 & -1 \end{bmatrix}$

6. $\begin{bmatrix} 5 & 0 & 2 \\ 0 & -1 & 3 \\ 0 & 0 & 1 \end{bmatrix}$

7. $\begin{bmatrix} -2 & 0 & 0 \\ 0 & -3 & 0 \\ 0 & 0 & -1 \end{bmatrix}$

8. $\begin{bmatrix} -15 & 0 & 4 \\ 3 & 0 & -5 \\ 12 & 0 & 6 \end{bmatrix}$

9. $\begin{bmatrix} -3 & 6 & 9 \\ 9 & 12 & -3 \\ 0 & 15 & -6 \end{bmatrix}$

10. $\begin{bmatrix} -15 & 0 & 3 \\ 3 & 9 & -6 \\ 12 & -3 & 6 \end{bmatrix}$

11. $\begin{bmatrix} 2 & 0 & -1 & 4 \\ -1 & 2 & 0 & 3 \\ 3 & 0 & 1 & 2 \\ -2 & 0 & 3 & 1 \end{bmatrix}$

12. $\begin{bmatrix} 2 & 0 & 0 & 0 \\ -3 & 1 & 0 & 0 \\ 4 & -1 & 3 & 0 \\ 5 & 2 & 1 & -1 \end{bmatrix}$

13. $\begin{bmatrix} -4 & 1 & 2 & 3 \\ 1 & -2 & 1 & 2 \\ 2 & -1 & 3 & 4 \\ 1 & 2 & 2 & -1 \end{bmatrix}$

14. $\begin{bmatrix} 3 & -1 & 2 & 1 \\ -2 & 0 & 1 & -3 \\ -1 & 2 & -3 & 4 \\ -2 & 1 & -2 & 1 \end{bmatrix}$

15. $\begin{bmatrix} -1 & 1 & -1 & 0 & 0 \\ 0 & 1 & -1 & 0 & 1 \\ 1 & 0 & 1 & -1 & 0 \\ 0 & -1 & 0 & 1 & -1 \\ 0 & 1 & 1 & -1 & 1 \end{bmatrix}$

16. $\begin{bmatrix} 1 & 2 & -1 & 3 & 4 \\ 2 & 3 & -1 & 2 & -2 \\ 1 & 2 & 0 & 1 & -1 \\ 1 & 0 & 2 & -1 & 0 \\ 0 & -1 & 1 & 0 & 2 \end{bmatrix}$

17. $\begin{bmatrix} -1 & 0 & 0 & 0 & 0 \\ 0 & -1 & 0 & 0 & 0 \\ 0 & 0 & -1 & 0 & 0 \\ 0 & 0 & 0 & -1 & 0 \\ 0 & 0 & 0 & 0 & -1 \end{bmatrix}$

18. $\begin{bmatrix} 0 & 0 & 0 & 0 & 2 \\ 0 & 0 & 0 & 2 & 0 \\ 0 & 0 & 2 & 0 & 0 \\ 0 & 2 & 0 & 0 & 0 \\ 2 & 0 & 0 & 0 & 0 \end{bmatrix}$

In Exercises 19 and 20, find (a) $|A|$, (b) $|B|$, (c) AB, and (d) $|AB|$. Then verify that $|A||B| = |AB|$.

19. $A = \begin{bmatrix} -1 & 2 \\ 0 & 1 \end{bmatrix}$, $B = \begin{bmatrix} 3 & 4 \\ 2 & 1 \end{bmatrix}$

20. $A = \begin{bmatrix} 1 & 2 & 3 \\ 4 & 5 & 6 \\ 7 & 8 & 0 \end{bmatrix}$, $B = \begin{bmatrix} 1 & 2 & 1 \\ 0 & -1 & 1 \\ 0 & 2 & 3 \end{bmatrix}$

In Exercises 21 and 22, find (a) $|A^T|$, (b) $|A^3|$, (c) $|A^TA|$, and (d) $|5A|$.

21. $A = \begin{bmatrix} -2 & 6 \\ 1 & 3 \end{bmatrix}$

22. $A = \begin{bmatrix} 3 & 0 & 1 \\ -1 & 0 & 0 \\ 2 & 1 & 2 \end{bmatrix}$

In Exercises 23 and 24, find (a) $|A|$ and (b) $|A^{-1}|$.

23. $A = \begin{bmatrix} 1 & 0 & -4 \\ 0 & 3 & 2 \\ -2 & 7 & 6 \end{bmatrix}$

24. $A = \begin{bmatrix} 2 & -1 & 4 \\ 5 & 0 & 3 \\ 1 & -2 & 0 \end{bmatrix}$

In Exercises 25–28, solve the system of linear equations by each of the following methods.

(a) Gaussian elimination with back-substitution
(b) Gauss-Jordan elimination
(c) Cramer's Rule

25. $\begin{aligned} x_1 + 2x_2 + x_3 &= 1 \\ 3x_1 + x_2 - x_3 &= 0 \\ 2x_1 + 2x_2 + 3x_3 &= 6 \end{aligned}$

26. $\begin{aligned} x_1 + 3x_2 + 2x_3 &= 5 \\ 2x_1 - x_2 - x_3 &= 0 \\ x_1 + 4x_2 + x_3 &= 3 \end{aligned}$

27. $\begin{aligned} x_1 + 2x_2 - x_3 &= 0 \\ 4x_1 + x_2 + 2x_3 &= 1 \\ x_1 - x_2 + 3x_3 &= 4 \end{aligned}$

28. $\begin{aligned} x_1 - x_2 + 2x_3 &= 3 \\ 3x_1 - 2x_2 + x_3 &= 1 \\ 2x_1 + x_2 + x_3 &= 0 \end{aligned}$

In Exercises 29–34, use the determinant of the coefficient matrix to determine whether the system of linear equations has a unique solution.

29. $\begin{aligned} 5x + 4y &= 2 \\ -x + y &= -22 \end{aligned}$

30. $\begin{aligned} 2x - 5y &= 2 \\ 3x - 7y &= 1 \end{aligned}$

31. $\begin{aligned} -x + y + 2z &= 1 \\ 2x + 3y + z &= -2 \\ 5x + 4y + 2z &= 4 \end{aligned}$

32. $\begin{aligned} 2x + 3y + z &= 10 \\ 2x - 3y - 3z &= 22 \\ 8x + 6y &= -2 \end{aligned}$

33. $\begin{aligned} x_1 + 2x_2 + 6x_3 &= 1 \\ 2x_1 + 5x_2 + 15x_3 &= 4 \\ 3x_1 + x_2 + 3x_3 &= -6 \end{aligned}$

34. $\begin{aligned} x_1 + 5x_2 + 3x_3 &= 14 \\ 4x_1 + 2x_2 + 5x_3 &= 3 \\ 3x_3 + 8x_4 + 6x_5 &= 16 \\ 2x_1 + 4x_2 - 2x_5 &= 0 \\ 2x_1 - x_3 &= 0 \end{aligned}$

True or False? In Exercises 35 and 36, determine whether each statement is true or false. If a statement is true, give a reason or cite an appropriate statement in the text. If a statement is false, provide an example that shows that the statement is not true in all cases or cite an appropriate statement in the text.

35. (a) The cofactor C_{22} of a given matrix is always a positive number.

(b) If a square matrix B is obtained from A by interchanging two rows, then $\det(B) = \det(A)$.

(c) If A is a square matrix of order n, then $\det(A) = -\det(A^T)$.

36. (a) If A and B are square matrices of order n such that $\det(AB) = -1$, then both A and B are nonsingular.

(b) If A is a 3×3 matrix with $\det(A) = 5$, then $\det(2A) = 10$.

(c) If A and B are square matrices of order n, then $\det(A + B) = \det(A) + \det(B)$.

37. If A is a 3×3 matrix such that $|A| = 2$, then what is the value of $|4A|$?

38. If A is a 4×4 matrix such that $|A| = -1$, then what is the value of $|2A|$?

39. Prove the following property.

$$\begin{vmatrix} a_{11} & a_{12} & a_{13} \\ a_{21} & a_{22} & a_{23} \\ a_{31} + c_{31} & a_{32} + c_{32} & a_{33} + c_{33} \end{vmatrix} = \begin{vmatrix} a_{11} & a_{12} & a_{13} \\ a_{21} & a_{22} & a_{23} \\ a_{31} & a_{32} & a_{33} \end{vmatrix} + \begin{vmatrix} a_{11} & a_{12} & a_{13} \\ a_{21} & a_{22} & a_{23} \\ c_{31} & c_{32} & c_{33} \end{vmatrix}$$

40. Illustrate the property given in Exercise 39 for the following.

$$A = \begin{bmatrix} 1 & 0 & 2 \\ 1 & -1 & 2 \\ 2 & 1 & -1 \end{bmatrix}, \quad c_{31} = 3, \quad c_{32} = 0, \quad c_{33} = 1$$

41. Find the determinant of the following $n \times n$ matrix.

$$\begin{bmatrix} 1 - n & 1 & 1 & \cdots & 1 \\ 1 & 1 - n & 1 & \cdots & 1 \\ \vdots & \vdots & \vdots & & \vdots \\ 1 & 1 & 1 & \cdots & 1 - n \end{bmatrix}$$

42. Show that

$$\begin{vmatrix} a & 1 & 1 & 1 \\ 1 & a & 1 & 1 \\ 1 & 1 & a & 1 \\ 1 & 1 & 1 & a \end{vmatrix} = (a + 3)(a - 1)^3.$$

In Exercises 43–46, find the eigenvalues and corresponding eigenvectors of the matrix.

43. $\begin{bmatrix} -3 & 10 \\ 5 & 2 \end{bmatrix}$

44. $\begin{bmatrix} 5 & 2 \\ 4 & 1 \end{bmatrix}$

45. $\begin{bmatrix} 1 & 0 & 0 \\ -2 & 3 & 0 \\ 0 & 0 & 4 \end{bmatrix}$

46. $\begin{bmatrix} -3 & 0 & 4 \\ 2 & 1 & 1 \\ -1 & 0 & 1 \end{bmatrix}$

Calculus In Exercises 47–50, find the Jacobians of the functions. If x, y, and z are continuous functions of u, v, and w with continuous first partial derivatives, the **Jacobians** $J(u, v)$ and $J(u, v, w)$ are given by

$$J(u, v) = \begin{vmatrix} \dfrac{\partial x}{\partial u} & \dfrac{\partial x}{\partial v} \\ \dfrac{\partial y}{\partial u} & \dfrac{\partial y}{\partial v} \end{vmatrix} \quad \text{and} \quad J(u, v, w) = \begin{vmatrix} \dfrac{\partial x}{\partial u} & \dfrac{\partial x}{\partial v} & \dfrac{\partial x}{\partial w} \\ \dfrac{\partial y}{\partial u} & \dfrac{\partial y}{\partial v} & \dfrac{\partial y}{\partial w} \\ \dfrac{\partial z}{\partial u} & \dfrac{\partial z}{\partial v} & \dfrac{\partial z}{\partial w} \end{vmatrix}.$$

47. $x = \frac{1}{2}(v - u), \quad y = \frac{1}{2}(v + u)$

48. $x = au + bv, \quad y = cu + dv$

49. $x = \frac{1}{2}(u + v), \quad y = \frac{1}{2}(u - v), \quad z = 2uvw$

50. $x = u - v + w, \quad y = 2uv, \quad z = u + v + w$

51. *Writing* Compare the various methods for calculating the determinant of a matrix. Which method requires the least amount of computation? Which method do you prefer if the matrix has very few zeros?

52. Prove that if $|A| = |B| \neq 0$ and A and B are of the same size, then there exists a matrix C such that $|C| = 1$ and $A = CB$.

The Adjoint of a Matrix

In Exercises 53 and 54, find the adjoint of the matrix.

53. $\begin{bmatrix} 1 & 0 & 0 \\ 2 & 1 & 0 \\ 2 & -1 & 1 \end{bmatrix}$

54. $\begin{bmatrix} 1 & 2 \\ -2 & 0 \end{bmatrix}$

Cramer's Rule

In Exercises 55–58, use the determinant of the coefficient matrix to determine whether the system of linear equations has a unique solution. If it does, use Cramer's Rule to find the solution.

55. $0.2x - 0.1y = 0.07$
$0.4x - 0.5y = -0.01$

56. $2x + y = 0.3$
$3x - y = -1.3$

57. $2x_1 + 3x_2 + 3x_3 = 3$
$6x_1 + 6x_2 + 12x_3 = 13$
$12x_1 + 9x_2 - x_3 = 2$

58. $4x_1 + 4x_2 + 4x_3 = 5$
$4x_1 - 2x_2 - 8x_3 = 1$
$8x_1 + 2x_2 - 4x_3 = 6$

Area, Volume, and Equations of Lines and Planes

In Exercises 59 and 60, use a determinant to find the area of the triangle with the given vertices.

59. $(1, 0), (5, 0), (5, 8)$

60. $(-4, 0), (4, 0), (0, 6)$

In Exercises 61 and 62, use a determinant to find an equation of the line passing through the given points.

61. $(-4, 0), (4, 4)$

62. $(2, 5), (6, -1)$

In Exercises 63 and 64, find an equation of the plane passing through the given points.

63. $(0, 0, 0), (1, 0, 3), (0, 3, 4)$

64. $(0, 0, 0), (2, -1, 1), (-3, 2, 5)$

True or False? In Exercises 65 and 66, determine whether each statement is true or false. If a statement is true, give a reason or cite an appropriate statement in the text. If a statement is false, provide an example that shows that the statement is not true in all cases or cite an appropriate statement in the text.

65. (a) In Cramer's Rule, the value of x_i is the quotient of two determinants, where the numerator is the determinant of the coefficient matrix.

(b) Three points (x_1, y_1), (x_2, y_2), and (x_3, y_3) are collinear if the determinant of the matrix that has the coordinates as entries in the first two columns and 1's as entries in the third column is nonzero.

66. (a) If A is a square matrix, then the matrix of cofactors of A is called the adjoint of A.

(b) In Cramer's Rule, the denominator is the determinant of the matrix formed by replacing the column corresponding to the variable being solved for with the column representing the constants.

CHAPTER 3 Projects

1 EIGENVALUES AND STOCHASTIC MATRICES

In Section 2.5, you studied a consumer preference model for competing cable television companies. The matrix representing the transition probabilities was

$$P = \begin{bmatrix} 0.70 & 0.15 & 0.15 \\ 0.20 & 0.80 & 0.15 \\ 0.10 & 0.05 & 0.70 \end{bmatrix}.$$

Given the initial state matrix X, you observed that the number of subscribers after 1 year is given by the product PX.

$$X = \begin{bmatrix} 15,000 \\ 20,000 \\ 65,000 \end{bmatrix} \longrightarrow PX = \begin{bmatrix} 0.70 & 0.15 & 0.15 \\ 0.20 & 0.80 & 0.15 \\ 0.10 & 0.05 & 0.70 \end{bmatrix} \begin{bmatrix} 15,000 \\ 20,000 \\ 65,000 \end{bmatrix} = \begin{bmatrix} 23,250 \\ 28,750 \\ 48,000 \end{bmatrix}$$

After ten years, the number of subscribers had nearly reached a **steady state.**

$$X = \begin{bmatrix} 15,000 \\ 20,000 \\ 65,000 \end{bmatrix} \longrightarrow P^{10}X = \begin{bmatrix} 33,287 \\ 47,147 \\ 19,566 \end{bmatrix}$$

That is, for large values of n, the product $P^n X$ approaches a limit $\overline{X}$, $P\overline{X} = \overline{X}$.

From your knowledge of eigenvalues, this means that 1 is an eigenvalue of P with corresponding eigenvector $\overline{X}$.

1. Use a computer or calculator to show that the eigenvalues and eigenvectors of P are as follows.

Eigenvalue	*Eigenvector*
$\lambda_1 = 1$	$\mathbf{x}_1 = \begin{bmatrix} 7 \\ 10 \\ 4 \end{bmatrix}$
$\lambda_2 = 0.65$	$\mathbf{x}_2 = \begin{bmatrix} 0 \\ -1 \\ 1 \end{bmatrix}$
$\lambda_3 = 0.55$	$\mathbf{x}_3 = \begin{bmatrix} -2 \\ 1 \\ 1 \end{bmatrix}$

2. Let S be the matrix whose columns are the eigenvectors of P. Show that $S^{-1}PS$ is a diagonal matrix D. What are the entries along the diagonal of D?

3. Show that $P^n = (SDS^{-1})^n = SD^nS^{-1}$. Use this result to calculate $P^{10}X$ and verify the result from Section 2.5.

2   THE CAYLEY-HAMILTON THEOREM

The **characteristic polynomial** of a square matrix A is given by the determinant $|\lambda I - A|$. If the order of A is n, then the characteristic polynomial $p(\lambda)$ is an nth-degree polynomial in the variable λ.

$$p(\lambda) = \det(\lambda I - A) = \lambda^n + c_{n-1}\lambda^{n-1} + \cdots + c_2\lambda^2 + c_1\lambda + c_0$$

The Cayley-Hamilton Theorem asserts that every square matrix satisfies its characteristic polynomial. That is, for the $n \times n$ matrix A,

$$p(A) = A^n + c_{n-1}A^{n-1} + \cdots + c_2A^2 + c_1A + c_0I = O.$$

Note that this is a matrix equation. The zero on the right is the $n \times n$ zero matrix, and the coefficient c_0 has been multiplied by the $n \times n$ identity matrix I.

1. Verify the Cayley-Hamilton Theorem for the matrix

$$\begin{bmatrix} 2 & -2 \\ -2 & -1 \end{bmatrix}.$$

2. Verify the Cayley-Hamilton Theorem for the matrix

$$\begin{bmatrix} 6 & 0 & 4 \\ -2 & 1 & 3 \\ 2 & 0 & 4 \end{bmatrix}.$$

3. Prove the Cayley-Hamilton Theorem for an arbitrary 2×2 matrix A,

$$A = \begin{bmatrix} a & b \\ c & d \end{bmatrix}.$$

4. If A is nonsingular and $p(A) = A^n + c_{n-1}A^{n-1} + \cdots + c_1A + c_0I = O$, show that

$$A^{-1} = \frac{1}{c_0}(-A^{n-1} - c_{n-1}A^{n-2} - \cdots - c_2A - c_1I).$$

Use this result to find the inverse of the matrix

$$A = \begin{bmatrix} 1 & 2 \\ 3 & 5 \end{bmatrix}.$$

5. The Cayley-Hamilton Theorem can be used to calculate powers A^n of the square matrix A. For example, the characteristic polynomial of the matrix

$$A = \begin{bmatrix} 3 & -1 \\ 2 & -1 \end{bmatrix}$$

is $p(\lambda) = \lambda^2 - 2\lambda - 1$.

The Cayley-Hamilton Theorem implies that

$$A^2 - 2A - I = O \qquad \text{or} \qquad A^2 = 2A + I.$$

Thus, A^2 is given in terms of lower powers of A.

$$A^2 = 2A + I = 2\begin{bmatrix} 3 & -1 \\ 2 & -1 \end{bmatrix} + \begin{bmatrix} 1 & 0 \\ 0 & 1 \end{bmatrix} = \begin{bmatrix} 7 & -2 \\ 4 & -1 \end{bmatrix}$$

Similarly, multiplying both sides of the equation $A^2 = 2A + I$ by A gives A^3 in terms of lower powers of A. Moreover, you can write A^3 in terms of just A and I after replacing A^2 by $2A + I$, as follows.

$$A^3 = 2A^2 + A = 2(2A + I) + A = 5A + 2I$$

(a) Use this method to find A^3 and A^4. (First write A^4 as a linear combination of A and I—that is, as a sum of scalar multiples of A and I.)

(b) Find A^5 for the matrix

$$\begin{bmatrix} 0 & 0 & 1 \\ 2 & 2 & -1 \\ 1 & 0 & 2 \end{bmatrix}.$$

(*Hint:* Find the characteristic polynomial of A, then use the Cayley-Hamilton Theorem to express A^3 as a linear combination of A^2, A, and I. Inductively express A^5 as a linear combination of A^2, A, and I.)

Take this test as you would take a test in class. When you are finished, check your work against the answers given in the back of the book.

1. Solve the system of linear equations.

$$\begin{aligned} 4x_1 + x_2 - 3x_3 &= 11 \\ 2x_1 - 3x_2 + 2x_3 &= 9 \\ x_1 + x_2 + x_3 &= -3 \end{aligned}$$

2. Find the solution set of the system of linear equations represented by the augmented matrix.

$$\begin{bmatrix} 0 & 1 & -1 & 0 & 2 \\ 1 & 0 & 2 & -1 & 0 \\ 1 & 2 & 0 & -1 & 4 \end{bmatrix}$$

3. Solve the homogeneous linear system corresponding to the following coefficient matrix.

$$\begin{bmatrix} 1 & 2 & 1 & -2 \\ 0 & 0 & 2 & -4 \\ -2 & -4 & 1 & -2 \end{bmatrix}$$

4. Find conditions on k such that the system is consistent.

$$\begin{aligned} x + 2y - z &= 3 \\ -x - y + z &= 2 \\ -x + y + z &= k \end{aligned}$$

5. A manufacturer produces three different models of a given product that are shipped to two different warehouses. The number of units of model i that are shipped to warehouse j is represented by a_{ij} in the matrix

$$A = \begin{bmatrix} 200 & 300 \\ 600 & 350 \\ 250 & 400 \end{bmatrix}.$$

The price of each model in dollars per unit is represented by the matrix

$$B = \begin{bmatrix} 12.50 & 9.00 & 21.50 \end{bmatrix}.$$

Find the product BA and state what each entry of the product represents.

6. Solve for x and y in the matrix equation $2A - B = I$ if

$$A = \begin{bmatrix} -1 & 1 \\ 2 & 3 \end{bmatrix} \quad \text{and} \quad B = \begin{bmatrix} x & 2 \\ y & 5 \end{bmatrix}.$$

7. Find $A^T A$ for the matrix

$$A = \begin{bmatrix} 1 & 2 & 3 \\ 4 & 5 & 6 \end{bmatrix}.$$

8. Find the inverses (if they exist) of the matrices.

(a) $\begin{bmatrix} -2 & 3 \\ 4 & 6 \end{bmatrix}$

(b) $\begin{bmatrix} -2 & 3 \\ 3 & 6 \end{bmatrix}$

9. Find the inverse of the matrix
$$\begin{bmatrix} 1 & 1 & 0 \\ -3 & 6 & 5 \\ 0 & 1 & 0 \end{bmatrix}.$$

10. Factor the matrix
$$A = \begin{bmatrix} 2 & -4 \\ 1 & 0 \end{bmatrix}$$
into a product of elementary matrices.

11. Find the determinant of the matrix
$$\begin{bmatrix} 5 & 1 & 2 & 4 \\ 1 & 0 & -2 & -3 \\ 1 & 1 & 6 & 1 \\ 1 & 0 & 0 & -4 \end{bmatrix}.$$

12. Find each determinant if
$$A = \begin{bmatrix} 1 & -3 \\ 4 & 2 \end{bmatrix} \quad \text{and} \quad B = \begin{bmatrix} -2 & 1 \\ 0 & 5 \end{bmatrix}.$$
    (a) $|A|$ (b) $|B|$ (c) $|AB|$ (d) $|A^{-1}|$

13. If $|A| = 7$ and A is of order 4, then find each determinant.
    (a) $|3A|$ (b) $|A^T|$ (c) $|A^{-1}|$ (d) $|A^3|$

14. Use linear equations to find the parabola $y = ax^2 + bx + c$ that passes through the points $(-1, 2)$, $(0, 1)$, and $(2, 6)$. Sketch the points and the parabola.

15. Let $\mathbf{x}_1, \mathbf{x}_2, \mathbf{x}_3$, and $\mathbf{b}$ be the following column matrices.
$$\mathbf{x}_1 = \begin{bmatrix} 1 \\ 0 \\ 1 \end{bmatrix} \quad \mathbf{x}_2 = \begin{bmatrix} 1 \\ 1 \\ 0 \end{bmatrix} \quad \mathbf{x}_3 = \begin{bmatrix} 0 \\ 1 \\ 1 \end{bmatrix} \quad \mathbf{b} = \begin{bmatrix} 1 \\ 2 \\ 3 \end{bmatrix}$$
    Find constants a, b, and c such that $a\mathbf{x}_1 + b\mathbf{x}_2 + c\mathbf{x}_3 = \mathbf{b}$.

16. Let A, B, and C be three nonzero $n \times n$ matrices such that $AC = BC$. Does it follow that $A = B$? Prove your answer.

17. For any matrix B, prove that the matrix $B^T B$ is symmetric.

18. Prove that if the matrix A has an inverse, then the inverse is unique.

19. (a) Define row equivalence of matrices.
    (b) Prove that if A is row-equivalent to B and B is row-equivalent to C, then A is row-equivalent to C.

20. A matrix A is **skew-symmetric** if $A^T = -A$. Prove that the determinant of an odd-order, skew-symmetric matrix is zero.

21. A matrix A is **idempotent** if $A^2 = A$.
    (a) Give three examples of idempotent 2×2 matrices.
    (b) Find the determinant of an idempotent matrix.

22. Let A and B be nonsingular matrices of the same order.
    (a) Is the sum $A + B$ nonsingular? Prove your answer.
    (b) Is the product AB nonsingular? Prove your answer.

23. Describe all possible 2×2 reduced row-echelon matrices.

WILLIAM ROWAN HAMILTON

1 8 0 5 – 1 8 6 5

WILLIAM ROWAN HAMILTON is considered to be Ireland's most famous mathematician. His early education was astonishing. By age 5 he could read and translate Latin, Greek, and Hebrew. By age 10 he was fluent in French and Italian and was studying Asian languages. At 17, he had a grasp of calculus and astronomy.

Hamilton's formal education began when he entered Trinity College in Dublin at age 18. He so impressed his teachers there that after 4 years in undergraduate school, he was elected to the position of professor of astronomy. In 1828, his first year on the faculty, he published an impressive work on optics entitled *A Theory of Systems of Rays*. This paper firmly established Hamilton's reputation, and it has become one of the classics of Western science. In it Hamilton included some of his own methods for working with systems of linear equations. He also introduced the notion of the characteristic equation of a matrix.

During the last 20 years of his life, Hamilton devoted most of his mathematical creativity to developing the theory of a special type of number he called a *quaternion*. This work paved the way for the development of the modern notion of a vector, and we still use Hamilton's **i**, **j**, **k** notation for the standard unit vectors in 3-space.

CHAPTER 4 VECTOR SPACES

4.1 | Vectors in R^n

You are probably already familiar with two- and three-dimensional vectors as they are used in physics and engineering. In those disciplines, a vector is characterized by two quantities (length and direction) and is represented by a directed line segment. In this chapter you will see that these are only two special types of vectors. They are, however, important here because their geometric representations can help you understand the more general definition of a vector.

This section begins with a short review of vectors in the plane, which is the way vectors were developed historically.

Vectors in the Plane

A **vector in the plane** is represented geometrically by a **directed line segment** whose **initial point** is the origin and whose **terminal point** is the point (x_1, x_2), as shown in Figure 4.1. This vector is represented by the same **ordered pair** used to represent its terminal point. That is,

$$\mathbf{x} = (x_1, x_2).$$

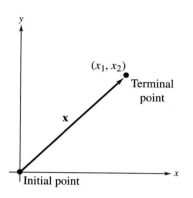

Figure 4.1

The coordinates x_1 and x_2 are called the **components** of the vector $\mathbf{x}$. Two vectors in the plane $\mathbf{u} = (u_1, u_2)$ and $\mathbf{v} = (v_1, v_2)$ are **equal** if and only if $u_1 = v_1$ and $u_2 = v_2$.

REMARK: The term *vector* derives from the Latin word *vectus*, meaning "to carry." The idea is that if you were to carry something from the origin to the point (x_1, x_2), the trip could be represented by the directed line segment from $(0, 0)$ to (x_1, x_2). Vectors are represented by lowercase letters set in boldface type (such as $\mathbf{u}$, $\mathbf{v}$, $\mathbf{w}$, and $\mathbf{x}$).

EXAMPLE 1   *Vectors in the Plane*

Use a directed line segment to represent each vector in the plane.
(a) $\mathbf{u} = (2, 3)$ (b) $\mathbf{v} = (-1, 2)$

Solution

To represent each vector, draw a directed line segment from the origin to the indicated terminal point, as shown in Figure 4.2.

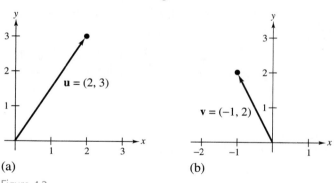

(a)   (b)

Figure 4.2

The first basic vector operation is **vector addition.** To add two vectors in the plane, add their corresponding components. That is, the **sum** of $\mathbf{u}$ and $\mathbf{v}$ is the vector given by

$$\mathbf{u} + \mathbf{v} = (u_1, u_2) + (v_1, v_2) = (u_1 + v_1, u_2 + v_2).$$

Geometrically, the sum of two vectors in the plane is represented as the diagonal of a parallelogram having $\mathbf{u}$ and $\mathbf{v}$ as its adjacent sides, as shown in Figure 4.3.

In the next example, one of the vectors you will add is the vector $(0, 0)$, called the **zero vector.** The zero vector is denoted by $\mathbf{0}$.

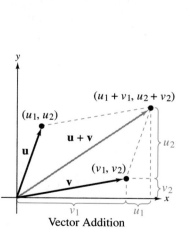

Vector Addition

Figure 4.3

EXAMPLE 2 *Adding Two Vectors in the Plane*

Find the sum of the vectors.

(a) $\mathbf{u} = (1, 4)$, $\mathbf{v} = (2, -2)$ (b) $\mathbf{u} = (3, -2)$, $\mathbf{v} = (-3, 2)$ (c) $\mathbf{u} = (2, 1)$, $\mathbf{v} = (0, 0)$

Solution

(a) $\mathbf{u} + \mathbf{v} = (1, 4) + (2, -2) = (3, 2)$
(b) $\mathbf{u} + \mathbf{v} = (3, -2) + (-3, 2) = (0, 0)$
(c) $\mathbf{u} + \mathbf{v} = (2, 1) + (0, 0) = (2, 1)$

Figure 4.4 gives the graphical representation of each sum.

Simulation

Explore this concept further with an electronic simulation available on the *Elementary Linear Algebra Learning Tools Student CD-ROM*.

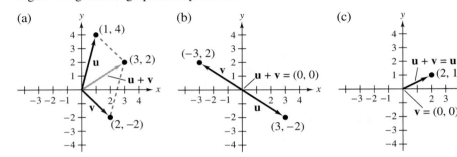

Figure 4.4

The second basic vector operation is called **scalar multiplication.** To multiply a vector $\mathbf{v}$ by a scalar c, multiply each of the components of $\mathbf{v}$ by c. That is,

$$c\mathbf{v} = c(v_1, v_2) = (cv_1, cv_2).$$

Recall from Chapter 2 that the word *scalar* is used to mean a real number. Historically, this usage arose from the fact that multiplying a vector by a real number changes the "scale" of the vector. For instance, if a vector $\mathbf{v}$ is multiplied by 2, the resulting vector $2\mathbf{v}$ is a vector having the same direction as $\mathbf{v}$ and twice the length. In general, for a scalar c, the vector $c\mathbf{v}$ will be $|c|$ times as long as $\mathbf{v}$. Moreover, if c is positive, then $c\mathbf{v}$ has the same direction as $\mathbf{v}$, and if c is negative, then $c\mathbf{v}$ and $\mathbf{v}$ have opposite directions. This is shown in Figure 4.5.

The product of a vector $\mathbf{v}$ with the scalar -1 is denoted by

$$-\mathbf{v} = (-1)\mathbf{v}.$$

The vector $-\mathbf{v}$ is called the **negative** of $\mathbf{v}$. The **difference** of $\mathbf{u}$ and $\mathbf{v}$ is defined to be

$$\mathbf{u} - \mathbf{v} = \mathbf{u} + (-\mathbf{v}),$$

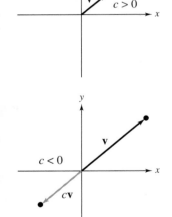

Figure 4.5

and you can say that $\mathbf{v}$ is **subtracted** from $\mathbf{u}$.

EXAMPLE 3 *Operations with Vectors in the Plane*

Given $\mathbf{v} = (-2, 5)$ and $\mathbf{u} = (3, 4)$, find each vector.

(a) $\frac{1}{2}\mathbf{v}$ (b) $\mathbf{u} - \mathbf{v}$ (c) $\frac{1}{2}\mathbf{v} + \mathbf{u}$

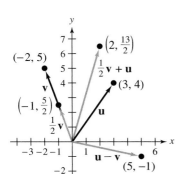

Figure 4.6

Solution   (a) Because $\mathbf{v} = (-2, 5)$, you have

$$\tfrac{1}{2}\mathbf{v} = \left(\tfrac{1}{2}(-2), \tfrac{1}{2}(5)\right) = \left(-1, \tfrac{5}{2}\right).$$

(b) By the definition of vector subtraction, you have

$$\mathbf{u} - \mathbf{v} = (3 - (-2), 4 - 5) = (5, -1).$$

(c) Using the result of part(a), you have

$$\tfrac{1}{2}\mathbf{v} + \mathbf{u} = \left(-1, \tfrac{5}{2}\right) + (3, 4) = \left(-1 + 3, \tfrac{5}{2} + 4\right) = \left(2, \tfrac{13}{2}\right).$$

Figure 4.6 gives a graphical representation of these vector operations.

Vector addition and scalar multiplication share many properties with matrix addition and scalar multiplication. The ten properties listed in the following theorem play a fundamental role in linear algebra. In fact, you will see in the next section that it is precisely these ten properties that are abstracted from vectors in the plane to define the general notion of a vector space.

Theorem 4.1

Properties of Vector Addition and Scalar Multiplication

Let $\mathbf{u}$, $\mathbf{v}$, and $\mathbf{w}$ be vectors in the plane, and let c and d be scalars.

1. $\mathbf{u} + \mathbf{v}$ is a vector in the plane.   **Closure under addition**
2. $\mathbf{u} + \mathbf{v} = \mathbf{v} + \mathbf{u}$ **Commutative property of addition**
3. $(\mathbf{u} + \mathbf{v}) + \mathbf{w} = \mathbf{u} + (\mathbf{v} + \mathbf{w})$ **Associative property of addition**
4. $\mathbf{u} + \mathbf{0} = \mathbf{u}$
5. $\mathbf{u} + (-\mathbf{u}) = \mathbf{0}$
6. $c\mathbf{u}$ is a vector in the plane.   **Closure under scalar multiplication**
7. $c(\mathbf{u} + \mathbf{v}) = c\mathbf{u} + c\mathbf{v}$ **Distributive property**
8. $(c + d)\mathbf{u} = c\mathbf{u} + d\mathbf{u}$ **Distributive property**
9. $c(d\mathbf{u}) = (cd)\mathbf{u}$
10. $1(\mathbf{u}) = \mathbf{u}$

Proof   The proof of each property is a straightforward application of the definition of vector addition and scalar multiplication combined with the corresponding properties of addition and multiplication of real numbers. For instance, to prove the associative property of vector addition, you can write

$$\begin{aligned}
(\mathbf{u} + \mathbf{v}) + \mathbf{w} &= [(u_1, u_2) + (v_1, v_2)] + (w_1, w_2) \\
&= (u_1 + v_1, u_2 + v_2) + (w_1, w_2) \\
&= ((u_1 + v_1) + w_1, (u_2 + v_2) + w_2) \\
&= (u_1 + (v_1 + w_1), u_2 + (v_2 + w_2)) \\
&= (u_1, u_2) + (v_1 + w_1, v_2 + w_2) \\
&= (u_1, u_2) + [(v_1, v_2) + (w_1, w_2)] \\
&= \mathbf{u} + (\mathbf{v} + \mathbf{w}).
\end{aligned}$$

Similarly, to prove the right distributive property of scalar multiplication over addition, you can write

$$
\begin{aligned}
(c + d)\mathbf{u} &= (c + d)(u_1, u_2) \\
&= ((c + d)u_1, (c + d)u_2) \\
&= (cu_1 + du_1, cu_2 + du_2) \\
&= (cu_1, cu_2) + (du_1, du_2) \\
&= c(u_1, u_2) + d(u_1, u_2) \\
&= c\mathbf{u} + d\mathbf{u}.
\end{aligned}
$$

The proofs of the other eight properties are left to you. (See Exercise 47.)

REMARK : Note that the associative property of vector addition allows you to write such expressions as $\mathbf{u} + \mathbf{v} + \mathbf{w}$ without ambiguity, because you obtain the same vector sum regardless of which addition is performed first.

Vectors in R^n

The discussion of vectors in the plane can now be extended to a discussion of vectors in n-space. A vector in n-space is represented by an **ordered n-tuple.** For instance, an ordered triple has the form (x_1, x_2, x_3), an ordered quadruple has the form (x_1, x_2, x_3, x_4), and a general ordered n-tuple has the form $(x_1, x_2, x_3, \ldots, x_n)$. The set of all n-tuples is called **n-space** and is denoted by R^n.

R^1 = 1-space = set of all real numbers

R^2 = 2-space = set of all ordered pairs of real numbers

R^3 = 3-space = set of all ordered triples of real numbers

R^4 = 4-space = set of all ordered quadruples of real numbers

$\vdots$

R^n = n-space = set of all ordered n-tuples of real numbers

The practice of using an ordered pair to represent either a point or a vector in R^2 continues in R^n. That is, an n-tuple $(x_1, x_2, x_3, \ldots, x_n)$ can be viewed as a **point** in R^n with the x_i's as its coordinates or as a **vector**

$$
\mathbf{x} = (x_1, x_2, x_3, \ldots, x_n) \qquad \text{Vector in } R^n
$$

with the x_i's as its components. As with vectors in the plane, two vectors in R^n are **equal** if and only if corresponding components are equal. [In the case of $n = 2$ or $n = 3$, the familiar (x, y) or (x, y, z) notation is occasionally used.]

The sum of two vectors in R^n and the scalar multiple of a vector in R^n are called the **standard operations in R^n** and are defined on the following page.

Definition of Vector Addition and Scalar Multiplication in R^n	Let $\mathbf{u} = (u_1, u_2, u_3, \ldots, u_n)$ and $\mathbf{v} = (v_1, v_2, v_3, \ldots, v_n)$ be vectors in R^n and let c be a real number. Then the **sum** of $\mathbf{u}$ and $\mathbf{v}$ is defined to be the vector $$\mathbf{u} + \mathbf{v} = (u_1 + v_1, u_2 + v_2, u_3 + v_3, \ldots, u_n + v_n),$$ and the **scalar multiple** of $\mathbf{u}$ by c is defined to be the vector $$c\mathbf{u} = (cu_1, cu_2, cu_3, \ldots, cu_n).$$

As with 2-space, the **negative** of a vector in R^n is defined as

$$-\mathbf{u} = (-u_1, -u_2, -u_3, \ldots, -u_n)$$

and the **difference** of two vectors in R^n is defined as

$$\mathbf{u} - \mathbf{v} = (u_1 - v_1, u_2 - v_2, u_3 - v_3, \ldots, u_n - v_n).$$

The **zero vector** in R^n is given by $\mathbf{0} = (0, 0, \ldots, 0)$.

EXAMPLE 4   *Vector Operations in R^3*

Given $\mathbf{u} = (-1, 0, 1)$ and $\mathbf{v} = (2, -1, 5)$ in R^3, find each vector.

(a) $\mathbf{u} + \mathbf{v}$ (b) $2\mathbf{u}$ (c) $\mathbf{v} - 2\mathbf{u}$

Solution   (a) To add two vectors, add their corresponding components as follows.

$$\mathbf{u} + \mathbf{v} = (-1, 0, 1) + (2, -1, 5) = (1, -1, 6)$$

(b) To multiply a vector by a scalar, multiply each component by the scalar as follows.

$$2\mathbf{u} = 2(-1, 0, 1) = (-2, 0, 2)$$

(c) Using the result of part (b), you have

$$\mathbf{v} - 2\mathbf{u} = (2, -1, 5) - (-2, 0, 2) = (4, -1, 3).$$

Figure 4.7 gives a graphical representation of these vector operations in R^3.

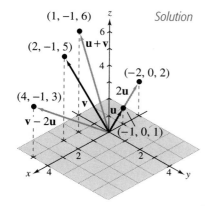

Figure 4.7

Technology Note

Some graphing calculators and computer software programs will perform vector addition and scalar multiplication. Using a TI-86, you can verify Example 4 as follows.

```
VECTOR:U          3
  e1=-1
  e2=0
  e3=1
```

```
VECTOR:V          3
  e1=2
  e2=-1
  e3=5
```

```
U+V
          [1  -1  6]
2U
          [-2  0  2]
V-2U
          [4  -1  3]
```

The following properties of vector addition and scalar multiplication for vectors in R^n are the same as those listed in Theorem 4.1 for vectors in a plane. Their proofs, based on the definitions of vector addition and scalar multiplication in R^n, are left as an exercise. (See Exercise 48.)

Theorem 4.2

Properties of Vector Addition and Scalar Multiplication

Let **u**, **v**, and **w** be vectors in R^n, and let c and d be scalars.
1. **u** + **v** is a vector in R^n.
2. **u** + **v** = **v** + **u**
3. (**u** + **v**) + **w** = **u** + (**v** + **w**)
4. **u** + **0** = **u**
5. **u** + (−**u**) = **0**
6. c**u** is a vector in R^n.
7. c(**u** + **v**) = c**u** + c**v**
8. $(c + d)$**u** = c**u** + d**u**
9. $c(d$**u**$) = (cd)$**u**
10. $1($**u**$) = $**u**

Using the ten properties given in Theorem 4.2, you can perform algebraic manipulations with vectors in R^n in much the same way as you do with real numbers, as demonstrated in the next example.

EXAMPLE 5   *Vector Operations in R^4*

Let **u** $= (2, -1, 5, 0)$, **v** $= (4, 3, 1, -1)$, and **w** $= (-6, 2, 0, 3)$ be vectors in R^4. Solve for **x** in each of the following.

(a) **x** $= 2$**u** $- ($**v** $+ 3$**w**$)$ (b) $3($**x** $+ $**w**$) = 2$**u** $- $**v** $+ $**x**

Solution   (a) Using the properties listed in Theorem 4.2, you have

$$\begin{aligned}
\mathbf{x} &= 2\mathbf{u} - (\mathbf{v} + 3\mathbf{w}) \\
&= 2\mathbf{u} - \mathbf{v} - 3\mathbf{w} \\
&= (4, -2, 10, 0) - (4, 3, 1, -1) - (-18, 6, 0, 9) \\
&= (4 - 4 + 18, -2 - 3 - 6, 10 - 1 - 0, 0 + 1 - 9) \\
&= (18, -11, 9, -8).
\end{aligned}$$

(b) Begin by solving for **x** as follows.

$$\begin{aligned}
3(\mathbf{x} + \mathbf{w}) &= 2\mathbf{u} - \mathbf{v} + \mathbf{x} \\
3\mathbf{x} + 3\mathbf{w} &= 2\mathbf{u} - \mathbf{v} + \mathbf{x} \\
3\mathbf{x} - \mathbf{x} &= 2\mathbf{u} - \mathbf{v} - 3\mathbf{w} \\
2\mathbf{x} &= 2\mathbf{u} - \mathbf{v} - 3\mathbf{w} \\
\mathbf{x} &= \tfrac{1}{2}(2\mathbf{u} - \mathbf{v} - 3\mathbf{w})
\end{aligned}$$

Using the result of part (a) produces

$$\mathbf{x} = \tfrac{1}{2}(18, -11, 9, -8)$$
$$= \left(9, -\tfrac{11}{2}, \tfrac{9}{2}, -4\right).$$

The zero vector $\mathbf{0}$ in R^n is called the **additive identity** in R^n. Similarly, the vector $-\mathbf{v}$ is called the **additive inverse** of $\mathbf{v}$. The following theorem summarizes several important properties of the additive identity and additive inverse in R^n.

Theorem 4.3

Properties of Additive Identity and Additive Inverse

Let $\mathbf{v}$ be a vector in R^n, and let c be a scalar. Then the following properties are true.
1. The additive identity is unique. That is, if $\mathbf{v} + \mathbf{u} = \mathbf{v}$, then $\mathbf{u} = \mathbf{0}$.
2. The additive inverse of $\mathbf{v}$ is unique. That is, if $\mathbf{v} + \mathbf{u} = \mathbf{0}$, then $\mathbf{u} = -\mathbf{v}$.
3. $0\mathbf{v} = \mathbf{0}$
4. $c\mathbf{0} = \mathbf{0}$
5. If $c\mathbf{v} = \mathbf{0}$, then $c = 0$ or $\mathbf{v} = \mathbf{0}$.
6. $-(-\mathbf{v}) = \mathbf{v}$

Proof

To prove the first property, assume that $\mathbf{v} + \mathbf{u} = \mathbf{v}$. Then the following steps are justified by Theorem 4.2.

$\mathbf{v} + \mathbf{u} = \mathbf{v}$	Given
$(\mathbf{v} + \mathbf{u}) + (-\mathbf{v}) = \mathbf{v} + (-\mathbf{v})$	Add $-\mathbf{v}$ to both sides.
$(\mathbf{v} + \mathbf{u}) + (-\mathbf{v}) = \mathbf{0}$	Additive inverse
$\mathbf{v} + [\mathbf{u} + (-\mathbf{v})] = \mathbf{0}$	Associative property
$\mathbf{v} + [(-\mathbf{v}) + \mathbf{u}] = \mathbf{0}$	Commutative property
$[\mathbf{v} + (-\mathbf{v})] + \mathbf{u} = \mathbf{0}$	Associative property
$\mathbf{0} + \mathbf{u} = \mathbf{0}$	Additive inverse
$\mathbf{u} + \mathbf{0} = \mathbf{0}$	Commutative property
$\mathbf{u} = \mathbf{0}$	Additive identity

As you gain experience in reading and writing proofs involving vector algebra, you will not need to list this many steps. For now, however, it's a good idea. The proofs of the other five properties are left as exercises. (See Exercises 49–53.)

REMARK: In Properties 3 and 5 of Theorem 4.3, note that two different zeros are used, the scalar 0 and the vector $\mathbf{0}$.

The next example illustrates an important type of problem in linear algebra—writing one vector $\mathbf{x}$ as the sum of scalar multiples of other vectors $\mathbf{v}_1, \mathbf{v}_2, \ldots,$ and $\mathbf{v}_n$. That is,

$$\mathbf{x} = c_1\mathbf{v}_1 + c_2\mathbf{v}_2 + \cdots + c_n\mathbf{v}_n.$$

The vector $\mathbf{x}$ is called a **linear combination** of the vectors $\mathbf{v}_1, \mathbf{v}_2, \ldots,$ and $\mathbf{v}_n$.

EXAMPLE 6 *Writing a Vector as a Linear Combination of Other Vectors*

Given $\mathbf{x} = (-1, -2, -2)$, $\mathbf{u} = (0, 1, 4)$, $\mathbf{v} = (-1, 1, 2)$, and $\mathbf{w} = (3, 1, 2)$ in R^3, find scalars a, b, and c such that

$$\mathbf{x} = a\mathbf{u} + b\mathbf{v} + c\mathbf{w}.$$

Solution By writing

$$\overbrace{(-1, -2, -2)}^{\mathbf{x}} = a\overbrace{(0, 1, 4)}^{\mathbf{u}} + b\overbrace{(-1, 1, 2)}^{\mathbf{v}} + c\overbrace{(3, 1, 2)}^{\mathbf{w}}$$
$$= (-b + 3c, a + b + c, 4a + 2b + 2c),$$

you can equate corresponding components so that they form the following system of three linear equations in a, b, and c.

$$-b + 3c = -1 \qquad \text{Equation from first component}$$
$$a + b + c = -2 \qquad \text{Equation from second component}$$
$$4a + 2b + 2c = -2 \qquad \text{Equation from third component}$$

Using the techniques of Chapter 1, solve for a, b, and c to get

$$a = 1, \quad b = -2, \quad \text{and} \quad c = -1.$$

Thus, $\mathbf{x}$ can be written as a linear combination of $\mathbf{u}$, $\mathbf{v}$, and $\mathbf{w}$.

$$\mathbf{x} = \mathbf{u} - 2\mathbf{v} - \mathbf{w}$$

Try using vector addition and scalar multiplication to check this result.

DISCOVERY

Is the vector (1, 1) a linear combination of the vectors (1, 2) and $(-2, -4)$? Graph these vectors in the plane and explain your answer geometrically. Similarly, determine whether the vector (1, 1) is a linear combination of the vectors (1, 2) and (2, 1). What is the geometric significance of these two questions? Is every vector in R^2 a linear combination of the vectors (1, 2) and (2, 1)? Give a geometric explanation for your answer.

You will often find it useful to represent a vector $\mathbf{u} = (u_1, u_2, \ldots, u_n)$ in R^n as either a $1 \times n$ row matrix (row vector),

$$\mathbf{u} = [u_1 \quad u_2 \quad \cdots \quad u_n],$$

or an $n \times 1$ column matrix (column vector),

$$\mathbf{u} = \begin{bmatrix} u_1 \\ u_2 \\ \vdots \\ u_n \end{bmatrix}.$$

This approach is valid because the matrix operations of addition and scalar multiplication give the same results as the corresponding vector operations. That is, the matrix sums

$$\mathbf{u} + \mathbf{v} = \begin{bmatrix} u_1 & u_2 & \cdots & u_n \end{bmatrix} + \begin{bmatrix} v_1 & v_2 & \cdots & v_n \end{bmatrix}$$
$$= \begin{bmatrix} u_1 + v_1 & u_2 + v_2 & \cdots & u_n + v_n \end{bmatrix}$$

and

$$\mathbf{u} + \mathbf{v} = \begin{bmatrix} u_1 \\ u_2 \\ \vdots \\ u_n \end{bmatrix} + \begin{bmatrix} v_1 \\ v_2 \\ \vdots \\ v_n \end{bmatrix} = \begin{bmatrix} u_1 + v_1 \\ u_2 + v_2 \\ \vdots \\ u_n + v_n \end{bmatrix}$$

yield the same results as the vector operation of addition,

$$\mathbf{u} + \mathbf{v} = (u_1, u_2, \ldots, u_n) + (v_1, v_2, \ldots, v_n) = (u_1 + v_1, u_2 + v_2, \ldots, u_n + v_n).$$

The same argument applies to scalar multiplication. The only difference in the three notations for vectors is how the components are displayed; the underlying operations are the same.

SECTION 4.1   Exercises

In Exercises 1–6, find the vector $\mathbf{v}$ and illustrate the indicated vector operations geometrically, where $\mathbf{u} = (-2, 3)$ and $\mathbf{w} = (-3, -2)$.

1. $\mathbf{v} = \frac{3}{2}\mathbf{u}$ **2.** $\mathbf{v} = \mathbf{u} + \mathbf{w}$ **3.** $\mathbf{v} = \mathbf{u} + 2\mathbf{w}$

4. $\mathbf{v} = -\mathbf{u} + \mathbf{w}$ **5.** $\mathbf{v} = \frac{1}{2}(3\mathbf{u} + \mathbf{w})$ **6.** $\mathbf{v} = \mathbf{u} - 2\mathbf{w}$

7. Given the vector $\mathbf{v} = (2, 1)$, sketch (a) $2\mathbf{v}$, (b) $-3\mathbf{v}$, and (c) $\frac{1}{2}\mathbf{v}$.

8. Given the vector $\mathbf{v} = (3, -2)$, sketch (a) $4\mathbf{v}$, (b) $-\frac{1}{2}\mathbf{v}$, and (c) $0\mathbf{v}$.

In Exercises 9–14, let $\mathbf{u} = (1, 2, 3)$, $\mathbf{v} = (2, 2, -1)$, and $\mathbf{w} = (4, 0, -4)$.

9. Find $\mathbf{u} - \mathbf{v}$ and $\mathbf{v} - \mathbf{u}$.

10. Find $\mathbf{u} - \mathbf{v} + 2\mathbf{w}$.

11. Find $2\mathbf{u} + 4\mathbf{v} - \mathbf{w}$.

12. Find $5\mathbf{u} - 3\mathbf{v} - \frac{1}{2}\mathbf{w}$.

13. Find $\mathbf{z}$, where $2\mathbf{z} - 3\mathbf{u} = \mathbf{w}$.

14. Find $\mathbf{z}$, where $2\mathbf{u} + \mathbf{v} - \mathbf{w} + 3\mathbf{z} = \mathbf{0}$.

15. Given the vector $\mathbf{v} = (1, 2, 2)$, sketch (a) $2\mathbf{v}$, (b) $-\mathbf{v}$, and (c) $\frac{1}{2}\mathbf{v}$.

16. Given the vector $\mathbf{v} = (2, 0, 1)$, sketch (a) $-\mathbf{v}$, (b) $2\mathbf{v}$, and (c) $\frac{1}{2}\mathbf{v}$.

17. Which of the vectors are scalar multiples of $\mathbf{z} = (3, 2, -5)$?
(a) $\mathbf{u} = (-6, -4, 10)$
(b) $\mathbf{v} = \left(2, \frac{4}{3}, -\frac{10}{3}\right)$
(c) $\mathbf{w} = (6, 4, 10)$

18. Which of the vectors are scalar multiples of $\mathbf{z} = \left(\frac{1}{2}, -\frac{2}{3}, \frac{3}{4}\right)$?
(a) $\mathbf{u} = (6, -4, 9)$
(b) $\mathbf{v} = \left(-1, \frac{4}{3}, -\frac{3}{2}\right)$
(c) $\mathbf{w} = (12, 0, 9)$

In Exercises 19–22, find (a) $\mathbf{u} - \mathbf{v}$ and (b) $2(\mathbf{u} + 3\mathbf{v})$.

19. $\mathbf{u} = (4, 0, -3, 5)$, $\mathbf{v} = (0, 2, 5, 4)$

20. $\mathbf{u} = (0, 4, 3, 4, 4)$, $\mathbf{v} = (6, 8, -3, 3, -5)$

21. $\mathbf{u} = (-7, 0, 0, 0, 9)$, $\mathbf{v} = (2, -3, -2, 3, 3)$

22. $\mathbf{u} = (6, -5, 4, 3)$, $\mathbf{v} = \left(-2, \frac{5}{3}, -\frac{4}{3}, -1\right)$

In Exercises 23 and 24, use a graphing utility with matrix capabilities to find the following, where $\mathbf{u} = (1, 2, -3, 1)$, $\mathbf{v} = (0, 2, -1, -2)$, and $\mathbf{w} = (2, -2, 1, 3)$.

23. (a) $\mathbf{u} + 2\mathbf{v}$
(b) $\mathbf{w} - 3\mathbf{u}$
(c) $4\mathbf{v} + \frac{1}{2}\mathbf{u} - \mathbf{w}$
(d) $\frac{1}{4}(3\mathbf{u} + 2\mathbf{v} - \mathbf{w})$

24. (a) $\mathbf{v} + 3\mathbf{w}$
(b) $2\mathbf{w} - \frac{1}{2}\mathbf{u}$
(c) $2\mathbf{u} + \mathbf{w} - 3\mathbf{v}$
(d) $\frac{1}{2}(4\mathbf{v} - 3\mathbf{u} + \mathbf{w})$

In Exercises 25 and 26, solve for $\mathbf{w}$ given that $\mathbf{u} = (1, -1, 0, 1)$ and $\mathbf{v} = (0, 2, 3, -1)$.

25. $2\mathbf{w} = \mathbf{u} - 3\mathbf{v}$ **26.** $\mathbf{w} + \mathbf{u} = -\mathbf{v}$

In Exercises 27–32, if possible, write **v** as a linear combination of **u** and **w**, where $\mathbf{u} = (1, 2)$ and $\mathbf{w} = (1, -1)$.

27. $\mathbf{v} = (2, 1)$

28. $\mathbf{v} = (0, 3)$

29. $\mathbf{v} = (3, 0)$

30. $\mathbf{v} = (1, -1)$

31. $\mathbf{v} = (-1, -2)$

32. $\mathbf{v} = (1, -4)$

In Exercises 33 and 34, find **w** such that $2\mathbf{u} + \mathbf{v} - 3\mathbf{w} = \mathbf{0}$.

33. $\mathbf{u} = (0, 2, 7, 5)$, $\mathbf{v} = (-3, 1, 4, -8)$

34. $\mathbf{u} = (0, 0, -8, 1)$, $\mathbf{v} = (1, -8, 0, 7)$

In Exercises 35–38, if possible, write **v** as a linear combination of $\mathbf{u}_1$, $\mathbf{u}_2$, and $\mathbf{u}_3$.

35. $\mathbf{u}_1 = (2, 3, 5)$, $\mathbf{u}_2 = (1, 2, 4)$, $\mathbf{u}_3 = (-2, 2, 3)$,
$\mathbf{v} = (10, 1, 4)$

36. $\mathbf{u}_1 = (1, 3, 5)$, $\mathbf{u}_2 = (2, -1, 3)$, $\mathbf{u}_3 = (-3, 2, -4)$,
$\mathbf{v} = (-1, 7, 2)$

37. $\mathbf{u}_1 = (1, 1, 2, 2)$, $\mathbf{u}_2 = (2, 3, 5, 6)$, $\mathbf{u}_3 = (-3, 1, -4, 2)$,
$\mathbf{v} = (0, 5, 3, 0)$

38. $\mathbf{u}_1 = (1, 3, 2, 1)$, $\mathbf{u}_2 = (2, -2, -5, 4)$, $\mathbf{u}_3 = (2, -1, 3, 6)$,
$\mathbf{v} = (2, 5, -4, 0)$

In Exercises 39 and 40, use a graphing utility with matrix capabilities to write **v** as a linear combination of $\mathbf{u}_1$, $\mathbf{u}_2$, $\mathbf{u}_3$, $\mathbf{u}_4$, $\mathbf{u}_5$, and $\mathbf{u}_6$. Then verify your solution.

39. $\mathbf{u}_1 = (1, 2, -3, 4, -1, 2)$
$\mathbf{u}_2 = (1, -2, 1, -1, 2, 1)$
$\mathbf{u}_3 = (0, 2, -1, 2, -1, -1)$
$\mathbf{u}_4 = (1, 0, 3, -4, 1, 2)$
$\mathbf{u}_5 = (1, -2, 1, -1, 2, -3)$
$\mathbf{u}_6 = (3, 2, 1, -2, 3, 0)$
$\mathbf{v} = (10, 30, -13, 14, -7, 27)$

40. $\mathbf{u}_1 = (1, -3, 4, -5, 2, -1)$
$\mathbf{u}_2 = (3, -2, 4, -3, -2, 1)$
$\mathbf{u}_3 = (1, 1, 1, -1, 4, -1)$
$\mathbf{u}_4 = (3, -1, 3, -4, 2, 3)$
$\mathbf{u}_5 = (1, -2, 1, 5, -3, 4)$
$\mathbf{u}_6 = (4, 2, -1, 3, -1, 1)$
$\mathbf{v} = (8, 17, -16, 26, 0, -4)$

True or False? In Exercises 41 and 42, determine whether each statement is true or false. If a statement is true, give a reason or cite an appropriate statement in the text. If a statement is false, provide an example that shows that the statement is not true in all cases or cite an appropriate statement in the text.

41. (a) Two vectors in R^n are equal if and only if their corresponding components are equal.

(b) For a nonzero scalar c, the vector $c\mathbf{v}$ is c times as long as **v** and has the same direction as **v** if $c > 0$ and the opposite direction if $c < 0$.

42. (a) To add two vectors in R^n, add their corresponding components.

(b) The zero vector **0** in R^n is defined as the additive inverse of a vector.

In Exercises 43 and 44, the zero vector $\mathbf{0} = (0, 0, 0)$ can be written as a linear combination of the vectors $\mathbf{v}_1$, $\mathbf{v}_2$, and $\mathbf{v}_3$ as $\mathbf{0} = 0\mathbf{v}_1 + 0\mathbf{v}_2 + 0\mathbf{v}_3$. This is called the *trivial* solution. Can you find a *nontrivial* way of writing **0** as a linear combination of the three vectors?

43. $\mathbf{v}_1 = (1, 0, 1)$, $\mathbf{v}_2 = (-1, 1, 2)$, $\mathbf{v}_3 = (0, 1, 4)$

44. $\mathbf{v}_1 = (1, 0, 1)$, $\mathbf{v}_2 = (-1, 1, 2)$, $\mathbf{v}_3 = (0, 1, 3)$

45. Illustrate properties 1–10 of Theorem 4.2 for $\mathbf{u} = (2, -1, 3, 6)$, $\mathbf{v} = (1, 4, 0, 1)$, $\mathbf{w} = (3, 0, 2, 0)$, $c = 5$, and $d = -2$.

46. Illustrate properties 1–10 of Theorem 4.2 for $\mathbf{u} = (2, -1, 3)$, $\mathbf{v} = (3, 4, 0)$, $\mathbf{w} = (7, 8, -4)$, $c = 2$, and $d = -1$.

47. Complete the proof of Theorem 4.1.

48. Prove each property of vector addition and scalar multiplication from Theorem 4.2.

(a) Property 1: $\mathbf{u} + \mathbf{v}$ is a vector in R^n.
(b) Property 2: $\mathbf{u} + \mathbf{v} = \mathbf{v} + \mathbf{u}$
(c) Property 3: $(\mathbf{u} + \mathbf{v}) + \mathbf{w} = \mathbf{u} + (\mathbf{v} + \mathbf{w})$
(d) Property 4: $\mathbf{u} + \mathbf{0} = \mathbf{u}$
(e) Property 5: $\mathbf{u} + (-\mathbf{u}) = \mathbf{0}$
(f) Property 6: $c\mathbf{u}$ is a vector in R^n.
(g) Property 7: $c(\mathbf{u} + \mathbf{v}) = c\mathbf{u} + c\mathbf{v}$
(h) Property 8: $(c + d)\mathbf{u} = c\mathbf{u} + d\mathbf{u}$
(i) Property 9: $c(d\mathbf{u}) = (cd)\mathbf{u}$
(j) Property 10: $1(\mathbf{u}) = \mathbf{u}$

In Exercises 49–53, complete the proofs of the remaining properties of Theorem 4.3 by supplying the justification for each step. Use the properties of vector addition and scalar multiplication from Theorem 4.2.

49. Property 2: The additive inverse of **v** is unique. That is, if $\mathbf{v} + \mathbf{u} = \mathbf{0}$, then $\mathbf{u} = -\mathbf{v}$.

$\mathbf{v} + \mathbf{u} = \mathbf{0}$	Given
$(-\mathbf{v}) + (\mathbf{v} + \mathbf{u}) = (-\mathbf{v}) + \mathbf{0}$	a. _____
$((-\mathbf{v}) + \mathbf{v}) + \mathbf{u} = -\mathbf{v}$	b. _____
$\mathbf{0} + \mathbf{u} = -\mathbf{v}$	c. _____
$\mathbf{u} = -\mathbf{v}$	d. _____

50. Property 3: $0\mathbf{v} = \mathbf{0}$

$$0\mathbf{v} = (0 + 0)\mathbf{v}$$ a. _____

$$0\mathbf{v} = 0\mathbf{v} + 0\mathbf{v}$$ b. _____

$$0\mathbf{v} + (-0\mathbf{v}) = (0\mathbf{v} + 0\mathbf{v}) + (-0\mathbf{v})$$ c. _____

$$\mathbf{0} = 0\mathbf{v} + (0\mathbf{v} + (-0\mathbf{v}))$$ d. _____

$$\mathbf{0} = 0\mathbf{v} + \mathbf{0}$$ e. _____

$$\mathbf{0} = 0\mathbf{v}$$ f. _____

51. Property 4: $c\mathbf{0} = \mathbf{0}$

$$c\mathbf{0} = c(\mathbf{0} + \mathbf{0})$$ a. _____

$$c\mathbf{0} = c\mathbf{0} + c\mathbf{0}$$ b. _____

$$c\mathbf{0} + (-c\mathbf{0}) = (c\mathbf{0} + c\mathbf{0}) + (-c\mathbf{0})$$ c. _____

$$\mathbf{0} = c\mathbf{0} + (c\mathbf{0} + (-c\mathbf{0}))$$ d. _____

$$\mathbf{0} = c\mathbf{0} + \mathbf{0}$$ e. _____

$$\mathbf{0} = c\mathbf{0}$$ f. _____

52. Property 5: If $c\mathbf{v} = \mathbf{0}$, then $c = 0$ or $\mathbf{v} = \mathbf{0}$. If $c = 0$, you are done. If $c \neq 0$, then c^{-1} exists, and you have

$$c^{-1}(c\mathbf{v}) = c^{-1}\mathbf{0}$$ a. _____

$$(c^{-1}c)\mathbf{v} = \mathbf{0}$$ b. _____

$$1\mathbf{v} = \mathbf{0}$$ c. _____

$$\mathbf{v} = \mathbf{0}.$$ d. _____

53. Property 6: $-(-\mathbf{v}) = \mathbf{v}$

$$-(-\mathbf{v}) + (-\mathbf{v}) = \mathbf{0} \text{ and } \mathbf{v} + (-\mathbf{v}) = \mathbf{0}$$ a. _____

$$-(-\mathbf{v}) + (-\mathbf{v}) = \mathbf{v} + (-\mathbf{v})$$ b. _____

$$-(-\mathbf{v}) + (-\mathbf{v}) + \mathbf{v} = \mathbf{v} + (-\mathbf{v}) + \mathbf{v}$$ c. _____

$$-(-\mathbf{v}) + ((-\mathbf{v}) + \mathbf{v}) = \mathbf{v} + ((-\mathbf{v}) + \mathbf{v})$$ d. _____

$$-(-\mathbf{v}) + \mathbf{0} = \mathbf{v} + \mathbf{0}$$ e. _____

$$-(-\mathbf{v}) = \mathbf{v}$$ f. _____

54. For each matrix, can you write the third column of the matrix as a linear combination of the first two columns?

(a) $\begin{bmatrix} 1 & 2 & 3 \\ 7 & 8 & 9 \\ 4 & 5 & 6 \end{bmatrix}$

(b) $\begin{bmatrix} 1 & 2 & 3 \\ 7 & 8 & 9 \\ 4 & 5 & 7 \end{bmatrix}$

55. *Writing*   Let $A\mathbf{x} = \mathbf{b}$ be a system of m linear equations in n unknowns. Designate the columns of A as $\mathbf{a}_1, \mathbf{a}_2, \ldots, \mathbf{a}_n$. If $\mathbf{b}$ is a linear combination of these n column vectors, explain why this implies that the linear system is consistent. Illustrate your answer with appropriate examples. What can you conclude about the linear system if $\mathbf{b}$ is not a linear combination of the columns of A?

56. *Writing*   How could you describe vector subtraction geometrically? What is the relationship between vector subtraction and the basic vector operations of addition and scalar multiplication?

4.2 | Vector Spaces

In Theorem 4.2, ten special properties of vector addition and scalar multiplication in R^n were listed. Suitable definitions of addition and scalar multiplication reveal that many other mathematical quantities (such as matrices, polynomials, and functions) also share these ten properties. *Any* set that satisfies these properties (or **axioms**) is called a **vector space,** and the objects in the set are called **vectors.**

It is important to realize that the following definition of a vector space is precisely that—a *definition.* You do not need to prove anything, because you are simply listing the axioms required of vector spaces. This type of definition is called an **abstraction** because you are abstracting a collection of properties from a particular setting R^n to form the axioms for a more general setting.

Definition of Vector Space   Let V be a set on which two operations (**vector addition** and **scalar multiplication**) are defined. If the following axioms are satisfied for every $\mathbf{u}$, $\mathbf{v}$, and $\mathbf{w}$ in V and every scalar (real number) c and d, then V is called a **vector space.**

Addition:

1. $\mathbf{u} + \mathbf{v}$ is in V.   Closure under addition
2. $\mathbf{u} + \mathbf{v} = \mathbf{v} + \mathbf{u}$ Commutative property
3. $\mathbf{u} + (\mathbf{v} + \mathbf{w}) = (\mathbf{u} + \mathbf{v}) + \mathbf{w}$ Associative property
4. V has a **zero vector 0** such that for   Additive identity
   every $\mathbf{u}$ in V, $\mathbf{u} + \mathbf{0} = \mathbf{u}$.
5. For every $\mathbf{u}$ in V, there is a vector   Additive inverse
   in V denoted by $-\mathbf{u}$ such that
   $\mathbf{u} + (-\mathbf{u}) = \mathbf{0}$.

Scalar Multiplication:

6. $c\mathbf{u}$ is in V.   Closure under scalar multiplication
7. $c(\mathbf{u} + \mathbf{v}) = c\mathbf{u} + c\mathbf{v}$ Distributive property
8. $(c + d)\mathbf{u} = c\mathbf{u} + d\mathbf{u}$ Distributive property
9. $c(d\mathbf{u}) = (cd)\mathbf{u}$ Associative property
10. $1(\mathbf{u}) = \mathbf{u}$ Scalar identity

It is important to realize that a vector space consists of four entities: a set of vectors, a set of scalars, and two operations. When you refer to a vector space V, be sure that all four entities are clearly stated or understood. Unless stated otherwise, assume that the set of scalars is the set of real numbers.

The first two examples of vector spaces are not surprising. They are, in fact, the models used to form the ten vector space axioms.

EXAMPLE 1   *R^2 with the Standard Operations Is a Vector Space*

The set of all ordered pairs of real numbers R^2 with the standard operations is a vector space. To verify this, look back at Theorem 4.1. Vectors in this space have the form

$$\mathbf{v} = (v_1, v_2).$$

EXAMPLE 2   *R^n with the Standard Operations Is a Vector Space*

The set of all ordered n-tuples of real numbers R^n with the standard operations is a vector space. This is verified by Theorem 4.2. Vectors in this space are of the form

$$\mathbf{v} = (v_1, v_2, v_3, \ldots, v_n).$$

REMARK: From Example 2 you can conclude that R^1, the set of real numbers (with the usual operations of addition and multiplication), is a vector space.

The next three examples describe vector spaces in which the basic set V does not consist of ordered n-tuples. Each example describes the set V and defines the two vector operations. To show that the set is a vector space, you must verify all ten axioms.

EXAMPLE 3   *The Vector Space of All 2 × 3 Matrices*

Show that the set of all 2×3 matrices with the operations of matrix addition and scalar multiplication is a vector space.

Solution   If A and B are 2×3 matrices and c is a scalar, then $A + B$ and cA are also 2×3 matrices. Therefore, the set is closed under matrix addition and scalar multiplication. Moreover, the other eight vector space axioms follow directly from Theorems 2.1 and 2.2 (see Section 2.2). Thus, you can conclude that the set is a vector space. Vectors in this space have the form

$$\mathbf{a} = A = \begin{bmatrix} a_{11} & a_{12} & a_{13} \\ a_{21} & a_{22} & a_{23} \end{bmatrix}.$$

REMARK: In the same way you are able to show that the set of all 2×3 matrices is a vector space, you can show that the set of all $m \times n$ matrices, denoted by $M_{m,n}$, is a vector space.

EXAMPLE 4   *The Vector Space of All Polynomials of Degree 2 or Less*

Let P_2 be the set of all polynomials of the form

$$p(x) = a_2 x^2 + a_1 x + a_0,$$

where a_0, a_1, and a_2 are real numbers. The *sum* of two polynomials $p(x) = a_2 x^2 + a_1 x + a_0$ and $q(x) = b_2 x^2 + b_1 x + b_0$ is defined in the usual way by

$$p(x) + q(x) = (a_2 + b_2)x^2 + (a_1 + b_1)x + (a_0 + b_0),$$

and the *scalar multiple* of $p(x)$ by the scalar c is defined by

$$cp(x) = ca_2 x^2 + ca_1 x + ca_0.$$

Show that P_2 is a vector space.

Solution   Verification of each of the ten vector space axioms is a straightforward application of the properties of real numbers. For instance, because the set of real numbers is closed under addition, it follows that $a_2 + b_2$, $a_1 + b_1$, and $a_0 + b_0$ are real numbers, and

$$p(x) + q(x) = (a_2 + b_2)x^2 + (a_1 + b_1)x + (a_0 + b_0)$$

is in the set P_2 because it is a polynomial of degree 2 or less. Thus, P_2 is closed under addition. Similarly, you can use the fact that the set of real numbers is closed under multiplication to show that P_2 is closed under scalar multiplication. To verify the commutative axiom of addition, write

$$p(x) + q(x) = (a_2x^2 + a_1x + a_0) + (b_2x^2 + b_1x + b_0)$$
$$= (a_2 + b_2)x^2 + (a_1 + b_1)x + (a_0 + b_0)$$
$$= (b_2 + a_2)x^2 + (b_1 + a_1)x + (b_0 + a_0)$$
$$= (b_2x^2 + b_1x + b_0) + (a_2x^2 + a_1x + a_0)$$
$$= q(x) + p(x).$$

Can you see where the commutative property of addition of real numbers was used? The zero vector in this space is the zero polynomial given by $\mathbf{0}(x) = 0x^2 + 0x + 0$, for all x. Try verifying the other vector space axioms. You may then conclude that P_2 is a vector space.

REMARK: Even though the zero polynomial $\mathbf{0}(x) = 0$ has no degree, P_2 is often described as the set of all polynomials of degree 2 *or less*.

P_n is defined to be the set of all polynomials of degree n or less (together with the zero polynomial). The procedure used to verify that P_2 is a vector space can be extended to show that P_n, with the usual operations of polynomial addition and scalar multiplication, is a vector space.

EXAMPLE 5   *The Vector Space of Continuous Functions (Calculus)*

Let $C(-\infty, \infty)$ be the set of all real-valued continuous functions defined on the entire real line. This set consists of all polynomial functions and all other functions that are continuous on the entire real line. For instance, $f(x) = \sin x$ and $g(x) = e^x$ are members of the set. Addition is defined by

$$(f + g)(x) = f(x) + g(x),$$

as shown in Figure 4.8. Scalar multiplication is defined by

$$(cf)(x) = c[f(x)].$$

Show that $C(-\infty, \infty)$ is a vector space.

Solution   To verify that the set $C(-\infty, \infty)$ is closed under addition and scalar multiplication, you can use a result from calculus—that the sum of two continuous functions is continuous and the product of a scalar and a continuous function is continuous. To verify that the set $C(-\infty, \infty)$ has an additive identity, consider the function f_0 that has a value of zero for all x. That is,

$$f_0(x) = 0, \quad x \text{ is any real number.}$$

This function is continuous on the entire real line (its graph is simply the line $y = 0$). Hence it is in the set $C(-\infty, \infty)$. Moreover, if f is any other function that is continuous on the entire real line, then

$$(f + f_0)(x) = f(x) + f_0(x) = f(x) + 0 = f(x).$$

Hence f_0 is the additive identity in $C(-\infty, \infty)$. The verification of the other vector space axioms is left to you.

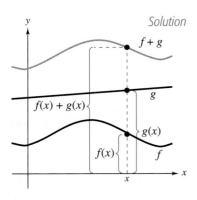

Figure 4.8

For convenience, the following summary lists some important vector spaces that will be referred to frequently in the remainder of this text. In each case the operations are the standard ones.

Summary of Important Vector Spaces

$$R = \text{set of all real numbers}$$
$$R^2 = \text{set of all ordered pairs}$$
$$R^3 = \text{set of all ordered triples}$$
$$R^n = \text{set of all } n\text{-tuples}$$
$$C(-\infty, \infty) = \text{set of all continuous functions defined on the real line}$$
$$C[a, b] = \text{set of all continuous functions defined on a closed interval } [a, b]$$
$$P = \text{set of all polynomials}$$
$$P_n = \text{set of all polynomials of degree } \leq n$$
$$M_{m,n} = \text{set of all } m \times n \text{ matrices}$$
$$M_{n,n} = \text{set of all } n \times n \text{ } square \text{ matrices}$$

You have seen the versatility of the concept of a vector space. For instance, a vector can be a real number, an n-tuple, a matrix, a polynomial, a continuous function, and so on. But what is the purpose? Why bother to define this abstraction? There are several reasons, but the most important is that this abstraction turns out to be mathematically efficient in the sense that general results that apply to all vector spaces can now be derived. Once a theorem has been proven for an abstract vector space, you need not give separate proofs for n-tuples, matrices, and polynomials. You can simply point out that the theorem is true for any vector space, regardless of the particular form the vectors happen to take. This process is illustrated in Theorem 4.4.

Theorem 4.4

Properties of Scalar Multiplication

Let $\mathbf{v}$ be any element of a vector space V, and let c be any scalar. Then the following properties are true.
1. $0\mathbf{v} = \mathbf{0}$ 2. $c\mathbf{0} = \mathbf{0}$
3. If $c\mathbf{v} = \mathbf{0}$, then $c = 0$ or $\mathbf{v} = \mathbf{0}$.   4. $(-1)\mathbf{v} = -\mathbf{v}$

Proof   To prove these properties, you are restricted to using the ten vector space axioms. For instance, to prove the second property, note from axiom 4 that $\mathbf{0} = \mathbf{0} + \mathbf{0}$. This allows you to write the following steps.

$c\mathbf{0} = c(\mathbf{0} + \mathbf{0})$	**Additive identity**
$c\mathbf{0} = c\mathbf{0} + c\mathbf{0}$	**Left distributive property**
$c\mathbf{0} + (-c\mathbf{0}) = (c\mathbf{0} + c\mathbf{0}) + (-c\mathbf{0})$	**Add $-c\mathbf{0}$ to both sides.**
$c\mathbf{0} + (-c\mathbf{0}) = c\mathbf{0} + [c\mathbf{0} + (-c\mathbf{0})]$	**Associative property**
$\mathbf{0} = c\mathbf{0} + \mathbf{0}$	**Additive inverse**
$\mathbf{0} = c\mathbf{0}$	**Additive identity**

To prove the third property, suppose that $c\mathbf{v} = \mathbf{0}$. To show that this implies either $c = 0$ or $\mathbf{v} = \mathbf{0}$, assume that $c \neq 0$. (If $c = 0$, you have nothing more to prove.) Now, because $c \neq 0$, you can use the reciprocal $1/c$ to show that $\mathbf{v} = \mathbf{0}$ as follows.

$$\mathbf{v} = 1\mathbf{v} = \left(\frac{1}{c}\right)(c)\mathbf{v} = \frac{1}{c}(c\mathbf{v}) = \frac{1}{c}(\mathbf{0}) = \mathbf{0}$$

Note that the last step uses property 2 (the one you just proved). The proofs of the first and fourth properties are left to you. (See Exercises 34 and 35.)

The remaining examples in this section describe some sets (with operations) that *do not* form vector spaces. To show that a set is not a vector space, you need only find one axiom that is not satisfied. For example, if you can find two members of V that do not commute ($\mathbf{u} + \mathbf{v} \neq \mathbf{v} + \mathbf{u}$), then regardless of how many other members of V do commute and how many of the other ten axioms are satisfied, you can still conclude that V is not a vector space.

EXAMPLE 6   *The Set of Integers Is Not a Vector Space*

The set of all integers (with the standard operations) does not form a vector space because it is not closed under scalar multiplication. For example,

$$\tfrac{1}{2}(1) = \tfrac{1}{2}.$$
$$\underset{Scalar}{\big\uparrow} \quad \underset{Integer}{} \quad \underset{Noninteger}{}$$

REMARK: In Example 6, notice that a single failure of one of the ten vector space axioms suffices to show that a set is not a vector space.

In Example 4 it was shown that the set of all polynomials of degree 2 or less forms a vector space. You will now see that the set of all polynomials whose degree is exactly 2 does not form a vector space.

EXAMPLE 7   *The Set of Second-Degree Polynomials Is Not a Vector Space*

The set of all second-degree polynomials is not a vector space because it is not closed under addition. To see this, consider the second-degree polynomials

$$p(x) = x^2 \qquad \text{and} \qquad q(x) = -x^2 + x + 1,$$

whose sum is the *first*-degree polynomial

$$p(x) + q(x) = x + 1.$$

The sets in Examples 6 and 7 are not vector spaces because they fail one or both closure axioms. In the next example you will look at a set that passes both tests for closure but still fails to be a vector space.

EXAMPLE 8 *A Set That Is Not a Vector Space*

Let $V = R^2$, the set of all ordered pairs of real numbers, with the standard operation of addition and the following *nonstandard* definition of scalar multiplication:

$$c(x_1, x_2) = (cx_1, 0).$$

Show that V is not a vector space.

Solution   In this example, the operation of scalar multiplication is not the standard one. For instance, the product of the scalar 2 times the ordered pair $(3, 4)$ does not equal $(6, 8)$. Instead, the second component of the product is 0,

$$2(3, 4) = (2 \cdot 3, 0) = (6, 0).$$

This example is interesting because it actually satisfies the first nine axioms of the definition of a vector space (try showing this). The tenth axiom is where you get into trouble. In attempting to verify that axiom, the nonstandard definition of scalar multiplication gives you

$$1(1, 1) = (1, 0) \neq (1, 1).$$

Hence, the tenth axiom is not verified, and the set (together with the two given operations) is not a vector space.

Do not be confused by the notation used for scalar multiplication in Example 8. In writing

$$c(x_1, x_2) = (cx_1, 0),$$

the scalar multiple of (x_1, x_2) by c is *defined* to be $(cx_1, 0)$. As it turns out, this nonstandard definition fails to satisfy the tenth vector space axiom.

SECTION 4.2 Exercises

In Exercises 1–6, describe the zero vector (the additive identity) of the vector space.

1. R^4 **2.** $C(-\infty, \infty)$ **3.** $M_{2,3}$

4. $M_{1,4}$ **5.** P_3 **6.** $M_{2,2}$

In Exercises 7–12, describe the additive inverse of a vector in the vector space.

7. R^4 **8.** $C(-\infty, \infty)$ **9.** $M_{2,3}$

10. $M_{1,4}$ **11.** P_3 **12.** $M_{2,2}$

In Exercises 13–24, determine whether the set, together with the indicated operations, is a vector space. If it is not, identify at least one of the ten vector space axioms that fails.

13. $M_{4,6}$ with the standard operations

14. $M_{1,1}$ with the standard operations

15. The set of all fifth-degree polynomials with the standard operations

16. P, the set of all polynomials, with the standard operations

17. The set $\{(x, x): x$ is a real number$\}$ with the standard operations

18. The set $\{(x, y): x \geq 0, y$ is a real number$\}$ with the standard operations in R^2

19. The set of all 2×2 matrices of the form

$$\begin{bmatrix} a & b \\ c & 0 \end{bmatrix}$$

with the standard operations

20. The set of all 2×2 matrices of the form

$$\begin{bmatrix} a & b \\ c & 1 \end{bmatrix}$$

with the standard operations

21. The set of all 2×2 singular matrices with the standard operations

22. The set of all 2×2 nonsingular matrices with the standard operations

23. The set of all 2×2 diagonal matrices with the standard operations

24. $C[0, 1]$, the set of all continuous functions defined on the interval $[0, 1]$, with the standard operations

25. Rather than use the standard definitions of addition and scalar multiplication in R^2, suppose these two operations are defined as follows.

(a) $(x_1, y_1) + (x_2, y_2) = (x_1 + x_2, y_1 + y_2)$
$$c(x, y) = (cx, y)$$
(b) $(x_1, y_1) + (x_2, y_2) = (x_1, 0)$
$$c(x, y) = (cx, cy)$$
(c) $(x_1, y_1) + (x_2, y_2) = (x_1 + x_2, y_1 + y_2)$
$$c(x, y) = \left(\sqrt{c}\,x, \sqrt{c}\,y\right)$$

With these new definitions, is R^2 a vector space? Justify your answers.

26. Rather than use the standard definitions of addition and scalar multiplication in R^3, suppose these two operations are defined as follows.

(a) $(x_1, y_1, z_1) + (x_2, y_2, z_2) = (x_1 + x_2, y_1 + y_2, z_1 + z_2)$
$$c(x, y, z) = (cx, cy, 0)$$
(b) $(x_1, y_1, z_1) + (x_2, y_2, z_2) = (0, 0, 0)$
$$c(x, y, z) = (cx, cy, cz)$$
(c) $(x_1, y_1, z_1) + (x_2, y_2, z_2)$
$$= (x_1 + x_2 + 1, y_1 + y_2 + 1, z_1 + z_2 + 1)$$
$$c(x, y, z) = (cx, cy, cz)$$
(d) $(x_1, y_1, z_1) + (x_2, y_2, z_2)$
$$= (x_1 + x_2 + 1, y_1 + y_2 + 1, z_1 + z_2 + 1)$$
$$c(x, y, z) = (cx + c - 1, cy + c - 1, cz + c - 1)$$

With these new definitions, is R^3 a vector space? Justify your answers.

27. Prove in full detail that $M_{2,2}$, with the standard operations, is a vector space.

28. Prove in full detail that the set $\{(x, 2x): x \text{ is a real number}\}$, with the standard operations in R^2, is a vector space.

29. Determine whether the set R^2, with the operations
$$(x_1, y_1) + (x_2, y_2) = (x_1 x_2, y_1 y_2)$$
and
$$c(x_1, y_1) = (cx_1, cy_1),$$
is a vector space. If it is, verify each vector space axiom; if not, state all vector space axioms that fail.

30. Let V be the set of all positive real numbers. Determine whether V is a vector space with the following operations.

$x + y = xy$	Addition
$cx = x^c$	Scalar multiplication

If it is, verify each vector space axiom; if not, state all vector space axioms that fail.

True or False?   In Exercises 31 and 32, determine whether each statement is true or false. If a statement is true, give a reason or cite an appropriate statement in the text. If a statement is false, provide an example that shows that the statement is not true in all cases or cite an appropriate statement in the text.

31. (a) A vector space consists of four entities: a set of vectors, a set of scalars, and two operations.

(b) The set of all integers with the standard operations is a vector space.

(c) The set of all pairs of real numbers of the form (x, y), where $y \geq 0$, with the standard operations on R^2 is a vector space.

32. (a) It is sufficient to show that a set is not a vector space by showing that just one axiom is not satisfied.

(b) The set of all first-degree polynomials with the standard operations is a vector space.

(c) The set of all pairs of real numbers of the form $(0, y)$, with the standard operations on R^2, is a vector space.

33. Complete the proof of the cancellation property of vector addition by supplying the justification for each step.

Prove that if $\mathbf{u}$, $\mathbf{v}$, and $\mathbf{w}$ are vectors in a vector space V such that $\mathbf{u} + \mathbf{w} = \mathbf{v} + \mathbf{w}$, then $\mathbf{u} = \mathbf{v}$.

$\mathbf{u} + \mathbf{w} = \mathbf{v} + \mathbf{w}$	Given
$(\mathbf{u} + \mathbf{w}) + (-\mathbf{w}) = (\mathbf{v} + \mathbf{w}) + (-\mathbf{w})$	a. _____
$\mathbf{u} + (\mathbf{w} + (-\mathbf{w})) = \mathbf{v} + (\mathbf{w} + (-\mathbf{w}))$	b. _____
$\mathbf{u} + \mathbf{0} = \mathbf{v} + \mathbf{0}$	c. _____
$\mathbf{u} = \mathbf{v}$	d. _____

34. Prove property 1 of Theorem 4.4.

35. Prove property 4 of Theorem 4.4.

36. Prove that in a given vector space V, the zero vector is unique.

37. Prove that in a given vector space V, the additive inverse of a vector is unique.

4.3 | Subspaces of Vector Spaces

In most important applications in linear algebra, vector spaces occur as **subspaces** of larger spaces. For instance, you will see that the solution set of a homogeneous system of linear equations in n variables is a subspace of R^n. (See Theorem 4.16.)

A subset of a vector space is a subspace if it is itself a vector space (with the *same* operations), as stated in the following definition.

Definition of Subspace of a Vector Space	A nonempty subset W of a vector space V is called a **subspace** of V if W is itself a vector space under the operations of addition and scalar multiplication defined in V.

REMARK: Note that if W is a subspace of V, it must be closed under the operations inherited from V.

EXAMPLE 1   *A Subspace of R^3*

Show that the set $W = \{(x_1, 0, x_3) : x_1 \text{ and } x_3 \text{ are real numbers}\}$ is a subspace of R^3 with the standard operations.

Solution   The set W is nonempty because it contains the zero vector $(0, 0, 0)$.

Graphically, the set W can be interpreted as simply the xz-plane, as shown in Figure 4.9. The set W is closed under addition because the sum of any two vectors in the xz-plane must also lie in the xz-plane. That is, if $(x_1, 0, x_3)$ and $(y_1, 0, y_3)$ are in W, then their sum $(x_1 + y_1, 0, x_3 + y_3)$ is also in W (because the second component is zero). Similarly, to see that W is closed under scalar multiplication, let $(x_1, 0, x_3)$ be in W and let c be a scalar. Then $c(x_1, 0, x_3) = (cx_1, 0, cx_3)$ has zero as its second component and must therefore be in W. The other eight vector space axioms can be verified as well, and these verifications are left to you.

Figure 4.9

To establish that a set W is a vector space, you must verify all ten vector space properties. However, if W is a subset of a larger vector space V (and the operations defined on W are the *same* as those defined on V), then most of the ten properties are *inherited* from the larger space and need no verification. The following theorem tells us that it is sufficient to test for closure in order to establish that a nonempty subset of a vector space is a subspace.

Theorem 4.5	*Test for a Subspace*

If W is a nonempty subset of a vector space V, then W is a subspace of V if and only if the following closure conditions hold.
1. If $\mathbf{u}$ and $\mathbf{v}$ are in W, then $\mathbf{u} + \mathbf{v}$ is in W.
2. If $\mathbf{u}$ is in W and c is any scalar, then $c\mathbf{u}$ is in W.

Proof   The proof of the theorem in one direction is straightforward. That is, if W is a subspace of V, then W is a vector space and must be closed under addition and scalar multiplication.

To prove the theorem in the other direction, assume that W is closed under addition and scalar multiplication. Note that if $\mathbf{u}$, $\mathbf{v}$, and $\mathbf{w}$ are in W, then they are also in V. Consequently, vector space axioms 2, 3, 7, 8, 9, and 10 are satisfied automatically. Furthermore, because W is closed under addition and scalar multiplication, it follows that for any $\mathbf{v}$ in W and scalar $c = 0$,

$$c\mathbf{v} = \mathbf{0}$$

and

$$(-1)\mathbf{v} = -\mathbf{v}.$$

Both lie in W, therefore satisfying the remaining axioms (4 and 5).

REMARK: Note that if W is a subspace of a vector space V, then both W and V must have the same zero vector $\mathbf{0}$. (See Exercise 27.)

Because a subspace of a vector space is itself a vector space, it must contain the zero vector. In fact, the simplest subspace of a vector space is the one consisting of only the zero vector,

$$W = \{\mathbf{0}\}.$$

This subspace is called the **zero subspace.** Another obvious subspace of V is V itself. Every vector space contains these two trivial subspaces, and subspaces other than these two are called **proper** (or nontrivial) subspaces.

EXAMPLE 2   *A Subspace of $M_{2,2}$*

Let W be the set of all 2×2 symmetric matrices. Show that W is a subspace of the vector space $M_{2,2}$, with the standard operations of matrix addition and scalar multiplication.

Solution   Recall that a matrix is called *symmetric* if it is equal to its own transpose. Because $M_{2,2}$ is a vector space, you need only show that W (a subset of $M_{2,2}$) satisfies the conditions of Theorem 4.5. Begin by observing that W is *nonempty*. W is closed under addition because $A_1 = A_1^T$ and $A_2 = A_2^T$ imply that

$$(A_1 + A_2)^T = A_1^T + A_2^T = A_1 + A_2.$$

So, if A_1 and A_2 are symmetric matrices of order 2, then so is $A_1 + A_2$. Similarly, W is closed under scalar multiplication because $A = A^T$ implies that

$$(cA)^T = cA^T = cA.$$

Hence, if A is a symmetric matrix of order 2, then so is cA.

The result of Example 2 can be generalized. That is, for any positive integer n, the set of symmetric matrices of order n is a subspace of the vector space $M_{n,n}$ with the standard operations. The next example describes a subset of $M_{n,n}$ that is *not* a subspace.

EXAMPLE 3 *The Set of Singular Matrices Is Not a Subspace of $M_{n,n}$*

Let W be the set of singular matrices of order 2. Show that W is not a subspace of $M_{2,2}$ with the standard operations.

Solution By Theorem 4.5, you can show that a subset W is not a subspace by showing that W is empty, that W is not closed under addition, or that W is not closed under scalar multiplication. For this particular set, W is nonempty and closed under scalar multiplication, but it is not closed under addition. To see this, let A and B be given by

$$A = \begin{bmatrix} 1 & 0 \\ 0 & 0 \end{bmatrix} \quad \text{and} \quad B = \begin{bmatrix} 0 & 0 \\ 0 & 1 \end{bmatrix}.$$

Then A and B are both singular (noninvertible), but their sum

$$A + B = \begin{bmatrix} 1 & 0 \\ 0 & 1 \end{bmatrix}$$

is nonsingular (invertible). Therefore W is not closed under addition, and by Theorem 4.5 you can conclude that it is not a subspace of $M_{2,2}$.

EXAMPLE 4 *The Set of First-Quadrant Vectors Is Not a Subspace of R^2*

Show that $W = \{(x_1, x_2): x_1 \geq 0 \text{ and } x_2 \geq 0\}$, with the standard operations, is not a subspace of R^2.

Solution This set is nonempty and closed under addition. It is not, however, closed under scalar multiplication. To see this, note that $(1, 1)$ is in W, but the scalar multiple

$$(-1)(1, 1) = (-1, -1)$$

is not in W. Therefore W is not a subspace of R^2.

You will often encounter sequences of subspaces that are nested within each other. For instance, consider the vector spaces

$$P_0, P_1, P_2, P_3, \ldots, \quad \text{and} \quad P_n,$$

where P_k is the set of all polynomials of degree less than or equal to k, with the standard operations. It is easy to show that if $j \leq k$, then P_j is a subspace of P_k. Thus you can write

$$P_0 \subset P_1 \subset P_2 \subset P_3 \subset \cdots \subset P_n.$$

Another nesting of subspaces is described in Example 5.

EXAMPLE 5   *Subspaces of Functions (Calculus)*

Let W_5 be the *vector space* of all functions defined on $[0, 1]$, and let W_1, W_2, W_3, and W_4 be defined as follows.

W_1 = set of all polynomial functions defined on the interval $[0, 1]$
W_2 = set of all functions that are differentiable on $[0, 1]$
W_3 = set of all functions that are continuous on $[0, 1]$
W_4 = set of all functions that are integrable on $[0, 1]$

Show that $W_1 \subset W_2 \subset W_3 \subset W_4 \subset W_5$ and that W_i is a subspace of W_j for $i \leq j$.

Solution   From calculus you know that every polynomial function is differentiable on $[0, 1]$. Therefore, $W_1 \subset W_2$. Moreover, $W_2 \subset W_3$ because every differentiable function is continuous, $W_3 \subset W_4$ because every continuous function is integrable, and $W_4 \subset W_5$ because every integrable function is, of course, a function. Therefore you have $W_1 \subset W_2 \subset W_3 \subset W_4 \subset W_5$, as shown in Figure 4.10. The verification that W_i is a subspace of W_j for $i \leq j$ is left to you.

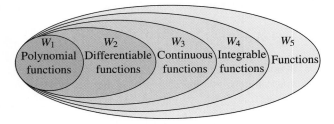

Figure 4.10

Note in Example 5 that if U, V, and W are vector spaces such that W is a subspace of V and V is a subspace of U, then W is also a subspace of U. This special case of the following theorem tells us that the intersection of two subspaces is a subspace, as shown in Figure 4.11.

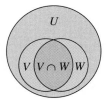

Figure 4.11   The intersection of two subspaces is a subspace.

Theorem 4.6   *The Intersection of Two Subspaces Is a Subspace*

If V and W are both subspaces of a vector space U, then the intersection of V and W (denoted by $V \cap W$) is also a subspace of U.

Proof   Because V and W are both subspaces of U, you know that both contain the zero vector. Thus $V \cap W$ is nonempty. To show that $V \cap W$ is closed under addition, let $\mathbf{v}_1$ and $\mathbf{v}_2$ be any two vectors in $V \cap W$. Then, because V and W are both subspaces of U, you know that both are closed under addition. Therefore, because $\mathbf{v}_1$ and $\mathbf{v}_2$ are both in V, their sum $\mathbf{v}_1 + \mathbf{v}_2$ must be in V. Similarly, $\mathbf{v}_1 + \mathbf{v}_2$ is in W because $\mathbf{v}_1$ and $\mathbf{v}_2$ are both in W. But this implies that $\mathbf{v}_1 + \mathbf{v}_2$ is in $V \cap W$, and it follows that $V \cap W$ is closed under addition. It is left to you to show (by a similar argument) that $V \cap W$ is closed under scalar multiplication. (See Exercise 32.)

REMARK: Theorem 4.6 states that the *intersection* of two subspaces is a subspace. In Exercise 28 you are asked to show that the *union* of two subspaces is not (in general) a subspace.

Subspace of R^n

R^n is a convenient source for examples of vector spaces, and the remainder of this section is devoted to looking at subspaces of R^n.

EXAMPLE 6   *Determining Subspaces of R^2*

Which of these two subsets is a subspace of R^2?
(a) The set of points on the line given by $x + 2y = 0$
(b) The set of points on the line given by $x + 2y = 1$

Solution   (a) Solving for x, you can see that a point in R^2 is on the line $x + 2y = 0$ if and only if it has the form $(-2t, t)$, where t is any real number. (See Figure 4.12.)

To show that this set is closed under addition, let

$$\mathbf{v}_1 = (-2t_1, t_1) \quad \text{and} \quad \mathbf{v}_2 = (-2t_2, t_2)$$

be any two points on the line. Then you have

$$\begin{aligned}
\mathbf{v}_1 + \mathbf{v}_2 &= (-2t_1, t_1) + (-2t_2, t_2) \\
&= (-2(t_1 + t_2), t_1 + t_2) \\
&= (-2t_3, t_3),
\end{aligned}$$

where $t_3 = t_1 + t_2$. Thus $\mathbf{v}_1 + \mathbf{v}_2$ lies on the line, and the set is closed under addition. In a similar way, you can show that the set is closed under scalar multiplication. Therefore, this set *is* a subspace of R^2.

(b) This subset of R^2 is *not* a subspace of R^2 because every subspace must contain the zero vector, and the zero vector $(0, 0)$ is not on the line. (See Figure 4.12.)

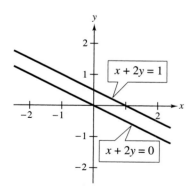

Figure 4.12

Of the two lines in Example 6, the one that is a subspace of R^2 is the one that passes through the origin. It can be shown that this is characteristic of subspaces of R^2. That is, if W is a subset of R^2, then it is a subspace if and only if one of the following is true.
1. W consists of the *single point* $(0, 0)$.
2. W consists of all points on a *line* that passes through the origin.
3. W consists of all of R^2.
These three possibilities are shown graphically in Figure 4.13.

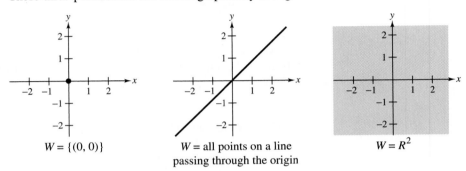

| $W = \{(0, 0)\}$ | W = all points on a line passing through the origin | $W = R^2$ |

Figure 4.13

EXAMPLE 7   *A Subset of R^2 That Is Not a Subspace*

Show that the subset of R^2 that consists of all points on the unit circle $x^2 + y^2 = 1$ is not a subspace.

Solution   This subset of R^2 is *not* a subspace because the points $(1, 0)$ and $(0, 1)$ are in the subset, but their sum $(1, 1)$ is not. (See Figure 4.14.) So, this subset is not closed under addition.

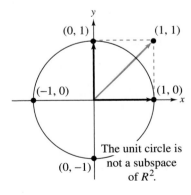

The unit circle is not a subspace of R^2.

Figure 4.14

REMARK: Another way you can tell that the subset shown in Figure 4.14 is not a subspace of R^2 is by noting that it does not contain the zero vector (the origin).

EXAMPLE 8   *Determining Subspaces of R^3*

Which of the following subsets is a subspace of R^3?
(a) $W = \{(x_1, x_2, 1): x_1 \text{ and } x_2 \text{ are real numbers}\}$
(b) $W = \{(x_1, x_1 + x_3, x_3): x_1 \text{ and } x_3 \text{ are real numbers}\}$

Solution   (a) Because $\mathbf{0} = (0, 0, 0)$ is not in W, you know that W is *not* a subspace of R^3.
(b) This set is nonempty because it contains the zero vector $(0, 0, 0)$.
Let $\mathbf{v} = (v_1, v_1 + v_3, v_3)$ and $\mathbf{u} = (u_1, u_1 + u_3, u_3)$ be two vectors in W, and let c be any real number. Show that W is closed under addition as follows.

$$\begin{aligned} \mathbf{v} + \mathbf{u} &= (v_1 + u_1, v_1 + v_3 + u_1 + u_3, v_3 + u_3) \\ &= (v_1 + u_1, (v_1 + u_1) + (v_3 + u_3), v_3 + u_3) \\ &= (x_1, x_1 + x_3, x_3) \end{aligned}$$

where $x_1 = v_1 + u_1$ and $x_3 = v_3 + u_3$. Hence $\mathbf{v} + \mathbf{u}$ is in W (because it is of the proper form). Similarly, W is closed under scalar multiplication because

$$\begin{aligned} c\mathbf{v} &= (cv_1, c(v_1 + v_3), cv_3) \\ &= (cv_1, cv_1 + cv_3, cv_3) \\ &= (x_1, x_1 + x_3, x_3), \end{aligned}$$

where $x_1 = cv_1$ and $x_3 = cv_3$. Hence $c\mathbf{v}$ is in W. Finally, because W is closed under addition and scalar multiplication, you can conclude that it *is* a subspace of R^3.

In Example 8, note that the graph of each subset is a plane in R^3. However, the only subset that is a *subspace* is the one represented by a plane that passes through the origin. (See Figure 4.15.)

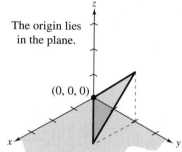

The origin does not lie in the plane.

The origin lies in the plane.

$(0, 0, 0)$

Figure 4.15

In general, you can show that a subset W of R^3 is a subspace of R^3 (with the standard operations) if and only if it has one of the following forms.
1. W consists of the *single point* $(0, 0, 0)$.
2. W consists of all points on a *line* that passes through the origin.
3. W consists of all points in a *plane* that passes through the origin.
4. W consists of all of R^3.

In Exercises 1–6, verify that W is a subspace of V. In each case, assume that V has the standard operations.

1. $W = \{(x_1, x_2, x_3, 0): x_1, x_2, \text{ and } x_3 \text{ are real numbers}\}$
   $V = R^4$

2. $W = \{(x, y, 2x - 3y): x \text{ and } y \text{ are real numbers}\}$
   $V = R^3$

3. W is the set of all 2×2 matrices of the form
   $$\begin{bmatrix} 0 & a \\ b & 0 \end{bmatrix}.$$
   $V = M_{2,2}$

4. W is the set of all 3×2 matrices of the form
   $$\begin{bmatrix} a & b \\ a + b & 0 \\ 0 & c \end{bmatrix}.$$
   $V = M_{3,2}$

5. *Calculus* W is the set of all functions that are continuous on $[0, 1]$. V is the set of all functions that are integrable on $[0, 1]$.

6. *Calculus* W is the set of all functions that are differentiable on $[0, 1]$. V is the set of all functions that are continuous on $[0, 1]$.

In Exercises 7–12, W is not a subspace of the vector space. Verify this by giving a specific example that violates the test for a vector subspace (Theorem 4.5).

7. W is the set of all vectors in R^3 whose third component is -1.

8. W is the set of all vectors in R^2 whose components are rational numbers.

9. W is the set of all nonnegative functions in $C(-\infty, \infty)$.

10. W is the set of all vectors in R^3 whose components are nonnegative.

11. W is the set of all matrices in $M_{n,n}$ with zero determinants.

12. W is the set of all matrices in $M_{n,n}$ such that $A^2 = A$.

13. Which of the following subsets of $C(-\infty, \infty)$ are subspaces of $C(-\infty, \infty)$?

    (a) The set of all nonnegative functions: $f(x) \geq 0$
    (b) The set of all even functions: $f(-x) = f(x)$
    (c) The set of all odd functions: $f(-x) = -f(x)$
    (d) The set of all constant functions: $f(x) = c$
    (e) The set of all functions such that $f(0) = 0$
    (f) The set of all functions such that $f(0) = 1$

14. Which of the following subsets of $M_{n,n}$ are subspaces of $M_{n,n}$ with the standard operations?

    (a) The set of all $n \times n$ upper triangular matrices
    (b) The set of all $n \times n$ matrices with integer entries
    (c) The set of all $n \times n$ matrices A that commute with a given matrix B
    (d) The set of all $n \times n$ singular matrices
    (e) The set of all $n \times n$ invertible matrices
    (f) The set of all $n \times n$ matrices whose entries add up to zero

In Exercises 15–20, determine whether the set W is a subspace of R^3 with the standard operations. Justify your answer.

15. $W = \{(x_1, 0, x_3): x_1 \text{ and } x_3 \text{ are real numbers}\}$

16. $W = \{(x_1, x_2, 4): x_1 \text{ and } x_2 \text{ are real numbers}\}$

17. $W = \{(a, b, a + 2b): a \text{ and } b \text{ are real numbers}\}$

18. $W = \{(s, s - t, t): s \text{ and } t \text{ are real numbers}\}$

19. $W = \{(x_1, x_2, x_1 x_2): x_1 \text{ and } x_2 \text{ are real numbers}\}$

20. $W = \{(x_1, 1/x_1, x_3): x_1 \text{ and } x_3 \text{ are real numbers, } x_1 \neq 0\}$

True or False?   In Exercises 21 and 22, determine whether each statement is true or false. If a statement is true, give a reason or cite an appropriate statement in the text. If a statement is false, provide an example that shows that the statement is not true in all cases or cite an appropriate statement in the text.

21. (a) Every vector space V contains at least one subspace that is the zero subspace.
    (b) If V and W are both subspaces of a vector space U, then the intersection of V and W is also a subspace.
    (c) If U, V, and W are vector spaces such that W is a subspace of V and U is a subspace of V, then $W = U$.

22. (a) Every vector space V contains two proper subspaces that are the zero subspace and itself.
    (b) If W is a subspace of R^2, then W must contain the vector $(0, 0)$.
    (c) If W and U are subspaces of a vector space V, then the union of W and U is a subspace of V.

23. *Guided Proof*   Prove that a nonempty set W is a subspace of a vector space V if and only if $ax + by$ is an element of W for all scalars a and b and all vectors x and y in W.

    Getting Started: In one direction, assume that W is a subspace, and show by using closure axioms that $ax + by$ lies in W. In the other direction, assume that $ax + by$ is an element of W for all real numbers a and b and elements x and y in W, and verify that W is closed under addition and scalar multiplication.

(i) If W is a subspace of V, then use scalar multiplication closure to show that $a\mathbf{x}$ and $b\mathbf{y}$ are in W. Now use additive closure to get the desired result.

(ii) Conversely, assume that $a\mathbf{x} + b\mathbf{y}$ is in W. By cleverly assigning specific values to a and b, show that W is closed under addition and scalar multiplication.

24. Let $\mathbf{x}$, $\mathbf{y}$, and $\mathbf{z}$ be vectors in a vector space V. Show that the set of all linear combinations of $\mathbf{x}$, $\mathbf{y}$, and $\mathbf{z}$

$$W = \{a\mathbf{x} + b\mathbf{y} + c\mathbf{z}: a, b, \text{ and } c \text{ are scalars}\}$$

is a subspace of V. This subspace is called the **span** of $\{\mathbf{x}, \mathbf{y}, \mathbf{z}\}$.

25. Let A be a fixed 2×3 matrix. Prove that the set

$$W = \left\{\mathbf{x} \in R^3 : A\mathbf{x} = \begin{bmatrix} 1 \\ 2 \end{bmatrix}\right\}$$

is not a subspace of R^3.

26. Let A be a fixed $m \times n$ matrix. Prove that the set

$$W = \{\mathbf{x} \in R^n : A\mathbf{x} = \mathbf{0}\}$$

is a subspace of R^n.

27. Let W be a subspace of the vector space V. Prove that the zero vector in V is also the zero vector in W.

28. Give an example showing that the union of two subspaces of a vector space V is not necessarily a subspace of V.

29. Let A be a fixed 2×2 matrix. Prove that the set

$$W = \{X : XA = AX\}$$

is a subspace of $M_{2,2}$.

30. *Calculus* Determine whether the set

$$S = \left\{f \in C[0, 1]: \int_0^1 f(x)\, dx = 0\right\}$$

is a subspace of $C[0, 1]$. Prove your answer.

31. Let V and W be two subspaces of a vector space U. Prove that the set

$$V + W = \{\mathbf{u} : \mathbf{u} = \mathbf{v} + \mathbf{w}, \text{ where } \mathbf{v} \in V \text{ and } \mathbf{w} \in W\}$$

is a subspace of U. Describe $V + W$ if V and W are the following subspaces of $U = R^2$:

$V = \{(x, 0) : x \text{ is a real number}\}$ and $W = \{(0, y) : y \text{ is a real number}\}$.

32. Complete the proof of Theorem 4.6 by showing that the intersection of two subspaces of a vector space is closed under scalar multiplication.

4.4 | Spanning Sets and Linear Independence

This section begins to develop procedures for representing each vector in a vector space as a **linear combination** of a select number of vectors in the space.

Definition of Linear Combination of Vectors

A vector $\mathbf{v}$ in a vector space V is called a **linear combination** of the vectors $\mathbf{u}_1, \mathbf{u}_2, \ldots, \mathbf{u}_k$ in V if $\mathbf{v}$ can be written in the form

$$\mathbf{v} = c_1\mathbf{u}_1 + c_2\mathbf{u}_2 + \cdots + c_k\mathbf{u}_k,$$

where $c_1, c_2, \ldots, c_k$ are scalars.

Often, one or more of the vectors in a given set can be written as linear combinations of other vectors in the set. Examples 1, 2, and 3 illustrate this possibility.

EXAMPLE 1 *Examples of Linear Combinations*

(a) For the set of vectors in R^3,

$$S = \{\overset{\mathbf{v}_1}{(1, 3, 1)}, \overset{\mathbf{v}_2}{(0, 1, 2)}, \overset{\mathbf{v}_3}{(1, 0, -5)}\},$$

v_1 is a linear combination of v_2 and v_3 because

$$v_1 = 3v_2 + v_3 = 3(0, 1, 2) + (1, 0, -5)$$
$$= (1, 3, 1).$$

(b) For the set of vectors in $M_{2,2}$,

$$S = \left\{ \overset{v_1}{\begin{bmatrix} 0 & 8 \\ 2 & 1 \end{bmatrix}}, \overset{v_2}{\begin{bmatrix} 0 & 2 \\ 1 & 0 \end{bmatrix}}, \overset{v_3}{\begin{bmatrix} -1 & 3 \\ 1 & 2 \end{bmatrix}}, \overset{v_4}{\begin{bmatrix} -2 & 0 \\ 1 & 3 \end{bmatrix}} \right\},$$

v_1 is a linear combination of v_2, v_3, and v_4 because

$$v_1 = v_2 + 2v_3 - v_4$$
$$= \begin{bmatrix} 0 & 2 \\ 1 & 0 \end{bmatrix} + 2\begin{bmatrix} -1 & 3 \\ 1 & 2 \end{bmatrix} - \begin{bmatrix} -2 & 0 \\ 1 & 3 \end{bmatrix}$$
$$= \begin{bmatrix} 0 & 8 \\ 2 & 1 \end{bmatrix}.$$

In Example 1 it was easy to verify that one of the vectors in the set S was a linear combination of the other vectors because you were given the appropriate coefficients to form the linear combination. In the next example, a procedure for finding the coefficients is demonstrated.

EXAMPLE 2   *Finding a Linear Combination*

Write the vector $w = (1, 1, 1)$ as a linear combination of vectors in the set S.

$$S = \{\overset{v_1}{(1, 2, 3)}, \overset{v_2}{(0, 1, 2)}, \overset{v_3}{(-1, 0, 1)}\}$$

Solution   You need to find scalars c_1, c_2, and c_3 such that

$$(1, 1, 1) = c_1(1, 2, 3) + c_2(0, 1, 2) + c_3(-1, 0, 1)$$
$$= (c_1 - c_3, 2c_1 + c_2, 3c_1 + 2c_2 + c_3).$$

Equating corresponding components yields the following system of linear equations.

$$\begin{aligned} c_1 \quad\quad - c_3 &= 1 \\ 2c_1 + c_2 \quad\quad &= 1 \\ 3c_1 + 2c_2 + c_3 &= 1 \end{aligned}$$

Using Gauss-Jordan elimination, you can show that this system has an infinite number of solutions, each of the form

$$c_1 = 1 + t, \quad\quad c_2 = -1 - 2t, \quad\quad c_3 = t.$$

To obtain one solution, you could let $t = 1$. Then $c_3 = 1$, $c_2 = -3$, and $c_1 = 2$, and you have

$$\mathbf{w} = 2\mathbf{v}_1 - 3\mathbf{v}_2 + \mathbf{v}_3.$$

Other choices for t would yield other ways to write $\mathbf{w}$ as a linear combination of $\mathbf{v}_1$, $\mathbf{v}_2$, and $\mathbf{v}_3$.

EXAMPLE 3   *Finding a Linear Combination*

If possible, write the vector $\mathbf{w} = (1, -2, 2)$ as a linear combination of vectors in the set S given in Example 2.

Solution   Following the procedure given in Example 2 results in the system

$$\begin{array}{rcrcrcr} c_1 & & & - & c_3 & = & 1 \\ 2c_1 & + & c_2 & & & = & -2 \\ 3c_1 & + & 2c_2 & + & c_3 & = & 2. \end{array}$$

The augmented matrix of this system row reduces to

$$\begin{bmatrix} 1 & 0 & -1 & 1 \\ 0 & 1 & 2 & -4 \\ 0 & 0 & 0 & 7 \end{bmatrix}.$$

From the third row you can conclude that the system of equations is inconsistent, and therefore there is no solution. Consequently, $\mathbf{w}$ *cannot* be written as a linear combination of $\mathbf{v}_1$, $\mathbf{v}_2$, and $\mathbf{v}_3$.

Spanning Sets

If every vector in a given vector space can be written as a linear combination of vectors in a given set S, then S is called a **spanning set** of the vector space.

Definition of Spanning Set of a Vector Space

Let $S = \{\mathbf{v}_1, \mathbf{v}_2, \ldots, \mathbf{v}_k\}$ be a subset of a vector space V. The set S is called a **spanning set** of V if *every* vector in V can be written as a linear combination of vectors in S. In such cases it is said that S **spans** V.

EXAMPLE 4   *Examples of Spanning Sets*

(a) The set $S = \{(1, 0, 0), (0, 1, 0), (0, 0, 1)\}$ spans R^3 because any vector $\mathbf{u} = (u_1, u_2, u_3)$ in R^3 can be written as

$$\mathbf{u} = u_1(1, 0, 0) + u_2(0, 1, 0) + u_3(0, 0, 1) = (u_1, u_2, u_3).$$

(b) The set $S = \{1, x, x^2\}$ spans P_2 because any polynomial $p(x) = a + bx + cx^2$ in P_2 can be written as

$$p(x) = a(1) + b(x) + c(x^2)$$
$$= a + bx + cx^2.$$

The spanning sets given in Example 4 are called the **standard spanning sets** of R^3 and P_2, respectively. (You will learn more about standard spanning sets in the next section.) In the next example you will look at a nonstandard spanning set of R^3.

EXAMPLE 5 *A Spanning Set for R^3*

Show that the set $S = \{(1, 2, 3), (0, 1, 2), (-2, 0, 1)\}$ spans R^3.

Solution Let $\mathbf{u} = (u_1, u_2, u_3)$ be *any* vector in R^3. You need to find scalars c_1, c_2, and c_3 such that

$$(u_1, u_2, u_3) = c_1(1, 2, 3) + c_2(0, 1, 2) + c_3(-2, 0, 1)$$
$$= (c_1 - 2c_3, 2c_1 + c_2, 3c_1 + 2c_2 + c_3).$$

This vector equation produces the system

$$\begin{array}{rcl} c_1 \quad\quad - 2c_3 &=& u_1 \\ 2c_1 + c_2 \quad\quad &=& u_2 \\ 3c_1 + 2c_2 + c_3 &=& u_3. \end{array}$$

The coefficient matrix for this system has a nonzero determinant, and it follows from the list of equivalent conditions given in Section 3.3 that the system has a unique solution. Therefore, any vector in R^3 can be written as a linear combination of the vectors in S, and you can conclude that the set S spans R^3.

EXAMPLE 6 *A Set That Does Not Span R^3*

From Example 3 you know that the set

$$S = \{(1, 2, 3), (0, 1, 2), (-1, 0, 1)\}$$

does not span R^3 because $\mathbf{w} = (1, -2, 2)$ is in R^3 and cannot be expressed as a linear combination of the vectors in S.

Comparing the sets of vectors in Examples 5 and 6, note that the sets are the same except for a seemingly insignificant difference in the third vector.

$$S_1 = \{(1, 2, 3), (0, 1, 2), (-2, 0, 1)\} \quad\quad \text{Example 5}$$
$$S_2 = \{(1, 2, 3), (0, 1, 2), (-1, 0, 1)\} \quad\quad \text{Example 6}$$

The difference, however, is significant, because the set S_1 spans R^3 whereas the set S_2 does not. The reason for this difference can be seen in Figure 4.16. The vectors in S_2 lie in a common plane; the vectors in S_1 do not.

$S_1 = \{(1, 2, 3), (0, 1, 2), (-2, 0, 1)\}$
The vectors in S_1 do not lie
in a common plane.

$S_2 = \{(1, 2, 3), (0, 1, 2), (-1, 0, 1)\}$
The vectors in S_2 lie in a
common plane.

Figure 4.16

Although the set S_2 does not span all of R^3, it does span a subspace of R^3—namely, the plane in which the three vectors of S_2 lie. This subspace is called the **span of S_2,** as indicated in the following definition.

Definition of the Span of a Set

If $S = \{\mathbf{v}_1, \mathbf{v}_2, \ldots, \mathbf{v}_k\}$ is a set of vectors in a vector space V, then the **span of S** is the set of all linear combinations of the vectors in S,

$$\text{span}(S) = \{c_1\mathbf{v}_1 + c_2\mathbf{v}_2 + \cdots + c_k\mathbf{v}_k : c_1, c_2, \ldots, c_k \text{ are real numbers}\}.$$

The span of S is denoted by $\text{span}(S)$ or $\text{span}\{\mathbf{v}_1, \mathbf{v}_2, \ldots, \mathbf{v}_k\}$. If $\text{span}(S) = V$, it is said that V is **spanned** by $\{\mathbf{v}_1, \mathbf{v}_2, \ldots, \mathbf{v}_k\}$, or that S **spans** V.

The following theorem tells you that the span of any finite nonempty subset of a vector space V is a subspace of V.

Theorem 4.7

Span(S) Is a Subspace of V

If $S = \{\mathbf{v}_1, \mathbf{v}_2, \ldots, \mathbf{v}_k\}$ is a set of vectors in a vector space V, then $\text{span}(S)$ is a subspace of V. Moreover, $\text{span}(S)$ is the smallest subspace of V that contains S, in the sense that every other subspace of V that contains S must contain $\text{span}(S)$.

Proof   To show that span(S), the set of all linear combinations of $\mathbf{v}_1, \mathbf{v}_2, \ldots, \mathbf{v}_k$, is a subspace of V, show that it is closed under addition and scalar multiplication. Consider any two vectors $\mathbf{u}$ and $\mathbf{v}$ in span(S),

$$\mathbf{u} = c_1\mathbf{v}_1 + c_2\mathbf{v}_2 + \cdots + c_k\mathbf{v}_k$$
$$\mathbf{v} = d_1\mathbf{v}_1 + d_2\mathbf{v}_2 + \cdots + d_k\mathbf{v}_k,$$

where $c_1, c_2, \ldots, c_k$ and $d_1, d_2, \ldots, d_k$ are scalars. Then

$$\mathbf{u} + \mathbf{v} = (c_1 + d_1)\mathbf{v}_1 + (c_2 + d_2)\mathbf{v}_2 + \cdots + (c_k + d_k)\mathbf{v}_k$$

and

$$c\mathbf{u} = (cc_1)\mathbf{v}_1 + (cc_2)\mathbf{v}_2 + \cdots + (cc_k)\mathbf{v}_k,$$

which means that $\mathbf{u} + \mathbf{v}$ and $c\mathbf{u}$ are also in span(S) because they can be written as linear combinations of vectors in S. Therefore, span(S) is a subspace of V. It is left to you to prove that span(S) is the smallest subspace of V that contains S. (See Exercise 40.)

Linear Dependence and Linear Independence

For a given set of vectors $S = \{\mathbf{v}_1, \mathbf{v}_2, \ldots, \mathbf{v}_k\}$ in a vector space V, the vector equation

$$c_1\mathbf{v}_1 + c_2\mathbf{v}_2 + \cdots + c_k\mathbf{v}_k = \mathbf{0}$$

always has the **trivial solution**

$$c_1 = 0, c_2 = 0, \ldots, c_k = 0.$$

Often, however, there are also **nontrivial** solutions. For instance, in Example 1(a) you saw that in the set

$$S = \{\overset{\mathbf{v}_1}{(1, 3, 1)}, \overset{\mathbf{v}_2}{(0, 1, 2)}, \overset{\mathbf{v}_3}{(1, 0, -5)}\},$$

the vector $\mathbf{v}_1$ can be written as a linear combination of the other two as follows.

$$\mathbf{v}_1 = 3\mathbf{v}_2 + \mathbf{v}_3$$

Hence the vector equation

$$c_1\mathbf{v}_1 + c_2\mathbf{v}_2 + c_3\mathbf{v}_3 = \mathbf{0}$$

has a nontrivial solution in which the coefficients are *not all zero:*

$$c_1 = 1, \qquad c_2 = -3, \qquad c_3 = -1.$$

This characteristic is described by saying that the set S is **linearly dependent.** Had the *only* solution been the trivial one ($c_1 = c_2 = c_3 = 0$), then the set S would have been **linearly independent.** This notion is essential to the study of linear algebra, and is stated formally in the following definition.

Definition of Linear Dependence and Linear Independence

A set of vectors $S = \{v_1, v_2, \ldots, v_k\}$ in a vector space V is called **linearly independent** if the vector equation

$$c_1v_1 + c_2v_2 + \cdots + c_kv_k = 0$$

has only the trivial solution, $c_1 = 0, c_2 = 0, \ldots, c_k = 0$. If there are also nontrivial solutions, then S is called **linearly dependent.**

EXAMPLE 7   *Examples of Linearly Dependent Sets*

(a) The set $S = \{(1, 2), (2, 4)\}$ in R^2 is linearly dependent because

$$-2(1, 2) + (2, 4) = (0, 0).$$

(b) The set $S = \{(1, 0), (0, 1), (-2, 5)\}$ in R^2 is linearly dependent because

$$2(1, 0) - 5(0, 1) + (-2, 5) = (0, 0).$$

(c) The set $S = \{(0, 0), (1, 2)\}$ in R^2 is linearly dependent because

$$1(0, 0) + 0(1, 2) = (0, 0).$$

The next example demonstrates a procedure for testing whether a set of vectors is linearly independent or linearly dependent.

EXAMPLE 8   *Testing for Linear Independence*

Determine whether the set of vectors in R^3 is linearly independent or linearly dependent.

$$S = \{\overset{v_1}{(1, 2, 3)}, \overset{v_2}{(0, 1, 2)}, \overset{v_3}{(-2, 0, 1)}\}$$

Solution   To test for linear independence or linear dependence, form the vector equation

$$c_1v_1 + c_2v_2 + c_3v_3 = 0.$$

If the only solution of this equation is

$$c_1 = c_2 = c_3 = 0,$$

then the set S is linearly independent. Otherwise, S is linearly dependent. Expanding this equation, you have

$$c_1(1, 2, 3) + c_2(0, 1, 2) + c_3(-2, 0, 1) = (0, 0, 0)$$
$$(c_1 - 2c_3, 2c_1 + c_2, 3c_1 + 2c_2 + c_3) = (0, 0, 0),$$

which yields the following homogeneous system of linear equations in c_1, c_2, and c_3.

$$\begin{aligned} c_1 \qquad\quad - 2c_3 &= 0 \\ 2c_1 + c_2 \qquad\quad &= 0 \\ 3c_1 + 2c_2 + c_3 &= 0 \end{aligned}$$

The augmented matrix of this system reduces by Gauss-Jordan elimination as follows.

$$
\begin{bmatrix}
1 & 0 & -2 & 0 \\
2 & 1 & 0 & 0 \\
3 & 2 & 1 & 0
\end{bmatrix}
\longrightarrow
\begin{bmatrix}
1 & 0 & 0 & 0 \\
0 & 1 & 0 & 0 \\
0 & 0 & 1 & 0
\end{bmatrix}
$$

This implies that the only solution is the trivial solution

$$c_1 = c_2 = c_3 = 0.$$

Therefore, S is linearly independent.

The steps shown in Example 8 are summarized as follows.

Testing for Linear Independence and Linear Dependence

Let $S = \{v_1, v_2, \ldots, v_k\}$ be a set of vectors in a vector space V. To determine whether S is linearly independent or linearly dependent, perform the following steps.
1. From the vector equation $c_1v_1 + c_2v_2 + \cdots + c_kv_k = 0$, write a homogeneous system of linear equations in the variables $c_1, c_2, \ldots,$ and c_k.
2. Use Gaussian elimination to determine whether the system has a unique solution.
3. If the system has only the trivial solution, $c_1 = 0, c_2 = 0, \ldots, c_k = 0$, then the set S is linearly independent. If the system also has nontrivial solutions, then S is linearly dependent.

EXAMPLE 9   *Testing for Linear Independence*

Determine whether the set of vectors in P_2 is linearly independent or linearly dependent.

$$S = \{\overset{v_1}{1 + x - 2x^2}, \overset{v_2}{2 + 5x - x^2}, \overset{v_3}{x + x^2}\}$$

Solution   Expanding the equation $c_1v_1 + c_2v_2 + c_3v_3 = 0$ produces

$$c_1(1 + x - 2x^2) + c_2(2 + 5x - x^2) + c_3(x + x^2) = 0 + 0x + 0x^2$$
$$(c_1 + 2c_2) + (c_1 + 5c_2 + c_3)x + (-2c_1 - c_2 + c_3)x^2 = 0 + 0x + 0x^2.$$

Equating corresponding coefficients of equal powers of x produces the following homogeneous system of linear equations in $c_1, c_2,$ and c_3.

$$
\begin{aligned}
c_1 + 2c_2 \quad\quad &= 0 \\
c_1 + 5c_2 + c_3 &= 0 \\
-2c_1 - c_2 + c_3 &= 0
\end{aligned}
$$

The augmented matrix of this system reduces by Gaussian elimination as follows.

$$
\begin{bmatrix}
1 & 2 & 0 & 0 \\
1 & 5 & 1 & 0 \\
-2 & -1 & 1 & 0
\end{bmatrix}
\longrightarrow
\begin{bmatrix}
1 & 2 & 0 & 0 \\
0 & 1 & \frac{1}{3} & 0 \\
0 & 0 & 0 & 0
\end{bmatrix}
$$

This implies that the system has an infinite number of solutions. Therefore, the system must have nontrivial solutions, and you can conclude that the set S is linearly dependent.

One nontrivial solution is

$$c_1 = 2, \quad c_2 = -1, \quad \text{and} \quad c_3 = 3,$$

which yields the nontrivial linear combination

$$(2)(1 + x - 2x^2) + (-1)(2 + 5x - x^2) + (3)(x + x^2) = 0.$$

EXAMPLE 10   *Testing for Linear Independence*

Determine whether the set of vectors in $M_{2,2}$ is linearly independent or linearly dependent.

$$S = \left\{ \overset{\mathbf{v}_1}{\begin{bmatrix} 2 & 1 \\ 0 & 1 \end{bmatrix}}, \overset{\mathbf{v}_2}{\begin{bmatrix} 3 & 0 \\ 2 & 1 \end{bmatrix}}, \overset{\mathbf{v}_3}{\begin{bmatrix} 1 & 0 \\ 2 & 0 \end{bmatrix}} \right\}$$

Solution   From the equation

$$c_1 \mathbf{v}_1 + c_2 \mathbf{v}_2 + c_3 \mathbf{v}_3 = \mathbf{0},$$

you have

$$c_1 \begin{bmatrix} 2 & 1 \\ 0 & 1 \end{bmatrix} + c_2 \begin{bmatrix} 3 & 0 \\ 2 & 1 \end{bmatrix} + c_3 \begin{bmatrix} 1 & 0 \\ 2 & 0 \end{bmatrix} = \begin{bmatrix} 0 & 0 \\ 0 & 0 \end{bmatrix},$$

which produces the following system of linear equations in c_1, c_2, and c_3.

$$\begin{aligned}
2c_1 + 3c_2 + c_3 &= 0 \\
c_1 \qquad\qquad &= 0 \\
2c_2 + 2c_3 &= 0 \\
c_1 + c_2 \qquad &= 0
\end{aligned}$$

Using Gaussian elimination, the augmented matrix of this system reduces as follows.

$$\begin{bmatrix} 2 & 3 & 1 & 0 \\ 1 & 0 & 0 & 0 \\ 0 & 2 & 2 & 0 \\ 1 & 1 & 0 & 0 \end{bmatrix} \longrightarrow \begin{bmatrix} 1 & 0 & 0 & 0 \\ 0 & 1 & 0 & 0 \\ 0 & 0 & 1 & 0 \\ 0 & 0 & 0 & 0 \end{bmatrix}$$

Thus, the system has only the trivial solution and you can conclude that the set S is linearly independent.

EXAMPLE 11 *Testing for Linear Independence*

Determine whether the set of vectors in $M_{4,1}$ is linearly independent or linearly dependent.

$$S = \left\{ \begin{matrix} \mathbf{v}_1 \\ \begin{bmatrix} 1 \\ 0 \\ -1 \\ 0 \end{bmatrix} \end{matrix}, \begin{matrix} \mathbf{v}_2 \\ \begin{bmatrix} 1 \\ 1 \\ 0 \\ 2 \end{bmatrix} \end{matrix}, \begin{matrix} \mathbf{v}_3 \\ \begin{bmatrix} 0 \\ 3 \\ 1 \\ -2 \end{bmatrix} \end{matrix}, \begin{matrix} \mathbf{v}_4 \\ \begin{bmatrix} 0 \\ 1 \\ -1 \\ 2 \end{bmatrix} \end{matrix} \right\}$$

Solution   From the equation

$$c_1 \mathbf{v}_1 + c_2 \mathbf{v}_2 + c_3 \mathbf{v}_3 + c_4 \mathbf{v}_4 = \mathbf{0},$$

you obtain

$$c_1 \begin{bmatrix} 1 \\ 0 \\ -1 \\ 0 \end{bmatrix} + c_2 \begin{bmatrix} 1 \\ 1 \\ 0 \\ 2 \end{bmatrix} + c_3 \begin{bmatrix} 0 \\ 3 \\ 1 \\ -2 \end{bmatrix} + c_4 \begin{bmatrix} 0 \\ 1 \\ -1 \\ 2 \end{bmatrix} = \begin{bmatrix} 0 \\ 0 \\ 0 \\ 0 \end{bmatrix}.$$

This equation produces the following system of linear equations in c_1, c_2, c_3, and c_4.

$$
\begin{aligned}
c_1 + c_2 \qquad\qquad &= 0 \\
c_2 + 3c_3 + c_4 &= 0 \\
-c_1 \qquad + c_3 - c_4 &= 0 \\
2c_2 - 2c_3 + 2c_4 &= 0
\end{aligned}
$$

Using Gaussian elimination, you can row reduce the augmented matrix of this system as follows.

$$
\begin{bmatrix}
1 & 1 & 0 & 0 & 0 \\
0 & 1 & 3 & 1 & 0 \\
-1 & 0 & 1 & -1 & 0 \\
0 & 2 & -2 & 2 & 0
\end{bmatrix}
\longrightarrow
\begin{bmatrix}
1 & 0 & 0 & 0 & 0 \\
0 & 1 & 0 & 0 & 0 \\
0 & 0 & 1 & 0 & 0 \\
0 & 0 & 0 & 1 & 0
\end{bmatrix}
$$

Thus, the system has only the trivial solution, and you can conclude that the set S is linearly independent.

If a set of vectors is linearly dependent, then by definition the equation

$$c_1 \mathbf{v}_1 + c_2 \mathbf{v}_2 + \cdots + c_k \mathbf{v}_k = \mathbf{0}$$

has a nontrivial solution (a solution for which not all the c_i's are zero). For instance, if $c_1 \neq 0$, then you can solve this equation for $\mathbf{v}_1$ and write $\mathbf{v}_1$ as a linear combination of the other vectors $\mathbf{v}_2$, $\mathbf{v}_3$, . . . , and $\mathbf{v}_k$. In other words, the vector $\mathbf{v}_1$ *depends* on the other vectors in the set. This property is characteristic of a linearly dependent set.

Theorem 4.8

A Property of Linearly Dependent Sets

A set $S = \{v_1, v_2, \ldots, v_k\}$, $k \geq 2$, is linearly dependent if and only if at least one of the vectors v_j can be written as a linear combination of the other vectors in S.

Proof   To prove the theorem in one direction, assume that S is a linearly dependent set. Then there exist scalars $c_1, c_2, c_3, \ldots, c_k$ (not all zero) such that

$$c_1v_1 + c_2v_2 + c_3v_3 + \cdots + c_kv_k = 0.$$

Because one of the coefficients must be nonzero, no generality is lost by assuming that $c_1 \neq 0$. Then solving for v_1 as a linear combination of the other vectors produces

$$c_1v_1 = -c_2v_2 - c_3v_3 - \cdots - c_kv_k$$

$$v_1 = -\frac{c_2}{c_1}v_2 - \frac{c_3}{c_1}v_3 - \cdots - \frac{c_k}{c_1}v_k.$$

Conversely, suppose the vector v_1 in S is a linear combination of the other vectors. That is,

$$v_1 = c_2v_2 + c_3v_3 + \cdots + c_kv_k.$$

Then the equation $-v_1 + c_2v_2 + c_3v_3 + \cdots + c_kv_k = 0$ has at least one coefficient, -1, that is nonzero, and you can conclude that S is linearly dependent.

EXAMPLE 12   *Writing a Vector as a Linear Combination of Other Vectors*

In Example 9, you determined that the set

$$S = \{\overset{v_1}{1 + x - 2x^2}, \overset{v_2}{2 + 5x - x^2}, \overset{v_3}{x + x^2}\}$$

is linearly dependent. Show that one of the vectors in this set can be written as a linear combination of the other two.

Solution   In Example 9, the equation $c_1v_1 + c_2v_2 + c_3v_3 = 0$ produced the system

$$\begin{aligned} c_1 + 2c_2 \quad\quad &= 0 \\ c_1 + 5c_2 + c_3 &= 0 \\ -2c_1 - c_2 + c_3 &= 0. \end{aligned}$$

This system has an infinite number of solutions given by $c_3 = 3t$, $c_2 = -t$, and $c_1 = 2t$. Letting $t = 1$ results in the equation $2v_1 - v_2 + 3v_3 = 0$. Therefore, v_2 can be written as a linear combination of v_1 and v_3 as follows.

$$v_2 = 2v_1 + 3v_3$$

A check yields

$$2 + 5x - x^2 = 2(1 + x - 2x^2) + 3(x + x^2)$$
$$= 2 + 2x - 4x^2 + 3x + 3x^2$$
$$= 2 + 5x - x^2.$$

Theorem 4.8 has a practical corollary that provides a simple test for determining whether *two* vectors are linearly dependent. In Exercise 59 you are asked to prove this corollary.

Corollary to Theorem 4.8   Two vectors $\mathbf{u}$ and $\mathbf{v}$ in a vector space V are linearly dependent if and only if one is a scalar multiple of the other.

REMARK: The zero vector is always a scalar multiple of another vector in a vector space.

EXAMPLE 13   *Testing for Linear Dependence of Two Vectors*

(a) The set

$$S = \{(\overset{\mathbf{v}_1}{1, 2, 0}), (\overset{\mathbf{v}_2}{-2, 2, 1})\}$$

is linearly independent because $\mathbf{v}_1$ and $\mathbf{v}_2$ are not scalar multiples of each other, as shown in Figure 4.17(a).

(b) The set

$$S = \{(\overset{\mathbf{v}_1}{4, -4, -2}), (\overset{\mathbf{v}_2}{-2, 2, 1})\}$$

is linearly dependent because $\mathbf{v}_1 = -2\mathbf{v}_2$, as shown in Figure 4.17(b).

(a)

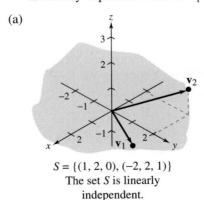

$S = \{(1, 2, 0), (-2, 2, 1)\}$
The set S is linearly
independent.

(b)

$S = \{(4, -4, -2), (-2, 2, 1)\}$
The set S is linearly
dependent because $\mathbf{v}_1 = -2\mathbf{v}_2$.

Figure 4.17

SECTION 4.4 Exercises

In Exercises 1–4, determine whether each vector can be written as a linear combination of the vectors in S.

1. $S = \{(2, -1, 3), (5, 0, 4)\}$
   (a) $\mathbf{u} = (1, 1, -1)$
   (b) $\mathbf{v} = \left(8, -\frac{1}{4}, \frac{27}{4}\right)$
   (c) $\mathbf{w} = (1, -8, 12)$

2. $S = \{(1, 2, -2), (2, -1, 1)\}$
   (a) $\mathbf{u} = (1, -5, -5)$
   (b) $\mathbf{v} = (-2, -6, 6)$
   (c) $\mathbf{w} = (-1, -22, 22)$

3. $S = \{(2, 0, 7), (2, 4, 5), (2, -12, 13)\}$
   (a) $\mathbf{u} = (-1, 5, -6)$
   (b) $\mathbf{v} = (-3, 15, 18)$
   (c) $\mathbf{w} = \left(\frac{1}{3}, \frac{4}{3}, \frac{1}{2}\right)$

4. $S = \{(6, -7, 8, 6), (4, 6, -4, 1)\}$
   (a) $\mathbf{u} = (-42, 113, -112, -60)$
   (b) $\mathbf{v} = \left(\frac{49}{2}, \frac{99}{4}, -14, \frac{19}{2}\right)$
   (c) $\mathbf{w} = \left(-4, -14, \frac{27}{2}, \frac{53}{8}\right)$

In Exercises 5–10, determine whether the set S spans R^2. If the set does not span R^2, give a geometric description of the subspace that it does span.

5. $S = \{(2, 1), (-1, 2)\}$
6. $S = \{(5, 0), (5, -4)\}$
7. $S = \{(-3, 5)\}$
8. $S = \{(1, 3), (-2, -6), (4, 12)\}$
9. $S = \{(-1, 2), (2, -4)\}$
10. $S = \{(-1, 4), (4, -1), (1, 1)\}$

In Exercises 11–16, determine whether the set S spans R^3. If the set does not span R^3, give a geometric description of the subspace that it does span.

11. $S = \{(4, 7, 3), (-1, 2, 6), (2, -3, 5)\}$
12. $S = \{(6, 7, 6), (3, 2, -4), (1, -3, 2)\}$
13. $S = \{(-2, 5, 0), (4, 6, 3)\}$
14. $S = \{(1, 0, 1), (1, 1, 0), (0, 1, 1)\}$
15. $S = \{(1, -2, 0), (0, 0, 1), (-1, 2, 0)\}$
16. $S = \{(1, 0, 3), (2, 0, -1), (4, 0, 5), (2, 0, 6)\}$

In Exercises 17–28, determine whether the set S is linearly independent or linearly dependent.

17. $S = \{(-2, 2), (3, 5)\}$ **18.** $S = \{(-2, 4), (1, -2)\}$
19. $S = \{(0, 0), (1, -1)\}$ **20.** $S = \{(1, 0), (1, 1), (2, -1)\}$
21. $S = \{(1, -4, 1), (6, 3, 2)\}$
22. $S = \{(6, 2, 1), (-1, 3, 2)\}$
23. $S = \{(1, 1, 1), (2, 2, 2), (3, 3, 3)\}$
24. $S = \left\{\left(\frac{3}{4}, \frac{5}{2}, \frac{3}{2}\right), \left(3, 4, \frac{7}{2}\right), \left(-\frac{3}{2}, 6, 2\right)\right\}$
25. $S = \{(-4, -3, 4), (1, -2, 3), (6, 0, 0)\}$
26. $S = \{(1, 0, 0), (0, 4, 0), (0, 0, -6), (1, 5, -3)\}$
27. $S = \{(4, -3, 6, 2), (1, 8, 3, 1), (3, -2, -1, 0)\}$
28. $S = \{(0, 0, 0, 1), (0, 0, 1, 1), (0, 1, 1, 1), (1, 1, 1, 1)\}$

In Exercises 29–32, show that the set is linearly dependent by finding a nontrivial linear combination (of vectors in the set) whose sum is the zero vector. Then express one of the vectors in the set as a linear combination of the other vectors in the set.

29. $S = \{(3, 4), (-1, 1), (2, 0)\}$
30. $S = \{(2, 4), (-1, -2), (0, 6)\}$
31. $S = \{(1, 1, 1), (1, 1, 0), (0, 1, 1), (0, 0, 1)\}$
32. $S = \{(1, 2, 3, 4), (1, 0, 1, 2), (1, 4, 5, 6)\}$

33. For which values of t is each set linearly independent?
   (a) $S = \{(t, 1, 1), (1, t, 1), (1, 1, t)\}$
   (b) $S = \{(t, 1, 1), (1, 0, 1), (1, 1, 3t)\}$

34. For which values of t is each set linearly independent?
   (a) $S = \{(t, 0, 0), (0, 1, 0), (0, 0, 1)\}$
   (b) $S = \{(t, t, t), (t, 1, 0), (t, 0, 1)\}$

35. Given the matrices

$$A = \begin{bmatrix} 2 & -3 \\ 4 & 1 \end{bmatrix} \quad \text{and} \quad B = \begin{bmatrix} 0 & 5 \\ 1 & -2 \end{bmatrix}$$

in $M_{2,2}$, determine which of the following are linear combinations of A and B.

   (a) $\begin{bmatrix} 6 & -19 \\ 10 & 7 \end{bmatrix}$ (b) $\begin{bmatrix} 6 & 2 \\ 9 & 11 \end{bmatrix}$

   (c) $\begin{bmatrix} -2 & 28 \\ 1 & -11 \end{bmatrix}$ (d) $\begin{bmatrix} 0 & 0 \\ 0 & 0 \end{bmatrix}$

36. Determine whether the matrices from $M_{2,2}$ form a linearly independent set.

$$A = \begin{bmatrix} 1 & -1 \\ 4 & 5 \end{bmatrix}, \quad B = \begin{bmatrix} 4 & 3 \\ -2 & 3 \end{bmatrix}, \quad C = \begin{bmatrix} 1 & -8 \\ 22 & 23 \end{bmatrix}$$

37. Determine whether each set in P_2 is linearly independent.
  (a) $S = \{2 - x, 2x - x^2, 6 - 5x + x^2\}$
  (b) $S = \{x^2 - 1, 2x + 5\}$
  (c) $S = \{x^2 + 3x + 1, 2x^2 + x - 1, 4x\}$

38. Determine whether the set $S = \{x^2 - 2x, x^3 + 8, x^3 - x^2, x^2 - 4\}$ spans P_3.

39. By inspection, determine why each of the sets is linearly dependent.
  (a) $S = \{(1, -2), (2, 3), (-2, 4)\}$
  (b) $S = \{(1, -6, 2), (2, -12, 4)\}$
  (c) $S = \{(0, 0), (1, 0)\}$

40. Complete the proof of Theorem 4.7.

In Exercises 41 and 42, determine whether the sets S_1 and S_2 span the same subspace of R^3.

41. $S_1 = \{(1, 2, -1), (0, 1, 1), (2, 5, -1)\}$
  $S_2 = \{(-2, -6, 0), (1, 1, -2)\}$

42. $S_1 = \{(0, 0, 1), (0, 1, 1), (2, 1, 1)\}$
  $S_2 = \{(1, 1, 1), (1, 1, 2), (2, 1, 1)\}$

True or False? In Exercises 43 and 44, determine whether each statement is true or false. If a statement is true, give a reason or cite an appropriate statement in the text. If a statement is false, provide an example that shows that the statement is not true in all cases or cite an appropriate statement in the text.

43. (a) A set of vectors $S = \{v_1, v_2, \ldots, v_k\}$ in a vector space is called linearly dependent if the vector equation $c_1 v_1 + c_2 v_2 + \cdots + c_k v_k = 0$ has only the trivial solution.
  (b) Two vectors u and v in a vector space V are linearly dependent if and only if one is a scalar multiple of the other.

44. (a) A set $S = \{v_1, v_2, \ldots, v_k\}$, $k \geq 2$, is linearly independent if and only if at least one of the vectors v_j can be written as a linear combination of the other vectors.
  (b) If a subset S spans a vector space V, then every vector in V can be written as a linear combination of the vectors in S.

In Exercises 45 and 46, prove that the set of vectors is linearly independent and spans R^3.

45. $B = \{(1, 1, 1), (1, 1, 0), (1, 0, 0)\}$
46. $B = \{(1, 2, 3), (3, 2, 1), (0, 0, 1)\}$

47. ***Guided Proof*** Prove that a nonempty subset of a finite set of linearly independent vectors is linearly independent.

Getting Started: You need to show that a subset of a linearly independent set of vectors cannot be linearly dependent.
  (i) Suppose S is a set of linearly independent vectors. Let T be a subset of S.
  (ii) If T is linearly dependent, then there exist constants not all zero satisfying the vector equation $c_1 v_1 + c_2 v_2 + \cdots + c_k v_k = 0$.
  (iii) Use this fact to derive a contradiction and thus conclude that T is linearly independent.

48. Prove that if S_1 is a nonempty subset of the finite set S_2 and S_1 is linearly dependent, then so is S_2.

49. Prove that any set of vectors containing the zero vector is linearly dependent.

50. Given that $\{u_1, u_2, \ldots, u_n\}$ is a linearly independent set of vectors and that the set $\{u_1, u_2, \ldots, u_n, v\}$ is linearly dependent, prove that v is a linear combination of the u_i's.

51. Let $\{v_1, v_2, \ldots, v_k\}$ be a linearly independent set of vectors in a vector space V. Delete the vector v_k from this set and prove that the set $\{v_1, v_2, \ldots, v_{k-1}\}$ cannot span V.

52. If V is spanned by $\{v_1, v_2, \ldots, v_k\}$ and one of these vectors can be written as a linear combination of the other $k - 1$ vectors, prove that the span of these $k - 1$ vectors is also V.

53. ***Writing*** The set $\{(1, 2, 3), (1, 0, -2), (-1, 0, 2)\}$ is linearly dependent, but $(1, 2, 3)$ cannot be written as a linear combination of $(1, 0, -2)$ and $(-1, 0, 2)$. Why does this statement not contradict Theorem 4.8?

54. ***Writing*** Under what conditions will a set consisting of a single vector be linearly independent?

55. Let $S = \{u, v\}$ be a linearly independent set. Prove that the set $\{u + v, u - v\}$ is linearly independent.

56. Let u, v, and w be any three vectors from a vector space V. Determine whether the set of vectors $\{v - u, w - v, u - w\}$ is linearly independent or dependent.

57. Let $f_1(x) = 3x$ and $f_2(x) = |x|$. Graph both functions on the interval $-2 \leq x \leq 2$. Show that these functions are linearly dependent in the vector space $C[0, 1]$, but linearly independent in $C[-1, 1]$.

58. ***Writing*** Let A be a nonsingular matrix of order 3. Prove that if $\{v_1, v_2, v_3\}$ is a linearly independent set in $M_{3,1}$, then the set $\{Av_1, Av_2, Av_3\}$ is also linearly independent. Explain, by means of an example, why this property is not true if A is singular.

59. Prove the corollary to Theorem 4.8: Two vectors u and v are linearly dependent if and only if one is a scalar multiple of the other.

4.5 | Basis and Dimension

In this section you will continue your study of spanning sets. In particular, you will look at spanning sets (in a vector space) that are both linearly independent *and* span the entire space. Such a set forms a **basis** for the vector space. (The plural of *basis* is *bases*.)

Definition of Basis

A set of vectors $S = \{\mathbf{v}_1, \mathbf{v}_2, \ldots, \mathbf{v}_n\}$ in a vector space V is called a **basis** for V if the following conditions are true.
1. S spans V.   2. S is linearly independent.

REMARK: This definition tells you that a basis has two features. A basis S must have *enough vectors* to span V, but *not so many vectors* that one of them could be written as a linear combination of the other vectors in S.

This definition does not imply that every vector space has a basis consisting of a finite number of vectors. In this text, however, the discussion of bases is restricted to those consisting of a finite number of vectors. Moreover, if a vector space V has a basis consisting of a finite number of vectors, then V is **finite dimensional.** Otherwise, V is called **infinite dimensional.** [The vector space P of *all* polynomials is infinite dimensional, as is the vector space $C(-\infty, \infty)$ of all continuous functions defined on the real line.] The vector space $V = \{\mathbf{0}\}$, consisting of the zero vector alone, is finite dimensional.

EXAMPLE 1   *The Standard Basis for R^3*

Show that the set is a basis for R^3.

$$S = \{(1, 0, 0), (0, 1, 0), (0, 0, 1)\}$$

Solution   Example 4(a) in Section 4.4 showed that S spans R^3. Furthermore, S is linearly independent because the vector equation

$$c_1(1, 0, 0) + c_2(0, 1, 0) + c_3(0, 0, 1) = (0, 0, 0)$$

has only the trivial solution $c_1 = c_2 = c_3 = 0$. (Try verifying this.) Therefore, S is a basis for R^3. (See Figure 4.18.)

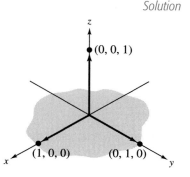

Figure 4.18

The basis $S = \{(1, 0, 0), (0, 1, 0), (0, 0, 1)\}$ is called the **standard basis** for R^3. This result can be generalized to n-space. That is, the vectors

$$\mathbf{e}_1 = (1, 0, \ldots, 0)$$
$$\mathbf{e}_2 = (0, 1, \ldots, 0)$$
$$\vdots$$
$$\mathbf{e}_n = (0, 0, \ldots, 1)$$

form a basis for R^n called the **standard basis** for R^n.

The next two examples describe nonstandard bases for R^2 and R^3.

EXAMPLE 2   *A Nonstandard Basis for R^2*

Show that the set

$$S = \{\overset{\mathbf{v}_1}{(1,\,1)},\, \overset{\mathbf{v}_2}{(1,\,-1)}\}$$

is a basis for R^2.

Solution   According to the definition of a basis for a vector space, you must show that S spans R^2 and that S is linearly independent.

To verify that S spans R^2, let

$$\mathbf{x} = (x_1, x_2)$$

represent an arbitrary vector in R^2. To show that $\mathbf{x}$ can be written as a linear combination of $\mathbf{v}_1$ and $\mathbf{v}_2$, consider the equation

$$c_1\mathbf{v}_1 + c_2\mathbf{v}_2 = \mathbf{x}$$
$$c_1(1,\,1) + c_2(1,\,-1) = (x_1, x_2)$$
$$(c_1 + c_2,\, c_1 - c_2) = (x_1, x_2).$$

Equating corresponding components yields the following system of linear equations.

$$c_1 + c_2 = x_1$$
$$c_1 - c_2 = x_2$$

Because the coefficient matrix of this system has a nonzero determinant, you know that the system has a unique solution. Therefore, you can conclude that S spans R^2.

To show that S is linearly independent, consider the following linear combination.

$$c_1\mathbf{v}_1 + c_2\mathbf{v}_2 = \mathbf{0}$$
$$c_1(1,\,1) + c_2(1,\,-1) = (0, 0)$$
$$(c_1 + c_2,\, c_1 - c_2) = (0, 0)$$

Equating corresponding components yields the following homogeneous system.

$$c_1 + c_2 = 0$$
$$c_1 - c_2 = 0$$

Because the coefficient matrix of this system has a nonzero determinant, you know that the system has only the trivial solution

$$c_1 = c_2 = 0.$$

Therefore, you can conclude that S is linearly independent.

Thus, you can conclude that S is a basis for R^2 because it is a linearly independent spanning set for R^2.

EXAMPLE 3   *A Nonstandard Basis for R^3*

From Examples 5 and 8 in the preceding section, you know that

$$S = \{(1, 2, 3), (0, 1, 2), (-2, 0, 1)\}$$

spans R^3 and is linearly independent. Therefore, S is a basis for R^3.

EXAMPLE 4   *A Basis for Polynomials*

Show that the vector space P_3 has the following basis.

$$S = \{1, x, x^2, x^3\}$$

Solution   It is clear that S spans P_3 because the span of S consists of all polynomials of the form

$$a_0 + a_1 x + a_2 x^2 + a_3 x^3, \quad a_0, a_1, a_2, \text{ and } a_3 \text{ are real,}$$

which is precisely the form of all polynomials in P_3.

To verify the linear independence of S, recall that the zero vector **0** in P_3 is the polynomial $\mathbf{0}(x) = 0$ for all x. Thus the test for linear independence yields the equation

$$a_0 + a_1 x + a_2 x^2 + a_3 x^3 = \mathbf{0}(x) = 0, \quad \text{for all } x.$$

This third-degree polynomial is said to be *identically equal to zero*. From algebra you know that for a polynomial to be identically equal to zero, all of its coefficients must be zero; that is,

$$a_0 = a_1 = a_2 = a_3 = 0.$$

Hence, S is linearly independent and is therefore a basis for P_3.

REMARK: The basis $S = \{1, x, x^2, x^3\}$ is called the **standard basis** for P_3. Similarly, the **standard basis** for P_n is

$$S = \{1, x, x^2, \ldots, x^n\}.$$

EXAMPLE 5   *A Basis for $M_{2,2}$*

It can be verified that the set

$$S = \left\{ \begin{bmatrix} 1 & 0 \\ 0 & 0 \end{bmatrix}, \begin{bmatrix} 0 & 1 \\ 0 & 0 \end{bmatrix}, \begin{bmatrix} 0 & 0 \\ 1 & 0 \end{bmatrix}, \begin{bmatrix} 0 & 0 \\ 0 & 1 \end{bmatrix} \right\}$$

is a basis for $M_{2,2}$. This set is called the **standard basis** for $M_{2,2}$. In a similar manner, the standard basis for the vector space $M_{m,n}$ consists of the mn distinct $m \times n$ matrices having a single 1 and all the other entries equal to zero.

Theorem 4.9	*Uniqueness of Basis Representation*

If $S = \{v_1, v_2, \ldots, v_n\}$ is a basis for a vector space V, then every vector in V can be written in one and only one way as a linear combination of vectors in S.

Proof   The existence portion of the proof is straightforward. That is, because S spans V, you know that an arbitrary vector $\mathbf{u}$ in V can be expressed as $\mathbf{u} = c_1 v_1 + c_2 v_2 + \cdots + c_n v_n$.

To prove uniqueness (that a given vector can be represented in only one way), suppose that $\mathbf{u}$ has another representation $\mathbf{u} = b_1 v_1 + b_2 v_2 + \cdots + b_n v_n$. Subtracting the second representation from the first produces

$$\mathbf{u} - \mathbf{u} = (c_1 - b_1)v_1 + (c_2 - b_2)v_2 + \cdots + (c_n - b_n)v_n = \mathbf{0}.$$

However, because S is linearly independent, the only solution to this equation is the trivial solution

$$c_1 - b_1 = 0, \qquad c_2 - b_2 = 0, \quad \ldots, \quad c_n - b_n = 0,$$

which means that $c_i = b_i$ for all $i = 1, 2, \ldots, n$. Hence, $\mathbf{u}$ has only one representation for the given basis S.

EXAMPLE 6   *Uniqueness of Basis Representation*

Let $\mathbf{u} = (u_1, u_2, u_3)$ be any vector in R^3. Show that the equation $\mathbf{u} = c_1 v_1 + c_2 v_2 + c_3 v_3$ has a unique solution for the basis $S = \{v_1, v_2, v_3\} = \{(1, 2, 3), (0, 1, 2), (-2, 0, 1)\}$.

Solution   From the equation

$$(u_1, u_2, u_3) = c_1(1, 2, 3) + c_2(0, 1, 2) + c_3(-2, 0, 1)$$
$$= (c_1 - 2c_3, 2c_1 + c_2, 3c_1 + 2c_2 + c_3),$$

the following system of linear equations is obtained.

$$\begin{matrix} c_1 & & -2c_3 = u_1 \\ 2c_1 & + c_2 & = u_2 \\ 3c_1 & + 2c_2 & + c_3 = u_3 \end{matrix} \qquad \underbrace{\begin{bmatrix} 1 & 0 & -2 \\ 2 & 1 & 0 \\ 3 & 2 & 1 \end{bmatrix}}_{A} \underbrace{\begin{bmatrix} c_1 \\ c_2 \\ c_3 \end{bmatrix}}_{\mathbf{c}} = \underbrace{\begin{bmatrix} u_1 \\ u_2 \\ u_3 \end{bmatrix}}_{\mathbf{u}}$$

Because the matrix A is invertible, you know that this system has a unique solution given by $\mathbf{c} = A^{-1}\mathbf{u}$. Solving for A^{-1} gives

$$A^{-1} = \begin{bmatrix} -1 & 4 & -2 \\ 2 & -7 & 4 \\ -1 & 2 & -1 \end{bmatrix},$$

which implies that

$$c_1 = -u_1 + 4u_2 - 2u_3$$
$$c_2 = 2u_1 - 7u_2 + 4u_3$$
$$c_3 = -u_1 + 2u_2 - u_3.$$

For instance, the vector $\mathbf{u} = (1, 0, 0)$ can be represented uniquely as a linear combination of $\mathbf{v}_1$, $\mathbf{v}_2$, and $\mathbf{v}_3$ as follows.

$$(1, 0, 0) = -\mathbf{v}_1 + 2\mathbf{v}_2 - \mathbf{v}_3$$

You will now study two major results concerning bases.

Theorem 4.10 **Bases and Linear Dependence**

If $S = \{\mathbf{v}_1, \mathbf{v}_2, \ldots, \mathbf{v}_n\}$ is a basis for a vector space V, then every set containing more than n vectors in V is linearly dependent.

Proof Let $S_1 = \{\mathbf{u}_1, \mathbf{u}_2, \ldots, \mathbf{u}_m\}$ be any set of m vectors in V, where $m > n$. To show that S_1 is linearly *dependent*, you need to find scalars $k_1, k_2, \ldots, k_m$ (not all zero) such that

$$k_1\mathbf{u}_1 + k_2\mathbf{u}_2 + \cdots + k_m\mathbf{u}_m = \mathbf{0} \qquad \text{Equation 1}$$

Because S is a basis for V, it follows that each $\mathbf{u}_i$ is a linear combination of vectors in S, and you can write

$$\mathbf{u}_1 = c_{11}\mathbf{v}_1 + c_{21}\mathbf{v}_2 + \cdots + c_{n1}\mathbf{v}_n$$
$$\mathbf{u}_2 = c_{12}\mathbf{v}_1 + c_{22}\mathbf{v}_2 + \cdots + c_{n2}\mathbf{v}_n$$
$$\vdots$$
$$\mathbf{u}_m = c_{1m}\mathbf{v}_1 + c_{2m}\mathbf{v}_2 + \cdots + c_{nm}\mathbf{v}_n.$$

Substituting each of these representations of $\mathbf{u}_i$ into Equation 1 and regrouping terms produces

$$d_1\mathbf{v}_1 + d_2\mathbf{v}_2 + \cdots + d_n\mathbf{v}_n = \mathbf{0},$$

where $d_i = c_{i1}k_1 + c_{i2}k_2 + \cdots + c_{im}k_m$. However, because the $\mathbf{v}_i$'s form a linearly independent set, you can conclude that each $d_i = 0$. Thus, the following system of equations is obtained.

$$c_{11}k_1 + c_{12}k_2 + \cdots + c_{1m}k_m = 0$$
$$c_{21}k_1 + c_{22}k_2 + \cdots + c_{2m}k_m = 0$$
$$\vdots$$
$$c_{n1}k_1 + c_{n2}k_2 + \cdots + c_{nm}k_m = 0$$

But this homogeneous system has fewer equations than variables $k_1, k_2, \ldots, k_m$, and from Theorem 1.1 you know that it must have *nontrivial* solutions. Consequently, S_1 is linearly dependent.

EXAMPLE 7 *Linearly Dependent Sets in R^3 and P_3*

(a) Because R^3 has a basis consisting of three vectors, the set
$$S = \{(1, 2, -1), (1, 1, 0), (2, 3, 0), (5, 9, -1)\}$$
must be linearly dependent.

(b) Because P_3 has a basis consisting of four vectors, the set
$$S = \{(1, 1 + x, 1 - x, 1 + x + x^2, 1 - x + x^2\}$$
must be linearly dependent.

Because R^n has the standard basis consisting of n vectors, it follows from Theorem 4.10 that every set of vectors in R^n containing more than n vectors must be linearly dependent. Another significant consequence of Theorem 4.10 is given in the following theorem.

Theorem 4.11

Number of Vectors in a Basis

If a vector space V has one basis with n vectors, then every basis for V has n vectors.

Proof Let
$$S_1 = \{\mathbf{v}_1, \mathbf{v}_2, \ldots, \mathbf{v}_n\}$$
be the given basis for V, and let
$$S_2 = \{\mathbf{u}_1, \mathbf{u}_2, \ldots, \mathbf{u}_m\}$$
be any other basis for V. Because S_1 is a basis and S_2 is linearly independent, Theorem 4.10 implies that $m \le n$. Similarly, $n \le m$ because S_1 is linearly independent and S_2 is a basis. Consequently, $n = m$.

EXAMPLE 8 *Spanning Sets and Bases*

Use Theorem 4.11 to explain why each of the following statements is true.
(a) The set $S_1 = \{(3, 2, 1), (7, -1, 4)\}$ is not a basis for R^3.
(b) The set
$$S_2 = \{x + 2, x^2, x^3 - 1, 3x + 1, x^2 - 2x + 3\}$$
is not a basis for P_3.

Solution (a) The standard basis for R^3 has three vectors, and S_1 has only two. Hence, by Theorem 4.11, S_1 cannot be a basis for R^3.
(b) The standard basis for P_3, $S = \{1, x, x^2, x^3\}$, has four elements. Therefore, by Theorem 4.11, the given set S_2 has too many elements to be a basis for P_3.

The Dimension of a Vector Space

The discussion of spanning sets, linear independence, and bases leads to an important notion in the study of vector spaces. By Theorem 4.11, you know that if a vector space V has a basis consisting of n vectors, then every other basis for the space also has n vectors. The number n is called the **dimension** of V.

Definition of Dimension of a Vector Space	If a vector space V has a basis consisting of n vectors, then the number n is called the **dimension** of V, denoted by $\dim(V) = n$. If V consists of the zero vector alone, the dimension of V is defined as zero.

This definition allows you to observe the following characteristics of the dimensions of some familiar vector spaces. In each case, the dimension is determined by simply counting the number of vectors in the standard basis.

1. The dimension of R^n with the standard operations is n.
2. The dimension of P_n with the standard operations is $n + 1$.
3. The dimension of $M_{m,n}$ with the standard operations is mn.

If W is a subspace of an n-dimensional vector space, then it can be shown that W is finite dimensional and that the dimension of W is less than or equal to n. (See Exercise 61.) In the next three examples, you will look at a technique for determining the dimension of a subspace. Basically, you determine the dimension by finding a set of linearly independent vectors that spans the subspace. This set is a basis for the subspace, and the dimension of the subspace is the number of vectors in the basis.

EXAMPLE 9   *Finding the Dimension of a Subspace*

Determine the dimension of each subspace of R^3.
(a) $W = \{(d, c - d, c): c \text{ and } d \text{ are real numbers}\}$
(b) $W = \{(2b, b, 0): b \text{ is a real number}\}$

Solution   The goal in each example is to find a set of linearly independent vectors that spans the subspace.
(a) By writing the representative vector $(d, c - d, c)$ as

$$(d, c - d, c) = (0, c, c) + (d, -d, 0)$$
$$= c(0, 1, 1) + d(1, -1, 0),$$

you can see that W is spanned by the set

$$S = \{(0, 1, 1), (1, -1, 0)\}.$$

Using the techniques described in the preceding section, you can show that this set is linearly independent. Therefore, it is a basis for W, and you can conclude that W is a two-dimensional subspace of R^3.

(b) By writing the representative vector $(2b, b, 0)$ as

$$(2b, b, 0) = b(2, 1, 0),$$

you can see that W is spanned by the set $S = \{(2, 1, 0)\}$. Therefore, W is a one-dimensional subspace of R^3.

REMARK: In Example 9(a), the subspace W is a two-dimensional plane in R^3 determined by the vectors $(0, 1, 1)$ and $(1, -1, 0)$. In Example 9(b), the subspace is a one-dimensional line.

EXAMPLE 10 *Finding the Dimension of a Subspace*

Find the dimension of the subspace W of R^4 spanned by

$$S = \{\overset{v_1}{(-1, 2, 5, 0)}, \overset{v_2}{(3, 0, 1, -2)}, \overset{v_3}{(-5, 4, 9, 2)}\}.$$

Solution   Although W is spanned by the set S, S is not a basis for W because S is a linearly dependent set. Specifically, v_3 can be written as a linear combination of v_1 and v_2 as follows.

$$v_3 = 2v_1 - v_2$$

This means that W is spanned by the set $S_1 = \{v_1, v_2\}$. Moreover, S_1 is linearly independent because neither vector is a scalar multiple of the other, and you can conclude that the dimension of W is 2.

EXAMPLE 11 *Finding the Dimension of a Subspace*

Let W be the subspace of all symmetric matrices in $M_{2,2}$. What is the dimension of W?

Solution   Every 2×2 symmetric matrix has the following form.

$$A = \begin{bmatrix} a & b \\ b & c \end{bmatrix} = \begin{bmatrix} a & 0 \\ 0 & 0 \end{bmatrix} + \begin{bmatrix} 0 & b \\ b & 0 \end{bmatrix} + \begin{bmatrix} 0 & 0 \\ 0 & c \end{bmatrix}$$

$$= a\begin{bmatrix} 1 & 0 \\ 0 & 0 \end{bmatrix} + b\begin{bmatrix} 0 & 1 \\ 1 & 0 \end{bmatrix} + c\begin{bmatrix} 0 & 0 \\ 0 & 1 \end{bmatrix}$$

Therefore, the set

$$S = \left\{ \begin{bmatrix} 1 & 0 \\ 0 & 0 \end{bmatrix}, \begin{bmatrix} 0 & 1 \\ 1 & 0 \end{bmatrix}, \begin{bmatrix} 0 & 0 \\ 0 & 1 \end{bmatrix} \right\}$$

spans W. Moreover, S can be shown to be linearly independent, and you can conclude that the dimension of W is 3.

Usually, to conclude that a set

$$S = \{v_1, v_2, \ldots, v_n\}$$

is a basis for a vector space V, you must show that S satisfies two conditions: S spans V and is linearly independent. However, if V is known to have a dimension of n, then the following theorem tells you that both conditions need not be checked: either one will suffice. The proof is left as an exercise. (See Exercise 62.)

Theorem 4.12

Basis Tests in an n-Dimensional Space

Let V be a vector space of dimension n.
1. If $S = \{v_1, v_2, \ldots, v_n\}$ is a linearly independent set of vectors in V, then S is a basis for V.
2. If $S = \{v_1, v_2, \ldots, v_n\}$ spans V, then S is a basis for V.

EXAMPLE 12   *Testing for a Basis in an n-Dimensional Space*

Show that the set of vectors is a basis for $M_{5,1}$.

$$S = \left\{ \begin{bmatrix} 1 \\ 2 \\ -1 \\ 3 \\ 4 \end{bmatrix}, \begin{bmatrix} 0 \\ 1 \\ 3 \\ -2 \\ 3 \end{bmatrix}, \begin{bmatrix} 0 \\ 0 \\ 2 \\ -1 \\ 5 \end{bmatrix}, \begin{bmatrix} 0 \\ 0 \\ 0 \\ 2 \\ -3 \end{bmatrix}, \begin{bmatrix} 0 \\ 0 \\ 0 \\ 0 \\ -2 \end{bmatrix} \right\}$$

Solution   Because S has five vectors and the dimension of $M_{5,1}$ is five, you can apply Theorem 4.12 to verify that S is a basis by showing either that S is linearly independent or that S spans $M_{5,1}$. To show the first of these, form the vector equation

$$c_1 v_1 + c_2 v_2 + c_3 v_3 + c_4 v_4 + c_5 v_5 = 0,$$

which yields the following homogeneous system of linear equations.

$$
\begin{aligned}
c_1 &&&&&&&&&= 0 \\
2c_1 &+ c_2 &&&&&&&&= 0 \\
-c_1 &+ 3c_2 &+ 2c_3 &&&&&&&= 0 \\
3c_1 &- 2c_2 &- c_3 &+ 2c_4 &&&&&= 0 \\
4c_1 &+ 3c_2 &+ 5c_3 &- 3c_4 &- 2c_5 &= 0
\end{aligned}
$$

Because this system has only the trivial solution, S must be linearly independent. Therefore, by Theorem 4.12, S is a basis for $M_{5,1}$.

In Exercises 1–4, write the standard basis for the vector space.

1. R^6
2. $M_{4,1}$
3. $M_{2,4}$
4. P_4

Writing In Exercises 5–8, explain why S is not a basis for R^2.

5. $S = \{(1, 2), (1, 0), (0, 1)\}$
6. $S = \{(-4, 5), (0, 0)\}$
7. $S = \{(6, -5), (12, -10)\}$
8. $S = \{(-3, 2)\}$

Writing In Exercises 9–12, explain why S is not a basis for R^3.

9. $S = \{(1, 3, 0), (4, 1, 2), (-2, 5, -2)\}$
10. $S = \{(7, 0, 3), (8, -4, 1)\}$
11. $S = \{(0, 0, 0), (1, 0, 0), (0, 1, 0)\}$
12. $S = \{(6, 4, 1), (3, -5, 1), (8, 13, 6), (0, 6, 9)\}$

Writing In Exercises 13–16, explain why S is not a basis for P_2.

13. $S = \{1, 2x, x^2 - 4, 5x\}$
14. $S = \{2, x, x + 3, 3x^2\}$
15. $S = \{1 - x, 1 - x^2, 3x^2 - 2x - 1\}$
16. $S = \{6x - 3, 3x^2, 1 - 2x - x^2\}$

Writing In Exercises 17 and 18, explain why S is not a basis for $M_{2,2}$.

17. $S = \left\{ \begin{bmatrix} 1 & 0 \\ 0 & 1 \end{bmatrix}, \begin{bmatrix} 0 & 1 \\ 1 & 0 \end{bmatrix} \right\}$

18. $S = \left\{ \begin{bmatrix} 1 & 0 \\ 0 & 0 \end{bmatrix}, \begin{bmatrix} 0 & 1 \\ 1 & 0 \end{bmatrix}, \begin{bmatrix} 1 & 0 \\ 0 & 1 \end{bmatrix}, \begin{bmatrix} 8 & -4 \\ -4 & 3 \end{bmatrix} \right\}$

In Exercises 19–22, determine whether the set $\{v_1, v_2\}$ is a basis for R^2.

19.

20.

21.

22.

In Exercises 23–28, determine whether S is a basis for the indicated vector space.

23. $S = \{(3, -2), (4, 5)\}$ for R^2
24. $S = \{(1, 5, 3), (0, 1, 2), (0, 0, 6)\}$ for R^3
25. $S = \{(0, 3, -2), (4, 0, 3), (-8, 15, -16)\}$ for R^3
26. $S = \{(0, 0, 0), (1, 5, 6), (6, 2, 1)\}$ for R^3
27. $S = \{(-1, 2, 0, 0), (2, 0, -1, 0), (3, 0, 0, 4), (0, 0, 5, 0)\}$ for R^4
28. $S = \{(1, 0, 0, 1), (0, 2, 0, 2), (1, 0, 1, 0), (0, 2, 2, 0)\}$ for R^4

In Exercises 29 and 30, determine whether S is a basis for $M_{2,2}$.

29. $S = \left\{ \begin{bmatrix} 2 & 0 \\ 0 & 3 \end{bmatrix}, \begin{bmatrix} 1 & 4 \\ 0 & 1 \end{bmatrix}, \begin{bmatrix} 0 & 1 \\ 3 & 2 \end{bmatrix}, \begin{bmatrix} 0 & 1 \\ 2 & 0 \end{bmatrix} \right\}$

30. $S = \left\{ \begin{bmatrix} 1 & 2 \\ -5 & 4 \end{bmatrix}, \begin{bmatrix} 2 & -7 \\ 6 & 2 \end{bmatrix}, \begin{bmatrix} 4 & -9 \\ 11 & 12 \end{bmatrix}, \begin{bmatrix} 12 & -16 \\ 17 & 42 \end{bmatrix} \right\}$

In Exercises 31–34, determine whether S is a basis for P_3.

31. $S = \{t^3 - 2t^2 + 1, t^2 - 4, t^3 + 2t, 5t\}$
32. $S = \{4t - t^2, 5 + t^3, 3t + 5, 2t^3 - 3t^2\}$
33. $S = \{4 - t, t^3, 6t^2, t^3 + 3t, 4t - 1\}$
34. $S = \{t^3 - 1, 2t^2, t + 3, 5 + 2t + 2t^2 + t^3\}$

In Exercises 35–38, determine whether S is a basis for R^3. If it is, write $u = (8, 3, 8)$ as a linear combination of the vectors in S.

35. $S = \{(4, 3, 2), (0, 3, 2), (0, 0, 2)\}$
36. $S = \{(0, 0, 0), (1, 3, 4), (6, 1, -2)\}$
37. $S = \left\{ \left(\frac{2}{3}, \frac{5}{2}, 1 \right), \left(1, \frac{3}{2}, 0 \right), (2, 12, 6) \right\}$
38. $S = \{(1, 4, 7), (3, 0, 1), (2, 1, 2)\}$

In Exercises 39–42, determine the dimension of the vector space.

39. R^6
40. R

41. P_7

42. $M_{2,3}$

43. Find a basis for $D_{3,3}$ (the vector space of all 3×3 diagonal matrices). What is the dimension of this vector space?

44. Find a basis for the vector space of all 3×3 symmetric matrices. What is the dimension of this vector space?

45. Find all subsets of the following set that form a basis for R^2.
$S = \{(1, 0), (0, 1), (1, 1)\}$

46. Find all subsets of the following set that form a basis for R^3.
$S = \{(1, 3, -2), (-4, 1, 1), (-2, 7, -3), (2, 1, 1)\}$

47. Find a basis for R^2 that includes the vector $(1, 1)$.

48. Find a basis for R^3 that includes the set $S = \{(1, 0, 2), (0, 1, 1)\}$.

In Exercises 49 and 50, (a) give a geometric description, (b) find a basis, and (c) determine the dimension of the subspace W of R^2.

49. $W = \{(2t, t): t$ is a real number$\}$

50. $W = \{(0, t): t$ is a real number$\}$

In Exercises 51 and 52, (a) give a geometric description, (b) find a basis, and (c) determine the dimension of the subspace W of R^3.

51. $W = \{(2t, t, -t): t$ is a real number$\}$

52. $W = \{(2s - t, s, t): s$ and t are real numbers$\}$

In Exercises 53–56, find (a) a basis for and (b) the dimension of the subspace W of R^4.

53. $W = \{(2s - t, s, t, s): s$ and t are real numbers$\}$

54. $W = \{(5t, -3t, t, t): t$ is a real number$\}$

55. $W = \{(0, 6t, t, -t): t$ is a real number$\}$

56. $W = \{(s + 4t, t, s, 2s - t): s$ and t are real numbers$\}$

True or False? In Exercises 57 and 58, determine whether each statement is true or false. If a statement is true, give a reason or cite an appropriate statement in the text. If a statement is false, provide an example that shows that the statement is not true in all cases or cite an appropriate statement in the text.

57. (a) If $\dim(V) = n$, then there exists a set of $n - 1$ vectors in V that will span V.
(b) If $\dim(V) = n$, then there exists a set of $n + 1$ vectors in V that will span V.

58. (a) If $\dim(V) = n$, then any set of $n + 1$ vectors in V must be dependent.

(b) If $\dim(V) = n$, then any set of $n - 1$ vectors in V must be independent.

59. Prove that if $S = \{\mathbf{v}_1, \mathbf{v}_2, \ldots, \mathbf{v}_n\}$ is a basis for a vector space V and c is a nonzero scalar, then the set $S_1 = \{c\mathbf{v}_1, c\mathbf{v}_2, \ldots, c\mathbf{v}_n\}$ is also a basis for V.

60. Prove that the vector space P of all polynomials is infinite dimensional.

61. Prove that if W is a subspace of a finite dimensional vector space V, then (dimension of W) $\le$ (dimension of V).

62. Prove Theorem 4.12.

63. ***Writing***
(a) Let $S_1 = \text{span}((1, 0, 0), (1, 1, 0))$ and $S_2 = \text{span}((0, 0, 1), (0, 1, 0))$ be subspaces of R^3. Find a basis for and the dimension of each of the subspaces $S_1, S_2, S_1 \cap S_2$, and $S_1 + S_2$. (See Exercise 31 in Section 4.3 for the definition of $S_1 + S_2$.)
(b) Let S_1 and S_2 be two-dimensional subspaces of R^3. Is it possible that $S_1 \cap S_2 = \{(0, 0, 0)\}$? Explain.

64. ***Guided Proof*** Let S be a spanning set for the finite dimensional vector space V. Prove that there exists a subset S' of S that forms a basis for V.

Getting Started: S is a spanning set, but it may not be a basis because it may be linearly dependent. You need to remove extra vectors so that a subset S' is a spanning set and is also linearly independent.

(i) If S is a linearly independent set, you are done. If not, remove some vector $\mathbf{v}$ from S that is a linear combination of the other vectors in S.

(ii) Call this set S_1. If S_1 is a linearly independent set, you are done. If not, continue to remove dependent vectors until you produce a linearly independent subset S'.

(iii) Thus, conclude that this subset is the minimal spanning set S'.

65. Let S be a linearly independent set of vectors from the finite dimensional vector space V. Prove that there exists a basis for V containing S.

66. Let V be a vector space of dimension n. Prove that any set of less than n vectors cannot span V.

4.6 | Rank of a Matrix and Systems of Linear Equations

In this section you will investigate the vector space spanned by the **row vectors** (or **column vectors**) of a matrix. Then you will see how these spaces relate to solutions of systems of linear equations.

To begin, you need to know some terminology. For an $m \times n$ matrix A, the n-tuples corresponding to the rows of A are called the **row vectors** of A.

$$
A = \begin{bmatrix} a_{11} & a_{12} & \cdots & a_{1n} \\ a_{21} & a_{22} & \cdots & a_{2n} \\ \vdots & \vdots & & \vdots \\ a_{m1} & a_{m2} & \cdots & a_{mn} \end{bmatrix}
\qquad
\begin{array}{l}
\textit{Row Vectors of A} \\
(a_{11}, a_{12}, \ldots, a_{1n}) \\
(a_{21}, a_{22}, \ldots, a_{2n}) \\
\vdots \\
(a_{m1}, a_{m2}, \ldots, a_{mn})
\end{array}
$$

Similarly, the columns of A are called the **column vectors** of A. You will find it useful to preserve the column notation for these column vectors.

$$
\begin{array}{cc}
 & \textit{Column Vectors of A} \\
A = \begin{bmatrix} a_{11} & a_{12} & \cdots & a_{1n} \\ a_{21} & a_{22} & \cdots & a_{2n} \\ \vdots & \vdots & & \vdots \\ a_{m1} & a_{m2} & \cdots & a_{mn} \end{bmatrix}
&
\begin{bmatrix} a_{11} \\ a_{21} \\ \vdots \\ a_{m1} \end{bmatrix}
\begin{bmatrix} a_{12} \\ a_{22} \\ \vdots \\ a_{m2} \end{bmatrix}
\cdots
\begin{bmatrix} a_{1n} \\ a_{2n} \\ \vdots \\ a_{mn} \end{bmatrix}
\end{array}
$$

EXAMPLE 1 *Row Vectors and Column Vectors*

For the matrix

$$
A = \begin{bmatrix} 0 & 1 & -1 \\ -2 & 3 & 4 \end{bmatrix},
$$

the row vectors are $(0, 1, -1)$ and $(-2, 3, 4)$ and the column vectors are

$$
\begin{bmatrix} 0 \\ -2 \end{bmatrix}, \quad \begin{bmatrix} 1 \\ 3 \end{bmatrix}, \quad \text{and} \quad \begin{bmatrix} -1 \\ 4 \end{bmatrix}.
$$

In Example 1, note that for an $m \times n$ matrix A, the row vectors are vectors in R^n and the column vectors are vectors in R^m. This leads to the following definitions of the **row space** and **column space** of a matrix.

Definition of Row Space and Column Space of a Matrix

Let A be an $m \times n$ matrix.
1. The **row space** of A is the subspace of R^n spanned by the row vectors of A.
2. The **column space** of A is the subspace of R^m spanned by the column vectors of A.

As it turns out, the row and column spaces of A share many properties. However, because of your familiarity with elementary row operations, you will begin by looking at the row space of a matrix. Recall that two matrices are row-equivalent if one can be obtained from the other by elementary row operations. The following theorem tells you that row-equivalent matrices have the same row space.

Theorem 4.13	*Row-Equivalent Matrices Have the Same Row Space*

If an $m \times n$ matrix A is row-equivalent to an $m \times n$ matrix B, then the row space of A is equal to the row space of B.

Proof   Because the rows of B can be obtained from the rows of A by elementary row operations (scalar multiplication and addition), it follows that the row vectors of B can be written as linear combinations of the row vectors of A. Hence, the row vectors of B lie in the row space of A, and the subspace spanned by the row vectors of B is contained in the row space of A. But it is also true that the rows of A can be obtained from the rows of B by elementary row operations. Thus, you can conclude that the two row spaces are subspaces of each other and are therefore equal.

REMARK: Note that Theorem 4.13 says that the row space of a matrix is not changed by elementary row operations. Elementary row operations can, however, change the *column* space.

If the matrix B is in row-echelon form, then its nonzero row vectors form a linearly independent set. (Try verifying this.) Consequently, they form a basis for the row space of B, and by Theorem 4.13 they also form a basis for the row space of A. This important result is stated in the next theorem.

Theorem 4.14	*Basis for the Row Space of a Matrix*

If a matrix A is row-equivalent to a matrix B in row-echelon form, then the nonzero row vectors of B form a basis for the row space of A.

EXAMPLE 2   *Finding a Basis for a Row Space*

Find a basis for the row space of

$$A = \begin{bmatrix} 1 & 3 & 1 & 3 \\ 0 & 1 & 1 & 0 \\ -3 & 0 & 6 & -1 \\ 3 & 4 & -2 & 1 \\ 2 & 0 & -4 & -2 \end{bmatrix}.$$

Solution   Using elementary *row* operations, rewrite A in row-echelon form as follows.

$$B = \begin{bmatrix} 1 & 3 & 1 & 3 \\ 0 & 1 & 1 & 0 \\ 0 & 0 & 0 & 1 \\ 0 & 0 & 0 & 0 \\ 0 & 0 & 0 & 0 \end{bmatrix} \begin{matrix} w_1 \\ w_2 \\ w_3 \\ \\ \end{matrix}$$

By Theorem 4.14, you can conclude that the nonzero row vectors of B,

$$\mathbf{w}_1 = (1, 3, 1, 3), \quad \mathbf{w}_2 = (0, 1, 1, 0), \quad \text{and} \quad \mathbf{w}_3 = (0, 0, 0, 1),$$

form a basis for the row space of A.

The technique used in Example 2 to find the row space of a matrix can be used to solve the following type of problem. Suppose you are asked to find a basis for the subspace spanned by the set $S = \{\mathbf{v}_1, \mathbf{v}_2, \ldots, \mathbf{v}_k\}$ in R^n. By using the vectors in S to form the rows of a matrix A, you can use elementary row operations to rewrite A in row-echelon form. The nonzero rows of this matrix will then form a basis for the subspace spanned by S. This is demonstrated in Example 3.

EXAMPLE 3   *Finding a Basis for a Subspace*

Find a basis for the subspace of R^3 spanned by

$$S = \{\overset{\mathbf{v}_1}{(-1, 2, 5)}, \overset{\mathbf{v}_2}{(3, 0, 3)}, \overset{\mathbf{v}_3}{(5, 1, 8)}\}.$$

Solution   Use $\mathbf{v}_1$, $\mathbf{v}_2$, and $\mathbf{v}_3$ to form the rows of a matrix A. Then write A in row-echelon form.

$$A = \begin{bmatrix} -1 & 2 & 5 \\ 3 & 0 & 3 \\ 5 & 1 & 8 \end{bmatrix} \begin{matrix} v_1 \\ v_2 \\ v_3 \end{matrix} \quad \longrightarrow \quad B = \begin{bmatrix} 1 & -2 & -5 \\ 0 & 1 & 3 \\ 0 & 0 & 0 \end{bmatrix} \begin{matrix} w_1 \\ w_2 \\ \end{matrix}$$

Therefore, the nonzero row vectors of B,

$$\mathbf{w}_1 = (1, -2, -5) \quad \text{and} \quad \mathbf{w}_2 = (0, 1, 3),$$

form a basis for the row space of A. That is, they form a basis for the subspace spanned by $S = \{\mathbf{v}_1, \mathbf{v}_2, \mathbf{v}_3\}$.

To find a basis for the column space of a matrix A, you have two options. On the one hand, you could use the fact that the column space of A is equal to the row space of A^T and apply the technique of Example 2 to the matrix A^T. On the other hand, observe that although row operations can change the column space of a matrix, they do not change the dependency relationships between columns. You are asked to prove this fact in Exercise 56.

For example, consider the row-equivalent matrices A and B from Example 2.

$$A = \begin{bmatrix} 1 & 3 & 1 & 3 \\ 0 & 1 & 1 & 0 \\ -3 & 0 & 6 & -1 \\ 3 & 4 & -2 & 1 \\ 2 & 0 & -4 & -2 \end{bmatrix} \qquad B = \begin{bmatrix} 1 & 3 & 1 & 3 \\ 0 & 1 & 1 & 0 \\ 0 & 0 & 0 & 1 \\ 0 & 0 & 0 & 0 \\ 0 & 0 & 0 & 0 \end{bmatrix}$$

$\quad\quad\;\; \mathbf{a}_1 \quad \mathbf{a}_2 \quad \mathbf{a}_3 \quad \mathbf{a}_4 \qquad\qquad\quad \mathbf{b}_1 \quad \mathbf{b}_2 \quad \mathbf{b}_3 \quad \mathbf{b}_4$

Notice that columns 1, 2, and 3 of matrix B satisfy the equation $\mathbf{b}_3 = -2\mathbf{b}_1 + \mathbf{b}_2$, and so do the corresponding columns of matrix A; that is,

$$\mathbf{a}_3 = -2\mathbf{a}_1 + \mathbf{a}_2.$$

Similarly, the column vectors $\mathbf{b}_1$, $\mathbf{b}_2$, and $\mathbf{b}_4$ of matrix B are linearly independent, and so are the corresponding columns of matrix A.

The following example shows how to find a basis for the column space of a matrix using both of these methods.

EXAMPLE 4 *Finding a Basis for the Column Space of a Matrix*

Find a basis for the column space of the matrix A given in Example 2.

$$A = \begin{bmatrix} 1 & 3 & 1 & 3 \\ 0 & 1 & 1 & 0 \\ -3 & 0 & 6 & -1 \\ 3 & 4 & -2 & 1 \\ 2 & 0 & -4 & -2 \end{bmatrix}$$

Solution 1 Take the transpose of A and use elementary row operations to write A^T in row-echelon form.

$$A^T = \begin{bmatrix} 1 & 0 & -3 & 3 & 2 \\ 3 & 1 & 0 & 4 & 0 \\ 1 & 1 & 6 & -2 & -4 \\ 3 & 0 & -1 & 1 & -2 \end{bmatrix} \longrightarrow \begin{bmatrix} 1 & 0 & -3 & 3 & 2 \\ 0 & 1 & 9 & -5 & -6 \\ 0 & 0 & 1 & -1 & -1 \\ 0 & 0 & 0 & 0 & 0 \end{bmatrix} \begin{matrix} \mathbf{w}_1 \\ \mathbf{w}_2 \\ \mathbf{w}_3 \\ \end{matrix}$$

Thus, $\mathbf{w}_1 = (1, 0, -3, 3, 2)$, $\mathbf{w}_2 = (0, 1, 9, -5, -6)$, and $\mathbf{w}_3 = (0, 0, 1, -1, -1)$ form a basis for the row space of A^T. This is equivalent to saying that the column vectors

$$\begin{bmatrix} 1 \\ 0 \\ -3 \\ 3 \\ 2 \end{bmatrix}, \quad \begin{bmatrix} 0 \\ 1 \\ 9 \\ -5 \\ -6 \end{bmatrix}, \quad \text{and} \quad \begin{bmatrix} 0 \\ 0 \\ 1 \\ -1 \\ -1 \end{bmatrix}$$

form a basis for the column space of A.

Solution 2   In Example 2, row operations were used on the original matrix A to obtain its row-echelon form B. It is easy to see that in the matrix B, the first, second, and fourth column vectors are linearly independent (these columns have the leading 1's). Hence, the corresponding columns of the matrix A are linearly independent, and a basis for the column space consists of the vectors

$$\begin{bmatrix} 1 \\ 0 \\ -3 \\ 3 \\ 2 \end{bmatrix}, \quad \begin{bmatrix} 3 \\ 1 \\ 0 \\ 4 \\ 0 \end{bmatrix}, \quad \text{and} \quad \begin{bmatrix} 3 \\ 0 \\ -1 \\ 1 \\ -2 \end{bmatrix}.$$

Notice that this is a different basis for the column space from that obtained in the first solution. Verify that these bases span the same subspace of R^5.

REMARK: Notice that in the second solution, the row-echelon form B indicates which columns of A form the basis of the column space. You do not use the column vectors of B to form the basis.

Notice in Examples 2 and 4 that both the row space and the column space of A have a dimension of 3 (because there are *three* vectors in both bases). This is generalized in the following theorem.

Theorem 4.15   Row and Column Spaces Have Equal Dimensions

If A is an $m \times n$ matrix, then the row space and column space of A have the same dimension.

Proof   Let $\mathbf{v}_1, \mathbf{v}_2, \ldots,$ and $\mathbf{v}_m$ be the row vectors and $\mathbf{u}_1, \mathbf{u}_2, \ldots,$ and $\mathbf{u}_n$ be the column vectors of the matrix

$$A = \begin{bmatrix} a_{11} & a_{12} & \cdots & a_{1n} \\ a_{21} & a_{22} & \cdots & a_{2n} \\ \vdots & \vdots & & \vdots \\ a_{m1} & a_{m2} & \cdots & a_{mn} \end{bmatrix}.$$

Suppose the row space of A has dimension r and basis $S = \{\mathbf{b}_1, \mathbf{b}_2, \ldots, \mathbf{b}_r\}$, where $\mathbf{b}_i = (b_{i1}, b_{i2}, \ldots, b_{in})$. Using this basis, you can write the row vectors of A as

$$\begin{aligned} \mathbf{v}_1 &= c_{11}\mathbf{b}_1 + c_{12}\mathbf{b}_2 + \cdots + c_{1r}\mathbf{b}_r \\ \mathbf{v}_2 &= c_{21}\mathbf{b}_1 + c_{22}\mathbf{b}_2 + \cdots + c_{2r}\mathbf{b}_r \\ &\ \ \vdots \\ \mathbf{v}_m &= c_{m1}\mathbf{b}_1 + c_{m2}\mathbf{b}_2 + \cdots + c_{mr}\mathbf{b}_r. \end{aligned}$$

Rewrite this system of vector equations as follows.

$$[a_{11}a_{12} \cdots a_{1n}] = c_{11}[b_{11}b_{12} \cdots b_{1n}] + c_{12}[b_{21}b_{22} \cdots b_{2n}] + \cdots + c_{1r}[b_{r1}b_{r2} \cdots b_{rn}]$$
$$[a_{21}a_{22} \cdots a_{2n}] = c_{21}[b_{11}b_{12} \cdots b_{1n}] + c_{22}[b_{21}b_{22} \cdots b_{2n}] + \cdots + c_{2r}[b_{r1}b_{r2} \cdots b_{rn}]$$
$$\vdots$$
$$[a_{m1}a_{m2} \cdots a_{mn}] = c_{m1}[b_{11}b_{12} \cdots b_{1n}] + c_{m2}[b_{21}b_{22} \cdots b_{2n}] + \cdots + c_{mr}[b_{r1}b_{r2} \cdots b_{rn}]$$

Now, take only entries corresponding to the first column of matrix A to obtain the following system of scalar equations.

$$a_{11} = c_{11}b_{11} + c_{12}b_{21} + c_{13}b_{31} + \cdots + c_{1r}b_{r1}$$
$$a_{21} = c_{21}b_{11} + c_{22}b_{21} + c_{23}b_{31} + \cdots + c_{2r}b_{r1}$$
$$a_{31} = c_{31}b_{11} + c_{32}b_{21} + c_{33}b_{31} + \cdots + c_{3r}b_{r1}$$
$$\vdots$$
$$a_{m1} = c_{m1}b_{11} + c_{m2}b_{21} + c_{m3}b_{31} + \cdots + c_{mr}b_{r1}$$

Similarly, for the entries of the jth column you can obtain the following system.

$$a_{1j} = c_{11}b_{1j} + c_{12}b_{2j} + c_{13}b_{3j} + \cdots + c_{1r}b_{rj}$$
$$a_{2j} = c_{21}b_{1j} + c_{22}b_{2j} + c_{23}b_{3j} + \cdots + c_{2r}b_{rj}$$
$$a_{3j} = c_{31}b_{1j} + c_{32}b_{2j} + c_{33}b_{3j} + \cdots + c_{3r}b_{rj}$$
$$\vdots$$
$$a_{mj} = c_{m1}b_{1j} + c_{m2}b_{2j} + c_{m3}b_{3j} + \cdots + c_{mr}b_{rj}$$

Now, let the vectors $\mathbf{c}_i = [c_{1i}c_{2i} \cdots c_{mi}]^T$. Then the system for the jth column can be rewritten in a vector form as

$$\mathbf{u}_j = b_{1j}\mathbf{c}_1 + b_{2j}\mathbf{c}_2 + \cdots + b_{rj}\mathbf{c}_r.$$

Put all column vectors together to obtain the following.

$$\mathbf{u}_1 = [a_{11} \quad a_{21} \cdots a_{m1}]^T = b_{11}\mathbf{c}_1 + b_{21}\mathbf{c}_2 + \cdots + b_{r1}\mathbf{c}_r$$
$$\mathbf{u}_2 = [a_{12} \quad a_{22} \cdots a_{m2}]^T = b_{12}\mathbf{c}_1 + b_{22}\mathbf{c}_2 + \cdots + b_{r2}\mathbf{c}_r$$
$$\vdots$$
$$\mathbf{u}_n = [a_{1n} \quad a_{2n} \cdots a_{mn}]^T = b_{1n}\mathbf{c}_1 + b_{2n}\mathbf{c}_2 + \cdots + b_{rn}\mathbf{c}_r$$

Because each column vector of A is a linear combination of r vectors, you know that the dimension of the column space of A is less than or equal to r (the dimension of the row space of A). That is,

dim(column space of A) $\leq$ dim(row space of A).

Repeating this procedure for A^T, you can conclude that the dimension of the column space of A^T is less than or equal to the dimension of the row space of A^T. But this implies that the dimension of the row space of A is less than or equal to the dimension of the column space of A. That is,

dim(row space of A) $\leq$ dim(column space of A).

Therefore, the two dimensions must be equal.

The dimension of the row (or column) space of a matrix has the following special name.

Definition of the Rank of a Matrix	The dimension of the row (or column) space of a matrix A is called the **rank** of A and is denoted by rank(A).

R E M A R K : Some texts distinguish between the *row rank* and the *column rank* of a matrix. But because these ranks are equal (Theorem 4.15), this text will not distinguish between them.

EXAMPLE 5   *Finding the Rank of a Matrix*

Find the rank of the matrix

$$A = \begin{bmatrix} 1 & -2 & 0 & 1 \\ 2 & 1 & 5 & -3 \\ 0 & 1 & 3 & 5 \end{bmatrix}.$$

Solution   Convert to row-echelon form as follows.

$$A = \begin{bmatrix} 1 & -2 & 0 & 1 \\ 2 & 1 & 5 & -3 \\ 0 & 1 & 3 & 5 \end{bmatrix} \longrightarrow B = \begin{bmatrix} 1 & -2 & 0 & 1 \\ 0 & 1 & 1 & -1 \\ 0 & 0 & 1 & 3 \end{bmatrix}$$

Because B has three nonzero rows, the rank of A is 3.

The Nullspace of a Matrix

The notions of row and column spaces and rank have some important applications to systems of linear equations. Consider first the homogeneous linear system

$$A\mathbf{x} = \mathbf{0}$$

where A is an $m \times n$ matrix, $\mathbf{x} = [x_1 \ x_2 \ \ldots \ x_n]^T$ is the column vector of unknowns, and $\mathbf{0} = [0 \ 0 \ \ldots \ 0]^T$ is the zero vector in R^m.

$$\begin{bmatrix} a_{11} & a_{12} & \cdots & a_{1n} \\ a_{21} & a_{22} & \cdots & a_{2n} \\ \vdots & \vdots & & \vdots \\ a_{m1} & a_{m2} & \cdots & a_{mn} \end{bmatrix} \begin{bmatrix} x_1 \\ x_2 \\ \vdots \\ x_n \end{bmatrix} = \begin{bmatrix} 0 \\ 0 \\ \vdots \\ 0 \end{bmatrix}$$

The following theorem tells you that the set of all solutions of this homogeneous system is a subspace of R^n.

Theorem 4.16	*Solutions of a Homogeneous System*

Solutions of a Homogeneous System

If A is an $m \times n$ matrix, then the set of all solutions of the homogeneous system of linear equations

$$A\mathbf{x} = \mathbf{0}$$

is a subspace of R^n called the **nullspace** of A and is denoted by $N(A)$. So,

$$N(A) = \{\mathbf{x} \in R^n : A\mathbf{x} = \mathbf{0}\}.$$

The dimension of the nullspace of A is called the **nullity** of A.

Proof   Because A is an $m \times n$ matrix, you know that $\mathbf{x}$ has size $n \times 1$. Thus, the set of all solutions of the system is a *subset* of R^n. This set is clearly nonempty, because $A\mathbf{0} = \mathbf{0}$. You can verify that it is a subspace by showing that it is closed under the operations of addition and scalar multiplication. Let $\mathbf{x}_1$ and $\mathbf{x}_2$ be two solution vectors of the system $A\mathbf{x} = \mathbf{0}$, and let c be a scalar. Because $A\mathbf{x}_1 = \mathbf{0}$ and $A\mathbf{x}_2 = \mathbf{0}$, you know that

$$A(\mathbf{x}_1 + \mathbf{x}_2) = A\mathbf{x}_1 + A\mathbf{x}_2 = \mathbf{0} + \mathbf{0} = \mathbf{0} \qquad \text{Addition}$$

and

$$A(c\mathbf{x}_1) = c(A\mathbf{x}_1) = c(\mathbf{0}) = \mathbf{0}. \qquad \text{Scalar multiplication}$$

Thus, both $(\mathbf{x}_1 + \mathbf{x}_2)$ and $c\mathbf{x}_1$ are solutions of $A\mathbf{x} = \mathbf{0}$, and you can conclude that the set of all solutions forms a subspace of R^n.

REMARK: The nullspace of A is also called the **solution space** of the system $A\mathbf{x} = \mathbf{0}$.

EXAMPLE 6   *Finding the Solution Space of a Homogeneous System*

Find the nullspace of the following matrix.

$$A = \begin{bmatrix} 1 & 2 & -2 & 1 \\ 3 & 6 & -5 & 4 \\ 1 & 2 & 0 & 3 \end{bmatrix}$$

Solution   The nullspace of A is the solution space of the homogeneous system $A\mathbf{x} = \mathbf{0}$.

To solve this system, you need to write the augmented matrix $[A \ \vdots \ \mathbf{0}]$ in reduced row-echelon form. However, because the system of equations is homogeneous, the right-hand column of the augmented matrix consists entirely of zeros and will not change as you do row operations. Hence, it is sufficient to find the reduced row-echelon form of A.

$$A = \begin{bmatrix} 1 & 2 & -2 & 1 \\ 3 & 6 & -5 & 4 \\ 1 & 2 & 0 & 3 \end{bmatrix} \longrightarrow \begin{bmatrix} 1 & 2 & 0 & 3 \\ 0 & 0 & 1 & 1 \\ 0 & 0 & 0 & 0 \end{bmatrix}$$

The system of equations corresponding to the reduced row-echelon form is

$$x_1 + 2x_2 \quad + 3x_4 = 0$$
$$x_3 + \quad x_4 = 0.$$

Choose x_2 and x_4 as free variables to represent the solutions in this parametric form.

$$x_1 = -2s - 3t, \quad x_2 = s, \quad x_3 = -t, \quad x_4 = t$$

This means that the solution space of $A\mathbf{x} = \mathbf{0}$ consists of all solution vectors $\mathbf{x}$ of the following form.

$$\mathbf{x} = \begin{bmatrix} x_1 \\ x_2 \\ x_3 \\ x_4 \end{bmatrix} = \begin{bmatrix} -2s - 3t \\ s \\ -t \\ t \end{bmatrix} = s\begin{bmatrix} -2 \\ 1 \\ 0 \\ 0 \end{bmatrix} + t\begin{bmatrix} -3 \\ 0 \\ -1 \\ 1 \end{bmatrix}$$

Thus, a basis for the nullspace of A consists of the vectors

$$\begin{bmatrix} -2 \\ 1 \\ 0 \\ 0 \end{bmatrix} \quad \text{and} \quad \begin{bmatrix} -3 \\ 0 \\ -1 \\ 1 \end{bmatrix}.$$

In other words, these two vectors are solutions of $A\mathbf{x} = \mathbf{0}$, and all solutions of this homogeneous system are linear combinations of these two vectors.

REMARK: For the basis in Example 6, it was not proven that the vectors were linearly independent, only that they spanned the solution set. It can be shown that when homogeneous systems are solved from the reduced row-echelon form, the spanning set is always independent.

In Example 6, the matrix A has four columns. Furthermore, the rank of the matrix is 2, and the dimension of the nullspace is 2. Therefore, you can see that

Number of columns = rank + nullity.

One way to see this is to look at the reduced row-echelon form of A.

$$\begin{bmatrix} 1 & 2 & 0 & 3 \\ 0 & 0 & 1 & 1 \\ 0 & 0 & 0 & 0 \end{bmatrix}$$

The columns with the leading 1's (columns 1 and 3) determine the rank of the matrix. The other columns (2 and 4) determine the nullity of the matrix because they correspond to the free variables. This relationship is generalized in the following theorem.

Theorem 4.17	*Dimension of the Solution Space*

If A is an $m \times n$ matrix of rank r, then the dimension of the solution space of $A\mathbf{x} = \mathbf{0}$ is $n - r$. That is,

$$n = \text{rank}(A) + \text{nullity}(A).$$

Proof   Because A has rank r, you know that it is row-equivalent to a reduced row-echelon matrix B with r nonzero rows. No generality is lost by assuming that the upper left corner of B has the form of the $r \times r$ identity matrix I_r. Moreover, because the zero rows of B contribute nothing to the solution, you can discard them to form the $r \times n$ matrix B', where $B' = [I_r \vdots C]$. The matrix C has $n - r$ columns corresponding to the variables $x_{r+1}, x_{r+2}, \ldots, x_n$. Thus, the solution space of $A\mathbf{x} = \mathbf{0}$ can be represented by the system

$$
\begin{aligned}
x_1 + & & c_{11}x_{r+1} + c_{12}x_{r+2} + \cdots + c_{1, n-r}x_n &= 0 \\
& x_2 + & c_{21}x_{r+1} + c_{22}x_{r+2} + \cdots + c_{2, n-r}x_n &= 0 \\
& & \vdots \qquad\qquad \vdots \qquad\qquad\quad \vdots & \quad\vdots \\
& x_r + & c_{r1}x_{r+1} + c_{r2}x_{r+2} + \cdots + c_{r, n-r}x_n &= 0.
\end{aligned}
$$

Solving for the first r variables in terms of the last $n - r$ variables produces $n - r$ vectors in the basis of the solution space. Consequently, the solution space has dimension $n - r$.

Example 7 illustrates this theorem and further explores the column space of a matrix.

EXAMPLE 7   *Rank and Nullity of a Matrix*

Let the column vectors of the matrix A be denoted by $\mathbf{a}_1, \mathbf{a}_2, \mathbf{a}_3, \mathbf{a}_4,$ and $\mathbf{a}_5$.

$$
A = \begin{bmatrix} 1 & 0 & -2 & 1 & 0 \\ 0 & -1 & -3 & 1 & 3 \\ -2 & -1 & 1 & -1 & 3 \\ 0 & 3 & 9 & 0 & -12 \end{bmatrix}
$$
$$
\quad\;\; \mathbf{a}_1 \;\;\; \mathbf{a}_2 \;\;\; \mathbf{a}_3 \;\;\; \mathbf{a}_4 \;\;\; \mathbf{a}_5
$$

(a) Find the rank and nullity of A.
(b) Find a subset of the column vectors of A that forms a basis for the column space of A.
(c) If possible, write the third column of A as a linear combination of the first two columns.

Solution   Let B be the reduced row-echelon form of A.

$$
A = \begin{bmatrix} 1 & 0 & -2 & 1 & 0 \\ 0 & -1 & -3 & 1 & 3 \\ -2 & -1 & 1 & -1 & 3 \\ 0 & 3 & 9 & 0 & -12 \end{bmatrix} \longrightarrow B = \begin{bmatrix} 1 & 0 & -2 & 0 & 1 \\ 0 & 1 & 3 & 0 & -4 \\ 0 & 0 & 0 & 1 & -1 \\ 0 & 0 & 0 & 0 & 0 \end{bmatrix}
$$
$$
\;\; \mathbf{a}_1 \;\;\; \mathbf{a}_2 \;\;\; \mathbf{a}_3 \;\;\; \mathbf{a}_4 \;\;\; \mathbf{a}_5 \qquad\qquad\qquad \mathbf{b}_1 \;\;\; \mathbf{b}_2 \;\;\; \mathbf{b}_3 \;\;\; \mathbf{b}_4 \;\;\; \mathbf{b}_5
$$

(a) Because B has three nonzero rows, the rank of A is 3. Furthermore, the number of columns of A is $n = 5$, which implies that the nullity of A is $n - \text{rank} = 5 - 3 = 2$.

(b) Because the first, second, and fourth column vectors of B are linearly independent, the corresponding column vectors of A,

$$
\mathbf{a}_1 = \begin{bmatrix} 1 \\ 0 \\ -2 \\ 0 \end{bmatrix}, \qquad \mathbf{a}_2 = \begin{bmatrix} 0 \\ -1 \\ -1 \\ 3 \end{bmatrix}, \qquad \text{and} \qquad \mathbf{a}_4 = \begin{bmatrix} 1 \\ 1 \\ -1 \\ 0 \end{bmatrix},
$$

form a basis for the column space of A.

(c) The third column of B is a linear combination of the first two columns: $\mathbf{b}_3 = -2\mathbf{b}_1 + 3\mathbf{b}_2$. Therefore, the same dependency relationship holds for the corresponding columns of matrix A.

$$
\mathbf{a}_3 = \begin{bmatrix} -2 \\ -3 \\ 1 \\ 9 \end{bmatrix} = -2 \begin{bmatrix} 1 \\ 0 \\ -2 \\ 0 \end{bmatrix} + 3 \begin{bmatrix} 0 \\ -1 \\ -1 \\ 3 \end{bmatrix} = -2\mathbf{a}_1 + 3\mathbf{a}_2
$$

Solutions of Systems of Linear Equations

You now know that the set of all solution vectors of the *homogeneous* linear system $A\mathbf{x} = \mathbf{0}$ is a subspace. Is this true also of the set of all solution vectors of the *nonhomogeneous* system $A\mathbf{x} = \mathbf{b}$, where $\mathbf{b} \neq \mathbf{0}$? The answer is "no," because the zero vector is never a solution of a nonhomogeneous system. There is a relationship, however, between the sets of solutions of the two systems $A\mathbf{x} = \mathbf{0}$ and $A\mathbf{x} = \mathbf{b}$. Specifically, if $\mathbf{x}_p$ is a *particular* solution of the nonhomogeneous system $A\mathbf{x} = \mathbf{b}$, then *every* solution of this system can be written in the form

$$
\mathbf{x} = \mathbf{x}_p + \mathbf{x}_h,
$$

Solution of $A\mathbf{x} = \mathbf{b}$	Solution of $A\mathbf{x} = \mathbf{0}$

where $\mathbf{x}_h$ is a solution of the corresponding homogeneous system $A\mathbf{x} = \mathbf{0}$. The following theorem states this important result.

Theorem 4.18 *Solutions of a Nonhomogeneous Linear System*

If $\mathbf{x}_p$ is a particular solution of the nonhomogeneous system $A\mathbf{x} = \mathbf{b}$, then every solution of this system can be written in the form $\mathbf{x} = \mathbf{x}_p + \mathbf{x}_h$, where $\mathbf{x}_h$ is a solution of the corresponding homogeneous system $A\mathbf{x} = \mathbf{0}$.

Proof Let **x** be any solution of $A\mathbf{x} = \mathbf{b}$. Then $(\mathbf{x} - \mathbf{x}_p)$ is a solution of the homogeneous system $A\mathbf{x} = \mathbf{0}$, because

$$A(\mathbf{x} - \mathbf{x}_p) = A\mathbf{x} - A\mathbf{x}_p = \mathbf{b} - \mathbf{b} = \mathbf{0}.$$

Letting $\mathbf{x}_h = \mathbf{x} - \mathbf{x}_p$, you have $\mathbf{x} = \mathbf{x}_p + \mathbf{x}_h$.

EXAMPLE 8 *Finding the Solution Set of a Nonhomogeneous System*

Find the set of all solution vectors of the system of linear equations.

$$\begin{array}{rcr} x_1 \quad\quad - 2x_3 + x_4 &=& 5 \\ 3x_1 + x_2 - 5x_3 \quad\quad &=& 8 \\ x_1 + 2x_2 \quad\quad - 5x_4 &=& -9 \end{array}$$

Solution The augmented matrix for the system $A\mathbf{x} = \mathbf{b}$ reduces as follows.

$$\begin{bmatrix} 1 & 0 & -2 & 1 & 5 \\ 3 & 1 & -5 & 0 & 8 \\ 1 & 2 & 0 & -5 & -9 \end{bmatrix} \longrightarrow \begin{bmatrix} 1 & 0 & -2 & 1 & 5 \\ 0 & 1 & 1 & -3 & -7 \\ 0 & 0 & 0 & 0 & 0 \end{bmatrix}$$

The system of linear equations corresponding to the reduced row-echelon matrix is

$$\begin{array}{rcr} x_1 \quad\quad - 2x_3 + x_4 &=& 5 \\ x_2 + x_3 - 3x_4 &=& -7. \end{array}$$

Letting $x_3 = s$ and $x_4 = t$, you can write a representative solution vector of $A\mathbf{x} = \mathbf{b}$ as follows.

$$\mathbf{x} = \begin{bmatrix} x_1 \\ x_2 \\ x_3 \\ x_4 \end{bmatrix} = \begin{bmatrix} 2s - t + 5 \\ -s + 3t - 7 \\ s + 0t + 0 \\ 0s + t + 0 \end{bmatrix} = s\begin{bmatrix} 2 \\ -1 \\ 1 \\ 0 \end{bmatrix} + t\begin{bmatrix} -1 \\ 3 \\ 0 \\ 1 \end{bmatrix} + \begin{bmatrix} 5 \\ -7 \\ 0 \\ 0 \end{bmatrix}$$

$$= s\mathbf{u}_1 + t\mathbf{u}_2 + \mathbf{x}_p$$

You can see that $\mathbf{x}_p$ is a *particular* solution vector of $A\mathbf{x} = \mathbf{b}$, and $\mathbf{x}_h = s\mathbf{u}_1 + t\mathbf{u}_2$ represents an arbitrary vector in the solution space of $A\mathbf{x} = \mathbf{0}$.

The final theorem in this section describes how the column space of a matrix can be used to determine whether a system of linear equations is consistent.

Theorem 4.19 *Solutions of a System of Linear Equations*

The system of linear equations $A\mathbf{x} = \mathbf{b}$ is consistent if and only if **b** is in the column space of A.

Proof Let

$$
A = \begin{bmatrix} a_{11} & a_{12} & \cdots & a_{1n} \\ a_{21} & a_{22} & \cdots & a_{2n} \\ \vdots & \vdots & & \vdots \\ a_{m1} & a_{m2} & \cdots & a_{mn} \end{bmatrix}, \quad \mathbf{x} = \begin{bmatrix} x_1 \\ x_2 \\ \vdots \\ x_n \end{bmatrix}, \quad \text{and} \quad \mathbf{b} = \begin{bmatrix} b_1 \\ b_2 \\ \vdots \\ b_m \end{bmatrix}
$$

be the coefficient matrix, the column matrix of unknowns, and the right-hand side, respectively, of the system $A\mathbf{x} = \mathbf{b}$. Then

$$
A\mathbf{x} = \begin{bmatrix} a_{11} & a_{12} & \cdots & a_{1n} \\ a_{21} & a_{22} & \cdots & a_{2n} \\ \vdots & \vdots & & \vdots \\ a_{m1} & a_{m2} & \cdots & a_{mn} \end{bmatrix} \begin{bmatrix} x_1 \\ x_2 \\ \vdots \\ x_n \end{bmatrix}
$$

$$
= \begin{bmatrix} a_{11}x_1 + a_{12}x_2 + \cdots + a_{1n}x_n \\ a_{21}x_1 + a_{22}x_2 + \cdots + a_{2n}x_n \\ \vdots & \vdots & & \vdots \\ a_{m1}x_1 + a_{m2}x_2 + \cdots + a_{mn}x_n \end{bmatrix}
$$

$$
= x_1 \begin{bmatrix} a_{11} \\ a_{21} \\ \vdots \\ a_{m1} \end{bmatrix} + x_2 \begin{bmatrix} a_{12} \\ a_{22} \\ \vdots \\ a_{m2} \end{bmatrix} + \cdots + x_n \begin{bmatrix} a_{1n} \\ a_{2n} \\ \vdots \\ a_{mn} \end{bmatrix}.
$$

Hence, $A\mathbf{x} = \mathbf{b}$ if and only if $\mathbf{b}$ is a linear combination of the columns of A. That is, the system is consistent if and only if $\mathbf{b}$ is in the subspace of R^m spanned by the columns of A.

EXAMPLE 9 *Consistency of a System of Linear Equations*

Consider the system of linear equations

$$
\begin{aligned}
x_1 + x_2 - x_3 &= -1 \\
x_1 \quad\quad + x_3 &= 3 \\
3x_1 + 2x_2 - x_3 &= 1.
\end{aligned}
$$

The rank of the coefficient matrix is equal to the rank of the augmented matrix.

$$
A = \begin{bmatrix} 1 & 1 & -1 \\ 1 & 0 & 1 \\ 3 & 2 & -1 \end{bmatrix} \longrightarrow \begin{bmatrix} 1 & 0 & 1 \\ 0 & 1 & -2 \\ 0 & 0 & 0 \end{bmatrix}
$$

$$
[A \;\vdots\; \mathbf{b}] = \begin{bmatrix} 1 & 1 & -1 & -1 \\ 1 & 0 & 1 & 3 \\ 3 & 2 & -1 & 1 \end{bmatrix} \longrightarrow \begin{bmatrix} 1 & 0 & 1 & 3 \\ 0 & 1 & -2 & -4 \\ 0 & 0 & 0 & 0 \end{bmatrix}
$$

Hence, $\mathbf{b}$ is in the column space of A, and the system of linear equations is consistent.

Systems of Linear Equations with Square Coefficient Matrices

The final result in this section summarizes several major results involving systems of linear equations, matrices, determinants, and vector spaces.

Summary of Equivalent Conditions for Square Matrices

If A is an $n \times n$ matrix, then the following conditions are equivalent.

1. A is invertible.
2. $A\mathbf{x} = \mathbf{b}$ has a unique solution for any $n \times 1$ matrix $\mathbf{b}$.
3. $A\mathbf{x} = \mathbf{0}$ has only the trivial solution.
4. A is row-equivalent to I_n.
5. $|A| \neq 0$.
6. Rank$(A) = n$.
7. The n row vectors of A are linearly independent.
8. The n column vectors of A are linearly independent.

SECTION 4.6 Exercises

In Exercises 1–8, find (a) the rank of the matrix, (b) a basis for the row space, and (c) a basis for the column space.

1. $\begin{bmatrix} 1 & 0 \\ 0 & 2 \end{bmatrix}$

2. $\begin{bmatrix} 1 & 2 & 3 \end{bmatrix}$

3. $\begin{bmatrix} 2 & 4 \\ 1 & 6 \end{bmatrix}$

4. $\begin{bmatrix} 1 & -3 & 2 \\ 4 & 2 & 1 \end{bmatrix}$

5. $\begin{bmatrix} 4 & 20 & 31 \\ 6 & -5 & -6 \\ 2 & -11 & -16 \end{bmatrix}$

6. $\begin{bmatrix} 2 & -3 & 1 \\ 5 & 10 & 6 \\ 8 & -7 & 5 \end{bmatrix}$

7. $\begin{bmatrix} -2 & -4 & 4 & 5 \\ 3 & 6 & -6 & -4 \\ -2 & -4 & 4 & 9 \end{bmatrix}$

8. $\begin{bmatrix} 2 & 4 & -3 & -6 \\ 7 & 14 & -6 & -3 \\ -2 & -4 & 1 & -2 \\ 2 & 4 & -2 & -2 \end{bmatrix}$

In Exercises 9–12, find a basis for the subspace of R^4 spanned by S.

9. $S = \{(2, 9, -2, 53), (-3, 2, 3, -2), (8, -3, -8, 17),$
$(0, -3, 0, 15)\}$

10. $S = \{(6, -3, 6, 34), (3, -2, 3, 19), (8, 3, -9, 6),$
$(-2, 0, 6, -5)\}$

11. $S = \{(-3, 2, 5, 28), (-6, 1, -8, -1), (14, -10, 12, -10),$
$(0, 5, 12, 50)\}$

12. $S = \{(2, 5, -3, -2), (-2, -3, 2, -5), (1, 3, -2, 2),$
$(-1, -5, 3, 5)\}$

In Exercises 13–20, find a basis for and the dimension of the solution space of $A\mathbf{x} = \mathbf{0}$.

13. $A = \begin{bmatrix} 2 & -1 \\ 1 & 3 \end{bmatrix}$

14. $A = \begin{bmatrix} 2 & -1 \\ -6 & 3 \end{bmatrix}$

15. $A = \begin{bmatrix} 1 & 2 & 3 \end{bmatrix}$

16. $A = \begin{bmatrix} 1 & 2 & 3 \\ 0 & 1 & 0 \end{bmatrix}$

17. $A = \begin{bmatrix} 1 & 2 & -3 \\ 2 & -1 & 4 \\ 4 & 3 & -2 \end{bmatrix}$

18. $A = \begin{bmatrix} 3 & -6 & 21 \\ -2 & 4 & -14 \\ 1 & -2 & 7 \end{bmatrix}$

19. $A = \begin{bmatrix} 1 & 3 & -2 & 4 \\ 0 & 1 & -1 & 2 \\ -2 & -6 & 4 & -8 \end{bmatrix}$

20. $A = \begin{bmatrix} 2 & 6 & 3 & 1 \\ 2 & 1 & 0 & -2 \\ 3 & -2 & 1 & 1 \\ 0 & 6 & 2 & 0 \end{bmatrix}$

In Exercises 21–26, find (a) a basis for and (b) the dimension of the solution space of the homogeneous system of linear equations.

21. $\begin{aligned} -x + y + z &= 0 \\ 3x - y &= 0 \\ 2x - 4y - 5z &= 0 \end{aligned}$

22. $\begin{aligned} 4x - y + 2z &= 0 \\ 2x + 3y - z &= 0 \\ 3x + y + z &= 0 \end{aligned}$

23. $\begin{aligned} x - 2y + 3z &= 0 \\ -3x + 6y - 9z &= 0 \end{aligned}$

24. $\begin{aligned} 3x_1 + 3x_2 + 15x_3 + 11x_4 &= 0 \\ x_1 - 3x_2 + x_3 + x_4 &= 0 \\ 2x_1 + 3x_2 + 11x_3 + 8x_4 &= 0 \end{aligned}$

25. $\begin{aligned} 9x_1 - 4x_2 - 2x_3 - 20x_4 &= 0 \\ 12x_1 - 6x_2 - 4x_3 - 29x_4 &= 0 \\ 3x_1 - 2x_2 \quad\quad - 7x_4 &= 0 \\ 3x_1 - 2x_2 - x_3 - 8x_4 &= 0 \end{aligned}$

26. $\begin{aligned} x_1 + 3x_2 + 2x_3 + 22x_4 + 13x_5 &= 0 \\ x_1 \quad\quad + x_3 - 2x_4 + x_5 &= 0 \\ 3x_1 + 6x_2 + 5x_3 + 42x_4 + 27x_5 &= 0 \end{aligned}$

In Exercises 27–32, (a) determine whether the nonhomogeneous system $A\mathbf{x} = \mathbf{b}$ is consistent and (b) if the system is consistent, write the solution in the form $\mathbf{x} = \mathbf{x}_h + \mathbf{x}_p$, where $\mathbf{x}_h$ is a solution of $A\mathbf{x} = \mathbf{0}$ and $\mathbf{x}_p$ is a particular solution of $A\mathbf{x} = \mathbf{b}$.

27. $\begin{aligned} x + 3y + 10z &= 18 \\ -2x + 7y + 32z &= 29 \\ -x + 3y + 14z &= 12 \\ x + y + 2z &= 8 \end{aligned}$

28. $\begin{aligned} 3x - 8y + 4z &= 19 \\ -6y + 2z + 4w &= 5 \\ 5x + 22z + w &= 29 \\ x - 2y + 2z &= 8 \end{aligned}$

29. $\begin{aligned} 3w - 2x + 16y - 2z &= -7 \\ -w + 5x - 14y + 18z &= 29 \\ 3w - x + 14y + 2z &= 1 \end{aligned}$

30. $\begin{aligned} 2x - 4y + 5z &= 8 \\ -7x + 14y + 4z &= -28 \\ 3x - 6y + z &= 12 \end{aligned}$

31. $\begin{aligned} x_1 + 2x_2 + x_3 + x_4 + 5x_5 &= 0 \\ -5x_1 - 10x_2 + 3x_3 + 3x_4 + 55x_5 &= -8 \\ x_1 + 2x_2 + 2x_3 - 3x_4 - 5x_5 &= 14 \\ -x_1 - 2x_2 + x_3 + x_4 + 15x_5 &= -2 \end{aligned}$

32. $\begin{aligned} 5x_1 - 4x_2 + 12x_3 - 33x_4 + 14x_5 &= -4 \\ -2x_1 + x_2 - 6x_3 + 12x_4 - 8x_5 &= 1 \\ 2x_1 - x_2 + 6x_3 - 12x_4 + 8x_5 &= -1 \end{aligned}$

In Exercises 33–36, determine whether $\mathbf{b}$ is in the column space of A. If it is, write $\mathbf{b}$ as a linear combination of the column vectors of A.

33. $A = \begin{bmatrix} -1 & 2 \\ 4 & 0 \end{bmatrix}, \quad \mathbf{b} = \begin{bmatrix} 3 \\ 4 \end{bmatrix}$

34. $A = \begin{bmatrix} -1 & 2 \\ 2 & -4 \end{bmatrix}, \quad \mathbf{b} = \begin{bmatrix} 2 \\ 4 \end{bmatrix}$

35. $A = \begin{bmatrix} 1 & 3 & 0 \\ -1 & 1 & 0 \\ 2 & 0 & 1 \end{bmatrix}, \quad \mathbf{b} = \begin{bmatrix} 1 \\ 2 \\ -3 \end{bmatrix}$

36. $A = \begin{bmatrix} 1 & 3 & 2 \\ -1 & 1 & 2 \\ 0 & 1 & 1 \end{bmatrix}, \quad \mathbf{b} = \begin{bmatrix} 1 \\ 1 \\ 0 \end{bmatrix}$

37. *Writing*   Explain why the row vectors of a 4×3 matrix form a linearly dependent set. (Assume all matrix entries are distinct.)

38. *Writing*   Explain why the column vectors of a 3×4 matrix form a linearly dependent set. (Assume all matrix entries are distinct.)

39. Prove that if A is not square, then either the row vectors of A or the column vectors of A form a linearly dependent set.

40. Give an example showing that the rank of the product of two matrices can be less than the rank of either matrix.

41. Give examples of matrices A and B of the same size such that

(a) rank$(A + B)$ < rank(A) and rank$(A + B)$ < rank(B)

(b) rank$(A + B)$ = rank(A) and rank$(A + B)$ = rank(B)

(c) rank$(A + B)$ > rank(A) and rank$(A + B)$ > rank(B).

42. Prove that the nonzero row vectors of a matrix in row-echelon form are linearly independent.

43. Let A be an $m \times n$ matrix (where $m < n$) whose rank is r.

(a) What is the largest value r can be?

(b) How many vectors are in a basis for the row space of A?

(c) How many vectors are in a basis for the column space of A?

(d) Which vector space R^k has the row space as a subspace?

(e) Which vector space R^k has the column space as a subspace?

44. Show that the three points (x_1, y_1), (x_2, y_2), and (x_3, y_3) in a plane are collinear if and only if the matrix

$$\begin{bmatrix} x_1 & y_1 & 1 \\ x_2 & y_2 & 1 \\ x_3 & y_3 & 1 \end{bmatrix}$$

has rank less than 3.

45. Given the matrices A and B, show that the row vectors of AB are in the row space of B and the column vectors of AB are in the column space of A.

46. Find the rank of the matrix

$$\begin{bmatrix} 1 & 2 & 3 & \cdots & n \\ n+1 & n+2 & n+3 & \cdots & 2n \\ 2n+1 & 2n+2 & 2n+3 & \cdots & 3n \\ \vdots & \vdots & \vdots & & \vdots \\ n^2-n+1 & n^2-n+2 & n^2-n+3 & \cdots & n^2 \end{bmatrix}$$

for $n = 2, 3$, and 4. Can you find a pattern in these ranks?

47. Prove each property of the system of linear equations in n variables $A\mathbf{x} = \mathbf{b}$.

(a) If $\text{rank}(A) = \text{rank}([A \ : \ \mathbf{b}]) = n$, then the system has a unique solution.

(b) If $\text{rank}(A) = \text{rank}([A \ : \ \mathbf{b}]) < n$, then the system has an infinite number of solutions.

(c) If $\text{rank}(A) < \text{rank}([A \ : \ \mathbf{b}])$, then the system is inconsistent.

True or False? In Exercises 48–50, determine whether each statement is true or false. If a statement is true, give a reason or cite an appropriate statement in the text. If a statement is false, provide an example that shows that the statement is not true in all cases or cite an appropriate statement in the text.

48. (a) If an $m \times n$ matrix A is row-equivalent to an $m \times n$ matrix B, then the row space of A is equivalent to the row space of B.

(b) If A is an $m \times n$ matrix of rank r, then the dimension of the solution space of $A\mathbf{x} = \mathbf{0}$ is $m - r$.

49. (a) If an $m \times n$ matrix B can be obtained from elementary row operations on an $m \times n$ matrix A, then the column space of B is equal to the column space of A.

(b) The system of linear equations $A\mathbf{x} = \mathbf{b}$ is inconsistent if and only if $\mathbf{b}$ is in the column space of A.

50. (a) The column space of a matrix A is equal to the row space of A^T.

(b) Row operations on a matrix A may change the dependency relationships among the columns of A.

In Exercises 51 and 52, use the fact that the matrices A and B are row-equivalent to answer the following about matrix A.

(a) Find the rank and nullity of A.

(b) Find a basis for the nullspace of A.

(c) Find a basis for the row space of A.

(d) Find a basis for the column space of A.

(e) Determine whether or not the rows of A are linearly independent.

(f) Let the columns of A be denoted by $\mathbf{a}_1$, $\mathbf{a}_2$, $\mathbf{a}_3$, $\mathbf{a}_4$, and $\mathbf{a}_5$. Which of the following sets are linearly independent?

(i) $\{\mathbf{a}_1, \mathbf{a}_2, \mathbf{a}_4\}$ (ii) $\{\mathbf{a}_1, \mathbf{a}_2, \mathbf{a}_3\}$ (iii) $\{\mathbf{a}_1, \mathbf{a}_3, \mathbf{a}_5\}$

51. $A = \begin{bmatrix} 1 & 2 & 1 & 0 & 0 \\ 2 & 5 & 1 & 1 & 0 \\ 3 & 7 & 2 & 2 & -2 \\ 4 & 9 & 3 & -1 & 4 \end{bmatrix}$

$B = \begin{bmatrix} 1 & 0 & 3 & 0 & -4 \\ 0 & 1 & -1 & 0 & 2 \\ 0 & 0 & 0 & 1 & -2 \\ 0 & 0 & 0 & 0 & 0 \end{bmatrix}$

52. $A = \begin{bmatrix} -2 & -5 & 8 & 0 & -17 \\ 1 & 3 & -5 & 1 & 5 \\ 3 & 11 & -19 & 7 & 1 \\ 1 & 7 & -13 & 5 & -3 \end{bmatrix}$

$B = \begin{bmatrix} 1 & 0 & 1 & 0 & 1 \\ 0 & 1 & -2 & 0 & 3 \\ 0 & 0 & 0 & 1 & -5 \\ 0 & 0 & 0 & 0 & 0 \end{bmatrix}$

53. Let A be an $m \times n$ matrix.

(a) Prove that the system of linear equations $A\mathbf{x} = \mathbf{b}$ is consistent for all column vectors $\mathbf{b}$ if and only if the rank of A is m.

(b) Prove that the homogeneous system of linear equations $A\mathbf{x} = \mathbf{0}$ has only the trivial solution if and only if the columns of A are linearly independent.

54. Let A and B be square matrices of order n satisfying $A\mathbf{x} = B\mathbf{x}$ for all $\mathbf{x} \in R^n$.

(a) Find the nullspace and nullity of $A - B$.

(b) Show that A and B must be identical.

55. Let A be an $m \times n$ matrix. Prove that $N(A) \subset N(A^TA)$.

56. Prove that row operations do not change the dependency relationships among columns of an $m \times n$ matrix.

4.7 | Coordinates and Change of Basis

In Section 4.5, you saw that if B is a basis for a vector space V, then every vector $\mathbf{x}$ in V can be expressed in one and only one way as a linear combination of vectors in B. The coefficients in the linear combination are the **coordinates of x relative to B.** In the context of coordinates, the order of the vectors in this basis is important, and this will sometimes be emphasized by referring to the basis B as an *ordered* basis.

Coordinate Representation Relative to a Basis

Let $B = \{\mathbf{v}_1, \mathbf{v}_2, \ldots, \mathbf{v}_n\}$ be an ordered basis for a vector space V and let $\mathbf{x}$ be a vector in V such that

$$\mathbf{x} = c_1\mathbf{v}_1 + c_2\mathbf{v}_2 + \cdots + c_n\mathbf{v}_n.$$

The scalars $c_1, c_2, \ldots, c_n$ are called the **coordinates of x relative to the basis B.** The **coordinate matrix** (or **coordinate vector**) **of x relative to B** is the column matrix in R^n whose components are the coordinates of $\mathbf{x}$.

$$[\mathbf{x}]_B = \begin{bmatrix} c_1 \\ c_2 \\ \vdots \\ c_n \end{bmatrix}$$

Coordinate Representation in R^n

In R^n, the notation for coordinate matrices conforms to the usual component notation, except that column notation is used for the coordinate matrix. In other words, writing a vector in R^n as $\mathbf{x} = (x_1, x_2, \ldots, x_n)$ means that the x_i's are the coordinates of $\mathbf{x}$ *relative to the standard basis S* in R^n. Thus, you have

$$[\mathbf{x}]_S = \begin{bmatrix} x_1 \\ x_2 \\ \vdots \\ x_n \end{bmatrix}.$$

EXAMPLE 1   *Coordinates and Components in R^n*

Find the coordinate matrix of $\mathbf{x} = (-2, 1, 3)$ in R^3 relative to the standard basis $S = \{(1, 0, 0), (0, 1, 0), (0, 0, 1)\}$.

Solution   Because $\mathbf{x}$ can be written as

$$\mathbf{x} = (-2, 1, 3) = -2(1, 0, 0) + 1(0, 1, 0) + 3(0, 0, 1),$$

you can see that the coordinate matrix of $\mathbf{x}$ relative to the standard basis is simply

$$[\mathbf{x}]_S = \begin{bmatrix} -2 \\ 1 \\ 3 \end{bmatrix}.$$

So, the components of $\mathbf{x}$ are the same as its coordinates relative to the standard basis.

EXAMPLE 2   *Finding a Coordinate Matrix Relative to a Standard Basis*

The coordinate matrix of $\mathbf{x}$ in R^2 relative to the (nonstandard) ordered basis $B = \{\mathbf{v}_1, \mathbf{v}_2\} = \{(1, 0), (1, 2)\}$ is

$$[\mathbf{x}]_B = \begin{bmatrix} 3 \\ 2 \end{bmatrix}.$$

Find the coordinates of $\mathbf{x}$ relative to the standard basis $B' = \{\mathbf{u}_1, \mathbf{u}_2\} = \{(1, 0), (0, 1)\}$.

Solution   Because $[\mathbf{x}]_B = \begin{bmatrix} 3 \\ 2 \end{bmatrix}$, you can write

$$\mathbf{x} = 3\mathbf{v}_1 + 2\mathbf{v}_2 = 3(1, 0) + 2(1, 2) = (5, 4).$$

Moreover, because $(5, 4) = 5(1, 0) + 4(0, 1)$, it follows that the coordinates of $\mathbf{x}$ relative to B' are given by

$$[\mathbf{x}]_{B'} = \begin{bmatrix} 5 \\ 4 \end{bmatrix}.$$

Figure 4.19 compares these two coordinate representations.

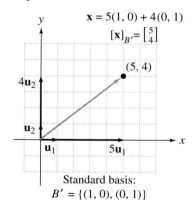

Figure 4.19

Example 2 shows that the procedure for finding the coordinate matrix relative to a *standard* basis is straightforward. The problem becomes a bit tougher, however, when you must find the coordinate matrix relative to a *nonstandard* basis. Here is an example.

EXAMPLE 3 *Finding a Coordinate Matrix Relative to a Nonstandard Basis*

Find the coordinate matrix of $x = (1, 2, -1)$ in R^3 relative to the (nonstandard) basis

$$B' = \{u_1, u_2, u_3\} = \{(1, 0, 1), (0, -1, 2), (2, 3, -5)\}.$$

Solution Begin by writing x as a linear combination of u_1, u_2, and u_3.

$$x = c_1u_1 + c_2u_2 + c_3u_3$$
$$(1, 2, -1) = c_1(1, 0, 1) + c_2(0, -1, 2) + c_3(2, 3, -5)$$

Equating corresponding components produces the following system of linear equations.

$$\begin{aligned} c_1 \quad\quad + 2c_3 &= 1 \\ -c_2 + 3c_3 &= 2 \\ c_1 + 2c_2 - 5c_3 &= -1 \end{aligned}$$

$$\begin{bmatrix} 1 & 0 & 2 \\ 0 & -1 & 3 \\ 1 & 2 & -5 \end{bmatrix} \begin{bmatrix} c_1 \\ c_2 \\ c_3 \end{bmatrix} = \begin{bmatrix} 1 \\ 2 \\ -1 \end{bmatrix}$$

The solution of this system is $c_1 = 5$, $c_2 = -8$, and $c_3 = -2$. Therefore,

$$x = 5(1, 0, 1) + (-8)(0, -1, 2) + (-2)(2, 3, -5),$$

and the coordinate matrix of x relative to B' is

$$[x]_{B'} = \begin{bmatrix} 5 \\ -8 \\ -2 \end{bmatrix}.$$

REMARK: Note that the solution in Example 3 is written as

$$[x]_{B'} = \begin{bmatrix} 5 \\ -8 \\ -2 \end{bmatrix}.$$

It would be incorrect to write the solution as

$$x = \begin{bmatrix} 5 \\ -8 \\ -2 \end{bmatrix}.$$

Do you see why?

Change of Basis in R^n

The procedure demonstrated in Examples 2 and 3 is called a **change of basis.** That is, you were given the coordinates of a vector relative to one basis B and were asked to find the coordinates relative to another basis B'.

For instance, if in Example 3 you let B be the standard basis, then the problem of finding the coordinate matrix of $\mathbf{x} = (1, 2, -1)$ relative to the basis B' becomes one of solving for c_1, c_2, and c_3 in the matrix equation

$$\underbrace{\begin{bmatrix} 1 & 0 & 2 \\ 0 & -1 & 3 \\ 1 & 2 & -5 \end{bmatrix}}_{P} \underbrace{\begin{bmatrix} c_1 \\ c_2 \\ c_3 \end{bmatrix}}_{[\mathbf{x}]_{B'}} = \underbrace{\begin{bmatrix} 1 \\ 2 \\ -1 \end{bmatrix}}_{[\mathbf{x}]_B}.$$

The matrix P is called the **transition matrix from B' to B,** where $[\mathbf{x}]_{B'}$ is the coordinate matrix of $\mathbf{x}$ relative to B', and $[\mathbf{x}]_B$ is the coordinate matrix of $\mathbf{x}$ relative to B. Multiplication by the transition matrix P changes a coordinate matrix relative to B' into a coordinate matrix relative to B. That is,

$$P[\mathbf{x}]_{B'} = [\mathbf{x}]_B. \qquad \text{Change of basis from } B' \text{ to } B$$

To perform a change of basis from B to B', use the matrix P^{-1} (the **transition matrix from B to B'**) and write

$$[\mathbf{x}]_{B'} = P^{-1}[\mathbf{x}]_B. \qquad \text{Change of basis from } B \text{ to } B'$$

This means that the change of basis problem in Example 3 can be represented by the matrix equation

$$\begin{bmatrix} c_1 \\ c_2 \\ c_3 \end{bmatrix} = \underbrace{\begin{bmatrix} -1 & 4 & 2 \\ 3 & -7 & -3 \\ 1 & -2 & -1 \end{bmatrix}}_{P^{-1}} \underbrace{\begin{bmatrix} 1 \\ 2 \\ -1 \end{bmatrix}}_{[\mathbf{x}]_B} = \underbrace{\begin{bmatrix} 5 \\ -8 \\ -2 \end{bmatrix}}_{[\mathbf{x}]_{B'}}.$$

This discussion generalizes as follows. Suppose that

$$B = \{\mathbf{v}_1, \mathbf{v}_2, \ldots, \mathbf{v}_n\} \qquad \text{and} \qquad B' = \{\mathbf{u}_1, \mathbf{u}_2, \ldots, \mathbf{u}_n\}$$

are two ordered bases for R^n. If $\mathbf{x}$ is a vector in R^n and

$$[\mathbf{x}]_B = \begin{bmatrix} c_1 \\ c_2 \\ \vdots \\ c_n \end{bmatrix} \qquad \text{and} \qquad [\mathbf{x}]_{B'} = \begin{bmatrix} d_1 \\ d_2 \\ \vdots \\ d_n \end{bmatrix}$$

are the coordinate matrices of $\mathbf{x}$ relative to B and B', then the **transition matrix P from B' to B** is the matrix P such that

$$[\mathbf{x}]_B = P[\mathbf{x}]_{B'}.$$

The following theorem tells you that the transition matrix P is invertible and that its inverse is the **transition matrix from B to B'.** That is,

$$[\mathbf{x}]_{B'} = P^{-1}[\mathbf{x}]_B.$$

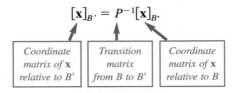

| Coordinate matrix of **x** relative to B' | Transition matrix from B to B' | Coordinate matrix of **x** relative to B |

Theorem 4.20   *The Inverse of a Transition Matrix*

If P is the transition matrix from a basis B' to a basis B in R^n, then P is invertible and the transition matrix from B to B' is given by P^{-1}.

Before proving Theorem 4.20, you will prove a preliminary lemma.

Lemma

Let $B = \{\mathbf{v}_1, \mathbf{v}_2, \ldots, \mathbf{v}_n\}$ and $B' = \{\mathbf{u}_1, \mathbf{u}_2, \ldots, \mathbf{u}_n\}$ be two bases for a vector space V. If

$$\mathbf{v}_1 = c_{11}\mathbf{u}_1 + c_{21}\mathbf{u}_2 + \cdots + c_{n1}\mathbf{u}_n$$
$$\mathbf{v}_2 = c_{12}\mathbf{u}_1 + c_{22}\mathbf{u}_2 + \cdots + c_{n2}\mathbf{u}_n$$
$$\vdots$$
$$\mathbf{v}_n = c_{1n}\mathbf{u}_1 + c_{2n}\mathbf{u}_2 + \cdots + c_{nn}\mathbf{u}_n,$$

then the transition matrix from B to B' is

$$Q = \begin{bmatrix} c_{11} & c_{12} & \cdots & c_{1n} \\ c_{21} & c_{22} & \cdots & c_{2n} \\ \vdots & \vdots & & \vdots \\ c_{n1} & c_{n2} & \cdots & c_{nn} \end{bmatrix}.$$

Proof (of lemma)   Let

$$\mathbf{v} = d_1\mathbf{v}_1 + d_2\mathbf{v}_2 + \cdots + d_n\mathbf{v}_n$$

be an arbitrary vector in V. The coordinate matrix of $\mathbf{v}$ with respect to the basis B is therefore

$$[\mathbf{v}]_B = \begin{bmatrix} d_1 \\ d_2 \\ \vdots \\ d_n \end{bmatrix}.$$

Then you have

$$Q[\mathbf{v}]_B = \begin{bmatrix} c_{11} & c_{12} & \cdots & c_{1n} \\ c_{21} & c_{22} & \cdots & c_{2n} \\ \vdots & \vdots & & \vdots \\ c_{n1} & c_{n2} & \cdots & c_{nn} \end{bmatrix} \begin{bmatrix} d_1 \\ d_2 \\ \vdots \\ d_n \end{bmatrix}$$

$$= \begin{bmatrix} c_{11}d_1 + c_{12}d_2 + \cdots + c_{1n}d_n \\ c_{21}d_1 + c_{22}d_2 + \cdots + c_{2n}d_n \\ \vdots & \vdots & & \vdots \\ c_{n1}d_1 + c_{n2}d_2 + \cdots + c_{nn}d_n \end{bmatrix}.$$

On the other hand, you can write

$$\mathbf{v} = d_1\mathbf{v}_1 + d_2\mathbf{v}_2 + \cdots + d_n\mathbf{v}_n$$
$$= d_1(c_{11}\mathbf{u}_1 + \cdots + c_{n1}\mathbf{u}_n) + \cdots + d_n(c_{1n}\mathbf{u}_1 + \cdots + c_{nn}\mathbf{u}_n)$$
$$= (d_1c_{11} + \cdots + d_nc_{1n})\mathbf{u}_1 + \cdots + (d_1c_{n1} + \cdots + d_nc_{nn})\mathbf{u}_n,$$

which implies that

$$[\mathbf{v}]_{B'} = \begin{bmatrix} c_{11}d_1 + c_{12}d_2 + \cdots + c_{1n}d_n \\ c_{21}d_1 + c_{22}d_2 + \cdots + c_{2n}d_n \\ \vdots & \vdots & & \vdots \\ c_{n1}d_1 + c_{n2}d_2 + \cdots + c_{nn}d_n \end{bmatrix}.$$

Therefore, $Q[\mathbf{v}]_B = [\mathbf{v}]_{B'}$ and you can conclude that Q is the transition matrix from B to B'.

Proof (of Theorem 4.20)   From the preceding lemma, let Q be the transition matrix from B to B'. Then

$$[\mathbf{v}]_B = P[\mathbf{v}]_{B'} \qquad \text{and} \qquad [\mathbf{v}]_{B'} = Q[\mathbf{v}]_B,$$

which implies that $[\mathbf{v}]_B = PQ[\mathbf{v}]_B$ for every vector $\mathbf{v}$ in R^n. From this it follows that $PQ = I$. Therefore, P is invertible and P^{-1} is equal to Q, the transition matrix from B to B'.

There is a nice way to use Gauss-Jordan elimination to find the transition matrix P^{-1}. First define two matrices B and B' whose columns correspond to the vectors in B and B'. That is,

$$B = \begin{bmatrix} v_{11} & v_{12} & \cdots & v_{1n} \\ v_{21} & v_{22} & \cdots & v_{2n} \\ \vdots & \vdots & & \vdots \\ v_{n1} & v_{n2} & \cdots & v_{nn} \end{bmatrix} \qquad \text{and} \qquad B' = \begin{bmatrix} u_{11} & u_{12} & \cdots & u_{1n} \\ u_{21} & u_{22} & \cdots & u_{2n} \\ \vdots & \vdots & & \vdots \\ u_{n1} & u_{n2} & \cdots & u_{nn} \end{bmatrix}.$$
$$\qquad\quad \mathbf{v}_1 \quad \mathbf{v}_2 \qquad \mathbf{v}_n \qquad\qquad\qquad\qquad \mathbf{u}_1 \quad \mathbf{u}_2 \qquad \mathbf{u}_n$$

Then, by reducing the $n \times 2n$ matrix $[B' \vdots B]$ so that the identity matrix I_n occurs in place of B', you obtain the matrix $[I_n \vdots P^{-1}]$. This procedure is stated formally in the following theorem.

| Theorem 4.21 | *Transition Matrix from B to B′* |

Let $B = \{\mathbf{v}_1, \mathbf{v}_2, \ldots, \mathbf{v}_n\}$ and $B' = \{\mathbf{u}_1, \mathbf{u}_2, \ldots, \mathbf{u}_n\}$ be two bases for R^n. Then the transition matrix P^{-1} from B to B' can be found by using Gauss-Jordan elimination on the $n \times 2n$ matrix $[B' \;\vdots\; B]$ as follows.

$$[B' \;\vdots\; B] \quad \longrightarrow \quad [I_n \;\vdots\; P^{-1}]$$

Proof To begin, let

$$\mathbf{v}_1 = c_{11}\mathbf{u}_1 + c_{21}\mathbf{u}_2 + \cdots + c_{n1}\mathbf{u}_n$$
$$\mathbf{v}_2 = c_{12}\mathbf{u}_1 + c_{22}\mathbf{u}_2 + \cdots + c_{n2}\mathbf{u}_n$$
$$\vdots$$
$$\mathbf{v}_n = c_{1n}\mathbf{u}_1 + c_{2n}\mathbf{u}_2 + \cdots + c_{nn}\mathbf{u}_n,$$

which implies that

$$c_{1i}\begin{bmatrix} u_{11} \\ u_{12} \\ \vdots \\ u_{1n} \end{bmatrix} + c_{2i}\begin{bmatrix} u_{21} \\ u_{22} \\ \vdots \\ u_{2n} \end{bmatrix} + \cdots + c_{ni}\begin{bmatrix} u_{n1} \\ u_{n2} \\ \vdots \\ u_{nn} \end{bmatrix} = \begin{bmatrix} v_{i1} \\ v_{i2} \\ \vdots \\ v_{in} \end{bmatrix}$$

for $i = 1, 2, \ldots, n$. From these vector equations you can write the following n systems of linear equations

$$u_{11}c_{1i} + u_{21}c_{2i} + \cdots + u_{n1}c_{ni} = v_{i1}$$
$$u_{12}c_{1i} + u_{22}c_{2i} + \cdots + u_{n2}c_{ni} = v_{i2}$$
$$\vdots$$
$$u_{1n}c_{1i} + u_{2n}c_{2i} + \cdots + u_{nn}c_{ni} = v_{in}$$

for $i = 1, 2, \ldots, n$. Because each of the n systems has the same coefficient matrix, you can reduce all n systems simultaneously using the following augmented matrix.

$$\left[\begin{array}{cccc:cccc} u_{11} & u_{21} & \cdots & u_{n1} & v_{11} & v_{21} & \cdots & v_{n1} \\ u_{12} & u_{22} & \cdots & u_{n2} & v_{12} & v_{22} & \cdots & v_{n2} \\ \vdots & \vdots & & \vdots & \vdots & \vdots & & \vdots \\ u_{1n} & u_{2n} & \cdots & u_{nn} & v_{1n} & v_{2n} & \cdots & v_{nn} \end{array}\right]$$

$$\underbrace{\qquad\qquad\qquad}_{B'} \qquad \underbrace{\qquad\qquad\qquad}_{B}$$

Applying Gauss-Jordan elimination to this matrix produces

$$\left[\begin{array}{cccc:cccc} 1 & 0 & \cdots & 0 & c_{11} & c_{12} & \cdots & c_{1n} \\ 0 & 1 & \cdots & 0 & c_{21} & c_{22} & \cdots & c_{2n} \\ \vdots & \vdots & & \vdots & \vdots & \vdots & & \vdots \\ 0 & 0 & \cdots & 1 & c_{n1} & c_{n2} & \cdots & c_{nn} \end{array}\right].$$

However, by the preceding lemma, the right-hand side of this matrix is $Q = P^{-1}$, which implies that the matrix has the form

$$[I \mathrel{\vdots} P^{-1}],$$

which proves the theorem.

In the following example, you will apply this procedure to the change of basis problem given in Example 3.

EXAMPLE 4   *Finding a Transition Matrix*

Find the transition matrix from B to B' for the following bases in R^3.

$$B = \{(1, 0, 0), (0, 1, 0), (0, 0, 1)\}$$

and

$$B' = \{(1, 0, 1), (0, -1, 2), (2, 3, -5)\}$$

Solution   First use the vectors in the two bases to form the matrices B and B'.

$$B = \begin{bmatrix} 1 & 0 & 0 \\ 0 & 1 & 0 \\ 0 & 0 & 1 \end{bmatrix} \quad \text{and} \quad B' = \begin{bmatrix} 1 & 0 & 2 \\ 0 & -1 & 3 \\ 1 & 2 & -5 \end{bmatrix}$$

Then form the matrix $[B' \mathrel{\vdots} B]$ and use Gauss-Jordan elimination to rewrite $[B' \mathrel{\vdots} B]$ as $[I_3 \mathrel{\vdots} P^{-1}]$.

$$\begin{bmatrix} 1 & 0 & 2 & \vdots & 1 & 0 & 0 \\ 0 & -1 & 3 & \vdots & 0 & 1 & 0 \\ 1 & 2 & -5 & \vdots & 0 & 0 & 1 \end{bmatrix} \longrightarrow \begin{bmatrix} 1 & 0 & 0 & \vdots & -1 & 4 & 2 \\ 0 & 1 & 0 & \vdots & 3 & -7 & -3 \\ 0 & 0 & 1 & \vdots & 1 & -2 & -1 \end{bmatrix}$$

From this you can conclude that the transition matrix from B to B' is

$$P^{-1} = \begin{bmatrix} -1 & 4 & 2 \\ 3 & -7 & -3 \\ 1 & -2 & -1 \end{bmatrix}.$$

Try multiplying P^{-1} by the coordinate matrix of $\mathbf{x} = (1, 2, -1)$ to see that the result is the same as the one obtained in Example 3.

DISCOVERY

Let $B = \{(1, 0), (1, 2)\}$ and $B' = \{(1, 0), (0, 1)\}$. Form the matrix $[B' \mathrel{\vdots} B]$. Make a conjecture about the necessity of using Gauss-Jordan elimination to obtain the transition matrix P^{-1} if the change of basis is from a nonstandard basis to a standard basis.

Note that when B is the standard basis, as in Example 4, the process of changing $[B' \ \vdots \ B]$ to $[I_n \ \vdots \ P^{-1}]$ becomes

$$[B' \ \vdots \ I_n] \ \longrightarrow \ [I_n \ \vdots \ P^{-1}].$$

But this is the same process used to find inverse matrices in Section 2.3. In other words, if B is the standard basis in R^n, then the transition matrix from B to B' is given by

$$P^{-1} = (B')^{-1}. \qquad \text{Standard basis to nonstandard basis}$$

The process is even simpler if B' is the standard basis, because the matrix $[B' \ \vdots \ B]$ is already in the form

$$[I_n \ \vdots \ B] = [I_n \ \vdots \ P^{-1}].$$

In this case, the transition matrix is simply

$$P^{-1} = B. \qquad \text{Nonstandard basis to standard basis}$$

For instance, the transition matrix in Example 2 from $B = \{(1, 0), (1, 2)\}$ to $B' = \{(1, 0), (0, 1)\}$ is

$$P^{-1} = B = \begin{bmatrix} 1 & 1 \\ 0 & 2 \end{bmatrix}.$$

EXAMPLE 5 *Finding a Transition Matrix*

Find the transition matrix from B to B' for the following bases for R^2.

$$B = \{(-3, 2), (4, -2)\}$$

and

$$B' = \{(-1, 2), (2, -2)\}$$

Solution Begin by forming the matrix

$$[B' \ \vdots \ B] = \begin{bmatrix} -1 & 2 & \vdots & -3 & 4 \\ 2 & -2 & \vdots & 2 & -2 \end{bmatrix}$$

and use Gauss-Jordan elimination to obtain the transition matrix P^{-1} from B to B':

$$[I_2 \ \vdots \ P^{-1}] = \begin{bmatrix} 1 & 0 & \vdots & -1 & 2 \\ 0 & 1 & \vdots & -2 & 3 \end{bmatrix}.$$

Thus, you have

$$P^{-1} = \begin{bmatrix} -1 & 2 \\ -2 & 3 \end{bmatrix}.$$

Technology   Note

Most graphing calculators and computer software programs have the capability to augment two matrices. After this has been done, you can use the *rref* (reduced row-echelon form) command to find the transition matrix P^{-1} from B to B'. For example, to find the transition matrix P^{-1} from B to B' in Example 5 using a TI-86, your screen would look like the following.

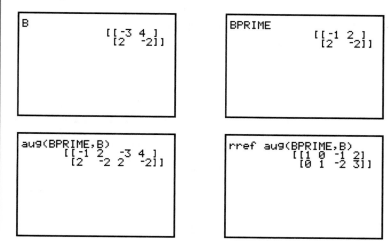

Use a graphing calculator with matrix capabilities to find the transition matrix P from B' to B.

In Example 5, if you had found the transition matrix from B' to B (rather than from B to B'), you would have obtained

$$[B \;\vdots\; B'] = \begin{bmatrix} -3 & 4 & \vdots & -1 & 2 \\ 2 & -2 & \vdots & 2 & -2 \end{bmatrix},$$

which reduces to

$$[I_2 \;\vdots\; P] = \begin{bmatrix} 1 & 0 & \vdots & 3 & -2 \\ 0 & 1 & \vdots & 2 & -1 \end{bmatrix}.$$

Therefore, the transition matrix from B' to B is

$$P = \begin{bmatrix} 3 & -2 \\ 2 & -1 \end{bmatrix}.$$

You can verify that this is the inverse of the transition matrix found in Example 5 by multiplication:

$$PP^{-1} = \begin{bmatrix} 3 & -2 \\ 2 & -1 \end{bmatrix}\begin{bmatrix} -1 & 2 \\ -2 & 3 \end{bmatrix} = \begin{bmatrix} 1 & 0 \\ 0 & 1 \end{bmatrix} = I_2.$$

Coordinate Representation in General n-Dimensional Spaces

One benefit of coordinate representation is that it enables you to represent vectors in an arbitrary n-dimensional space using the same notation used in R^n. For instance, in Example 6, note that the coordinate matrix of a vector in P_3 is a vector in R^4.

EXAMPLE 6 *Coordinate Representation in P_3*

Find the coordinate matrix of $p = 3x^3 - 2x^2 + 4$ relative to the standard basis in P_3,

$$S = \{1, x, x^2, x^3\}.$$

Solution First write p as a linear combination of the basis vectors (in the given order).

$$p = 4(1) + 0(x) + (-2)(x^2) + 3(x^3)$$

This tells you that the coordinate matrix of p relative to S is

$$[p]_S = \begin{bmatrix} 4 \\ 0 \\ -2 \\ 3 \end{bmatrix}.$$

In the preceding section, you saw that it is sometimes convenient to represent $n \times 1$ matrices as n-tuples. The next example adds some justification to this practice.

EXAMPLE 7 *Coordinate Representation in $M_{3,1}$*

Find the coordinate matrix of

$$X = \begin{bmatrix} -1 \\ 4 \\ 3 \end{bmatrix}$$

relative to the standard basis in $M_{3,1}$,

$$S = \left\{ \begin{bmatrix} 1 \\ 0 \\ 0 \end{bmatrix}, \begin{bmatrix} 0 \\ 1 \\ 0 \end{bmatrix}, \begin{bmatrix} 0 \\ 0 \\ 1 \end{bmatrix} \right\}.$$

Solution Because X can be written as

$$X = \begin{bmatrix} -1 \\ 4 \\ 3 \end{bmatrix} = (-1)\begin{bmatrix} 1 \\ 0 \\ 0 \end{bmatrix} + 4\begin{bmatrix} 0 \\ 1 \\ 0 \end{bmatrix} + 3\begin{bmatrix} 0 \\ 0 \\ 1 \end{bmatrix},$$

the coordinate matrix of X relative to S is

$$[X]_S = \begin{bmatrix} -1 \\ 4 \\ 3 \end{bmatrix}.$$

REMARK: In Section 6.2 you will learn more about the use of R^n to represent an arbitrary n-dimensional vector space.

Theorems 4.20 and 4.21 can be generalized to cover arbitrary n-dimensional spaces. This text, however, does not cover this.

SECTION 4.7   Exercises

In Exercises 1–6, you are given the coordinate matrix of **x** relative to a (nonstandard) basis B. Find the coordinate vector of **x** relative to the standard basis in R^n.

1. $B = \{(2, -1), (0, 1)\}$,
   $[\mathbf{x}]_B = [4, 1]^T$
2. $B = \{(-1, 4), (4, -1)\}$,
   $[\mathbf{x}]_B = [-2, 3]^T$
3. $B = \{(1, 0, 1), (1, 1, 0), (0, 1, 1)\}$,
   $[\mathbf{x}]_B = [2, 3, 1]^T$
4. $B = \left\{\left(\frac{3}{4}, \frac{5}{2}, \frac{3}{2}\right), \left(3, 4, \frac{7}{2}\right), \left(-\frac{3}{2}, 6, 2\right)\right\}$,
   $[\mathbf{x}]_B = [2, 0, 4]^T$
5. $B = \{(0, 0, 0, 1), (0, 0, 1, 1), (0, 1, 1, 1), (1, 1, 1, 1)\}$,
   $[\mathbf{x}]_B = [1, -2, 3, -1]^T$
6. $B = \{(4, 0, 7, 3), (0, 5, -1, -1), (-3, 4, 2, 1), (0, 1, 5, 0)\}$,
   $[\mathbf{x}]_B = [-2, 3, 4, 1]^T$

In Exercises 7–12, find the coordinate matrix of **x** in R^n relative to the basis B.

7. $B = \{(4, 0), (0, 3)\}$,
   $\mathbf{x} = (12, 6)$
8. $B = \{(-6, 7), (4, -3)\}$,
   $\mathbf{x} = (-26, 32)$
9. $B = \{(8, 11, 0), (7, 0, 10), (1, 4, 6)\}$,
   $\mathbf{x} = (3, 19, 2)$
10. $B = \left\{\left(\frac{3}{2}, 4, 1\right), \left(\frac{3}{4}, \frac{5}{2}, 0\right), \left(1, \frac{1}{2}, 2\right)\right\}$,
    $\mathbf{x} = \left(3, -\frac{1}{2}, 8\right)$

11. $B = \{(4, 3, 3), (-11, 0, 11), (0, 9, 2)\}$,
    $\mathbf{x} = (11, 18, -7)$
12. $B = \{(9, -3, 15, 4), (3, 0, 0, 1), (0, -5, 6, 8), (3, -4, 2, -3)\}$,
    $\mathbf{x} = (0, -20, 7, 15)$

In Exercises 13–22, find the transition matrix from B to B'.

13. $B = \{(1, 0), (0, 1)\}$,
    $B' = \{(2, 4), (1, 3)\}$
14. $B = \{(1, 0), (0, 1)\}$,
    $B' = \{(1, 1), (5, 6)\}$
15. $B = \{(2, 4), (-1, 3)\}$,
    $B' = \{(1, 0), (0, 1)\}$
16. $B = \{(1, 1), (1, 0)\}$,
    $B' = \{(1, 0), (0, 1)\}$
17. $B = \{(1, 0, 0), (0, 1, 0), (0, 0, 1)\}$,
    $B' = \{(1, 0, 0), (0, 2, 8), (6, 0, 12)\}$
18. $B = \{(1, 0, 0), (0, 1, 0), (0, 0, 1)\}$,
    $B' = \{(1, 3, -1), (2, 7, -4), (2, 9, -7)\}$
19. $B = \{(1, 3, 3), (1, 5, 6), (1, 4, 5)\}$,
    $B' = \{(1, 0, 0), (0, 1, 0), (0, 0, 1)\}$
20. $B = \{(2, -1, 4), (0, 2, 1), (-3, 2, 1)\}$,
    $B' = \{(1, 0, 0), (0, 1, 0), (0, 0, 1)\}$
21. $B = \{(1, 0, 0, 0), (0, 1, 0, 0), (0, 0, 1, 0), (0, 0, 0, 1)\}$,
    $B' = \{(1, 3, 2, -1), (-2, -5, -5, 4), (-1, -2, -2, 4),$
    $(-2, -3, -5, 11)\}$
22. $B = \{(1, 1, 1, 1), (0, 1, 1, 1), (0, 0, 1, 1), (0, 0, 0, 1)\}$,
    $B' = \{(1, 0, 0, 0), (0, 1, 0, 0), (0, 0, 1, 0), (0, 0, 0, 1)\}$

In Exercises 23–26, use Theorem 4.21 to (a) find the transition matrix from B to B', (b) find the transition matrix from B' to B, (c) verify that the two transition matrices are inverses of each other, and (d) find $[\mathbf{x}]_B$ given $[\mathbf{x}]_{B'}$.

23. $B = \{(1, 3), (-2, -2)\}, \quad B' = \{(-12, 0), (-4, 4)\}$,

$$[\mathbf{x}]_{B'} = \begin{bmatrix} -1 \\ 3 \end{bmatrix}$$

24. $B = \{(2, -2), (6, 3)\}, \quad B' = \{(1, 1), (32, 31)\}$,

$$[\mathbf{x}]_{B'} = \begin{bmatrix} 2 \\ -1 \end{bmatrix}$$

25. $B = \{(1, 0, 2), (0, 1, 3), (1, 1, 1)\}$,
$B' = \{(2, 1, 1), (1, 0, 0), (0, 2, 1)\}$,

$$[\mathbf{x}]_{B'} = \begin{bmatrix} 1 \\ 2 \\ -1 \end{bmatrix}$$

26. $B = \{(1, 1, 1), (1, -1, 1), (0, 0, 1)\}$,
$B' = \{(2, 2, 0), (0, 1, 1), (1, 0, 1)\}$,

$$[\mathbf{x}]_{B'} = \begin{bmatrix} 2 \\ 3 \\ 1 \end{bmatrix}$$

In Exercises 27 and 28, use a graphing utility with matrix capabilities to (a) find the transition matrix from B to B', (b) find the transition matrix from B' to B, (c) verify that the two transition matrices are inverses of one another, and (d) find $[\mathbf{x}]_B$ given $[\mathbf{x}]_{B'}$.

27. $B = \{(4, 2, -4), (6, -5, -6), (2, -1, 8)\}$,
$B' = \{(1, 0, 4), (4, 2, 8), (2, 5, -2)\}$,

$$[\mathbf{x}]_{B'} = \begin{bmatrix} 1 \\ -1 \\ 2 \end{bmatrix}$$

28. $B = \{(1, 3, 4), (2, -5, 2), (-4, 2, -6)\}$,
$B' = \{(1, 2, -2), (4, 1, -4), (-2, 5, 8)\}$,

$$[\mathbf{x}]_{B'} = \begin{bmatrix} -1 \\ 0 \\ 2 \end{bmatrix}$$

In Exercises 29 and 30, find the coordinate matrix of p relative to the standard basis in P_2.

29. $p = x^2 + 11x + 4$

30. $p = 3x^2 + 114x + 13$

In Exercises 31 and 32, find the coordinate matrix of X relative to the standard basis in $M_{3,1}$.

31. $X = \begin{bmatrix} 0 \\ 3 \\ 2 \end{bmatrix}$

32. $X = \begin{bmatrix} 2 \\ -1 \\ 4 \end{bmatrix}$

True or False?   In Exercises 33 and 34, determine whether each statement is true or false. If a statement is true, give a reason or cite an appropriate statement in the text. If a statement is false, provide an example that shows that the statement is not true in all cases or cite an appropriate statement in the text.

33. (a) If P is the transition matrix from the basis B to B', then the equation $P[\mathbf{x}]_{B'} = [\mathbf{x}]_B$ represents the change of basis from B to B'.

(b) If B is the standard basis in R^n, then the transition matrix from B to B' is $P^{-1} = (B')^{-1}$.

34. (a) If P is the transition matrix used to perform a change of basis from B' to B, then P^{-1} is the transition matrix from B to B'.

(b) To perform the change of basis from a nonstandard basis B' to the standard basis B, the transition matrix P^{-1} is simply B.

35. Let P be the transition matrix from B'' to B', and let Q be the transition matrix from B' to B. What is the transition matrix from B'' to B?

36. Let P be the transition matrix from B'' to B', and let Q be the transition matrix from B' to B. What is the transition matrix from B to B''?

37. *Writing*   Let B and B' be two bases for the vector space R^n. Discuss the nature of the transition matrix from B to B' if one of the bases is the standard basis.

38. *Writing*   Is it possible for a transition matrix to equal the identity matrix? Illustrate your answer with appropriate examples.

4.8 | Applications of Vector Spaces

Linear Differential Equations (Calculus)

A **linear differential equation of order** n is of the form

$$y^{(n)} + g_{n-1}(x)y^{(n-1)} + \cdots + g_1(x)y' + g_0(x)y = f(x),$$

where $g_0, g_1, \ldots, g_{n-1}$ and f are functions of x with a common domain. If $f(x) = 0$, the equation is **homogeneous.** Otherwise it is **nonhomogeneous.** A function y is called a **solution** of the linear differential equation if the equation is satisfied when y and its first n derivatives are substituted into the equation.

EXAMPLE 1   *A Second-Order Linear Differential Equation*

Show that both $y_1 = e^x$ and $y_2 = e^{-x}$ are solutions of the second-order linear differential equation $y'' - y = 0$.

Solution   For the function $y_1 = e^x$, you have $y_1' = e^x$ and $y_1'' = e^x$. Therefore,

$$y_1'' - y_1 = e^x - e^x = 0,$$

and $y_1 = e^x$ is a solution of the given linear differential equation. Similarly, for $y_2 = e^{-x}$, you have

$$y_2' = -e^{-x} \quad \text{and} \quad y_2'' = e^{-x}.$$

This implies that

$$y_2'' - y_2 = e^{-x} - e^{-x} = 0.$$

Therefore, $y_2 = e^{-x}$ is also a solution of the given linear differential equation.

There are two important observations that you can make about Example 1. The first is that in the vector space $C''(-\infty, \infty)$ of all twice differentiable functions defined on the entire real line, the two solutions $y_1 = e^x$ and $y_2 = e^{-x}$ are *linearly independent*. This means that the only solution of

$$C_1 y_1 + C_2 y_2 = 0$$

that is valid for all x is $C_1 = C_2 = 0$. The second observation is that every *linear combination* of y_1 and y_2 is also a solution of the given linear differential equation. To see this, let $y = C_1 y_1 + C_2 y_2$. Then

$$y = C_1 e^x + C_2 e^{-x}$$
$$y' = C_1 e^x - C_2 e^{-x}$$
$$y'' = C_1 e^x + C_2 e^{-x}.$$

Substituting into the differential equation $y'' - y = 0$ produces

$$y'' - y = (C_1 e^x + C_2 e^{-x}) - (C_1 e^x + C_2 e^{-x}) = 0.$$

Thus, $y = C_1 e^x + C_2 e^{-x}$ is a solution.

These two observations are generalized in the following theorem, which is stated without proof.

Solutions of a Linear Homogeneous Differential Equation

Every nth-order linear homogeneous differential equation

$$y^{(n)} + g_{n-1}(x)y^{(n-1)} + \cdots + g_1(x)y' + g_0(x)y = 0$$

has n linearly independent solutions. Moreover, if $\{y_1, y_2, \ldots, y_n\}$ is a set of linearly independent solutions, then every solution is of the form

$$y = C_1 y_1 + C_2 y_2 + \cdots + C_n y_n,$$

where $C_1, C_2, \ldots,$ and C_n are real numbers.

REMARK: The solution $y = C_1 y_1 + C_2 y_2 + \cdots + C_n y_n$ is called the **general solution** of the given differential equation.

In light of the preceding theorem, you can see the importance of being able to determine whether a set of solutions is linearly independent. Before describing a way of testing for linear independence, you are given a preliminary definition.

Definition of the Wronskian of a Set of Functions

Let $\{y_1, y_2, \ldots, y_n\}$ be a set of functions, each of which has $n - 1$ derivatives on an interval I. The determinant

$$W(y_1, y_2, \ldots, y_n) = \begin{vmatrix} y_1 & y_2 & \cdots & y_n \\ y_1' & y_2' & \cdots & y_n' \\ \vdots & \vdots & & \vdots \\ y_1^{(n-1)} & y_2^{(n-1)} & \cdots & y_n^{(n-1)} \end{vmatrix}$$

is called the **Wronskian** of the given set of functions.

REMARK: The Wronskian of a set of functions is named after the Polish mathematician Josef Maria Wronski (1778–1853).

EXAMPLE 2 *Finding the Wronskian of a Set of Functions*

(a) The Wronskian of the set $\{1 - x, 1 + x, 2 - x\}$ is

$$W = \begin{vmatrix} 1 - x & 1 + x & 2 - x \\ -1 & 1 & -1 \\ 0 & 0 & 0 \end{vmatrix} = 0.$$

(b) The Wronskian of the set $\{x, x^2, x^3\}$ is

$$W = \begin{vmatrix} x & x^2 & x^3 \\ 1 & 2x & 3x^2 \\ 0 & 2 & 6x \end{vmatrix} = 2x^3.$$

The Wronskian in part (a) of Example 2 is said to be **identically equal to zero,** because it is zero for any value of x. The Wronskian in part (b) is not identically equal to zero because values of x exist for which this Wronskian is nonzero.

The following theorem shows how the Wronskian of a set of functions can be used to test for linear independence.

Wronskian Test for Linear Independence   Let $\{y_1, y_2, \ldots, y_n\}$ be a set of n solutions of an nth-order linear homogeneous differential equation. This set is linearly independent if and only if the Wronskian is not identically equal to zero.

REMARK: This test does *not* apply to an arbitrary set of functions. Each of the functions $y_1, y_2, \ldots,$ and y_n must be a solution of the same linear homogeneous differential equation of order n.

EXAMPLE 3   *Testing a Set of Solutions for Linear Independence*

Determine whether $\{1, \cos x, \sin x\}$ is a set of linearly independent solutions of the linear homogeneous differential equation

$$y''' + y' = 0.$$

Solution   Begin by observing that each of the given functions is a solution of $y''' + y' = 0$. (Try checking this.) Next, testing for linear independence produces the Wronskian of the three functions as follows.

$$W = \begin{vmatrix} 1 & \cos x & \sin x \\ 0 & -\sin x & \cos x \\ 0 & -\cos x & -\sin x \end{vmatrix}$$
$$= \sin^2 x + \cos^2 x$$
$$= 1$$

Because W is not identically equal to zero, you can conclude that the set $\{1, \cos x, \sin x\}$ is linearly independent. Moreover, because this set consists of three linearly independent solutions of a third-order linear homogeneous differential equation, you can conclude that the general solution is

$$y = C_1 + C_2 \cos x + C_3 \sin x.$$

EXAMPLE 4   *Testing a Set of Solutions for Linear Independence*

Determine whether $\{e^x, xe^x, (x + 1)e^x\}$ is a set of linearly independent solutions of the linear homogeneous differential equation

$$y''' - 3y'' + 3y' - y = 0.$$

Solution   As in Example 3, begin by verifying that each of the given functions is actually a solution of $y''' - 3y'' + 3y' - y = 0$. (This verification is left to you.) Testing for linear independence produces the Wronskian of the three functions as follows.

$$W = \begin{vmatrix} e^x & xe^x & (x + 1)e^x \\ e^x & (x + 1)e^x & (x + 2)e^x \\ e^x & (x + 2)e^x & (x + 3)e^x \end{vmatrix}$$

$$= 0$$

So, the set $\{e^x, xe^x, (x + 1)e^x\}$ is linearly dependent.

In Example 4, the Wronskian was used to determine that the set

$$\{e^x, xe^x, (x + 1)e^x\}$$

is linearly dependent. Another way to determine the linear dependence of this set is to observe that the third function is a linear combination of the first two. That is,

$$(x + 1)e^x = e^x + xe^x.$$

Try showing that the different set

$$\{e^x, xe^x, x^2e^x\}$$

forms a linearly independent set of solutions of the differential equation

$$y''' - 3y'' + 3y' - y = 0.$$

Conic Sections and Rotation

Every conic section in the xy-plane has an equation of the form

$$ax^2 + bxy + cy^2 + dx + ey + f = 0.$$

Identifying the graph of this equation is fairly simple as long as b, the coefficient of the xy-term, is zero. In such cases the conic axes are parallel to the coordinate axes, and the identification is accomplished by writing the equation in standard (completed square) form. The standard forms of the equations of the four basic conics are given in the following summary. For circles, ellipses, and hyperbolas, the point (h, k) is the center. For parabolas, the point (h, k) is the vertex.

Standard Forms of
Equations of Conics

Circle $(r = \text{radius})$: $(x - h)^2 + (y - k)^2 = r^2$

Ellipse $(2\alpha = \text{major axis length}, 2\beta = \text{minor axis length})$:

Hyperbola $(2\alpha = \text{transverse axis length}, 2\beta = \text{minor axis length})$:

Parabola $(p = \text{directed distance from vertex to focus})$:

EXAMPLE 5 *Identifying Conic Sections*

(a) The standard form of $x^2 - 2x + 4y - 3 = 0$ is

$$(x - 1)^2 = 4(-1)(y - 1).$$

Thus, the graph of this equation is a parabola with the vertex at $(h, k) = (1, 1)$. The axis of the parabola is vertical. Because $p = -1$, the focus is the point $(1, 0)$. Finally, because the focus lies below the vertex, the parabola opens downward, as shown in Figure 4.20(a).

(b) The standard form of $x^2 + 4y^2 + 6x - 8y + 9 = 0$ is

$$\frac{(x + 3)^2}{2^2} + \frac{(y - 1)^2}{1^2} = 1.$$

Thus, the graph of this equation is an ellipse with its center at $(h, k) = (-3, 1)$. The major axis is horizontal, and its length is $2\alpha = 4$. The length of the minor axis is $2\beta = 2$. The vertices of this ellipse occur at $(-5, 1)$ and $(-1, 1)$, and the endpoints of the minor axis occur at $(-3, 2)$ and $(-3, 0)$, as shown in Figure 4.20(b).

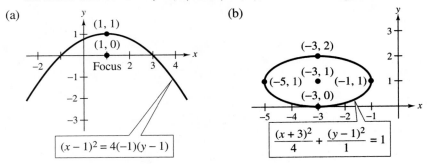

Figure 4.20

The equations of the conics in Example 5 have no xy-term. Consequently, the axes of the corresponding conics are parallel to the coordinate axes. For second-degree polynomial equations that have an xy-term, the axes of the corresponding conics are not parallel to the coordinate axes. In such cases it is helpful to *rotate* the standard axes to form a new x'-axis and y'-axis. The required rotation angle θ (measured counterclockwise) is given by $\cot 2\theta = (a - c)/b$. With this rotation, the standard basis in the plane

$$B = \{(1, 0), (0, 1)\}$$

is rotated to form the new basis

$$B' = \{(\cos \theta, \sin \theta), (-\sin \theta, \cos \theta)\},$$

as shown in Figure 4.21.

To find the coordinates of a point (x, y) relative to this new basis, you can use a transition matrix, as demonstrated in Example 6.

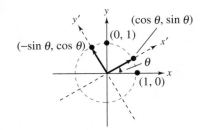

Figure 4.21

EXAMPLE 6   *A Transition Matrix for Rotation in the Plane*

Find the coordinates of a point (x, y) in R^2 relative to the basis

$$B' = \{(\cos \theta, \sin \theta), (-\sin \theta, \cos \theta)\}.$$

Solution   By Theorem 4.21 you have

$$[B' \vdots B] = \begin{bmatrix} \cos \theta & -\sin \theta & \vdots & 1 & 0 \\ \sin \theta & \cos \theta & \vdots & 0 & 1 \end{bmatrix}.$$

Because B is the standard basis in R^2, P^{-1} is given by $(B')^{-1}$. You can use Theorem 3.10 to find $(B')^{-1}$. This results in the following.

$$[I \vdots P^{-1}] = \begin{bmatrix} 1 & 0 & \vdots & \cos \theta & \sin \theta \\ 0 & 1 & \vdots & -\sin \theta & \cos \theta \end{bmatrix}$$

By letting $[x', y']^T$ be the coordinates of (x, y) relative to B', you can use the transition matrix P^{-1} as follows.

$$\begin{bmatrix} \cos \theta & \sin \theta \\ -\sin \theta & \cos \theta \end{bmatrix} \begin{bmatrix} x \\ y \end{bmatrix} = \begin{bmatrix} x' \\ y' \end{bmatrix}$$

Thus, the x'- and y'-coordinates are given by

$$x' = x \cos \theta + y \sin \theta$$
$$y' = -x \sin \theta + y \cos \theta.$$

The last two equations in Example 6 give the $x'y'$-coordinates in terms of the xy-coordinates. To perform a rotation of axes for a second-degree polynomial equation, it is helpful to express the xy-coordinates in terms of the $x'y'$-coordinates. To do this, solve the last two equations in Example 6 for x and y to obtain

$$x = x' \cos \theta - y' \sin \theta \qquad \text{and} \qquad y = x' \sin \theta + y' \cos \theta.$$

Substituting these expressions for x and y into the given second-degree equation produces a second-degree polynomial equation in x' and y' that has no $x'y'$-term.

Rotation of Axes

The second-degree equation $ax^2 + bxy + cy^2 + dx + ey + f = 0$ can be written in the form

$$a'(x')^2 + c'(y')^2 + d'x' + e'y' + f' = 0$$

by rotating the coordinate axes counterclockwise through the angle θ, where θ is defined by $\cot 2\theta = (a - c)/b$. The coefficients of the new equation are obtained from the substitutions

$$x = x' \cos \theta - y' \sin \theta$$
$$y = x' \sin \theta + y' \cos \theta.$$

REMARK: When you solve for $\sin \theta$ and $\cos \theta$, the trigonometric identity $\cot 2\theta = (\cot^2 \theta - 1)/(2 \cot \theta)$ is often useful.

Example 7 demonstrates how to identify the graph of a second-degree polynomial by rotating the coordinate axes.

EXAMPLE 7   *Rotation of a Conic Section*

Perform a rotation of axes to eliminate the xy-term in

$$5x^2 - 6xy + 5y^2 + 14\sqrt{2}x - 2\sqrt{2}y + 18 = 0,$$

and sketch the graph of the resulting equation in the $x'y'$-plane.

Solution   The angle of rotation is given by

$$\cot 2\theta = \frac{a - c}{b} = 0.$$

This implies that $\theta = \pi/4$. Thus,

$$\sin \theta = \frac{1}{\sqrt{2}} \quad \text{and} \quad \cos \theta = \frac{1}{\sqrt{2}}.$$

By substituting

$$x = x' \cos \theta - y' \sin \theta = \frac{1}{\sqrt{2}}(x' - y')$$

and

$$y = x' \sin \theta + y' \cos \theta = \frac{1}{\sqrt{2}}(x' + y')$$

into the original equation and simplifying, you obtain

$$(x')^2 + 4(y')^2 + 6x' - 8y' + 9 = 0.$$

Finally, by completing the square, you find the standard form of this equation to be

$$\frac{(x' + 3)^2}{2^2} + \frac{(y' - 1)^2}{1^2} = 1,$$

which is the equation of an ellipse, as shown in Figure 4.22.

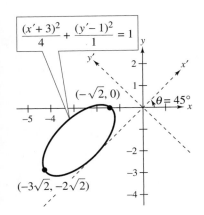

Figure 4.22

In Example 7 the new (rotated) basis for R^2 is

$$B' = \left\{ \left(\frac{1}{\sqrt{2}}, \frac{1}{\sqrt{2}} \right), \left(-\frac{1}{\sqrt{2}}, \frac{1}{\sqrt{2}} \right) \right\},$$

and the coordinates of the vertices of the ellipse relative to B' are $[-5, 1]^T$ and $[-1, 1]^T$.

To find the coordinates of the vertices relative to the standard basis $B = \{(1, 0), (0, 1)\}$, use the equations

$$x = \frac{1}{\sqrt{2}}(x' - y') \quad \text{and} \quad y = \frac{1}{\sqrt{2}}(x' + y')$$

to obtain

$$\left(-3\sqrt{2}, -2\sqrt{2}\right) \quad \text{and} \quad \left(-\sqrt{2}, 0\right),$$

as shown in Figure 4.22.

SECTION 4.8	**Exercises**

Linear Differential Equations (Calculus)

In Exercises 1–4, determine which functions are solutions of the linear differential equation.

1. $y'' + y = 0$

   (a) e^x (b) $\sin x$ (c) $\cos x$ (d) $\sin x - \cos x$

2. $y''' + 3y'' + 3y' + y = 0$

   (a) x (b) e^x (c) e^{-x} (d) xe^{-x}

3. $y'' + 4y' + 4y = 0$

   (a) e^{-2x} (b) xe^{-2x} (c) $x^2 e^{-2x}$ (d) $(x + 2)e^{-2x}$

4. $y'''' - 2y''' + y'' = 0$

   (a) 1 (b) x (c) x^2 (d) e^x

In Exercises 5–8, find the Wronskian for the set of functions.

5. $\{e^x, e^{-x}\}$

6. $\{x, \sin x, \cos x\}$

7. $\{e^{-x}, xe^{-x}, (x + 3)e^{-x}\}$

8. $\{1, e^x, e^{2x}\}$

In Exercises 9–16, test the given set of solutions for linear independence.

Differential Equation	*Solutions*
9. $y'' + y = 0$	$\{\sin x, \cos x\}$
10. $y'' + 4y' + 4y = 0$	$\{e^{-2x}, xe^{-2x}\}$
11. $y''' + 4y'' + 4y' = 0$	$\{e^{-2x}, xe^{-2x}, (2x + 1)e^{-2x}\}$
12. $y''' + y' = 0$	$\{1, \sin x, \cos x\}$
13. $y''' + y' = 0$	$\{2, -1 + 2\sin x, 1 + \sin x\}$
14. $y''' + 3y'' + 3y' + y = 0$	$\{e^{-x}, xe^{-x}, x^2 e^{-x}\}$
15. $y''' + 3y'' + 3y' + y = 0$	$\{e^{-x}, xe^{-x}, e^{-x} + xe^{-x}\}$
16. $y'''' - 2y''' + y'' = 0$	$\{1, x, e^x, xe^x\}$

17. Find the general solution of the differential equation given in Exercise 9.

18. Find the general solution of the differential equation given in Exercise 10.

19. Find the general solution of the differential equation given in Exercise 12.

20. Find the general solution of the differential equation given in Exercise 16.

21. Prove that $y = C_1 \cos ax + C_2 \sin ax$ is the general solution of $y'' + a^2 y = 0$, $a \neq 0$.

22. Prove that the set $\{e^{ax}, e^{bx}\}$ is linearly independent if and only if $a \neq b$.

23. Prove that the set $\{e^{ax}, xe^{ax}\}$ is linearly independent.

24. Prove that the set $\{e^{ax} \cos bx, e^{ax} \sin bx\}$, where $b \neq 0$, is linearly independent.

25. *Writing*   Is the sum of two solutions of a nonhomogeneous linear differential equation also a solution? Explain your answer.

26. *Writing*   Is the scalar multiple of a solution of a nonhomogeneous linear differential equation also a solution? Explain your answer.

Conic Sections and Rotation

In Exercises 27–36, identify and sketch the graph.

27. $y^2 + x = 0$

28. $5x^2 + 3y^2 - 15 = 0$

29. $\dfrac{x^2}{9} - \dfrac{y^2}{16} - 1 = 0$

30. $x^2 - 2x + 8y + 17 = 0$

31. $9x^2 + 25y^2 - 36x - 50y + 61 = 0$

32. $9x^2 - y^2 + 54x + 10y + 55 = 0$

33. $x^2 + 4y^2 + 4x + 32y + 64 = 0$

34. $2x^2 - y^2 + 4x + 10y - 22 = 0$

35. $x^2 + 4x + 6y - 2 = 0$

36. $y^2 + 8x + 6y + 25 = 0$

In Exercises 37–44, perform a rotation of axes to eliminate the xy-term, and sketch the graph of the conic.

37. $xy + 1 = 0$

38. $4x^2 + 2xy + 4y^2 - 15 = 0$

39. $5x^2 - 2xy + 5y^2 - 24 = 0$

40. $x^2 + 2xy + y^2 - 8x + 8y = 0$

41. $13x^2 + 6\sqrt{3}xy + 7y^2 - 16 = 0$

42. $3x^2 - 2\sqrt{3}xy + y^2 + 2x + 2\sqrt{3}\,y = 0$

43. $x^2 + 2\sqrt{3}xy + 3y^2 - 2\sqrt{3}x + 2y + 16 = 0$

44. $7x^2 - 2\sqrt{3}xy + 5y^2 = 16$

In Exercises 45 and 46, perform a rotation of axes to eliminate the xy-term, and sketch the graph of the "degenerate" conic.

45. $x^2 - 2xy + y^2 = 0$

46. $5x^2 - 2xy + 5y^2 = 0$

47. Prove that a rotation of $\theta = \pi/4$ will eliminate the xy-term from the equation
$$ax^2 + bxy + ay^2 + dx + ey + f = 0.$$

48. Prove that a rotation of θ, where $\cot 2\theta = (a - c)/b$, will eliminate the xy-term from the equation
$$ax^2 + bxy + cy^2 + dx + ey + f = 0.$$

49. For the equation $ax^2 + bxy + cy^2 = 0$, define the matrix A to be
$$A = \begin{bmatrix} a & b/2 \\ b/2 & c \end{bmatrix}.$$
Prove that if $|A| \neq 0$, then the graph of $ax^2 + bxy + cy^2 = 0$ is two intersecting lines.

50. In Exercise 49, let $|A| = 0$ and describe the graph of $ax^2 + bxy + cy^2 = 0$.

CHAPTER 4 Summary

After studying this chapter, you should have acquired the following skills.

■ Perform vector operations (vector addition and scalar multiplication) on vectors in R^n. Recognize and use the properties of these operations.

■ Determine whether a given set with two operations is a vector space.

■ Recognize standard examples of vector spaces: R^n, $M_{m,n}$, P_n, P, $C(-\infty, \infty)$, $C[a, b]$.

■ Determine whether a given subset W of a vector space V is a subspace.

■ Recognize subspaces of R^2 and R^3 and understand their geometric interpretations.

■ Determine whether a vector **x** is a linear combination of a given finite set of vectors in V and be able to write this linear combination.

■ Determine whether a given set S of vectors in a vector space V is a spanning set for V—that is, whether $V = \text{span}(S)$.

■ Determine whether a given finite set of vectors in a vector space V is linearly independent.

■ Determine whether a given finite set of vectors in V forms a basis for V and recognize standard bases in the vector spaces R^n, $M_{m,n}$, and P_n.

■ Find the dimension of a subspace of R^n, $M_{m,n}$, or P_n.

■ Use the tests for bases in an n-dimensional vector space.

■ Find a basis for the column or row space of a matrix.

■ Find the rank of a given matrix.

■ Find a basis for and the dimension of the nullspace (nullity) of a matrix.

■ Find a general solution of a consistent system $A\mathbf{x} = \mathbf{b}$ in the form $\mathbf{x} = \mathbf{x}_p + \mathbf{x}_h$.

■ Know and use the equivalent conditions for an invertible matrix A also viewed as a matrix of coefficients for a square linear system.

■ Use the definition of the coordinate matrix $[\mathbf{x}]_B$ for a vector $\mathbf{x}$ relative to the basis B to find $[\mathbf{x}]_B$ in R^n, P_n, and $M_{m,n}$.

■ Find the transition matrix from the basis B to the basis B' in R^n.

■ Find the coordinate matrix $[\mathbf{x}]_B$ relative to the basis B in R^n using the transition matrix from the standard basis to B.

■ Find $[\mathbf{x}]_{B'}$ for a vector $\mathbf{x}$ in R^n, given the transition matrix from B to B' and the coordinate matrix $[\mathbf{x}]_B$.

■ Find the transition matrix from B to B' given that P is the transition matrix from B' to B.

■ Determine whether a function is a solution of a given differential equation.

■ Find the Wronskian for a given set of functions.

■ Test a given set of solutions for linear independence.

■ Find the general solution of a given differential equation.

■ Identify and sketch the graph of a given conic section.

■ Perform a rotation of axes to eliminate the xy-term in the equation of a conic and sketch the graph of a conic or a degenerate conic.

CHAPTER 4   Review Exercises

In Exercises 1–4, find (a) $\mathbf{u} + \mathbf{v}$, (b) $2\mathbf{v}$, (c) $\mathbf{u} - \mathbf{v}$, and (d) $3\mathbf{u} - 2\mathbf{v}$.

1. $\mathbf{u} = (-1, 2, 3)$, $\mathbf{v} = (1, 0, 2)$

2. $\mathbf{u} = (-1, 2, 1)$, $\mathbf{v} = (0, 1, 1)$

3. $\mathbf{u} = (3, -1, 2, 3)$, $\mathbf{v} = (0, 2, 2, 1)$

4. $\mathbf{u} = (0, 1, -1, 2)$, $\mathbf{v} = (1, 0, 0, 2)$

In Exercises 5 and 6, solve for $\mathbf{x}$ given that $\mathbf{u} = (1, -1, 2)$, $\mathbf{v} = (0, 2, 3)$, and $\mathbf{w} = (0, 1, 1)$.

5. $2\mathbf{x} - \mathbf{u} + 3\mathbf{v} + \mathbf{w} = \mathbf{0}$ **6.** $3\mathbf{x} + 2\mathbf{u} - \mathbf{v} + 2\mathbf{w} = \mathbf{0}$

In Exercises 7 and 8, write $\mathbf{v}$ as a linear combination of $\mathbf{u}_1$, $\mathbf{u}_2$, and $\mathbf{u}_3$, if possible.

7. $\mathbf{v} = (1, 2, 3, 5)$, $\mathbf{u}_1 = (1, 2, 3, 4)$, $\mathbf{u}_2 = (-1, -2, -3, 4)$, $\mathbf{u}_3 = (0, 0, 1, 1)$

8. $\mathbf{v} = (4, 4, 5)$, $\mathbf{u}_1 = (1, 2, 3)$, $\mathbf{u}_2 = (-2, 0, 1)$, $\mathbf{u}_3 = (1, 0, 0)$

In Exercises 9 and 10, describe the zero vector and the additive inverse of a vector in the given vector space.

9. $M_{3,4}$ **10.** P_8

In Exercises 11–16, determine whether W is a subspace of the vector space.

11. $W = \{(x, y): x = 2y\}$, $V = R^2$

12. $W = \{(x, y): x - y = 1\}$, $V = R^2$

13. $W = \{(x, 2x, 3x): x$ is a real number$\}$, $V = R^3$

14. $W = \{(x, y, z): x \geq 0\}$, $V = R^3$

15. $W = \{f: f(0) = -1\}$, $V = C[-1, 1]$

16. $W = \{f: f(-1) = 0\}$, $V = C[-1, 1]$

17. Which of the following subsets of R^3 is a subspace of R^3?

(a) $W = \{(x_1, x_2, x_3): x_1^2 + x_2^2 + x_3^2 = 0\}$

(b) $W = \{(x_1, x_2, x_3): x_1^2 + x_2^2 + x_3^2 = 1\}$

18. Which of the following subsets of R^3 is a subspace of R^3?

(a) $W = \{(x_1, x_2, x_3): x_1 + x_2 + x_3 = 0\}$

(b) $W = \{(x_1, x_2, x_3): x_1 + x_2 + x_3 = 1\}$

In Exercises 19–22, determine whether S (a) spans R^3, (b) is linearly independent, and (c) is a basis for R^3.

19. $S = \{(1, -5, 4), (11, 6, -1), (2, 3, 5)\}$

20. $S = \{(4, 0, 1), (0, -3, 2), (5, 10, 0)\}$

21. $S = \left\{\left(-\frac{1}{2}, \frac{3}{4}, -1\right), (5, 2, 3), (-4, 6, -8)\right\}$

22. $S = \{(1, 0, 0), (0, 1, 0), (0, 0, 1), (-1, 2, -3)\}$

23. Determine whether $S = \{1 - t, 2t + 3t^2, t^2 - 2t^3, 2 + t^3\}$ is a basis for P_3.

24. Determine whether the set is a basis for $M_{2,2}$.

$$S = \left\{\begin{bmatrix} 1 & 0 \\ 2 & 3 \end{bmatrix}, \begin{bmatrix} -2 & 1 \\ -1 & 0 \end{bmatrix}, \begin{bmatrix} 3 & 4 \\ 2 & 3 \end{bmatrix}, \begin{bmatrix} -3 & -3 \\ 1 & 3 \end{bmatrix}\right\}$$

In Exercises 25–28, find (a) a basis for and (b) the dimension of the solution space of the homogeneous system of equations.

25.
$$\begin{aligned} 2x_1 + 4x_2 + 3x_3 - 6x_4 &= 0 \\ x_1 + 2x_2 + 2x_3 - 5x_4 &= 0 \\ 3x_1 + 6x_2 + 5x_3 - 11x_4 &= 0 \end{aligned}$$

26.
$$\begin{aligned} 3x_1 + 8x_2 + 2x_3 + 3x_4 &= 0 \\ 4x_1 + 6x_2 + 2x_3 - x_4 &= 0 \\ 3x_1 + 4x_2 + x_3 - 3x_4 &= 0 \end{aligned}$$

27.
$$\begin{aligned} x_1 - 3x_2 + x_3 + x_4 &= 0 \\ 2x_1 + x_2 - x_3 + 2x_4 &= 0 \\ x_1 + 4x_2 - 2x_3 + x_4 &= 0 \\ 5x_1 - 8x_2 + 2x_3 + 5x_4 &= 0 \end{aligned}$$

28.
$$\begin{aligned} -x_1 + 2x_2 - x_3 + 2x_4 &= 0 \\ -2x_1 + 2x_2 + x_3 + 4x_4 &= 0 \\ 3x_1 + 2x_2 + 2x_3 + 5x_4 &= 0 \\ -3x_1 + 8x_2 + 5x_3 + 17x_4 &= 0 \end{aligned}$$

In Exercises 29–32, find a basis for the solution space of $Ax = 0$. Then verify that rank(A) + nullity(A) = n.

29. $A = \begin{bmatrix} 5 & -8 \\ -10 & 16 \end{bmatrix}$

30. $A = \begin{bmatrix} 1 & 4 \\ 3 & 2 \end{bmatrix}$

31. $A = \begin{bmatrix} 2 & -3 & -6 & -4 \\ 1 & 5 & -3 & 11 \\ 2 & 7 & -6 & 16 \end{bmatrix}$

32. $A = \begin{bmatrix} 1 & 3 & 2 \\ 4 & -1 & -18 \\ -1 & 3 & 10 \\ 1 & 2 & 0 \end{bmatrix}$

In Exercises 33–36, find (a) the rank and (b) a basis for the row space of the given matrix.

33. $\begin{bmatrix} 1 & 2 \\ -4 & 3 \\ 6 & 1 \end{bmatrix}$

34. $\begin{bmatrix} 2 & -1 & 4 \\ 1 & 5 & 6 \\ 1 & 16 & 14 \end{bmatrix}$

35. $\begin{bmatrix} 7 & 0 & 2 \\ 4 & 1 & 6 \\ -1 & 16 & 14 \end{bmatrix}$

36. $\begin{bmatrix} 1 & -4 & 0 & 4 \end{bmatrix}$

In Exercises 37–40, find the coordinate matrix of **x** relative to the standard basis.

37. $B = \{(1, 1), (-1, 1)\}, \quad [\mathbf{x}]_B = [3, 5]^T$

38. $B = \{(2, 0), (3, 3)\}, \quad [\mathbf{x}]_B = [1, 1]^T$

39. $B = \{(1, 0, 0), (1, 1, 0), (0, 1, 1)\}, \quad [\mathbf{x}]_B = [2, 0, -1]^T$

40. $B = \left\{\left(\frac{1}{2}, \frac{1}{2}\right), (1, 0)\right\}, \quad [\mathbf{x}]_B = \left[\frac{1}{2}, \frac{1}{2}\right]^T$

In Exercises 41–44, find the coordinate matrix of **x** relative to the given (nonstandard) basis for R^n.

41. $B' = \{(5, 0), (0, -8)\}, \quad \mathbf{x} = (2, 2)$

42. $B' = \{(1, 2, 3), (1, 2, 0), (0, -6, 2)\}, \quad \mathbf{x} = (3, -3, 0)$

43. $B' = \{(9, -3, 15, 4), (-3, 0, 0, -1), (0, -5, 6, 8), (-3, 4, -2, 3)\}, \quad \mathbf{x} = (21, -5, 43, 14)$

44. $B' = \{(1, 0, 0), (0, 1, 0), (1, 1, 1)\}, \quad \mathbf{x} = (4, -2, 9)$

In Exercises 45 and 46, find the coordinate matrix of **x** relative to the basis B'.

45. $B = \{(1, 1), (-1, 1)\}, \quad B' = \{(0, 1), (1, 2)\}, \quad [\mathbf{x}]_B = [3, -3]^T$

46. $B = \{(1, 0, 0), (1, 1, 0), (1, 1, 1)\}, \quad B' = \{(0, 0, 1), (0, 1, 1), (1, 1, 1)\}, \quad [\mathbf{x}]_B = [-1, 2, -3]^T$

In Exercises 47–50, find the transition matrix from B to B'.

47. $B = \{(1, -1), (3, 1)\}, \quad B' = \{(1, 0), (0, 1)\}$

48. $B = \{(1, 0, 0), (0, 1, 0), (0, 0, 1)\}, \quad B' = \{(0, 0, 1), (0, 1, 0), (1, 0, 0)\}$

49. $B = \{(1, -1), (3, 1)\}, \quad B' = \{(1, 2), (-1, 0)\}$

50. $B = \{(1, 1, 1), (1, 1, 0), (1, 0, 0)\}, \quad B' = \{(1, 2, 3), (0, 1, 0), (1, 0, 1)\}$

51. Let W be the subspace of P_3 of all polynomials such that $p(0) = 0$, and let U be the subspace of all polynomials such that $p(1) = 0$. Find a basis for W, a basis for U, and a basis for their intersection $W \cap U$.

52. *Calculus* Let $V = C'(-\infty, \infty)$, the vector space of all continuously differentiable functions on the real line.

(a) Prove that $W = \{f: f' = 3f\}$ is a subspace of V.

(b) Prove that $U = \{f: f' = f + 1\}$ is not a subspace of V.

53. *Writing* Let $B = \{p_1(x), p_2(x), \ldots, p_n(x), p_{n+1}(x)\}$ be a basis for P_n. Must B contain a polynomial of each degree $0, 1, 2, \ldots, n$? Explain your reasoning.

54. Let A and B be $n \times n$ matrices with $A \neq O$ and $B \neq O$. Prove that if A is symmetric and B is skew-symmetric ($B^T = -B$), then $\{A, B\}$ is a linearly independent set.

55. Let $V = P_5$ and consider the set W of all polynomials of the form $(x^3 + x)p(x)$, where $p(x)$ is in P_2. Is W a subspace of V? Prove your answer.

56. Let $\mathbf{v}_1$, $\mathbf{v}_2$, and $\mathbf{v}_3$ be three linearly independent vectors in a vector space V. Is the set $\{\mathbf{v}_1 - \mathbf{v}_2, \mathbf{v}_2 - \mathbf{v}_3, \mathbf{v}_3 - \mathbf{v}_1\}$ linearly dependent or linearly independent?

57. Let A be an $n \times n$ square matrix. Prove that the row vectors of A are linearly dependent if and only if the column vectors of A are linearly dependent.

58. Let A be an $n \times n$ square matrix, and let λ be a scalar. Prove that the set

$$S = \{\mathbf{x}: A\mathbf{x} = \lambda\mathbf{x}\}$$

is a subspace of R^n. Determine the dimension of S if $\lambda = 3$ and

$$A = \begin{bmatrix} 3 & 1 & 0 \\ 0 & 3 & 0 \\ 0 & 0 & 1 \end{bmatrix}.$$

59. Let $f(x) = x$ and $g(x) = |x|$.

(a) Show that f and g are linearly independent in $C[-1, 1]$.

(b) Show that f and g are linearly dependent in $C[0, 1]$.

60. Given a set of functions, describe how their domain can influence whether the set is linearly independent or dependent.

True or False? In Exercises 61–64, determine whether each statement is true or false. If a statement is true, give a reason or cite an appropriate statement in the text. If a statement is false, provide an example that shows that the statement is not true in all cases or cite an appropriate statement in the text.

61. (a) The standard operations in R^n are vector addition and scalar multiplication.

(b) The additive inverse of a given vector is not unique.

(c) A vector space consists of four entities: a set of vectors, a set of scalars, and two operations.

62. (a) The set $W = \{(0, x^2, x^3): x^2 \text{ and } x^3 \text{ are real numbers}\}$ is a subspace of R^3.

(b) A linearly independent spanning set S is called a basis of a vector space V.

(c) If A is an invertible $n \times n$ matrix, then the n row vectors of A are linearly dependent.

63. (a) The set of all n-tuples is called n-space and is denoted by R^n.

(b) The additive identity of a given vector is not unique.

(c) Once a theorem has been proved for an abstract vector space, you need not give separate proofs for n-tuples, matrices, and polynomials.

64. (a) The set of points on the line given by $x + y = 0$ is a subspace of R^2.

(b) A set of vectors $S = \{\mathbf{v}_1, \mathbf{v}_2, \ldots, \mathbf{v}_n\}$ in a vector space V is linearly independent if the vector equation $c_1\mathbf{v}_1 + c_2\mathbf{v}_2 + \cdots + c_n\mathbf{v}_n = \mathbf{0}$ has only the trivial solution.

(c) Elementary row operations preserve the column space of the matrix A.

Linear Differential Equations (Calculus)

In Exercises 65 and 66, determine whether each function is a solution of the linear differential equation.

65. $y'' - y' - 6y = 0$

(a) e^{3x} (b) e^{2x} (c) e^{-3x} (d) e^{-2x}

66. $y'''' - y = 0$

(a) e^x (b) e^{-x} (c) $\cos x$ (d) $\sin x$

In Exercises 67 and 68, find the Wronskian for the set of functions.

67. $\{1, x, e^x\}$ **68.** $\{1, x, 2 + x\}$

In Exercises 69–72, test the given set of solutions for linear independence.

Differential Equation	*Solutions*
69. $y'' + 6y' + 9y = 0$	$\{e^{-3x}, xe^{-3x}\}$
70. $y'' + 6y' + 9y = 0$	$\{e^{-3x}, 3e^{-3x}\}$
71. $y''' - 6y'' + 11y' - 6y = 0$	$\{e^x, e^{2x}, e^x - e^{2x}\}$
72. $y'' + 4y = 0$	$\{\sin 2x, \cos 2x\}$

Conic Sections and Rotation

In Exercises 73–76, identify and sketch the graph of the equation.

73. $x^2 + y^2 - 4x + 2y - 4 = 0$

74. $x^2 - y^2 + 2x - 3 = 0$

75. $2x^2 - 20x - y + 46 = 0$

76. $4x^2 + y^2 + 32x + 4y + 63 = 0$

In Exercises 77 and 78, perform a rotation of axes to eliminate the xy-term, and sketch the graph of the conic.

77. $xy = 3$

78. $9x^2 + 4xy + 9y^2 - 20 = 0$

CHAPTER 4 **Projects**

1 SOLUTIONS OF LINEAR SYSTEMS

Write a short paragraph to answer each of the following questions about solutions of systems of linear equations. You should not perform any calculations, but instead base your explanations on the appropriate properties from the text.

1. One solution of the homogeneous linear system

$$x + 2y + z + 3w = 0$$
$$x - y \quad\quad + w = 0$$
$$y - z + 2w = 0$$

is $x = -2$, $y = -1$, $z = 1$, and $w = 1$. Explain why $x = 4$, $y = 2$, $z = -2$, and $w = -2$ must also be a solution. Do not perform any row operations.

2. The vectors $\mathbf{x}_1$ and $\mathbf{x}_2$ are solutions of the homogeneous linear system $A\mathbf{x} = \mathbf{0}$. Explain why the vector $2\mathbf{x}_1 - 3\mathbf{x}_2$ must also be a solution.

3. Consider the two systems represented by the augmented matrices.

$$\begin{bmatrix} 1 & 1 & -5 & 3 \\ 1 & 0 & -2 & 1 \\ 2 & -1 & -1 & 0 \end{bmatrix} \quad \begin{bmatrix} 1 & 1 & -5 & -9 \\ 1 & 0 & -2 & -3 \\ 2 & -1 & -1 & 0 \end{bmatrix}$$

If the first system is known to be consistent, explain why the second system is also consistent. Do not perform any row operations.

4. The vectors $\mathbf{x}_1$ and $\mathbf{x}_2$ are solutions of the linear system $A\mathbf{x} = \mathbf{b}$. Is the vector $2\mathbf{x}_1 - 3\mathbf{x}_2$ also a solution? Why or why not?

5. The linear systems $A\mathbf{x} = \mathbf{b}_1$ and $A\mathbf{x} = \mathbf{b}_2$ are consistent. Is the system $A\mathbf{x} = \mathbf{b}_1 + \mathbf{b}_2$ necessarily consistent? Why or why not?

6. Consider the linear system $A\mathbf{x} = \mathbf{b}$. If the rank of A equals the rank of the augmented matrix for the system, explain why the system must be consistent. Contrast this to the case in which the rank of A is less than the rank of the augmented matrix.

2 DIRECT SUM

Let U and W be subspaces of the vector space V. You learned in Section 4.3 that the intersection $U \cap W$ is also a subspace of V, whereas the union $U \cup W$ is, in general, not a subspace. In this project you will explore the **sum** and **direct sum** of subspaces, focusing especially on their geometric interpretation in R^n.

1. Define the sum of the subspaces U and W as follows.

    $$U + W = \{\mathbf{u} + \mathbf{w} : \mathbf{u} \in U, \mathbf{w} \in W\}$$

    Prove that $U + W$ is a subspace of V.

2. Consider the following subspaces of $V = R^3$.

    $$U = \{(x, y, x - y) : x, y \in R\}$$
    $$W = \{(x, 0, x) : x \in R\}$$
    $$Z = \{(x, x, x) : x \in R\}$$

    Find $U + W$, $U + Z$, and $W + Z$.

3. If U and W are subspaces of V such that $V = U + W$ and $U \cap W = \{\mathbf{0}\}$, prove that every vector in V has a *unique* representation of the form $\mathbf{u} + \mathbf{w}$, where $\mathbf{u}$ is in U and $\mathbf{w}$ is in W. In this case, we say that V is the **direct sum** of U and W, and write

    $$V = U \oplus W. \qquad \text{Direct sum}$$

    Which of the sums in part (2) of this project are direct sums?

4. Let $V = U \oplus W$ and suppose that $\{\mathbf{u}_1, \mathbf{u}_2, \ldots, \mathbf{u}_k\}$ is a basis for the subspace U and $\{\mathbf{w}_1, \mathbf{w}_2, \ldots, \mathbf{w}_m\}$ is a basis for the subspace W. Prove that the set

    $$\{\mathbf{u}_1, \ldots, \mathbf{u}_k, \mathbf{w}_1, \ldots, \mathbf{w}_m\}$$

    is a basis for V.

5. Consider the following subspaces of $V = R^3$.

    $$U = \{(x, 0, y) : x, y \in R\}$$
    $$W = \{(0, x, y) : x, y \in R\}$$

    Show that $R^3 = U + W$. Is R^3 the *direct* sum of U and W? What are the dimensions of U, W, $U \cap W$, and $U + W$? In general, formulate a conjecture that relates the dimensions of U, W, $U \cap W$, and $U + W$.

6. Can you find two 2-dimensional subspaces of R^3 whose intersection is just the zero vector? Why or why not?

**JEAN-BAPTISTE
JOSEPH FOURIER**

1 7 6 8 – 1 8 3 0

JEAN-BAPTISTE JOSEPH FOURIER
was born in Auxerre, France. He was orphaned at a young age, and later entered a local military college run by Benedictines. Around the time of the French Revolution, he returned to the military academy at Auxerre as a professor of mathematics, gaining a reputation as a gifted teacher and mathematician.

When Napoleon created the École Normale in 1794, Fourier was nominated to study there. Fourier later began teaching at the École Centrale (now called the Polytechnique). His innovative approach to education brought the school rapid success, and before long the college was graduating talented scientists, engineers, and mathematicians.

In 1798 Fourier accompanied Napoleon as the French invaded Egypt. He remained in Egypt, founding schools and teaching, until 1801, when he returned to France to accept a government position at

Grenoble. There, in addition to doing his administrative work, Fourier conducted research in mathematical physics and wrote one of his most famous papers, The Mathematical Theory of Heat (1811). This paper was awarded the Grand Prize by the Academy of Sciences of Paris.

In 1822 Fourier published a classic text that has the same title as his earlier paper on heat. This work contains many important results dealing with differential equations, eigenvalues, and representation of functions by trigonometric series (later called Fourier series). Fourier's work with Fourier series forced mathematicians to reconsider the then-accepted, but narrow, definition of a function.

During this time, Fourier was appointed secretary of the Academy of Sciences, a position he held until his death in 1830.

5 INNER PRODUCT SPACES

CHAPTER

5.1 | Length and Dot Product in R^n

Section 4.1 mentioned that vectors in the plane can be characterized as directed line segments having a certain *length* and *direction*. In this section R^2 will be used as a model to define these and other geometric properties (such as distance and angle) of vectors in R^n. Then, in the next section, these ideas will be extended to general vector spaces.

You will begin by reviewing the definition of the length of a vector in R^2. If $\mathbf{v} = (v_1, v_2)$ is a vector in the plane, then the **length,** or **magnitude,** of $\mathbf{v}$, denoted by $\|\mathbf{v}\|$, is defined as

$$\|\mathbf{v}\| = \sqrt{v_1^2 + v_2^2}.$$

This definition corresponds to the usual notion of length in Euclidean geometry. That is, the vector **v** is thought of as the hypotenuse of a right triangle whose sides have lengths of $|v_1|$ and $|v_2|$, as shown in Figure 5.1. Applying the Pythagorean Theorem produces

$$\|\mathbf{v}\|^2 = |v_1|^2 + |v_2|^2 = v_1^2 + v_2^2.$$

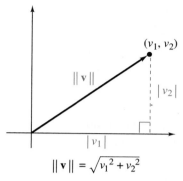

$$\|\mathbf{v}\| = \sqrt{v_1^2 + v_2^2}$$

Figure 5.1

Using R^2 as a model, the length of a vector in R^n is defined as follows.

Definition of Length of a Vector in R^n

The **length,** or **magnitude,** of a vector $\mathbf{v} = (v_1, v_2, \ldots, v_n)$ in R^n is given by

$$\|\mathbf{v}\| = \sqrt{v_1^2 + v_2^2 + \cdots + v_n^2}.$$

REMARK: The length of a vector is also called its **norm.** If $\|\mathbf{v}\| = 1$, then the vector **v** is called a **unit vector.**

This definition shows that the length of a vector cannot be negative. That is, $\|\mathbf{v}\| \geq 0$. Moreover, $\|\mathbf{v}\| = 0$ if and only if **v** is the zero vector **0**.

Technology Note

You can use a graphing calculator or computer software program to find the length, or norm, of a vector. For example, using a TI-86, the length of the vector $\mathbf{v} = (2, -1, -2)$ can be found as follows.

Verify the length of **v** in Example 1(a) on the next page.

EXAMPLE 1 *The Length of a Vector in R^n*

(a) In R^5, the length of $\mathbf{v} = (0, -2, 1, 4, -2)$ is given by

$$\|\mathbf{v}\| = \sqrt{0^2 + (-2)^2 + 1^2 + 4^2 + (-2)^2} = \sqrt{25} = 5.$$

(b) In R^3, the length of $\mathbf{v} = \left(2/\sqrt{17}, -2/\sqrt{17}, 3/\sqrt{17}\right)$ is given by

$$\|\mathbf{v}\| = \sqrt{\left(\frac{2}{\sqrt{17}}\right)^2 + \left(-\frac{2}{\sqrt{17}}\right)^2 + \left(\frac{3}{\sqrt{17}}\right)^2} = \sqrt{\frac{17}{17}} = 1.$$

Because its length is 1, $\mathbf{v}$ is a unit vector, as shown in Figure 5.2

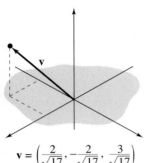

$\mathbf{v} = \left(\dfrac{2}{\sqrt{17}}, -\dfrac{2}{\sqrt{17}}, \dfrac{3}{\sqrt{17}}\right)$

Figure 5.2

Each vector in the standard basis for R^n has length 1 and is called a **standard unit vector** in R^n. In physics and engineering it is common to denote the standard unit vectors in R^2 and R^3 as follows.

$$\{\mathbf{i}, \mathbf{j}\} = \{(1, 0), (0, 1)\}$$

and

$$\{\mathbf{i}, \mathbf{j}, \mathbf{k}\} = \{(1, 0, 0), (0, 1, 0), (0, 0, 1)\}$$

Two nonzero vectors $\mathbf{u}$ and $\mathbf{v}$ in R^n are **parallel** if one is a scalar multiple of the other—that is, $\mathbf{u} = c\mathbf{v}$. Moreover, if $c > 0$, then $\mathbf{u}$ and $\mathbf{v}$ have the **same direction,** and if $c < 0$, $\mathbf{u}$ and $\mathbf{v}$ have **opposite directions.** The following theorem gives a formula for finding the length of a scalar multiple of a vector.

Theorem 5.1

Length of a Scalar Multiple

Let $\mathbf{v}$ be a vector in R^n and c be a scalar. Then

$$\|c\mathbf{v}\| = |c|\,\|\mathbf{v}\|,$$

where $|c|$ is the absolute value of c.

Proof Because $c\mathbf{v} = (cv_1, cv_2, \ldots, cv_n)$, it follows that

$$
\begin{aligned}
\|c\mathbf{v}\| &= \|(cv_1, cv_2, \ldots, cv_n)\| \\
&= \sqrt{(cv_1)^2 + (cv_2)^2 + \cdots + (cv_n)^2} \\
&= \sqrt{c^2(v_1^2 + v_2^2 + \cdots + v_n^2)} \\
&= |c|\sqrt{v_1^2 + v_2^2 + \cdots + v_n^2} \\
&= |c|\,\|\mathbf{v}\|.
\end{aligned}
$$

One important use of Theorem 5.1 is in finding a unit vector having the same direction as a given vector. The following theorem provides a procedure for doing this.

Theorem 5.2	*Unit Vector in the Direction of* **v**

If **v** is a nonzero vector in R^n, then the vector

$$\mathbf{u} = \frac{\mathbf{v}}{\|\mathbf{v}\|}$$

has length 1 and has the same direction as **v**. This vector **u** is called the **unit vector in the direction of v**.

Proof   Because **v** is nonzero, you know that $\|\mathbf{v}\| \neq 0$. Thus $1/\|\mathbf{v}\|$ is positive, and you can write **u** as a positive scalar multiple of **v**.

$$\mathbf{u} = \left(\frac{1}{\|\mathbf{v}\|}\right)\mathbf{v}$$

Therefore, it follows that **u** has the same direction as **v**. Finally, **u** has length 1 because

$$\|\mathbf{u}\| = \left\| \frac{\mathbf{v}}{\|\mathbf{v}\|} \right\| = \frac{1}{\|\mathbf{v}\|}\|\mathbf{v}\| = 1.$$

The process of finding the unit vector in the direction of **v** is called **normalizing** the vector **v**. The procedure is demonstrated in the next example.

EXAMPLE 2   *Finding a Unit Vector*

Find the unit vector in the direction of $\mathbf{v} = (3, -1, 2)$, and verify that this vector has length 1.

Solution   The unit vector in the direction of **v** is

$$\frac{\mathbf{v}}{\|\mathbf{v}\|} = \frac{(3, -1, 2)}{\sqrt{3^2 + (-1)^2 + 2^2}}$$

$$= \frac{1}{\sqrt{14}}(3, -1, 2)$$

$$= \left(\frac{3}{\sqrt{14}}, \frac{-1}{\sqrt{14}}, \frac{2}{\sqrt{14}}\right),$$

which is a unit vector because

$$\sqrt{\left(\frac{3}{\sqrt{14}}\right)^2 + \left(\frac{-1}{\sqrt{14}}\right)^2 + \left(\frac{2}{\sqrt{14}}\right)^2} = \sqrt{\frac{14}{14}} = 1.$$

(See Figure 5.3.)

Figure 5.3

Technology Note

You can use a graphing calculator or computer software program to find the unit vector for a given vector. For example, you can use a TI-86 to find the unit vector for $\mathbf{v} = (-3, 4)$ as follows.

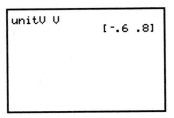

Distance Between Two Vectors in R^n

To define the **distance between two vectors** in R^n, once again R^2 will be used as the model. The Distance Formula from analytic geometry tells you that the distance d between two points in the plane, (u_1, u_2) and (v_1, v_2), is given by

$$d = \sqrt{(u_1 - v_1)^2 + (u_2 - v_2)^2}.$$

In vector terminology, this distance can be viewed as the length of $\mathbf{u} - \mathbf{v}$, where $\mathbf{u} = (u_1, u_2)$ and $\mathbf{v} = (v_1, v_2)$, as shown in Figure 5.4. That is,

$$\|\mathbf{u} - \mathbf{v}\| = \sqrt{(u_1 - v_1)^2 + (u_2 - v_2)^2},$$

which leads to the following definition.

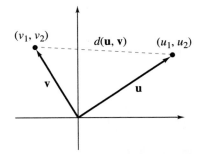

$$d(\mathbf{u}, \mathbf{v}) = \|\mathbf{u} - \mathbf{v}\| = \sqrt{(u_1 - v_1)^2 + (u_2 - v_2)^2}$$

Figure 5.4

<table>
<tr><td>

Definition of Distance Between Two Vectors

</td><td>

The **distance between two vectors u** and **v** in R^n is

$$d(\mathbf{u}, \mathbf{v}) = \|\mathbf{u} - \mathbf{v}\|.$$

</td></tr>
</table>

You can easily verify the following properties of distance.

1. $d(\mathbf{u}, \mathbf{v}) \geq 0$
2. $d(\mathbf{u}, \mathbf{v}) = 0$ if and only if $\mathbf{u} = \mathbf{v}$.
3. $d(\mathbf{u}, \mathbf{v}) = d(\mathbf{v}, \mathbf{u})$

EXAMPLE 3   *Finding the Distance Between Two Vectors*

The distance between $\mathbf{u} = (0, 2, 2)$ and $\mathbf{v} = (2, 0, 1)$ is

$$d(\mathbf{u}, \mathbf{v}) = \|\mathbf{u} - \mathbf{v}\| = \|(0 - 2, 2 - 0, 2 - 1)\|$$
$$= \sqrt{(-2)^2 + 2^2 + 1^2} = 3.$$

Dot Product and the Angle Between Two Vectors

|| v − u ||

|| u ||

θ

|| v ||

Angle Between Two Vectors

Figure 5.5

To find the angle θ ($0 \leq \theta \leq \pi$) between two nonzero vectors $\mathbf{u} = (u_1, u_2)$ and $\mathbf{v} = (v_1, v_2)$ in R^2, the Law of Cosines can be applied to the triangle shown in Figure 5.5 to obtain

$$\|\mathbf{v} - \mathbf{u}\|^2 = \|\mathbf{u}\|^2 + \|\mathbf{v}\|^2 - 2\|\mathbf{u}\| \|\mathbf{v}\| \cos \theta.$$

Expanding and solving for $\cos \theta$ yields

$$\cos \theta = \frac{u_1 v_1 + u_2 v_2}{\|\mathbf{u}\| \|\mathbf{v}\|}.$$

The numerator of the above quotient is defined to be the **dot product** of $\mathbf{u}$ and $\mathbf{v}$ and is denoted by

$$\mathbf{u} \cdot \mathbf{v} = u_1 v_1 + u_2 v_2.$$

This definition is generalized to R^n as follows.

<table>
<tr><td>

Definition of Dot Product in R^n

</td><td>

The **dot product** of $\mathbf{u} = (u_1, u_2, \ldots, u_n)$ and $\mathbf{v} = (v_1, v_2, \ldots, v_n)$ is the *scalar* quantity

$$\mathbf{u} \cdot \mathbf{v} = u_1 v_1 + u_2 v_2 + \cdots + u_n v_n.$$

</td></tr>
</table>

REMARK: Be sure you see that the dot product of two vectors is a scalar, not another vector.

EXAMPLE 4   *Finding the Dot Product of Two Vectors*

The dot product of $\mathbf{u} = (1, 2, 0, -3)$ and $\mathbf{v} = (3, -2, 4, 2)$ is

$$\mathbf{u} \cdot \mathbf{v} = (1)(3) + (2)(-2) + (0)(4) + (-3)(2) = -7.$$

Technology Note

You can use a graphing calculator or computer software program to find the dot product of two vectors. Using a TI-86, you can verify Example 4 as follows.

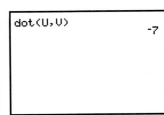

Theorem 5.3   *Properties of the Dot Product*

If $\mathbf{u}$, $\mathbf{v}$, and $\mathbf{w}$ are vectors in R^n and c is a scalar, then the following properties are true.
1. $\mathbf{u} \cdot \mathbf{v} = \mathbf{v} \cdot \mathbf{u}$
2. $\mathbf{u} \cdot (\mathbf{v} + \mathbf{w}) = \mathbf{u} \cdot \mathbf{v} + \mathbf{u} \cdot \mathbf{w}$
3. $c(\mathbf{u} \cdot \mathbf{v}) = (c\mathbf{u}) \cdot \mathbf{v} = \mathbf{u} \cdot (c\mathbf{v})$
4. $\mathbf{v} \cdot \mathbf{v} = \|\mathbf{v}\|^2$
5. $\mathbf{v} \cdot \mathbf{v} \geq 0$, and $\mathbf{v} \cdot \mathbf{v} = 0$ if and only if $\mathbf{v} = \mathbf{0}$.

Proof   The proofs of the properties follow easily from the definition of dot product. For example, to prove the first property, you can write

$$\mathbf{u} \cdot \mathbf{v} = u_1 v_1 + u_2 v_2 + \cdots + u_n v_n$$
$$= v_1 u_1 + v_2 u_2 + \cdots + v_n u_n$$
$$= \mathbf{v} \cdot \mathbf{u}.$$

In Section 4.1, R^n was defined to be the *set* of all ordered n-tuples of real numbers. When R^n is combined with the standard operations of vector addition, scalar multiplication, vector length, and the dot product, the resulting vector space is called **Euclidean n-space.** In the remainder of this text, unless stated otherwise, you may assume R^n to have the standard Euclidean operations.

EXAMPLE 5   *Finding Dot Products*

Given $\mathbf{u} = (2, -2)$, $\mathbf{v} = (5, 8)$, and $\mathbf{w} = (-4, 3)$, find each of the following.

(a) $\mathbf{u} \cdot \mathbf{v}$ (b) $(\mathbf{u} \cdot \mathbf{v})\mathbf{w}$ (c) $\mathbf{u} \cdot (2\mathbf{v})$ (d) $\|\mathbf{w}\|^2$ (e) $\mathbf{u} \cdot (\mathbf{v} - 2\mathbf{w})$

Solution   (a) By definition, you have

$$\mathbf{u} \cdot \mathbf{v} = 2(5) + (-2)(8) = -6.$$

(b) Using the result in part (a), you have

$$(\mathbf{u} \cdot \mathbf{v})\mathbf{w} = -6\mathbf{w} = -6(-4, 3) = (24, -18).$$

(c) By property 3 of Theorem 5.3, you have

$$\mathbf{u} \cdot (2\mathbf{v}) = 2(\mathbf{u} \cdot \mathbf{v}) = 2(-6) = -12.$$

(d) By property 4 of Theorem 5.3, you have

$$\|\mathbf{w}\|^2 = \mathbf{w} \cdot \mathbf{w} = (-4)(-4) + (3)(3) = 25.$$

(e) Because $2\mathbf{w} = (-8, 6)$, you have

$$\mathbf{v} - 2\mathbf{w} = (5 - (-8), 8 - 6) = (13, 2).$$

Consequently,

$$\mathbf{u} \cdot (\mathbf{v} - 2\mathbf{w}) = 2(13) + (-2)(2) = 26 - 4 = 22.$$

EXAMPLE 6   *Using Properties of the Dot Product*

Given two vectors $\mathbf{u}$ and $\mathbf{v}$ in R^n such that $\mathbf{u} \cdot \mathbf{u} = 39$, $\mathbf{u} \cdot \mathbf{v} = -3$, and $\mathbf{v} \cdot \mathbf{v} = 79$, evaluate $(\mathbf{u} + 2\mathbf{v}) \cdot (3\mathbf{u} + \mathbf{v})$.

Solution   Using Theorem 5.3, rewrite the given dot product as

$$\begin{aligned}
(\mathbf{u} + 2\mathbf{v}) \cdot (3\mathbf{u} + \mathbf{v}) &= \mathbf{u} \cdot (3\mathbf{u} + \mathbf{v}) + (2\mathbf{v}) \cdot (3\mathbf{u} + \mathbf{v}) \\
&= \mathbf{u} \cdot (3\mathbf{u}) + \mathbf{u} \cdot \mathbf{v} + (2\mathbf{v}) \cdot (3\mathbf{u}) + (2\mathbf{v}) \cdot \mathbf{v} \\
&= 3(\mathbf{u} \cdot \mathbf{u}) + \mathbf{u} \cdot \mathbf{v} + 6(\mathbf{v} \cdot \mathbf{u}) + 2(\mathbf{v} \cdot \mathbf{v}) \\
&= 3(\mathbf{u} \cdot \mathbf{u}) + 7(\mathbf{u} \cdot \mathbf{v}) + 2(\mathbf{v} \cdot \mathbf{v}) \\
&= 3(39) + 7(-3) + 2(79) = 254.
\end{aligned}$$

DISCOVERY

How does the dot product of two vectors compare with the product of their lengths? For instance, let $\mathbf{u} = (1, 1)$ and $\mathbf{v} = (-4, -3)$. Calculate $\mathbf{u} \cdot \mathbf{v}$ and $\|\mathbf{u}\| \, \|\mathbf{v}\|$. Repeat this experiment with other choices for u and v. Formulate a conjecture about the relationship between $\mathbf{u} \cdot \mathbf{v}$ and $\|\mathbf{u}\| \, \|\mathbf{v}\|$.

To define the angle θ between two vectors $\mathbf{u}$ and $\mathbf{v}$ in R^n, you can use the formula in R^2

$$\cos \theta = \frac{\mathbf{u} \cdot \mathbf{v}}{\|\mathbf{u}\| \, \|\mathbf{v}\|}.$$

For such a definition to make sense, however, the value of the right-hand side of this formula cannot exceed 1 in absolute value. This fact comes from a famous theorem named after the French mathematician Augustin-Louis Cauchy (1789–1857) and the German mathematician Hermann Schwarz (1843–1921).

Theorem 5.4	*The Cauchy-Schwarz Inequality*

If $\mathbf{u}$ and $\mathbf{v}$ are vectors in R^n, then

$$|\mathbf{u} \cdot \mathbf{v}| \leq \|\mathbf{u}\| \, \|\mathbf{v}\|,$$

where $|\mathbf{u} \cdot \mathbf{v}|$ denotes the *absolute value* of $\mathbf{u} \cdot \mathbf{v}$.

Proof   *Case 1.* If $\mathbf{u} = \mathbf{0}$, then it follows that

$$|\mathbf{u} \cdot \mathbf{v}| = |\mathbf{0} \cdot \mathbf{v}| = 0 \quad \text{and} \quad \|\mathbf{u}\| \, \|\mathbf{v}\| = 0\|\mathbf{v}\| = 0.$$

Hence, the theorem is true if $\mathbf{u} = \mathbf{0}$.

Case 2. If $\mathbf{u} \neq \mathbf{0}$, let t be any real number and consider the vector $t\mathbf{u} + \mathbf{v}$. Because $(t\mathbf{u} + \mathbf{v}) \cdot (t\mathbf{u} + \mathbf{v}) \geq 0$, it follows that

$$(t\mathbf{u} + \mathbf{v}) \cdot (t\mathbf{u} + \mathbf{v}) = t^2(\mathbf{u} \cdot \mathbf{u}) + 2t(\mathbf{u} \cdot \mathbf{v}) + \mathbf{v} \cdot \mathbf{v} \geq 0.$$

Now, let $a = \mathbf{u} \cdot \mathbf{u}$, $b = 2(\mathbf{u} \cdot \mathbf{v})$, and $c = \mathbf{v} \cdot \mathbf{v}$ to obtain the quadratic inequality $at^2 + bt + c \geq 0$. Because this quadratic is never negative, it has either no real roots or a single repeated real root. But by the Quadratic Formula, this implies that the discriminant, $b^2 - 4ac$, is less than or equal to zero. Thus

$$b^2 - 4ac \leq 0$$
$$b^2 \leq 4ac$$
$$4(\mathbf{u} \cdot \mathbf{v})^2 \leq 4(\mathbf{u} \cdot \mathbf{u})(\mathbf{v} \cdot \mathbf{v})$$
$$(\mathbf{u} \cdot \mathbf{v})^2 \leq (\mathbf{u} \cdot \mathbf{u})(\mathbf{v} \cdot \mathbf{v}).$$

Taking the square root of both sides produces

$$|\mathbf{u} \cdot \mathbf{v}| \leq \sqrt{\mathbf{u} \cdot \mathbf{u}} \, \sqrt{\mathbf{v} \cdot \mathbf{v}} = \|\mathbf{u}\| \, \|\mathbf{v}\|.$$

EXAMPLE 7   *An Example of the Cauchy-Schwarz Inequality*

Verify the Cauchy-Schwarz Inequality for $\mathbf{u} = (1, -1, 3)$ and $\mathbf{v} = (2, 0, -1)$.

Solution   Because $\mathbf{u} \cdot \mathbf{v} = -1$, $\mathbf{u} \cdot \mathbf{u} = 11$, and $\mathbf{v} \cdot \mathbf{v} = 5$, you have

$$|\mathbf{u} \cdot \mathbf{v}| = |-1| = 1$$

and

$$\|\mathbf{u}\|\|\mathbf{v}\| = \sqrt{\mathbf{u} \cdot \mathbf{u}}\sqrt{\mathbf{v} \cdot \mathbf{v}} = \sqrt{11}\sqrt{5} = \sqrt{55}.$$

Thus, the inequality holds, and you have $|\mathbf{u} \cdot \mathbf{v}| \leq \|\mathbf{u}\|\|\mathbf{v}\|$.

The Cauchy-Schwarz Inequality leads to the definition of the angle between two nonzero vectors in R^n.

Definition of the Angle Between Two Vectors in R^n

The **angle** θ between two nonzero vectors in R^n is given by

$$\cos \theta = \frac{\mathbf{u} \cdot \mathbf{v}}{\|\mathbf{u}\|\|\mathbf{v}\|}, \quad 0 \leq \theta \leq \pi.$$

REMARK: The angle between the zero vector and another vector is not defined.

EXAMPLE 8   *Finding the Angle Between Two Vectors*

The angle between $\mathbf{u} = (-4, 0, 2, -2)$ and $\mathbf{v} = (2, 0, -1, 1)$ is given by

$$\cos \theta = \frac{\mathbf{u} \cdot \mathbf{v}}{\|\mathbf{u}\|\|\mathbf{v}\|} = \frac{-12}{\sqrt{24}\sqrt{6}} = -\frac{12}{\sqrt{144}} = -1.$$

Consequently, $\theta = \pi$. It makes sense that $\mathbf{u}$ and $\mathbf{v}$ should have opposite directions, because $\mathbf{u} = -2\mathbf{v}$.

Note that because $\|\mathbf{u}\|$ and $\|\mathbf{v}\|$ are always positive, $\mathbf{u} \cdot \mathbf{v}$ and $\cos \theta$ will always have the same sign. Moreover, because the cosine is positive in the first quadrant and negative in the second quadrant, the sign of the dot product of two vectors can be used to determine whether the angle between them is acute or obtuse, as shown in Figure 5.6.

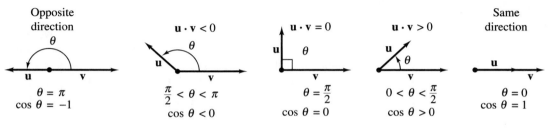

Figure 5.6

Figure 5.6 shows that two nonzero vectors meet at a right angle if and only if their dot product is zero. Two such vectors are said to be **orthogonal** (or perpendicular).

Definition of Orthogonal Vectors	Two vectors **u** and **v** in R^n are **orthogonal** if $$\mathbf{u} \cdot \mathbf{v} = 0.$$

REMARK: Even though the angle between the zero vector and another vector is not defined, it is convenient to extend the definition of orthogonality to include the zero vector. In other words, the vector **0** is said to be orthogonal to every vector.

EXAMPLE 9   *Orthogonal Vectors in R^n*

(a) The vectors $\mathbf{u} = (1, 0, 0)$ and $\mathbf{v} = (0, 1, 0)$ are orthogonal because

$$\mathbf{u} \cdot \mathbf{v} = (1)(0) + (0)(1) + (0)(0) = 0.$$

(b) The vectors $\mathbf{u} = (3, 2, -1, 4)$ and $\mathbf{v} = (1, -1, 1, 0)$ are orthogonal because

$$\mathbf{u} \cdot \mathbf{v} = (3)(1) + (2)(-1) + (-1)(1) + (4)(0) = 0.$$

EXAMPLE 10   *Finding Orthogonal Vectors*

Determine all vectors in R^2 that are orthogonal to $\mathbf{u} = (4, 2)$.

Solution   Let $\mathbf{v} = (v_1, v_2)$ be orthogonal to **u**. Then

$$\mathbf{u} \cdot \mathbf{v} = (4, 2) \cdot (v_1, v_2) = 4v_1 + 2v_2 = 0,$$

which implies that $2v_2 = -4v_1$ and $v_2 = -2v_1$. Therefore, every vector that is orthogonal to $(4, 2)$ is of the form

$$\mathbf{v} = (t, -2t) = t(1, -2),$$

where t is a real number. (See Figure 5.7.)

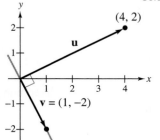

Figure 5.7

The Cauchy-Schwarz Inequality can be used to prove another well-known inequality called the **Triangle Inequality.** The name "Triangle Inequality" is derived from the interpretation of the theorem in R^2, illustrated for the vectors **u** and **v** in Figure 5.8(a). If you consider $\|\mathbf{u}\|$ and $\|\mathbf{v}\|$ to be the lengths of two sides of a triangle, you see that the length of the third side is $\|\mathbf{u} + \mathbf{v}\|$. Moreover, because the length of any side of a triangle cannot be greater than the sum of the lengths of the other two sides, you have

$$\|\mathbf{u} + \mathbf{v}\| \leq \|\mathbf{u}\| + \|\mathbf{v}\|.$$

Figure 5.8(b) illustrates the Triangle Inequality for the vectors **u** and **v** in R^3.

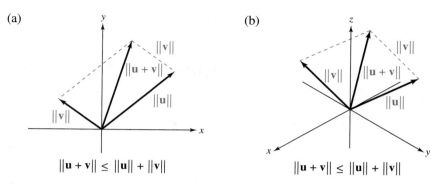

Figure 5.8

These results can be generalized to R^n in the following theorem.

| Theorem 5.5 | *The Triangle Inequality* |

If **u** and **v** are vectors in R^n, then

$$\|\mathbf{u} + \mathbf{v}\| \le \|\mathbf{u}\| + \|\mathbf{v}\|.$$

Proof   Using the properties of the dot product, you have

$$\begin{aligned}
\|\mathbf{u} + \mathbf{v}\|^2 &= (\mathbf{u} + \mathbf{v}) \cdot (\mathbf{u} + \mathbf{v}) \\
&= \mathbf{u} \cdot (\mathbf{u} + \mathbf{v}) + \mathbf{v} \cdot (\mathbf{u} + \mathbf{v}) \\
&= \mathbf{u} \cdot \mathbf{u} + 2(\mathbf{u} \cdot \mathbf{v}) + \mathbf{v} \cdot \mathbf{v} \\
&= \|\mathbf{u}\|^2 + 2(\mathbf{u} \cdot \mathbf{v}) + \|\mathbf{v}\|^2 \\
&\le \|\mathbf{u}\|^2 + 2|\mathbf{u} \cdot \mathbf{v}| + \|\mathbf{v}\|^2.
\end{aligned}$$

Now, by the Cauchy-Schwarz Inequality, $|\mathbf{u} \cdot \mathbf{v}| \le \|\mathbf{u}\|\,\|\mathbf{v}\|$, and you can write

$$\begin{aligned}
\|\mathbf{u} + \mathbf{v}\|^2 &\le \|\mathbf{u}\|^2 + 2|\mathbf{u} \cdot \mathbf{v}| + \|\mathbf{v}\|^2 \\
&\le \|\mathbf{u}\|^2 + 2\|\mathbf{u}\|\,\|\mathbf{v}\| + \|\mathbf{v}\|^2 \\
&= (\|\mathbf{u}\| + \|\mathbf{v}\|)^2.
\end{aligned}$$

Because both $\|\mathbf{u} + \mathbf{v}\|$ and $(\|\mathbf{u}\| + \|\mathbf{v}\|)$ are nonnegative, taking the square root of both sides yields

$$\|\mathbf{u} + \mathbf{v}\| \le \|\mathbf{u}\| + \|\mathbf{v}\|.$$

REMARK: Equality occurs in the Triangle Inequality if and only if the vectors **u** and **v** have the same direction. (See Exercise 86.)

From the proof of the Triangle Inequality you have

$$\|\mathbf{u} + \mathbf{v}\|^2 = \|\mathbf{u}\|^2 + 2(\mathbf{u} \cdot \mathbf{v}) + \|\mathbf{v}\|^2.$$

If $\mathbf{u}$ and $\mathbf{v}$ are orthogonal, then $\mathbf{u} \cdot \mathbf{v} = 0$, and you have the following extension of the **Pythagorean Theorem** to R^n.

Theorem 5.6

The Pythagorean Theorem

If $\mathbf{u}$ and $\mathbf{v}$ are vectors in R^n, then $\mathbf{u}$ and $\mathbf{v}$ are orthogonal if and only if

$$\|\mathbf{u} + \mathbf{v}\|^2 = \|\mathbf{u}\|^2 + \|\mathbf{v}\|^2.$$

This result is illustrated graphically for R^2 and R^3 in Figure 5.9.

The Dot Product and Matrix Multiplication

It is often useful to represent a vector $\mathbf{u} = (u_1, u_2, \ldots, u_n)$ in R^n as an $n \times 1$ column matrix. In this notation, the dot product of two vectors

$$\mathbf{u} = \begin{bmatrix} u_1 \\ u_2 \\ \vdots \\ u_n \end{bmatrix} \quad \text{and} \quad \mathbf{v} = \begin{bmatrix} v_1 \\ v_2 \\ \vdots \\ v_n \end{bmatrix}$$

can be represented as the matrix product of the transpose of $\mathbf{u}$ times $\mathbf{v}$.

$$\mathbf{u} \cdot \mathbf{v} = \mathbf{u}^T \mathbf{v} = [u_1 \ u_2 \ \ldots \ u_n] \begin{bmatrix} v_1 \\ v_2 \\ \vdots \\ v_n \end{bmatrix} = [u_1 v_1 + u_2 v_2 + \cdots + u_n v_n]$$

For example, the dot product of the vectors

$$\mathbf{u} = \begin{bmatrix} 1 \\ 2 \\ -1 \end{bmatrix} \quad \text{and} \quad \mathbf{v} = \begin{bmatrix} 3 \\ -2 \\ 4 \end{bmatrix}$$

is given by

$$\mathbf{u} \cdot \mathbf{v} = \mathbf{u}^T \mathbf{v} = [1 \ \ 2 \ \ -1] \begin{bmatrix} 3 \\ -2 \\ 4 \end{bmatrix} = [(1)(3) + (2)(-2) + (-1)(4)] = -5$$

In this light, many of the properties of the dot product are direct consequences of the corresponding properties of matrix multiplication.

Figure 5.9

SECTION 5.1   **Exercises**

In Exercises 1–6, find the length of the vector.

1. $\mathbf{v} = (4, 3)$

2. $\mathbf{v} = (0, 1)$

3. $\mathbf{v} = (1, 2, 2)$

4. $\mathbf{v} = (2, 0, 6)$

5. $\mathbf{v} = (2, 0, -5, 5)$

6. $\mathbf{v} = (2, -4, 5, -1, 1)$

In Exercises 7–10, find (a) $\|\mathbf{u}\|$, (b) $\|\mathbf{v}\|$, and (c) $\|\mathbf{u} + \mathbf{v}\|$.

7. $\mathbf{u} = (0, 4, 3), \quad \mathbf{v} = (1, -2, 1)$

8. $\mathbf{u} = \left(1, \frac{1}{2}\right), \quad \mathbf{v} = \left(2, -\frac{1}{2}\right)$

9. $\mathbf{u} = (0, 1, -1, 2), \quad \mathbf{v} = (1, 1, 3, 0)$

10. $\mathbf{u} = (1, 0, 0, 0), \quad \mathbf{v} = (0, 1, 0, 0)$

In Exercises 11–14, find a unit vector (a) in the direction of $\mathbf{u}$ and (b) in the direction opposite that of $\mathbf{u}$.

11. $\mathbf{u} = (-5, 12)$

12. $\mathbf{u} = (1, -1)$

13. $\mathbf{u} = (3, 2, -5)$

14. $\mathbf{u} = (1, 0, 2, 2)$

15. For what values of c is $\|c(1, 2, 3)\| = 1$?

16. For what values of c is $\|c(2, 2, -1)\| = 3$?

In Exercises 17–20, find the vector $\mathbf{v}$ with the given length that has the same direction as the vector $\mathbf{u}$.

17. $\|\mathbf{v}\| = 4, \quad \mathbf{u} = (1, 1)$

18. $\|\mathbf{v}\| = 4, \quad \mathbf{u} = (-1, 1)$

19. $\|\mathbf{v}\| = 2, \quad \mathbf{u} = \left(\sqrt{3}, 3, 0\right)$

20. $\|\mathbf{v}\| = 3, \quad \mathbf{u} = (0, 2, 1, -1)$

21. Given the vector $\mathbf{v} = (8, 8, 6)$, find $\mathbf{u}$ such that

  (a) $\mathbf{u}$ has the same direction as $\mathbf{v}$ and one-half its length.

  (b) $\mathbf{u}$ has the direction opposite that of $\mathbf{v}$ and one-fourth its length.

22. Given the vector $\mathbf{v} = (-1, 3, 0, 4)$, find $\mathbf{u}$ such that

  (a) $\mathbf{u}$ has the same direction as $\mathbf{v}$ and one-half its length.

  (b) $\mathbf{u}$ has the direction opposite that of $\mathbf{v}$ and one-fourth its length.

In Exercises 23–26, find the distance between $\mathbf{u}$ and $\mathbf{v}$.

23. $\mathbf{u} = (1, -1), \quad \mathbf{v} = (-1, 1)$

24. $\mathbf{u} = (3, 4), \quad \mathbf{v} = (7, 1)$

25. $\mathbf{u} = (1, 1, 2), \quad \mathbf{v} = (-1, 3, 0)$

26. $\mathbf{u} = (0, 1, 2, 3), \quad \mathbf{v} = (1, 0, 4, -1)$

In Exercises 27–30, find (a) $\mathbf{u} \cdot \mathbf{v}$, (b) $\mathbf{u} \cdot \mathbf{u}$, (c) $\|\mathbf{u}\|^2$, (d) $(\mathbf{u} \cdot \mathbf{v})\mathbf{v}$, and (e) $\mathbf{u} \cdot (5\mathbf{v})$.

27. $\mathbf{u} = (3, 4), \quad \mathbf{v} = (2, -3)$

28. $\mathbf{u} = (-1, 1, -2), \quad \mathbf{v} = (1, -3, -2)$

29. $\mathbf{u} = (4, 0, -3, 5), \quad \mathbf{v} = (0, 2, 5, 4)$

30. $\mathbf{u} = (0, 4, 3, 4, 4), \quad \mathbf{v} = (6, 8, -3, 3, -5)$

31. Find $(\mathbf{u} + \mathbf{v}) \cdot (2\mathbf{u} - \mathbf{v})$, given that $\mathbf{u} \cdot \mathbf{u} = 4$, $\mathbf{u} \cdot \mathbf{v} = -5$, and $\mathbf{v} \cdot \mathbf{v} = 10$.

32. Find $(3\mathbf{u} - \mathbf{v}) \cdot (\mathbf{u} - 3\mathbf{v})$, given that $\mathbf{u} \cdot \mathbf{u} = 8$, $\mathbf{u} \cdot \mathbf{v} = 7$, and $\mathbf{v} \cdot \mathbf{v} = 6$.

In Exercises 33–44, use a graphing utility with vector capabilities to find each of the following. Round your answer to four decimal places, if necessary.

  (a) Norm of $\mathbf{u}$ and $\mathbf{v}$

  (b) A unit vector in the direction of $\mathbf{v}$

  (c) A unit vector in the opposite direction of $\mathbf{u}$

  (d) $\mathbf{u} \cdot \mathbf{v}$

  (e) $\mathbf{u} \cdot \mathbf{u}$

  (f) $\mathbf{v} \cdot \mathbf{v}$

33. $\mathbf{u} = (5, -12), \quad \mathbf{v} = (-8, -15)$

34. $\mathbf{u} = (3, -4), \quad \mathbf{v} = (5, 12)$

35. $\mathbf{u} = (5, 12), \quad \mathbf{v} = (-12, 5)$

36. $\mathbf{u} = (3, -4), \quad \mathbf{v} = (4, 3)$

37. $\mathbf{u} = (10, -24), \quad \mathbf{v} = (-5, -12)$

38. $\mathbf{u} = (9, 12), \quad \mathbf{v} = (-7, 24)$

39. $\mathbf{u} = (0, 5, 12), \quad \mathbf{v} = (0, -5, -12)$

40. $\mathbf{u} = (3, 0, -4), \quad \mathbf{v} = (-3, -4, 0)$

41. $\mathbf{u} = (0, 2, 2, -1, 1, -2),$
    $\mathbf{v} = (2, 0, 1, 1, 2, -2)$

42. $\mathbf{u} = (1, 2, 3, -2, -1, -3),$
    $\mathbf{v} = (-1, 0, 2, 1, 2, -3)$

43. $\mathbf{u} = (-1, 1, 2, -1, 1, 1, -2, 1),$
    $\mathbf{v} = (-1, 0, 1, 2, -2, 1, 1, -2)$

44. $\mathbf{u} = (3, -1, 2, 1, 0, 1, 2, -1),$
    $\mathbf{v} = (1, 2, 0, -1, 2, -2, 1, 0)$

In Exercises 45 and 46, verify the Cauchy-Schwarz Inequality for the vectors.

45. $\mathbf{u} = (3, 4), \quad \mathbf{v} = (2, -3)$

46. $\mathbf{u} = (1, 1, -2), \quad \mathbf{v} = (1, -3, -2)$

In Exercises 47–52, find the angle θ between the vectors.

47. $\mathbf{u} = (3, 1), \quad \mathbf{v} = (-2, 4)$

48. $\mathbf{u} = \left(\cos \frac{\pi}{6}, \sin \frac{\pi}{6}\right), \quad \mathbf{v} = \left(\cos \frac{3\pi}{4}, \sin \frac{3\pi}{4}\right)$

49. $\mathbf{u} = (1, 1, 1), \quad \mathbf{v} = (2, 1, -1)$

50. $\mathbf{u} = (2, 3, 1), \quad \mathbf{v} = (-3, 2, 0)$

51. $\mathbf{u} = (0, 1, 0, 1)$, $\quad \mathbf{v} = (3, 3, 3, 3)$

52. $\mathbf{u} = (1, 3, -1, 2, 0)$, $\quad \mathbf{v} = (-1, 4, 5, -3, 2)$

In Exercises 53–58, determine all vectors $\mathbf{v}$ that are orthogonal to $\mathbf{u}$.

53. $\mathbf{u} = (0, 5)$ **54.** $\mathbf{u} = (2, 7)$

55. $\mathbf{u} = (-3, 2)$ **56.** $\mathbf{u} = (0, 0)$

57. $\mathbf{u} = (4, -1, 0)$ **58.** $\mathbf{u} = (0, 1, 0, 0, 0)$

In Exercises 59–64, determine whether $\mathbf{u}$ and $\mathbf{v}$ are orthogonal, parallel, or neither.

59. $\mathbf{u} = (2, 18)$, $\quad \mathbf{v} = \left(\frac{3}{2}, -\frac{1}{6}\right)$ **60.** $\mathbf{u} = (4, 3)$, $\quad \mathbf{v} = \left(\frac{1}{2}, -\frac{2}{3}\right)$

61. $\mathbf{u} = \left(-\frac{1}{3}, \frac{2}{3}\right)$, $\quad \mathbf{v} = (2, -4)$

62. $\mathbf{u} = (0, 1, 6)$, $\quad \mathbf{v} = (1, -2, -1)$

63. $\mathbf{u} = (\cos\theta, \sin\theta, -1)$, $\quad \mathbf{v} = (\sin\theta, -\cos\theta, 0)$

64. $\mathbf{u} = \left(4, \frac{3}{2}, -1, \frac{1}{2}\right)$, $\quad \mathbf{v} = \left(-2, -\frac{3}{4}, \frac{1}{2}, -\frac{1}{4}\right)$

In Exercises 65–68, use a graphing utility with vector capabilities to determine whether $\mathbf{u}$ and $\mathbf{v}$ are orthogonal, parallel, or neither.

65. $\mathbf{u} = \left(-2, \frac{1}{2}, -1, 3\right)$, $\quad \mathbf{v} = \left(\frac{3}{2}, 1, -\frac{5}{2}, 0\right)$

66. $\mathbf{u} = \left(-\frac{21}{2}, \frac{43}{2}, -12, \frac{3}{2}\right)$, $\quad \mathbf{v} = \left(0, 6, \frac{21}{2}, -\frac{9}{2}\right)$

67. $\mathbf{u} = \left(-\frac{3}{4}, \frac{3}{2}, -\frac{9}{2}, -6\right)$, $\quad \mathbf{v} = \left(\frac{3}{8}, -\frac{3}{4}, \frac{9}{8}, 3\right)$

68. $\mathbf{u} = \left(-\frac{4}{3}, \frac{8}{3}, -4, -\frac{32}{3}\right)$, $\quad \mathbf{v} = \left(-\frac{16}{3}, -2, \frac{4}{3}, -\frac{2}{3}\right)$

True or False? In Exercises 69 and 70, determine whether each statement is true or false. If a statement is true, give a reason or cite an appropriate statement in the text. If a statement is false, provide an example that shows that the statement is not true in all cases or cite an appropriate statement in the text.

69. (a) The length or norm of a vector is given by
$\|\mathbf{v}\| = |v_1 + v_2 + v_3 + \cdots + v_n|$.

    (b) The dot product of two vectors $\mathbf{u}$ and $\mathbf{v}$ is another vector given by $\mathbf{u} \cdot \mathbf{v} = (u_1v_1, u_2v_2, u_3v_3, \ldots, u_nv_n)$.

70. (a) If $\mathbf{v}$ is a nonzero vector in R^n, the unit vector is given by $\mathbf{u} = \mathbf{v}/|\mathbf{v}|$.

    (b) If $\mathbf{u} \cdot \mathbf{v} < 0$, then the angle θ between $\mathbf{u}$ and $\mathbf{v}$ is acute.

Writing In Exercises 71 and 72, write a short paragraph explaining why each expression is meaningless. Assume that $\mathbf{u}$ and $\mathbf{v}$ are vectors in R^n, and c is a scalar.

71. (a) $\|\mathbf{u} \cdot \mathbf{v}\|$ (b) $\mathbf{u} + (\mathbf{u} \cdot \mathbf{v})$

72. (a) $(\mathbf{u} \cdot \mathbf{v}) \cdot \mathbf{u}$ (b) $c \cdot (\mathbf{u} \cdot \mathbf{v})$

In Exercises 73 and 74, verify the Triangle Inequality for the vectors $\mathbf{u}$ and $\mathbf{v}$.

73. $\mathbf{u} = (4, 0)$, $\mathbf{v} = (1, 1)$ **74.** $\mathbf{u} = (1, 1, 1)$, $\mathbf{v} = (0, 1, -2)$

In Exercises 75 and 76, verify the Pythagorean Theorem for the vectors $\mathbf{u}$ and $\mathbf{v}$.

75. $\mathbf{u} = (1, -1)$, $\quad \mathbf{v} = (1, 1)$

76. $\mathbf{u} = (3, 4, -2)$, $\quad \mathbf{v} = (4, -3, 0)$

77. ***Writing*** What is known about θ, the angle between $\mathbf{u}$ and $\mathbf{v}$, if

    (a) $\mathbf{u} \cdot \mathbf{v} = 0$ (b) $\mathbf{u} \cdot \mathbf{v} > 0$ (c) $\mathbf{u} \cdot \mathbf{v} < 0$?

78. Let $\mathbf{v} = (v_1, v_2)$ be a vector in R^2. Show that $(v_2, -v_1)$ is orthogonal to $\mathbf{v}$, and use this fact to find two unit vectors orthogonal to $(12, 5)$.

79. Find the angle between the diagonal of a cube and one of its edges.

80. Find the angle between the diagonal of a cube and the diagonal of one of its sides.

81. Prove that if $\mathbf{u}$, $\mathbf{v}$, and $\mathbf{w}$ are vectors in R^n, then $(\mathbf{u} + \mathbf{v}) \cdot \mathbf{w} = \mathbf{u} \cdot \mathbf{w} + \mathbf{v} \cdot \mathbf{w}$.

82. ***Guided Proof*** Prove that if $\mathbf{u}$ is orthogonal to $\mathbf{v}$ and $\mathbf{w}$, then $\mathbf{u}$ is orthogonal to $c\mathbf{v} + d\mathbf{w}$ for any scalars c and d.

    Getting Started: To prove that $\mathbf{u}$ is orthogonal to $c\mathbf{v} + d\mathbf{w}$, you need to show that the dot product of $\mathbf{u}$ and $c\mathbf{v} + d\mathbf{w}$ is 0.

    (i) Rewrite the dot product of $\mathbf{u}$ and $c\mathbf{v} + d\mathbf{w}$ as a linear combination of $(\mathbf{u} \cdot \mathbf{v})$ and $(\mathbf{u} \cdot \mathbf{w})$ using Properties 2 and 3 of Theorem 5.3.

    (ii) Use the fact that $\mathbf{u}$ is orthogonal to $\mathbf{v}$ and $\mathbf{w}$ and the result of part (i) to lead to the conclusion that $\mathbf{u}$ is orthogonal to $c\mathbf{v} + d\mathbf{w}$.

83. Prove that if $\mathbf{u}$ and $\mathbf{v}$ are vectors in R^n, then
$\mathbf{u} \cdot \mathbf{v} = \frac{1}{4}\|\mathbf{u} + \mathbf{v}\|^2 - \frac{1}{4}\|\mathbf{u} - \mathbf{v}\|^2$.

84. Prove that if $\mathbf{u}$ and $\mathbf{v}$ are vectors in R^n, then
$\|\mathbf{u} + \mathbf{v}\|^2 + \|\mathbf{u} - \mathbf{v}\|^2 = 2\|\mathbf{u}\|^2 + 2\|\mathbf{v}\|^2$.

85. Prove that the vectors $\mathbf{u} = (\cos\theta)\mathbf{i} - (\sin\theta)\mathbf{j}$ and $\mathbf{v} = (\sin\theta)\mathbf{i} + (\cos\theta)\mathbf{j}$ are orthogonal unit vectors for any value of θ. Graph $\mathbf{u}$ and $\mathbf{v}$ for $\theta = \pi/3$.

86. Prove that $\|\mathbf{u} + \mathbf{v}\| = \|\mathbf{u}\| + \|\mathbf{v}\|$ if and only if $\mathbf{u}$ and $\mathbf{v}$ have the same direction.

87. ***Writing*** The vector $\mathbf{u} = (3240, 1450, 2235)$ gives the number of bushels of corn, oats, and wheat raised by a farmer in a certain year. The vector $\mathbf{v} = (2.22, 1.85, 3.25)$ gives the price in dollars per bushel of each of the crops. Find the dot product $\mathbf{u} \cdot \mathbf{v}$, and explain what information it gives.

88. ***Writing*** Let $\mathbf{x}$ be a solution to the $m \times n$ homogeneous linear system of equations $A\mathbf{x} = \mathbf{0}$. Explain why $\mathbf{x}$ is orthogonal to the row vectors of A.

89. Use the matrix multiplication interpretation of the dot product, $\mathbf{u} \cdot \mathbf{v} = \mathbf{u}^T\mathbf{v}$, to prove the first three properties of Theorem 5.3.

5.2 | Inner Product Spaces

In Section 5.1 the concepts of length, distance, and angle were extended from R^2 to R^n. This section extends the concepts one step further—to general vector spaces. This is accomplished by using the notion of an **inner product** of two vectors.

You already have one example of an inner product: the *dot product* in R^n. The dot product, called the **Euclidean inner product,** is only one of several inner products that can be defined on R^n. To distinguish between the standard inner product and other possible inner products, use the following notation.

$\mathbf{u} \cdot \mathbf{v} =$ dot product (Euclidean inner product for R^n)

$\langle \mathbf{u}, \mathbf{v} \rangle =$ general inner product for vector space V

A general inner product is defined in much the same way that a general vector space is defined—that is, in order for a function to qualify as an inner product, it must satisfy a set of axioms. The axioms parallel Properties 1, 2, 3, and 5 of the dot product given in Theorem 5.3.

Definition of Inner Product

Let $\mathbf{u}$, $\mathbf{v}$, and $\mathbf{w}$ be vectors in a vector space V, and let c be any scalar. An **inner product** on V is a function that associates a real number $\langle \mathbf{u}, \mathbf{v} \rangle$ with each pair of vectors $\mathbf{u}$ and $\mathbf{v}$ and satisfies the following axioms.

1. $\langle \mathbf{u}, \mathbf{v} \rangle = \langle \mathbf{v}, \mathbf{u} \rangle$
2. $\langle \mathbf{u}, \mathbf{v} + \mathbf{w} \rangle = \langle \mathbf{u}, \mathbf{v} \rangle + \langle \mathbf{u}, \mathbf{w} \rangle$
3. $c\langle \mathbf{u}, \mathbf{v} \rangle = \langle c\mathbf{u}, \mathbf{v} \rangle$
4. $\langle \mathbf{v}, \mathbf{v} \rangle \geq 0$, and $\langle \mathbf{v}, \mathbf{v} \rangle = 0$ if and only if $\mathbf{v} = \mathbf{0}$.

R E M A R K : A vector space V with an inner product is called an **inner product space.** Whenever an inner product space is referred to, assume that the set of scalars is the set of real numbers.

EXAMPLE 1   *The Euclidean Inner Product for R^n*

Show that the dot product in R^n satisfies the four axioms of an inner product.

Solution   In R^n, the dot product of two vectors $\mathbf{u} = (u_1, u_2, \ldots, u_n)$ and $\mathbf{v} = (v_1, v_2, \ldots, v_n)$ is given by

$$\mathbf{u} \cdot \mathbf{v} = u_1v_1 + u_2v_2 + \cdots + u_nv_n.$$

By Theorem 5.3, you know that this dot product satisfies the required four axioms, and thus it is an inner product on R^n.

The Euclidean inner product is not the only inner product that can be defined on R^n. A different inner product is illustrated in Example 2. Note that to show that a function is an inner product, you must show that the four inner product axioms are satisfied.

EXAMPLE 2   *A Different Inner Product for R^2*

Show that the function defines an inner product on R^2, where $\mathbf{u} = (u_1, u_2)$ and $\mathbf{v} = (v_1, v_2)$.

$$\langle \mathbf{u}, \mathbf{v} \rangle = u_1 v_1 + 2u_2 v_2$$

Solution   1. Because the product of real numbers is commutative,

$$\langle \mathbf{u}, \mathbf{v} \rangle = u_1 v_1 + 2u_2 v_2 = v_1 u_1 + 2v_2 u_2 = \langle \mathbf{v}, \mathbf{u} \rangle.$$

2. Let $\mathbf{w} = (w_1, w_2)$. Then

$$\begin{aligned}
\langle \mathbf{u}, \mathbf{v} + \mathbf{w} \rangle &= u_1(v_1 + w_1) + 2u_2(v_2 + w_2) \\
&= u_1 v_1 + u_1 w_1 + 2u_2 v_2 + 2u_2 w_2 \\
&= (u_1 v_1 + 2u_2 v_2) + (u_1 w_1 + 2u_2 w_2) \\
&= \langle \mathbf{u}, \mathbf{v} \rangle + \langle \mathbf{u}, \mathbf{w} \rangle.
\end{aligned}$$

3. If c is any scalar, then

$$c\langle \mathbf{u}, \mathbf{v} \rangle = c(u_1 v_1 + 2u_2 v_2) = (cu_1)v_1 + 2(cu_2)v_2 = \langle c\mathbf{u}, \mathbf{v} \rangle.$$

4. Because the square of a real number is nonnegative,

$$\langle \mathbf{v}, \mathbf{v} \rangle = v_1^2 + 2v_2^2 \geq 0.$$

Moreover, this expression is equal to zero if and only if $\mathbf{v} = \mathbf{0}$ (that is, if and only if $v_1 = v_2 = 0$).

Example 2 can be generalized to show that

$$\langle \mathbf{u}, \mathbf{v} \rangle = c_1 u_1 v_1 + c_2 u_2 v_2 + \cdots + c_n u_n v_n, \quad c_i > 0$$

is an inner product on R^n. The positive constants $c_1, \ldots, c_n$ are called **weights.** If any c_i is negative or 0, then this function does not define an inner product.

EXAMPLE 3   *A Function That Is Not an Inner Product*

Show that the function is not an inner product on R^3, where $\mathbf{u} = (u_1, u_2, u_3)$ and $\mathbf{v} = (v_1, v_2, v_3)$.

$$\langle \mathbf{u}, \mathbf{v} \rangle = u_1 v_1 - 2u_2 v_2 + u_3 v_3$$

Solution   Observe that Axiom 4 is not satisfied. For example, let $\mathbf{v} = (1, 2, 1)$. Then $\langle \mathbf{v}, \mathbf{v} \rangle = (1)(1) - 2(2)(2) + (1)(1) = -6$, which is less than zero.

EXAMPLE 4   *An Inner Product on $M_{2,2}$*

Let

$$A = \begin{bmatrix} a_{11} & a_{12} \\ a_{21} & a_{22} \end{bmatrix} \quad \text{and} \quad B = \begin{bmatrix} b_{11} & b_{12} \\ b_{21} & b_{22} \end{bmatrix}$$

be matrices in the vector space $M_{2,2}$. The function given by

$$\langle A, B \rangle = a_{11}b_{11} + a_{21}b_{21} + a_{12}b_{12} + a_{22}b_{22}$$

is an inner product on $M_{2,2}$. The verification of the four inner product axioms is left to you.

You obtain the inner product described in the next example from calculus. The verification of the inner product properties depends on the properties of the definite integral.

EXAMPLE 5   *An Inner Product Defined by a Definite Integral (Calculus)*

Let f and g be real-valued continuous functions in the vector space $C[a, b]$. Show that

$$\langle f, g \rangle = \int_a^b f(x)g(x)\, dx$$

defines an inner product on $C[a, b]$.

Solution   You can use familiar properties from calculus to verify the four parts of the definition.

1. $\langle f, g \rangle = \int_a^b f(x)g(x)\, dx = \int_a^b g(x)f(x)\, dx = \langle g, f \rangle$

2. $\langle f, g + h \rangle = \int_a^b f(x)[g(x) + h(x)]\, dx = \int_a^b [f(x)g(x) + f(x)h(x)]\, dx$

$$= \int_a^b f(x)g(x)\, dx + \int_a^b f(x)h(x)\, dx = \langle f, g \rangle + \langle f, h \rangle$$

3. $c\langle f, g \rangle = c \int_a^b f(x)g(x)\, dx = \int_a^b cf(x)g(x)\, dx = \langle cf, g \rangle$

4. Because $[f(x)]^2 \geq 0$ for all x, you know from calculus that

$$\langle f, f \rangle = \int_a^b [f(x)]^2\, dx \geq 0$$

with

$$\langle f, f \rangle = \int_a^b [f(x)]^2\, dx = 0$$

if and only if f is the zero function in $C[a, b]$, or if $a = b$.

The following theorem lists some easily verified properties of inner products.

Theorem 5.7

Properties of Inner Products

Let **u**, **v**, and **w** be vectors in an inner product space V, and let c be any real number.
1. $\langle \mathbf{0}, \mathbf{v} \rangle = \langle \mathbf{v}, \mathbf{0} \rangle = 0$
2. $\langle \mathbf{u} + \mathbf{v}, \mathbf{w} \rangle = \langle \mathbf{u}, \mathbf{w} \rangle + \langle \mathbf{v}, \mathbf{w} \rangle$
3. $\langle \mathbf{u}, c\mathbf{v} \rangle = c\langle \mathbf{u}, \mathbf{v} \rangle$

Proof   The proof of the first property follows. The proofs of the other two properties are left as exercises. (See Exercises 57 and 58.) From the definition of an inner product, you know that $\langle \mathbf{0}, \mathbf{v} \rangle = \langle \mathbf{v}, \mathbf{0} \rangle$, so you need show only one of these to be zero. Using the fact that $0(\mathbf{v}) = \mathbf{0}$,

$$\langle \mathbf{0}, \mathbf{v} \rangle = \langle 0(\mathbf{v}), \mathbf{v} \rangle$$
$$= 0\langle \mathbf{v}, \mathbf{v} \rangle$$
$$= 0.$$

The definitions of norm (or length), distance, and angle for general inner product spaces closely parallel those given for Euclidean n-space. Note that the definition of the angle θ between **u** and **v** presumes that

$$-1 \leq \frac{\langle \mathbf{u}, \mathbf{v} \rangle}{\|\mathbf{u}\| \, \|\mathbf{v}\|} \leq 1$$

for a general inner product, which follows from the Cauchy-Schwarz Inequality given later in Theorem 5.8.

Definition of Norm, Distance, and Angle

Let **u** and **v** be vectors in an inner product space V.
1. The **norm** (or **length**) of **u** is $\|\mathbf{u}\| = \sqrt{\langle \mathbf{u}, \mathbf{u} \rangle}$.
2. The **distance** between **u** and **v** is $d(\mathbf{u}, \mathbf{v}) = \|\mathbf{u} - \mathbf{v}\|$.
3. The **angle** between two nonzero vectors **u** and **v** is given by

$$\cos \theta = \frac{\langle \mathbf{u}, \mathbf{v} \rangle}{\|\mathbf{u}\| \, \|\mathbf{v}\|}, \quad 0 \leq \theta \leq \pi.$$

4. **u** and **v** are **orthogonal** if $\langle \mathbf{u}, \mathbf{v} \rangle = 0$.

REMARK: If $\|\mathbf{v}\| = 1$, then **v** is called a **unit vector**. Moreover, if **v** is any nonzero vector in an inner product space V, then the vector $\mathbf{u} = \mathbf{v}/\|\mathbf{v}\|$ is a unit vector and is called the **unit vector in the direction of v**.

EXAMPLE 6   *Finding Inner Products*

For polynomials $p = a_0 + a_1 x + \cdots + a_n x^n$ and $q = b_0 + b_1 x + \cdots + b_n x^n$ in the vector space P_n, the function given by

$$\langle p, q \rangle = a_0 b_0 + a_1 b_1 + \cdots + a_n b_n$$

is an inner product. Let $p(x) = 1 - 2x^2$, $q(x) = 4 - 2x + x^2$, and $r(x) = x + 2x^2$ be polynomials in P_2, and determine each of the following.

(a) $\langle p, q \rangle$ (b) $\langle q, r \rangle$ (c) $\|q\|$ (d) $d(p, q)$

Solution   (a) The inner product of p and q is given by

$$\langle p, q \rangle = a_0 b_0 + a_1 b_1 + a_2 b_2$$
$$= (1)(4) + (0)(-2) + (-2)(1) = 2.$$

(b) The inner product of q and r is given by

$$\langle q, r \rangle = (4)(0) + (-2)(1) + (1)(2) = 0.$$

(c) The norm of q is given by

$$\|q\| = \sqrt{\langle q, q \rangle} = \sqrt{4^2 + (-2)^2 + 1^2} = \sqrt{21}.$$

(d) The distance between p and q is given by

$$d(p, q) = \|p - q\| = \|(1 - 2x^2) - (4 - 2x + x^2)\|$$
$$= \|-3 + 2x - 3x^2\|$$
$$= \sqrt{(-3)^2 + 2^2 + (-3)^2} = \sqrt{22}.$$

Notice that the vectors q and r are orthogonal.

Orthogonality depends on the particular inner product used. That is, two vectors may be orthogonal with respect to one inner product but not another. Try reworking Example 6 using the inner product $\langle p, q \rangle = a_0 b_0 + a_1 b_1 + 2 a_2 b_2$. With this inner product the only orthogonal pair is p and q.

EXAMPLE 7   *Using the Inner Product on C[0, 1] (Calculus)*

Use the inner product defined in Example 5 and the functions $f(x) = x$ and $g(x) = x^2$ in $C[0, 1]$ to find each of the following.

(a) $\|f\|$ (b) $d(f, g)$

Solution   (a) Because $f(x) = x$, you have

$$\|f\|^2 = \langle f, f \rangle = \int_0^1 (x)(x)\, dx = \int_0^1 x^2\, dx = \left[\frac{x^3}{3} \right]_0^1 = \frac{1}{3}.$$

Therefore, $\|f\| = \dfrac{1}{\sqrt{3}}$.

(b) To find $d(f, g)$, write

$$[d(f, g)]^2 = \langle f - g, f - g \rangle$$

$$= \int_0^1 [f(x) - g(x)]^2 \, dx$$

$$= \int_0^1 [x - x^2]^2 \, dx$$

$$= \int_0^1 [x^2 - 2x^3 + x^4] \, dx$$

$$= \left[\frac{x^3}{3} - \frac{x^4}{2} + \frac{x^5}{5} \right]_0^1$$

$$= \frac{1}{30}.$$

Therefore, $d(f, g) = \dfrac{1}{\sqrt{30}}$.

Technology   Note

Many graphing calculators and computer software programs have built-in routines for approximating definite integrals. For example, on the TI-86, you can use the *fnInt* command from the *calc* menu to verify Example 7(b) as shown below.

```
√(fnInt((x-x²)²,x,0,1
))
         .182574185835
```

The result should be approximately $0.182574 \approx \dfrac{1}{\sqrt{30}}$.

In Example 7, you found that the distance between the functions $f(x) = x$ and $g(x) = x^2$ in $C[0, 1]$ is $1/\sqrt{30} \approx 0.183$. In practice, the *actual* distance between a pair of vectors is not as useful as the *relative* distance between several pairs. For instance, the distance between $g(x) = x^2$ and $h(x) = x^2 + 1$ in $C[0, 1]$ is 1. From Figure 5.10, this seems reasonable. That is, whatever norm is defined on $C[0, 1]$, it seems reasonable that you would want to say that f and g are closer than g and h.

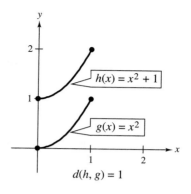

Figure 5.10

The properties of length and distance that were listed for R^n in the preceding section also hold for general inner product spaces. For instance, if $\mathbf{u}$ and $\mathbf{v}$ are vectors in an inner product space, then the following properties are true.

Properties of Norm
1. $\|\mathbf{u}\| \geq 0$
2. $\|\mathbf{u}\| = 0$ if and only if $\mathbf{u} = \mathbf{0}$.
3. $\|c\mathbf{u}\| = |c|\,\|\mathbf{u}\|$

Properties of Distance
1. $d(\mathbf{u}, \mathbf{v}) \geq 0$
2. $d(\mathbf{u}, \mathbf{v}) = 0$ if and only if $\mathbf{u} = \mathbf{v}$.
3. $d(\mathbf{u}, \mathbf{v}) = d(\mathbf{v}, \mathbf{u})$

Theorem 5.8 lists the general inner product space versions of the Cauchy-Schwarz Inequality, the Triangle Inequality, and the general Pythagorean Theorem.

Theorem 5.8

Let $\mathbf{u}$ and $\mathbf{v}$ be vectors in an inner product space V.
1. Cauchy-Schwarz Inequality: $|\langle \mathbf{u}, \mathbf{v} \rangle| \leq \|\mathbf{u}\|\,\|\mathbf{v}\|$
2. Triangle Inequality: $\|\mathbf{u} + \mathbf{v}\| \leq \|\mathbf{u}\| + \|\mathbf{v}\|$
3. Pythagorean Theorem: $\mathbf{u}$ and $\mathbf{v}$ are orthogonal if and only if

$$\|\mathbf{u} + \mathbf{v}\|^2 = \|\mathbf{u}\|^2 + \|\mathbf{v}\|^2.$$

The proofs of these three results parallel those given for Theorems 5.4, 5.5, and 5.6. Simply substitute $\langle \mathbf{u}, \mathbf{v} \rangle$ for the Euclidean inner product $\mathbf{u} \cdot \mathbf{v}$.

EXAMPLE 8   *An Example of the Cauchy-Schwarz Inequality (Calculus)*

Let $f(x) = 1$ and $g(x) = x$ be functions in the vector space $C[0, 1]$, with the inner product defined in Example 5. Verify that

$$|\langle f, g \rangle| \leq \|f\|\,\|g\|.$$

Solution   For the left side of this inequality you have

$$\langle f, g \rangle = \int_0^1 f(x)g(x)\,dx = \int_0^1 x\,dx = \frac{x^2}{2}\Big]_0^1 = \frac{1}{2}.$$

For the right side of the inequality you have

$$\|f\|^2 = \int_0^1 f(x)f(x)\,dx = \int_0^1 dx = x\Big]_0^1 = 1$$

and

$$\|g\|^2 = \int_0^1 g(x)g(x)\,dx = \int_0^1 x^2\,dx = \frac{x^3}{3}\Big]_0^1 = \frac{1}{3}.$$

Therefore,

$$\|f\|\,\|g\| = \sqrt{(1)\left(\frac{1}{3}\right)} = \frac{1}{\sqrt{3}} \approx 0.577,$$

and it follows that $|\langle f, g \rangle| \le \|f\|\,\|g\|$.

Orthogonal Projections in Inner Product Spaces

Let **u** and **v** be vectors in the plane. If **v** is nonzero, then **u** can be orthogonally projected onto **v**, as shown in Figure 5.11. This projection is denoted by $\text{proj}_{\mathbf{v}}\mathbf{u}$. Because $\text{proj}_{\mathbf{v}}\mathbf{u}$ is a scalar multiple of **v**, you can write

$$\text{proj}_{\mathbf{v}}\mathbf{u} = a\mathbf{v}.$$

If $a > 0$, as shown in Figure 5.11(a), then $\cos\theta > 0$ and the length of $\text{proj}_{\mathbf{v}}\mathbf{u}$ is

$$\|a\mathbf{v}\| = |a|\,\|\mathbf{v}\| = a\|\mathbf{v}\| = \|\mathbf{u}\|\cos\theta = \frac{\|\mathbf{u}\|\,\|\mathbf{v}\|\cos\theta}{\|\mathbf{v}\|} = \frac{\mathbf{u}\cdot\mathbf{v}}{\|\mathbf{v}\|},$$

which implies that $a = (\mathbf{u}\cdot\mathbf{v})/\|\mathbf{v}\|^2 = (\mathbf{u}\cdot\mathbf{v})/(\mathbf{v}\cdot\mathbf{v})$. Therefore,

$$\text{proj}_{\mathbf{v}}\mathbf{u} = \frac{\mathbf{u}\cdot\mathbf{v}}{\mathbf{v}\cdot\mathbf{v}}\mathbf{v}.$$

If $a < 0$, as shown in Figure 5.11(b), then it can be shown that the orthogonal projection of **u** onto **v** is given by the same formula.

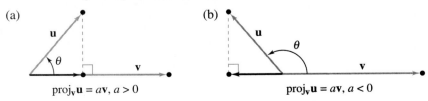

(a)   **u** θ **v** $\text{proj}_{\mathbf{v}}\mathbf{u} = a\mathbf{v},\ a > 0$

(b)   **u** θ **v** $\text{proj}_{\mathbf{v}}\mathbf{u} = a\mathbf{v},\ a < 0$

Figure 5.11

EXAMPLE 9   *Finding the Orthogonal Projection of u onto v*

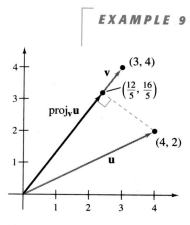

Figure 5.12

In R^2, the orthogonal projection of $\mathbf{u} = (4, 2)$ onto $\mathbf{v} = (3, 4)$ is given by

$$\text{proj}_\mathbf{v}\mathbf{u} = \frac{\mathbf{u} \cdot \mathbf{v}}{\mathbf{v} \cdot \mathbf{v}}\mathbf{v} = \frac{(4, 2) \cdot (3, 4)}{(3, 4) \cdot (3, 4)}(3, 4)$$

$$= \frac{20}{25}(3, 4)$$

$$= \left(\frac{12}{5}, \frac{16}{5}\right),$$

as shown in Figure 5.12.

The notion of orthogonal projection extends naturally to a general inner product space in the following way.

Definition of Orthogonal Projection

Let $\mathbf{u}$ and $\mathbf{v}$ be vectors in an inner product space V, such that $\mathbf{v} \neq \mathbf{0}$. Then the **orthogonal projection** of $\mathbf{u}$ onto $\mathbf{v}$ is given by

$$\text{proj}_\mathbf{v}\mathbf{u} = \frac{\langle \mathbf{u}, \mathbf{v} \rangle}{\langle \mathbf{v}, \mathbf{v} \rangle}\mathbf{v}.$$

REMARK: If $\mathbf{v}$ is a unit vector, then $\langle \mathbf{v}, \mathbf{v} \rangle = \|\mathbf{v}\|^2 = 1$, and the formula for the orthogonal projection of $\mathbf{u}$ onto $\mathbf{v}$ takes the following simpler form.

$$\text{proj}_\mathbf{v}\mathbf{u} = \langle \mathbf{u}, \mathbf{v} \rangle\mathbf{v}$$

EXAMPLE 10   *Finding an Orthogonal Projection in R^3*

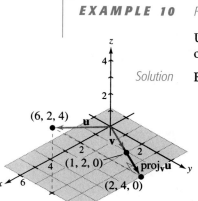

Figure 5.13

Use the Euclidean inner product in R^3 to find the orthogonal projection of $\mathbf{u} = (6, 2, 4)$ onto $\mathbf{v} = (1, 2, 0)$.

Solution   Because $\mathbf{u} \cdot \mathbf{v} = 10$ and $\|\mathbf{v}\|^2 = \mathbf{v} \cdot \mathbf{v} = 5$, the orthogonal projection of $\mathbf{u}$ onto $\mathbf{v}$ is

$$\text{proj}_\mathbf{v}\mathbf{u} = \frac{\mathbf{u} \cdot \mathbf{v}}{\mathbf{v} \cdot \mathbf{v}}\mathbf{v}$$

$$= \frac{10}{5}(1, 2, 0)$$

$$= 2(1, 2, 0)$$

$$= (2, 4, 0),$$

as shown in Figure 5.13.

REMARK: Notice in Example 10 that $\mathbf{u} - \text{proj}_\mathbf{v}\mathbf{u} = (6, 2, 4) - (2, 4, 0) = (4, -2, 4)$ is orthogonal to $\mathbf{v} = (1, 2, 0)$. This is true in general: if $\mathbf{u}$ and $\mathbf{v}$ are nonzero vectors in an inner product space, then $\mathbf{u} - \text{proj}_\mathbf{v}\mathbf{u}$ is orthogonal to $\mathbf{v}$. (See Exercise 56.)

An important property of orthogonal projections that will be used in approximation problems (see Section 5.4) is given in the following theorem. It states that, of all possible scalar multiples of a vector $\mathbf{v}$, the orthogonal projection of $\mathbf{u}$ onto $\mathbf{v}$ is the one that is closest to $\mathbf{u}$, as shown in Figure 5.14. For instance, in Example 10, this theorem implies that, of all the scalar multiples of the vector $\mathbf{v} = (1, 2, 0)$, the vector $\text{proj}_\mathbf{v}\mathbf{u} = (2, 4, 0)$ is closest to $\mathbf{u} = (6, 2, 4)$. You are asked to prove this explicitly in Exercise 62.

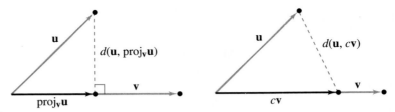

Figure 5.14

Theorem 5.9   *Orthogonal Projection and Distance*

Let $\mathbf{u}$ and $\mathbf{v}$ be two vectors in an inner product space V, such that $\mathbf{v} \neq \mathbf{0}$. Then

$$d(\mathbf{u}, \text{proj}_\mathbf{v}\mathbf{u}) < d(\mathbf{u}, c\mathbf{v}), \quad c \neq \frac{\langle \mathbf{u}, \mathbf{v} \rangle}{\langle \mathbf{v}, \mathbf{v} \rangle}.$$

Proof   Let $b = \langle \mathbf{u}, \mathbf{v} \rangle / \langle \mathbf{v}, \mathbf{v} \rangle$. Then you can write

$$\|\mathbf{u} - c\mathbf{v}\|^2 = \|(\mathbf{u} - b\mathbf{v}) + (b - c)\mathbf{v}\|^2,$$

where $(\mathbf{u} - b\mathbf{v})$ and $(b - c)\mathbf{v}$ are orthogonal. You can verify this by using the inner product axioms to show that

$$\langle (\mathbf{u} - b\mathbf{v}), (b - c)\mathbf{v} \rangle = 0.$$

Now, by the Pythagorean Theorem you can write

$$\|(\mathbf{u} - b\mathbf{v}) + (b - c)\mathbf{v}\|^2 = \|\mathbf{u} - b\mathbf{v}\|^2 + \|(b - c)\mathbf{v}\|^2,$$

which implies that

$$\|\mathbf{u} - c\mathbf{v}\|^2 = \|\mathbf{u} - b\mathbf{v}\|^2 + (b - c)^2\|\mathbf{v}\|^2.$$

Because $b \neq c$ and $\mathbf{v} \neq \mathbf{0}$, you know that $(b - c)^2 \|\mathbf{v}\|^2 > 0$. Therefore,

$$\|\mathbf{u} - b\mathbf{v}\|^2 < \|\mathbf{u} - c\mathbf{v}\|^2,$$

and it follows that

$$d(\mathbf{u}, b\mathbf{v}) < d(\mathbf{u}, c\mathbf{v}).$$

The next example discusses a type of orthogonal projection in the inner product space $C[a, b]$.

EXAMPLE 11   *Finding an Orthogonal Projection in $C[a, b]$ (Calculus)*

Let $f(x) = 1$ and $g(x) = x$ be functions in $C[0, 1]$. Use the inner product defined in Example 5 to find the orthogonal projection of f onto g.

Solution   From Example 8 you know that

$$\langle f, g \rangle = \frac{1}{2} \quad \text{and} \quad \|g\|^2 = \langle g, g \rangle = \frac{1}{3}.$$

Therefore, the orthogonal projection of f onto g is

$$\text{proj}_g f = \frac{\langle f, g \rangle}{\langle g, g \rangle} g$$

$$= \frac{1/2}{1/3} x$$

$$= \frac{3}{2} x.$$

SECTION 5.2   Exercises

In Exercises 1–8, find each of the following for the given inner product defined in R^n.

(a) $\langle \mathbf{u}, \mathbf{v} \rangle$
(b) $\|\mathbf{u}\|$
(c) $\|\mathbf{v}\|$
(d) $d(\mathbf{u}, \mathbf{v})$

1. $\mathbf{u} = (3, 4), \quad \mathbf{v} = (5, -12), \quad \langle \mathbf{u}, \mathbf{v} \rangle = \mathbf{u} \cdot \mathbf{v}$

2. $\mathbf{u} = (1, 1), \quad \mathbf{v} = (7, 9), \quad \langle \mathbf{u}, \mathbf{v} \rangle = \mathbf{u} \cdot \mathbf{v}$

3. $\mathbf{u} = (-4, 3), \quad \mathbf{v} = (0, 5), \quad \langle \mathbf{u}, \mathbf{v} \rangle = 3u_1 v_1 + u_2 v_2$

4. $\mathbf{u} = (0, -6), \quad \mathbf{v} = (-1, 1), \quad \langle \mathbf{u}, \mathbf{v} \rangle = u_1 v_1 + 2u_2 v_2$

5. $\mathbf{u} = (0, 9, 4), \quad \mathbf{v} = (9, -2, -4), \quad \langle \mathbf{u}, \mathbf{v} \rangle = \mathbf{u} \cdot \mathbf{v}$

6. $\mathbf{u} = (8, 0, -8), \quad \mathbf{v} = (8, 3, 16),$
$\langle \mathbf{u}, \mathbf{v} \rangle = 2u_1 v_1 + 3u_2 v_2 + u_3 v_3$

7. $\mathbf{u} = (1, 1, 1), \quad \mathbf{v} = (2, 5, 2),$
$\langle \mathbf{u}, \mathbf{v} \rangle = u_1 v_1 + 2u_2 v_2 + u_3 v_3$

8. $\mathbf{u} = (2, 0, 1, -1), \quad \mathbf{v} = (2, 2, 0, 1),$
$\langle \mathbf{u}, \mathbf{v} \rangle = \mathbf{u} \cdot \mathbf{v}$

Calculus In Exercises 9–12, use the functions f and g in $C[-1, 1]$ to find (a) $\langle f, g \rangle$, (b) $\|f\|$, (c) $\|g\|$, and (d) $d(f, g)$ for the inner product given by

$$\langle f, g \rangle = \int_{-1}^{1} f(x)g(x)\, dx.$$

9. $f(x) = x^2$, $g(x) = x^2 + 1$

10. $f(x) = -x$, $g(x) = x^2 - x + 2$

11. $f(x) = x$, $g(x) = e^x$

12. $f(x) = 1$, $g(x) = 3x^2 - 1$

In Exercises 13 and 14, use the inner product $\langle A, B \rangle = 2a_{11}b_{11} + a_{12}b_{12} + a_{21}b_{21} + 2a_{22}b_{22}$ to find (a) $\langle A, B \rangle$, (b) $\|A\|$, (c) $\|B\|$, and (d) $d(A, B)$ for the matrices in $M_{2,2}$.

13. $A = \begin{bmatrix} -1 & 3 \\ 4 & -2 \end{bmatrix}$, $B = \begin{bmatrix} 0 & -2 \\ 1 & 1 \end{bmatrix}$

14. $A = \begin{bmatrix} 1 & 0 \\ 0 & 1 \end{bmatrix}$, $B = \begin{bmatrix} 0 & 1 \\ 1 & 0 \end{bmatrix}$

In Exercises 15 and 16, use the inner product $\langle p, q \rangle = a_0b_0 + a_1b_1 + a_2b_2$ to find (a) $\langle p, q \rangle$, (b) $\|p\|$, (c) $\|q\|$, and (d) $d(p, q)$ for the polynomials in P_2.

15. $p(x) = 1 - x + 3x^2$, $q(x) = x - x^2$

16. $p(x) = 1 + x^2$, $q(x) = 1 - x^2$

In Exercises 17–20, prove that the indicated function is an inner product.

17. $\langle \mathbf{u}, \mathbf{v} \rangle$ as given in Exercise 3

18. $\langle \mathbf{u}, \mathbf{v} \rangle$ as given in Exercise 6

19. $\langle A, B \rangle$ as given in Exercises 13 and 14

20. $\langle p, q \rangle$ as given in Exercises 15 and 16

Writing In Exercises 21–24, state why $\langle \mathbf{u}, \mathbf{v} \rangle$ is not an inner product for $\mathbf{u} = (u_1, u_2)$ and $\mathbf{v} = (v_1, v_2)$ in R^2.

21. $\langle \mathbf{u}, \mathbf{v} \rangle = u_1v_1$

22. $\langle \mathbf{u}, \mathbf{v} \rangle = u_1v_1 - u_2v_2$

23. $\langle \mathbf{u}, \mathbf{v} \rangle = u_1^2v_1^2 + u_2^2v_2^2$

24. $\langle \mathbf{u}, \mathbf{v} \rangle = 3u_1v_2 - u_2v_1$

In Exercises 25–30, find the angle between the vectors.

25. $\mathbf{u} = (3, 4)$, $\mathbf{v} = (5, -12)$, $\langle \mathbf{u}, \mathbf{v} \rangle = \mathbf{u} \cdot \mathbf{v}$

26. $\mathbf{u} = (-4, 3)$, $\mathbf{v} = (0, 5)$, $\langle \mathbf{u}, \mathbf{v} \rangle = 3u_1v_1 + u_2v_2$

27. $\mathbf{u} = (1, 1, 1)$, $\mathbf{v} = (2, -2, 2)$, $\langle \mathbf{u}, \mathbf{v} \rangle = u_1v_1 + 2u_2v_2 + u_3v_3$

28. $p(x) = 1 - x + x^2$, $q(x) = 1 + x + x^2$, $\langle p, q \rangle = a_0b_0 + a_1b_1 + a_2b_2$

29. *Calculus* $f(x) = x$, $g(x) = x^2$,
$$\langle f, g \rangle = \int_{-1}^{1} f(x)g(x)\, dx$$

30. *Calculus* $f(x) = 1$, $g(x) = x^2$,
$$\langle f, g \rangle = \int_{-1}^{1} f(x)g(x)\, dx$$

In Exercises 31–36, verify

    (a) the Cauchy-Schwarz Inequality and

    (b) the Triangle Inequality.

31. $\mathbf{u} = (5, 12)$, $\mathbf{v} = (3, 4)$, $\langle \mathbf{u}, \mathbf{v} \rangle = \mathbf{u} \cdot \mathbf{v}$

32. $\mathbf{u} = (1, 0, 4)$, $\mathbf{v} = (-5, 4, 1)$, $\langle \mathbf{u}, \mathbf{v} \rangle = \mathbf{u} \cdot \mathbf{v}$

33. $p(x) = 2x$, $q(x) = 3x^2 + 1$, $\langle p, q \rangle = a_0b_0 + a_1b_1 + a_2b_2$

34. $A = \begin{bmatrix} 0 & 3 \\ 2 & 1 \end{bmatrix}$, $B = \begin{bmatrix} -3 & 1 \\ 4 & 3 \end{bmatrix}$,
$\langle A, B \rangle = a_{11}b_{11} + a_{12}b_{12} + a_{21}b_{21} + a_{22}b_{22}$

35. *Calculus* $f(x) = \sin x$, $g(x) = \cos x$,
$$\langle f, g \rangle = \int_{-\pi}^{\pi} f(x)g(x)\, dx$$

36. *Calculus* $f(x) = 1$, $g(x) = \cos \pi x$,
$$\langle f, g \rangle = \int_{0}^{2} f(x)g(x)\, dx$$

Calculus In Exercises 37–40, show that f and g are orthogonal in the inner product space $C[a, b]$ with the inner product given by

$$\langle f, g \rangle = \int_{a}^{b} f(x)g(x)\, dx.$$

37. $C[-\pi, \pi]$, $f(x) = \cos x$, $g(x) = \sin x$

38. $C[-1, 1]$, $f(x) = x$, $g(x) = \frac{1}{2}(3x^2 - 1)$

39. $C[-1, 1]$, $f(x) = x$, $g(x) = \frac{1}{2}(5x^3 - 3x)$

40. $C[0, \pi]$, $f(x) = 1$, $g(x) = \cos(2nx)$, $n = 1, 2, 3, \ldots$

In Exercises 41 and 42, (a) find $\text{proj}_\mathbf{v}\mathbf{u}$, (b) find $\text{proj}_\mathbf{u}\mathbf{v}$, and (c) sketch a graph of both results.

41. $\mathbf{u} = (1, 2)$, $\mathbf{v} = (2, 1)$

42. $\mathbf{u} = (-1, 3)$, $\mathbf{v} = (4, 4)$

In Exercises 43 and 44, find (a) $\text{proj}_\mathbf{v}\mathbf{u}$ and (b) $\text{proj}_\mathbf{u}\mathbf{v}$.

43. $\mathbf{u} = (1, 3, -2)$, $\mathbf{v} = (0, -1, 1)$

44. $\mathbf{u} = (0, 1, 3, -6)$, $\mathbf{v} = (-1, 1, 2, 2)$

Calculus   In Exercises 45–50, find the orthogonal projection of f onto g. Use the inner product in $C[a, b]$ given by

$$\langle f, g \rangle = \int_a^b f(x)g(x)\, dx.$$

45. $C[-1, 1]$, $f(x) = x$, $g(x) = 1$

46. $C[-1, 1]$, $f(x) = x^3 - x$, $g(x) = 2x - 1$

47. $C[0, 1]$, $f(x) = x$, $g(x) = e^x$

48. $C[-\pi, \pi]$, $f(x) = \sin x$, $g(x) = \cos x$

49. $C[-\pi, \pi]$, $f(x) = 1$, $g(x) = \sin 3x$

50. $C[-\pi, \pi]$, $f(x) = \sin 2x$, $g(x) = \sin 3x$

True or False?   In Exercises 51 and 52, determine whether each statement is true or false. If a statement is true, give a reason or cite an appropriate statement in the text. If a statement is false, provide an example that shows that the statement is not true in all cases or cite an appropriate statement in the text.

51. (a) The dot product is the only inner product that can be defined in R^n.

(b) Of all the possible scalar multiples of a vector $\mathbf{v}$, the orthogonal projection of $\mathbf{u}$ onto $\mathbf{v}$ is the vector closest to $\mathbf{u}$.

52. (a) The norm of the vector $\mathbf{u}$ is defined as the angle between the vector $\mathbf{u}$ and the positive x-axis.

(b) The angle θ between a vector $\mathbf{v}$ and the projection of $\mathbf{u}$ onto $\mathbf{v}$ is obtuse if the scalar $a < 0$ and acute if $a > 0$, where $a\mathbf{v} = \text{proj}_\mathbf{v}\mathbf{u}$.

53. Let $\mathbf{u} = (4, 2)$ and $\mathbf{v} = (2, -2)$ be vectors in R^2 with the inner product $\langle \mathbf{u}, \mathbf{v} \rangle = u_1 v_1 + 2u_2 v_2$.

(a) Show that $\mathbf{u}$ and $\mathbf{v}$ are orthogonal.

(b) Sketch the vectors $\mathbf{u}$ and $\mathbf{v}$. Are they orthogonal in the Euclidean sense?

54. Prove that $\|\mathbf{u} + \mathbf{v}\|^2 + \|\mathbf{u} - \mathbf{v}\|^2 = 2\|\mathbf{u}\|^2 + 2\|\mathbf{v}\|^2$ for any vectors $\mathbf{u}$ and $\mathbf{v}$ in an inner product space V.

55. Prove that the following function is an inner product for R^n.

$$\langle \mathbf{u}, \mathbf{v} \rangle = c_1 u_1 v_1 + c_2 u_2 v_2 + \cdots + c_n u_n v_n, \quad c_i > 0$$

56. Let $\mathbf{u}$ and $\mathbf{v}$ be nonzero vectors in an inner product space V. Prove that $\mathbf{u} - \text{proj}_\mathbf{v}\mathbf{u}$ is orthogonal to $\mathbf{v}$.

57. Prove property 2 of Theorem 5.7: If $\mathbf{u}$, $\mathbf{v}$, and $\mathbf{w}$ are vectors in an inner product space, then $\langle \mathbf{u} + \mathbf{v}, \mathbf{w} \rangle = \langle \mathbf{u}, \mathbf{w} \rangle + \langle \mathbf{v}, \mathbf{w} \rangle$.

58. Prove property 3 of Theorem 5.7: If $\mathbf{u}$ and $\mathbf{v}$ are vectors in an inner product space and c is a scalar, then $\langle \mathbf{u}, c\mathbf{v} \rangle = c\langle \mathbf{u}, \mathbf{v} \rangle$.

59. ***Guided Proof***   Let W be a subspace of the inner product space V. Prove that the following set $W^\perp$ is a subspace of V.

$$W^\perp = \{\mathbf{v} \in V : \langle \mathbf{v}, \mathbf{w} \rangle = 0 \text{ for all } \mathbf{w} \in W\}$$

Getting Started: To prove that $W^\perp$ is a subspace of V, you must show that $W^\perp$ is nonempty and that the closure conditions for a subspace hold (Theorem 4.5).

(i) Find an obvious vector in $W^\perp$ to conclude that it is nonempty.

(ii) To show the closure of $W^\perp$ under addition, you need to show that $\langle \mathbf{v}_1 + \mathbf{v}_2, \mathbf{w} \rangle = 0$ for all $\mathbf{w} \in W$ and for any $\mathbf{v}_1, \mathbf{v}_2 \in W^\perp$. Use the properties of inner products and the fact that $\langle \mathbf{v}_1, \mathbf{w} \rangle$ and $\langle \mathbf{v}_2, \mathbf{w} \rangle$ are both zero to show this.

(iii) To show closure under multiplication by a scalar, proceed similarly to part (ii). You need to use the properties of inner products and the condition of belonging to $W^\perp$.

60. Use the result of Exercise 59 to find $W^\perp$ if W is a span of $(1, 2, 3)$ in $V = R^3$.

61. ***Guided Proof***   Let $\langle \mathbf{u}, \mathbf{v} \rangle$ be the Euclidean inner product on R^n. Use the fact that $\langle \mathbf{u}, \mathbf{v} \rangle = \mathbf{u}^T \mathbf{v}$ to prove that for any $n \times n$ matrix A

(a) $\langle A^T \mathbf{u}, \mathbf{v} \rangle = \langle \mathbf{u}, A\mathbf{v} \rangle$ and (b) $\langle A^T A \mathbf{u}, \mathbf{u} \rangle = \|A\mathbf{u}\|^2$.

Getting Started: To prove (a) and (b), you can make use of both the properties of transposes (Theorem 2.6) and the properties of dot products (Theorem 5.3).

(i) To prove part (a), you can make repeated use of the given property $\langle \mathbf{u}, \mathbf{v} \rangle = \mathbf{u}^T \mathbf{v}$ and Property 4 of Theorem 2.6.

(ii) To prove part (b), you can make use of the given property $\langle \mathbf{u}, \mathbf{v} \rangle = \mathbf{u}^T \mathbf{v}$, Property 4 of Theorem 2.6, and Property 4 of Theorem 5.3.

62. Let $\mathbf{u} = (6, 2, 4)$ and $\mathbf{v} = (1, 2, 0)$ be the two vectors in Example 10. Without using Theorem 5.9, show that among all the scalar multiples $c\mathbf{v}$ of the vector $\mathbf{v}$, the projection of $\mathbf{u}$ onto $\mathbf{v}$ is the vector closest to $\mathbf{u}$—that is, show that $d(\mathbf{u}, \text{proj}_\mathbf{v}\mathbf{u})$ is a minimum.

5.3 | Orthonormal Bases: Gram-Schmidt Process

You saw in Section 4.7 that a vector space can have many different bases. While studying that section, you should have noticed that certain bases are more convenient than others. For example, R^3 has the convenient standard basis $B = \{(1, 0, 0), (0, 1, 0), (0, 0, 1)\}$. This set is the *standard* basis for R^3 because it has special characteristics that are particularly useful. One important characteristic is that the three vectors in the basis are *mutually orthogonal*. That is,

$$(1, 0, 0) \cdot (0, 1, 0) = 0$$
$$(1, 0, 0) \cdot (0, 0, 1) = 0$$
$$(0, 1, 0) \cdot (0, 0, 1) = 0.$$

A second important characteristic is that each vector in the basis is a *unit vector*.

This section identifies some advantages of bases consisting of mutually orthogonal unit vectors and develops a procedure, known as the *Gram-Schmidt orthonormalization process*, for constructing such bases.

Definition of Orthogonal and Orthonormal Sets

A set S of vectors in an inner product space V is called **orthogonal** if every pair of vectors in S is orthogonal. If, in addition, each vector in the set is a unit vector, then S is called **orthonormal.**

For $S = \{\mathbf{v}_1, \mathbf{v}_2, \ldots, \mathbf{v}_n\}$, this definition has the following form.

Orthogonal	*Orthonormal*
1. $\langle \mathbf{v}_i, \mathbf{v}_j \rangle = 0, \ i \neq j$	1. $\langle \mathbf{v}_i, \mathbf{v}_j \rangle = 0, \ i \neq j$
	2. $\|\mathbf{v}_i\| = 1, \ i = 1, 2, \ldots, n$

If S is a *basis*, then it is called an **orthogonal basis** or an **orthonormal basis,** respectively.

The standard basis for R^n is orthonormal, but it is not the only orthonormal basis for R^n. For instance, a nonstandard orthonormal basis for R^3 can be formed by rotating the standard basis about the z-axis to form

$$B = \{(\cos \theta, \sin \theta, 0), (-\sin \theta, \cos \theta, 0), (0, 0, 1)\},$$

as shown in Figure 5.15. Try verifying that the dot product of any two distinct vectors in B is zero, and that each vector in B is a unit vector. Example 1 describes another nonstandard orthonormal basis for R^3.

Figure 5.15

EXAMPLE 1   *A Nonstandard Orthonormal Basis for R^3*

Show that the set is an orthonormal basis for R^3.

$$S = \left\{ \overset{\mathbf{v}_1}{\left(\frac{1}{\sqrt{2}}, \frac{1}{\sqrt{2}}, 0 \right)}, \overset{\mathbf{v}_2}{\left(-\frac{\sqrt{2}}{6}, \frac{\sqrt{2}}{6}, \frac{2\sqrt{2}}{3} \right)}, \overset{\mathbf{v}_3}{\left(\frac{2}{3}, -\frac{2}{3}, \frac{1}{3} \right)} \right\}$$

Solution   First show that the three vectors are mutually orthogonal.

$$\mathbf{v}_1 \cdot \mathbf{v}_2 = -\frac{1}{6} + \frac{1}{6} + 0 = 0$$

$$\mathbf{v}_1 \cdot \mathbf{v}_3 = \frac{2}{3\sqrt{2}} - \frac{2}{3\sqrt{2}} + 0 = 0$$

$$\mathbf{v}_2 \cdot \mathbf{v}_3 = -\frac{\sqrt{2}}{9} - \frac{\sqrt{2}}{9} + \frac{2\sqrt{2}}{9} = 0$$

Furthermore, each vector is of length 1 because

$$\|\mathbf{v}_1\| = \sqrt{\mathbf{v}_1 \cdot \mathbf{v}_1} = \sqrt{\tfrac{1}{2} + \tfrac{1}{2} + 0} = 1$$

$$\|\mathbf{v}_2\| = \sqrt{\mathbf{v}_2 \cdot \mathbf{v}_2} = \sqrt{\tfrac{2}{36} + \tfrac{2}{36} + \tfrac{8}{9}} = 1$$

$$\|\mathbf{v}_3\| = \sqrt{\mathbf{v}_3 \cdot \mathbf{v}_3} = \sqrt{\tfrac{4}{9} + \tfrac{4}{9} + \tfrac{1}{9}} = 1.$$

Therefore, S is an orthonormal set. Because the three vectors do not lie in the same plane (see Figure 5.16), you know that they span R^3. Hence, by Theorem 4.12, they form a (nonstandard) orthonormal basis for R^3.

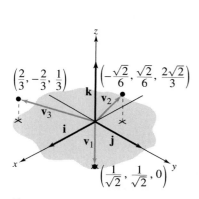

$\left(\frac{2}{3}, -\frac{2}{3}, \frac{1}{3}\right)$

$\left(-\frac{\sqrt{2}}{6}, \frac{\sqrt{2}}{6}, \frac{2\sqrt{2}}{3}\right)$

$\left(\frac{1}{\sqrt{2}}, \frac{1}{\sqrt{2}}, 0\right)$

Figure 5.16

EXAMPLE 2   *An Orthonormal Basis for P_3*

In P_3, with the inner product

$$\langle p, q \rangle = a_0 b_0 + a_1 b_1 + a_2 b_2 + a_3 b_3,$$

the standard basis $B = \{1, x, x^2, x^3\}$ is orthonormal. The verification of this is left as an exercise. (See Exercise 11.)

The orthogonal set in the following example is used to construct Fourier approximations of continuous functions. (See Section 5.5.)

EXAMPLE 3   *An Orthogonal Set in $C[0, 2\pi]$ (Calculus)*

In $C[0, 2\pi]$, with the inner product

$$\langle f, g \rangle = \int_0^{2\pi} f(x)g(x)\, dx,$$

show that the set

$$S = \{1, \sin x, \cos x, \sin 2x, \cos 2x, \ldots, \sin nx, \cos nx\}$$

is orthogonal.

Solution   To show that this set is orthogonal, you need to verify the following inner products, where m and n are positive integers.

$$\langle 1, \sin nx \rangle = \int_0^{2\pi} \sin nx \, dx = 0$$

$$\langle 1, \cos nx \rangle = \int_0^{2\pi} \cos nx \, dx = 0$$

$$\langle \sin mx, \cos nx \rangle = \int_0^{2\pi} \sin mx \cos nx \, dx = 0$$

$$\langle \sin mx, \sin nx \rangle = \int_0^{2\pi} \sin mx \sin nx \, dx = 0, \quad m \neq n$$

$$\langle \cos mx, \cos nx \rangle = \int_0^{2\pi} \cos mx \cos nx \, dx = 0, \quad m \neq n$$

One of these products is verified, and the others are left to you. If $m \neq n$, then the formula for rewriting a product of trigonometric functions as a sum can be used to obtain

$$\int_0^{2\pi} \sin mx \cos nx \, dx = \frac{1}{2} \int_0^{2\pi} [\sin(m + n)x + \sin(m - n)x] \, dx$$

$$= -\frac{1}{2} \left[\frac{\cos(m + n)x}{m + n} + \frac{\cos(m - n)x}{m - n} \right]_0^{2\pi}$$

$$= 0.$$

If $m = n$, then

$$\int_0^{2\pi} \sin mx \cos mx \, dx = \frac{1}{2m} \left[\sin^2 mx \right]_0^{2\pi}$$

$$= 0.$$

Note that Example 3 shows only that the set S is orthogonal. This particular set is not orthonormal. An orthonormal set can be formed, however, by normalizing each vector in S. That is, because

$$\|1\|^2 = \int_0^{2\pi} dx = 2\pi$$

$$\|\sin nx\|^2 = \int_0^{2\pi} \sin^2 nx \, dx = \pi$$

$$\|\cos nx\|^2 = \int_0^{2\pi} \cos^2 nx \, dx = \pi,$$

it follows that the set

$$\left\{ \frac{1}{\sqrt{2\pi}}, \frac{1}{\sqrt{\pi}} \sin x, \frac{1}{\sqrt{\pi}} \cos x, \ldots, \frac{1}{\sqrt{\pi}} \sin nx, \frac{1}{\sqrt{\pi}} \cos nx \right\}$$

is orthonormal.

Each set in Examples 1, 2, and 3 is linearly independent. Linear independence is a characteristic of any orthogonal set of nonzero vectors, as stated in the following theorem.

Theorem 5.10

Orthogonal Sets Are Linearly Independent

If $S = \{\mathbf{v}_1, \mathbf{v}_2, \ldots, \mathbf{v}_n\}$ is an orthogonal set of *nonzero* vectors in an inner product space V, then S is linearly independent.

Proof   **You** need to show that the vector equation

$$c_1 \mathbf{v}_1 + c_2 \mathbf{v}_2 + \cdots + c_n \mathbf{v}_n = \mathbf{0}$$

implies that $c_1 = c_2 = \cdots = c_n = 0$. To do this, form the inner product of the left side of the equation with each vector in S. That is, for each i,

$$\langle (c_1 \mathbf{v}_1 + c_2 \mathbf{v}_2 + \cdots + c_i \mathbf{v}_i + \cdots + c_n \mathbf{v}_n), \mathbf{v}_i \rangle = \langle \mathbf{0}, \mathbf{v}_i \rangle$$
$$c_1 \langle \mathbf{v}_1, \mathbf{v}_i \rangle + c_2 \langle \mathbf{v}_2, \mathbf{v}_i \rangle + \cdots + c_i \langle \mathbf{v}_i, \mathbf{v}_i \rangle + \cdots + c_n \langle \mathbf{v}_n, \mathbf{v}_i \rangle = 0.$$

Now, because S is orthogonal, $\langle \mathbf{v}_i, \mathbf{v}_j \rangle = 0$ for $j \neq i$, and thus the equation reduces to

$$c_i \langle \mathbf{v}_i, \mathbf{v}_i \rangle = 0.$$

But because each vector in S is nonzero, you know that

$$\langle \mathbf{v}_i, \mathbf{v}_i \rangle = \|\mathbf{v}_i\|^2 \neq 0.$$

So every c_i must be zero and the set must be linearly independent.

As a consequence of Theorems 4.12 and 5.10, you have the following result.

Corollary to Theorem 5.10

If V is an inner product space of dimension n, then any orthogonal set of n nonzero vectors is a basis for V.

EXAMPLE 4   *Using Orthogonality to Test for a Basis*

Show that the set is a basis for R^4.

$$S = \{ \overset{\mathbf{v}_1}{(2, 3, 2, -2)}, \overset{\mathbf{v}_2}{(1, 0, 0, 1)}, \overset{\mathbf{v}_3}{(-1, 0, 2, 1)}, \overset{\mathbf{v}_4}{(-1, 2, -1, 1)} \}$$

Solution The set S has four nonzero vectors. Thus, by the corollary to Theorem 5.10, you can show that S is a basis for R^4 by showing that it is an orthogonal set, as follows.

$$\mathbf{v}_1 \cdot \mathbf{v}_2 = 2 + 0 + 0 - 2 = 0$$
$$\mathbf{v}_1 \cdot \mathbf{v}_3 = -2 + 0 + 4 - 2 = 0$$
$$\mathbf{v}_1 \cdot \mathbf{v}_4 = -2 + 6 - 2 - 2 = 0$$
$$\mathbf{v}_2 \cdot \mathbf{v}_3 = -1 + 0 + 0 + 1 = 0$$
$$\mathbf{v}_2 \cdot \mathbf{v}_4 = -1 + 0 + 0 + 1 = 0$$
$$\mathbf{v}_3 \cdot \mathbf{v}_4 = 1 + 0 - 2 + 1 = 0$$

Thus S is orthogonal, and by the corollary to Theorem 5.10 it is a basis for R^4.

Section 4.7 discussed a technique for finding a coordinate representation relative to a nonstandard basis. If the basis is *orthonormal*, this procedure can be streamlined.

Before presenting this result, you will look at an example in R^2. Figure 5.17 shows that $\mathbf{i} = (1, 0)$ and $\mathbf{j} = (0, 1)$ form an orthonormal basis for R^2. Any vector $\mathbf{w}$ in R^2 can be represented as $\mathbf{w} = \mathbf{w}_1 + \mathbf{w}_2$, where $\mathbf{w}_1 = \text{proj}_\mathbf{i}\mathbf{w}$ and $\mathbf{w}_2 = \text{proj}_\mathbf{j}\mathbf{w}$. Because $\mathbf{i}$ and $\mathbf{j}$ are unit vectors, it follows that $\mathbf{w}_1 = (\mathbf{w} \cdot \mathbf{i})\mathbf{i}$ and $\mathbf{w}_2 = (\mathbf{w} \cdot \mathbf{j})\mathbf{j}$. Consequently,

$$\begin{aligned} \mathbf{w} &= \mathbf{w}_1 + \mathbf{w}_2 \\ &= (\mathbf{w} \cdot \mathbf{i})\mathbf{i} + (\mathbf{w} \cdot \mathbf{j})\mathbf{j} \\ &= c_1\mathbf{i} + c_2\mathbf{j}, \end{aligned}$$

which shows that the coefficients c_1 and c_2 are simply the dot products of $\mathbf{w}$ with the respective basis vectors. This result is generalized in the following theorem.

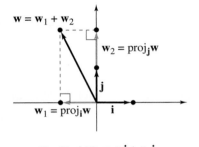

$\mathbf{w} = \mathbf{w}_1 + \mathbf{w}_2$

$\mathbf{w}_2 = \text{proj}_\mathbf{j}\mathbf{w}$

$\mathbf{j}$

$\mathbf{w}_1 = \text{proj}_\mathbf{i}\mathbf{w}$ $\quad \mathbf{i}$

$\mathbf{w} = \mathbf{w}_1 + \mathbf{w}_2 = c_1\mathbf{i} + c_2\mathbf{j}$

Figure 5.17

Theorem 5.11 *Coordinates Relative to an Orthonormal Basis*

If $B = \{\mathbf{v}_1, \mathbf{v}_2, \ldots, \mathbf{v}_n\}$ is an orthonormal basis for an inner product space V, then the coordinate representation of a vector $\mathbf{w}$ with respect to B is

$$\mathbf{w} = \langle \mathbf{w}, \mathbf{v}_1 \rangle \mathbf{v}_1 + \langle \mathbf{w}, \mathbf{v}_2 \rangle \mathbf{v}_2 + \cdots + \langle \mathbf{w}, \mathbf{v}_n \rangle \mathbf{v}_n.$$

Proof Because B is a basis for V, there must exist unique scalars $c_1, c_2, \ldots, c_n$ such that

$$\mathbf{w} = c_1\mathbf{v}_1 + c_2\mathbf{v}_2 + \cdots + c_n\mathbf{v}_n.$$

Taking the inner product (with $\mathbf{v}_i$) of both sides of this equation, you have

$$\begin{aligned} \langle \mathbf{w}, \mathbf{v}_i \rangle &= \langle (c_1\mathbf{v}_1 + c_2\mathbf{v}_2 + \cdots + c_n\mathbf{v}_n), \mathbf{v}_i \rangle \\ &= c_1\langle \mathbf{v}_1, \mathbf{v}_i \rangle + c_2\langle \mathbf{v}_2, \mathbf{v}_i \rangle + \cdots + c_n\langle \mathbf{v}_n, \mathbf{v}_i \rangle \end{aligned}$$

and by the orthogonality of B this equation reduces to

$$\langle \mathbf{w}, \mathbf{v}_i \rangle = c_i\langle \mathbf{v}_i, \mathbf{v}_i \rangle.$$

Because B is orthonormal, you have $\langle \mathbf{v}_i, \mathbf{v}_i \rangle = \|\mathbf{v}_i\|^2 = 1$, and it follows that $\langle \mathbf{w}, \mathbf{v}_i \rangle = c_i$.

In Theorem 5.11 the coordinates of $\mathbf{w}$ relative to the *orthonormal* basis B are called the **Fourier coefficients** of $\mathbf{w}$ relative to B, after the French mathematician Jean-Baptiste Joseph Fourier (1768–1830). The corresponding coordinate matrix of $\mathbf{w}$ relative to B is

$$[\mathbf{w}]_B = [c_1 \ c_2 \ \cdots \ c_n]^T$$
$$= [\langle \mathbf{w}, \mathbf{v}_1 \rangle \ \langle \mathbf{w}, \mathbf{v}_2 \rangle \ \cdots \ \langle \mathbf{w}, \mathbf{v}_n \rangle]^T.$$

EXAMPLE 5   *Representing Vectors Relative to an Orthonormal Basis*

Find the coordinates of $\mathbf{w} = (5, -5, 2)$ relative to the following orthonormal basis for R^3.

$$B = \left\{ \overset{\mathbf{v}_1}{\left(\tfrac{3}{5}, \tfrac{4}{5}, 0\right)}, \overset{\mathbf{v}_2}{\left(-\tfrac{4}{5}, \tfrac{3}{5}, 0\right)}, \overset{\mathbf{v}_3}{\left(0, 0, 1\right)} \right\}$$

Solution   Because B is orthonormal, you can use Theorem 5.11 to find the required coordinates.

$$\mathbf{w} \cdot \mathbf{v}_1 = (5, -5, 2) \cdot \left(\tfrac{3}{5}, \tfrac{4}{5}, 0\right) = -1$$
$$\mathbf{w} \cdot \mathbf{v}_2 = (5, -5, 2) \cdot \left(-\tfrac{4}{5}, \tfrac{3}{5}, 0\right) = -7$$
$$\mathbf{w} \cdot \mathbf{v}_3 = (5, -5, 2) \cdot (0, 0, 1) = 2$$

Thus, the coordinate matrix relative to B is

$$[\mathbf{w}]_B = \begin{bmatrix} -1 \\ -7 \\ 2 \end{bmatrix}.$$

Gram-Schmidt Orthonormalization Process

Having seen one of the advantages of orthonormal bases (the straightforwardness of coordinate representation), you will now look at a procedure for finding such a basis. This procedure is called the **Gram-Schmidt orthonormalization process,** after the Danish mathematician Jorgen Pederson Gram (1850–1916) and the German mathematician Erhardt Schmidt (1876–1959). It has three steps.

1. Begin with a basis for the inner product space. It need not be orthogonal nor consist of unit vectors.
2. Convert the given basis to an orthogonal basis.
3. Normalize each vector in the orthogonal basis to form an orthonormal basis.

REMARK: The Gram-Schmidt orthonormalization process leads to a matrix factorization similar to the *LU*-factorization you studied in Chapter 2. You are asked to investigate this *QR*-**factorization** in Project 1 at the end of this chapter.

Theorem 5.12   *Gram-Schmidt Orthonormalization Process*

1. Let $B = \{\mathbf{v}_1, \mathbf{v}_2, \ldots, \mathbf{v}_n\}$ be a basis for an inner product space V.
2. Let $B' = \{\mathbf{w}_1, \mathbf{w}_2, \ldots, \mathbf{w}_n\}$, where $\mathbf{w}_i$ is given by

$$\mathbf{w}_1 = \mathbf{v}_1$$

$$\mathbf{w}_2 = \mathbf{v}_2 - \frac{\langle \mathbf{v}_2, \mathbf{w}_1 \rangle}{\langle \mathbf{w}_1, \mathbf{w}_1 \rangle} \mathbf{w}_1$$

$$\mathbf{w}_3 = \mathbf{v}_3 - \frac{\langle \mathbf{v}_3, \mathbf{w}_1 \rangle}{\langle \mathbf{w}_1, \mathbf{w}_1 \rangle} \mathbf{w}_1 - \frac{\langle \mathbf{v}_3, \mathbf{w}_2 \rangle}{\langle \mathbf{w}_2, \mathbf{w}_2 \rangle} \mathbf{w}_2$$

$$\vdots$$

$$\mathbf{w}_n = \mathbf{v}_n - \frac{\langle \mathbf{v}_n, \mathbf{w}_1 \rangle}{\langle \mathbf{w}_1, \mathbf{w}_1 \rangle} \mathbf{w}_1 - \frac{\langle \mathbf{v}_n, \mathbf{w}_2 \rangle}{\langle \mathbf{w}_2, \mathbf{w}_2 \rangle} \mathbf{w}_2 - \cdots - \frac{\langle \mathbf{v}_n, \mathbf{w}_{n-1} \rangle}{\langle \mathbf{w}_{n-1}, \mathbf{w}_{n-1} \rangle} \mathbf{w}_{n-1}.$$

Then B' is an *orthogonal* basis for V.

3. Let $\mathbf{u}_i = \dfrac{\mathbf{w}_i}{\|\mathbf{w}_i\|}$. Then the set $B'' = \{\mathbf{u}_1, \mathbf{u}_2, \ldots, \mathbf{u}_n\}$ is an *orthonormal* basis for V.
Moreover, span $\{\mathbf{v}_1, \mathbf{v}_2, \ldots, \mathbf{v}_k\}$ = span$\{\mathbf{u}_1, \mathbf{u}_2, \ldots, \mathbf{u}_k\}$ for $k = 1, 2, \ldots, n$.

Rather than give a general proof of this theorem, it seems more instructive to discuss a special case for which you can use a geometric model. Let $\{\mathbf{v}_1, \mathbf{v}_2\}$ be a basis for R^2, as shown in Figure 5.18. To determine an orthogonal basis for R^2, first choose one of the original vectors, say $\mathbf{v}_1$. Now you want to find a second vector that is orthogonal to $\mathbf{v}_1$. Figure 5.19 shows that $\mathbf{v}_2 - \text{proj}_{\mathbf{v}_1} \mathbf{v}_2$ has this property.

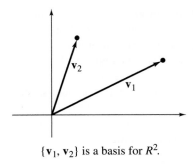

$\{\mathbf{v}_1, \mathbf{v}_2\}$ is a basis for R^2.

Figure 5.18

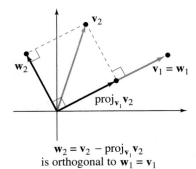

$\mathbf{w}_2 = \mathbf{v}_2 - \text{proj}_{\mathbf{v}_1} \mathbf{v}_2$
is orthogonal to $\mathbf{w}_1 = \mathbf{v}_1$

Figure 5.19

Thus, by letting

$$\mathbf{w}_1 = \mathbf{v}_1 \qquad \text{and} \qquad \mathbf{w}_2 = \mathbf{v}_2 - \text{proj}_{\mathbf{v}_1}\mathbf{v}_2 = \mathbf{v}_2 - \frac{\mathbf{v}_2 \cdot \mathbf{w}_1}{\mathbf{w}_1 \cdot \mathbf{w}_1}\mathbf{w}_1,$$

you can conclude that the set $\{\mathbf{w}_1, \mathbf{w}_2\}$ is orthogonal. By the corollary to Theorem 5.10, it is therefore a basis for R^2. Finally, by normalizing $\mathbf{w}_1$ and $\mathbf{w}_2$, you obtain the following orthonormal basis for R^2.

$$\{\mathbf{u}_1, \mathbf{u}_2\} = \left\{\frac{\mathbf{w}_1}{\|\mathbf{w}_1\|}, \frac{\mathbf{w}_2}{\|\mathbf{w}_2\|}\right\}$$

EXAMPLE 6   *Applying the Gram-Schmidt Orthonormalization Process*

Apply the Gram-Schmidt orthonormalization process to the following basis for R^2.

$$B = \{\overset{\mathbf{v}_1}{(1, 1)}, \overset{\mathbf{v}_2}{(0, 1)}\}$$

Solution   The Gram-Schmidt orthonormalization process produces

$$\mathbf{w}_1 = \mathbf{v}_1 = (1, 1)$$

$$\mathbf{w}_2 = \mathbf{v}_2 - \frac{\mathbf{v}_2 \cdot \mathbf{w}_1}{\mathbf{w}_1 \cdot \mathbf{w}_1}\mathbf{w}_1$$

$$= (0, 1) - \tfrac{1}{2}(1, 1)$$

$$= \left(-\tfrac{1}{2}, \tfrac{1}{2}\right).$$

The set $B' = \{\mathbf{w}_1, \mathbf{w}_2\}$ is an orthogonal basis for R^2. By normalizing each vector in B', you obtain

$$\mathbf{u}_1 = \frac{\mathbf{w}_1}{\|\mathbf{w}_1\|} = \frac{1}{\sqrt{2}}(1, 1)$$

$$= \frac{\sqrt{2}}{2}(1, 1)$$

$$= \left(\frac{\sqrt{2}}{2}, \frac{\sqrt{2}}{2}\right)$$

$$\mathbf{u}_2 = \frac{\mathbf{w}_2}{\|\mathbf{w}_2\|} = \frac{1}{1/\sqrt{2}}\left(-\frac{1}{2}, \frac{1}{2}\right)$$

$$= \sqrt{2}\left(-\frac{1}{2}, \frac{1}{2}\right)$$

$$= \left(-\frac{\sqrt{2}}{2}, \frac{\sqrt{2}}{2}\right).$$

Thus, $B'' = \{\mathbf{u}_1, \mathbf{u}_2\}$ is an orthonormal basis for R^2. See Figure 5.20 on the next page.

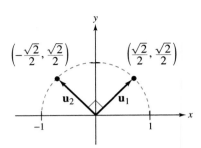

Given basis: $B = \{\mathbf{v}_1, \mathbf{v}_2\}$ Orthonormal basis: $B'' = \{\mathbf{u}_1, \mathbf{u}_2\}$

Figure 5.20

REMARK: An orthonormal set derived by the Gram-Schmidt orthonormalization process depends on the order of the vectors in the basis. For instance, try reworking Example 6 with the original basis ordered as $\{\mathbf{v}_2, \mathbf{v}_1\}$ rather than $\{\mathbf{v}_1, \mathbf{v}_2\}$.

EXAMPLE 7 *Applying the Gram-Schmidt Orthonormalization Process*

Apply the Gram-Schmidt orthonormalization process to the following basis for R^3.

$$B = \{(1, 1, 0), (1, 2, 0), (0, 1, 2)\}$$

Solution   Applying the Gram-Schmidt orthonormalization process produces

$$\mathbf{w}_1 = \mathbf{v}_1 = (1, 1, 0)$$

$$\mathbf{w}_2 = \mathbf{v}_2 - \frac{\mathbf{v}_2 \cdot \mathbf{w}_1}{\mathbf{w}_1 \cdot \mathbf{w}_1} \mathbf{w}_1$$

$$= (1, 2, 0) - \frac{3}{2}(1, 1, 0) = \left(-\frac{1}{2}, \frac{1}{2}, 0\right)$$

$$\mathbf{w}_3 = \mathbf{v}_3 - \frac{\mathbf{v}_3 \cdot \mathbf{w}_1}{\mathbf{w}_1 \cdot \mathbf{w}_1} \mathbf{w}_1 - \frac{\mathbf{v}_3 \cdot \mathbf{w}_2}{\mathbf{w}_2 \cdot \mathbf{w}_2} \mathbf{w}_2$$

$$= (0, 1, 2) - \frac{1}{2}(1, 1, 0) - \frac{1/2}{1/2}\left(-\frac{1}{2}, \frac{1}{2}, 0\right)$$

$$= (0, 0, 2).$$

The set $B' = \{\mathbf{w}_1, \mathbf{w}_2, \mathbf{w}_3\}$ is an orthogonal basis for R^3. Normalizing each vector in B' produces

$$\mathbf{u}_1 = \frac{\mathbf{w}_1}{\|\mathbf{w}_1\|} = \frac{1}{\sqrt{2}}(1, 1, 0) = \left(\frac{\sqrt{2}}{2}, \frac{\sqrt{2}}{2}, 0\right)$$

$$\mathbf{u}_2 = \frac{\mathbf{w}_2}{\|\mathbf{w}_2\|} = \frac{1}{1/\sqrt{2}}\left(-\frac{1}{2}, \frac{1}{2}, 0\right) = \left(-\frac{\sqrt{2}}{2}, \frac{\sqrt{2}}{2}, 0\right)$$

$$\mathbf{u}_3 = \frac{\mathbf{w}_3}{\|\mathbf{w}_3\|} = \frac{1}{2}(0, 0, 2) = (0, 0, 1).$$

Thus, $B'' = \{\mathbf{u}_1, \mathbf{u}_2, \mathbf{u}_3\}$ is an orthonormal basis for R^3.

Examples 6 and 7 applied the Gram-Schmidt orthonormalization process to bases for R^2 and R^3. The process works equally well for a subspace of an inner product space. The procedure is demonstrated in the next example.

EXAMPLE 8   *Applying the Gram-Schmidt Orthonormalization Process*

The vectors $\mathbf{v}_1 = (0, 1, 0)$ and $\mathbf{v}_2 = (1, 1, 1)$ span a plane in R^3. Find an orthonormal basis for this subspace.

Solution   Applying the Gram-Schmidt orthonormalization process produces

$$\mathbf{w}_1 = \mathbf{v}_1 = (0, 1, 0)$$

$$\mathbf{w}_2 = \mathbf{v}_2 - \frac{\mathbf{v}_2 \cdot \mathbf{w}_1}{\mathbf{w}_1 \cdot \mathbf{w}_1}\mathbf{w}_1$$

$$= (1, 1, 1) - \frac{1}{1}(0, 1, 0)$$

$$= (1, 0, 1).$$

Normalizing $\mathbf{w}_1$ and $\mathbf{w}_2$ produces the orthonormal set

$$\mathbf{u}_1 = \frac{\mathbf{w}_1}{\|\mathbf{w}_1\|} = (0, 1, 0)$$

$$\mathbf{u}_2 = \frac{\mathbf{w}_2}{\|\mathbf{w}_2\|}$$

$$= \frac{1}{\sqrt{2}}(1, 0, 1)$$

$$= \left(\frac{\sqrt{2}}{2}, 0, \frac{\sqrt{2}}{2}\right).$$

See Figure 5.21.

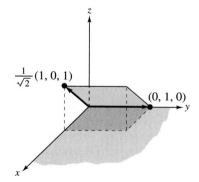

Figure 5.21

EXAMPLE 9   *Applying the Gram-Schmidt Orthonormalization Process (Calculus)*

Apply the Gram-Schmidt orthonormalization process to the basis $B = \{1, x, x^2\}$ in P_2, using the inner product

$$\langle p, q \rangle = \int_{-1}^{1} p(x)q(x)\, dx.$$

Solution   Let $B = \{1, x, x^2\} = \{\mathbf{v}_1, \mathbf{v}_2, \mathbf{v}_3\}$. Then you have

$$\mathbf{w}_1 = \mathbf{v}_1 = 1$$

$$\mathbf{w}_2 = \mathbf{v}_2 - \frac{\langle \mathbf{v}_2, \mathbf{w}_1 \rangle}{\langle \mathbf{w}_1, \mathbf{w}_1 \rangle} \mathbf{w}_1$$

$$= x - \frac{0}{2}(1) = x$$

$$\mathbf{w}_3 = \mathbf{v}_3 - \frac{\langle \mathbf{v}_3, \mathbf{w}_1 \rangle}{\langle \mathbf{w}_1, \mathbf{w}_1 \rangle} \mathbf{w}_1 - \frac{\langle \mathbf{v}_3, \mathbf{w}_2 \rangle}{\langle \mathbf{w}_2, \mathbf{w}_2 \rangle} \mathbf{w}_2$$

$$= x^2 - \frac{2/3}{2}(1) - \frac{0}{2/3}(x)$$

$$= x^2 - \frac{1}{3}.$$

(In Exercises 33–36 you are asked to verify these calculations.) Now, by normalizing $B' = \{\mathbf{w}_1, \mathbf{w}_2, \mathbf{w}_3\}$, you have

$$\mathbf{u}_1 = \frac{\mathbf{w}_1}{\|\mathbf{w}_1\|} = \frac{1}{\sqrt{2}}(1) = \frac{1}{\sqrt{2}}$$

$$\mathbf{u}_2 = \frac{\mathbf{w}_2}{\|\mathbf{w}_2\|} = \frac{1}{\sqrt{2/3}}(x) = \frac{\sqrt{3}}{\sqrt{2}} x$$

$$\mathbf{u}_3 = \frac{\mathbf{w}_3}{\|\mathbf{w}_3\|} = \frac{1}{\sqrt{8/45}}\left(x^2 - \frac{1}{3}\right) = \frac{\sqrt{5}}{2\sqrt{2}}(3x^2 - 1).$$

REMARK: The polynomials $\mathbf{u}_1, \mathbf{u}_2$, and $\mathbf{u}_3$ in Example 9 are called the first three **normalized Legendre polynomials,** after the French mathematician Adrien-Marie Legendre (1752–1833).

The computations in the Gram-Schmidt orthonormalization process are sometimes simpler when each vector $\mathbf{w}_i$ is normalized *before* it is used to determine the next vector. This **alternative form of the Gram-Schmidt orthonormalization process** has the following steps.

$$\mathbf{u}_1 = \frac{\mathbf{w}_1}{\|\mathbf{w}_1\|} = \frac{\mathbf{v}_1}{\|\mathbf{v}_1\|}$$

$$\mathbf{u}_2 = \frac{\mathbf{w}_2}{\|\mathbf{w}_2\|}, \text{ where } \mathbf{w}_2 = \mathbf{v}_2 - \langle \mathbf{v}_2, \mathbf{u}_1 \rangle \mathbf{u}_1$$

$$\mathbf{u}_3 = \frac{\mathbf{w}_3}{\|\mathbf{w}_3\|}, \text{ where } \mathbf{w}_3 = \mathbf{v}_3 - \langle \mathbf{v}_3, \mathbf{u}_1 \rangle \mathbf{u}_1 - \langle \mathbf{v}_3, \mathbf{u}_2 \rangle \mathbf{u}_2$$

$$\vdots$$

$$\mathbf{u}_n = \frac{\mathbf{w}_n}{\|\mathbf{w}_n\|}, \text{ where } \mathbf{w}_n = \mathbf{v}_n - \langle \mathbf{v}_n, \mathbf{u}_1 \rangle \mathbf{u}_1 - \cdots - \langle \mathbf{v}_n, \mathbf{u}_{n-1} \rangle \mathbf{u}_{n-1}$$

EXAMPLE 10   *Alternative Form of Gram-Schmidt Orthonormalization Process*

Find an orthonormal basis for the solution space of the homogeneous system of linear equations.

$$x_1 + x_2 \qquad + 7x_4 = 0$$
$$2x_1 + x_2 + 2x_3 + 6x_4 = 0$$

Solution   The augmented matrix for this system reduces as follows.

$$\begin{bmatrix} 1 & 1 & 0 & 7 & 0 \\ 2 & 1 & 2 & 6 & 0 \end{bmatrix} \longrightarrow \begin{bmatrix} 1 & 0 & 2 & -1 & 0 \\ 0 & 1 & -2 & 8 & 0 \end{bmatrix}$$

If you let $x_3 = s$ and $x_4 = t$, each solution of the system has the form

$$\begin{bmatrix} x_1 \\ x_2 \\ x_3 \\ x_4 \end{bmatrix} = \begin{bmatrix} -2s + t \\ 2s - 8t \\ s \\ t \end{bmatrix} = s\begin{bmatrix} -2 \\ 2 \\ 1 \\ 0 \end{bmatrix} + t\begin{bmatrix} 1 \\ -8 \\ 0 \\ 1 \end{bmatrix}.$$

Therefore, one basis for the solution space is

$$B = \{\mathbf{v}_1, \mathbf{v}_2\} = \{(-2, 2, 1, 0), (1, -8, 0, 1)\}.$$

To find an orthonormal basis $B' = \{\mathbf{u}_1, \mathbf{u}_2\}$, use the alternative form of the Gram-Schmidt orthonormalization process as follows.

$$\mathbf{u}_1 = \frac{\mathbf{v}_1}{\|\mathbf{v}_1\|}$$

$$= \frac{1}{3}(-2, 2, 1, 0)$$

$$= \left(-\frac{2}{3}, \frac{2}{3}, \frac{1}{3}, 0\right)$$

$$\mathbf{w}_2 = \mathbf{v}_2 - \langle \mathbf{v}_2, \mathbf{u}_1 \rangle \mathbf{u}_1$$

$$= (1, -8, 0, 1) - \left[(1, -8, 0, 1) \cdot \left(-\frac{2}{3}, \frac{2}{3}, \frac{1}{3}, 0\right)\right]\left(-\frac{2}{3}, \frac{2}{3}, \frac{1}{3}, 0\right)$$

$$= (-3, -4, 2, 1)$$

$$\mathbf{u}_2 = \frac{\mathbf{w}_2}{\|\mathbf{w}_2\|}$$

$$= \frac{1}{\sqrt{30}}(-3, -4, 2, 1)$$

$$= \left(-\frac{3}{\sqrt{30}}, -\frac{4}{\sqrt{30}}, \frac{2}{\sqrt{30}}, \frac{1}{\sqrt{30}}\right)$$

In Exercises 1–10, determine whether the set of vectors in R^n is orthogonal, orthonormal, or neither.

1. $\{(-4, 6), (5, 0)\}$

2. $\{(11, 4), (8, -3)\}$

3. $\{\left(\frac{3}{5}, \frac{4}{5}\right), \left(-\frac{4}{5}, \frac{3}{5}\right)\}$

4. $\{(1, 2), \left(-\frac{2}{5}, \frac{1}{5}\right)\}$

5. $\{(4, -1, 1), (-1, 0, 4), (-4, -17, -1)\}$

6. $\left\{\left(\frac{\sqrt{2}}{2}, 0, \frac{\sqrt{2}}{2}\right), \left(-\frac{\sqrt{6}}{6}, \frac{\sqrt{6}}{3}, \frac{\sqrt{6}}{6}\right), \left(\frac{\sqrt{3}}{3}, \frac{\sqrt{3}}{3}, -\frac{\sqrt{3}}{3}\right)\right\}$

7. $\left\{\left(\frac{\sqrt{2}}{3}, 0, -\frac{\sqrt{2}}{6}\right), \left(0, \frac{2\sqrt{5}}{5}, -\frac{\sqrt{5}}{5}\right), \left(\frac{\sqrt{5}}{5}, 0, \frac{1}{2}\right)\right\}$

8. $\{(-6, 3, 2, 1), (2, 0, 6, 0)\}$

9. $\left\{\left(\frac{\sqrt{2}}{2}, 0, 0, \frac{\sqrt{2}}{2}\right), \left(0, \frac{\sqrt{2}}{2}, \frac{\sqrt{2}}{2}, 0\right), \left(-\frac{1}{2}, \frac{1}{2}, -\frac{1}{2}, \frac{1}{2}\right)\right\}$

10. $\left\{\left(\frac{\sqrt{10}}{10}, 0, 0, \frac{3\sqrt{10}}{10}\right), (0, 0, 1, 0), (0, 1, 0, 0),\right.$
$\left.\left(-\frac{3\sqrt{10}}{10}, 0, 0, \frac{\sqrt{10}}{10}\right)\right\}$

11. Complete Example 2 by verifying that $\{1, x, x^2, x^3\}$ is an orthonormal basis for P_3 with the inner product $\langle p, q \rangle = a_0 b_0 + a_1 b_1 + a_2 b_2 + a_3 b_3$.

12. Verify that $\{(\sin\theta, \cos\theta), (\cos\theta, -\sin\theta)\}$ is an orthonormal basis for R^2.

In Exercises 13–18, find the coordinates of **x** relative to the orthonormal basis B in R^n.

13. $B = \left\{\left(-\frac{2\sqrt{13}}{13}, \frac{3\sqrt{13}}{13}\right), \left(\frac{3\sqrt{13}}{13}, \frac{2\sqrt{13}}{13}\right)\right\}$, $\mathbf{x} = (1, 2)$

14. $B = \left\{\left(\frac{\sqrt{5}}{5}, \frac{2\sqrt{5}}{5}\right), \left(-\frac{2\sqrt{5}}{5}, \frac{\sqrt{5}}{5}\right)\right\}$, $\mathbf{x} = (-3, 4)$

15. $B = \left\{\left(\frac{\sqrt{10}}{10}, 0, \frac{3\sqrt{10}}{10}\right), (0, 1, 0), \left(-\frac{3\sqrt{10}}{10}, 0, \frac{\sqrt{10}}{10}\right)\right\}$,
$\mathbf{x} = (2, -2, 1)$

16. $B = \{(1, 0, 0), (0, 1, 0), (0, 0, 1)\}$,
$\mathbf{x} = (3, -5, 11)$

17. $B = \left\{\left(\frac{3}{5}, \frac{4}{5}, 0\right), \left(-\frac{4}{5}, \frac{3}{5}, 0\right), (0, 0, 1)\right\}$,
$\mathbf{x} = (5, 10, 15)$

18. $B = \left\{\left(\frac{5}{13}, 0, \frac{12}{13}, 0\right), (0, 1, 0, 0), \left(-\frac{12}{13}, 0, \frac{5}{13}, 0\right), (0, 0, 0, 1)\right\}$,
$\mathbf{x} = (2, -1, 4, 3)$

In Exercises 19–26, use the Gram-Schmidt orthonormalization process to transform the given basis for R^n into an orthonormal basis. Use the Euclidean inner product for R^n and use the vectors in the order in which they are given.

19. $B = \{(3, 4), (1, 0)\}$

20. $B = \{(0, 1), (2, 5)\}$

21. $B = \{(1, -2, 2), (2, 2, 1), (2, -1, -2)\}$

22. $B = \{(1, 0, 0), (1, 1, 1), (1, 1, -1)\}$

23. $B = \{(4, -3, 0), (1, 2, 0), (0, 0, 4)\}$

24. $B = \{(0, 1, 2), (2, 0, 0), (1, 1, 1)\}$

25. $B = \{(0, 1, 1), (1, 1, 0), (1, 0, 1)\}$

26. $B = \{(3, 4, 0, 0), (-1, 1, 0, 0), (2, 1, 0, -1), (0, 1, 1, 0)\}$

In Exercises 27–32, use the Gram-Schmidt orthonormalization process to transform the given basis for a subspace of R^n into an orthonormal basis for the subspace. Use the Euclidean inner product for R^n and use the vectors in the order in which they are given.

27. $B = \{(-8, 3, 5)\}$

28. $B = \{(4, -7, 6)\}$

29. $B = \{(3, 4, 0), (1, 0, 0)\}$

30. $B = \{(1, 2, 0), (2, 0, -2)\}$

31. $B = \{(1, 2, -1, 0), (2, 2, 0, 1)\}$

32. $B = \{(7, 24, 0, 0), (0, 0, 1, 1), (0, 0, 1, -2)\}$

Calculus In Exercises 33–36, let $B = \{1, x, x^2\}$ be a basis for P_2 with the inner product

$$\langle p, q \rangle = \int_{-1}^{1} p(x)q(x)\, dx.$$

Complete Example 9 by verifying the indicated inner products.

33. $\langle x, 1 \rangle = 0$

34. $\langle 1, 1 \rangle = 2$

35. $\langle x^2, 1 \rangle = \frac{2}{3}$

36. $\langle x^2, x \rangle = 0$

True or False? In Exercises 37 and 38, determine whether each statement is true or false. If a statement is true, give a reason or cite an appropriate statement in the text. If a statement is false, provide an example that shows that the statement is not true in all cases or cite an appropriate statement in the text.

37. (a) A set S of vectors in an inner product space V is orthogonal if every pair of vectors in S is orthogonal.

(b) To show that a set of nonzero vectors is a basis for R^n, it is sufficient to show that the set is an orthogonal set.

(c) An orthonormal basis derived by the Gram-Schmidt orthonormalization process does not depend on the order of the vectors in the basis.

38. (a) A set S of vectors in an inner product space V is orthonormal if every vector is a unit vector and each pair of vectors is orthogonal.

(b) If a given set of nonzero vectors S in an inner product space V is orthogonal, then S is linearly independent.

(c) The Gram-Schmidt orthonormalization process is a procedure for finding an orthonormal basis for an inner product space V.

In Exercises 39–42, find an orthonormal basis for the solution space of the homogeneous system of linear equations.

39. $2x_1 + x_2 - 6x_3 + 2x_4 = 0$
$x_1 + 2x_2 - 3x_3 + 4x_4 = 0$
$x_1 + x_2 - 3x_3 + 2x_4 = 0$

40. $x_1 + x_2 - 3x_3 - 2x_4 = 0$
$2x_1 - x_2 \qquad - x_4 = 0$
$3x_1 + x_2 - 5x_3 - 4x_4 = 0$

41. $x_1 + x_2 - x_3 - x_4 = 0$
$2x_1 + x_2 - 2x_3 - 2x_4 = 0$

42. $x + 3y - 3z = 0$

In Exercises 43–48, let $p(x) = a_0 + a_1x + a_2x^2$ and $q(x) = b_0 + b_1x + b_2x^2$ be vectors in P_2 with

$$\langle p, q \rangle = a_0b_0 + a_1b_1 + a_2b_2.$$

Determine whether the given second-degree polynomials form an orthonormal set, and if not, use the Gram-Schmidt orthonormalization process to form an orthonormal set.

43. $\left\{ \dfrac{x^2 + 1}{\sqrt{2}}, \dfrac{x^2 + x - 1}{\sqrt{3}} \right\}$

44. $\{ \sqrt{2}(x^2 - 1), \sqrt{2}(x^2 + x + 2) \}$

45. $\{ x^2, x^2 + 2x, x^2 + 2x + 1 \}$

46. $\{ 1, x, x^2 \}$

47. $\{ x^2 - 1, x - 1 \}$

48. $\left\{ \dfrac{3x^2 + 4x}{5}, \dfrac{-4x^2 + 3x}{5}, 1 \right\}$

49. Use the inner product $\langle u, v \rangle = 2u_1v_1 + u_2v_2$ in R^2 and the Gram-Schmidt orthonormalization process to transform $\{(2, -1), (-2, 10)\}$ into an orthonormal basis.

50. *Writing* Explain why the result of Exercise 49 is not an orthonormal basis when the Euclidean inner product on R^2 is used.

51. Let $\{u_1, u_2, \ldots, u_n\}$ be an orthonormal basis for R^n. Prove that $\|v\|^2 = |v \cdot u_1|^2 + |v \cdot u_2|^2 + \cdots + |v \cdot u_n|^2$ for any vector v in R^n. This equation is called **Parseval's equality**.

52. *Guided Proof* Prove that if w is orthogonal to each vector in $S = \{v_1, v_2, \ldots, v_n\}$, then w is orthogonal to every linear combination of vectors in S.

Getting Started: To prove that w is orthogonal to every linear combination of vectors in S, you need to show that their dot product is 0.

(i) Write v as a linear combination of vectors, with arbitrary scalars $c_1, \ldots, c_n$, in S.
(ii) Form the inner product of w and v.
(iii) Use the properties of inner products to rewrite the inner product $\langle w, v \rangle$ as a linear combination of the inner products $\langle w, v_i \rangle, i = 1, \ldots, n$.
(iv) Use the fact that w is orthogonal to each vector in S to lead to the conclusion that w is orthogonal to v.

53. Let P be an $n \times n$ matrix. Prove that the following conditions are equivalent.

(a) $P^{-1} = P^T$. (Such a matrix is called *orthogonal*.)
(b) The row vectors of P form an orthonormal basis for R^n.
(c) The column vectors of P form an orthonormal basis for R^n.

54. Use each matrix to illustrate the result of Exercise 53.

(a) $P = \begin{bmatrix} -1 & 0 & 0 \\ 0 & 0 & 1 \\ 0 & -1 & 0 \end{bmatrix}$

(b) $P = \begin{bmatrix} 1/\sqrt{2} & 1/\sqrt{2} & 0 \\ 1/\sqrt{2} & -1/\sqrt{2} & 0 \\ 0 & 0 & 1 \end{bmatrix}$

55. Find an orthonormal basis for R^4 that includes the vectors

$$v_1 = \left(\frac{1}{\sqrt{2}}, 0, \frac{1}{\sqrt{2}}, 0 \right) \quad \text{and} \quad v_2 = \left(0, -\frac{1}{\sqrt{2}}, 0, \frac{1}{\sqrt{2}} \right).$$

56. Let W be a subspace of R^n. Prove that the following set is a subspace of R^n.

$$W^\perp = \{ v: w \cdot v = 0 \text{ for every } w \text{ in } W \}$$

Then prove that the intersection of W and $W^\perp$ is $\{0\}$.

In Exercises 57–60, find bases for the following four **fundamental subspaces** of the given matrix A.

$N(A) = $ nullspace of A $\qquad N(A^T) = $ nullspace of A^T
$R(A) = $ column space of A $\qquad R(A^T) = $ column space of A^T

Then, show that $N(A) = R(A^T)^\perp$ and $N(A^T) = R(A)^\perp$.

57. $\begin{bmatrix} 1 & 1 & -1 \\ 0 & 2 & 1 \\ 1 & 3 & 0 \end{bmatrix}$

58. $\begin{bmatrix} 0 & 1 & -1 \\ 0 & -2 & 2 \\ 0 & -1 & 1 \end{bmatrix}$

59. $\begin{bmatrix} 1 & 0 & 1 \\ 1 & 1 & 1 \end{bmatrix}$

60. $\begin{bmatrix} 0 & 0 & 1 & 2 & 0 \\ 1 & -2 & 0 & 2 & 0 \\ -1 & 2 & 1 & 0 & 0 \\ 0 & 0 & 1 & 2 & 1 \end{bmatrix}$

61. Let A be an $m \times n$ matrix.
  (a) Explain why $R(A^T)$ is the same as the row space of A.
  (b) Prove that $N(A) \subset R(A^T)^{\perp}$.
  (c) Prove that $N(A) = R(A^T)^{\perp}$.
  (d) Prove that $N(A^T) = R(A)^{\perp}$.

5.4 | Mathematical Models and Least Squares Analysis

In this section, you will look at *inconsistent* systems of linear equations, and show how to find the "best possible solution" of such a system. The necessity of "solving" inconsistent systems arises in the computation of the least squares regression line, as illustrated in Example 1.

EXAMPLE 1 *Least Squares Regression Line*

Let $(1, 0)$, $(2, 1)$, and $(3, 3)$ be three points in the plane, as shown in Figure 5.22. How can you find the line $y = c_0 + c_1 x$ that "best fits" these points? One way is to note that if the three points were collinear, then the following system of equations would be consistent.

$$c_0 + c_1 = 0$$
$$c_0 + 2c_1 = 1$$
$$c_0 + 3c_1 = 3$$

This system can be written in the matrix form $A\mathbf{x} = \mathbf{b}$, where

$$A = \begin{bmatrix} 1 & 1 \\ 1 & 2 \\ 1 & 3 \end{bmatrix}, \quad \mathbf{b} = \begin{bmatrix} 0 \\ 1 \\ 3 \end{bmatrix}, \quad \text{and} \quad \mathbf{x} = \begin{bmatrix} c_0 \\ c_1 \end{bmatrix}.$$

However, because the points are not collinear, the system is inconsistent. Although it is impossible to find $\mathbf{x}$ such that $A\mathbf{x} = \mathbf{b}$, you can look for an $\mathbf{x}$ that *minimizes* the norm of the error $\|A\mathbf{x} - \mathbf{b}\|$. The solution

$$\mathbf{x} = \begin{bmatrix} c_0 \\ c_1 \end{bmatrix}$$

of this minimization problem is called the **least squares regression line** $y = c_0 + c_1 x$.

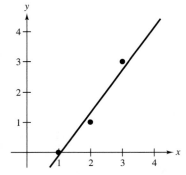

Figure 5.22

In Section 2.5, you briefly studied the least squares regression line and how to calculate it using matrices. Now, you will combine the ideas of orthogonality and projection to develop this concept in more generality. To begin, consider the linear system $A\mathbf{x} = \mathbf{b}$, where A is an $m \times n$ matrix and $\mathbf{b}$ is a column vector in R^m. You already know how to use

Gaussian elimination with back-substitution to solve for **x** if the system is consistent. However, if the system is inconsistent, it is still useful to find the "best possible" solution; that is, the value of **x** for which the difference between $A\mathbf{x}$ and **b** is smallest. One way to define "best possible" is to require that the norm of $A\mathbf{x} - \mathbf{b}$ be minimized. This definition is the heart of the **least squares problem.**

Least Squares Problem

Given an $m \times n$ matrix A and a vector **b** in R^m, the **least squares problem** is to find **x** in R^n such that $\|A\mathbf{x} - \mathbf{b}\|^2$ is minimized.

REMARK: The term **least squares** comes from the fact that minimizing $\|A\mathbf{x} - \mathbf{b}\|$ is equivalent to minimizing $\|A\mathbf{x} - \mathbf{b}\|^2$, which is a sum of squares.

Orthogonal Subspaces

To solve the least squares problem, you first need to develop the concept of orthogonal subspaces. Two subspaces of R^n are said to be **orthogonal** if the vectors in each subspace are orthogonal to the vectors in the other subspace.

Definition of Orthogonal Subspaces

The subspaces S_1 and S_2 of R^n are **orthogonal** if $\mathbf{v}_1 \cdot \mathbf{v}_2 = 0$ for all $\mathbf{v}_1$ in S_1 and all $\mathbf{v}_2$ in S_2.

EXAMPLE 2   *Orthogonal Subspaces*

The subspaces

$$S_1 = \text{span}\left(\begin{bmatrix} 1 \\ 0 \\ 1 \end{bmatrix}, \begin{bmatrix} 1 \\ 1 \\ 0 \end{bmatrix} \right) \quad \text{and} \quad S_2 = \text{span}\left(\begin{bmatrix} -1 \\ 1 \\ 1 \end{bmatrix} \right)$$

are orthogonal because the dot product of any vector in S_1 times any vector in S_2 is zero.

Notice in Example 2 that the zero vector is the only vector common to both S_1 and S_2. This is true in general. If S_1 and S_2 are orthogonal subspaces of R^n, then their intersection consists only of the zero vector. You are asked to prove this fact in Exercise 35.

Given a subspace S of R^n, the set of all vectors orthogonal to every vector in S is called the **orthogonal complement** of S, as given in the following definition.

Definition of Orthogonal Complement

If S is a subspace of R^n, then the **orthogonal complement of S** is the set

$$S^\perp = \{\mathbf{u} \in R^n : \mathbf{v} \cdot \mathbf{u} = 0 \text{ for all vectors } \mathbf{v} \in S\}.$$

The orthogonal complement of the trivial subspace $\{0\}$ is all of R^n, and, conversely, the orthogonal complement of R^n is the trivial subspace $\{0\}$. In Example 2, the subspace S_1 is the orthogonal complement of S_2, and the subspace S_2 is the orthogonal complement of S_1. In general, the orthogonal complement of a subspace of R^n is itself a subspace of R^n (see Exercise 36). You can find the orthogonal complement of a subspace of R^n by finding the nullspace of a matrix, as illustrated in the next example.

EXAMPLE 3   *Finding the Orthogonal Complement*

Find the orthogonal complement of the subspace S of R^4 spanned by the two column vectors $\mathbf{v}_1$ and $\mathbf{v}_2$ of the matrix A.

$$A = \begin{bmatrix} 1 & 0 \\ 2 & 0 \\ 1 & 0 \\ 0 & 1 \end{bmatrix}$$
$$\qquad \mathbf{v}_1 \quad \mathbf{v}_2$$

Solution   A vector $\mathbf{u} \in R^4$ will be in the orthogonal complement of S if its dot product with the two columns of A, $\mathbf{v}_1$ and $\mathbf{v}_2$, is zero. If you take the transpose of A, then you will see that the orthogonal complement of S consists of all the vectors $\mathbf{u}$ such that $A^T\mathbf{u} = \mathbf{0}$.

$$A^T\mathbf{u} = \mathbf{0}$$

$$\begin{bmatrix} 1 & 2 & 1 & 0 \\ 0 & 0 & 0 & 1 \end{bmatrix} \begin{bmatrix} x_1 \\ x_2 \\ x_3 \\ x_4 \end{bmatrix} = \begin{bmatrix} 0 \\ 0 \end{bmatrix}$$

That is, the orthogonal complement of S is the nullspace of the matrix A^T:

$$S^\perp = N(A^T).$$

Using the techniques for solving homogeneous linear systems, you can find that a basis for the orthogonal complement consists of the two vectors

$$\mathbf{u}_1 = \begin{bmatrix} -2 \\ 1 \\ 0 \\ 0 \end{bmatrix} \quad \text{and} \quad \mathbf{u}_2 = \begin{bmatrix} -1 \\ 0 \\ 1 \\ 0 \end{bmatrix}.$$

Notice in Example 3 that R^4 is split up into two subspaces, $S = \text{span}(\mathbf{v}_1, \mathbf{v}_2)$ and $S^\perp = \text{span}(\mathbf{u}_1, \mathbf{u}_2)$. In fact, the four vectors $\mathbf{v}_1$, $\mathbf{v}_2$, $\mathbf{u}_1$, and $\mathbf{u}_2$ form a basis for R^4. Each vector in R^4 can be *uniquely* written as a sum of a vector from S and a vector from $S^\perp$. This concept is generalized in the following definition.

Definition of Direct Sum	Let S_1 and S_2 be two subspaces of R^n. If each vector $\mathbf{x} \in R^n$ can be uniquely written as a sum of a vector $\mathbf{s}_1$ from S_1 and a vector $\mathbf{s}_2$ from S_2, $\mathbf{x} = \mathbf{s}_1 + \mathbf{s}_2$, then R^n is the **direct sum** of S_1 and S_2, and you can write $R^n = S_1 \oplus S_2$.

EXAMPLE 4   *Direct Sum*

(a) From Example 2, you can see that R^3 is the direct sum of the subspaces

$$S_1 = \text{span}\left(\begin{bmatrix} 1 \\ 0 \\ 1 \end{bmatrix}, \begin{bmatrix} 1 \\ 1 \\ 0 \end{bmatrix}\right) \quad \text{and} \quad S_2 = \text{span}\left(\begin{bmatrix} -1 \\ 1 \\ 1 \end{bmatrix}\right).$$

(b) From Example 3 you can see that $R^4 = S \oplus S^\perp$, where

$$S = \text{span}\left(\begin{bmatrix} 1 \\ 2 \\ 1 \\ 0 \end{bmatrix}, \begin{bmatrix} 0 \\ 0 \\ 0 \\ 1 \end{bmatrix}\right) \quad \text{and} \quad S^\perp = \text{span}\left(\begin{bmatrix} -2 \\ 1 \\ 0 \\ 0 \end{bmatrix}, \begin{bmatrix} -1 \\ 0 \\ 1 \\ 0 \end{bmatrix}\right).$$

The following theorem collects some important facts about orthogonal complements and direct sums.

Theorem 5.13	*Properties of Orthogonal Subspaces*
	Let S be a subspace of R^n. Then the following properties are true. 1. $\dim(S) + \dim(S^\perp) = n$ 2. $R^n = S \oplus S^\perp$ 3. $(S^\perp)^\perp = S$

Proof   1. If $S = R^n$ or $S = \{\mathbf{0}\}$, then Property 1 is trivial. So let $\{\mathbf{v}_1, \mathbf{v}_2, \ldots, \mathbf{v}_t\}$ be a basis for S, $0 < t < n$. Let A be the $n \times t$ matrix whose columns are the basis vectors $\mathbf{v}_i$. Then $S = R(A)$, which implies that $S^\perp = N(A^T)$, where A^T is a $t \times n$ matrix of rank t (see Section 5.3, Exercise 61). Because the dimension of $N(A^T)$ is $n - t$, you have shown that

$$\dim(S) + \dim(S^\perp) = t + (n - t) = n.$$

2. If $S = R^n$ or $S = \{\mathbf{0}\}$, then Property 2 is trivial. So let $\{\mathbf{v}_1, \mathbf{v}_2, \ldots, \mathbf{v}_t\}$ be a basis for S and let $\{\mathbf{v}_{t+1}, \mathbf{v}_{t+2}, \ldots, \mathbf{v}_n\}$ be a basis for $S^\perp$. It can be shown that the set $\{\mathbf{v}_1, \mathbf{v}_2, \ldots, \mathbf{v}_t, \mathbf{v}_{t+1}, \ldots, \mathbf{v}_n\}$ is linearly independent, and hence forms a basis for R^n. Let $\mathbf{x} \in R^n$, $\mathbf{x} = c_1\mathbf{v}_1 + \cdots + c_t\mathbf{v}_t + c_{t+1}\mathbf{v}_{t+1} + \cdots + c_n\mathbf{v}_n$. If you write $\mathbf{v} = c_1\mathbf{v}_1 + \cdots + c_t\mathbf{v}_t$ and $\mathbf{w} = c_{t+1}\mathbf{v}_{t+1} + \cdots + c_n\mathbf{v}_n$, then you have expressed an arbitrary vector $\mathbf{x}$ as the sum of a vector from S and a vector from $S^\perp$, $\mathbf{x} = \mathbf{v} + \mathbf{w}$.

To show the uniqueness of this representation, assume that $\mathbf{x} = \mathbf{v} + \mathbf{w} = \hat{\mathbf{v}} + \hat{\mathbf{w}}$. This implies that $\hat{\mathbf{v}} - \mathbf{v} = \mathbf{w} - \hat{\mathbf{w}}$. Thus, the two vectors $\hat{\mathbf{v}} - \mathbf{v}$ and $\mathbf{w} - \hat{\mathbf{w}}$ are in both S and $S^{\perp}$. Since $S \cap S^{\perp} = \{\mathbf{0}\}$, you must have $\hat{\mathbf{v}} = \mathbf{v}$ and $\mathbf{w} = \hat{\mathbf{w}}$.

3. Let $\mathbf{v} \in S$. Then $\mathbf{v} \cdot \mathbf{u} = 0$ for all $\mathbf{u} \in S^{\perp}$, which implies that $\mathbf{v} \in (S^{\perp})^{\perp}$. On the other hand, if $\mathbf{v} \in (S^{\perp})^{\perp}$, then, because $R^n = S \oplus S^{\perp}$, you can write $\mathbf{v}$ as the unique sum of the vector from S and a vector from $S^{\perp}$, $\mathbf{v} = \mathbf{s} + \mathbf{w}$, $\mathbf{s} \in S$, $\mathbf{w} \in S^{\perp}$. Because $\mathbf{w}$ is in $S^{\perp}$, it is orthogonal to every vector in S, and in particular to $\mathbf{v}$. Thus,

$$
\begin{aligned}
0 = \mathbf{w} \cdot \mathbf{v} &= \mathbf{w} \cdot (\mathbf{s} + \mathbf{w}) \\
&= \mathbf{w} \cdot \mathbf{s} + \mathbf{w} \cdot \mathbf{w} \\
&= \mathbf{w} \cdot \mathbf{w}.
\end{aligned}
$$

This implies that $\mathbf{w} = \mathbf{0}$ and $\mathbf{v} = \mathbf{s} + \mathbf{w} = \mathbf{s} \in S$.

You studied the projection of one vector onto another in Section 5.3. This is now generalized to projections of a vector $\mathbf{v}$ onto a subspace S. Because $R^n = S \oplus S^{\perp}$, every vector $\mathbf{v}$ in R^n can be uniquely written as a sum of a vector from S and a vector from $S^{\perp}$:

$$\mathbf{v} = \mathbf{v}_1 + \mathbf{v}_2, \qquad \mathbf{v}_1 \in S, \qquad \mathbf{v}_2 \in S^{\perp}.$$

The vector $\mathbf{v}_1$ is called the **projection** of $\mathbf{v}$ onto the subspace S, and is denoted $\mathbf{v}_1 = \text{proj}_S \mathbf{v}$. Furthermore, $\mathbf{v}_2 = \mathbf{v} - \mathbf{v}_1 = \mathbf{v} - \text{proj}_S \mathbf{v}$, which implies that the vector $\mathbf{v} - \text{proj}_S \mathbf{v}$ is orthogonal to the subspace S.

Given a subspace S of R^n, you can use the Gram-Schmidt orthonormalization process to calculate an orthonormal basis for S. It is then an easy matter to compute the projection of a vector $\mathbf{v}$ onto S using the following theorem. You are asked to prove this theorem in Exercise 37.

Theorem 5.14   *Projection onto a Subspace*

If $\{\mathbf{u}_1, \mathbf{u}_2, \dots, \mathbf{u}_t\}$ is an orthonormal basis for the subspace S of R^n, and $\mathbf{v} \in R^n$, then

$$\text{proj}_S \mathbf{v} = (\mathbf{v} \cdot \mathbf{u}_1)\mathbf{u}_1 + (\mathbf{v} \cdot \mathbf{u}_2)\mathbf{u}_2 + \cdots + (\mathbf{v} \cdot \mathbf{u}_t)\mathbf{u}_t.$$

EXAMPLE 5   *Projection onto a Subspace*

Find the projection of the vector $\mathbf{v} = \begin{bmatrix} 1 \\ 1 \\ 3 \end{bmatrix}$ onto the subspace S of R^3 spanned by the vectors

$$\mathbf{w}_1 = \begin{bmatrix} 0 \\ 3 \\ 1 \end{bmatrix} \quad \text{and} \quad \mathbf{w}_2 = \begin{bmatrix} 2 \\ 0 \\ 0 \end{bmatrix}.$$

Solution   By normalizing $\mathbf{w}_1$ and $\mathbf{w}_2$, you obtain an orthonormal basis for S.

$$\{\mathbf{u}_1, \mathbf{u}_2\} = \left\{\frac{1}{\sqrt{10}}\mathbf{w}_1, \frac{1}{2}\mathbf{w}_2\right\}$$

$$= \left\{\begin{bmatrix} 0 \\ 3/\sqrt{10} \\ 1/\sqrt{10} \end{bmatrix}, \begin{bmatrix} 1 \\ 0 \\ 0 \end{bmatrix}\right\}$$

Use Theorem 5.14 to find the projection of $\mathbf{v}$ onto S.

$$\operatorname{proj}_S\mathbf{v} = (\mathbf{v} \cdot \mathbf{u}_1)\mathbf{u}_1 + (\mathbf{v} \cdot \mathbf{u}_2)\mathbf{u}_2$$

$$= \frac{6}{\sqrt{10}}\begin{bmatrix} 0 \\ 3/\sqrt{10} \\ 1/\sqrt{10} \end{bmatrix} + 1\begin{bmatrix} 1 \\ 0 \\ 0 \end{bmatrix}$$

$$= \begin{bmatrix} 1 \\ \frac{9}{5} \\ \frac{3}{5} \end{bmatrix}$$

The projection of $\mathbf{v}$ onto the plane S is illustrated in Figure 5.23.

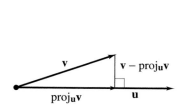

Figure 5.23

Theorem 5.9 said that among all the scalar multiples of a vector $\mathbf{u}$, the orthogonal projection of $\mathbf{v}$ onto $\mathbf{u}$ is the one that is closest to $\mathbf{v}$. Example 5 suggests that this property is also true for projections onto subspaces. That is, among all the vectors in the subspace S, the vector $\operatorname{proj}_S\mathbf{v}$ is the closest vector to $\mathbf{v}$. These two results are illustrated in Figure 5.24.

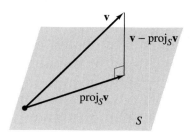

Figure 5.24

Theorem 5.15   *Orthogonal Projection and Distance*

Let S be a subspace of R^n and $\mathbf{v} \in R^n$. Then for all $\mathbf{u} \in S$, $\mathbf{u} \neq \operatorname{proj}_S\mathbf{v}$,

$$\|\mathbf{v} - \operatorname{proj}_S\mathbf{v}\| < \|\mathbf{v} - \mathbf{u}\|.$$

Proof Let $\mathbf{u} \in S$, $\mathbf{u} \neq \text{proj}_S\mathbf{v}$. By adding and subtracting the same quantity $\text{proj}_S\mathbf{v}$ to and from the vector $\mathbf{v} - \mathbf{u}$, you obtain

$$\mathbf{v} - \mathbf{u} = (\mathbf{v} - \text{proj}_S\mathbf{v}) + (\text{proj}_S\mathbf{v} - \mathbf{u}).$$

Observe that $(\text{proj}_S\mathbf{v} - \mathbf{u})$ is in S and $(\mathbf{v} - \text{proj}_S\mathbf{v})$ is orthogonal to S. Hence, $(\mathbf{v} - \text{proj}_S\mathbf{v})$ and $(\text{proj}_S\mathbf{v} - \mathbf{u})$ are orthogonal vectors, and you can use the Pythagorean Theorem (Theorem 5.6) to obtain

$$\|\mathbf{v} - \mathbf{u}\|^2 = \|\mathbf{v} - \text{proj}_S\mathbf{v}\|^2 + \|\text{proj}_S\mathbf{v} - \mathbf{u}\|^2.$$

Because $\mathbf{u} \neq \text{proj}_S\mathbf{v}$, the second term on the right is positive, and you have

$$\|\mathbf{v} - \text{proj}_S\mathbf{v}\| < \|\mathbf{v} - \mathbf{u}\|.$$

Fundamental Subspaces of a Matrix

You need to develop one more idea before solving the least squares problem. Recall that if A is an $m \times n$ matrix, the column space of A is a subspace of R^m consisting of all vectors of the form $A\mathbf{x}$, $\mathbf{x} \in R^n$. The four **fundamental subspaces** of the matrix A are defined as follows (see Exercises 57–61 in Section 5.3).

$$N(A) = \text{nullspace of } A \qquad N(A^T) = \text{nullspace of } A^T$$
$$R(A) = \text{column space of } A \qquad R(A^T) = \text{column space of } A^T$$

It turns out that these subspaces play a crucial role in the solution of the least squares problem.

EXAMPLE 6 *Fundamental Subspaces*

Find the four fundamental subspaces of the matrix

$$A = \begin{bmatrix} 1 & 2 & 0 \\ 0 & 0 & 1 \\ 0 & 0 & 0 \\ 0 & 0 & 0 \end{bmatrix}.$$

Solution The column space of A is simply the span of the first and third columns, because the second column is a scalar multiple of the first.

$$R(A) = \text{span}\left(\begin{bmatrix} 1 \\ 0 \\ 0 \\ 0 \end{bmatrix}, \begin{bmatrix} 0 \\ 1 \\ 0 \\ 0 \end{bmatrix} \right)$$

The column space of A^T is equivalent to the row space of A, which is spanned by the first two rows.

$$R(A^T) = \text{span}\left(\begin{bmatrix} 1 \\ 2 \\ 0 \end{bmatrix}, \begin{bmatrix} 0 \\ 0 \\ 1 \end{bmatrix} \right)$$

The nullspace of A is the solution space of the homogeneous system $A\mathbf{x} = \mathbf{0}$.

$$N(A) = \text{span}\left(\begin{bmatrix} -2 \\ 1 \\ 0 \end{bmatrix} \right)$$

Finally, the nullspace of A^T is the solution space of the homogeneous system whose coefficient matrix is A^T.

$$N(A^T) = \text{span}\left(\begin{bmatrix} 0 \\ 0 \\ 1 \\ 0 \end{bmatrix}, \begin{bmatrix} 0 \\ 0 \\ 0 \\ 1 \end{bmatrix} \right)$$

Observe in Example 6 that $R(A)$ and $N(A^T)$ are orthogonal subspaces of R^4, and $R(A^T)$ and $N(A)$ are orthogonal subspaces of R^3. In fact, even more about these subspaces is true, as stated in the following theorem.

Theorem 5.16

Fundamental Subspaces of a Matrix

If A is an $m \times n$ matrix, then
1. $R(A)$ and $N(A^T)$ are orthogonal subspaces of R^m.
2. $R(A^T)$ and $N(A)$ are orthogonal subspaces of R^n.
3. $R(A) \oplus N(A^T) = R^m$.
4. $R(A^T) \oplus N(A) = R^n$.

Proof   To prove Property 1, let $\mathbf{v} \in R(A)$ and $\mathbf{u} \in N(A^T)$. Because the column space of A is equal to the row space of A^T, you can see that $A^T\mathbf{u} = \mathbf{0}$ implies that $\mathbf{u} \cdot \mathbf{v} = 0$. Property 2 follows from applying Property 1 to A^T.

To prove Property 3, observe that $R(A)^\perp = N(A^T)$. Hence,

$$R^m = R(A) \oplus R(A)^\perp = R(A) \oplus N(A^T).$$

A similar argument applied to $R(A^T)$ proves Property 4.

Least Squares

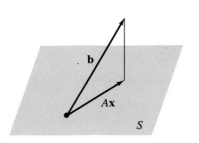

Figure 5.25

You have now developed all the tools needed to solve the least squares problem. Recall that you are attempting to find a vector **x** that minimizes $\|A\mathbf{x} - \mathbf{b}\|$, where A is an $m \times n$ matrix and **b** is a vector in R^m. Let S be the column space of A: $S = R(A)$. You can assume that **b** is not in S, because otherwise the system $A\mathbf{x} = \mathbf{b}$ would be consistent. Hence, you are looking for a vector $A\mathbf{x}$ in S that is as close as possible to **b**, as indicated in Figure 5.25.

From Theorem 5.15 you know that the desired vector is the projection of **b** onto S. Letting $A\hat{\mathbf{x}} = \text{proj}_S \mathbf{b}$ be that projection, you can see that $A\hat{\mathbf{x}} - \mathbf{b} = \text{proj}_S \mathbf{b} - \mathbf{b}$ is orthogonal to $S = R(A)$. But this implies that $A\hat{\mathbf{x}} - \mathbf{b}$ is in $R(A)^\perp$, which equals $N(A^T)$ according to Theorem 5.16. This is the crucial observation: $A\hat{\mathbf{x}} - \mathbf{b}$ is in the nullspace of A^T. Thus, you have the following:

$$A^T(A\hat{\mathbf{x}} - \mathbf{b}) = \mathbf{0}$$
$$A^TA\hat{\mathbf{x}} - A^T\mathbf{b} = \mathbf{0}$$
$$A^TA\hat{\mathbf{x}} = A^T\mathbf{b}.$$

Therefore, the solution of the least squares problem comes down to solving the $n \times n$ linear system of equations

$$A^TA\mathbf{x} = A^T\mathbf{b}.$$

These equations are called the **normal equations** of the least squares problem $A\mathbf{x} = \mathbf{b}$.

EXAMPLE 7   *Solving the Normal Equations*

Find the solution of the least squares problem

$$A\mathbf{x} = \mathbf{b}$$

$$\begin{bmatrix} 1 & 1 \\ 1 & 2 \\ 1 & 3 \end{bmatrix} \begin{bmatrix} c_0 \\ c_1 \end{bmatrix} = \begin{bmatrix} 0 \\ 1 \\ 3 \end{bmatrix}$$

that was presented in Example 1.

Solution   Begin by calculating the following matrix products.

$$A^TA = \begin{bmatrix} 1 & 1 & 1 \\ 1 & 2 & 3 \end{bmatrix} \begin{bmatrix} 1 & 1 \\ 1 & 2 \\ 1 & 3 \end{bmatrix} = \begin{bmatrix} 3 & 6 \\ 6 & 14 \end{bmatrix}$$

$$A^T\mathbf{b} = \begin{bmatrix} 1 & 1 & 1 \\ 1 & 2 & 3 \end{bmatrix} \begin{bmatrix} 0 \\ 1 \\ 3 \end{bmatrix} = \begin{bmatrix} 4 \\ 11 \end{bmatrix}$$

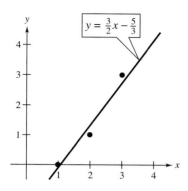

Figure 5.26

The normal equations are

$$A^T A \mathbf{x} = A^T \mathbf{b}$$

$$\begin{bmatrix} 3 & 6 \\ 6 & 14 \end{bmatrix} \begin{bmatrix} c_0 \\ c_1 \end{bmatrix} = \begin{bmatrix} 4 \\ 11 \end{bmatrix}.$$

The solution of this system of equations is $\mathbf{x} = \begin{bmatrix} -\frac{5}{3} \\ \frac{3}{2} \end{bmatrix}$, which implies that the least

squares regression line for the data is $y = \frac{3}{2}x - \frac{5}{3}$, as indicated in Figure 5.26.

Technology   Note

Many graphing calculators and computers have built-in programs for finding the least squares regression line for a set of data points. If you have access to such a program, try verifying the result of Example 7.

REMARK: For an $m \times n$ matrix A, the normal equations form an $n \times n$ system of linear equations. This system is always consistent, but it may have an infinite number of solutions. It can be shown, however, that there is a unique solution if the rank of A is n.

The following example illustrates how to solve the projection problem from Example 5 using normal equations.

EXAMPLE 8   *Orthogonal Projection onto a Subspace*

Find the orthogonal projection of the vector $\mathbf{b} = \begin{bmatrix} 1 \\ 1 \\ 3 \end{bmatrix}$ onto the column space S of the matrix

$$A = \begin{bmatrix} 0 & 2 \\ 3 & 0 \\ 1 & 0 \end{bmatrix}.$$

Solution   To find the orthogonal projection of $\mathbf{b}$ onto S, first solve the least squares problem $A\mathbf{x} = \mathbf{b}$. As in Example 7, calculate the following matrix products.

$$A^T A = \begin{bmatrix} 0 & 3 & 1 \\ 2 & 0 & 0 \end{bmatrix} \begin{bmatrix} 0 & 2 \\ 3 & 0 \\ 1 & 0 \end{bmatrix} = \begin{bmatrix} 10 & 0 \\ 0 & 4 \end{bmatrix}$$

$$A^T \mathbf{b} = \begin{bmatrix} 0 & 3 & 1 \\ 2 & 0 & 0 \end{bmatrix} \begin{bmatrix} 1 \\ 1 \\ 3 \end{bmatrix} = \begin{bmatrix} 6 \\ 2 \end{bmatrix}$$

The normal equations are

$$A^T A \mathbf{x} = A^T \mathbf{b}$$

$$\begin{bmatrix} 10 & 0 \\ 0 & 4 \end{bmatrix} \begin{bmatrix} x_1 \\ x_2 \end{bmatrix} = \begin{bmatrix} 6 \\ 2 \end{bmatrix}.$$

The solution of these equations is easily seen to be

$$\mathbf{x} = \begin{bmatrix} x_1 \\ x_2 \end{bmatrix} = \begin{bmatrix} \frac{3}{5} \\ \frac{1}{2} \end{bmatrix}.$$

Finally, the projection of **b** onto S is given by

$$A\mathbf{x} = \begin{bmatrix} 0 & 2 \\ 3 & 0 \\ 1 & 0 \end{bmatrix} \begin{bmatrix} \frac{3}{5} \\ \frac{1}{2} \end{bmatrix} = \begin{bmatrix} 1 \\ \frac{9}{5} \\ \frac{3}{5} \end{bmatrix},$$

which agrees with the solution obtained in Example 5.

Mathematical Modeling

Least squares problems play a fundamental role in mathematical modeling of real-life phenomena. The next example shows how to model the world population using a least squares quadratic polynomial.

EXAMPLE 9   *World Population*

Table 5.1 shows the world population in billions for five different years. (Source: U.S. Census Bureau)

TABLE 5.1

Year	1980	1985	1990	1995	2000
Population (y)	4.5	4.9	5.3	5.7	6.1

Let $x = 0$ represent the year 1980. Find the least squares regression quadratic polynomial $y = c_0 + c_1 x + c_2 x^2$ for these data and use the model to estimate the population for the year 2010.

Solution   By substituting the data points $(0, 4.5)$, $(5, 4.9)$, $(10, 5.3)$, $(15, 5.7)$, and $(20, 6.1)$ into the quadratic polynomial $y = c_0 + c_1 x + c_2 x^2$, you obtain the following system of linear equations.

$$
\begin{aligned}
c_0 & & & & & = 4.5 \\
c_0 & + & 5c_1 & + & 25c_2 & = 4.9 \\
c_0 & + & 10c_1 & + & 100c_2 & = 5.3 \\
c_0 & + & 15c_1 & + & 225c_2 & = 5.7 \\
c_0 & + & 20c_1 & + & 400c_2 & = 6.1
\end{aligned}
$$

This produces the least squares problem

$$A\mathbf{x} = \mathbf{b}$$

$$
\begin{bmatrix}
1 & 0 & 0 \\
1 & 5 & 25 \\
1 & 10 & 100 \\
1 & 15 & 225 \\
1 & 20 & 400
\end{bmatrix}
\begin{bmatrix}
c_0 \\
c_1 \\
c_2
\end{bmatrix}
=
\begin{bmatrix}
4.5 \\
4.9 \\
5.3 \\
5.7 \\
6.1
\end{bmatrix}.
$$

The normal equations are

$$A^T A \mathbf{x} = A^T \mathbf{b}$$

$$
\begin{bmatrix}
5 & 50 & 750 \\
50 & 750 & 12{,}500 \\
750 & 12{,}500 & 221{,}250
\end{bmatrix}
\begin{bmatrix}
c_0 \\
c_1 \\
c_2
\end{bmatrix}
=
\begin{bmatrix}
26.5 \\
285 \\
4375
\end{bmatrix}
$$

and their solution is

$$
\mathbf{x} =
\begin{bmatrix}
c_0 \\
c_1 \\
c_2
\end{bmatrix}
\approx
\begin{bmatrix}
4.5 \\
0.08 \\
0
\end{bmatrix}.
$$

Note that $c_2 = 0$. Hence, the least squares polynomial for these data is the following linear polynomial.

$$y = 4.5 + 0.08x$$

Evaluating this polynomial at $x = 30$ gives the estimate of the world population for the year 2010:

$$y = 4.5 + 0.08(30) \approx 6.9 \text{ billion.}$$

Least squares models can arise in many other contexts. Section 5.5 explores some applications of least squares to approximation of functions. In the final example of this section, a nonlinear model is used to find a relationship between the period of a planet and its mean distance from the sun.

EXAMPLE 10 *Application to Astronomy*

Table 5.2 shows the mean distances x and the periods y of the six planets that are closest to the sun. The mean distance is given in terms of astronomical units (where the Earth's mean distance is defined to be 1.0), and the period is given in years. Find a model for these data.

TABLE 5.2

Planet	Mercury	Venus	Earth	Mars	Jupiter	Saturn
Distance, x	0.387	0.723	1.0	1.523	5.203	9.541
Period, y	0.241	0.615	1.0	1.881	11.861	29.457

If you plot the data as given, they do not seem to lie in a straight line. However, by taking the logarithm of each coordinate, you obtain points of the form $(\ln x, \ln y)$, as shown in Table 5.3.

TABLE 5.3

Planet	Mercury	Venus	Earth	Mars	Jupiter	Saturn
$\ln x$	−0.949	−0.324	0.0	0.421	1.649	2.256
$\ln y$	−1.423	−0.486	0.0	0.632	2.473	3.383

Figure 5.27 shows a plot of the transformed points and suggests that the least squares regression line would be a good fit. Using the techniques of this section, you can find that the line is

$$\ln y = \tfrac{3}{2} \ln x \qquad \text{or} \qquad y = x^{3/2}.$$

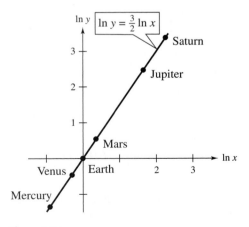

Technology Note

You can use a computer or graphing calculator with a built-in power regression program to verify the result of Example 10. For example, using the data in Table 5.2 and a TI-86 calculator, the power fit program from the *stat* menu gives $y \approx 1.00042x^{1.49954}$.

Figure 5.27

In Exercises 1–4, determine whether the sets are orthogonal.

1. $S_1 = \text{span}\left\{\begin{bmatrix} 2 \\ 1 \\ -1 \end{bmatrix}, \begin{bmatrix} 0 \\ 1 \\ 1 \end{bmatrix}\right\}$ $S_2 = \text{span}\left\{\begin{bmatrix} -1 \\ 2 \\ 0 \end{bmatrix}\right\}$

2. $S_1 = \text{span}\left\{\begin{bmatrix} -3 \\ 0 \\ 1 \end{bmatrix}\right\}$ $S_2 = \text{span}\left\{\begin{bmatrix} 2 \\ 1 \\ 6 \end{bmatrix}, \begin{bmatrix} 0 \\ 1 \\ 0 \end{bmatrix}\right\}$

3. $S_1 = \text{span}\left\{\begin{bmatrix} 1 \\ 1 \\ 1 \\ 1 \end{bmatrix}\right\}$ $S_2 = \text{span}\left\{\begin{bmatrix} -1 \\ 1 \\ -1 \\ 1 \end{bmatrix}, \begin{bmatrix} 0 \\ 2 \\ -2 \\ 0 \end{bmatrix}\right\}$

4. $S_1 = \text{span}\left\{\begin{bmatrix} 0 \\ 0 \\ 2 \\ 1 \end{bmatrix}, \begin{bmatrix} 0 \\ 0 \\ 1 \\ -2 \end{bmatrix}\right\}$ $S_2 = \text{span}\left\{\begin{bmatrix} 3 \\ 2 \\ 0 \\ 0 \end{bmatrix}, \begin{bmatrix} 0 \\ 1 \\ -2 \\ 2 \end{bmatrix}\right\}$

In Exercises 5–8, find the orthogonal complement $S^\perp$.

5. S is the subspace of R^3 consisting of the xz-plane.

6. S is the subspace of R^5 consisting of all vectors whose third and fourth components are zero.

7. $S = \text{span}\left\{\begin{bmatrix} 1 \\ 2 \\ 0 \\ 0 \end{bmatrix}, \begin{bmatrix} 0 \\ 1 \\ 0 \\ 1 \end{bmatrix}\right\}$ **8.** $S = \text{span}\left\{\begin{bmatrix} 0 \\ 1 \\ -1 \\ 1 \end{bmatrix}\right\}$

9. Find the orthogonal complement of the *solution* of Exercise 7.

10. Find the orthogonal complement of the *solution* of Exercise 8.

In Exercises 11–14, find the projection of the vector **v** onto the subspace S.

11. $S = \text{span}\left\{\begin{bmatrix} 0 \\ 0 \\ -1 \\ 1 \end{bmatrix}, \begin{bmatrix} 0 \\ 1 \\ 1 \\ 1 \end{bmatrix}\right\}$, $\mathbf{v} = \begin{bmatrix} 1 \\ 0 \\ 1 \\ 1 \end{bmatrix}$

12. $S = \text{span}\left\{\begin{bmatrix} -1 \\ 2 \\ 0 \\ 0 \end{bmatrix}, \begin{bmatrix} 0 \\ 0 \\ 1 \\ 0 \end{bmatrix}, \begin{bmatrix} 0 \\ 0 \\ 0 \\ 1 \end{bmatrix}\right\}$, $\mathbf{v} = \begin{bmatrix} 1 \\ 1 \\ 1 \\ 1 \end{bmatrix}$

13. $S = \text{span}\left\{\begin{bmatrix} 1 \\ 0 \\ 1 \end{bmatrix}, \begin{bmatrix} 0 \\ 1 \\ 1 \end{bmatrix}\right\}$, $\mathbf{v} = \begin{bmatrix} 2 \\ 3 \\ 4 \end{bmatrix}$

14. $S = \text{span}\left\{\begin{bmatrix} 1 \\ 1 \\ 1 \\ 1 \end{bmatrix}, \begin{bmatrix} 0 \\ 1 \\ -1 \\ 0 \end{bmatrix}, \begin{bmatrix} 0 \\ 1 \\ 1 \\ 0 \end{bmatrix}\right\}$, $\mathbf{v} = \begin{bmatrix} 1 \\ 2 \\ 3 \\ 4 \end{bmatrix}$

In Exercises 15 and 16, find the orthogonal projection of $\mathbf{b} = [2 \;\; -2 \;\; 1]^T$ onto the column space of the matrix A.

15. $A = \begin{bmatrix} 1 & 2 \\ 0 & 1 \\ 1 & 1 \end{bmatrix}$ **16.** $A = \begin{bmatrix} 0 & 2 \\ 1 & 1 \\ 1 & 3 \end{bmatrix}$

In Exercises 17–20, find bases for the four fundamental subspaces of the matrix A.

17. $A = \begin{bmatrix} 1 & 2 & 3 \\ 0 & 1 & 0 \end{bmatrix}$ **18.** $A = \begin{bmatrix} 0 & -1 & 1 \\ 1 & 2 & 0 \\ 1 & 1 & 1 \end{bmatrix}$

19. $A = \begin{bmatrix} 1 & 0 & 0 & 1 \\ 0 & 1 & 1 & 1 \\ 1 & 1 & 1 & 2 \\ 1 & 2 & 2 & 3 \end{bmatrix}$

20. $A = \begin{bmatrix} 1 & 0 & -1 \\ 0 & -1 & 1 \\ 1 & 1 & 0 \\ 1 & 0 & 1 \end{bmatrix}$

In Exercises 21–24, find the least squares solution of the system $A\mathbf{x} = \mathbf{b}$.

21. $A = \begin{bmatrix} 2 & 1 \\ 1 & 2 \\ 1 & 1 \end{bmatrix}$ $\mathbf{b} = \begin{bmatrix} 2 \\ 0 \\ -3 \end{bmatrix}$

22. $A = \begin{bmatrix} 0 & 1 \\ 1 & 0 \\ 1 & 2 \end{bmatrix}$ $\mathbf{b} = \begin{bmatrix} -1 \\ -1 \\ 3 \end{bmatrix}$

23. $A = \begin{bmatrix} 1 & 0 & 1 \\ 1 & 1 & 1 \\ 0 & 1 & 1 \\ 1 & 1 & 0 \end{bmatrix}$ $\mathbf{b} = \begin{bmatrix} 4 \\ -1 \\ 0 \\ 1 \end{bmatrix}$

24. $A = \begin{bmatrix} 0 & 2 & 0 \\ 1 & 1 & 0 \\ -1 & 0 & 1 \\ 1 & 1 & 1 \\ 0 & 0 & 1 \end{bmatrix}$ $\mathbf{b} = \begin{bmatrix} 0 \\ 1 \\ 0 \\ 0 \\ 1 \end{bmatrix}$

In Exercises 25–28, find the least squares regression line for the data points. Graph the points and the line on the same set of axes.

25. $(-1, 1), (1, 0), (3, -3)$
26. $(-2, -1), (-1, 0), (1, 0), (2, 2)$
27. $(-3, -3), (-2, -2), (0, 0), (1, 2)$
28. $(-2, 1), (-1, 2), (0, 1), (1, 2), (2, 1)$

In Exercises 29 and 30, find the least squares quadratic polynomial for the data points.

29. $(-2, 0), (-1, 0), (0, 1), (1, 2), (2, 5)$
30. $(0, 0), (2, 2), (3, 6), (4, 12)$

31. The table shows the annual sales, in millions of dollars, for Target Corporation and Wal-Mart Stores for 1992 through 2000. Find an appropriate model for each company and use the model to predict sales for the year 2005. (Source: Target Corporation and Wal-Mart Stores)

Year	1992	1993	1994	1995	1996
Target	17,927	19,233	21,311	23,516	25,371
Wal-Mart	55,484	67,345	82,494	93,627	104,859

Year	1997	1998	1999	2000
Target	27,487	30,662	33,702	36,903
Wal-Mart	117,958	137,634	165,013	191,329

32. The table shows the total domestic catch of fish x, in millions of pounds, and the portion of the total domestic catch of fish for human consumption y, in millions of pounds, for the years 1989 to 1999. Find an appropriate model for the relationship between total catch and the portion of total catch for human consumption. (Source: United States National Oceanic and Atmospheric Administration)

Year	1989	1990	1991	1992	1993	1994
x	8,463	9,404	9,484	9,637	10,467	10,461
y	6,204	7,041	7,031	7,618	8,214	7,936

Year	1995	1996	1997	1998	1999
x	9,788	9,565	9,845	9,194	9,339
y	7,667	7,476	7,248	7,174	6,832

True or False? In Exercises 33 and 34, determine whether each statement is true or false. If a statement is true, give a reason or cite an appropriate statement in the text. If a statement is false, provide an example that shows that the statement is not true in all cases or cite an appropriate statement in the text.

33. (a) If S_1 and S_2 are orthogonal subspaces of R^n, then their intersection is an empty set.
 (b) If each vector $\mathbf{v} \in R^n$ can be uniquely written as a sum of a vector $\mathbf{s}_1$ from S_1 and a vector $\mathbf{s}_2$ from S_2, then R^n is called the direct sum of S_1 and S_2.
 (c) The solution of the least squares problem consists essentially of solving the normal equations—that is, solving the $n \times n$ linear system of equations $A^T A \mathbf{x} = A^T \mathbf{b}$.

34. (a) If A is an $m \times n$ matrix, then $R(A)$ and $N(A^T)$ are orthogonal subspaces of R^n.
 (b) The set of all vectors orthogonal to every vector in a subspace S is called the orthogonal complement of S.
 (c) Given an $m \times n$ matrix A and a vector $\mathbf{b}$ in R^m, the least squares problem is to find $\mathbf{x}$ in R^n such that $\|A\mathbf{x} - \mathbf{b}\|^2$ is minimized.

35. Prove that if S_1 and S_2 are orthogonal subspaces of R^n, then their intersection consists only of the zero vector.

36. Prove that the orthogonal complement of a subspace of R^n is itself a subspace of R^n.

37. Prove Theorem 5.14.

38. Prove that if S_1 and S_2 are subspaces of R^n and $R^n = S_1 \oplus S_2$, then $S_1 \cap S_2 = \{\mathbf{0}\}$.

39. ***Writing*** Describe the normal equations for the least squares problem if the $m \times n$ matrix A has orthonormal columns.

5.5 | Applications of Inner Product Spaces

The Cross Product of Two Vectors in Space

Many problems in linear algebra involve finding a vector that is orthogonal to each vector in a given set. Here you will look at a vector product that yields a vector in R^3 that is orthogonal to two given vectors. This vector product is called the **cross product,** and it is most conveniently defined and calculated with vectors written in standard unit vector form.

$$\mathbf{v} = (v_1, v_2, v_3) = v_1\mathbf{i} + v_2\mathbf{j} + v_3\mathbf{k}$$

Definition of Cross Product of Two Vectors

Let $\mathbf{u} = u_1\mathbf{i} + u_2\mathbf{j} + u_3\mathbf{k}$ and $\mathbf{v} = v_1\mathbf{i} + v_2\mathbf{j} + v_3\mathbf{k}$ be vectors in R^3. The **cross product** of $\mathbf{u}$ and $\mathbf{v}$ is the vector

$$\mathbf{u} \times \mathbf{v} = (u_2v_3 - u_3v_2)\mathbf{i} - (u_1v_3 - u_3v_1)\mathbf{j} + (u_1v_2 - u_2v_1)\mathbf{k}.$$

REMARK: The cross product is defined only for vectors in R^3. The cross product of two vectors in R^2 or of vectors in R^n, $n > 3$, is not defined here.

A convenient way to remember the formula for the cross product $\mathbf{u} \times \mathbf{v}$ is to use the following determinant form.

$$\mathbf{u} \times \mathbf{v} = \begin{vmatrix} \mathbf{i} & \mathbf{j} & \mathbf{k} \\ u_1 & u_2 & u_3 \\ v_1 & v_2 & v_3 \end{vmatrix} \qquad \longleftarrow \quad \text{Components of u} \\ \longleftarrow \quad \text{Components of v}$$

Technically this is not a determinant because the entries are not all real numbers. Nevertheless, it is useful because it provides an easy way to remember the cross product formula. Using cofactor expansion along the first row, you obtain

$$\mathbf{u} \times \mathbf{v} = \begin{vmatrix} u_2 & u_3 \\ v_2 & v_3 \end{vmatrix}\mathbf{i} - \begin{vmatrix} u_1 & u_3 \\ v_1 & v_3 \end{vmatrix}\mathbf{j} + \begin{vmatrix} u_1 & u_2 \\ v_1 & v_2 \end{vmatrix}\mathbf{k}$$
$$= (u_2v_3 - u_3v_2)\mathbf{i} - (u_1v_3 - u_3v_1)\mathbf{j} + (u_1v_2 - u_2v_1)\mathbf{k},$$

which yields the formula given in the definition. Be sure to note that the **j**-component is preceded by a minus sign.

EXAMPLE 1   *Finding the Cross Product of Two Vectors*

Given $\mathbf{u} = \mathbf{i} - 2\mathbf{j} + \mathbf{k}$ and $\mathbf{v} = 3\mathbf{i} + \mathbf{j} - 2\mathbf{k}$, find each of the following.

(a) $\mathbf{u} \times \mathbf{v}$ (b) $\mathbf{v} \times \mathbf{u}$ (c) $\mathbf{v} \times \mathbf{v}$

Solution (a) $\mathbf{u} \times \mathbf{v} = \begin{vmatrix} \mathbf{i} & \mathbf{j} & \mathbf{k} \\ 1 & -2 & 1 \\ 3 & 1 & -2 \end{vmatrix}$

$= \begin{vmatrix} -2 & 1 \\ 1 & -2 \end{vmatrix} \mathbf{i} - \begin{vmatrix} 1 & 1 \\ 3 & -2 \end{vmatrix} \mathbf{j} + \begin{vmatrix} 1 & -2 \\ 3 & 1 \end{vmatrix} \mathbf{k}$

$= 3\mathbf{i} + 5\mathbf{j} + 7\mathbf{k}$

Simulation

Explore this concept further with an electronic simulation available on the *Elementary Linear Algebra* Learning Tools Student CD-ROM.

(b) $\mathbf{v} \times \mathbf{u} = \begin{vmatrix} \mathbf{i} & \mathbf{j} & \mathbf{k} \\ 3 & 1 & -2 \\ 1 & -2 & 1 \end{vmatrix}$

$= -3\mathbf{i} - 5\mathbf{j} - 7\mathbf{k}$

Note that this result is the negative of that in part (a).

(c) $\mathbf{v} \times \mathbf{v} = \begin{vmatrix} \mathbf{i} & \mathbf{j} & \mathbf{k} \\ 3 & 1 & -2 \\ 3 & 1 & -2 \end{vmatrix}$

$= \begin{vmatrix} 1 & -2 \\ 1 & -2 \end{vmatrix} \mathbf{i} - \begin{vmatrix} 3 & -2 \\ 3 & -2 \end{vmatrix} \mathbf{j} + \begin{vmatrix} 3 & 1 \\ 3 & 1 \end{vmatrix} \mathbf{k}$

$= 0\mathbf{i} + 0\mathbf{j} + 0\mathbf{k} = \mathbf{0}$

Technology   Note

Some graphing calculators and computer software programs have vector capabilities that include finding a cross product. For example, on a TI-86, you can verify $\mathbf{u} \times \mathbf{v}$ in Example 1(a) as follows.

```
cross(U,V)
            [3 5 7]
```

The results obtained in Example 1 suggest some interesting *algebraic* properties of the cross product. For instance,

$$\mathbf{u} \times \mathbf{v} = -(\mathbf{v} \times \mathbf{u}) \qquad \text{and} \qquad \mathbf{v} \times \mathbf{v} = \mathbf{0}.$$

These properties, along with several others, are given in Theorem 5.17.

Theorem 5.17

Algebraic Properties of the Cross Product

If **u**, **v**, and **w** are vectors in R^3 and c is a scalar, then the following properties are true.

1. $\mathbf{u} \times \mathbf{v} = -(\mathbf{v} \times \mathbf{u})$
2. $\mathbf{u} \times (\mathbf{v} + \mathbf{w}) = (\mathbf{u} \times \mathbf{v}) + (\mathbf{u} \times \mathbf{w})$
3. $c(\mathbf{u} \times \mathbf{v}) = c\mathbf{u} \times \mathbf{v} = \mathbf{u} \times c\mathbf{v}$
4. $\mathbf{u} \times \mathbf{0} = \mathbf{0} \times \mathbf{u} = \mathbf{0}$
5. $\mathbf{u} \times \mathbf{u} = \mathbf{0}$
6. $\mathbf{u} \cdot (\mathbf{v} \times \mathbf{w}) = (\mathbf{u} \times \mathbf{v}) \cdot \mathbf{w}$

Proof   The proof of the first property is given here. The proofs of the other properties are left to you. (See Exercises 23–27.) Let **u** and **v** be given by

$$\mathbf{u} = u_1\mathbf{i} + u_2\mathbf{j} + u_3\mathbf{k}$$

and

$$\mathbf{v} = v_1\mathbf{i} + v_2\mathbf{j} + v_3\mathbf{k}.$$

Then $\mathbf{u} \times \mathbf{v}$ is

$$\mathbf{u} \times \mathbf{v} = \begin{vmatrix} \mathbf{i} & \mathbf{j} & \mathbf{k} \\ u_1 & u_2 & u_3 \\ v_1 & v_2 & v_3 \end{vmatrix}$$

$$= (u_2v_3 - u_3v_2)\mathbf{i} - (u_1v_3 - u_3v_1)\mathbf{j} + (u_1v_2 - u_2v_1)\mathbf{k},$$

and $\mathbf{v} \times \mathbf{u}$ is

$$\mathbf{v} \times \mathbf{u} = \begin{vmatrix} \mathbf{i} & \mathbf{j} & \mathbf{k} \\ v_1 & v_2 & v_3 \\ u_1 & u_2 & u_3 \end{vmatrix}$$

$$= (v_2u_3 - v_3u_2)\mathbf{i} - (v_1u_3 - v_3u_1)\mathbf{j} + (v_1u_2 - v_2u_1)\mathbf{k}$$

$$= -(u_2v_3 - u_3v_2)\mathbf{i} + (u_1v_3 - u_3v_1)\mathbf{j} - (u_1v_2 - u_2v_1)\mathbf{k}$$

$$= -(\mathbf{u} \times \mathbf{v}).$$

Property 1 of Theorem 5.17 tells you that the vectors $\mathbf{u} \times \mathbf{v}$ and $\mathbf{v} \times \mathbf{u}$ have equal lengths but opposite directions. The geometric implication of this result will be discussed after some geometric properties of the cross product of two vectors are established.

| Theorem 5.18 | Geometric Properties of the Cross Product |

If **u** and **v** are nonzero vectors in R^3, then the following properties are true.

1. **u** × **v** is orthogonal to both **u** and **v**.
2. The angle θ between **u** and **v** is given by

$$\|\mathbf{u} \times \mathbf{v}\| = \|\mathbf{u}\|\,\|\mathbf{v}\| \sin \theta.$$

3. **u** and **v** are parallel if and only if **u** × **v** = **0**.
4. The parallelogram having **u** and **v** as adjacent sides has an area of $\|\mathbf{u} \times \mathbf{v}\|$.

Proof   The proof of Property 4 follows. The proofs of the other properties are left to you. (See Exercises 28–30.) To prove Property 4, let **u** and **v** represent adjacent sides of a parallelogram, as shown in Figure 5.28. By Property 2, the area of the parallelogram is given by

$$\text{Area} = \overbrace{\|\mathbf{u}\|}^{Base}\ \overbrace{\|\mathbf{v}\| \sin \theta}^{Height} = \|\mathbf{u} \times \mathbf{v}\|.$$

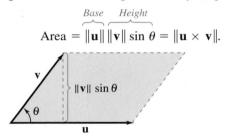

Figure 5.28

Property 1 states that the vector **u** × **v** is orthogonal to both **u** and **v**. This implies that **u** × **v** (and **v** × **u**) is orthogonal to the plane determined by **u** and **v**. One way to remember the orientation of the vectors **u**, **v**, and **u** × **v** is to compare them with the unit vectors **i**, **j**, and **k**, as shown in Figure 5.29. The three vectors **u**, **v**, and **u** × **v** form a *right-handed system*, whereas the three vectors **u**, **v**, and **v** × **u** form a *left-handed system.*

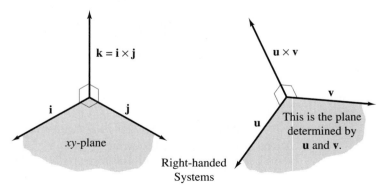

Figure 5.29

EXAMPLE 2   *Finding a Vector Orthogonal to Two Given Vectors*

Find a unit vector that is orthogonal to both

$$\mathbf{u} = \mathbf{i} - 4\mathbf{j} + \mathbf{k}$$

and

$$\mathbf{v} = 2\mathbf{i} + 3\mathbf{j}.$$

Solution   From Property 1 of Theorem 5.18, you know that the cross product

$$\mathbf{u} \times \mathbf{v} = \begin{vmatrix} \mathbf{i} & \mathbf{j} & \mathbf{k} \\ 1 & -4 & 1 \\ 2 & 3 & 0 \end{vmatrix}$$

$$= -3\mathbf{i} + 2\mathbf{j} + 11\mathbf{k}$$

is orthogonal to both **u** and **v**, as shown in Figure 5.30. Then, by dividing by the length of **u** × **v**,

$$\|\mathbf{u} \times \mathbf{v}\| = \sqrt{(-3)^2 + 2^2 + 11^2}$$

$$= \sqrt{134},$$

you obtain the unit vector

$$\frac{\mathbf{u} \times \mathbf{v}}{\|\mathbf{u} \times \mathbf{v}\|} = -\frac{3}{\sqrt{134}}\mathbf{i} + \frac{2}{\sqrt{134}}\mathbf{j} + \frac{11}{\sqrt{134}}\mathbf{k},$$

which is orthogonal to both **u** and **v** as follows.

$$\left(-\frac{3}{\sqrt{134}}, \frac{2}{\sqrt{134}}, \frac{11}{\sqrt{134}}\right) \cdot (1, -4, 1) = 0$$

$$\left(-\frac{3}{\sqrt{134}}, \frac{2}{\sqrt{134}}, \frac{11}{\sqrt{134}}\right) \cdot (2, 3, 0) = 0$$

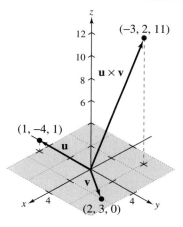

Figure 5.30

EXAMPLE 3   *Finding the Area of a Parallelogram*

Find the area of the parallelogram that has

$$\mathbf{u} = -3\mathbf{i} + 4\mathbf{j} + \mathbf{k}$$

and

$$\mathbf{v} = -2\mathbf{j} + 6\mathbf{k}$$

as adjacent sides, as shown in Figure 5.31.

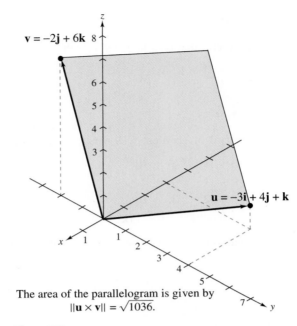

$\mathbf{v} = -2\mathbf{j} + 6\mathbf{k}$

$\mathbf{u} = -3\mathbf{i} + 4\mathbf{j} + \mathbf{k}$

The area of the parallelogram is given by
$$\|\mathbf{u} \times \mathbf{v}\| = \sqrt{1036}.$$

Figure 5.31

Solution From Property 4 of Theorem 5.18, you know that the area of this parallelogram is given by $\|\mathbf{u} \times \mathbf{v}\|$. Because

$$\mathbf{u} \times \mathbf{v} = \begin{vmatrix} \mathbf{i} & \mathbf{j} & \mathbf{k} \\ -3 & 4 & 1 \\ 0 & -2 & 6 \end{vmatrix} = 26\mathbf{i} + 18\mathbf{j} + 6\mathbf{k},$$

the area of the parallelogram is

$$\|\mathbf{u} \times \mathbf{v}\| = \sqrt{26^2 + 18^2 + 6^2} = \sqrt{1036} \approx 32.19.$$

Least Squares Approximations (Calculus)

Many problems in the physical sciences and engineering involve an approximation of a function f by another function g. If f is in $C[a, b]$ (the inner product space of all continuous functions on $[a, b]$), then usually g is chosen from a given subspace W of $C[a, b]$. For instance, to approximate the function

$$f(x) = e^x, \quad 0 \le x \le 1,$$

you could choose one of the following forms of g.

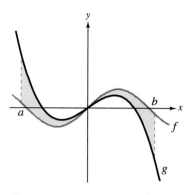

Figure 5.32

1. $g(x) = a_0 + a_1 x, \quad 0 \le x \le 1$ Linear
2. $g(x) = a_0 + a_1 x + a_2 x^2, \quad 0 \le x \le 1$ Quadratic
3. $g(x) = a_0 + a_1 \cos x + a_2 \sin x, \quad 0 \le x \le 1$ Trigonometric

Before discussing ways of finding the function g, you must define how one function can "best" approximate another function. One natural way would be to require that the area bounded by the graphs of f and g on the interval $[a, b]$,

$$\text{Area} = \int_a^b |f(x) - g(x)| \, dx,$$

be a minimum with respect to other functions in the subspace W, as shown in Figure 5.32. But because integrands involving absolute value are often difficult to evaluate, it is more common to square the integrand to obtain

$$\int_a^b [f(x) - g(x)]^2 \, dx.$$

With this criterion, the function g is called the **least squares approximation** of f with respect to the inner product space W.

Definition of Least Squares Approximation

Let f be continuous on $[a, b]$, and let W be a subspace of $C[a, b]$. A function g in W is called a **least squares approximation** of f with respect to W if the value of

$$I = \int_a^b [f(x) - g(x)]^2 \, dx$$

is a minimum with respect to all other functions in W.

REMARK: If the subspace W in this definition is the entire space $C[a, b]$, then $g(x) = f(x)$, which implies that $I = 0$.

EXAMPLE 4   *Finding a Least Squares Approximation*

Find the least squares approximation $g(x) = a_0 + a_1 x$ for

$$f(x) = e^x, \quad 0 \le x \le 1.$$

Solution   For this approximation you need to find the constants a_0 and a_1 that minimize the value of

$$I = \int_0^1 [f(x) - g(x)]^2 \, dx$$

$$= \int_0^1 (e^x - a_0 - a_1 x)^2 \, dx.$$

Evaluating this integral, you have

$$I = \int_0^1 (e^x - a_0 - a_1 x)^2 \, dx$$

$$= \int_0^1 (e^{2x} - 2a_0 e^x - 2a_1 x e^x + a_0^2 + 2a_0 a_1 x + a_1^2 x^2) \, dx$$

$$= \left[\frac{1}{2} e^{2x} - 2a_0 e^x - 2a_1 e^x (x - 1) + a_0^2 x + a_0 a_1 x^2 + a_1^2 \frac{x^3}{3} \right]_0^1$$

$$= \frac{1}{2}(e^2 - 1) - 2a_0(e - 1) - 2a_1 + a_0^2 + a_0 a_1 + \frac{1}{3} a_1^2.$$

Now, considering I to be a function of the variables a_0 and a_1, use calculus to determine the values of a_0 and a_1 that minimize I. Specifically, by setting the partial derivatives

$$\frac{\partial I}{\partial a_0} = 2a_0 - 2e + 2 + a_1$$

$$\frac{\partial I}{\partial a_1} = a_0 + \frac{2}{3} a_1 - 2$$

equal to zero, you obtain the following two linear equations in a_0 and a_1.

$$2a_0 + a_1 = 2(e - 1)$$
$$3a_0 + 2a_1 = 6$$

The solution of this system is

$$a_0 = 4e - 10 \approx 0.873 \qquad \text{and} \qquad a_1 = 18 - 6e \approx 1.690.$$

Therefore, the best *linear approximation* for $f(x) = e^x$ on the interval $[0, 1]$ is given by

$$g(x) = 4e - 10 + (18 - 6e)x$$
$$\approx 0.873 + 1.690x.$$

Figure 5.33 shows the graphs of f and g on $[0, 1]$.

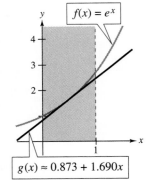

Figure 5.33

Of course, the approximation obtained in Example 4 depends on the definition of the best approximation. For instance, if the definition of "best" had been the *Taylor polynomial of degree one* centered at 0.5, then the approximating function g would have been

$$g(x) = f(0.5) + f'(0.5)(x - 0.5)$$
$$= e^{0.5} + e^{0.5}(x - 0.5)$$
$$\approx 0.824 + 1.649x.$$

Moreover, the function g obtained in Example 4 is only the best *linear* approximation of f (according to the least squares criterion). In Example 5 you will find the best *quadratic* approximation.

EXAMPLE 5   *Finding a Least Squares Approximation*

Find the least squares approximation $g(x) = a_0 + a_1x + a_2x^2$ for

$$f(x) = e^x, \quad 0 \le x \le 1.$$

Solution   For this approximation you need to find the values of a_0, a_1, and a_2 that minimize the value of

$$I = \int_0^1 [f(x) - g(x)]^2 \, dx$$

$$= \int_0^1 (e^x - a_0 - a_1x - a_2x^2)^2 \, dx.$$

Integrating and then setting the partial derivatives of I (with respect to a_0, a_1, and a_2) equal to zero produces the following system of linear equations.

$$
\begin{aligned}
6a_0 + 3a_1 + 2a_2 &= 6(e - 1) \\
6a_0 + 4a_1 + 3a_2 &= 12 \\
20a_0 + 15a_1 + 12a_2 &= 60(e - 2)
\end{aligned}
$$

The solution of this system is

$$
\begin{aligned}
a_0 &= -105 + 39e \approx 1.013 \\
a_1 &= 588 - 216e \approx 0.851 \\
a_2 &= -570 + 210e \approx 0.839.
\end{aligned}
$$

Thus, the approximating function g is

$$g(x) \approx 1.013 + 0.851x + 0.839x^2.$$

The graphs of f and g are shown in Figure 5.34.

$g(x) \approx 1.013 + 0.851x + 0.839x^2$

$f(x) = e^x$

Figure 5.34

The integral I (given in the definition of the least squares approximation) can be expressed in vector form. To do this, use the inner product defined in Example 5 in Section 5.2:

$$\langle f, g \rangle = \int_a^b f(x)g(x) \, dx.$$

With this inner product you have

$$I = \int_a^b [f(x) - g(x)]^2 \, dx = \langle f - g, f - g \rangle = \|f - g\|^2.$$

This means that the least squares approximating function g is the function that minimizes $\|f - g\|^2$ or, equivalently, minimizes $\|f - g\|$. In other words, the least squares approximation of a function f is the function g (in the subspace W) closest to f in terms of the inner product $\langle f, g \rangle$. The following theorem gives you a way of determining the function g.

Theorem 5.19	Least Squares Approximation

Let f be continuous on $[a, b]$, and let W be a finite-dimensional subspace of $C[a, b]$. The least squares approximating function of f with respect to W is given by

$$g = \langle f, \mathbf{w}_1 \rangle \mathbf{w}_1 + \langle f, \mathbf{w}_2 \rangle \mathbf{w}_2 + \cdots + \langle f, \mathbf{w}_n \rangle \mathbf{w}_n,$$

where $B = \{\mathbf{w}_1, \mathbf{w}_2, \ldots, \mathbf{w}_n\}$ is an orthonormal basis for W.

Proof To show that g is the least squares approximating function of f, prove that the inequality

$$\|f - g\| \le \|f - \mathbf{w}\|$$

is true for any vector $\mathbf{w}$ in W. By writing $f - g$ as

$$f - g = f - \langle f, \mathbf{w}_1 \rangle \mathbf{w}_1 - \langle f, \mathbf{w}_2 \rangle \mathbf{w}_2 - \cdots - \langle f, \mathbf{w}_n \rangle \mathbf{w}_n,$$

you can see that $f - g$ is orthogonal to each $\mathbf{w}_i$, which in turn implies that it is orthogonal to each vector in W. In particular, $f - g$ is orthogonal to $g - \mathbf{w}$. This allows you to apply the Pythagorean Theorem to the vector sum

$$f - \mathbf{w} = (f - g) + (g - \mathbf{w})$$

to conclude that

$$\|f - \mathbf{w}\|^2 = \|f - g\|^2 + \|g - \mathbf{w}\|^2.$$

Therefore, it follows that

$$\|f - g\|^2 \le \|f - \mathbf{w}\|^2,$$

which then implies that

$$\|f - g\| \le \|f - \mathbf{w}\|.$$

Now observe how Theorem 5.19 can be used to produce the least squares approximation obtained in Example 4. First apply the Gram-Schmidt orthonormalization process to the standard basis $\{1, x\}$ to obtain the orthonormal basis $B = \{1, \sqrt{3}(2x - 1)\}$. Then, by Theorem 5.19, the least squares approximation for e^x in the subspace of all linear functions is given by

$$g(x) = \langle e^x, 1 \rangle(1) + \langle e^x, \sqrt{3}(2x - 1) \rangle \sqrt{3}(2x - 1)$$

$$= \int_0^1 e^x \, dx + \sqrt{3}(2x - 1) \int_0^1 \sqrt{3} e^x (2x - 1) \, dx$$

$$= \int_0^1 e^x \, dx + 3(2x - 1) \int_0^1 e^x (2x - 1) \, dx$$

$$= (e - 1) + 3(2x - 1)(3 - e)$$

$$= 4e - 10 + (18 - 6e)x,$$

which agrees with the result obtained in Example 4.

EXAMPLE 6   *Finding a Least Squares Approximation*

Find the least squares approximation for $f(x) = \sin x$, $0 \le x \le \pi$, with respect to the subspace W of quadratic functions.

Solution   To use Theorem 5.19, apply the Gram-Schmidt orthonormalization process to the standard basis for W, $\{1, x, x^2\}$, to obtain the orthonormal basis

$$B = \{\mathbf{w}_1, \mathbf{w}_2, \mathbf{w}_3\}$$

$$= \left\{ \frac{1}{\sqrt{\pi}}, \frac{\sqrt{3}}{\pi\sqrt{\pi}}(2x - \pi), \frac{\sqrt{5}}{\pi^2\sqrt{\pi}}(6x^2 - 6\pi x + \pi^2) \right\}.$$

The least squares approximating function g is given by

$$g(x) = \langle f, \mathbf{w}_1 \rangle \mathbf{w}_1 + \langle f, \mathbf{w}_2 \rangle \mathbf{w}_2 + \langle f, \mathbf{w}_3 \rangle \mathbf{w}_3,$$

and you have

$$\langle f, \mathbf{w}_1 \rangle = \frac{1}{\sqrt{\pi}} \int_0^\pi \sin x \, dx = \frac{2}{\sqrt{\pi}}$$

$$\langle f, \mathbf{w}_2 \rangle = \frac{\sqrt{3}}{\pi\sqrt{\pi}} \int_0^\pi \sin x (2x - \pi) \, dx = 0$$

$$\langle f, \mathbf{w}_3 \rangle = \frac{\sqrt{5}}{\pi^2\sqrt{\pi}} \int_0^\pi \sin x (6x^2 - 6\pi x + \pi^2) \, dx$$

$$= \frac{2\sqrt{5}}{\pi^2\sqrt{\pi}}(\pi^2 - 12).$$

Therefore, g is given by

$$g(x) = \frac{2}{\pi} + \frac{10(\pi^2 - 12)}{\pi^5}(6x^2 - 6\pi x + \pi^2)$$

$$\approx -0.4177x^2 + 1.3122x - 0.0505.$$

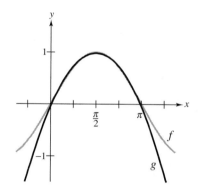

Figure 5.35

The graphs of f and g are shown in Figure 5.35.

Fourier Approximations (Calculus)

You will now look at a special type of least squares approximation called a **Fourier approximation.** For this approximation, consider functions of the form

$$g(x) = \frac{a_0}{2} + a_1 \cos x + \cdots + a_n \cos nx + b_1 \sin x + \cdots + b_n \sin nx$$

in the subspace W of $C[0, 2\pi]$ spanned by the basis

$$S = \{1, \cos x, \cos 2x, \ldots, \cos nx, \sin x, \sin 2x, \ldots, \sin nx\}.$$

These $2n + 1$ vectors are *orthogonal* in the inner product space $C[0, 2\pi]$ because

$$\langle f, g \rangle = \int_0^{2\pi} f(x)g(x)\, dx = 0, \quad f \neq g,$$

as demonstrated in Example 3 in Section 5.3. Moreover, by normalizing each function in this basis, you obtain the orthonormal basis

$$B = \{\mathbf{w}_0, \mathbf{w}_1, \ldots, \mathbf{w}_n, \mathbf{w}_{n+1}, \ldots, \mathbf{w}_{2n}\}$$

$$= \left\{ \frac{1}{\sqrt{2\pi}}, \frac{1}{\sqrt{\pi}} \cos x, \ldots, \frac{1}{\sqrt{\pi}} \cos nx, \frac{1}{\sqrt{\pi}} \sin x, \ldots, \frac{1}{\sqrt{\pi}} \sin nx \right\}.$$

With this orthonormal basis, you can apply Theorem 5.19 to write

$$g(x) = \langle f, \mathbf{w}_0 \rangle \mathbf{w}_0 + \langle f, \mathbf{w}_1 \rangle \mathbf{w}_1 + \cdots + \langle f, \mathbf{w}_{2n} \rangle \mathbf{w}_{2n}.$$

The coefficients $a_0, a_1, \ldots, a_n, b_1, \ldots, b_n$ for $g(x)$ in the equation

$$g(x) = \frac{a_0}{2} + a_1 \cos x + \cdots + a_n \cos nx + b_1 \sin x + \cdots + b_n \sin nx$$

are given by the following integrals.

$$a_0 = \langle f, \mathbf{w}_0 \rangle \frac{2}{\sqrt{2\pi}} = \frac{2}{\sqrt{2\pi}} \int_0^{2\pi} f(x) \frac{1}{\sqrt{2\pi}}\, dx = \frac{1}{\pi} \int_0^{2\pi} f(x)\, dx$$

$$a_1 = \langle f, \mathbf{w}_1 \rangle \frac{1}{\sqrt{\pi}} = \frac{1}{\sqrt{\pi}} \int_0^{2\pi} f(x) \frac{1}{\sqrt{\pi}} \cos x\, dx = \frac{1}{\pi} \int_0^{2\pi} f(x) \cos x\, dx$$

$$\vdots$$

$$a_n = \langle f, \mathbf{w}_n \rangle \frac{1}{\sqrt{\pi}} = \frac{1}{\sqrt{\pi}} \int_0^{2\pi} f(x) \frac{1}{\sqrt{\pi}} \cos nx\, dx = \frac{1}{\pi} \int_0^{2\pi} f(x) \cos nx\, dx$$

$$b_1 = \langle f, \mathbf{w}_{n+1} \rangle \frac{1}{\sqrt{\pi}} = \frac{1}{\sqrt{\pi}} \int_0^{2\pi} f(x) \frac{1}{\sqrt{\pi}} \sin x\, dx = \frac{1}{\pi} \int_0^{2\pi} f(x) \sin x\, dx$$

$$\vdots$$

$$b_n = \langle f, \mathbf{w}_{2n} \rangle \frac{1}{\sqrt{\pi}} = \frac{1}{\sqrt{\pi}} \int_0^{2\pi} f(x) \frac{1}{\sqrt{\pi}} \sin nx\, dx = \frac{1}{\pi} \int_0^{2\pi} f(x) \sin nx\, dx$$

The function $g(x)$ is called the **nth-order Fourier approximation** of f on the interval $[0, 2\pi]$. Like Fourier coefficients, this function is named after the French mathematician Jean-Baptiste Joseph Fourier (1768–1830). This brings you to Theorem 5.20.

Theorem 5.20

Fourier Approximation

On the interval $[0, 2\pi]$, the least squares approximation of a continuous function f with respect to the vector space spanned by $\{1, \cos x, \ldots, \cos nx, \sin x, \ldots, \sin nx\}$ is given by

$$g(x) = \frac{a_0}{2} + a_1 \cos x + \cdots + a_n \cos nx + b_1 \sin x + \cdots + b_n \sin nx,$$

where the **Fourier coefficients** $a_0, a_1, \ldots, a_n, b_1, \ldots, b_n$ are

$$a_0 = \frac{1}{\pi} \int_0^{2\pi} f(x)\, dx$$

$$a_j = \frac{1}{\pi} \int_0^{2\pi} f(x) \cos jx\, dx, \quad j = 1, 2, \ldots, n$$

$$b_j = \frac{1}{\pi} \int_0^{2\pi} f(x) \sin jx\, dx, \quad j = 1, 2, \ldots, n.$$

EXAMPLE 7   Finding a Fourier Approximation

Find the third-order Fourier approximation of $f(x) = x$, $\quad 0 \le x \le 2\pi$.

Solution   Using Theorem 5.20, you have

$$g(x) = \frac{a_0}{2} + a_1 \cos x + a_2 \cos 2x + a_3 \cos 3x + b_1 \sin x + b_2 \sin 2x + b_3 \sin 3x,$$

where

$$a_0 = \frac{1}{\pi} \int_0^{2\pi} x\, dx = \frac{1}{\pi} 2\pi^2 = 2\pi$$

$$a_j = \frac{1}{\pi} \int_0^{2\pi} x \cos jx\, dx = \left[\frac{1}{\pi j^2} \cos jx + \frac{x}{\pi j} \sin jx \right]_0^{2\pi} = 0$$

$$b_j = \frac{1}{\pi} \int_0^{2\pi} x \sin jx\, dx = \left[\frac{1}{\pi j^2} \sin jx - \frac{x}{\pi j} \cos jx \right]_0^{2\pi} = -\frac{2}{j}.$$

This implies that $b_1 = -2$, $b_2 = -\frac{2}{2} = -1$, and $b_3 = -\frac{2}{3}$. Therefore, you have

$$g(x) = \frac{2\pi}{2} - 2 \sin x - \sin 2x - \frac{2}{3} \sin 3x$$

$$= \pi - 2 \sin x - \sin 2x - \frac{2}{3} \sin 3x.$$

The graphs of f and g are compared in Figure 5.36.

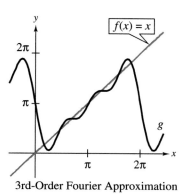

3rd-Order Fourier Approximation

Figure 5.36

In Example 7 the general pattern for the Fourier coefficients appears to be $a_0 = 2\pi$, $a_1 = a_2 = \cdots = a_n = 0$, and

$$b_1 = -\frac{2}{1}, \; b_2 = -\frac{2}{2}, \; \ldots, \; b_n = -\frac{2}{n}.$$

Thus the nth-order Fourier approximation of $f(x) = x$ is

$$g(x) = \pi - 2\left(\sin x + \frac{1}{2}\sin 2x + \frac{1}{3}\sin 3x + \cdots + \frac{1}{n}\sin nx\right).$$

As n increases, the Fourier approximation improves. For instance, Figure 5.37 shows the fourth- and fifth-order Fourier approximations of $f(x) = x$, $0 \le x \le 2\pi$.

4th-Order Fourier Approximation

5th-Order Fourier Approximation

Figure 5.37

In advanced courses it is shown that as $n \to \infty$, the approximation error $\|f - g\|$ approaches zero for all x in the interval $(0, 2\pi)$. The infinite *series* for $g(x)$ is called a **Fourier series.**

EXAMPLE 8   *Finding a Fourier Approximation*

Find the fourth-order Fourier approximation of $f(x) = |x - \pi|$, $0 \le x \le 2\pi$.

Solution   Using Theorem 5.20, find the Fourier coefficients as follows.

$$a_0 = \frac{1}{\pi}\int_0^{2\pi} |x - \pi| \, dx = \pi$$

$$a_j = \frac{1}{\pi}\int_0^{2\pi} |x - \pi| \cos jx \, dx$$

$$= \frac{2}{\pi}\int_0^{\pi} (\pi - x) \cos jx \, dx$$

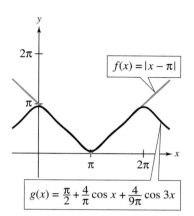

$$= \frac{2}{\pi j^2}(1 - \cos j\pi)$$

$$b_j = \frac{1}{\pi} \int_0^{2\pi} |x - \pi| \sin jx \, dx = 0$$

Thus, $a_1 = 4/\pi$, $a_2 = 0$, $a_3 = 4/9\pi$, and $a_4 = 0$, which means that the fourth-order Fourier approximation of f is

$$g(x) = \frac{\pi}{2} + \frac{4}{\pi} \cos x + \frac{4}{9\pi} \cos 3x.$$

The graphs of f and g are compared in Figure 5.38.

Figure 5.38

SECTION 5.5 Exercises

The Cross Product of Two Vectors in Space

In Exercises 1–8, find $\mathbf{u} \times \mathbf{v}$ and show that it is orthogonal to both $\mathbf{u}$ and $\mathbf{v}$.

1. $\mathbf{u} = (1, 0, 0)$, $\mathbf{v} = (0, 1, 0)$
2. $\mathbf{u} = (0, 1, 0)$, $\mathbf{v} = (1, 0, 0)$
3. $\mathbf{u} = \mathbf{j}$, $\mathbf{v} = \mathbf{k}$
4. $\mathbf{u} = \mathbf{k}$, $\mathbf{v} = \mathbf{j}$
5. $\mathbf{u} = \mathbf{i} + \mathbf{j} + \mathbf{k}$, $\mathbf{v} = 2\mathbf{i} + \mathbf{j} - \mathbf{k}$
6. $\mathbf{u} = \mathbf{i} - 2\mathbf{j} + \mathbf{k}$, $\mathbf{v} = -\mathbf{i} + 3\mathbf{j} - 2\mathbf{k}$
7. $\mathbf{u} = (2, -3, 1)$, $\mathbf{v} = (1, -2, 1)$
8. $\mathbf{u} = (4, 1, 0)$, $\mathbf{v} = (3, 2, -2)$

In Exercises 9–12, use a graphing utility with vector capabilities to find $\mathbf{u} \times \mathbf{v}$ and then show that it is orthogonal to both $\mathbf{u}$ and $\mathbf{v}$.

9. $\mathbf{u} = (1, 2, -1)$, $\mathbf{v} = (2, 1, 2)$
10. $\mathbf{u} = (0, 1, -1)$, $\mathbf{v} = (1, 2, 0)$
11. $\mathbf{u} = 2\mathbf{i} - \mathbf{j} + \mathbf{k}$, $\mathbf{v} = \mathbf{i} - 2\mathbf{j} + \mathbf{k}$
12. $\mathbf{u} = 2\mathbf{i} + \mathbf{j} - \mathbf{k}$, $\mathbf{v} = \mathbf{i} - \mathbf{j} + 2\mathbf{k}$

In Exercises 13–16, find the area of the parallelogram that has the vectors as adjacent sides.

13. $\mathbf{u} = \mathbf{j}$, $\mathbf{v} = \mathbf{j} + \mathbf{k}$
14. $\mathbf{u} = \mathbf{i} + \mathbf{j} + \mathbf{k}$, $\mathbf{v} = \mathbf{j} + \mathbf{k}$

15. $\mathbf{u} = (3, 2, -1)$, $\mathbf{v} = (1, 2, 3)$
16. $\mathbf{u} = (2, -1, 0)$, $\mathbf{v} = (-1, 2, 0)$

In Exercises 17–19, find $\mathbf{u} \cdot (\mathbf{v} \times \mathbf{w})$. This quantity is called the **triple scalar product** of $\mathbf{u}$, $\mathbf{v}$, and $\mathbf{w}$.

17. $\mathbf{u} = \mathbf{i}$, $\mathbf{v} = \mathbf{j}$, $\mathbf{w} = \mathbf{k}$
18. $\mathbf{u} = (1, 1, 1)$, $\mathbf{v} = (2, 1, 0)$, $\mathbf{w} = (0, 0, 1)$
19. $\mathbf{u} = (2, 0, 1)$, $\mathbf{v} = (0, 3, 0)$, $\mathbf{w} = (0, 0, 1)$

20. Show that the volume of a parallelepiped having $\mathbf{u}$, $\mathbf{v}$, and $\mathbf{w}$ as adjacent sides is given by $|\mathbf{u} \cdot (\mathbf{v} \times \mathbf{w})|$.

21. Use the result of Exercise 20 to find the volume of the parallelepiped with $\mathbf{u} = \mathbf{i} + \mathbf{j}$, $\mathbf{v} = \mathbf{j} + \mathbf{k}$, and $\mathbf{w} = \mathbf{i} + 2\mathbf{k}$ as adjacent edges. (See Figure 5.39.)

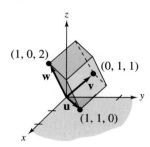

Figure 5.39

22. Find the volume of the parallelepiped shown in Figure 5.40, with **u**, **v**, and **w** as adjacent sides.

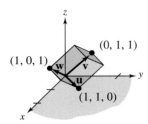

Figure 5.40

23. Prove that $c\mathbf{u} \times \mathbf{v} = c(\mathbf{u} \times \mathbf{v}) = \mathbf{u} \times c\mathbf{v}$.

24. Prove that $\mathbf{u} \times (\mathbf{v} + \mathbf{w}) = (\mathbf{u} \times \mathbf{v}) + (\mathbf{u} \times \mathbf{w})$.

25. Prove that $\mathbf{u} \times \mathbf{0} = \mathbf{0} \times \mathbf{u} = \mathbf{0}$.

26. Prove that $\mathbf{u} \times \mathbf{u} = \mathbf{0}$.

27. Prove that $\mathbf{u} \cdot (\mathbf{v} \times \mathbf{w}) = (\mathbf{u} \times \mathbf{v}) \cdot \mathbf{w}$.

28. Prove that $\mathbf{u} \times \mathbf{v}$ is orthogonal to both **u** and **v**.

29. Prove that the angle θ between **u** and **v** is given by $\|\mathbf{u} \times \mathbf{v}\| = \|\mathbf{u}\|\, \|\mathbf{v}\| \sin \theta$.

30. Prove that $\mathbf{u} \times \mathbf{v} = \mathbf{0}$ if and only if **u** and **v** are parallel.

31. Prove **Lagrange's Identity:**
$\|\mathbf{u} \times \mathbf{v}\|^2 = \|\mathbf{u}\|^2 \|\mathbf{v}\|^2 - (\mathbf{u} \cdot \mathbf{v})^2$.

32. (a) Prove that $\mathbf{u} \times (\mathbf{v} \times \mathbf{w}) = (\mathbf{u} \cdot \mathbf{w})\mathbf{v} - (\mathbf{u} \cdot \mathbf{v})\mathbf{w}$.
(b) Find an example for which $\mathbf{u} \times (\mathbf{v} \times \mathbf{w}) \neq (\mathbf{u} \times \mathbf{v}) \times \mathbf{w}$.

Least Squares Approximations (Calculus)

In Exercises 33–38, (a) find the *linear* least squares approximating function g for the function f and (b) sketch the graphs of f and g.

33. $f(x) = x^2, \quad 0 \le x \le 1$

34. $f(x) = \sqrt{x}, \quad 1 \le x \le 4$

35. $f(x) = e^{2x}, \quad 0 \le x \le 1$

36. $f(x) = \cos x, \quad 0 \le x \le \pi$

37. $f(x) = \sin x, \quad 0 \le x \le \pi/2$

38. $f(x) = \sin x, \quad -\pi/2 \le x \le \pi/2$

In Exercises 39–42, (a) find the *quadratic* least squares approximating function g for the function f and (b) graph f and g.

39. $f(x) = x^3, \quad 0 \le x \le 1$

40. $f(x) = \sqrt{x}, \quad 1 \le x \le 4$

41. $f(x) = \sin x, \quad 0 \le x \le \pi$

42. $f(x) = \sin x, \quad -\pi/2 \le x \le \pi/2$

Fourier Approximations (Calculus)

In Exercises 43–52, find the Fourier approximation of specified order for the function on the interval $[0, 2\pi]$.

43. $f(x) = \pi - x, \quad$ third order

44. $f(x) = \pi - x, \quad$ fourth order

45. $f(x) = (x - \pi)^2, \quad$ third order

46. $f(x) = (x - \pi)^2, \quad$ fourth order

47. $f(x) = e^{-x}, \quad$ first order

48. $f(x) = e^{-x}, \quad$ second order

49. $f(x) = 1 + x, \quad$ third order

50. $f(x) = 1 + x, \quad$ fourth order

51. $f(x) = 2 \sin x \cos x, \quad$ fourth order

52. $f(x) = \sin^2 x, \quad$ fourth order

53. Use the results of Exercises 43 and 44 to find the nth-order Fourier approximation of $f(x) = \pi - x$ on the interval $[0, 2\pi]$.

54. Use the results of Exercises 45 and 46 to find the nth-order Fourier approximation of $f(x) = (x - \pi)^2$ on the interval $[0, 2\pi]$.

CHAPTER 5 **Summary**

After studying this chapter, you should have acquired the following skills.

◼ For a given vector **v**, find the length of **v**, a vector **u** with the same length in the same direction as **v**, and a unit vector in the same or opposite direction as **v**.

◼ For two given vectors **u** and **v**, find the distance between **u** and **v**, the dot product of **u** and **v**, and the angle θ between **u** and **v**.

◼ For two given vectors **u** and **v**, verify the Cauchy-Schwarz Inequality for **u** and **v**, verify the Triangle Inequality for **u** and **v**, and verify the Pythagorean Theorem for **u** and **v**.

■ Determine all vectors that are orthogonal to a given vector.

■ Determine whether two given vectors are orthogonal, parallel, or neither.

■ Determine whether a function defines an inner product on R^n, $M_{n,n}$, or P_n, and find the inner product as defined for two given vectors $\langle \mathbf{u}, \mathbf{v} \rangle$ in R^n, $M_{n,n}$, or P_n.

■ Find the projection of a vector onto a given vector or subspace.

■ Determine whether a given set of vectors in R^n is orthogonal, orthonormal, or neither.

■ Find the coordinates of $\mathbf{x}$ relative to the orthonormal basis in R^n.

■ Apply the Gram-Schmidt orthonormalization process to find an orthonormal basis for a given basis, subspace, or inner product space.

■ Find an orthonormal basis for the solution space of a homogeneous system of linear equations.

■ Determine whether given subspaces are orthogonal.

■ Find the orthogonal complement of a given subspace.

■ Find the least squares solution of a system $A\mathbf{x} = \mathbf{b}$.

■ Find the least squares regression line for a given set of data points.

■ For two given vectors $\mathbf{u}$ and $\mathbf{v}$, find the cross product of $\mathbf{u}$ and $\mathbf{v}$, and show that the cross product of $\mathbf{u}$ and $\mathbf{v}$ is orthogonal to both vectors.

■ Find the linear or quadratic least squares approximating function for a given function.

■ Find the nth-order Fourier approximation for a given function.

CHAPTER 5   Review Exercises

In Exercises 1–4, find (a) $\|\mathbf{u}\|$, (b) $\|\mathbf{v}\|$, (c) $\mathbf{u} \cdot \mathbf{v}$, and (d) $d(\mathbf{u}, \mathbf{v})$.

1. $\mathbf{u} = (1, 2)$, $\mathbf{v} = (4, 1)$

2. $\mathbf{u} = (1, -1, 2)$, $\mathbf{v} = (2, 3, 1)$

3. $\mathbf{u} = (1, -2, 2, 0)$, $\mathbf{v} = (2, -1, 0, 2)$

4. $\mathbf{u} = (1, -1, 0, 1, 1)$, $\mathbf{v} = (0, 1, -2, 2, 1)$

In Exercises 5 and 6, find $\|\mathbf{v}\|$ and find a unit vector in the direction of $\mathbf{v}$.

5. $\mathbf{v} = (5, 3, -2)$ **6.** $\mathbf{v} = (1, -2, 1)$

In Exercises 7–10, find the angle between $\mathbf{u}$ and $\mathbf{v}$.

7. $\mathbf{u} = (2, 2)$, $\mathbf{v} = (-3, 3)$

8. $\mathbf{u} = \left(\cos \dfrac{3\pi}{4}, \sin \dfrac{3\pi}{4} \right)$, $\mathbf{v} = \left(\cos \dfrac{2\pi}{3}, \sin \dfrac{2\pi}{3} \right)$

9. $\mathbf{u} = (10, -5, 15)$, $\mathbf{v} = (-2, 1, -3)$

10. $\mathbf{u} = (1, 0, -3, 0)$, $\mathbf{v} = (2, -2, 1, 1)$

In Exercises 11–14, find $\text{proj}_{\mathbf{v}}\mathbf{u}$.

11. $\mathbf{u} = (2, 4)$, $\mathbf{v} = (1, -5)$

12. $\mathbf{u} = (2, 3)$, $\mathbf{v} = (0, 4)$

13. $\mathbf{u} = (0, -1, 2)$, $\mathbf{v} = (3, 2, 4)$

14. $\mathbf{u} = (1, 2, -1)$, $\mathbf{v} = (0, 2, 3)$

15. For $\mathbf{u} = \left(2, -\frac{1}{2}, 1 \right)$ and $\mathbf{v} = \left(\frac{3}{2}, 2, -1 \right)$, (a) find the inner product given by $\langle \mathbf{u}, \mathbf{v} \rangle = u_1 v_1 + 2u_2 v_2 + 3u_3 v_3$, and (b) use this inner product to find the distance between $\mathbf{u}$ and $\mathbf{v}$.

16. For $\mathbf{u} = \left(0, 3, \frac{1}{3} \right)$ and $\mathbf{v} = \left(\frac{4}{3}, 1, -3 \right)$, (a) find the inner product given by $\langle \mathbf{u}, \mathbf{v} \rangle = 2u_1 v_1 + u_2 v_2 + 2u_3 v_3$, and (b) use this inner product to find the distance between $\mathbf{u}$ and $\mathbf{v}$.

17. Verify the Triangle Inequality and the Cauchy-Schwarz Inequality for $\mathbf{u}$ and $\mathbf{v}$ given in Exercise 15. (Use the inner product given in Exercise 15.)

18. Verify the Triangle Inequality and the Cauchy-Schwarz Inequality for $\mathbf{u}$ and $\mathbf{v}$ given in Exercise 16. (Use the inner product given in Exercise 16.)

In Exercises 19 and 20, find all vectors that are orthogonal to **u**.

19. $\mathbf{u} = (0, -4, 3)$ **20.** $\mathbf{u} = (1, -2, 2, 1)$

In Exercises 21–24, use the Gram-Schmidt orthonormalization process to transform the basis into an orthonormal basis. (Use the Euclidean inner product.)

21. $B = \{(1, 1), (0, 1)\}$

22. $B = \{(3, 4), (1, 2)\}$

23. $B = \{(0, 3, 4), (1, 0, 0), (1, 1, 0)\}$

24. $B = \{(0, 0, 2), (0, 1, 1), (1, 1, 1)\}$

25. Let $B = \{(0, 2, -2), (1, 0, -2)\}$ be a basis for a subspace of R^3, and let $\mathbf{x} = (-1, 4, -2)$ be a vector in the subspace.

    (a) Write **x** as a linear combination of the vectors in B. That is, find the coordinates of **x** relative to B.

    (b) Use the Gram-Schmidt orthonormalization process to transform B into an orthonormal set B'.

    (c) Write **x** as a linear combination of the vectors in B'. That is, find the coordinates of **x** relative to B'.

26. Repeat Exercise 25 for $B = \{(-1, 2, 2), (1, 0, 0)\}$ and $\mathbf{x} = (-3, 4, 4)$.

Calculus   In Exercises 27–30, let f and g be functions in the vector space $C[a, b]$ with inner product

$$\langle f, g \rangle = \int_a^b f(x)g(x)\, dx.$$

27. Let $f(x) = x$ and $g(x) = x^2$ be vectors in $C[0, 1]$.

    (a) Find $\langle f, g \rangle$.

    (b) Find $\|g\|$.

    (c) Find $d(f, g)$.

    (d) Orthonormalize the set $B = \{f, g\}$.

28. Let $f(x) = x + 2$ and $g(x) = 15x - 8$ be vectors in $C[0, 1]$.

    (a) Find $\langle f, g \rangle$.

    (b) Find $\langle -4f, g \rangle$.

    (c) Find $\|f\|$.

    (d) Orthonormalize the set $B = \{f, g\}$.

29. Show that $f(x) = \sqrt{1 - x^2}$ and $g(x) = 2x\sqrt{1 - x^2}$ are orthogonal in $C[-1, 1]$.

30. Apply the Gram-Schmidt orthonormalization process to the following set in $C[-\pi, \pi]$.

$$S = \{1, \cos x, \sin x, \cos 2x, \sin 2x, \ldots, \cos nx, \sin nx\}$$

31. Find an orthonormal basis for the following subspace of Euclidean 3-space.

$$W = \{(x_1, x_2, x_3): x_1 + x_2 + x_3 = 0\}$$

32. Find an orthonormal basis for the solution space of the homogeneous system of linear equations.

$$x + y - z + w = 0$$
$$2x - y + z + 2w = 0$$

Calculus   In Exercises 33 and 34, (a) find the inner product, (b) determine whether the vectors are orthogonal, and (c) verify the Cauchy-Schwarz Inequality for the vectors.

33. $f(x) = x$, $g(x) = \dfrac{1}{x^2 + 1}$, $\langle f, g \rangle = \displaystyle\int_{-1}^1 f(x)g(x)\, dx$

34. $f(x) = x$, $g(x) = 4x^2$, $\langle f, g \rangle = \displaystyle\int_0^1 f(x)g(x)\, dx$

35. Prove that if **u** and **v** are vectors in an inner product space such that $\|\mathbf{u}\| \le 1$ and $\|\mathbf{v}\| \le 1$, then $|\langle \mathbf{u}, \mathbf{v} \rangle| \le 1$.

36. Prove that if **u** and **v** are vectors in an inner product space V, then

$$\big| \|\mathbf{u}\| - \|\mathbf{v}\| \big| \le \|\mathbf{u} \pm \mathbf{v}\|.$$

37. Let V be an m-dimensional subspace of R^n such that $m < n$. Prove that any vector **u** in R^n can be uniquely written in the form $\mathbf{u} = \mathbf{v} + \mathbf{w}$, where **v** is in V and **w** is orthogonal to every vector in V.

38. Let V be the two-dimensional subspace of R^4 spanned by $(0, 1, 0, 1)$ and $(0, 2, 0, 0)$. Write the vector $\mathbf{u} = (1, 1, 1, 1)$ in the form $\mathbf{u} = \mathbf{v} + \mathbf{w}$, where **v** is in V and **w** is orthogonal to every vector in V.

39. Let $\{\mathbf{u}_1, \mathbf{u}_2, \ldots, \mathbf{u}_m\}$ be an orthonormal subset of R^n, and let **v** be any vector in R^n. Prove that

$$\|\mathbf{v}\|^2 \ge \sum_{i=1}^m (\mathbf{v} \cdot \mathbf{u}_i)^2.$$

(This inequality is called **Bessel's Inequality.**)

40. Let $\{x_1, x_2, \ldots, x_n\}$ be a set of real numbers. Use the Cauchy-Schwarz Inequality to prove that

$$(x_1 + x_2 + \cdots + x_n)^2 \le n(x_1^2 + x_2^2 + \cdots + x_n^2).$$

41. Let **u** and **v** be vectors in an inner product space V. Prove that $\|\mathbf{u} + \mathbf{v}\| = \|\mathbf{u} - \mathbf{v}\|$ if and only if **u** and **v** are orthogonal.

42. *Writing*   Let $\{\mathbf{u}_1, \mathbf{u}_2, \ldots, \mathbf{u}_n\}$ be a dependent set of vectors in an inner product space V. Describe the result of applying the Gram-Schmidt orthonormalization process to this set.

43. Find the orthogonal complement $S^\perp$ of the subspace S of R^3 spanned by the two column vectors of the matrix

$$A = \begin{bmatrix} 1 & 2 \\ 2 & 1 \\ 0 & -1 \end{bmatrix}.$$

44. Find the projection of the vector $\mathbf{v} = \begin{bmatrix} 1 & 0 & -2 \end{bmatrix}^T$ onto the subspace

$$S = \text{span}\left\{ \begin{bmatrix} 0 \\ -1 \\ 1 \end{bmatrix}, \begin{bmatrix} 0 \\ 1 \\ 1 \end{bmatrix} \right\}.$$

45. Find bases for the four fundamental subspaces of the matrix

$$A = \begin{bmatrix} 0 & 1 & 0 \\ 0 & -3 & 0 \\ 1 & 0 & 1 \end{bmatrix}.$$

46. Find the least squares regression line for the set of data points $\{(-2, 2), (-1, 1), (0, 1), (1, 3)\}$.

Graph the points and the line on the same set of axes.

Mathematical Modeling

47. The table shows the average retail prices, in dollars per pound, of selected food items from 1993 to 2000. Find an appropriate least squares regression model for each item. Let $x = 3$ correspond to 1993. Use the models to find the expected food prices in 2005. (Source: U.S. Census Bureau)

Year	1993	1994	1995	1996
Ice Cream, y_I	2.59	2.62	2.68	2.94
T-Bone Steak, y_T	5.77	5.86	5.92	5.87
Ground Beef, y_G	2.25	2.14	2.04	2.05

Year	1997	1998	1999	2000
Ice Cream, y_I	3.02	3.30	3.40	3.66
T-Bone Steak, y_T	6.07	6.40	6.71	6.82
Ground Beef, y_G	2.06	2.08	2.15	2.33

48. The table shows the average annual salary of a Major League baseball player from 1990 to 1999. Find an appropriate least squares regression model and use the model to predict the average annual salary in 2005. Let $x = 0$ correspond to 1990. (Source: Major League Baseball Players Association)

Year	1990	1991	1992	1993	1994
Salary (in millions of dollars), y	0.60	0.85	1.03	1.08	1.17

Year	1995	1996	1997	1998	1999
Salary (in millions of dollars), y	1.11	1.12	1.34	1.40	1.61

The Cross Product of Two Vectors in Space

In Exercises 49 and 50, find $\mathbf{u} \times \mathbf{v}$ and show that it is orthogonal to both $\mathbf{u}$ and $\mathbf{v}$.

49. $\mathbf{u} = (1, 1, 1), \quad \mathbf{v} = (1, 0, 0)$

50. $\mathbf{u} = \mathbf{j} + 6\mathbf{k}, \quad \mathbf{v} = \mathbf{i} - 2\mathbf{j} + \mathbf{k}$

51. Find the area of the parallelogram that has $\mathbf{u} = (1, 3, 0)$ and $\mathbf{v} = (-1, 0, 2)$ as adjacent sides.

52. Prove that $\|\mathbf{u} \times \mathbf{v}\| = \|\mathbf{u}\| \|\mathbf{v}\|$ if and only if $\mathbf{u}$ and $\mathbf{v}$ are orthogonal.

In Exercises 53 and 54, the volume of the parallelepiped having $\mathbf{u}$, $\mathbf{v}$, and $\mathbf{w}$ as adjacent sides is given by the **triple scalar product** $|\mathbf{u} \cdot (\mathbf{v} \times \mathbf{w})|$. Find the volume of the parallelepiped having the three vectors as adjacent sides.

53. $\mathbf{u} = (1, 0, 0), \quad \mathbf{v} = (0, 0, 1), \quad \mathbf{w} = (0, 1, 0)$

54. $\mathbf{u} = (1, 2, 1), \quad \mathbf{v} = (-1, -1, 0), \quad \mathbf{w} = (3, 4, -1)$

Least Squares Approximations (Calculus)

In Exercises 55–58, find the *linear* least squares approximating function g for the function f. Then sketch the graphs of f and g.

55. $f(x) = x^3, \quad -1 \le x \le 1$

56. $f(x) = x^3, \quad 0 \le x \le 2$

57. $f(x) = \sin 2x, \quad 0 \le x \le \pi/2$

58. $f(x) = \sin x \cos x, \quad 0 \le x \le \pi$

In Exercises 59 and 60, find the *quadratic* least squares approximating function g for the function f. Then sketch the graphs of f and g.

59. $f(x) = \sqrt{x}, 0 \le x \le 1$ **60.** $f(x) = 1/x, 1 \le x \le 2$

Fourier Approximations (Calculus)

In Exercises 61 and 62, find the nth-order Fourier approximation of the function.

61. $f(x) = x^2, \quad -\pi \le x \le \pi$, first order

62. $f(x) = x, \quad -\pi \le x \le \pi$, second order

True or False? In Exercises 63 and 64, determine whether each statement is true or false. If a statement is true, give a reason or cite an appropriate statement in the text. If a statement is false, provide an example that shows that the statement is not true in all cases or cite an appropriate statement in the text.

63. (a) The cross product of two nonzero vectors in R^3 yields a vector that is orthogonal to the two given vectors that produced it.

    (b) The cross product of two nonzero vectors in R^3 is commutative.

(c) The least squares approximation of a function f is the function g (in the subspace W) closest to f in terms of the inner product $\langle f, g \rangle$.

64. (a) The vectors in R^3 $\mathbf{u} \times \mathbf{v}$ and $\mathbf{v} \times \mathbf{u}$ have equal lengths but opposite directions.

    (b) If $\mathbf{u}$ and $\mathbf{v}$ are two nonzero vectors in R^3, then $\mathbf{u}$ and $\mathbf{v}$ are parallel if and only if $\mathbf{u} \times \mathbf{v} = \mathbf{0}$.

    (c) A special type of least squares approximation, the Fourier approximation, is spanned by the basis $S = \{1, \cos x, \cos 2x, \ldots, \cos nx, \sin x, \sin 2x, \ldots, \sin nx\}$.

CHAPTER 5 | **Projects**

1 THE QR-FACTORIZATION

The Gram-Schmidt orthonormalization process leads to an important factorization of matrices called the **QR-factorization.** If A is an $m \times n$ matrix of rank n, then A can be expressed as the product $A = QR$ of an $m \times n$ matrix Q and an $n \times n$ matrix R, where Q has orthonormal columns and R is upper triangular.

The columns of A can be considered a basis for a subspace of R^m, and the columns of Q are the result of applying the Gram-Schmidt orthonormalization process to this set of column vectors.

Recall that in Example 7, Section 5.3, the Gram-Schmidt orthonormalization process was used on the column vectors of the matrix

$$A = \begin{bmatrix} 1 & 1 & 0 \\ 1 & 2 & 1 \\ 0 & 0 & 2 \end{bmatrix}.$$

An orthonormal basis for R^3 was obtained, which is labeled here as $\mathbf{q}_1, \mathbf{q}_2, \mathbf{q}_3$.

$$\mathbf{q}_1 = \left(\sqrt{2}/2, \sqrt{2}/2, 0 \right)$$
$$\mathbf{q}_2 = \left(-\sqrt{2}/2, \sqrt{2}/2, 0 \right)$$
$$\mathbf{q}_3 = (0, 0, 1)$$

These vectors form the columns of the matrix Q.

$$Q = \begin{bmatrix} \sqrt{2}/2 & -\sqrt{2}/2 & 0 \\ \sqrt{2}/2 & \sqrt{2}/2 & 0 \\ 0 & 0 & 1 \end{bmatrix}$$

The upper triangular matrix R consists of the following dot products.

$$\begin{bmatrix} \mathbf{v}_1 \cdot \mathbf{q}_1 & \mathbf{v}_2 \cdot \mathbf{q}_1 & \mathbf{v}_3 \cdot \mathbf{q}_1 \\ 0 & \mathbf{v}_2 \cdot \mathbf{q}_2 & \mathbf{v}_3 \cdot \mathbf{q}_2 \\ 0 & 0 & \mathbf{v}_3 \cdot \mathbf{q}_3 \end{bmatrix} = \begin{bmatrix} \sqrt{2} & 3\sqrt{2}/2 & \sqrt{2}/2 \\ 0 & \sqrt{2}/2 & \sqrt{2}/2 \\ 0 & 0 & 2 \end{bmatrix}$$

It is now an easy exercise to verify that $A = QR$.

In general, if A is an $m \times n$ matrix of rank n, then the QR-factorization of A can be constructed if you keep track of the dot products used in the Gram-Schmidt orthonormalization process as applied to the columns of A. The columns of the $m \times n$ matrix Q are the orthonormal vectors that result from the Gram-Schmidt orthonormalization process. The $n \times n$ upper triangular matrix R consists of certain dot products of the original column vectors $\mathbf{v}_i$ and the orthonormal column vectors $\mathbf{q}_i$. If the columns of the matrix A are denoted $\mathbf{v}_1, \mathbf{v}_2, \ldots, \mathbf{v}_n$ and the columns of Q are denoted $\mathbf{q}_1, \mathbf{q}_2, \ldots, \mathbf{q}_n$, then the QR-factorization of A is as follows.

$$A = QR$$

$$[\mathbf{v}_1 \quad \mathbf{v}_2 \quad \cdots \quad \mathbf{v}_n] = [\mathbf{q}_1 \quad \mathbf{q}_2 \quad \cdots \quad \mathbf{q}_n] \begin{bmatrix} \mathbf{v}_1 \cdot \mathbf{q}_1 & \mathbf{v}_2 \cdot \mathbf{q}_1 & \cdots & \mathbf{v}_n \cdot \mathbf{q}_1 \\ 0 & \mathbf{v}_2 \cdot \mathbf{q}_2 & \cdots & \mathbf{v}_n \cdot \mathbf{q}_2 \\ \vdots & \vdots & & \vdots \\ 0 & 0 & \cdots & \mathbf{v}_n \cdot \mathbf{q}_n \end{bmatrix}$$

1. Verify the matrix equation $A = QR$ for the preceding example.
2. Find the QR-factorization of each matrix.

(a) $A = \begin{bmatrix} 1 & 1 \\ 0 & 1 \\ 1 & 0 \end{bmatrix}$ (b) $A = \begin{bmatrix} 1 & 0 \\ 0 & 0 \\ 1 & 1 \\ 1 & 2 \end{bmatrix}$

(c) $A = \begin{bmatrix} 1 & 0 & 0 \\ 1 & 1 & 0 \\ 1 & 1 & 1 \end{bmatrix}$ (d) $A = \begin{bmatrix} 1 & 0 & -1 \\ 1 & 2 & 0 \\ 1 & 2 & 0 \\ 1 & 0 & 0 \end{bmatrix}$

3. Let $A = QR$ be the QR-factorization of the $m \times n$ matrix A of rank n. Show how the least squares problem can be solved using just matrix multiplication and back-substitution.
4. Use the result of Part 3 to solve the least squares problem $A\mathbf{x} = \mathbf{b}$ if A is the matrix from Part 2(a) and $\mathbf{b} = [-1 \quad 1 \quad -1]^T$.

The QR-factorization of a matrix forms the basis for many algorithms of linear algebra. Computer routines for the computation of eigenvalues (see Chapter 7) are based on this factorization, as are algorithms for computing the least squares regression line for a set of data points. However, it should be mentioned that, in practice, techniques other than the Gram-Schmidt orthonormalization process are actually used to compute the *QR*-factorization of a matrix.

2 ORTHOGONAL MATRICES AND CHANGE OF BASIS

Let $B = \{\mathbf{v}_1, \mathbf{v}_2, \ldots, \mathbf{v}_n\}$ be an ordered basis for the vector space V. Recall that the coordinate matrix of a vector $\mathbf{x} = c_1\mathbf{v}_1 + c_2\mathbf{v}_2 + \cdots + c_n\mathbf{v}_n$ in V is the column vector

$$[\mathbf{x}]_B = \begin{bmatrix} c_1 \\ c_2 \\ \vdots \\ c_n \end{bmatrix}.$$

If B' is another basis for V, then the transition matrix P from B' to B changes a coordinate matrix relative to B' into a coordinate matrix relative to B,

$$P[\mathbf{x}]_{B'} = [\mathbf{x}]_B.$$

The question you will explore now is whether there are transition matrices P that preserve the length of the coordinate matrix—that is, given $P[\mathbf{x}]_{B'} = [\mathbf{x}]_B$, does $\|[\mathbf{x}]_{B'}\| = \|[\mathbf{x}]_B\|$?

For example, consider the transition matrix from Example 5 in Section 4.7,

$$P = \begin{bmatrix} 3 & -2 \\ 2 & -1 \end{bmatrix}$$

relative to the bases for R^2,

$$B = \{(-3, 2), (4, -2)\} \qquad \text{and} \qquad B' = \{(-1, 2), (2, -2)\}.$$

If $\mathbf{x} = (-1, 2)$, then $[\mathbf{x}]_{B'} = [1 \quad 0]^T$ and $[\mathbf{x}]_B = P[\mathbf{x}]_{B'} = [3 \quad 2]^T$. Thus, using the Euclidean norm for R^2,

$$\|[\mathbf{x}]_{B'}\| = 1 \neq \sqrt{13} = \|[\mathbf{x}]_B\|.$$

You will see in this project that if the transition matrix P is **orthogonal,** then the norm of the coordinate vector will remain unchanged.

Definition of
Orthogonal Matrix

The square matrix P is **orthogonal** if it is invertible and $P^{-1} = P^T$.

1. Show that the matrix P defined previously is *not* orthogonal.
2. Show that for any real number θ, the matrix $\begin{bmatrix} \cos \theta & -\sin \theta \\ \sin \theta & \cos \theta \end{bmatrix}$ is orthogonal.
3. Show that a matrix is orthogonal if and only if its columns are pairwise orthogonal.
4. Prove that the inverse of an orthogonal matrix is orthogonal.
5. Is the sum of orthogonal matrices orthogonal? Is the product of orthogonal matrices orthogonal? Illustrate your answers with appropriate examples.
6. What is the determinant of an orthogonal matrix?
7. Prove that if P is an $m \times n$ orthogonal matrix, then $\|P\mathbf{x}\| = \|\mathbf{x}\|$ for all vectors $\mathbf{x}$ in R^n.
8. Verify the result of Part 7 using the bases $B = \{(1, 0), (0, 1)\}$ and

$$B' = \left\{ \left(\frac{-2}{\sqrt{5}}, \frac{1}{\sqrt{5}} \right), \left(\frac{1}{\sqrt{5}}, \frac{2}{\sqrt{5}} \right) \right\}.$$

CHAPTERS 4 & 5 Cumulative Test

Take this test as you would take a test in class. After you are done, check your work against the answers given in the back of the book.

1. Given the vectors $\mathbf{v} = (-1, 2)$ and $\mathbf{w} = (3, -1)$, sketch each vector.
    (a) $\mathbf{v} + \mathbf{w}$
    (b) $3\mathbf{v}$
    (c) $2\mathbf{v} - 4\mathbf{w}$

2. If possible, write $\mathbf{w} = (3, 2, 1)$ as a linear combination of the vectors $\mathbf{v}_1$, $\mathbf{v}_2$, and $\mathbf{v}_3$.

    $\mathbf{v}_1 = (1, 2, 0)$, $\mathbf{v}_2 = (-1, 0, 1)$, $\mathbf{v}_3 = (0, 3, 0)$

3. Prove that the set of all singular 2×2 matrices is not a vector space.

4. Determine whether the set is a subspace of R^4.

    $\{(x, x + y, y, -y): x, y \in R\}$

5. Determine whether the set is a subspace of R^3.

    $\{(x, xy, y): x, y \in R\}$

6. Determine whether the columns of the matrix A span R^4.

$$A = \begin{bmatrix} 1 & 2 & -1 & 0 \\ 1 & 3 & 0 & 2 \\ 0 & 0 & 1 & -1 \\ 1 & 0 & 0 & 1 \end{bmatrix}$$

7. (a) Define what it means to say that a set of vectors is *linearly independent.*
   (b) Determine whether the set S is linearly dependent or independent.
   $$S = \{(1, 0, 1, 0), (0, 3, 0, 1), (1, 1, 2, 2), (3, -4, 2, -3)\}$$

8. Find the dimension of and a basis for the subspace of $M_{3,3}$ consisting of all the 3×3 symmetric matrices.

9. (a) Define *basis* of a vector space.
   (b) Determine if the set is a basis for R^3.
   $$\{(1, 2, 1), (0, 1, 2), (2, 1, -3)\}$$

10. Find a basis for the solution space of $A\mathbf{x} = \mathbf{0}$ if

$$A = \begin{bmatrix} 1 & 1 & 0 & 0 \\ -2 & -2 & 0 & 0 \\ 0 & 0 & 1 & 1 \\ -1 & -1 & 0 & 0 \end{bmatrix}.$$

11. Find the coordinates $[\mathbf{v}]_B$ of the vector $\mathbf{v} = (1, 2, -3)$ relative to the basis
    $$B = \{(0, 1, 1), (1, 1, 0), (1, 0, 1)\}.$$

12. Find the transition matrix from the basis $B = \{(2, 1, 0), (1, 0, 0), (0, 1, 1)\}$ to the basis
    $$B' = \{(1, 0, 2), (1, 1, 1), (0, 1, 3)\}.$$

13. Let $\mathbf{u} = (1, 0, 2)$ and $\mathbf{v} = (-2, 1, 3)$.
    (a) Find $\|\mathbf{u}\|$.
    (b) Find the distance between $\mathbf{u}$ and $\mathbf{v}$.
    (c) Find $\mathbf{u} \cdot \mathbf{v}$.

14. Find the inner product of $f(x) = x^2$ and $g(x) = x - 2$ from $C[0, 1]$ using the integral inner product
    $$\langle f, g \rangle = \int_0^1 f(x)g(x)\,dx.$$

15. Use the Gram-Schmidt orthonormalization process to transform the following set of vectors into an orthonormal basis for R^3.
    $$\{(2, 0, 0), (1, 1, 1), (0, 1, 2)\}$$

16. Let $\mathbf{u} = (1, 2)$ and $\mathbf{v} = (-3, 2)$. Find $\text{proj}_{\mathbf{v}}\mathbf{u}$, and graph $\mathbf{u}$, $\mathbf{v}$, and $\text{proj}_{\mathbf{v}}\mathbf{u}$ on the same set of coordinate axes.

17. Find the four fundamental subspaces of the matrix

$$A = \begin{bmatrix} 0 & 1 & 1 & 0 \\ -1 & 0 & 0 & 1 \\ 1 & 1 & 1 & 1 \end{bmatrix}.$$

18. Find the orthogonal complement $S^\perp$ of the set

$$S = \text{span}\left(\begin{bmatrix} 1 \\ 0 \\ 1 \end{bmatrix}, \begin{bmatrix} -1 \\ 1 \\ 0 \end{bmatrix} \right).$$

19. Use the axioms for a vector space to prove that $0\mathbf{v} = \mathbf{0}$ for all vectors $\mathbf{v} \in V$.

20. Suppose that $\mathbf{x}_1, \ldots, \mathbf{x}_n$ are linearly independent vectors and $\mathbf{y}$ is a vector not in their span. Prove that the vectors $\mathbf{x}_1, \ldots, \mathbf{x}_n$ and $\mathbf{y}$ are linearly independent.

21. Let W be a subspace of an inner product space V. Prove that the set

    $W^\perp = \{\mathbf{v} \in V: \langle \mathbf{v}, \mathbf{w} \rangle = 0 \quad \text{for all} \quad \mathbf{w} \in W\}$

    is a subspace of V.

22. Find the least squares regression line for the points $\{(1, 1), (3, 0), (5, -3)\}$. Graph the points and the line.

23. The two matrices A and B are row-equivalent.

$$A = \begin{bmatrix} 2 & -4 & 0 & 1 & 7 & 11 \\ 1 & -2 & -1 & 1 & 9 & 12 \\ -1 & 2 & 1 & 3 & -5 & 16 \\ 4 & -8 & 1 & -1 & 6 & -2 \end{bmatrix}$$

$$B = \begin{bmatrix} 1 & -2 & 0 & 0 & 3 & 2 \\ 0 & 0 & 1 & 0 & -5 & -3 \\ 0 & 0 & 0 & 1 & 1 & 7 \\ 0 & 0 & 0 & 0 & 0 & 0 \end{bmatrix}$$

    (a) Find the rank of A.
    (b) Find a basis for the row space of A.
    (c) Find a basis for the column space of A.
    (d) Find a basis for the nullspace of A.
    (e) Is the last column of A in the span of the first three columns?
    (f) Are the first three columns of A linearly independent?
    (g) Is the last column of A in the span of columns 1, 3, and 4?
    (h) Are columns 1, 3, and 4 linearly dependent?

24. Let $\mathbf{u}$ and $\mathbf{v}$ be vectors in an inner product space V. Prove that $\|\mathbf{u} + \mathbf{v}\| = \|\mathbf{u} - \mathbf{v}\|$ if and only if $\mathbf{u}$ and $\mathbf{v}$ are orthogonal.

EMMY NOETHER

1 8 8 2 – 1 9 3 5

EMMY NOETHER, generally recognized as the leading woman mathematician in recent history, was born in Erlangen, Germany, where her father was a professor of mathematics.

At the beginning of the twentieth century it was still considered inappropriate for a woman to be interested in mathematics. Emmy Noether's talent for abstract reasoning, however, was so unusual that she was able to overcome the obstacles placed in her way. In 1907 she became the first woman to be granted a doctoral degree from a German university.

A few years after receiving her doctorate, Noether was invited to the University of Göttingen to work with Felix Klein and David Hilbert on the general theory of relativity. In spite of considerable opposition to the idea of granting a professorship to a woman, in 1922 Noether was finally made a professor at the university.

During the 1920s Noether published several papers on axiomatic systems, differential operators, and the representation of noncommutative algebras as linear transformations. Albert Einstein said of her, "In the realm of algebra in which the most gifted mathematicians have been busy for centuries, she discovered methods which have proved of enormous importance."

The rise of the National Socialist Party forced Noether to leave Germany in 1933. She emigrated to the United States, where she accepted a position as professor of mathematics at Bryn Mawr College in Philadelphia. During her year and a half at Bryn Mawr, Noether often traveled to Princeton University, where she delivered lectures at the Institute for Advanced Study.

CHAPTER 6 LINEAR TRANSFORMATIONS

6.1 | Introduction to Linear Transformations

In this chapter you will learn about functions that **map** a vector space V into a vector space W. This type of function is denoted by

$$T: V \to W.$$

The standard function terminology is used for such functions. For instance, V is called the **domain** of T, and W is called the **codomain** of T. If $\mathbf{v}$ is in V and $\mathbf{w}$ is in W such that

$$T(\mathbf{v}) = \mathbf{w},$$

then $\mathbf{w}$ is called the **image** of $\mathbf{v}$ under T. The set of all images of vectors in V is called the **range** of T, and the set of all $\mathbf{v}$ in V such that $T(\mathbf{v}) = \mathbf{w}$ is called the **preimage** of $\mathbf{w}$. (See Figure 6.1 on the following page.)

REMARK: For a vector
$\mathbf{v} = (v_1, v_2, \ldots, v_n)$ in R^n, it
would be technically correct to
use double parentheses to denote
$T(\mathbf{v})$ as $T(\mathbf{v}) = T((v_1, v_2, \ldots, v_n))$.
For convenience, however, one
set of parentheses is dropped,
producing

$$T(\mathbf{v}) = T(v_1, v_2, \ldots, v_n).$$

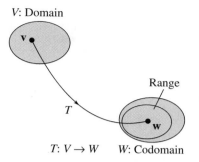

Figure 6.1

EXAMPLE 1   *A Function from R^2 into R^2*

For any vector $\mathbf{v} = (v_1, v_2)$ in R^2, let $T: R^2 \rightarrow R^2$ be defined by

$$T(v_1, v_2) = (v_1 - v_2, v_1 + 2v_2).$$

(a) Find the image of $\mathbf{v} = (-1, 2)$.
(b) Find the preimage of $\mathbf{w} = (-1, 11)$.

Solution   (a) For $\mathbf{v} = (-1, 2)$ you have

$$T(-1, 2) = (-1 - 2, -1 + 2(2)) = (-3, 3).$$

(b) If $T(\mathbf{v}) = (v_1 - v_2, v_1 + 2v_2) = (-1, 11)$, then

$$v_1 - v_2 = -1$$
$$v_1 + 2v_2 = 11.$$

This system of equations has the unique solution $v_1 = 3$ and $v_2 = 4$. Thus, the preimage of $(-1, 11)$ is the set in R^2 consisting of the single vector $(3, 4)$.

This chapter centers on functions (from one vector space to another) that preserve the operations of vector addition and scalar multiplication. Such functions are called **linear transformations.**

Definition of a Linear Transformation   Let V and W be vector spaces. The function $T: V \rightarrow W$ is called a **linear transformation** of V into W if the following two properties are true for all $\mathbf{u}$ and $\mathbf{v}$ in V and for any scalar c.
1. $T(\mathbf{u} + \mathbf{v}) = T(\mathbf{u}) + T(\mathbf{v})$
2. $T(c\mathbf{u}) = cT(\mathbf{u})$

A linear transformation is said to be *operation preserving*, because the same result occurs whether the operations of addition and scalar multiplication are performed before or after the linear transformation is applied. Although the same symbols are used to denote the vector operations in both V and W, you should note that the operations may be different, as indicated in the following diagram.

| Addition in V | Addition in W | Scalar multiplication in V | Scalar multiplication in W |

$$T(\mathbf{u} + \mathbf{v}) = T(\mathbf{u}) + T(\mathbf{v}) \qquad T(c\mathbf{u}) = cT(\mathbf{u})$$

EXAMPLE 2   *Verifying a Linear Transformation from R^2 into R^2*

Show that the function given in Example 1 is a linear transformation from R^2 into R^2.

$$T(v_1, v_2) = (v_1 - v_2, v_1 + 2v_2)$$

Solution   To show that the function T is a linear transformation, you must show that it preserves vector addition and scalar multiplication. To do this, let $\mathbf{v} = (v_1, v_2)$ and $\mathbf{u} = (u_1, u_2)$ be vectors in R^2 and let c be any real number. Then, using the properties of vector addition and scalar multiplication, you have the following.

1. Because $\mathbf{u} + \mathbf{v} = (u_1, u_2) + (v_1, v_2) = (u_1 + v_1, u_2 + v_2)$, you have

$$
\begin{aligned}
T(\mathbf{u} + \mathbf{v}) &= T(u_1 + v_1, u_2 + v_2) \\
&= ((u_1 + v_1) - (u_2 + v_2), (u_1 + v_1) + 2(u_2 + v_2)) \\
&= ((u_1 - u_2) + (v_1 - v_2), (u_1 + 2u_2) + (v_1 + 2v_2)) \\
&= (u_1 - u_2, u_1 + 2u_2) + (v_1 - v_2, v_1 + 2v_2) \\
&= T(\mathbf{u}) + T(\mathbf{v}).
\end{aligned}
$$

2. Because $c\mathbf{u} = c(u_1, u_2) = (cu_1, cu_2)$, you have

$$
\begin{aligned}
T(c\mathbf{u}) &= T(cu_1, cu_2) \\
&= (cu_1 - cu_2, cu_1 + 2cu_2) \\
&= c(u_1 - u_2, u_1 + 2u_2) \\
&= cT(\mathbf{u}).
\end{aligned}
$$

REMARK: A linear transformation $T: V \rightarrow V$ from a vector space into *itself* (as in Example 2) is called a **linear operator.**

Therefore, T is a linear transformation.

Most of the common functions studied in calculus are not linear transformations.

EXAMPLE 3   *Some Functions That Are Not Linear Transformations*

(a) $f(x) = \sin x$ is not a linear transformation from R into R because, in general,

$$\sin(x_1 + x_2) \neq \sin x_1 + \sin x_2.$$

For instance, $\sin[(\pi/2) + (\pi/3)] \neq \sin(\pi/2) + \sin(\pi/3)$.

(b) $f(x) = x^2$ is not a linear transformation from R into R because, in general,

$$(x_1 + x_2)^2 \neq x_1^2 + x_2^2.$$

For instance, $(1 + 2)^2 \neq 1^2 + 2^2.$

(c) $f(x) = x + 1$ is not a linear transformation from R into R because

$$f(x_1 + x_2) = x_1 + x_2 + 1$$

whereas

$$f(x_1) + f(x_2) = (x_1 + 1) + (x_2 + 1) = x_1 + x_2 + 2.$$

So $f(x_1 + x_2) \neq f(x_1) + f(x_2).$

REMARK: The function in Example 3(c) points out two uses of the term *linear*. In calculus, $f(x) = x + 1$ is called a linear function because its graph is a line. It is not a linear transformation from the vector space R into R, however, because it preserves neither vector addition nor scalar multiplication.

Two simple linear transformations are the **zero transformation** and the **identity transformation,** which are defined as follows.

1. $T(\mathbf{v}) = \mathbf{0}$, for all $\mathbf{v}$ Zero transformation $(T: V \rightarrow W)$
2. $T(\mathbf{v}) = \mathbf{v}$, for all $\mathbf{v}$ Identity transformation $(T: V \rightarrow V)$

The verifications of the linearity of these two transformations are left as exercises. (See Exercises 58 and 59.)

Note that the linear transformation in Example 2 has the property that the zero vector is mapped to itself. That is, $T(\mathbf{0}) = \mathbf{0}$. (Try checking this.) This property is true for all linear transformations, as stated in the next theorem.

| Theorem 6.1 | *Properties of Linear Transformations* |

Let T be a linear transformation from V into W, where $\mathbf{u}$ and $\mathbf{v}$ are in V. Then the following properties are true.

1. $T(\mathbf{0}) = \mathbf{0}$
2. $T(-\mathbf{v}) = -T(\mathbf{v})$
3. $T(\mathbf{u} - \mathbf{v}) = T(\mathbf{u}) - T(\mathbf{v})$
4. If $\mathbf{v} = c_1\mathbf{v}_1 + c_2\mathbf{v}_2 + \cdots + c_n\mathbf{v}_n,$

then

$$T(\mathbf{v}) = T(c_1\mathbf{v}_1 + c_2\mathbf{v}_2 + \cdots + c_n\mathbf{v}_n)$$
$$= c_1T(\mathbf{v}_1) + c_2T(\mathbf{v}_2) + \cdots + c_nT(\mathbf{v}_n).$$

Proof   To prove the first property, note that $0\mathbf{v} = \mathbf{0}$. Then it follows that

$$T(\mathbf{0}) = T(0\mathbf{v}) = 0T(\mathbf{v}) = \mathbf{0}.$$

The second property follows from $-\mathbf{v} = (-1)\mathbf{v}$, which implies that

$$T(-\mathbf{v}) = T[(-1)\mathbf{v}] = (-1)T(\mathbf{v}) = -T(\mathbf{v}).$$

The third property follows from $\mathbf{u} - \mathbf{v} = \mathbf{u} + (-\mathbf{v})$, which implies that

$$T(\mathbf{u} - \mathbf{v}) = T[\mathbf{u} + (-1)\mathbf{v}] = T(\mathbf{u}) + (-1)T(\mathbf{v}) = T(\mathbf{u}) - T(\mathbf{v}).$$

The proof of the fourth property is left to you.

Property 4 of Theorem 6.1 tells you that a linear transformation $T: V \to W$ is determined completely by its action on a basis of V. In other words, if $\{\mathbf{v}_1, \mathbf{v}_2, \ldots, \mathbf{v}_n\}$ is a basis for the vector space V and if $T(\mathbf{v}_1), T(\mathbf{v}_2), \ldots, T(\mathbf{v}_n)$ are given, then $T(\mathbf{v})$ is determined for *any* $\mathbf{v}$ in V. The use of this property is demonstrated in Example 4.

EXAMPLE 4   *Linear Transformations and Bases*

Let $T: R^3 \to R^3$ be a linear transformation such that

$$T(1, 0, 0) = (2, -1, 4)$$
$$T(0, 1, 0) = (1, 5, -2)$$
$$T(0, 0, 1) = (0, 3, 1).$$

Find $T(2, 3, -2)$.

Solution   Because $(2, 3, -2)$ can be written as

$$(2, 3, -2) = 2(1, 0, 0) + 3(0, 1, 0) - 2(0, 0, 1),$$

you can use Property 4 of Theorem 6.1 to write

$$\begin{aligned} T(2, 3, -2) &= 2T(1, 0, 0) + 3T(0, 1, 0) - 2T(0, 0, 1) \\ &= 2(2, -1, 4) + 3(1, 5, -2) - 2(0, 3, 1) \\ &= (7, 7, 0). \end{aligned}$$

Another advantage of Theorem 6.1 is that it provides a quick way to spot functions that are not linear transformations. That is, because all four conditions of the theorem must be true of a linear transformation, it follows that if any one of the properties is not satisfied for a function T, then the function is not a linear transformation. For example, the function given by

$$T(x_1, x_2) = (x_1 + 1, x_2)$$

is not a linear transformation from R^2 to R^2 because $T(0, 0) \neq (0, 0)$.

In the next example a matrix is used to define a linear transformation from R^2 into R^3. The vector $\mathbf{v} = (v_1, v_2)$ is written in the matrix form

$$\mathbf{v} = \begin{bmatrix} v_1 \\ v_2 \end{bmatrix}$$

so that it can be multiplied *on the left* by a matrix of order 3×2.

EXAMPLE 5   *A Linear Transformation Defined by a Matrix*

The function $T: R^2 \rightarrow R^3$ is defined as follows.

$$T(\mathbf{v}) = A\mathbf{v} = \begin{bmatrix} 3 & 0 \\ 2 & 1 \\ -1 & -2 \end{bmatrix} \begin{bmatrix} v_1 \\ v_2 \end{bmatrix}$$

(a) Find $T(\mathbf{v})$, where $\mathbf{v} = (2, -1)$.

(b) Show that T is a linear transformation from R^2 into R^3.

Solution   (a) Because $\mathbf{v} = (2, -1)$, you have

$$T(\mathbf{v}) = A\mathbf{v} = \begin{bmatrix} 3 & 0 \\ 2 & 1 \\ -1 & -2 \end{bmatrix} \begin{bmatrix} 2 \\ -1 \end{bmatrix} = \begin{bmatrix} 6 \\ 3 \\ 0 \end{bmatrix}.$$

$\underset{\text{Vector in } R^2}{\uparrow} \qquad \underset{\text{Vector in } R^3}{\uparrow}$

Therefore, you have $T(2, -1) = (6, 3, 0)$.

(b) Begin by observing that T does map a vector in R^2 to a vector in R^3. To show that T is a linear transformation, use the properties of matrix multiplication, as given in Theorem 2.3. For any vectors $\mathbf{u}$ and $\mathbf{v}$ in R^2, the distributive property of matrix multiplication over addition produces

$$T(\mathbf{u} + \mathbf{v}) = A(\mathbf{u} + \mathbf{v}) = A\mathbf{u} + A\mathbf{v} = T(\mathbf{u}) + T(\mathbf{v}).$$

Similarly, for any vector $\mathbf{u}$ in R^2 and any scalar c, the commutative property of scalar multiplication with matrix multiplication produces

$$T(c\mathbf{u}) = A(c\mathbf{u}) = c(A\mathbf{u}) = cT(\mathbf{u}).$$

Example 5 illustrates an important result regarding the representation of linear transformations from R^n into R^m. This result is presented in two stages. Theorem 6.2 on the next page states that every $m \times n$ matrix represents a linear transformation from R^n into R^m. Then, in Section 6.3, you will see the converse—that every linear transformation from R^n into R^m can be represented by an $m \times n$ matrix.

Note that in part (b) of Example 5 no reference is made to the specific matrix A. This verification therefore serves as a general proof that the function defined by any $m \times n$ matrix is a linear transformation from R^n into R^m.

Theorem 6.2	*The Linear Transformation Given by a Matrix*

Let A be an $m \times n$ matrix. The function T defined by

$$T(\mathbf{v}) = A\mathbf{v}$$

is a linear transformation from R^n into R^m. In order to conform to matrix multiplication with an $m \times n$ matrix, the vectors in R^n are represented by $n \times 1$ matrices and the vectors in R^m are represented by $m \times 1$ matrices.

REMARK: The $m \times n$ zero matrix corresponds to the zero transformation from R^n into R^m, and the $n \times n$ identity matrix I_n corresponds to the identity transformation from R^n into R^n.

Be sure you see that an $m \times n$ matrix A defines a linear transformation from R^n into R^m.

$$A\mathbf{v} = \begin{bmatrix} a_{11} & a_{12} & \cdots & a_{1n} \\ a_{21} & a_{22} & \cdots & a_{2n} \\ \vdots & \vdots & & \vdots \\ a_{m1} & a_{m2} & \cdots & a_{mn} \end{bmatrix} \begin{bmatrix} v_1 \\ v_2 \\ \vdots \\ v_n \end{bmatrix} = \begin{bmatrix} a_{11}v_1 + a_{12}v_2 + \cdots + a_{1n}v_n \\ a_{21}v_1 + a_{22}v_2 + \cdots + a_{2n}v_n \\ \vdots & \vdots & & \vdots \\ a_{m1}v_1 + a_{m2}v_2 + \cdots + a_{mn}v_n \end{bmatrix}$$

Vector
in R^n

Vector
in R^m

EXAMPLE 6 *Linear Transformations Given by Matrices*

The linear transformation $T: R^n \rightarrow R^m$ is defined by $T(\mathbf{v}) = A\mathbf{v}$. Find the dimensions of R^n and R^m for the linear transformation given by each matrix.

(a) $A = \begin{bmatrix} 0 & 1 & -1 \\ 2 & 3 & 0 \\ 4 & 2 & 1 \end{bmatrix}$

(b) $A = \begin{bmatrix} 2 & -3 \\ -5 & 0 \\ 0 & -2 \end{bmatrix}$

(c) $A = \begin{bmatrix} 1 & 0 & -1 & 2 \\ 3 & 1 & 0 & 0 \end{bmatrix}$

Solution   (a) Because the size of this matrix is 3×3, it defines a linear transformation from R^3 into R^3.

$$A\mathbf{v} = \begin{bmatrix} 0 & 1 & -1 \\ 2 & 3 & 0 \\ 4 & 2 & 1 \end{bmatrix} \begin{bmatrix} v_1 \\ v_2 \\ v_3 \end{bmatrix} = \begin{bmatrix} u_1 \\ u_2 \\ u_3 \end{bmatrix}$$

| Vector in R^3 | Vector in R^3 |

(b) Because the size of this matrix is 3×2, it defines a linear transformation from R^2 into R^3.

(c) Because the size of this matrix is 2×4, it defines a linear transformation from R^4 into R^2.

In the next example a common type of linear transformation from R^2 into R^2 is discussed.

EXAMPLE 7   *Rotation in the Plane*

Show that the linear transformation $T: R^2 \rightarrow R^2$ given by the matrix

$$A = \begin{bmatrix} \cos \theta & -\sin \theta \\ \sin \theta & \cos \theta \end{bmatrix}$$

has the property that it rotates every vector in R^2 counterclockwise about the origin through the angle θ.

Solution   From Theorem 6.2 you know that T is a linear transformation. To show that it rotates every vector in R^2 counterclockwise through the angle θ, let $\mathbf{v} = (x, y)$ be a vector in R^2. Using polar coordinates, you can write $\mathbf{v}$ as

$$\mathbf{v} = (x, y) = (r \cos \alpha, r \sin \alpha),$$

where r is the length of $\mathbf{v}$ and α is the angle from the positive x-axis counterclockwise to the vector $\mathbf{v}$. Now, applying the linear transformation T to $\mathbf{v}$ produces

$$\begin{aligned} T(\mathbf{v}) = A\mathbf{v} &= \begin{bmatrix} \cos \theta & -\sin \theta \\ \sin \theta & \cos \theta \end{bmatrix} \begin{bmatrix} x \\ y \end{bmatrix} \\ &= \begin{bmatrix} \cos \theta & -\sin \theta \\ \sin \theta & \cos \theta \end{bmatrix} \begin{bmatrix} r \cos \alpha \\ r \sin \alpha \end{bmatrix} \\ &= \begin{bmatrix} r \cos \theta \cos \alpha - r \sin \theta \sin \alpha \\ r \sin \theta \cos \alpha + r \cos \theta \sin \alpha \end{bmatrix} \\ &= \begin{bmatrix} r \cos(\theta + \alpha) \\ r \sin(\theta + \alpha) \end{bmatrix}. \end{aligned}$$

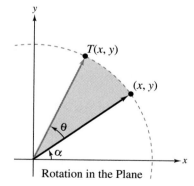

Rotation in the Plane

Figure 6.2

Therefore, the vector $T(\mathbf{v})$ has the same length as $\mathbf{v}$. Furthermore, because the angle from the positive x-axis to $T(\mathbf{v})$ is $\theta + \alpha$, $T(\mathbf{v})$ is the vector that results from rotating the vector $\mathbf{v}$ counterclockwise through the angle θ, as shown in Figure 6.2 on the previous page.

REMARK: The linear transformation in Example 7 is called a **rotation** in R^2. Rotations in R^2 preserve both vector length and the angle between two vectors. That is, the angle between $\mathbf{u}$ and $\mathbf{v}$ is equal to the angle between $T(\mathbf{u})$ and $T(\mathbf{v})$.

EXAMPLE 8   *A Projection in R^3*

The linear transformation $T: R^3 \to R^3$ given by

$$A = \begin{bmatrix} 1 & 0 & 0 \\ 0 & 1 & 0 \\ 0 & 0 & 0 \end{bmatrix}$$

is called a **projection** in R^3. If $\mathbf{v} = (x, y, z)$ is a vector in R^3, then $T(\mathbf{v}) = (x, y, 0)$. In other words, T maps every vector in R^3 to its orthogonal projection in the xy-plane, as shown in Figure 6.3.

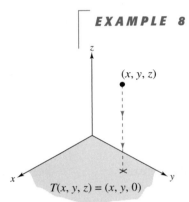

$T(x, y, z) = (x, y, 0)$

Projection onto xy-plane

Figure 6.3

So far only linear transformations from R^n into R^m have been discussed. In the remainder of this section, some linear transformations involving vector spaces other than R^n will be considered.

EXAMPLE 9   *A Linear Transformation from $M_{m,n}$ into $M_{n,m}$*

Let $T: M_{m,n} \to M_{n,m}$ be the function that maps an $m \times n$ matrix A to its transpose. That is,

$$T(A) = A^T.$$

Show that T is a linear transformation.

Solution   Let A and B be $m \times n$ matrices. From Theorem 2.6 you have

$$\begin{aligned} T(A + B) &= (A + B)^T \\ &= A^T + B^T \\ &= T(A) + T(B) \end{aligned}$$

and

$$\begin{aligned} T(cA) &= (cA)^T \\ &= c(A^T) \\ &= cT(A). \end{aligned}$$

Therefore, T is a linear transformation from $M_{m,n}$ into $M_{n,m}$.

EXAMPLE 10   *The Differential Operator (Calculus)*

Let $C'[a, b]$ be the set of all functions whose derivatives are continuous on $[a, b]$. Show that the differential operator D_x defines a linear transformation from $C'[a, b]$ into $C[a, b]$.

Solution   Using operator notation, you can write

$$D_x(f) = \frac{d}{dx}[f],$$

where f is in $C'[a, b]$. To show that D_x is a linear transformation, you must use calculus. Specifically, because the derivative of the sum of two functions is equal to the sum of their derivatives and because the sum of two continuous functions is continuous, you have

$$D_x(f + g) = \frac{d}{dx}[f + g] = \frac{d}{dx}[f] + \frac{d}{dx}[g]$$
$$= D_x(f) + D_x(g).$$

Similarly, because the derivative of a scalar multiple of a function is equal to the scalar multiple of the derivative and because the scalar multiple of a continuous function is continuous, you have

$$D_x(cf) = \frac{d}{dx}[cf] = c\left(\frac{d}{dx}[f]\right)$$
$$= cD_x(f).$$

Therefore, D_x is a linear transformation from $C'[a, b]$ into $C[a, b]$.

The linear transformation D_x in Example 10 is called the **differential operator.** For polynomials, the differential operator is a linear transformation from P_n into P_{n-1}, because the derivative of a polynomial function of degree n is a polynomial function of degree $n - 1$ or less. That is,

$$D_x(a_n x^n + \cdot\cdot\cdot + a_1 x + a_0) = na_n x^{n-1} + \cdot\cdot\cdot + a_1.$$

The next example describes a linear transformation from the vector space of polynomial functions P into the vector space of real numbers R.

EXAMPLE 11   *The Definite Integral as a Linear Transformation (Calculus)*

Let $T: P \rightarrow R$ be defined by

$$T(p) = \int_a^b p(x)\, dx.$$

Show that T is a linear transformation from P, the vector space of polynomial functions, into R, the vector space of real numbers.

Solution Using properties of definite integrals, you can write

$$T(p + q) = \int_a^b [p(x) + q(x)]\, dx$$

$$= \int_a^b p(x)\, dx + \int_a^b q(x)\, dx$$

$$= T(p) + T(q)$$

and

$$T(cp) = \int_a^b c[p(x)]\, dx = c \int_a^b p(x)\, dx = cT(p).$$

Therefore, T is a linear transformation.

SECTION 6.1 Exercises

In Exercises 1–6, use the function to find (a) the image of **v** and (b) the preimage of **w**.

1. $T(v_1, v_2) = (v_1 + v_2, v_1 - v_2)$, $\mathbf{v} = (3, -4)$, $\mathbf{w} = (3, 19)$

2. $T(v_1, v_2) = (2v_2 - v_1, v_1, v_2)$, $\mathbf{v} = (0, 6)$, $\mathbf{w} = (3, 1, 2)$

3. $T(v_1, v_2, v_3) = (v_2 - v_1, v_1 + v_2, 2v_1)$, $\mathbf{v} = (2, 3, 0)$, $\mathbf{w} = (-11, -1, 10)$

4. $T(v_1, v_2, v_3) = (2v_1 + v_2, 2v_2 - 3v_1, v_1 - v_3)$, $\mathbf{v} = (-4, 5, 1)$, $\mathbf{w} = (4, 1, -1)$

5. $T(v_1, v_2, v_3) = (4v_2 - v_1, 4v_1 + 5v_2)$, $\mathbf{v} = (2, -3, -1)$, $\mathbf{w} = (3, 9)$

6. $T(v_1, v_2) = \left(\dfrac{\sqrt{2}}{2} v_1 - \dfrac{\sqrt{2}}{2} v_2, v_1 + v_2, 2v_1 - v_2 \right)$, $\mathbf{v} = (1, 1)$, $\mathbf{w} = \left(-5\sqrt{2}, -2, -16 \right)$

In Exercises 7–14, determine whether the function is a linear transformation.

7. $T: R^2 \rightarrow R^2$, $T(x, y) = (x, 1)$

8. $T: R^3 \rightarrow R^3$, $T(x, y, z) = (x + y, x - y, z)$

9. $T: R^2 \rightarrow R^3$, $T(x, y) = \left(\sqrt{x}, xy, \sqrt{y} \right)$

10. $T: M_{2,2} \rightarrow R$, $T(A) = |A|$

11. $T: M_{3,3} \rightarrow M_{3,3}$, $T(A) = \begin{bmatrix} 0 & 0 & 1 \\ 0 & 1 & 0 \\ 1 & 0 & 0 \end{bmatrix} A$

12. $T: M_{2,2} \rightarrow M_{2,2}$, $T(A) = A^T$

13. $T: P_2 \rightarrow P_2$, $T(a_0 + a_1 x + a_2 x^2) = (a_0 + a_1 + a_2) + (a_1 + a_2)x + a_2 x^2$

14. $T: P_2 \rightarrow P_2$, $T(a_0 + a_1 x + a_2 x^2) = a_1 + 2a_2 x$

In Exercises 15–18, let $T: R^3 \rightarrow R^3$ be a linear transformation such that $T(1, 0, 0) = (2, 4, -1)$, $T(0, 1, 0) = (1, 3, -2)$, and $T(0, 0, 1) = (0, -2, 2)$. Find the following.

15. $T(0, 3, -1)$ **16.** $T(2, -1, 0)$

17. $T(2, -4, 1)$ **18.** $T(-2, 4, -1)$

In Exercises 19–23, the linear transformation $T: R^n \rightarrow R^m$ is defined by $T(\mathbf{v}) = A\mathbf{v}$. Find the dimensions of R^n and R^m.

19. $A = \begin{bmatrix} 0 & 1 & -2 & 1 \\ -1 & 4 & 5 & 0 \\ 0 & 1 & 3 & 1 \end{bmatrix}$

20. $A = \begin{bmatrix} 1 & 2 \\ -2 & 4 \\ -2 & 2 \end{bmatrix}$

21. $A = \begin{bmatrix} -1 & 2 & 1 & 3 & 4 \\ 0 & 0 & 2 & -1 & 0 \end{bmatrix}$

22. $A = \begin{bmatrix} -1 & 0 & 0 & 0 \\ 0 & 1 & 0 & 0 \\ 0 & 0 & 2 & 0 \\ 0 & 0 & 0 & 1 \end{bmatrix}$

23. $A = \begin{bmatrix} 0 & -1 \\ -1 & 0 \end{bmatrix}$

24. For the linear transformation given in Exercise 19, find (a) $T(1, 0, 2, 3)$ and (b) the preimage of $(0, 0, 0)$.

25. *Writing* For the linear transformation given in Exercise 20, find (a) $T(2, 4)$ and (b) the preimage of $(-1, 2, 2)$. (c) Then explain why the vector $(1, 1, 1)$ has no preimage under this transformation.

26. For the linear transformation given in Exercise 21, find (a) $T(1, 0, -1, 3, 0)$ and (b) the preimage of $(-1, 8)$.

27. For the linear transformation given in Exercise 22, find (a) $T(1, 1, 1, 1)$ and (b) the preimage of $(1, 1, 1, 1)$.

28. For the linear transformation given in Exercise 23, find (a) $T(1, 1)$, (b) the preimage of $(1, 1)$, and (c) the preimage of $(0, 0)$.

29. Let T be the linear transformation from R^2 into R^2 given by
$$T(x, y) = (x \cos \theta - y \sin \theta, x \sin \theta + y \cos \theta).$$
Find (a) $T(4, 4)$ for $\theta = 45°$, (b) $T(4, 4)$ for $\theta = 30°$, and (c) $T(5, 0)$ for $\theta = 120°$.

30. For the linear transformation given in Exercise 29, let $\theta = 45°$ and find the preimage of $\mathbf{v} = (1, 1)$.

31. *Calculus* Let D_x be the linear transformation from $C'[a, b]$ into $C[a, b]$ given in Example 10. Decide whether each statement is true or false. Explain your reasoning.

(a) $D_x(e^{x^2} + 2x) = D_x(e^{x^2}) + 2D_x(x)$

(b) $D_x(x^2 - \ln x) = D_x(x^2) - D_x(\ln x)$

(c) $D_x(\sin 2x) = 2D_x(\sin x)$

32. *Calculus* For the linear transformation given in Example 10, find the preimage of each function.

(a) $f(x) = 2x + 1$ (b) $f(x) = e^x$ (c) $f(x) = \sin x$

33. *Calculus* Let T be the linear transformation from P into R given by
$$T(p) = \int_0^1 p(x) \, dx.$$
Find (a) $T(3x^2 - 2)$, (b) $T(x^3 - x^5)$, and (c) $T(4x - 6)$.

34. *Calculus* Let T be the linear transformation from P_2 into R given by the integral in Exercise 33. Find the preimage of 1. That is, find the polynomial function(s) of degree two or less such that $T(p) = 1$.

35. Let T be a linear transformation from R^2 into R^2 such that $T(1, 1) = (1, 0)$ and $T(1, -1) = (0, 1)$. Find $T(1, 0)$ and $T(0, 2)$.

36. Let T be a linear transformation from R^2 into R^2 such that $T(1, 0) = (1, 1)$ and $T(0, 1) = (-1, 1)$. Find $T(1, 4)$ and $T(-2, 1)$.

37. Let T be a linear transformation from P_2 into P_2 such that $T(1) = x$, $T(x) = 1 + x$, and $T(x^2) = 1 + x + x^2$. Find $T(2 - 6x + x^2)$.

38. Let T be a linear transformation from $M_{2,2}$ into $M_{2,2}$ such that
$$T\left(\begin{bmatrix} 1 & 0 \\ 0 & 0 \end{bmatrix}\right) = \begin{bmatrix} 1 & -1 \\ 0 & 2 \end{bmatrix}, \quad T\left(\begin{bmatrix} 0 & 1 \\ 0 & 0 \end{bmatrix}\right) = \begin{bmatrix} 0 & 2 \\ 1 & 1 \end{bmatrix},$$
$$T\left(\begin{bmatrix} 0 & 0 \\ 1 & 0 \end{bmatrix}\right) = \begin{bmatrix} 1 & 2 \\ 0 & 1 \end{bmatrix}, \quad T\left(\begin{bmatrix} 0 & 0 \\ 0 & 1 \end{bmatrix}\right) = \begin{bmatrix} 3 & -1 \\ 1 & 0 \end{bmatrix}.$$
Find $T\left(\begin{bmatrix} 1 & 3 \\ -1 & 4 \end{bmatrix}\right)$.

True or False? In Exercises 39 and 40, determine whether each statement is true or false. If a statement is true, give a reason or cite an appropriate statement in the text. If a statement is false, provide an example that shows that the statement is not true in all cases or cite an appropriate statement in the text.

39. (a) Linear transformations are functions from one vector space to another that preserve the operations of vector addition and scalar multiplication.

(b) The function $f(x) = \cos x$ is a linear transformation from R into R.

(c) For polynomials, the differential operator D_x is a linear transformation from P_n into P_{n-1}.

40. (a) A linear transformation is operation preserving if the same result occurs whether the operations of addition and scalar multiplication are performed before or after the linear transformation is applied.

(b) The function $g(x) = x^3$ is a linear transformation from R into R.

(c) Any linear function of the form $f(x) = ax + b$ is a linear transformation from R into R.

41. *Writing* Suppose $T: R^2 \to R^2$ such that $T(1, 0) = (1, 0)$ and $T(0, 1) = (0, 0)$.

(a) Determine $T(x, y)$ for (x, y) in R^2.

(b) Give a geometric description of T.

42. *Writing* Suppose $T: R^2 \to R^2$ such that $T(1, 0) = (0, 1)$ and $T(0, 1) = (1, 0)$.

(a) Determine $T(x, y)$ for (x, y) in R^2.

(b) Give a geometric description of T.

In Exercises 43–48, let T be the function from R^2 into R^2 such that $T(\mathbf{u}) = \text{proj}_\mathbf{v}\mathbf{u}$, where $\mathbf{v} = (1, 1)$.

43. Find $T(x, y)$. **44.** Find $T(5, 0)$.

45. Prove that $T(\mathbf{u} + \mathbf{w}) = T(\mathbf{u}) + T(\mathbf{w})$ for every $\mathbf{u}$ and $\mathbf{w}$ in R^2.

46. Prove that $T(c\mathbf{u}) = cT(\mathbf{u})$ for every $\mathbf{u}$ in R^2. This result and the result in Exercise 45 prove that T is a linear transformation from R^2 into R^2.

47. *Writing* Find $T(3, 4)$ and $T(T(3, 4))$ and give a geometric description of the result.

48. Show that T is given by the matrix

$$A = \begin{bmatrix} \frac{1}{2} & \frac{1}{2} \\ \frac{1}{2} & \frac{1}{2} \end{bmatrix}.$$

In Exercises 49–52, use the concept of a fixed point of a linear transformation $T: V \to V$. A vector $\mathbf{u}$ is a **fixed point** if $T(\mathbf{u}) = \mathbf{u}$.

49. Prove that $\mathbf{0}$ is a fixed point of any linear transformation $T: V \to V$.

50. Prove that the set of fixed points of a linear transformation $T: V \to V$ is a subspace of V.

51. Determine all fixed points of the linear transformation $T: R^2 \to R^2$ given by $T(x, y) = (x, 2y)$.

52. Determine all fixed points of the linear transformation $T: R^2 \to R^2$ given by $T(x, y) = (y, x)$.

Exercises 53–55 deal with translations in the plane. A **translation** is a function of the form $T(x, y) = (x - h, y - k)$, where at least one of the constants h and k is nonzero.

53. Show that a translation in the plane is not a linear transformation.

54. For the translation $T(x, y) = (x - 2, y + 1)$, determine the images of $(0, 0)$, $(2, -1)$, and $(5, 4)$.

55. Show that a translation in the plane has no fixed points.

56. Let $S = \{\mathbf{v}_1, \mathbf{v}_2, \ldots, \mathbf{v}_n\}$ be a set of linearly dependent vectors in V, and let T be a linear transformation from V into V. Prove that the set

$$\{T(\mathbf{v}_1), T(\mathbf{v}_2), \ldots, T(\mathbf{v}_n)\}$$

is linearly dependent.

57. Let $S = \{\mathbf{v}_1, \mathbf{v}_2, \mathbf{v}_3\}$ be a set of linearly independent vectors in R^3. Find a linear transformation T from R^3 into R^3 such that the set $\{T(\mathbf{v}_1), T(\mathbf{v}_2), T(\mathbf{v}_3)\}$ is linearly dependent.

58. Prove that the zero transformation $T: V \to W$ is a linear transformation.

59. Prove that the identity transformation $T: V \to V$ is a linear transformation.

60. Let V be an inner product space. For a fixed vector $\mathbf{v}_0$ in V, define $T: V \to R$ by $T(\mathbf{v}) = \langle \mathbf{v}, \mathbf{v}_0 \rangle$. Prove that T is a linear transformation.

61. Let $T: M_{n,n} \to R$ be defined by $T(A) = a_{11} + a_{22} + \cdots + a_{nn}$ (the trace of A). Prove that T is a linear transformation.

62. Let V be an inner product space with a subspace W having $B = \{\mathbf{w}_1, \mathbf{w}_2, \ldots, \mathbf{w}_n\}$ as an orthonormal basis. Show that the function $T: V \to W$ given by

$$T(\mathbf{v}) = \langle \mathbf{v}, \mathbf{w}_1 \rangle \mathbf{w}_1 + \langle \mathbf{v}, \mathbf{w}_2 \rangle \mathbf{w}_2 + \cdots + \langle \mathbf{v}, \mathbf{w}_n \rangle \mathbf{w}_n$$

is a linear transformation. T is called the **orthogonal projection of V onto W**.

63. *Guided Proof* Let $\{\mathbf{v}_1, \mathbf{v}_2, \ldots, \mathbf{v}_n\}$ be a basis for a vector space V. Prove that if a linear transformation $T: V \to V$ satisfies $T(\mathbf{v}_i) = \mathbf{0}$ for $i = 1, 2, \ldots, n$, then T is the zero transformation.

Getting Started: To prove that T is the zero transformation, you need to show that $T(\mathbf{v}) = \mathbf{0}$ for every vector $\mathbf{v}$ in V.

(i) Let $\mathbf{v}$ be an arbitrary vector in V such that $\mathbf{v} = c_1\mathbf{v}_1 + c_2\mathbf{v}_2 + \cdots + c_n\mathbf{v}_n$.

(ii) Use the definition and properties of linear transformations to rewrite $T(\mathbf{v})$ as a linear combination of $T(\mathbf{v}_i)$.

(iii) Use the fact that $T(\mathbf{v}_i) = \mathbf{0}$ to conclude that $T(\mathbf{v}) = \mathbf{0}$ and that T therefore is the zero transformation.

64. *Guided Proof* Prove that $T: V \to W$ is a linear transformation if and only if $T(a\mathbf{u} + b\mathbf{v}) = aT(\mathbf{u}) + bT(\mathbf{v})$ for all vectors $\mathbf{u}$ and $\mathbf{v}$ and all scalars a and b.

Getting Started: Because this is an "if and only if" statement, you need to prove the statement in both directions. To prove that T is a linear transformation, you need to show that the function satisfies the definition of a linear transformation. In the other direction, suppose that T is a linear transformation. You can use the definition and properties of a linear transformation to prove that $T(a\mathbf{u} + b\mathbf{v}) = aT(\mathbf{u}) + bT(\mathbf{v})$.

(i) Suppose that $T(a\mathbf{u} + b\mathbf{v}) = aT(\mathbf{u}) + bT(\mathbf{v})$. Show that T preserves the properties of vector addition and scalar multiplication by choosing appropriate values for a and b.

(ii) To prove the statement in the other direction, assume that T is a linear transformation. Use the properties and definition of a linear transformation to show that $T(a\mathbf{u} + b\mathbf{v}) = aT(\mathbf{u}) + bT(\mathbf{v})$.

6.2 | The Kernel and Range of a Linear Transformation

You know from Theorem 6.1 that for any linear transformation $T: V \rightarrow W$, the zero vector in V is mapped to the zero vector in W. That is,

$$T(\mathbf{0}) = \mathbf{0}.$$

The question you will consider first in this section is whether there are *other* vectors $\mathbf{v}$ such that $T(\mathbf{v}) = \mathbf{0}$. The collection of all such elements is called the **kernel** of T. Note that the symbol $\mathbf{0}$ is used to represent the zero vector in both V and W, although these two zero vectors are often different.

Definition of Kernel of a Linear Transformation

Let $T: V \rightarrow W$ be a linear transformation. Then the set of all vectors $\mathbf{v}$ in V that satisfy $T(\mathbf{v}) = \mathbf{0}$ is called the **kernel** of T and is denoted by $\ker(T)$.

Sometimes the kernel of a transformation is obvious and can be found by inspection, as demonstrated in Examples 1, 2, and 3.

EXAMPLE 1   *Finding the Kernel of a Linear Transformation*

Let $T: M_{3,2} \rightarrow M_{2,3}$ be the linear transformation that maps a 3×2 matrix A to its transpose. That is,

$$T(A) = A^T.$$

Find the kernel of T.

Solution   For this linear transformation, the 3×2 zero matrix is clearly the only matrix in $M_{3,2}$ whose transpose is the zero matrix in $M_{2,3}$.

Zero Matrix in $M_{3,2}$ *Zero Matrix in $M_{2,3}$*

$$\mathbf{0} = \begin{bmatrix} 0 & 0 \\ 0 & 0 \\ 0 & 0 \end{bmatrix} \qquad \mathbf{0} = \begin{bmatrix} 0 & 0 & 0 \\ 0 & 0 & 0 \end{bmatrix}$$

Therefore, the kernel of T consists of a single element: the zero matrix in $M_{3,2}$.

EXAMPLE 2   *The Kernel of the Zero and Identity Transformations*

(a) The kernel of the zero transformation $T: V \rightarrow W$ consists of all of V because $T(\mathbf{v}) = \mathbf{0}$ for every $\mathbf{v}$ in V. That is,

$$\ker(T) = V.$$

(b) The kernel of the identity transformation $T: V \rightarrow V$ consists of the single element **0**. That is,

$$\ker(T) = \{\mathbf{0}\}.$$

EXAMPLE 3 *Finding the Kernel of a Linear Transformation*

Find the kernel of the projection $T: R^3 \rightarrow R^3$ given by

$$T(x, y, z) = (x, y, 0).$$

Solution   This linear transformation projects the vector (x, y, z) in R^3 to the vector $(x, y, 0)$ in the xy-plane. Therefore the kernel consists of all vectors lying on the z-axis. That is,

$$\ker(T) = \{(0, 0, z): z \text{ is a real number}\}.$$

(See Figure 6.4.)

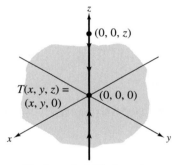

The kernel of T is the set of all vectors on the z-axis.

Figure 6.4

Finding the kernels of the linear transformations in Examples 1, 2, and 3 was fairly easy. Generally the kernel of a linear transformation is not so obvious, and finding it requires a little work, as illustrated in the next two examples.

EXAMPLE 4 *Finding the Kernel of a Linear Transformation*

Find the kernel of the linear transformation $T: R^2 \rightarrow R^3$ given by

$$T(x_1, x_2) = (x_1 - 2x_2, 0, -x_1).$$

Solution   To find $\ker(T)$, you need to find all $\mathbf{x} = (x_1, x_2)$ in R^2 such that

$$T(x_1, x_2) = (x_1 - 2x_2, 0, -x_1) = (0, 0, 0).$$

This leads to the homogeneous system

$$x_1 - 2x_2 = 0$$
$$0 = 0$$
$$-x_1 \qquad = 0,$$

which has only the trivial solution $(x_1, x_2) = (0, 0)$. Thus, you have

$$\ker(T) = \{(0, 0)\} = \{\mathbf{0}\}.$$

EXAMPLE 5   *Finding the Kernel of a Linear Transformation*

Find the kernel of the linear transformation $T: R^3 \rightarrow R^2$ defined by $T(\mathbf{x}) = A\mathbf{x}$, where

$$A = \begin{bmatrix} 1 & -1 & -2 \\ -1 & 2 & 3 \end{bmatrix}.$$

Solution   The kernel of T is the set of all $\mathbf{x} = (x_1, x_2, x_3)$ in R^3 such that

$$T(x_1, x_2, x_3) = (0, 0).$$

From this equation you can write the following homogeneous system.

$$\begin{bmatrix} 1 & -1 & -2 \\ -1 & 2 & 3 \end{bmatrix} \begin{bmatrix} x_1 \\ x_2 \\ x_3 \end{bmatrix} = \begin{bmatrix} 0 \\ 0 \end{bmatrix} \quad \longrightarrow \quad \begin{array}{r} x_1 - x_2 - 2x_3 = 0 \\ -x_1 + 2x_2 + 3x_3 = 0 \end{array}$$

Writing the augmented matrix of this system in reduced row-echelon form produces

$$\begin{bmatrix} 1 & 0 & -1 & 0 \\ 0 & 1 & 1 & 0 \end{bmatrix} \quad \longrightarrow \quad \begin{array}{r} x_1 = x_3 \\ x_2 = -x_3. \end{array}$$

Using the parameter $t = x_3$ produces the family of solutions

$$\begin{bmatrix} x_1 \\ x_2 \\ x_3 \end{bmatrix} = \begin{bmatrix} t \\ -t \\ t \end{bmatrix} = t \begin{bmatrix} 1 \\ -1 \\ 1 \end{bmatrix}.$$

Thus, the kernel of T is given by

$$\ker(T) = \{t(1, -1, 1): t \text{ is a real number}\}$$
$$= \text{span}\{(1, -1, 1)\}.$$

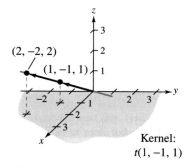

(2, −2, 2)

(1, −1, 1)

Kernel:
$t(1, -1, 1)$

Figure 6.5

Note that in Example 5 the kernel of T contains an infinite number of vectors. Of course, the zero vector is in $\ker(T)$, but the kernel also contains such nonzero vectors as $(1, -1, 1)$ and $(2, -2, 2)$, as shown in Figure 6.5. Figure 6.5 shows that this particular kernel is a line passing through the origin, which implies that it is a subspace of R^3. In Theorem 6.3 you will see that the kernel of every linear transformation $T: V \rightarrow W$ is a subspace of V.

Theorem 6.3	The Kernel Is a Subspace of V
	The kernel of a linear transformation $T: V \rightarrow W$ is a subspace of the domain V.

Proof From Theorem 6.1 you know that $\ker(T)$ is a nonempty subset of V. Therefore, by Theorem 4.5, you can show that $\ker(T)$ is a subspace of V by showing that it is closed under vector addition and scalar multiplication. To do so, let $\mathbf{u}$ and $\mathbf{v}$ be vectors in the kernel of T. Then

$$T(\mathbf{u} + \mathbf{v}) = T(\mathbf{u}) + T(\mathbf{v})$$
$$= \mathbf{0} + \mathbf{0}$$
$$= \mathbf{0},$$

which implies that $\mathbf{u} + \mathbf{v}$ is in the kernel. Moreover, if c is any scalar, then

$$T(c\mathbf{u}) = cT(\mathbf{u})$$
$$= c\mathbf{0}$$
$$= \mathbf{0},$$

which implies that $c\mathbf{u}$ is in the kernel.

REMARK: As a result of Theorem 6.3, the kernel of T is sometimes called the **nullspace** of T.

The next example shows how to find a basis for the kernel of a transformation defined by a matrix.

EXAMPLE 6 *Finding a Basis for the Kernel*

Let $T: R^5 \rightarrow R^4$ be defined by $T(\mathbf{x}) = A\mathbf{x}$, where $\mathbf{x}$ is in R^5 and

$$A = \begin{bmatrix} 1 & 2 & 0 & 1 & -1 \\ 2 & 1 & 3 & 1 & 0 \\ -1 & 0 & -2 & 0 & 1 \\ 0 & 0 & 0 & 2 & 8 \end{bmatrix}.$$

Find a basis for $\ker(T)$ as a subspace of R^5.

Solution Following the procedure shown in Example 5, reduce the augmented matrix $[A : \mathbf{0}]$ to echelon form as follows.

$$\begin{bmatrix} 1 & 0 & 2 & 0 & -1 & 0 \\ 0 & 1 & -1 & 0 & -2 & 0 \\ 0 & 0 & 0 & 1 & 4 & 0 \\ 0 & 0 & 0 & 0 & 0 & 0 \end{bmatrix} \longrightarrow \begin{aligned} x_1 &= -2x_3 + x_5 \\ x_2 &= x_3 + 2x_5 \\ x_4 &= -4x_5 \end{aligned}$$

DISCOVERY

What is the rank of the matrix A in Example 6? Formulate a conjecture relating the dimension of the kernel, the rank, and the number of columns of A. Verify your conjecture for the matrix in Example 5.

Letting $x_3 = s$ and $x_5 = t$, you have

$$\mathbf{x} = \begin{bmatrix} x_1 \\ x_2 \\ x_3 \\ x_4 \\ x_5 \end{bmatrix} = \begin{bmatrix} -2s + t \\ s + 2t \\ s + 0t \\ 0s - 4t \\ 0s + t \end{bmatrix} = s\begin{bmatrix} -2 \\ 1 \\ 1 \\ 0 \\ 0 \end{bmatrix} + t\begin{bmatrix} 1 \\ 2 \\ 0 \\ -4 \\ 1 \end{bmatrix}.$$

So one basis for the kernel of T is given by

$$B = \{(-2, 1, 1, 0, 0), (1, 2, 0, -4, 1)\}.$$

In the solution given in Example 6, a basis for the kernel of T was found by solving the homogeneous system given by $A\mathbf{x} = \mathbf{0}$. This procedure is a familiar one—it is the same procedure used to find the *solution space* of $A\mathbf{x} = \mathbf{0}$. In other words, the kernel of T is the nullspace of the matrix A, as given in the following corollary to Theorem 6.3.

Corollary to Theorem 6.3

Let $T: R^n \rightarrow R^m$ be the linear transformation given by $T(\mathbf{x}) = A\mathbf{x}$. Then the kernel of T is equal to the solution space of $A\mathbf{x} = \mathbf{0}$.

The Range of a Linear Transformation

The kernel is one of two critical subspaces associated with a linear transformation. The second is the **range** of T, denoted by range(T). Recall from Section 6.1 that the range of $T: V \rightarrow W$ is the set of all vectors $\mathbf{w}$ in W that are images of vectors in V. That is, range(T) = $\{T(\mathbf{v}): \mathbf{v}$ is in $V\}$.

Theorem 6.4

The Range of T is a Subspace of W

The range of a linear transformation $T: V \rightarrow W$ is a subspace of W.

Proof   The range of T is nonempty because $T(\mathbf{0}) = \mathbf{0}$ implies that the range contains the zero vector. To show that it is closed under vector addition, let $T(\mathbf{u})$ and $T(\mathbf{v})$ be vectors in the range of T. Because $\mathbf{u}$ and $\mathbf{v}$ are in V, it follows that $\mathbf{u} + \mathbf{v}$ is also in V. Hence, the sum $T(\mathbf{u}) + T(\mathbf{v}) = T(\mathbf{u} + \mathbf{v})$ is in the range of T.

To show closure under scalar multiplication, let $T(\mathbf{u})$ be a vector in the range of T and let c be a scalar. Because $\mathbf{u}$ is in V, it follows that $c\mathbf{u}$ is also in V. Hence, the scalar multiple $cT(\mathbf{u}) = T(c\mathbf{u})$ is in the range of T.

Domain   Kernel

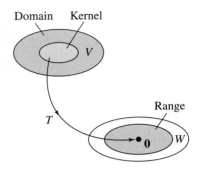

T

Range

W

0

Figure 6.6

Note that the kernel and range of a linear transformation $T: V \to W$ are subspaces of V and W, respectively, as illustrated in Figure 6.6.

To find a basis for the range of a linear transformation defined by $T(\mathbf{x}) = A\mathbf{x}$, observe that the range consists of all vectors $\mathbf{b}$ such that the system $A\mathbf{x} = \mathbf{b}$ is consistent. By writing the system

$$\begin{bmatrix} a_{11} & a_{12} & \cdots & a_{1n} \\ a_{21} & a_{22} & \cdots & a_{2n} \\ \vdots & \vdots & & \vdots \\ a_{m1} & a_{m2} & \cdots & a_{mn} \end{bmatrix} \begin{bmatrix} x_1 \\ x_2 \\ \vdots \\ x_n \end{bmatrix} = \begin{bmatrix} b_1 \\ b_2 \\ \vdots \\ b_m \end{bmatrix}$$

in the form

$$A\mathbf{x} = x_1 \begin{bmatrix} a_{11} \\ a_{21} \\ \vdots \\ a_{m1} \end{bmatrix} + x_2 \begin{bmatrix} a_{12} \\ a_{22} \\ \vdots \\ a_{m2} \end{bmatrix} + \cdots + x_n \begin{bmatrix} a_{1n} \\ a_{2n} \\ \vdots \\ a_{mn} \end{bmatrix} = \begin{bmatrix} b_1 \\ b_2 \\ \vdots \\ b_m \end{bmatrix} = \mathbf{b}$$

you can see that $\mathbf{b}$ is in the range of T if and only if $\mathbf{b}$ is a linear combination of the column vectors of A. Thus *the column space of the matrix A is the same as the range of T.*

Corollary to Theorem 6.4   Let $T: R^n \to R^m$ be the linear transformation given by $T(\mathbf{x}) = A\mathbf{x}$. Then the column space of A is equal to the range of T.

In Example 4 in Section 4.6, you saw two procedures for finding a basis for the column space of a matrix. In the following example, the second procedure given in that example will be used to find a basis for the range of a linear transformation defined by a matrix.

EXAMPLE 7   *Finding a Basis for the Range of a Linear Transformation*

Let $T: R^5 \to R^4$ be the linear transformation given in Example 6. Find a basis for the range of T.

Solution   The echelon form of A was calculated in Example 6.

$$A = \begin{bmatrix} 1 & 2 & 0 & 1 & -1 \\ 2 & 1 & 3 & 1 & 0 \\ -1 & 0 & -2 & 0 & 1 \\ 0 & 0 & 0 & 2 & 8 \end{bmatrix} \Rightarrow \begin{bmatrix} 1 & 0 & 2 & 0 & -1 \\ 0 & 1 & -1 & 0 & -2 \\ 0 & 0 & 0 & 1 & 4 \\ 0 & 0 & 0 & 0 & 0 \end{bmatrix}$$

Because the leading 1's appear in columns 1, 2, and 4 of the reduced matrix on the right, the corresponding column vectors of A form a basis for the column space of A. Therefore, one basis for the range of T is

$$B = \{(1, 2, -1, 0), (2, 1, 0, 0), (1, 1, 0, 2)\}.$$

The dimensions of the kernel and range of a linear transformation are given the following names.

Definition of Rank and Nullity of a Linear Transformation

Let $T: V \rightarrow W$ be a linear transformation. The dimension of the kernel of T is called the **nullity** of T and is denoted by **nullity**(T). The dimension of the range of T is called the **rank** of T and is denoted by **rank**(T).

REMARK: If T is given by a matrix A, then the rank of T is equal to the rank of A, as defined in Section 4.6.

In Examples 6 and 7, the nullity and rank of T are related to the dimension of the domain as follows.

$$\text{rank}(T) + \text{nullity}(T) = 3 + 2 = 5 = \text{dimension of domain}$$

This relationship is true for any linear transformation from a finite-dimensional vector space, as stated in the following theorem.

Theorem 6.5

Sum of Rank and Nullity

Let $T: V \rightarrow W$ be a linear transformation from an n-dimensional vector space V into a vector space W. Then the sum of the dimensions of the range and kernel is equal to the dimension of the domain. That is,

$$\text{rank}(T) + \text{nullity}(T) = n$$

or

$$\text{dim(range)} + \text{dim(kernel)} = \text{dim(domain)}.$$

Proof   The proof given here covers the case in which T is represented by an $m \times n$ matrix A. The general case will follow in the next section, where you will see that any linear transformation from an n-dimensional space to an m-dimensional space can be represented by a matrix. To prove this theorem, assume that the matrix A has a rank of r. Then you have

$$\text{rank}(T) = \text{dim(range of } T) = \text{dim(column space)} = \text{rank}(A) = r.$$

From Theorem 4.17, however, you know that

$$\text{nullity}(T) = \text{dim(kernel of } T) = \text{dim(solution space)} = n - r.$$

Thus, it follows that

$$\text{rank}(T) + \text{nullity}(T) = r + (n - r) = n.$$

EXAMPLE 8 *Finding the Rank and Nullity of a Linear Transformation*

Find the rank and nullity of the linear transformation $T: R^3 \to R^3$ defined by the matrix

$$A = \begin{bmatrix} 1 & 0 & -2 \\ 0 & 1 & 1 \\ 0 & 0 & 0 \end{bmatrix}.$$

Solution Because A is in row-echelon form and has two nonzero rows, it has a rank of 2. Thus, the rank of T is 2, and the nullity is dim(domain) $-$ rank $= 3 - 2 = 1$.

REMARK: One way to visualize the relationship between the rank and the nullity of a linear transformation given by a matrix is to observe that the rank is determined by the number of leading 1's, and the nullity by the number of free variables (columns without leading 1's). Their sum must be the total number of columns of the matrix, which is the dimension of the domain. In Example 8, the first two columns have leading 1's, indicating that the rank is 2. The third column corresponds to a free variable, indicating that the nullity is 1.

EXAMPLE 9 *Finding the Rank and Nullity of a Linear Transformation*

Let $T: R^5 \to R^7$ be a linear transformation.

(a) Find the dimension of the kernel of T if the dimension of the range is 2.
(b) Find the rank of T if the nullity of T is 4.
(c) Find the rank of T if ker$(T) = \{0\}$.

Solution (a) By Theorem 6.5, with $n = 5$, you have

$$\dim(\text{kernel}) = n - \dim(\text{range}) = 5 - 2 = 3.$$

(b) Again by Theorem 6.5, you have

$$\text{rank}(T) = n - \text{nullity}(T) = 5 - 4 = 1.$$

(c) In this case, the nullity of T is 0. Hence

$$\text{rank}(T) = n - \text{nullity}(T) = 5 - 0 = 5.$$

One-to-One and Onto Linear Transformations

This section began with a question: how many vectors in the domain of a linear transformation are mapped to the zero vector? Theorem 6.6 shows that if the zero vector is the only vector $\mathbf{v}$ such that $T(\mathbf{v}) = \mathbf{0}$, then T is one-to-one. A function $T: V \to W$ is called **one-to-one** if the preimage of every $\mathbf{w}$ in the range consists of a single vector, as shown in Figure 6.7. This is equivalent to saying that T is one-to-one if and only if, for all $\mathbf{u}$ and $\mathbf{v}$ in V, $T(\mathbf{u}) = T(\mathbf{v})$ implies that $\mathbf{u} = \mathbf{v}$.

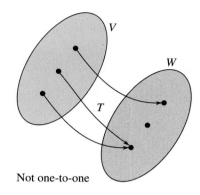

One-to-one   Not one-to-one

Figure 6.7

Theorem 6.6	*One-to-One Linear Transformations*

Let $T: V \to W$ be a linear transformation. Then T is one-to-one if and only if $\ker(T) = \{\mathbf{0}\}$.

Proof   Suppose T is one-to-one. Then $T(\mathbf{v}) = \mathbf{0}$ can have only one solution: $\mathbf{v} = \mathbf{0}$. In that case, $\ker(T) = \{\mathbf{0}\}$. Conversely, suppose $\ker(T) = \{\mathbf{0}\}$ and $T(\mathbf{u}) = T(\mathbf{v})$. Because T is a linear transformation, it follows that

$$T(\mathbf{u} - \mathbf{v}) = T(\mathbf{u}) - T(\mathbf{v}) = \mathbf{0}.$$

This implies that the vector $\mathbf{u} - \mathbf{v}$ lies in the kernel of T and must therefore equal $\mathbf{0}$. Hence, $\mathbf{u} - \mathbf{v} = \mathbf{0}$ and $\mathbf{u} = \mathbf{v}$, and you can conclude that T is one-to-one.

EXAMPLE 10   *One-to-One and Not One-to-One Linear Transformations*

(a) The linear transformation $T: M_{m,n} \to M_{n,m}$ given by $T(A) = A^T$ is one-to-one because its kernel consists of only the $m \times n$ zero matrix.

(b) The zero transformation $T: R^3 \to R^3$ is not one-to-one because its kernel is all of R^3.

A function $T: V \to W$ is said to be **onto** if every element in W has a preimage in V. In other words, T is onto W when W is equal to the range of T; hence, the following theorem.

Theorem 6.7	*Onto Linear Transformations*

Let $T: V \to W$ be a linear transformation, where W is finite dimensional. Then T is onto if and only if the rank of T is equal to the dimension of W.

For vector spaces of equal dimensions, you can combine the results of Theorems 6.5, 6.6, and 6.7 to obtain the following theorem relating the concepts of one-to-one and onto.

Theorem 6.8	One-to-One and Onto Linear Transformations

Let $T: V \to W$ be a linear transformation with vector spaces V and W *both* of dimension n. Then T is one-to-one if and only if it is onto.

Proof If T is one-to-one, then by Theorem 6.6 $\ker(T) = \{\mathbf{0}\}$, and $\dim(\ker(T)) = 0$. In that case, Theorem 6.5 produces

$$\dim(\text{range of } T) = n - \dim(\ker(T))$$
$$= n$$
$$= \dim(W).$$

Consequently, by Theorem 6.7, T is onto. Similarly, if T is onto, then

$$\dim(\text{range of } T) = \dim(W) = n,$$

which by Theorem 6.5 implies that $\dim(\ker(T)) = 0$. Therefore, by Theorem 6.6, T is one-to-one.

The next example brings together several ideas related to the kernel and range of a linear transformation.

EXAMPLE 11 *Summarizing Several Results*

The linear transformation $T: R^n \to R^m$ is given by $T(\mathbf{x}) = A\mathbf{x}$. Find the nullity and rank of T and determine whether T is one-to-one, onto, or neither.

(a) $A = \begin{bmatrix} 1 & 2 & 0 \\ 0 & 1 & 1 \\ 0 & 0 & 1 \end{bmatrix}$ (b) $A = \begin{bmatrix} 1 & 2 \\ 0 & 1 \\ 0 & 0 \end{bmatrix}$

(c) $A = \begin{bmatrix} 1 & 2 & 0 \\ 0 & 1 & -1 \end{bmatrix}$ (d) $A = \begin{bmatrix} 1 & 2 & 0 \\ 0 & 1 & 1 \\ 0 & 0 & 0 \end{bmatrix}$

Solution Note that each matrix is already in echelon form, so that its rank can be determined by inspection.

$T: R^n \to R^m$	Dim(domain)	Dim(range) Rank(T)	Dim(kernel) Nullity(T)	One-to-One	Onto
(a) $T: R^3 \to R^3$	3	3	0	Yes	Yes
(b) $T: R^2 \to R^3$	2	2	0	Yes	No
(c) $T: R^3 \to R^2$	3	2	1	No	Yes
(d) $T: R^3 \to R^3$	3	2	1	No	No

Isomorphisms of Vector Spaces

This section ends with a very important result that can be a great aid in your understanding of vector spaces. The result describes a way to think of distinct vector spaces as being "essentially the same"—at least with respect to the operations of vector addition and scalar multiplication. For example, the vector spaces R^3 and $M_{3,1}$ are essentially the same with respect to their standard operations. Such spaces are said to be **isomorphic** to each other. (The Greek word *isos* means "equal.")

Definition of Isomorphism

A linear transformation $T: V \to W$ that is one-to-one and onto is called an **isomorphism.** Moreover, if V and W are vector spaces such that there exists an isomorphism from V to W, then V and W are said to be **isomorphic** to each other.

One way in which isomorphic spaces are "essentially the same" is that they have the same dimensions, as stated in the following theorem. In fact, the theorem goes even further, stating that if two vector spaces have the same finite dimension, then they must be isomorphic.

Theorem 6.9

Isomorphic Spaces and Dimension

Two finite-dimensional vector spaces V and W are isomorphic if and only if they are of the same dimension.

Proof   Assume that V is isomorphic to W, where V has dimension n. By the definition of isomorphic spaces, you know that there exists a linear transformation $T: V \to W$ that is one-to-one and onto. Because T is one-to-one, it follows that $\dim(\text{kernel}) = 0$, which also implies that

$$\dim(\text{range}) = \dim(\text{domain}) = n.$$

In addition, because T is onto, you can conclude that

$$\dim(\text{range}) = \dim(W) = n.$$

To prove the theorem in the other direction, assume that V and W both have dimension n. Let $B = \{\mathbf{v}_1, \mathbf{v}_2, \ldots, \mathbf{v}_n\}$ be a basis of V, and let $B' = \{\mathbf{w}_1, \mathbf{w}_2, \ldots, \mathbf{w}_n\}$ be a basis of W. Then an arbitrary vector in V can be represented as

$$\mathbf{v} = c_1\mathbf{v}_1 + c_2\mathbf{v}_2 + \cdots + c_n\mathbf{v}_n,$$

and you can define a linear transformation $T: V \to W$ as follows.

$$T(\mathbf{v}) = c_1\mathbf{w}_1 + c_2\mathbf{w}_2 + \cdots + c_n\mathbf{w}_n$$

It can be shown that this linear transformation is both one-to-one and onto. Thus, V and W are isomorphic.

Our study of vector spaces has given much greater coverage to R^n than to other vector spaces. This preference for R^n stems from its notational convenience and from the geometric models available for R^2 and R^3. Theorem 6.9 tells you that R^n is a perfect model for every n-dimensional vector space. Example 12 lists some vector spaces that are isomorphic to R^4.

EXAMPLE 12   *Isomorphic Vector Spaces*

The following vector spaces are isomorphic to each other.

(a) $R^4 = $ 4-space
(b) $M_{4,1} = $ space of all 4×1 matrices
(c) $M_{2,2} = $ space of all 2×2 matrices
(d) $P_3 = $ space of all polynomials of degree 3 or less
(e) $V = \{(x_1, x_2, x_3, x_4, 0): x_i \text{ is a real number}\}$ (subspace of R^5)

Example 12 tells you that the elements in these spaces behave the same way *as vectors* even though they are distinct mathematical entities. Thus, the convention of using the notation for an n-tuple and an $n \times 1$ matrix interchangeably is justified.

SECTION 6.2   Exercises

In Exercises 1–6, find the kernel of the linear transformation.

1. $T: R^3 \to R^3$, $T(x, y, z) = (0, 0, 0)$
2. $T: R^4 \to R^4$, $T(x, y, z, w) = (y, x, w, z)$
3. $T: P_3 \to R$, $T(a_0 + a_1 x + a_2 x^2 + a_3 x^3) = a_0$
4. $T: P_2 \to P_1$, $T(a_0 + a_1 x + a_2 x^2) = a_1 + 2a_2 x$
5. $T: R^2 \to R^2$, $T(x, y) = (x + 2y, y - x)$
6. $T: R^2 \to R^2$, $T(x, y) = (x - y, y - x)$

In Exercises 7–12, the linear transformation T is given by $T(\mathbf{v}) = A\mathbf{v}$. Find a basis for (a) the kernel of T and (b) the range of T.

7. $A = \begin{bmatrix} 1 & 2 \\ 3 & 4 \end{bmatrix}$
8. $A = \begin{bmatrix} 1 & 2 \\ -2 & -4 \end{bmatrix}$

9. $A = \begin{bmatrix} 1 & -1 & 2 \\ 0 & 1 & 2 \end{bmatrix}$
10. $A = \begin{bmatrix} 1 & 2 \\ -1 & -2 \\ 1 & 1 \end{bmatrix}$

11. $A = \begin{bmatrix} 1 & 2 & -1 & 4 \\ 3 & 1 & 2 & -1 \\ -4 & -3 & -1 & -3 \\ -1 & -2 & 1 & 1 \end{bmatrix}$

12. $A = \begin{bmatrix} -1 & 3 & 2 & 1 & 4 \\ 2 & 3 & 5 & 0 & 0 \\ 2 & 1 & 2 & 1 & 0 \end{bmatrix}$

In Exercises 13–24, the linear transformation T is given by $T(\mathbf{x}) = A\mathbf{x}$. Find (a) ker($T$), (b) nullity($T$), (c) range($T$), and (d) rank($T$).

13. $A = \begin{bmatrix} -1 & 1 \\ 1 & 1 \end{bmatrix}$
14. $A = \begin{bmatrix} 3 & 2 \\ -9 & -6 \end{bmatrix}$

15. $A = \begin{bmatrix} 5 & -3 \\ 1 & 1 \\ 1 & -1 \end{bmatrix}$
16. $A = \begin{bmatrix} 4 & 1 \\ 0 & 0 \\ 2 & -3 \end{bmatrix}$

17. $A = \begin{bmatrix} 0 & -2 & 3 \\ 4 & 0 & 11 \end{bmatrix}$
18. $A = \begin{bmatrix} 1 & 1 & 0 & 0 \\ 0 & 0 & 1 & 1 \end{bmatrix}$

19. $A = \begin{bmatrix} \frac{9}{10} & \frac{3}{10} \\ \frac{3}{10} & \frac{1}{10} \end{bmatrix}$
20. $A = \begin{bmatrix} \frac{1}{26} & -\frac{5}{26} \\ -\frac{5}{26} & \frac{25}{26} \end{bmatrix}$

21. $A = \begin{bmatrix} 2 & 2 & -3 & 1 & 13 \\ 1 & 1 & 1 & 1 & -1 \\ 3 & 3 & -5 & 0 & 14 \\ 6 & 6 & -2 & 4 & 16 \end{bmatrix}$

Data

22. $A = \begin{bmatrix} 3 & -2 & 6 & -1 & 15 \\ 4 & 3 & 8 & 10 & -14 \\ 2 & -3 & 4 & -4 & 20 \end{bmatrix}$

Data

23. $A = \begin{bmatrix} \frac{4}{9} & -\frac{4}{9} & \frac{2}{9} \\ -\frac{4}{9} & \frac{4}{9} & -\frac{2}{9} \\ \frac{2}{9} & -\frac{2}{9} & \frac{1}{9} \end{bmatrix}$

24. $A = \begin{bmatrix} 1 & 0 & 0 \\ 0 & 0 & 0 \\ 0 & 0 & 1 \end{bmatrix}$

In Exercises 25–32, let $T: R^3 \to R^3$ be a linear transformation. Use the given information to find the nullity of T and give a geometric description of the kernel and range of T.

25. rank(T) = 2   **26.** rank(T) = 1
27. rank(T) = 0   **28.** rank(T) = 3

29. T is the counterclockwise rotation of $45°$ about the z-axis:

$$T(x, y, z) = \left(\frac{\sqrt{2}}{2}x - \frac{\sqrt{2}}{2}y, \frac{\sqrt{2}}{2}x + \frac{\sqrt{2}}{2}y, z \right)$$

30. T is the reflection through the yz-coordinate plane:
$T(x, y, z) = (-x, y, z)$

31. T is the projection onto the vector $\mathbf{v} = (1, 2, 2)$:

$$T(x, y, z) = \frac{x + 2y + 2z}{9}(1, 2, 2)$$

32. T is the projection onto the xy-coordinate plane:
$T(x, y, z) = (x, y, 0)$

In Exercises 33–36, find the nullity of T.

33. $T: R^4 \to R^2$, rank(T) = 2   **34.** $T: R^5 \to R^2$, rank(T) = 2
35. $T: R^4 \to R^4$, rank(T) = 0   **36.** $T: P_3 \to P_1$, rank(T) = 2

37. Identify the zero element and standard basis for each of the isomorphic vector spaces in Example 12.

38. Which vector spaces are isomorphic to R^6?
(a) $M_{2,3}$ (b) P_6 (c) $C[0, 6]$
(d) $M_{6,1}$ (e) P_5
(f) $\{(x_1, x_2, x_3, 0, x_5, x_6, x_7): x_i \text{ is a real number}\}$

39. Calculus Let $T: P_4 \to P_3$ be given by $T(p) = p'$. What is the kernel of T?

40. Calculus Let $T: P_2 \to R$ be given by

$$T(p) = \int_0^1 p(x) \, dx.$$

What is the kernel of T?

41. Let $T: R^3 \to R^3$ be the linear transformation that projects $\mathbf{u}$ onto $\mathbf{v} = (2, -1, 1)$.
(a) Find the rank and nullity of T.
(b) Find a basis for the kernel of T.

42. Repeat Exercise 41 for $\mathbf{v} = (3, 0, 4)$.

In Exercises 43–46, verify that the matrix defines a linear function T that is one-to-one and onto.

43. $A = \begin{bmatrix} -1 & 0 \\ 0 & 1 \end{bmatrix}$ **44.** $A = \begin{bmatrix} 1 & 0 \\ 0 & -1 \end{bmatrix}$

45. $A = \begin{bmatrix} 1 & 0 & 0 \\ 0 & 0 & 1 \\ 0 & 1 & 0 \end{bmatrix}$ **46.** $A = \begin{bmatrix} 1 & 2 & 3 \\ -1 & 2 & 4 \\ 0 & 4 & 1 \end{bmatrix}$

True or False? In Exercises 47 and 48, determine whether each statement is true or false. If a statement is true, give a reason or cite an appropriate statement in the text. If a statement is false, provide an example that shows that the statement is not true in all cases or cite an appropriate statement in the text.

47. (a) The set of all vectors mapped from a vector space V to another vector space W by a linear transformation T is known as the kernel of T.
(b) The range of a linear transformation from a vector space V to a vector space W is a subspace of the vector space V.
(c) A linear transformation T from V to W is called one-to-one if and only if for all $\mathbf{u}$ and $\mathbf{v}$ in V, $T(\mathbf{u}) = T(\mathbf{v})$ implies that $\mathbf{u} = \mathbf{v}$.
(d) The vector spaces R^3 and $M_{3,1}$ are isomorphic to each other.

48. (a) The kernel of a linear transformation T from a vector space V to a vector space W is a subspace of the vector space V.
(b) The dimension of a linear transformation T from a vector space V to a vector space W is called the rank of T.
(c) A linear transformation T from V to W is one-to-one if the preimage of every $\mathbf{w}$ in the range consists of a single vector $\mathbf{v}$.
(d) The vector spaces R^2 and P_1 are isomorphic to each other.

49. For the transformation $T: R^n \to R^n$ given by $T(\mathbf{v}) = A\mathbf{v}$, what can be said about the rank of T if (a) det(A) $\neq 0$ and (b) det(A) = 0?

50. Let $T: M_{n,n} \to M_{n,n}$ be given by $T(A) = A - A^T$. Show that the kernel of T is the set of $n \times n$ symmetric matrices.

51. Determine a relationship among m, n, j, and k such that $M_{m,n}$ is isomorphic to $M_{j,k}$.

52. *Guided Proof* Let B be an invertible $n \times n$ matrix. Prove that the linear transformation $T: M_{n,n} \to M_{n,n}$ given by $T(A) = AB$ is an isomorphism.

Getting Started: To show that the linear transformation is an isomorphism, you need to show that T is both onto and one-to-one.

(i) Because T is a linear transformation with vector spaces of equal dimension, then by Theorem 6.8 you only need to show that T is one-to-one.

(ii) To show that T is one-to-one, you need to determine the kernel of T and show that it is $\{\mathbf{0}\}$ (Theorem 6.6). Use the fact that B is an invertible $n \times n$ matrix and $T(A) = AB$.

(iii) Conclude that T is an isomorphism.

53. Let $T: V \to W$ be a linear transformation. Prove that T is one-to-one if and only if the rank of T equals the dimension of V.

54. Let $T: V \to W$ be a linear transformation, and let U be a subspace of W. Prove that the set $T^{-1}(U) = \{\mathbf{v} \in V : T(\mathbf{v}) \in U\}$ is a subspace of V. What is $T^{-1}(U)$ if $U = \{\mathbf{0}\}$?

55. *Writing* Are the vector spaces R^4, $M_{2,2}$, and $M_{1,4}$ exactly the same? Describe their similarities and differences.

56. *Writing* Let $T: R^m \to R^n$ be a linear transformation. Explain the differences between the concepts of one-to-one and onto. What can you say about m and n if T is onto? What can you say about m and n if T is one-to-one?

6.3 | Matrices for Linear Transformations

Which representation of $T: R^3 \to R^3$ is better,

$$T(x_1, x_2, x_3) = (2x_1 + x_2 - x_3, -x_1 + 3x_2 - 2x_3, 3x_2 + 4x_3)$$

or

$$T(\mathbf{x}) = A\mathbf{x} = \begin{bmatrix} 2 & 1 & -1 \\ -1 & 3 & -2 \\ 0 & 3 & 4 \end{bmatrix} \begin{bmatrix} x_1 \\ x_2 \\ x_3 \end{bmatrix}?$$

The second representation is better than the first for at least three reasons: it is simpler to write, simpler to read, and more easily adapted for computer use. Later you will see that matrix representation of a linear transformation also has some theoretical advantages. In this section you will see that for linear transformations involving finite-dimensional vector spaces, matrix representation is always possible.

The key to representing a linear transformation $T: V \to W$ by a matrix is to determine how it acts on a basis of V. Once you know the image of every vector in the basis, you can use the properties of linear transformations to determine $T(\mathbf{v})$ for any $\mathbf{v}$ in V.

For convenience, the first three theorems in this section are stated in terms of linear transformations from R^n into R^m, relative to the standard bases in R^n and R^m. At the end of the section these results are generalized to include nonstandard bases and general vector spaces.

Recall that the standard basis for R^n, written in column vector notation, is given by

$$B = \{\mathbf{e}_1, \mathbf{e}_2, \ldots, \mathbf{e}_n\} = \left\{ \begin{bmatrix} 1 \\ 0 \\ \vdots \\ 0 \end{bmatrix}, \begin{bmatrix} 0 \\ 1 \\ \vdots \\ 0 \end{bmatrix}, \ldots, \begin{bmatrix} 0 \\ 0 \\ \vdots \\ 1 \end{bmatrix} \right\}.$$

| Theorem 6.10 | *Standard Matrix for a Linear Transformation* |

Let $T: R^n \to R^m$ be a linear transformation such that

$$T(\mathbf{e}_1) = \begin{bmatrix} a_{11} \\ a_{21} \\ \vdots \\ a_{m1} \end{bmatrix}, \quad T(\mathbf{e}_2) = \begin{bmatrix} a_{12} \\ a_{22} \\ \vdots \\ a_{m2} \end{bmatrix}, \dots, T(\mathbf{e}_n) = \begin{bmatrix} a_{1n} \\ a_{2n} \\ \vdots \\ a_{mn} \end{bmatrix}.$$

Then the $m \times n$ matrix whose n columns correspond to $T(\mathbf{e}_i)$,

$$A = \begin{bmatrix} a_{11} & a_{12} & \cdots & a_{1n} \\ a_{21} & a_{22} & \cdots & a_{2n} \\ \vdots & \vdots & & \vdots \\ a_{m1} & a_{m2} & \cdots & a_{mn} \end{bmatrix},$$

is such that $T(\mathbf{v}) = A\mathbf{v}$ for every $\mathbf{v}$ in R^n. A is called the **standard matrix** for T.

Proof   To show that $T(\mathbf{v}) = A\mathbf{v}$ for any $\mathbf{v}$ in R^n, you can write

$$\mathbf{v} = \begin{bmatrix} v_1 \\ v_2 \\ \vdots \\ v_n \end{bmatrix} = v_1\mathbf{e}_1 + v_2\mathbf{e}_2 + \cdots + v_n\mathbf{e}_n.$$

Because T is a linear transformation, you have

$$\begin{aligned} T(\mathbf{v}) &= T(v_1\mathbf{e}_1 + v_2\mathbf{e}_2 + \cdots + v_n\mathbf{e}_n) \\ &= T(v_1\mathbf{e}_1) + T(v_2\mathbf{e}_2) + \cdots + T(v_n\mathbf{e}_n) \\ &= v_1 T(\mathbf{e}_1) + v_2 T(\mathbf{e}_2) + \cdots + v_n T(\mathbf{e}_n). \end{aligned}$$

On the other hand, the matrix product $A\mathbf{v}$ is given by

$$A\mathbf{v} = \begin{bmatrix} a_{11} & a_{12} & \cdots & a_{1n} \\ a_{21} & a_{22} & \cdots & a_{2n} \\ \vdots & \vdots & & \vdots \\ a_{m1} & a_{m2} & \cdots & a_{mn} \end{bmatrix} \begin{bmatrix} v_1 \\ v_2 \\ \vdots \\ v_n \end{bmatrix} = \begin{bmatrix} a_{11}v_1 + a_{12}v_2 + \cdots + a_{1n}v_n \\ a_{21}v_1 + a_{22}v_2 + \cdots + a_{2n}v_n \\ \vdots & & \vdots \\ a_{m1}v_1 + a_{m2}v_2 + \cdots + a_{mn}v_n \end{bmatrix}$$

$$= v_1 \begin{bmatrix} a_{11} \\ a_{21} \\ \vdots \\ a_{m1} \end{bmatrix} + v_2 \begin{bmatrix} a_{12} \\ a_{22} \\ \vdots \\ a_{m2} \end{bmatrix} + \cdots + v_n \begin{bmatrix} a_{1n} \\ a_{2n} \\ \vdots \\ a_{mn} \end{bmatrix}$$

$$= v_1 T(\mathbf{e}_1) + v_2 T(\mathbf{e}_2) + \cdots + v_n T(\mathbf{e}_n).$$

Therefore, $T(\mathbf{v}) = A\mathbf{v}$ for each $\mathbf{v}$ in R^n.

EXAMPLE 1   *Finding the Standard Matrix of a Linear Transformation*

Find the standard matrix for the linear transformation $T: R^3 \to R^2$ defined by

$$T(x, y, z) = (x - 2y, 2x + y).$$

Solution   Begin by finding the images of $e_1, e_2,$ and e_3.

Vector Notation   *Matrix Notation*

$$T(e_1) = T(1, 0, 0) = (1, 2) \qquad T(e_1) = T\left(\begin{bmatrix} 1 \\ 0 \\ 0 \end{bmatrix}\right) = \begin{bmatrix} 1 \\ 2 \end{bmatrix}$$

$$T(e_2) = T(0, 1, 0) = (-2, 1) \qquad T(e_2) = T\left(\begin{bmatrix} 0 \\ 1 \\ 0 \end{bmatrix}\right) = \begin{bmatrix} -2 \\ 1 \end{bmatrix}$$

$$T(e_3) = T(0, 0, 1) = (0, 0) \qquad T(e_3) = T\left(\begin{bmatrix} 0 \\ 0 \\ 1 \end{bmatrix}\right) = \begin{bmatrix} 0 \\ 0 \end{bmatrix}$$

By Theorem 6.10, the columns of A consist of $T(e_1)$, $T(e_2)$, and $T(e_3)$, and you have

$$A = [T(e_1) \;\vdots\; T(e_2) \;\vdots\; T(e_3)] = \begin{bmatrix} 1 & -2 & 0 \\ 2 & 1 & 0 \end{bmatrix}.$$

As a check, note that

$$A \begin{bmatrix} x \\ y \\ z \end{bmatrix} = \begin{bmatrix} 1 & -2 & 0 \\ 2 & 1 & 0 \end{bmatrix} \begin{bmatrix} x \\ y \\ z \end{bmatrix} = \begin{bmatrix} x - 2y \\ 2x + y \end{bmatrix},$$

which is equivalent to $T(x, y, z) = (x - 2y, 2x + y)$.

A little practice will enable you to determine the standard matrix for a linear transformation, such as the one in Example 1, by inspection. For instance, the standard matrix for the linear transformation given by

$$T(x_1, x_2, x_3) = (x_1 - 2x_2 + 5x_3, 2x_1 + 3x_3, 4x_1 + x_2 - 2x_3)$$

is found by using the coefficients of $x_1, x_2,$ and x_3 to form the rows of A as follows.

$$A = \begin{bmatrix} 1 & -2 & 5 \\ 2 & 0 & 3 \\ 4 & 1 & -2 \end{bmatrix} \begin{matrix} \longleftarrow & 1x_1 - 2x_2 + 5x_3 \\ \longleftarrow & 2x_1 + 0x_2 + 3x_3 \\ \longleftarrow & 4x_1 + 1x_2 - 2x_3 \end{matrix}$$

EXAMPLE 2   *Finding the Standard Matrix of a Linear Transformation*

The linear transformation $T: R^2 \to R^2$ is given by projecting each point in R^2 onto the x-axis, as shown in Figure 6.8. Find the standard matrix for T.

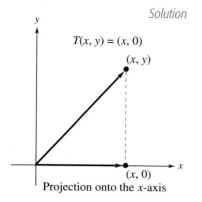

y

$T(x, y) = (x, 0)$

(x, y)

$(x, 0)$ x

Projection onto the x-axis

Figure 6.8

Solution   This linear transformation is given by

$$T(x, y) = (x, 0).$$

Therefore, the standard matrix for T is

$$A = [T(1, 0) \;\vdots\; T(0, 1)]$$

$$= \begin{bmatrix} 1 & 0 \\ 0 & 0 \end{bmatrix}.$$

The standard matrix for the zero transformation from R^n into R^m is the $m \times n$ zero matrix, and the standard matrix for the identity transformation from R^n into R^n is I_n.

Composition of Linear Transformations

The **composition,** T, of $T_1: R^n \rightarrow R^m$ with $T_2: R^m \rightarrow R^p$ is defined by

$$T(\mathbf{v}) = T_2(T_1(\mathbf{v})),$$

where $\mathbf{v}$ is a vector in R^n. This composition is denoted by

$$T = T_2 \circ T_1.$$

The domain of T is defined to be the domain of T_1. Moreover, the composition is not defined unless the range of T_1 lies within the domain of T_2, as shown in Figure 6.9.

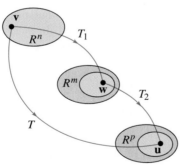

Composition of Transformations

Figure 6.9

The following theorem emphasizes the usefulness of matrices for representing linear transformations. The theorem not only states that the composition of two linear transformations is a linear transformation, but also says that the standard matrix of the composition is the product of the standard matrices of the two original linear transformations.

Theorem 6.11	*Composition of Linear Transformations*

Let $T_1: R^n \to R^m$ and $T_2: R^m \to R^p$ be linear transformations with standard matrices A_1 and A_2. The **composition** $T: R^n \to R^p$, defined by $T(\mathbf{v}) = T_2(T_1(\mathbf{v}))$, is a linear transformation. Moreover, the standard matrix A for T is given by the matrix product

$$A = A_2 A_1.$$

Proof To show that T is a linear transformation, let $\mathbf{u}$ and $\mathbf{v}$ be vectors in R^n and let c be any scalar. Then, because T_1 and T_2 are linear transformations, you can write

$$\begin{aligned} T(\mathbf{u} + \mathbf{v}) &= T_2(T_1(\mathbf{u} + \mathbf{v})) \\ &= T_2(T_1(\mathbf{u}) + T_1(\mathbf{v})) \\ &= T_2(T_1(\mathbf{u})) + T_2(T_1(\mathbf{v})) = T(\mathbf{u}) + T(\mathbf{v}) \end{aligned}$$

$$\begin{aligned} T(c\mathbf{v}) &= T_2(T_1(c\mathbf{v})) \\ &= T_2(cT_1(\mathbf{v})) \\ &= cT_2(T_1(\mathbf{v})) = cT(\mathbf{v}). \end{aligned}$$

REMARK: Theorem 6.11 can be generalized to cover the composition of n linear transformations. That is, if the standard matrices of $T_1, T_2, \ldots, T_n$ are $A_1, A_2, \ldots, A_n$, then the standard matrix of the composition T is given by

$$A = A_n A_{n-1} \cdots A_2 A_1.$$

Now, to show that $A_2 A_1$ is the standard matrix for T, use the associative property of matrix multiplication to write

$$T(\mathbf{v}) = T_2(T_1(\mathbf{v})) = T_2(A_1\mathbf{v}) = A_2(A_1\mathbf{v}) = (A_2 A_1)\mathbf{v}.$$

Because matrix multiplication is not commutative, order is important when the compositions of linear transformations are formed. In general, the composition $T_2 \circ T_1$ is not the same as $T_1 \circ T_2$, as demonstrated in the next example.

EXAMPLE 3 *The Standard Matrix of a Composition*

Let T_1 and T_2 be linear transformations from R^3 into R^3 such that

$$T_1(x, y, z) = (2x + y, 0, x + z) \qquad \text{and} \qquad T_2(x, y, z) = (x - y, z, y).$$

Find the standard matrices for the compositions $T = T_2 \circ T_1$ and $T' = T_1 \circ T_2$.

Solution The standard matrices for T_1 and T_2 are

$$A_1 = \begin{bmatrix} 2 & 1 & 0 \\ 0 & 0 & 0 \\ 1 & 0 & 1 \end{bmatrix} \qquad \text{and} \qquad A_2 = \begin{bmatrix} 1 & -1 & 0 \\ 0 & 0 & 1 \\ 0 & 1 & 0 \end{bmatrix}.$$

Thus, by Theorem 6.11, the standard matrix for T is given by

$$A = A_2 A_1 = \begin{bmatrix} 1 & -1 & 0 \\ 0 & 0 & 1 \\ 0 & 1 & 0 \end{bmatrix} \begin{bmatrix} 2 & 1 & 0 \\ 0 & 0 & 0 \\ 1 & 0 & 1 \end{bmatrix} = \begin{bmatrix} 2 & 1 & 0 \\ 1 & 0 & 1 \\ 0 & 0 & 0 \end{bmatrix},$$

and the standard matrix for T' is given by

$$A' = A_1 A_2 = \begin{bmatrix} 2 & 1 & 0 \\ 0 & 0 & 0 \\ 1 & 0 & 1 \end{bmatrix} \begin{bmatrix} 1 & -1 & 0 \\ 0 & 0 & 1 \\ 0 & 1 & 0 \end{bmatrix} = \begin{bmatrix} 2 & -2 & 1 \\ 0 & 0 & 0 \\ 1 & 0 & 0 \end{bmatrix}.$$

Another benefit of matrix representation is that it can represent the **inverse** of a linear transformation. Before seeing how this works, consider the following definition.

Definition of Inverse Linear Transformation

If $T_1: R^n \rightarrow R^n$ and $T_2: R^n \rightarrow R^n$ are linear transformations such that for every $\mathbf{v}$ in R^n

$$T_2(T_1(\mathbf{v})) = \mathbf{v} \quad \text{and} \quad T_1(T_2(\mathbf{v})) = \mathbf{v},$$

then T_2 is called the **inverse** of T_1 and T_1 is said to be **invertible.**

Not every linear transformation has an inverse. However, if the transformation T_1 is invertible, then the inverse is unique and denoted by T_1^{-1}.

Just as the inverse of a function of a real variable can be thought of as undoing what the function did, the inverse of a linear transformation T can be thought of as undoing the mapping done by T. For instance, if T is a linear transformation from R^3 onto R^3 such that

$$T(1, 4, -5) = (2, 3, 1)$$

and if T^{-1} exists, then T^{-1} maps $(2, 3, 1)$ back to its preimage under T. That is,

$$T^{-1}(2, 3, 1) = (1, 4, -5).$$

The following theorem states that a linear transformation is invertible if and only if it is an isomorphism (one-to-one and onto). You are asked to prove this theorem in Exercise 62.

Theorem 6.12

Existence of an Inverse Transformation

Let $T: R^n \rightarrow R^n$ be a linear transformation with standard matrix A. Then the following conditions are equivalent.

1. T is invertible.
2. T is an isomorphism.
3. A is invertible.

And, if T is invertible with standard matrix A, then the standard matrix for T^{-1} is A^{-1}.

REMARK: Several other conditions are equivalent to the three given in Theorem 6.12; see the summary of equivalent conditions given in Section 4.6.

EXAMPLE 4 *Finding the Inverse of a Linear Transformation*

The linear transformation $T: R^3 \rightarrow R^3$ is defined by

$$T(x_1, x_2, x_3) = (2x_1 + 3x_2 + x_3, 3x_1 + 3x_2 + x_3, 2x_1 + 4x_2 + x_3).$$

Show that T is invertible, and find its inverse.

Solution The standard matrix for T is

$$A = \begin{bmatrix} 2 & 3 & 1 \\ 3 & 3 & 1 \\ 2 & 4 & 1 \end{bmatrix}.$$

Using the techniques for matrix inversion (see Section 2.3), you can find that A is invertible and its inverse is given by

$$A^{-1} = \begin{bmatrix} -1 & 1 & 0 \\ -1 & 0 & 1 \\ 6 & -2 & -3 \end{bmatrix}.$$

Therefore, T is invertible and its standard matrix is A^{-1}.

Using the standard matrix for the inverse, you can find the rule for T^{-1} by computing the image of an arbitrary vector $\mathbf{v} = (x_1, x_2, x_3)$.

$$A^{-1}\mathbf{v} = \begin{bmatrix} -1 & 1 & 0 \\ -1 & 0 & 1 \\ 6 & -2 & -3 \end{bmatrix} \begin{bmatrix} x_1 \\ x_2 \\ x_3 \end{bmatrix} = \begin{bmatrix} -x_1 + x_2 \\ -x_1 + x_3 \\ 6x_1 - 2x_2 - 3x_3 \end{bmatrix}$$

In other words,

$$T^{-1}(x_1, x_2, x_3) = (-x_1 + x_2, -x_1 + x_3, 6x_1 - 2x_2 - 3x_3).$$

Nonstandard Bases and General Vector Spaces

You will now consider the more general problem of finding a matrix for a linear transformation $T: V \rightarrow W$, where B and B' are ordered bases for V and W, respectively. Recall that the coordinate matrix of $\mathbf{v}$ relative to B is denoted by $[\mathbf{v}]_B$. In order to represent the linear transformation T, A must be multiplied by a *coordinate matrix relative to B*. The result of the multiplication will be a *coordinate matrix relative to B'*. That is,

$$[T(\mathbf{v})]_{B'} = A[\mathbf{v}]_B.$$

A is called the **matrix of T relative to the bases B and B'.**

To find the matrix A, you will use a procedure similar to the one used to find the standard matrix for T. That is, the images of the vectors in B are written as coordinate matrices relative to the basis B'. These coordinate matrices form the columns of A.

Transformation Matrix for Nonstandard Bases

Let V and W be finite-dimensional vector spaces with bases B and B', respectively, where

$$B = \{\mathbf{v}_1, \mathbf{v}_2, \ldots, \mathbf{v}_n\}.$$

If $T: V \to W$ is a linear transformation such that

$$[T(\mathbf{v}_1)]_{B'} = \begin{bmatrix} a_{11} \\ a_{21} \\ \vdots \\ a_{m1} \end{bmatrix}, [T(\mathbf{v}_2)]_{B'} = \begin{bmatrix} a_{12} \\ a_{22} \\ \vdots \\ a_{m2} \end{bmatrix}, \ldots, [T(\mathbf{v}_n)]_{B'} = \begin{bmatrix} a_{1n} \\ a_{2n} \\ \vdots \\ a_{mn} \end{bmatrix},$$

then the $m \times n$ matrix whose n columns correspond to $[T(\mathbf{v}_i)]_{B'}$,

$$A = \begin{bmatrix} a_{11} & a_{12} & \cdots & a_{1n} \\ a_{21} & a_{22} & \cdots & a_{2n} \\ \vdots & \vdots & & \vdots \\ a_{m1} & a_{m2} & \cdots & a_{mn} \end{bmatrix},$$

is such that $[T(\mathbf{v})]_{B'} = A[\mathbf{v}]_B$ for every $\mathbf{v}$ in V.

EXAMPLE 5   *Finding a Matrix Relative to Nonstandard Bases*

Let $T: R^2 \to R^2$ be a linear transformation defined by

$$T(x_1, x_2) = (x_1 + x_2, 2x_1 - x_2).$$

Find the matrix of T relative to the bases

$$B = \{\overset{\mathbf{v}_1}{(1, 2)}, \overset{\mathbf{v}_2}{(-1, 1)}\} \quad \text{and} \quad B' = \{\overset{\mathbf{w}_1}{(1, 0)}, \overset{\mathbf{w}_2}{(0, 1)}\}.$$

Solution   By definition of T, you have

$$T(\mathbf{v}_1) = T(1, 2) = (3, 0) = 3\mathbf{w}_1 + 0\mathbf{w}_2$$
$$T(\mathbf{v}_2) = T(-1, 1) = (0, -3) = 0\mathbf{w}_1 - 3\mathbf{w}_2.$$

Therefore, the coordinate matrices of $T(\mathbf{v}_1)$ and $T(\mathbf{v}_2)$ relative to B' are

$$[T(\mathbf{v}_1)]_{B'} = \begin{bmatrix} 3 \\ 0 \end{bmatrix} \quad \text{and} \quad [T(\mathbf{v}_2)]_{B'} = \begin{bmatrix} 0 \\ -3 \end{bmatrix}.$$

The matrix for T relative to B and B' is formed by using these coordinate matrices as columns to produce

$$A = \begin{bmatrix} 3 & 0 \\ 0 & -3 \end{bmatrix}.$$

EXAMPLE 6   *Using a Matrix to Represent a Linear Transformation*

For the linear transformation $T: R^2 \rightarrow R^2$ given in Example 5, use the matrix A to find $T(\mathbf{v})$, where $\mathbf{v} = (2, 1)$.

Solution   Using the basis $B = \{(1, 2), (-1, 1)\}$, you find that

$$\mathbf{v} = (2, 1) = 1(1, 2) - 1(-1, 1),$$

which implies that

$$[\mathbf{v}]_B = \begin{bmatrix} 1 \\ -1 \end{bmatrix}.$$

Therefore, $[T(\mathbf{v})]_{B'}$ is given by

$$A[\mathbf{v}]_B = \begin{bmatrix} 3 & 0 \\ 0 & -3 \end{bmatrix} \begin{bmatrix} 1 \\ -1 \end{bmatrix} = \begin{bmatrix} 3 \\ 3 \end{bmatrix} = [T(\mathbf{v})]_{B'}.$$

Finally, because $B' = \{(1, 0), (0, 1)\}$, it follows that

$$T(\mathbf{v}) = 3(1, 0) + 3(0, 1) = (3, 3).$$

You can check this result by directly calculating $T(\mathbf{v})$ using the definition of T given in Example 5:

$$T(2, 1) = (2 + 1, 2(2) - 1) = (3, 3).$$

In the special case when $V = W$ and $B = B'$, the matrix A is called the **matrix of T relative to the basis B.** In such cases the matrix of the identity transformation is simply I_n. To see this, let $B = \{\mathbf{v}_1, \mathbf{v}_2, \ldots, \mathbf{v}_n\}$. Because the identity transformation maps each $\mathbf{v}_i$ to itself, you have

$$[T(\mathbf{v}_1)]_B = \begin{bmatrix} 1 \\ 0 \\ \vdots \\ 0 \end{bmatrix}, [T(\mathbf{v}_2)]_B = \begin{bmatrix} 0 \\ 1 \\ \vdots \\ 0 \end{bmatrix}, \ldots, [T(\mathbf{v}_n)]_B = \begin{bmatrix} 0 \\ 0 \\ \vdots \\ 1 \end{bmatrix},$$

and it follows that $A = I_n$.

In the next example you will construct a matrix representing the differential operator discussed in Example 10 in Section 6.1.

EXAMPLE 7   *A Matrix for the Differential Operator (Calculus)*

Let $D_x: P_2 \rightarrow P_1$ be the differential operator that maps a quadratic polynomial p onto its derivative p'. Find the matrix for D_x using the bases

$$B = \{1, x, x^2\} \quad \text{and} \quad B' = \{1, x\}.$$

Solution   The derivatives of the basis vectors are

$$D_x(1) = 0 = 0(1) + 0(x)$$
$$D_x(x) = 1 = 1(1) + 0(x)$$
$$D_x(x^2) = 2x = 0(1) + 2(x).$$

Thus, the coordinate matrices relative to B' are

$$[D_x(1)]_{B'} = \begin{bmatrix} 0 \\ 0 \end{bmatrix}, \quad [D_x(x)]_{B'} = \begin{bmatrix} 1 \\ 0 \end{bmatrix}, \quad [D_x(x^2)]_{B'} = \begin{bmatrix} 0 \\ 2 \end{bmatrix},$$

and the matrix for D_x is

$$A = \begin{bmatrix} 0 & 1 & 0 \\ 0 & 0 & 2 \end{bmatrix}.$$

Note that this matrix *does* produce the derivative of a quadratic polynomial $p(x) = a + bx + cx^2$.

$$Ap = \begin{bmatrix} 0 & 1 & 0 \\ 0 & 0 & 2 \end{bmatrix} \begin{bmatrix} a \\ b \\ c \end{bmatrix} = \begin{bmatrix} b \\ 2c \end{bmatrix} \implies b + 2cx = D_x[a + bx + cx^2]$$

SECTION 6.3   Exercises

In Exercises 1–8, find the standard matrix for the linear transformation T.

1. $T(x, y) = (x + 2y, x - 2y)$
2. $T(x, y) = (3x + 2y, 2y - x)$
3. $T(x, y) = (2x - 3y, x - y, y - 4x)$
4. $T(x, y) = (4x + y, 0, 2x - 3y)$
5. $T(x, y, z) = (x + y, x - y, z - x)$
6. $T(x, y, z) = (5x - 3y + z, 2z + 4y, 5x + 3y)$
7. $T(x, y, z) = (3z - 2y, 4x + 11z)$
8. $T(x_1, x_2, x_3, x_4) = (0, 0, 0, 0)$

In Exercises 9–12, use the standard matrix for the linear transformation T to find the image of the vector $\mathbf{v}$.

9. $T(x, y, z) = (13x - 9y + 4z, 6x + 5y - 3z)$, $\mathbf{v} = (1, -2, 1)$
10. $T(x, y) = (x + y, x - y, 2x, 2y)$, $\mathbf{v} = (3, -3)$
11. $T(x_1, x_2, x_3, x_4) = (x_1 + x_2, x_3 + x_4)$, $\mathbf{v} = (1, -1, 1, -1)$
12. $T(x_1, x_2, x_3, x_4) = (2x_1 - x_3, 3x_2 - 4x_4, 4x_3 - x_1, x_2 + x_4)$, $\mathbf{v} = (1, 2, 3, -2)$

In Exercises 13–24, (a) find the standard matrix A for the linear transformation T, (b) use A to find the image of the vector $\mathbf{v}$, and (c) sketch the graph of $\mathbf{v}$ and its image.

13. T is the reflection through the origin in R^2: $T(x, y) = (-x, -y)$, $\mathbf{v} = (3, 4)$.
14. T is the reflection in the line $y = x$ in R^2: $T(x, y) = (y, x)$, $\mathbf{v} = (3, 4)$.
15. T is the counterclockwise rotation of $135°$ in R^2, $\mathbf{v} = (4, 4)$.
16. T is the clockwise rotation (θ is negative) of $60°$ in R^2, $\mathbf{v} = (1, 2)$.
17. T is the reflection through the xy-coordinate plane in R^3: $T(x, y, z) = (x, y, -z)$, $\mathbf{v} = (3, 2, 2)$.
18. T is the reflection through the yz-coordinate plane in R^3: $T(x, y, z) = (-x, y, z)$, $\mathbf{v} = (2, 3, 4)$.
19. T is the counterclockwise rotation of $180°$ in R^2, $\mathbf{v} = (1, 2)$.
20. T is the counterclockwise rotation of $45°$ in R^2, $\mathbf{v} = (2, 2)$.
21. T is the projection onto the vector $\mathbf{w} = (3, 1)$ in R^2: $T(\mathbf{v}) = \text{proj}_{\mathbf{w}} \mathbf{v}$, $\mathbf{v} = (1, 4)$.

22. T is the projection onto the vector $\mathbf{w} = (-1, 5)$ in R^2: $T(\mathbf{v}) = \text{proj}_\mathbf{w}\mathbf{v}$, $\mathbf{v} = (2, -3)$.

23. T is the reflection through the vector $\mathbf{w} = (3, 1)$ in R^2, $\mathbf{v} = (1, 4)$. [The reflection of a vector $\mathbf{v}$ through $\mathbf{w}$ is given by $T(\mathbf{v}) = 2\,\text{proj}_\mathbf{w}\mathbf{v} - \mathbf{v}$.]

24. Repeat Exercise 23 for $\mathbf{w} = (4, -2)$ and $\mathbf{v} = (5, 0)$.

In Exercises 25–28, (a) find the standard matrix A for the linear transformation T, (b) use A to find the image of the vector $\mathbf{v}$, and (c) use a graphing utility and A to verify your result from part (b).

25. $T(x, y, z) = (2x + 3y - z, 3x - 2z, 2x - y + z)$, $\mathbf{v} = (1, 2, -1)$

26. $T(x, y, z) = (3x - 2y + z, 2x - 3y, y - 4z)$, $\mathbf{v} = (2, -1, -1)$

27. $T(x_1, x_2, x_3, x_4) = (x_1 - x_2, x_3, x_1 + 2x_2 - x_4, x_4)$, $\mathbf{v} = (1, 0, 1, -1)$

28. $T(x_1, x_2, x_3, x_4) = (x_1 + 2x_2, x_2 - x_1, 2x_3 - x_4, x_1)$, $\mathbf{v} = (0, 1, -1, 1)$

In Exercises 29–34, find the standard matrices for $T = T_2 \circ T_1$ and $T' = T_1 \circ T_2$.

29. $T_1: R^2 \to R^2$, $T_1(x, y) = (x - 2y, 2x + 3y)$
$T_2: R^2 \to R^2$, $T_2(x, y) = (2x, x - y)$

30. $T_1: R^2 \to R^2$, $T_1(x, y) = (x - 2y, 2x + 3y)$
$T_2: R^2 \to R^2$, $T_2(x, y) = (y, 0)$

31. $T_1: R^3 \to R^3$, $T_1(x, y, z) = (x, y, z)$
$T_2: R^3 \to R^3$, $T_2(x, y, z) = (0, x, 0)$

32. $T_1: R^3 \to R^3$, $T_1(x, y, z) = (x + 2y, y - z, -2x + y + 2z)$
$T_2: R^3 \to R^3$, $T_2(x, y, z) = (y + z, x + z, 2y - 2z)$

33. $T_1: R^2 \to R^3$, $T_1(x, y) = (-x + 2y, x + y, x - y)$
$T_2: R^3 \to R^2$, $T_2(x, y, z) = (x - 3y, z + 3x)$

34. $T_1: R^2 \to R^3$, $T_1(x, y) = (x, y, y)$
$T_2: R^3 \to R^2$, $T_2(x, y, z) = (y, z)$

In Exercises 35–40, determine whether the linear transformation is invertible. If it is, find its inverse.

35. $T(x, y) = (x + y, x - y)$

36. $T(x_1, x_2, x_3) = (x_1, x_1 + x_2, x_1 + x_2 + x_3)$

37. $T(x, y) = (2x, 0)$

38. $T(x, y) = (x + y, 3x + 3y)$

39. $T(x, y) = (5x, 5y)$

40. $T(x_1, x_2, x_3, x_4) = (x_1 - 2x_2, x_2, x_3 + x_4, x_3)$

In Exercises 41–48, find $T(\mathbf{v})$ by using (a) the standard matrix and (b) the matrix relative to B and B'.

41. $T: R^2 \to R^3$, $T(x, y) = (x + y, x, y)$, $\mathbf{v} = (5, 4)$, $B = \{(1, -1), (0, 1)\}$, $B' = \{(1, 1, 0), (0, 1, 1), (1, 0, 1)\}$

42. $T: R^2 \to R^3$, $T(x, y) = (x - y, 0, x + y)$, $\mathbf{v} = (-3, 2)$, $B = \{(1, 2), (1, 1)\}$, $B' = \{(1, 1, 1), (1, 1, 0), (0, 1, 1)\}$

43. $T: R^3 \to R^2$, $T(x, y, z) = (x - y, y - z)$, $\mathbf{v} = (1, 2, -3)$, $B = \{(1, 1, 1), (1, 1, 0), (0, 1, 1)\}$, $B' = \{(1, 2), (1, 1)\}$

44. $T: R^3 \to R^2$, $T(x, y, z) = (2x - z, y - 2x)$, $\mathbf{v} = (0, -5, 7)$, $B = \{(2, 0, 1), (0, 2, 1), (1, 2, 1)\}$, $B' = \{(1, 1), (2, 0)\}$

45. $T: R^3 \to R^4$, $T(x, y, z) = (2x, x + y, y + z, x + z)$, $\mathbf{v} = (1, -5, 2)$, $B = \{(2, 0, 1), (0, 2, 1), (1, 2, 1)\}$, $B' = \{(1, 0, 0, 1), (0, 1, 0, 1), (1, 0, 1, 0), (1, 1, 0, 0)\}$

46. $T: R^4 \to R^2$, $T(x_1, x_2, x_3, x_4) = (x_1 + x_2 + x_3 + x_4, x_4 - x_1)$, $\mathbf{v} = (4, -3, 1, 1)$, $B = \{(1, 0, 0, 1), (0, 1, 0, 1), (1, 0, 1, 0), (1, 1, 0, 0)\}$, $B' = \{(1, 1), (2, 0)\}$

47. $T: R^3 \to R^3$, $T(x, y, z) = (x + y + z, 2z - x, 2y - z)$, $\mathbf{v} = (4, -5, 10)$, $B = \{(2, 0, 1), (0, 2, 1), (1, 2, 1)\}$, $B' = \{(1, 1, 1), (1, 1, 0), (0, 1, 1)\}$

48. $T: R^2 \to R^2$, $T(x, y) = (2x - 12y, x - 5y)$, $\mathbf{v} = (10, 5)$, $B = B' = \{(4, 1), (3, 1)\}$

49. Let $T: P_2 \to P_3$ be given by $T(p) = xp$. Find the matrix of T relative to the bases $B = \{1, x, x^2\}$ and $B' = \{1, x, x^2, x^3\}$.

50. Let $T: P_2 \to P_4$ be given by $T(p) = x^2p$. Find the matrix of T relative to the bases $B = \{1, x, x^2\}$ and $B' = \{1, x, x^2, x^3, x^4\}$.

51. *Calculus* Let $B = \{1, x, e^x, xe^x\}$ be a basis of a subspace W of the space of continuous functions, and let D_x be the differential operator on W. Find the matrix for D_x relative to the basis B.

52. *Calculus* Repeat Exercise 51 for $B = \{e^{2x}, xe^{2x}, x^2e^{2x}\}$.

53. *Calculus* Use the matrix from Exercise 51 to evaluate $D_x[3x - 2xe^x]$.

54. *Calculus* Use the matrix from Exercise 52 to evaluate $D_x[5e^{2x} - 3xe^{2x} + x^2e^{2x}]$.

55. *Calculus* Let $B = \{1, x, x^2, x^3\}$ be a basis for P_3, and let $T: P_3 \to P_4$ be the linear transformation given by

$$T(x^k) = \int_0^x t^k\,dt.$$

(a) Find the matrix A for T with respect to B and the standard basis for P_4.
(b) Use A to integrate $p(x) = 6 - 2x + 3x^3$.

True or False? In Exercises 56 and 57, determine whether each statement is true or false. If a statement is true, give a reason or cite an appropriate statement in the text. If a statement is false, provide an example that shows that the statement is not true in all cases or cite an appropriate statement in the text.

56. (a) If T is a linear transformation R^n to R^m, the $m \times n$ matrix A is called the standard matrix such that $T(\mathbf{v}) = A\mathbf{v}$ for every $\mathbf{v}$ in R^n.
(b) The composition T of linear transformations T_1 and T_2, defined by $T(\mathbf{v}) = T_2(T_1(\mathbf{v}))$, has the standard matrix A given by the matrix product $A = A_2 A_1$.
(c) All linear transformations T have a unique inverse T^{-1}.

57. (a) The composition T of linear transformations T_1 and T_2, given by $T(\mathbf{v}) = T_2(T_1(\mathbf{v}))$, is defined if the range of T_1 lies within the domain of T_2.
(b) In general, the compositions $T_2 \circ T_1$ and $T_1 \circ T_2$ have the same standard matrix A.
(c) If $T: R^n \to R^n$ is an invertible linear transformation with standard matrix A, then T^{-1} has the same standard matrix A.

58. Let T be a linear transformation such that $T(\mathbf{v}) = k\mathbf{v}$ for $\mathbf{v}$ in R^n. Find the standard matrix for T.

59. Let $T: M_{2,3} \to M_{3,2}$ be given by $T(A) = A^T$. Find the matrix for T relative to the standard bases for $M_{2,3}$ and $M_{3,2}$.

60. Show that the linear transformation T given in Exercise 59 is an isomorphism and find the matrix for the inverse of T.

61. ***Guided Proof*** Let $T_1: V \to V$ and $T_2: V \to V$ be one-to-one linear transformations. Prove that the composition $T = T_2 \circ T_1$ is one-to-one and that T^{-1} exists and is equal to $T_1^{-1} \circ T_2^{-1}$.

Getting Started: To show that T is one-to-one, you can use the definition of a one-to-one transformation and show that $T(\mathbf{u}) = T(\mathbf{v})$ implies $\mathbf{u} = \mathbf{v}$. For the second statement, you first need to use Theorems 6.8 and 6.12 to show that T is invertible, and then show that $T \circ (T_1^{-1} \circ T_2^{-1})$ and $(T_1^{-1} \circ T_2^{-1}) \circ T$ are identity transformations.

(i) Let $T(\mathbf{u}) = T(\mathbf{v})$. Recall that $(T_2 \circ T_1)(\mathbf{v}) = T_2(T_1(\mathbf{v}))$ for all vectors $\mathbf{v}$. Now use the fact that T_2 and T_1 are one-to-one to conclude that $\mathbf{u} = \mathbf{v}$.
(ii) Use Theorems 6.8 and 6.12 to show that T_1, T_2, and T are all invertible transformations. Therefore T_1^{-1} and T_2^{-1} exist.
(iii) Form the composition $T' = T_1^{-1} \circ T_2^{-1}$. It is a linear transformation from V to V. To show that it is the inverse of T, you need to determine whether the composition of T with T' on both sides gives an identity transformation.

62. Prove Theorem 6.12.

63. ***Writing*** Is it always preferable to use the standard basis for R^n? Discuss the advantages and disadvantages of using different bases.

64. ***Writing*** Look back at Theorem 4.19 and rephrase it in terms of what you have learned in this chapter.

6.4 | Transition Matrices and Similarity

In Section 6.3 you saw that the matrix of a linear transformation $T: V \to V$ depends on the basis of V. In other words, the matrix of T relative to a basis B is different from the matrix of T relative to another basis B'.

A classical problem in linear algebra is this: Is it possible to find a basis B such that the matrix of T relative to B is diagonal? The solution of this problem is discussed in Chapter 7. This section lays a foundation for solving the problem. You will see how the matrices of a linear transformation relative to two different bases are related. In this section A, A', P, and P^{-1} represent the following four *square* matrices.

1. Matrix of T relative to B: A
2. Matrix of T relative to B': A'
3. Transition matrix from B' to B: P
4. Transition matrix from B to B': P^{-1}

Note that in Figure 6.10 there are two ways to get from the coordinate matrix $[\mathbf{v}]_{B'}$ to the coordinate matrix $[T(\mathbf{v})]_{B'}$. One way is direct, using the matrix A' to obtain

$$A'[\mathbf{v}]_{B'} = [T(\mathbf{v})]_{B'}.$$

The other way is indirect, using the matrices P, A, and P^{-1} to obtain

$$P^{-1}AP[\mathbf{v}]_{B'} = [T(\mathbf{v})]_{B'}.$$

But by the definition of the matrix of a linear transformation relative to a basis, this implies

$$A' = P^{-1}AP.$$

This relationship is demonstrated in Example 1.

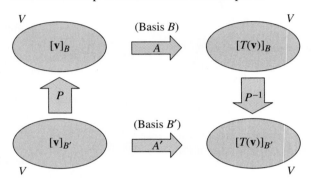

Figure 6.10

EXAMPLE 1 *Finding a Matrix for a Linear Transformation*

Find the matrix A' for $T: R^2 \to R^2$,

$$T(x_1, x_2) = (2x_1 - 2x_2, -x_1 + 3x_2),$$

relative to the basis $B' = \{(1, 0), (1, 1)\}$.

Solution   The standard matrix for T is

$$A = \begin{bmatrix} 2 & -2 \\ -1 & 3 \end{bmatrix}.$$

Furthermore, using the techniques of Section 4.7, you can find that the transition matrix from B' to the standard basis $B = \{(1, 0), (0, 1)\}$ is

$$P = \begin{bmatrix} 1 & 1 \\ 0 & 1 \end{bmatrix}.$$

The inverse of this matrix is the transition matrix from B to B',

$$P^{-1} = \begin{bmatrix} 1 & -1 \\ 0 & 1 \end{bmatrix}.$$

Therefore, the matrix for T relative to B' is

$$A' = P^{-1}AP = \begin{bmatrix} 1 & -1 \\ 0 & 1 \end{bmatrix} \begin{bmatrix} 2 & -2 \\ -1 & 3 \end{bmatrix} \begin{bmatrix} 1 & 1 \\ 0 & 1 \end{bmatrix} = \begin{bmatrix} 3 & -2 \\ -1 & 2 \end{bmatrix}.$$

In Example 1 the basis B is the standard basis for R^2. In the next example both B and B' are nonstandard bases.

EXAMPLE 2   *Finding a Matrix for a Linear Transformation*

Let

$$B = \{(-3, 2), (4, -2)\} \quad \text{and} \quad B' = \{(-1, 2), (2, -2)\}$$

be bases for R^2, and let

$$A = \begin{bmatrix} -2 & 7 \\ -3 & 7 \end{bmatrix}$$

be the matrix for $T: R^2 \to R^2$ relative to B. Find A', the matrix of T relative to B'.

Solution   In Example 5 in Section 4.7 you found that

$$P = \begin{bmatrix} 3 & -2 \\ 2 & -1 \end{bmatrix}$$

and

$$P^{-1} = \begin{bmatrix} -1 & 2 \\ -2 & 3 \end{bmatrix}.$$

Therefore, the matrix of T relative to B' is given by

$$A' = P^{-1}AP = \begin{bmatrix} -1 & 2 \\ -2 & 3 \end{bmatrix} \begin{bmatrix} -2 & 7 \\ -3 & 7 \end{bmatrix} \begin{bmatrix} 3 & -2 \\ 2 & -1 \end{bmatrix} = \begin{bmatrix} 2 & 1 \\ -1 & 3 \end{bmatrix}.$$

The diagram in Figure 6.10 should help you to remember the roles of the matrices $A, A', P,$ and P^{-1}.

EXAMPLE 3   *Using a Matrix for a Linear Transformation*

For the linear transformation $T: R^2 \to R^2$ given in Example 2, find $[\mathbf{v}]_B$, $[T(\mathbf{v})]_B$, and $[T(\mathbf{v})]_{B'}$ for the vector $\mathbf{v}$ whose coordinate matrix is

$$[\mathbf{v}]_{B'} = \begin{bmatrix} -3 \\ -1 \end{bmatrix}.$$

Solution   To find $[\mathbf{v}]_B$, use the transition matrix P from B' to B.

$$[\mathbf{v}]_B = P[\mathbf{v}]_{B'} = \begin{bmatrix} 3 & -2 \\ 2 & -1 \end{bmatrix} \begin{bmatrix} -3 \\ -1 \end{bmatrix} = \begin{bmatrix} -7 \\ -5 \end{bmatrix}$$

To find $[T(\mathbf{v})]_B$, multiply $[\mathbf{v}]_B$ by the matrix A to obtain

$$[T(\mathbf{v})]_B = A[\mathbf{v}]_B = \begin{bmatrix} -2 & 7 \\ -3 & 7 \end{bmatrix} \begin{bmatrix} -7 \\ -5 \end{bmatrix} = \begin{bmatrix} -21 \\ -14 \end{bmatrix}.$$

To find $[T(\mathbf{v})]_{B'}$, multiply $[T(\mathbf{v})]_B$ by P^{-1} to obtain

$$[T(\mathbf{v})]_{B'} = P^{-1}[T(\mathbf{v})]_B = \begin{bmatrix} -1 & 2 \\ -2 & 3 \end{bmatrix} \begin{bmatrix} -21 \\ -14 \end{bmatrix} = \begin{bmatrix} -7 \\ 0 \end{bmatrix}$$

or multiply $[\mathbf{v}]_{B'}$ by A' to obtain

$$[T(\mathbf{v})]_{B'} = A'[\mathbf{v}]_{B'} = \begin{bmatrix} 2 & 1 \\ -1 & 3 \end{bmatrix} \begin{bmatrix} -3 \\ -1 \end{bmatrix} = \begin{bmatrix} -7 \\ 0 \end{bmatrix}.$$

REMARK: It is instructive to note that the transformation T in Examples 2 and 3 is given by the rule $T(x, y) = \left(x - \frac{3}{2}y, 2x + 4y\right)$. Verify the results of Example 3 by showing that $\mathbf{v} = (1, -4)$ and $T(\mathbf{v}) = (7, -14)$.

Similar Matrices

Two square matrices A and A' that are related by an equation $A' = P^{-1}AP$ are called **similar** matrices, as indicated in the following definition.

Definition of Similar Matrices

For square matrices A and A' of order n, A' is said to be **similar** to A if there exists an invertible matrix P such that $A' = P^{-1}AP$.

If A' is similar to A, then it is also true that A is similar to A', as stated in the following theorem. Thus, it makes sense to say simply that **A and A' are similar.**

Theorem 6.13

Properties of Similar Matrices

Let A, B, and C be square matrices of order n. Then the following properties are true.
1. A is similar to A.
2. If A is similar to B, then B is similar to A.
3. If A is similar to B and B is similar to C, then A is similar to C.

Proof   The first property follows from the fact that $A = I_n A I_n$. To prove the second property, write

$$A = P^{-1} B P$$
$$P A P^{-1} = P(P^{-1} B P) P^{-1}$$
$$P A P^{-1} = B$$
$$Q^{-1} A Q = B, \text{ where } Q = P^{-1}.$$

The proof of the third property is left to you. (See Exercise 19.)

From the definition of similarity it follows that any two matrices that represent the same linear transformation $T: V \to V$ with respect to different bases must be similar.

EXAMPLE 4   *Similar Matrices*

(a) From Example 1, the matrices

$$A = \begin{bmatrix} 2 & -2 \\ -1 & 3 \end{bmatrix} \quad \text{and} \quad A' = \begin{bmatrix} 3 & -2 \\ -1 & 2 \end{bmatrix}$$

are similar because $A' = P^{-1} A P$, where $P = \begin{bmatrix} 1 & 1 \\ 0 & 1 \end{bmatrix}$.

(b) From Example 2, the matrices

$$A = \begin{bmatrix} -2 & 7 \\ -3 & 7 \end{bmatrix} \quad \text{and} \quad A' = \begin{bmatrix} 2 & 1 \\ -1 & 3 \end{bmatrix}$$

are similar because $A' = P^{-1} A P$, where $P = \begin{bmatrix} 3 & -2 \\ 2 & -1 \end{bmatrix}$.

You have seen that the matrix for a linear transformation $T: V \to V$ depends on the basis used for V. This observation leads naturally to the question: What choice of basis will make the matrix for T as simple as possible? Is it always the *standard* basis? Not necessarily, as the following example demonstrates.

EXAMPLE 5   *A Comparison of Two Matrices for a Linear Transformation*

Suppose

$$A = \begin{bmatrix} 1 & 3 & 0 \\ 3 & 1 & 0 \\ 0 & 0 & -2 \end{bmatrix}$$

is the matrix for $T: R^3 \to R^3$ relative to the standard basis. Find the matrix for T relative to the basis

$$B' = \{(1, 1, 0), (1, -1, 0), (0, 0, 1)\}.$$

Solution The transition matrix from B' to the standard matrix has columns consisting of the vectors in B',

$$P = \begin{bmatrix} 1 & 1 & 0 \\ 1 & -1 & 0 \\ 0 & 0 & 1 \end{bmatrix},$$

and it follows that

$$P^{-1} = \begin{bmatrix} \frac{1}{2} & \frac{1}{2} & 0 \\ \frac{1}{2} & -\frac{1}{2} & 0 \\ 0 & 0 & 1 \end{bmatrix}.$$

Therefore, the matrix for T relative to B' is

$$A' = P^{-1}AP = \begin{bmatrix} \frac{1}{2} & \frac{1}{2} & 0 \\ \frac{1}{2} & -\frac{1}{2} & 0 \\ 0 & 0 & 1 \end{bmatrix} \begin{bmatrix} 1 & 3 & 0 \\ 3 & 1 & 0 \\ 0 & 0 & -2 \end{bmatrix} \begin{bmatrix} 1 & 1 & 0 \\ 1 & -1 & 0 \\ 0 & 0 & 1 \end{bmatrix}$$

$$= \begin{bmatrix} 4 & 0 & 0 \\ 0 & -2 & 0 \\ 0 & 0 & -2 \end{bmatrix}.$$

Note that matrix A' is diagonal.

Diagonal matrices have many computational advantages over nondiagonal ones. For instance, for the diagonal matrix

$$D = \begin{bmatrix} d_1 & 0 & \cdots & 0 \\ 0 & d_2 & \cdots & 0 \\ \vdots & \vdots & & \vdots \\ 0 & 0 & \cdots & d_n \end{bmatrix},$$

the kth power is given as follows.

$$D^k = \begin{bmatrix} d_1^k & 0 & \cdots & 0 \\ 0 & d_2^k & \cdots & 0 \\ \vdots & \vdots & & \vdots \\ 0 & 0 & \cdots & d_n^k \end{bmatrix}$$

A diagonal matrix is its own transpose. Moreover, if all the diagonal elements are nonzero, then the inverse of a diagonal matrix is the matrix whose main diagonal elements are the reciprocals of corresponding elements in the original matrix. Given such computational advantages, it is important to find ways (if possible) to choose a basis for V such that the transformation matrix is diagonal, as it is in Example 5. You will pursue this problem in the next chapter.

In Exercises 1–6, (a) find the matrix A' for T relative to the basis B' and (b) show that A' is similar to A, the standard matrix for T.

1. $T: R^2 \to R^2$, $T(x, y) = (2x - y, y - x)$, $B' = \{(1, -2), (0, 3)\}$
2. $T: R^2 \to R^2$, $T(x, y) = (x + y, 4y)$, $B' = \{(-4, 1), (1, -1)\}$
3. $T: R^3 \to R^3$, $T(x, y, z) = (x, y, z)$,
   $B' = \{(1, 1, 0), (1, 0, 1), (0, 1, 1)\}$
4. $T: R^3 \to R^3$, $T(x, y, z) = (0, 0, 0)$,
   $B' = \{(1, 1, 0), (1, 0, 1), (0, 1, 1)\}$
5. $T: R^3 \to R^3$, $T(x, y, z) = (x - y + 2z, 2x + y - z, x + 2y + z)$,
   $B' = \{(1, 0, 1), (0, 2, 2), (1, 2, 0)\}$
6. $T: R^3 \to R^3$, $T(x, y, z) = (x, x + 2y, x + y + 3z)$,
   $B' = \{(1, -1, 0), (0, 0, 1), (0, 1, -1)\}$

7. Let $B = \{(1, 3), (-2, -2)\}$ and $B' = \{(-12, 0), (-4, 4)\}$ be bases for R^2, and let

$$A = \begin{bmatrix} 3 & 2 \\ 0 & 4 \end{bmatrix}$$

   be the matrix for $T: R^2 \to R^2$ relative to B.
   (a) Find the transition matrix P from B' to B.
   (b) Use the matrices A and P to find $[\mathbf{v}]_B$ and $[T(\mathbf{v})]_B$, where

   $$[\mathbf{v}]_{B'} = \begin{bmatrix} -1 \\ 2 \end{bmatrix}.$$

   (c) Find A' (the matrix of T relative to B') and P^{-1}.
   (d) Find $[T(\mathbf{v})]_{B'}$ in two ways: first as $P^{-1}[T(\mathbf{v})]_B$ and then as $A'[\mathbf{v}]_{B'}$.

8. Repeat Exercise 7 for $B = \{(1, 1), (-2, 3)\}$, $B' = \{(1, -1), (0, 1)\}$, and

$$[\mathbf{v}]_{B'} = \begin{bmatrix} 1 \\ -3 \end{bmatrix}.$$

9. Let $B = \{(1, 1, 0), (1, 0, 1), (0, 1, 1)\}$ and $B' = \{(1, 0, 0), (0, 1, 0), (0, 0, 1)\}$ be bases for R^3, and let

$$A = \begin{bmatrix} \frac{3}{2} & -1 & -\frac{1}{2} \\ -\frac{1}{2} & 2 & \frac{1}{2} \\ \frac{1}{2} & 1 & \frac{5}{2} \end{bmatrix}$$

   be the matrix for $T: R^3 \to R^3$ relative to B.
   (a) Find the transition matrix P from B' to B.
   (b) Use the matrices A and P to find $[\mathbf{v}]_B$ and $[T(\mathbf{v})]_B$, where

   $$[\mathbf{v}]_{B'} = \begin{bmatrix} 1 \\ 0 \\ -1 \end{bmatrix}.$$

   (c) Find A' (the matrix of T relative to B') and P^{-1}.
   (d) Find $[T(\mathbf{v})]_{B'}$: first as $P^{-1}[T(\mathbf{v})]_B$ and then as $A'[\mathbf{v}]_{B'}$.

10. Repeat Exercise 9 for $B = \{(1, 0, 0), (0, 1, 0), (0, 0, 1)\}$, $B' = \{(1, 1, -1), (1, -1, 1), (-1, 1, 1)\}$, and

$$[\mathbf{v}]_{B'} = \begin{bmatrix} 2 \\ 1 \\ 1 \end{bmatrix}.$$

11. Prove that if A and B are similar, then $|A| = |B|$. Is the converse true?

12. Illustrate the result of Exercise 11 using the matrices

$$A = \begin{bmatrix} 1 & 0 & 0 \\ 0 & -2 & 0 \\ 0 & 0 & 3 \end{bmatrix}, \quad B = \begin{bmatrix} 11 & 7 & 10 \\ 10 & 8 & 10 \\ -18 & -12 & -17 \end{bmatrix},$$

$$P = \begin{bmatrix} -1 & 1 & 0 \\ 2 & 1 & 2 \\ 1 & 1 & 1 \end{bmatrix},$$

$$P^{-1} = \begin{bmatrix} -1 & -1 & 2 \\ 0 & -1 & 2 \\ 1 & 2 & -3 \end{bmatrix}, \text{ where } B = P^{-1}AP.$$

13. Let A and B be similar matrices.
    (a) Prove that A^T and B^T are similar.
    (b) Prove that if A is nonsingular, then B is also nonsingular and A^{-1} and B^{-1} are similar.
    (c) Prove that there exists a matrix P such that $B^k = P^{-1}A^kP$.

14. Use the result of Exercise 13 to find B^4, where $B = P^{-1}AP$ for the matrices

$$A = \begin{bmatrix} 1 & 0 \\ 0 & 2 \end{bmatrix}, \quad B = \begin{bmatrix} -4 & -15 \\ 2 & 7 \end{bmatrix},$$

$$P = \begin{bmatrix} 2 & 5 \\ 1 & 3 \end{bmatrix}, \quad P^{-1} = \begin{bmatrix} 3 & -5 \\ -1 & 2 \end{bmatrix}.$$

15. Determine all $n \times n$ matrices that are similar to I_n.

16. Prove that if A is idempotent and B is similar to A, then B is idempotent. (An $n \times n$ matrix A is idempotent if $A = A^2$.)

17. Let A be an $n \times n$ matrix such that $A^2 = O$. Prove that if B is similar to A, then $B^2 = O$.

18. Let $B = P^{-1}AP$. Prove that if $A\mathbf{x} = \mathbf{x}$, then $PBP^{-1}\mathbf{x} = \mathbf{x}$.

19. Complete the proof of Theorem 6.13 by proving that if A is similar to B and B is similar to C, then A is similar to C.

20. *Writing* Suppose that A and B are similar. Explain why they have the same rank.

21. Prove that if A and B are similar, then A^2 is similar to B^2.

22. Prove that if A and B are similar, then A^k is similar to B^k for any positive integer k.

23. Let $A = CD$, where C is an invertible $n \times n$ matrix. Prove that the matrix DC is similar to A.

24. Let $B = P^{-1}AP$, where B is a diagonal matrix with main diagonal entries $b_{11}, b_{22}, \ldots, b_{nn}$. Prove that

$$\begin{bmatrix} a_{11} & a_{12} & \cdots & a_{1n} \\ a_{21} & a_{22} & \cdots & a_{2n} \\ \vdots & \vdots & & \vdots \\ a_{n1} & a_{n2} & \cdots & a_{nn} \end{bmatrix} \begin{bmatrix} p_{1i} \\ p_{2i} \\ \vdots \\ p_{ni} \end{bmatrix} = b_{ii} \begin{bmatrix} p_{1i} \\ p_{2i} \\ \vdots \\ p_{ni} \end{bmatrix},$$

for $i = 1, 2, \ldots, n$.

25. *Writing* Let $B = \{\mathbf{v}_1, \mathbf{v}_2, \ldots, \mathbf{v}_n\}$ be a basis for the vector space V, let B' be the standard basis, and consider the identity transformation $I: V \rightarrow V$. What can you say about the matrix of I relative to both B and B'? What can you say about the matrix of I relative to B? Relative to B'?

True or False? In Exercises 26 and 27, determine whether each statement is true or false. If a statement is true, give a reason or cite an appropriate statement in the text. If a statement is false, provide an example that shows that the statement is not true in all cases or cite an appropriate statement in the text.

26. (a) The matrix of a linear transformation A' relative to the basis B' is equal to the product $P^{-1}AP$, where P^{-1} is the transition matrix from B to B', A is the matrix of the linear transformation relative to basis B, and P is the transition matrix from B' to B.

(b) Two matrices that represent the same linear transformation $T: V \rightarrow V$ with respect to different bases are not necessarily similar.

27. (a) The matrix of a linear transformation A relative to the basis B is equal to the product $PA'P^{-1}$, where P is the transition matrix from B' to B, A' is the matrix of the linear transformation relative to basis B', and P^{-1} is the transition matrix from B to B'.

(b) The standard basis for R^n will always make the coordinate matrix for the linear transformation T the simplest matrix possible.

6.5 | Applications of Linear Transformations

The Geometry of Linear Transformations in the Plane

This section gives geometric interpretations of linear transformations represented by 2×2 elementary matrices. A summary of the various types of 2×2 elementary matrices is followed by examples in which each type of matrix is examined in more detail.

Elementary Matrices for Linear Transformations in the Plane

Reflection in y-Axis
$$A = \begin{bmatrix} -1 & 0 \\ 0 & 1 \end{bmatrix}$$

Reflection in x-Axis
$$A = \begin{bmatrix} 1 & 0 \\ 0 & -1 \end{bmatrix}$$

Reflection in Line $y = x$
$$A = \begin{bmatrix} 0 & 1 \\ 1 & 0 \end{bmatrix}$$

Horizontal Expansion $(k > 1)$ or Contraction $(0 < k < 1)$
$$A = \begin{bmatrix} k & 0 \\ 0 & 1 \end{bmatrix}$$

Vertical Expansion $(k > 1)$ or Contraction $(0 < k < 1)$
$$A = \begin{bmatrix} 1 & 0 \\ 0 & k \end{bmatrix}$$

Horizontal Shear
$$A = \begin{bmatrix} 1 & k \\ 0 & 1 \end{bmatrix}$$

Vertical Shear
$$A = \begin{bmatrix} 1 & 0 \\ k & 1 \end{bmatrix}$$

EXAMPLE 1   *Reflections in the Plane*

The transformations defined by the following matrices are called **reflections.** Reflections have the effect of mapping a point in the *xy*-plane to its "mirror image" with respect to one of the coordinate axes or the line given by $y = x$, as shown in Figure 6.11.

(a) Reflection in the *y*-axis:

$$T(x, y) = (-x, y)$$

$$\begin{bmatrix} -1 & 0 \\ 0 & 1 \end{bmatrix} \begin{bmatrix} x \\ y \end{bmatrix} = \begin{bmatrix} -x \\ y \end{bmatrix}$$

(b) Reflection in the *x*-axis:

$$T(x, y) = (x, -y)$$

$$\begin{bmatrix} 1 & 0 \\ 0 & -1 \end{bmatrix} \begin{bmatrix} x \\ y \end{bmatrix} = \begin{bmatrix} x \\ -y \end{bmatrix}$$

(c) Reflection in the line $y = x$:

$$T(x, y) = (y, x)$$

$$\begin{bmatrix} 0 & 1 \\ 1 & 0 \end{bmatrix} \begin{bmatrix} x \\ y \end{bmatrix} = \begin{bmatrix} y \\ x \end{bmatrix}$$

  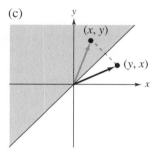

Reflection in the Plane

Figure 6.11

EXAMPLE 2   *Expansions and Contractions in the Plane*

The transformations defined by the following matrices are called **expansions** or **contractions,** depending on the value of the positive scalar *k*.

(a) Horizontal contractions and expansions:

$$T(x, y) = (kx, y)$$

$$\begin{bmatrix} k & 0 \\ 0 & 1 \end{bmatrix} \begin{bmatrix} x \\ y \end{bmatrix} = \begin{bmatrix} kx \\ y \end{bmatrix}$$

(b) Vertical contractions and expansions:

$$T(x, y) = (x, ky)$$

$$\begin{bmatrix} 1 & 0 \\ 0 & k \end{bmatrix} \begin{bmatrix} x \\ y \end{bmatrix} = \begin{bmatrix} x \\ ky \end{bmatrix}$$

Note that in Figures 6.12 and 6.13, the distance the point (x, y) is moved by a contraction or an expansion is proportional to its x- or y-coordinate. For instance, under the transformation given by $T(x, y) = (2x, y)$, the point $(1, 3)$ would be moved one unit to the right, but the point $(4, 3)$ would be moved four units to the right.

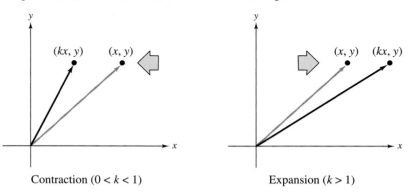

Contraction $(0 < k < 1)$ Expansion $(k > 1)$

Figure 6.12

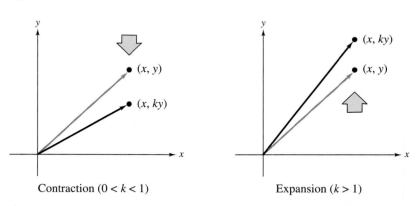

Contraction $(0 < k < 1)$ Expansion $(k > 1)$

Figure 6.13

The third type of linear transformation in the plane corresponding to an elementary matrix is called a **shear,** as described in Example 3.

EXAMPLE 3   *Shears in the Plane*

The transformations defined by the following matrices are shears.

$$T(x, y) = (x + ky, y) \qquad\qquad T(x, y) = (x, y + kx)$$

$$\begin{bmatrix} 1 & k \\ 0 & 1 \end{bmatrix}\begin{bmatrix} x \\ y \end{bmatrix} = \begin{bmatrix} x + ky \\ y \end{bmatrix} \qquad \begin{bmatrix} 1 & 0 \\ k & 1 \end{bmatrix}\begin{bmatrix} x \\ y \end{bmatrix} = \begin{bmatrix} x \\ kx + y \end{bmatrix}$$

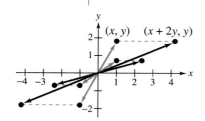

Figure 6.14

(a) The horizontal shear given by $T(x, y) = (x + 2y, y)$ is shown in Figure 6.14. Under this transformation, points in the upper half-plane are "sheared" to the right by amounts proportional to their y-coordinates. Points in the lower half-plane are "sheared" to the left by amounts proportional to the absolute values of their y-coordinates. Points on the x-axis are unmoved by this transformation.

(b) The vertical shear given by $T(x, y) = (x, y + 2x)$ is shown in Figure 6.15. Here, points in the right half-plane are "sheared" upward by amounts proportional to their x-coordinates. Points in the left half-plane are "sheared" downward by amounts proportional to the absolute values of their x-coordinates. Points on the y-axis are unmoved.

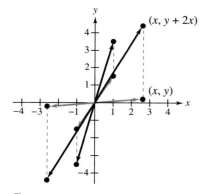

Figure 6.15

Computer Graphics

Linear transformations are useful in computer graphics. In Example 7 in Section 6.1 you saw how a linear transformation could be used to rotate figures in the plane. Here you will see how linear transformations can be used to rotate figures in three-dimensional space.

Suppose you want to rotate the point (x, y, z) counterclockwise about the z-axis through an angle θ, as shown in Figure 6.16. Letting the coordinates of the rotated point be (x', y', z'), you have

Figure 6.16

$$\begin{bmatrix} x' \\ y' \\ z' \end{bmatrix} = \begin{bmatrix} \cos\theta & -\sin\theta & 0 \\ \sin\theta & \cos\theta & 0 \\ 0 & 0 & 1 \end{bmatrix}\begin{bmatrix} x \\ y \\ z \end{bmatrix} = \begin{bmatrix} x\cos\theta - y\sin\theta \\ x\sin\theta + y\cos\theta \\ z \end{bmatrix}.$$

Example 4 shows how to use this matrix to rotate a figure in three-dimensional space.

EXAMPLE 4   *Rotation About the z-Axis*

The eight vertices of a rectangular box having sides of lengths 1, 2, and 3 are as follows.

$$V_1 = (0, 0, 0), \qquad V_2 = (1, 0, 0), \qquad V_3 = (1, 2, 0), \qquad V_4 = (0, 2, 0),$$

$$V_5 = (0, 0, 3), \qquad V_6 = (1, 0, 3), \qquad V_7 = (1, 2, 3), \qquad V_8 = (0, 2, 3)$$

Find the coordinates of the box when it is rotated counterclockwise about the z-axis through each angle.

(a) $\theta = 60°$ (b) $\theta = 90°$ (c) $\theta = 120°$

Solution   The original box is shown in Figure 6.17.

(a) The matrix that yields a rotation of $60°$ is

$$A = \begin{bmatrix} \cos 60° & -\sin 60° & 0 \\ \sin 60° & \cos 60° & 0 \\ 0 & 0 & 1 \end{bmatrix} = \begin{bmatrix} 1/2 & -\sqrt{3}/2 & 0 \\ \sqrt{3}/2 & 1/2 & 0 \\ 0 & 0 & 1 \end{bmatrix}.$$

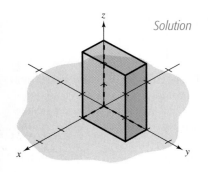

Figure 6.17

Multiplying this matrix by the eight vertices produces the following rotated vertices.

Original Vertex	*Rotated Vertex*
$V_1 = (0, 0, 0)$	$(0, 0, 0)$
$V_2 = (1, 0, 0)$	$(0.5, 0.87, 0)$
$V_3 = (1, 2, 0)$	$(-1.23, 1.87, 0)$
$V_4 = (0, 2, 0)$	$(-1.73, 1, 0)$
$V_5 = (0, 0, 3)$	$(0, 0, 3)$
$V_6 = (1, 0, 3)$	$(0.5, 0.87, 3)$
$V_7 = (1, 2, 3)$	$(-1.23, 1.87, 3)$
$V_8 = (0, 2, 3)$	$(-1.73, 1, 3)$

(a)

A computer-generated graph of the rotated box is shown in Figure 6.18(a). Note that in this graph line segments representing the sides of the box are drawn between images of pairs of vertices that are connected in the original box. For instance, because V_1 and V_2 are connected in the original box, the computer is told to connect the images of V_1 and V_2 in the rotated box.

(b) The matrix that yields a rotation of $90°$ is

$$A = \begin{bmatrix} \cos 90° & -\sin 90° & 0 \\ \sin 90° & \cos 90° & 0 \\ 0 & 0 & 1 \end{bmatrix} = \begin{bmatrix} 0 & -1 & 0 \\ 1 & 0 & 0 \\ 0 & 0 & 1 \end{bmatrix},$$

(b)

and the graph of the rotated box is shown in Figure 6.18(b).

(c) The matrix that yields a rotation of $120°$ is

$$A = \begin{bmatrix} \cos 120° & -\sin 120° & 0 \\ \sin 120° & \cos 120° & 0 \\ 0 & 0 & 1 \end{bmatrix} = \begin{bmatrix} -1/2 & -\sqrt{3}/2 & 0 \\ \sqrt{3}/2 & -1/2 & 0 \\ 0 & 0 & 1 \end{bmatrix},$$

(c)

Figure 6.18

and the graph of the rotated box is shown in Figure 6.18(c).

In Example 4, matrices were used to perform rotations about the z-axis. Similarly, you can use matrices to rotate figures about the x- or y-axis. All three types of rotations are summarized as follows.

Rotation About the x-Axis

$$\begin{bmatrix} 1 & 0 & 0 \\ 0 & \cos\theta & -\sin\theta \\ 0 & \sin\theta & \cos\theta \end{bmatrix}$$

Rotation About the y-Axis

$$\begin{bmatrix} \cos\theta & 0 & \sin\theta \\ 0 & 1 & 0 \\ -\sin\theta & 0 & \cos\theta \end{bmatrix}$$

Rotation About the z-Axis

$$\begin{bmatrix} \cos\theta & -\sin\theta & 0 \\ \sin\theta & \cos\theta & 0 \\ 0 & 0 & 1 \end{bmatrix}$$

In each case the rotation is oriented counterclockwise relative to a person facing the negative direction of the indicated axis, as shown in Figure 6.19.

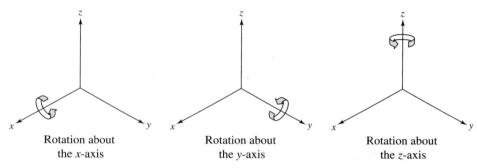

Rotation about
the x-axis

Rotation about
the y-axis

Rotation about
the z-axis

Figure 6.19

EXAMPLE 5   *Rotations About the x-Axis and y-Axis*

(a) The matrix that yields a rotation of 90° about the x-axis is

$$A = \begin{bmatrix} 1 & 0 & 0 \\ 0 & \cos\theta & -\sin\theta \\ 0 & \sin\theta & \cos\theta \end{bmatrix} = \begin{bmatrix} 1 & 0 & 0 \\ 0 & 0 & -1 \\ 0 & 1 & 0 \end{bmatrix},$$

and the graph of the rotated box from Example 4 is shown in Figure 6.20(a) on the following page.

(b) The matrix that yields a rotation of 90° about the y-axis is

$$A = \begin{bmatrix} \cos\theta & 0 & \sin\theta \\ 0 & 1 & 0 \\ -\sin\theta & 0 & \cos\theta \end{bmatrix} = \begin{bmatrix} 0 & 0 & 1 \\ 0 & 1 & 0 \\ -1 & 0 & 0 \end{bmatrix},$$

and the graph of the rotated box from Example 4 is shown in Figure 6.20(b) on the following page.

Simulation

Explore this concept further with an electronic simulation available on the *Elementary Linear Algebra* Learning Tools Student CD-ROM.

(a)

90°

(b)

90°

Figure 6.20

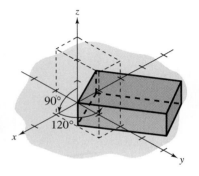

90°

120°

Figure 6.21

Rotations about the coordinate axes can be combined to produce any desired view of a figure. For instance, Figure 6.21 shows the rotation produced by first rotating the box (from Example 4) 90° about the y-axis, then further rotating the box 120° about the z-axis.

The use of computer graphics has become common among designers in many fields. Simply by entering into a computer the coordinates that form the outline of an object, a designer can see the object before it is created. As a simple example, the images of the toy boat shown in Figure 6.22 were created using only 27 points in space. Once the points have been stored in the computer, the boat can be viewed from any perspective.

Figure 6.22

The Geometry of Linear Transformations in the Plane

1. Let $T: R^2 \rightarrow R^2$ be a reflection in the x-axis. Find the image of each vector.

(a) $(3, 5)$ (b) $(2, -1)$ (c) $(a, 0)$

(d) $(0, b)$ (e) $(-c, d)$ (f) $(f, -g)$

2. Let $T: R^2 \rightarrow R^2$ be a reflection in the line $y = x$. Find the image of each vector.

(a) $(0, 1)$ (b) $(-1, 3)$ (c) $(a, 0)$

(d) $(0, b)$ (e) $(-c, d)$ (f) $(f, -g)$

3. Let $T(1, 0) = (0, 1)$ and $T(0, 1) = (1, 0)$.

(a) Determine $T(x, y)$ for any (x, y).

(b) Give a geometric description of T.

4. Let $T(1, 0) = (2, 0)$ and $T(0, 1) = (0, 1)$.

(a) Determine $T(x, y)$ for any (x, y).

(b) Give a geometric description of T.

In Exercises 5–8, (a) identify the transformation and (b) graphically represent the transformation for an arbitrary vector in the plane.

5. $T(x, y) = (x, y/2)$

6. $T(x, y) = (4x, y)$

7. $T(x, y) = (x + 3y, y)$

8. $T(x, y) = (x, 2x + y)$

In Exercises 9–12, find all fixed points of the linear transformation. The vector $\mathbf{v}$ is a fixed point of T if $T(\mathbf{v}) = \mathbf{v}$.

9. A reflection in the y-axis

10. A reflection in the line $y = x$

11. A vertical contraction

12. A horizontal shear

In Exercises 13–18, sketch the image of the unit square with vertices at $(0, 0)$, $(1, 0)$, $(1, 1)$, and $(0, 1)$ under the given transformation.

13. T is a reflection in the x-axis.

14. T is a reflection in the line $y = x$.

15. T is the contraction given by $T(x, y) = (x/2, y)$.

16. T is the expansion given by $T(x, y) = (x, 3y)$.

17. T is the shear given by $T(x, y) = (x + 2y, y)$.

18. T is the shear given by $T(x, y) = (x, y + 3x)$.

In Exercises 19–24, sketch the image of the rectangle with vertices at $(0, 0)$, $(0, 2)$, $(1, 2)$, and $(1, 0)$ under the given transformation.

19. T is a reflection in the y-axis.

20. T is a reflection in the line $y = x$.

21. T is the contraction given by $T(x, y) = (x, y/2)$.

22. T is the expansion given by $T(x, y) = (2x, y)$.

23. T is the shear given by $T(x, y) = (x + y, y)$.

24. T is the shear given by $T(x, y) = (x, y + 2x)$.

25. The linear transformation defined by a diagonal matrix with positive main diagonal elements is called a **magnification**. Find the images of $(1, 0)$, $(0, 1)$, and $(2, 2)$ under the linear transformation A and graphically interpret your result.

$$A = \begin{bmatrix} 2 & 0 \\ 0 & 3 \end{bmatrix}$$

26. Repeat Exercise 25 for the linear transformation defined by

$$A = \begin{bmatrix} 3 & 0 \\ 0 & 3 \end{bmatrix}.$$

In Exercises 27–32, give a geometric description of the linear transformation defined by the elementary matrix.

27. $A = \begin{bmatrix} 2 & 0 \\ 0 & 1 \end{bmatrix}$ **28.** $A = \begin{bmatrix} 1 & 0 \\ 2 & 1 \end{bmatrix}$

29. $A = \begin{bmatrix} 0 & 1 \\ 1 & 0 \end{bmatrix}$ **30.** $A = \begin{bmatrix} 1 & 3 \\ 0 & 1 \end{bmatrix}$

31. $A = \begin{bmatrix} 1 & 0 \\ 0 & -2 \end{bmatrix}$ **32.** $A = \begin{bmatrix} -1 & 0 \\ 0 & 1 \end{bmatrix}$

In Exercises 33 and 34, give a geometric description of the linear transformation defined by the matrix product.

33. $A = \begin{bmatrix} 2 & 0 \\ 2 & 1 \end{bmatrix} = \begin{bmatrix} 2 & 0 \\ 0 & 1 \end{bmatrix} \begin{bmatrix} 1 & 0 \\ 2 & 1 \end{bmatrix}$

34. $A = \begin{bmatrix} 0 & 3 \\ 1 & 0 \end{bmatrix} = \begin{bmatrix} 0 & 1 \\ 1 & 0 \end{bmatrix} \begin{bmatrix} 1 & 0 \\ 0 & 3 \end{bmatrix}$

Computer Graphics

In Exercises 35–38, find the matrix that will produce the indicated rotation.

35. $30°$ about the z-axis   **36.** $60°$ about the x-axis

37. $60°$ about the y-axis   **38.** $120°$ about the x-axis

In Exercises 39–42, find the image of the vector $(1, 1, 1)$ for the indicated rotation.

39. $30°$ about the z-axis
40. $60°$ about the x-axis
41. $60°$ about the y-axis
42. $120°$ about the x-axis

In Exercises 43–48, determine which single counterclockwise rotation about the x-, y-, or z-axis will produce the indicated tetrahedron. The original tetrahedron position is illustrated in Figure 6.23.

Figure 6.23

43.

44.

45.

46.

47.

48.

In Exercises 49–52, determine the matrix that will produce the indicated pair of rotations. Then find the image of the line segment from $(0, 0, 0)$ to $(1, 1, 1)$ under this composition.

49. $90°$ about the x-axis followed by $90°$ about the y-axis
50. $45°$ about the y-axis followed by $90°$ about the z-axis
51. $30°$ about the z-axis followed by $60°$ about the y-axis
52. $45°$ about the z-axis followed by $135°$ about the x-axis

CHAPTER 6 Summary

After studying this chapter, you should have acquired the following skills.

■ Find the image and preimage of a given function.

■ Determine whether a function from one vector space to another is a linear transformation.

■ Find the kernel and range of a given linear transformation.

■ Determine whether a given linear transformation is one-to-one or onto.

■ Find the basis for the kernel and range of a given linear transformation T and determine the nullity and rank of T.

■ Verify that a given matrix defines a linear function that is one-to-one and onto.

■ Determine whether two vector spaces are isomorphic.

■ Find the standard matrix for a given linear transformation and use this matrix to find the image of a given vector. Sketch the graph of the vector and its image.

■ Find the standard matrix of the composition of linear transformations.

■ Determine whether a given linear transformation is invertible and find its inverse if it exists.

■ Find the matrix of a given linear transformation relative to a nonstandard basis.

■ Know and use the definition and properties of similar matrices.

■ Identify linear transformations defined by reflections, expansions, contractions, shears, or rotations.

CHAPTER 6   Review Exercises

In Exercises 1 and 2, find (a) the image of **v** and (b) the preimage of **w** for the linear transformation.

1. $T: R^2 \rightarrow R^2$, $T(v_1, v_2) = (v_1, v_1 + 2v_2)$, $\mathbf{v} = (2, -3)$,
$\mathbf{w} = (4, 12)$

2. $T: R^3 \rightarrow R^3$, $T(v_1, v_2, v_3) = (0, v_1 + v_2, v_2 + v_3)$,
$\mathbf{v} = (-3, 2, 5)$, $\mathbf{w} = (0, 2, 5)$

In Exercises 3–8, determine whether the function is a linear transformation. If it is, find its standard matrix A.

3. $T: R^2 \rightarrow R^2$, $T(x_1, x_2) = (x_1 + 2x_2, -x_1 - x_2)$

4. $T: R^2 \rightarrow R^2$, $T(x_1, x_2) = (x_1 + 3, x_2)$

5. $T: R^2 \rightarrow R^2$, $T(x, y) = (x + h, y + k)$, $h \neq 0$ or $k \neq 0$
(translation in the plane)

6. $T: R^2 \rightarrow R^2$, $T(x, y) = (|x|, |y|)$

7. $T: R^3 \rightarrow R^3$, $T(x_1, x_2, x_3) = (x_1 - x_2, x_2 - x_3, x_3 - x_1)$

8. $T: R^3 \rightarrow R^3$, $T(x, y, z) = (z, y, x)$

9. Let T be a linear transformation from R^2 to R^2 such that $T(2, 0) = (1, 1)$ and $T(0, 3) = (3, 3)$. Find $T(1, 1)$ and $T(0, 1)$.

10. Let T be a linear transformation from R^3 to R such that $T(1, 1, 1) = 1$, $T(1, 1, 0) = 2$, and $T(1, 0, 0) = 3$. Find $T(0, 1, 1)$.

In Exercises 11–14, find the indicated power of A, the standard matrix for T.

11. $T: R^3 \rightarrow R^3$, reflection in the xy-plane. Find A^2.

12. $T: R^3 \rightarrow R^3$, projection onto the xy-plane. Find A^2.

13. $T: R^2 \rightarrow R^2$, counterclockwise rotation through the angle θ. Find A^3.

14. *Calculus* $T: P_3 \rightarrow P_3$, differential operator. Find A^2.

In Exercises 15–18, the linear transformation $T: R^n \rightarrow R^m$ is defined by $T(\mathbf{v}) = A\mathbf{v}$. For each matrix A, (a) determine the dimensions of R^n and R^m, (b) find the image $T(\mathbf{v})$ of the given vector $\mathbf{v}$, and (c) find the preimage of the given vector $\mathbf{w}$.

15. $A = \begin{bmatrix} 0 & 1 & 2 \\ -2 & 0 & 0 \end{bmatrix}$, $\mathbf{v} = (6, 1, 1)$, $\mathbf{w} = (3, 5)$

16. $A = [1, 1]$, $\mathbf{v} = (2, 3)$, $\mathbf{w} = (4)$

17. $A = \begin{bmatrix} 1 & 1 & 1 \\ 0 & 1 & 1 \\ 0 & 0 & 1 \end{bmatrix}$, $\mathbf{v} = (2, 1, -5)$, $\mathbf{w} = (6, 4, 2)$

18. $A = \begin{bmatrix} 4 & 0 \\ 0 & 5 \\ 1 & 1 \end{bmatrix}$, $\mathbf{v} = (2, 2)$, $\mathbf{w} = (4, -5, 0)$

In Exercises 19 and 20, find a basis for (a) ker(T) and (b) range(T).

19. $T: R^4 \rightarrow R^3$, $T(w, x, y, z) =$
$(2w + 4x + 6y + 5z, -w - 2x + 2y, 8y + 4z)$

20. $T: R^3 \rightarrow R^3$, $T(x, y, z) = (x + 2y, y + 2z, z + 2x)$

In Exercises 21 and 22, the linear transformation T is given by $T(\mathbf{v}) = A\mathbf{v}$. Find a basis for (a) the kernel of T and (b) the range of T, and then find (c) the rank of T and (d) the nullity of T.

21. $A = \begin{bmatrix} 1 & 2 \\ -1 & 0 \\ 1 & 1 \end{bmatrix}$ **22.** $A = \begin{bmatrix} 2 & 1 & 3 \\ 1 & 1 & 0 \\ 0 & 1 & -3 \end{bmatrix}$

23. Given $T: R^5 \rightarrow R^3$ and nullity(T) = 2, find rank(T).

24. Given $T: P_5 \rightarrow P_3$ and nullity(T) = 4, find rank(T).

25. Given $T: P_4 \rightarrow R^5$ and rank(T) = 3, find nullity(T).

26. Given $T: M_{2,2} \rightarrow M_{2,2}$ and rank(T) = 3, find nullity(T).

In Exercises 27–30, determine whether the transformation T has an inverse. If it does, find A and A^{-1}.

27. $T: R^2 \to R^2$, $T(x, y) = (2x, y)$

28. $T: R^2 \to R^2$, $T(x, y) = (x \cos \theta - y \sin \theta, x \sin \theta + y \cos \theta)$

29. $T: R^3 \to R^3$, $T(x, y, z) = (x, y, 0)$

30. $T: R^3 \to R^2$, $T(x, y, z) = (x + y, y - z)$

In Exercises 31 and 32, find the standard matrices for $T = T_1 \circ T_2$ and $T' = T_2 \circ T_1$.

31. $T_1: R^2 \to R^3$, $T_1(x, y) = (x, x + y, y)$
$T_2: R^3 \to R^2$, $T_2(x, y, z) = (0, y)$

32. $T_1: R \to R^2$, $T(x) = (x, 3x)$
$T_2: R^2 \to R$, $T(x, y) = (y + 2x)$

33. Use the standard matrix for counterclockwise rotation in R^2 to rotate the triangle with vertices $(3, 5)$, $(5, 3)$, and $(3, 0)$ counterclockwise $90°$ about the origin. Graph the triangles.

34. Rotate the triangle in Exercise 33 counterclockwise $90°$ about the point $(5, 3)$. Graph the triangles.

In Exercises 35–38, determine whether the linear transformation represented by the matrix A is (a) one-to-one, (b) onto, and (c) invertible.

35. $A = \begin{bmatrix} 2 & 0 \\ 0 & 3 \end{bmatrix}$

36. $A = \begin{bmatrix} 1 & 1 & 1 \\ 0 & 1 & 1 \end{bmatrix}$

37. $A = \begin{bmatrix} 1 & \frac{1}{4} \\ 0 & 1 \end{bmatrix}$

38. $A = \begin{bmatrix} 4 & 0 & 7 \\ 5 & 5 & 1 \\ 0 & 0 & 2 \end{bmatrix}$

In Exercises 39 and 40, find $T(\mathbf{v})$ by using (a) the standard matrix and (b) the matrix relative to B and B'.

39. $T: R^2 \to R^3$, $T(x, y) = (-x, y, x + y)$, $\mathbf{v} = (0, 1)$,
$B = \{(1, 1), (1, -1)\}$, $B' = \{(0, 1, 0), (0, 0, 1), (1, 0, 0)\}$

40. $T: R^2 \to R^2$, $T(x, y) = (2y, 0)$, $\mathbf{v} = (-1, 3)$,
$B = \{(2, 1), (-1, 0)\}$, $B' = \{(-1, 0), (2, 2)\}$

In Exercises 41 and 42, find the matrix A' for T relative to the basis B' and show that A' is similar to A, the standard matrix for T.

41. $T: R^2 \to R^2$, $T(x, y) = (x - 3y, y - x)$, $B' = \{(1, -1), (1, 1)\}$

42. $T: R^3 \to R^3$, $T(x, y, z) = (x + 3y, 3x + y, -2z)$,
$B' = \{(1, 1, 0), (1, -1, 0), (0, 0, 1)\}$

43. Let $T: R^3 \to R^3$ be given by $T(\mathbf{v}) = \text{proj}_{\mathbf{u}}\mathbf{v}$, where $\mathbf{u} = (0, 1, 2)$.

(a) Find A, the standard matrix for T.

(b) Let S be the linear transformation represented by $I - A$. Show that S is of the form $S(\mathbf{v}) = \text{proj}_{\mathbf{w}_1}\mathbf{v} + \text{proj}_{\mathbf{w}_2}\mathbf{v}$, where $\mathbf{w}_1$ and $\mathbf{w}_2$ are fixed vectors in R^3.

(c) Show that the kernel of T is equal to the range of S.

44. Let $T: R^2 \to R^2$ be given by $T(\mathbf{v}) = \text{proj}_{\mathbf{u}}\mathbf{v}$, where $\mathbf{u} = (4, 3)$.

(a) Find A, the standard matrix for T, and show that $A^2 = A$.

(b) Show that $(I - A)^2 = I - A$.

(c) Find $A\mathbf{v}$ and $(I - A)\mathbf{v}$ for $\mathbf{v} = (5, 0)$.

(d) Sketch the graph of $\mathbf{u}$, $\mathbf{v}$, $A\mathbf{v}$, and $(I - A)\mathbf{v}$.

45. Let S and T be linear transformations from V into W. Show that $S + T$ and kT are both linear transformations, where $(S + T)(\mathbf{v}) = S(\mathbf{v}) + T(\mathbf{v})$ and $(kT)(\mathbf{v}) = kT(\mathbf{v})$.

46. Suppose A and B are similar matrices and A is invertible.

(a) Prove that B is invertible.

(b) Prove that A^{-1} and B^{-1} are similar.

In Exercises 47 and 48, the sum $S + T$ of two linear transformations $S: V \to W$ and $T: V \to W$ is defined as $(S + T)(\mathbf{v}) = S(\mathbf{v}) + T(\mathbf{v})$.

47. Prove that $\text{rank}(S + T) \leq \text{rank}(S) + \text{rank}(T)$.

48. Give an example for each of the following.

(a) $\text{Rank}(S + T) = \text{rank}(S) + \text{rank}(T)$

(b) $\text{Rank}(S + T) < \text{rank}(S) + \text{rank}(T)$

49. Let $T: P_3 \to R$ such that $T(a_0 + a_1x + a_2x^2 + a_3x^3) = a_0 + a_1 + a_2 + a_3$.

(a) Prove that T is linear.

(b) Find the rank and nullity of T.

(c) Find a basis for the kernel of T.

50. Let $T: V \to U$ and $S: U \to W$ be linear transformations.

(a) Prove that if S and T are both one-to-one, then so is $S \circ T$.

(b) Prove that the kernel of T is contained in the kernel of $S \circ T$.

(c) Prove that if $S \circ T$ is onto, then so is S.

51. Let V be an inner product space. For a fixed nonzero vector $\mathbf{v}_0$ in V, let $T: V \to R$ be the linear transformation $T(\mathbf{v}) = \langle \mathbf{v}, \mathbf{v}_0 \rangle$. Find the kernel, range, rank, and nullity of T.

52. *Calculus* Let $B = \{1, x, \sin x, \cos x\}$ be a basis for a subspace W of the space of continuous functions, and let D_x be the differential operator on W. Find the matrix for D_x relative to the basis B. Find the range and kernel of D_x.

53. *Writing* Under what conditions are the spaces $M_{m,n}$ and $M_{p,q}$ isomorphic? Describe an isomorphism T in this case.

54. *Calculus*   Let $T: P_3 \to P_3$ be given by $T(p) = p(x) + p'(x)$. Find the rank and nullity of T.

The Geometry of Linear Transformations in the Plane

In Exercises 55–58, (a) identify the transformation and (b) graphically represent the transformation for an arbitrary vector in the plane.

55. $T(x, y) = (x, 2y)$

56. $T(x, y) = (x + y, y)$

57. $T(x, y) = (x, y + 3x)$

58. $T(x, y) = (5x, y)$

In Exercises 59–62, sketch the image of the triangle with vertices $(0, 0)$, $(1, 0)$, and $(0, 1)$ under the given transformation.

59. T is a reflection in the x-axis.

60. T is the expansion given by $T(x, y) = (2x, y)$.

61. T is the shear given by $T(x, y) = (x + 3y, y)$.

62. T is the shear given by $T(x, y) = (x, y + 2x)$.

In Exercises 63 and 64, give a geometric description of the linear transformation defined by the matrix product.

63. $\begin{bmatrix} 0 & 2 \\ 1 & 0 \end{bmatrix} = \begin{bmatrix} 2 & 0 \\ 0 & 1 \end{bmatrix} \begin{bmatrix} 0 & 1 \\ 1 & 0 \end{bmatrix}$

64. $\begin{bmatrix} 1 & 0 \\ 6 & 2 \end{bmatrix} = \begin{bmatrix} 1 & 0 \\ 0 & 2 \end{bmatrix} \begin{bmatrix} 1 & 0 \\ 3 & 1 \end{bmatrix}$

Computer Graphics

In Exercises 65 and 66, find the matrix that will produce the indicated rotation and then find the image of the vector $(1, -1, 1)$.

65. 45° about the z-axis

66. 90° about the x-axis

In Exercises 67 and 68, determine the matrix that will produce the indicated pair of rotations.

67. 60° about the x-axis followed by 30° about the z-axis

68. 120° about the y-axis followed by 45° about the z-axis

In Exercises 69 and 70, find the image of the unit cube with vertices $(0, 0, 0)$, $(1, 0, 0)$, $(1, 1, 0)$, $(0, 1, 0)$, $(0, 0, 1)$, $(1, 0, 1)$, $(1, 1, 1)$, and $(0, 1, 1)$ when it is rotated by the given angle.

69. 45° about the z-axis

70. 90° about the x-axis

True or False?   In Exercises 71–74, determine whether each statement is true or false. If a statement is true, give a reason or cite an appropriate statement in the text. If a statement is false, provide an example that shows that the statement is not true in all cases or cite an appropriate statement in the text.

71. (a) Linear transformations called reflections that map a point in the xy-plane to its mirror image across the line $y = x$ are defined by the standard matrix $\begin{bmatrix} 1 & 0 \\ 0 & 1 \end{bmatrix}$ in $M_{2,2}$.

  (b) The linear transformations called horizontal expansions or contractions are defined by the matrix $\begin{bmatrix} k & 0 \\ 0 & 1 \end{bmatrix}$ in $M_{2,2}$.

  (c) The matrix $\begin{bmatrix} 1 & 0 & 0 \\ 0 & 1/2 & -\sqrt{3}/2 \\ 0 & \sqrt{3}/2 & 1/2 \end{bmatrix}$ would rotate a point 60° about the x-axis.

72. (a) Linear transformations called reflections that map a point in the xy-plane to its mirror image across the x-axis are defined by the matrix $\begin{bmatrix} 1 & 0 \\ 0 & -1 \end{bmatrix}$ in $M_{2,2}$.

  (b) The linear transformations called vertical expansions or contractions are defined by the matrix $\begin{bmatrix} 1 & 0 \\ 0 & k \end{bmatrix}$ in $M_{2,2}$.

  (c) The matrix $\begin{bmatrix} \sqrt{3}/2 & 0 & 1/2 \\ 0 & 1 & 0 \\ -1/2 & 0 & \sqrt{3}/2 \end{bmatrix}$ would rotate a point 30° about the y-axis.

73. (a) In calculus, any linear function is also a linear transformation from R^2 to R^2.

  (b) A linear transformation is said to be onto if and only if for all **u** and **v** in V, $T(\mathbf{u}) = T(\mathbf{v})$ implies $\mathbf{u} = \mathbf{v}$.

  (c) Given the computational advantages, it is best to choose a basis for V such that the transformation matrix is diagonal.

74. (a) For polynomials, the differential operator D_x is a linear transformation from P_n to P_{n-1}.

  (b) The set of all vectors **v** in V that satisfy $T(\mathbf{v}) = \mathbf{v}$ is called the kernel of T.

  (c) The standard matrix A of the composition of two linear transformations $T(\mathbf{v}) = T_2(T_1(\mathbf{v}))$ is the product of the standard matrix for T_2 and the standard matrix for T_1.

CHAPTER 6   Projects

1   REFLECTIONS IN THE PLANE (I)

Let ℓ be the line $ax + by = 0$ in the plane. The linear transformation $L: R^2 \to R^2$ that sends a point (x, y) to its mirror image in ℓ is called the **reflection** in ℓ. (See Figure 6.24.)

The goal of these two projects is to find the matrix of this reflection relative to the standard basis. The first project is based on transition matrices, and the second project uses projections.

1. Find the standard matrix of L for the line $x = 0$.
2. Find the standard matrix of L for the line $y = 0$.
3. Find the standard matrix of L for the line $x - y = 0$.
4. Consider the line ℓ given by $x - 2y = 0$. Find a vector $\mathbf{v}$ parallel to ℓ and another vector $\mathbf{w}$ orthogonal to ℓ. Determine the matrix A of the reflection in ℓ relative to the ordered basis $\{\mathbf{v}, \mathbf{w}\}$. Finally, use the appropriate transition matrix to find the matrix of the reflection relative to the standard basis. Use this matrix to find $L(2, 1)$, $L(-1, 2)$, and $L(5, 0)$.
5. Consider the general line $ax + by = 0$. Let $\mathbf{v}$ be a vector parallel to ℓ, and let $\mathbf{w}$ be a vector orthogonal to ℓ. Determine the matrix A of the reflection in ℓ relative to the ordered basis $\{\mathbf{v}, \mathbf{w}\}$. Finally, use the appropriate transition matrix to find the matrix of L relative to the standard basis.
6. Find the standard matrix of the reflection in the line $3x + 4y = 0$. Use this matrix to find the images of the points $(3, 4)$, $(-4, 3)$, and $(0, 5)$.

Figure 6.24

2   REFLECTIONS IN THE PLANE (II)

In this second project, you will use projections to determine the standard matrix for the reflection L in the line $ax + by = 0$. (See Figure 6.25.) Recall that the projection of the vector $\mathbf{u}$ onto the vector $\mathbf{v}$ is given by

$$\text{proj}_{\mathbf{v}}\mathbf{u} = \frac{\mathbf{u} \cdot \mathbf{v}}{\mathbf{v} \cdot \mathbf{v}}\mathbf{v}.$$

1. Find the standard matrix for the projection onto the y-axis. That is, find the standard matrix for $\text{proj}_{\mathbf{v}}\mathbf{u}$ if $\mathbf{v} = (0, 1)$.
2. Find the standard matrix for the projection onto the x-axis.
3. Consider the line ℓ given by $x - 2y = 0$. Find a vector $\mathbf{v}$ parallel to ℓ and another vector $\mathbf{w}$ orthogonal to ℓ. Determine the matrix A of the projection onto ℓ relative to the ordered basis $\{\mathbf{v}, \mathbf{w}\}$. Finally, use the appropriate transition matrix to find the matrix of the projection relative to the standard basis. Use this matrix to find $\text{proj}_{\mathbf{v}}\mathbf{u}$ for the cases $\mathbf{u} = (2, 1)$, $\mathbf{u} = (-1, 2)$, and $\mathbf{u} = (5, 0)$.
4. Consider the general line $ax + by = 0$. Let $\mathbf{v}$ be a vector parallel to ℓ, and let $\mathbf{w}$ be a vector orthogonal to ℓ. Determine the matrix A of the projection onto ℓ relative to the ordered basis $\{\mathbf{v}, \mathbf{w}\}$. Finally, use the appropriate transition matrix to find the matrix of the projection relative to the standard basis.
5. Use Figure 6.26 to show that

   $$\text{proj}_{\mathbf{v}}\mathbf{u} = \tfrac{1}{2}(\mathbf{u} + L(\mathbf{u})),$$

   where L is the reflection in the line ℓ. Solve this equation for L and compare your answer with the formula from the first project.

Figure 6.25

Figure 6.26

**JAMES JOSEPH
SYLVESTER**

1 8 1 4 – 1 8 9 7

JAMES JOSEPH SYLVESTER, founder of the *American Journal of Mathematics*, was born in London. Although little is known of his early life, by the age of 14 he was studying at the University of London under Augustus DeMorgan, a well-known English mathematician. At 15, Sylvester entered the Royal Institute at Liverpool, where he won several prizes in mathematics.

After receiving his degree from Trinity College in 1837, Sylvester embarked on a series of careers in England and the United States. These included a professorship at the University of Virginia, an actuarial position in England, enrollment at a law school (with subsequent work as a lawyer), a professorship at a military academy, a professorship of mathematics at Johns Hopkins University, and finally a professorship of

mathematics at Oxford University, a position he held until his death in 1897.

In 1850 Sylvester met Arthur Cayley and the two began a life-long friendship. Sylvester attributed his "restoration to the enjoyment of mathematical life" to this meeting, and during the next half century he contributed much to the theory of determinants. His mathematical papers tended to be emotional discourses rather than systematic developments of theory, and he often claimed results to be "self-evident." Although his exuberance occasionally led to incorrect conjectures, Sylvester's intuition was usually right, probably because, as he said, "I really love my subject."

CHAPTER 7

EIGENVALUES AND EIGENVECTORS

7.1 | Eigenvalues and Eigenvectors

This section presents one of the most important problems in linear algebra, the **eigenvalue problem.** Its central question can be stated as follows. If A is an $n \times n$ matrix, do there exist nonzero vectors $\mathbf{x}$ in R^n such that $A\mathbf{x}$ is a scalar multiple of $\mathbf{x}$? The scalar, denoted by the Greek letter lambda (λ), is called an **eigenvalue** of the matrix A, and the nonzero vector $\mathbf{x}$ is called an **eigenvector** of A corresponding to λ. The terms *eigenvalue* and *eigenvector* are derived from the German word *Eigenwert*, meaning "proper value." Thus, you have

Eigenvalue

$$A\mathbf{x} = \lambda\mathbf{x}.$$

Eigenvector

Although you briefly looked at the eigenvalue problem in Section 3.4, the approach in this chapter will not depend on that material.

Eigenvalues and eigenvectors have many important applications, some of which are discussed in Section 7.4. For now you will consider a geometric interpretation of the problem in R^2. If λ is an eigenvalue of a matrix A and $\mathbf{x}$ is an eigenvector of A corresponding to λ, then multiplication of $\mathbf{x}$ by the matrix A produces a vector $\lambda\mathbf{x}$ that is parallel to $\mathbf{x}$, as shown in Figure 7.1.

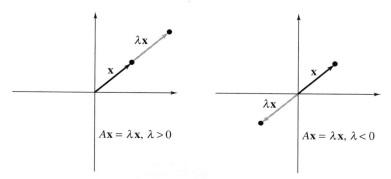

Figure 7.1

Definition of Eigenvalue and Eigenvector

Let A be an $n \times n$ matrix. The scalar λ is called an **eigenvalue** of A if there is a *nonzero* vector $\mathbf{x}$ such that

$$A\mathbf{x} = \lambda\mathbf{x}.$$

The vector $\mathbf{x}$ is called an **eigenvector** of A corresponding to λ.

Only real eigenvalues are presented in this chapter.

REMARK: Note that an eigen*vector* cannot be zero. Allowing $\mathbf{x}$ to be the zero vector would render the definition meaningless, because $A\mathbf{0} = \lambda\mathbf{0}$ is true for all real values of λ. An eigen*value* of $\lambda = 0$, however, is possible. (See Example 2.)

A matrix can have more than one eigenvalue, as demonstrated in Examples 1 and 2.

EXAMPLE 1   *Verifying Eigenvalues and Eigenvectors*

For the matrix

$$A = \begin{bmatrix} 2 & 0 \\ 0 & -1 \end{bmatrix},$$

verify that $\mathbf{x}_1 = (1, 0)$ is an eigenvector of A corresponding to the eigenvalue $\lambda_1 = 2$, and that $\mathbf{x}_2 = (0, 1)$ is an eigenvector of A corresponding to the eigenvalue $\lambda_2 = -1$.

Solution   Multiplying $\mathbf{x}_1$ by A produces

$$A\mathbf{x}_1 = \begin{bmatrix} 2 & 0 \\ 0 & -1 \end{bmatrix} \begin{bmatrix} 1 \\ 0 \end{bmatrix}$$

$$= \begin{bmatrix} 2 \\ 0 \end{bmatrix}$$

$$= 2 \begin{bmatrix} 1 \\ 0 \end{bmatrix}.$$

Eigenvalue   Eigenvector

Thus, $\mathbf{x}_1 = (1, 0)$ is an eigenvector of A corresponding to the eigenvalue $\lambda_1 = 2$. Similarly, multiplying $\mathbf{x}_2$ by A produces

$$A\mathbf{x}_2 = \begin{bmatrix} 2 & 0 \\ 0 & -1 \end{bmatrix} \begin{bmatrix} 0 \\ 1 \end{bmatrix}$$

$$= \begin{bmatrix} 0 \\ -1 \end{bmatrix}$$

$$= -1 \begin{bmatrix} 0 \\ 1 \end{bmatrix}.$$

Thus, $\mathbf{x}_2 = (0, 1)$ is an eigenvector of A corresponding to the eigenvalue $\lambda_2 = -1$.

EXAMPLE 2   *Verifying Eigenvalues and Eigenvectors*

For the matrix

$$A = \begin{bmatrix} 1 & -2 & 1 \\ 0 & 0 & 0 \\ 0 & 1 & 1 \end{bmatrix},$$

verify that

$$\mathbf{x}_1 = (-3, -1, 1) \quad \text{and} \quad \mathbf{x}_2 = (1, 0, 0)$$

are eigenvectors of A and find their corresponding eigenvalues.

Solution   Multiplying $\mathbf{x}_1$ by A produces

$$A\mathbf{x}_1 = \begin{bmatrix} 1 & -2 & 1 \\ 0 & 0 & 0 \\ 0 & 1 & 1 \end{bmatrix} \begin{bmatrix} -3 \\ -1 \\ 1 \end{bmatrix} = \begin{bmatrix} 0 \\ 0 \\ 0 \end{bmatrix} = 0 \begin{bmatrix} -3 \\ -1 \\ 1 \end{bmatrix}.$$

Thus, $\mathbf{x}_1 = (-3, -1, 1)$ is an eigenvector of A corresponding to the eigenvalue $\lambda_1 = 0$.

DISCOVERY

In Example 2, $\lambda_2 = 1$ is an eigenvalue of the matrix A. Calculate the determinant of the matrix $A - \lambda_2 I$, where I is the 3×3 identity matrix. Repeat this experiment for the other eigenvalue, $\lambda_1 = 0$. In general, if λ is an eigenvalue of the matrix A, what is the value of $|A - \lambda I|$?

Similarly, multiplying $\mathbf{x}_2$ by A produces

$$Ax_2 = \begin{bmatrix} 1 & -2 & 1 \\ 0 & 0 & 0 \\ 0 & 1 & 1 \end{bmatrix}\begin{bmatrix} 1 \\ 0 \\ 0 \end{bmatrix} = \begin{bmatrix} 1 \\ 0 \\ 0 \end{bmatrix} = 1\begin{bmatrix} 1 \\ 0 \\ 0 \end{bmatrix}.$$

Thus, $\mathbf{x}_2 = (1, 0, 0)$ is an eigenvector of A corresponding to the eigenvalue $\lambda_2 = 1$.

Eigenspaces

Although Examples 1 and 2 list only one eigenvector for each eigenvalue, each of the four eigenvalues in Examples 1 and 2 has an infinite number of eigenvectors. For instance, in Example 1 the vectors $(2, 0)$ and $(-3, 0)$ are eigenvectors of A corresponding to the eigenvalue 2. In fact, if A is an $n \times n$ matrix with an eigenvalue λ and a corresponding eigenvector $\mathbf{x}$, then every nonzero scalar multiple of $\mathbf{x}$ is also an eigenvector of A. This may be seen by letting c be a nonzero scalar, which then produces

$$A(c\mathbf{x}) = c(A\mathbf{x}) = c(\lambda\mathbf{x}) = \lambda(c\mathbf{x}).$$

It is also true that if $\mathbf{x}_1$ and $\mathbf{x}_2$ are eigenvectors corresponding to the *same* eigenvalue λ, then their sum is also an eigenvector corresponding to λ, because

$$A(\mathbf{x}_1 + \mathbf{x}_2) = A\mathbf{x}_1 + A\mathbf{x}_2 = \lambda\mathbf{x}_1 + \lambda\mathbf{x}_2 = \lambda(\mathbf{x}_1 + \mathbf{x}_2).$$

In other words, the set of all eigenvectors of a given eigenvalue λ, together with the zero vector, is a subspace of R^n. This special subspace of R^n is called the **eigenspace** of λ.

Theorem 7.1   *Eigenvectors of λ Form a Subspace*

If A is an $n \times n$ matrix with an eigenvalue λ, then the set of all eigenvectors of λ, together with the zero vector

$$\{\mathbf{0}\} \cup \{\mathbf{x}: \mathbf{x} \text{ is an eigenvector of } \lambda\},$$

is a subspace of R^n. This subspace is called the **eigenspace** of λ.

Determining the eigenvalues and corresponding eigenspaces for a matrix can be difficult. Occasionally, however, you can find eigenvalues and eigenspaces by simple inspection, as demonstrated in Example 3.

EXAMPLE 3   *An Example of Eigenspaces in the Plane*

Find the eigenvalues and corresponding eigenspaces of

$$A = \begin{bmatrix} -1 & 0 \\ 0 & 1 \end{bmatrix}.$$

Solution

Geometrically, multiplying a vector (x, y) in R^2 by the matrix A corresponds to a reflection in the y-axis. That is, if $\mathbf{v} = (x, y)$, then

$$A\mathbf{v} = \begin{bmatrix} -1 & 0 \\ 0 & 1 \end{bmatrix} \begin{bmatrix} x \\ y \end{bmatrix} = \begin{bmatrix} -x \\ y \end{bmatrix}.$$

Figure 7.2 illustrates that the only vectors reflected onto scalar multiples of themselves are those lying on either the x-axis or the y-axis.

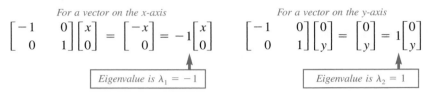

For a vector on the x-axis

$$\begin{bmatrix} -1 & 0 \\ 0 & 1 \end{bmatrix} \begin{bmatrix} x \\ 0 \end{bmatrix} = \begin{bmatrix} -x \\ 0 \end{bmatrix} = -1 \begin{bmatrix} x \\ 0 \end{bmatrix}$$

Eigenvalue is $\lambda_1 = -1$

For a vector on the y-axis

$$\begin{bmatrix} -1 & 0 \\ 0 & 1 \end{bmatrix} \begin{bmatrix} 0 \\ y \end{bmatrix} = \begin{bmatrix} 0 \\ y \end{bmatrix} = 1 \begin{bmatrix} 0 \\ y \end{bmatrix}$$

Eigenvalue is $\lambda_2 = 1$

Thus, the eigenvectors corresponding to $\lambda_1 = -1$ are the nonzero vectors on the x-axis, and the eigenvectors corresponding to $\lambda_2 = 1$ are the nonzero vectors on the y-axis. This implies that the eigenspace corresponding to $\lambda_1 = -1$ is the x-axis, and the eigenspace corresponding to $\lambda_2 = 1$ is the y-axis.

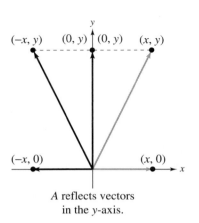

A reflects vectors in the y-axis.

Figure 7.2

Finding Eigenvalues and Eigenvectors

The geometric solution given in Example 3 is not typical of the general eigenvalue problem. A general approach will now be described.

To find the eigenvalues and eigenvectors for an $n \times n$ matrix A, let I be the $n \times n$ identity matrix. Writing the equation $A\mathbf{x} = \lambda\mathbf{x}$ in the form $\lambda I\mathbf{x} = A\mathbf{x}$ then produces

$$(\lambda I - A)\mathbf{x} = \mathbf{0}.$$

This homogeneous system of equations has nonzero solutions if and only if the coefficient matrix $(\lambda I - A)$ is *not* invertible—that is, if and only if the determinant of $(\lambda I - A)$ is zero. This is formally stated in the following theorem.

Theorem 7.2

Eigenvalues and Eigenvectors of a Matrix

Let A be an $n \times n$ matrix.
1. An eigenvalue of A is a scalar λ such that

$$\det(\lambda I - A) = 0.$$

2. The eigenvectors of A corresponding to λ are the nonzero solutions of

$$(\lambda I - A)\mathbf{x} = \mathbf{0}.$$

The equation $\det(\lambda I - A) = 0$ is called the **characteristic equation** of A. Moreover, when expanded to polynomial form, the polynomial

$$|\lambda I - A| = \lambda^n + c_{n-1}\lambda^{n-1} + \cdots + c_1\lambda + c_0$$

is called the **characteristic polynomial** of A. This definition tells you that the eigenvalues of an $n \times n$ matrix A correspond to the roots of the characteristic polynomial of A. Because the characteristic polynomial of A is of degree n, A can have at most n distinct eigenvalues.

REMARK: The Fundamental Theorem of Algebra states that an nth-degree polynomial has precisely n roots. These n roots, however, include both repeated and complex roots. In this chapter you will be concerned only with the real roots of characteristic polynomials—that is, real eigenvalues.

EXAMPLE 4   *Finding Eigenvalues and Eigenvectors*

Find the eigenvalues and corresponding eigenvectors of

$$A = \begin{bmatrix} 2 & -12 \\ 1 & -5 \end{bmatrix}.$$

Solution   The characteristic polynomial of A is

$$\begin{aligned}
|\lambda I - A| &= \begin{vmatrix} \lambda - 2 & 12 \\ -1 & \lambda + 5 \end{vmatrix} \\
&= (\lambda - 2)(\lambda + 5) - (-12) \\
&= \lambda^2 + 3\lambda - 10 + 12 \\
&= \lambda^2 + 3\lambda + 2 \\
&= (\lambda + 1)(\lambda + 2).
\end{aligned}$$

Thus, the characteristic equation is $(\lambda + 1)(\lambda + 2) = 0$, which gives $\lambda_1 = -1$ and $\lambda_2 = -2$ as the eigenvalues of A. To find the corresponding eigenvectors, use Gauss-Jordan elimination to solve the homogeneous linear system given by $(\lambda I - A)\mathbf{x} = \mathbf{0}$ twice: first for $\lambda = \lambda_1 = -1$, and then for $\lambda = \lambda_2 = -2$. For $\lambda_1 = -1$, the coefficient matrix is

$$(-1)I - A = \begin{bmatrix} -1 - 2 & 12 \\ -1 & -1 + 5 \end{bmatrix} = \begin{bmatrix} -3 & 12 \\ -1 & 4 \end{bmatrix},$$

which row reduces to

$$\begin{bmatrix} 1 & -4 \\ 0 & 0 \end{bmatrix}.$$

Therefore, $x_1 - 4x_2 = 0$. Letting $x_2 = t$, you can conclude that every eigenvector of λ_1 is of the form

$$\mathbf{x} = \begin{bmatrix} x_1 \\ x_2 \end{bmatrix} = \begin{bmatrix} 4t \\ t \end{bmatrix} = t \begin{bmatrix} 4 \\ 1 \end{bmatrix}, \quad t \neq 0.$$

For $\lambda_2 = -2$, you have

$$(-2)I - A = \begin{bmatrix} -2 - 2 & 12 \\ -1 & -2 + 5 \end{bmatrix} = \begin{bmatrix} -4 & 12 \\ -1 & 3 \end{bmatrix} \longrightarrow \begin{bmatrix} 1 & -3 \\ 0 & 0 \end{bmatrix}.$$

Letting $x_2 = t$, you can conclude that every eigenvector of λ_2 is of the form

$$\mathbf{x} = \begin{bmatrix} x_1 \\ x_2 \end{bmatrix} = \begin{bmatrix} 3t \\ t \end{bmatrix} = t \begin{bmatrix} 3 \\ 1 \end{bmatrix}, \quad t \neq 0.$$

Try checking that $A\mathbf{x} = \lambda_i \mathbf{x}$ for the eigenvalues and eigenvectors in this example.

The homogeneous systems that arise when you are finding eigenvectors will always row reduce to a matrix having at least one row of zeros, because the systems must have non-trivial solutions. The steps used to find the eigenvalues and corresponding eigenvectors of a matrix are summarized as follows.

Finding Eigenvalues and Eigenvectors

Let A be an $n \times n$ matrix.
1. Form the characteristic equation $|\lambda I - A| = 0$. It will be a polynomial equation of degree n in the variable λ.
2. Find the real roots of the characteristic equation. These are the eigenvalues of A.
3. For each eigenvalue λ_i, find the eigenvectors corresponding to λ_i by solving the homogeneous system $(\lambda_i I - A)\mathbf{x} = \mathbf{0}$. This requires row reducing an $n \times n$ matrix. The resulting reduced row-echelon form must have at least one row of zeros.

Finding the eigenvalues for an $n \times n$ matrix can be difficult because it involves the factorization of an nth-degree polynomial. Once an eigenvalue has been found, however, finding the corresponding eigenvectors is a straightforward application of Gauss-Jordan reduction.

EXAMPLE 5   *Finding Eigenvalues and Eigenvectors*

Find the eigenvalues and corresponding eigenvectors for

$$A = \begin{bmatrix} 2 & 1 & 0 \\ 0 & 2 & 0 \\ 0 & 0 & 2 \end{bmatrix}.$$

What is the dimension of the eigenspace of each eigenvalue?

Solution   The characteristic polynomial of A is

$$|\lambda I - A| = \begin{vmatrix} \lambda - 2 & -1 & 0 \\ 0 & \lambda - 2 & 0 \\ 0 & 0 & \lambda - 2 \end{vmatrix} = (\lambda - 2)^3.$$

So the characteristic equation is $(\lambda - 2)^3 = 0$.

Thus, the only eigenvalue is $\lambda = 2$. To find the eigenvectors for $\lambda = 2$, solve the homogeneous linear system represented by $(2I - A)\mathbf{x} = \mathbf{0}$.

$$2I - A = \begin{bmatrix} 0 & -1 & 0 \\ 0 & 0 & 0 \\ 0 & 0 & 0 \end{bmatrix}$$

This implies that $x_2 = 0$. Using the parameters $s = x_1$ and $t = x_3$, you can find that the eigenvectors of $\lambda = 2$ are of the form

$$\mathbf{x} = \begin{bmatrix} x_1 \\ x_2 \\ x_3 \end{bmatrix} = \begin{bmatrix} s \\ 0 \\ t \end{bmatrix} = s \begin{bmatrix} 1 \\ 0 \\ 0 \end{bmatrix} + t \begin{bmatrix} 0 \\ 0 \\ 1 \end{bmatrix}, \quad s \text{ and } t \text{ not both zero.}$$

Because $\lambda = 2$ has two linearly independent eigenvectors, the dimension of its eigenspace is 2.

If an eigenvalue λ_1 occurs as a *multiple root* (k times) of the characteristic polynomial, then λ_1 has **multiplicity** k. This implies that $(\lambda - \lambda_1)^k$ is a factor of the characteristic polynomial and $(\lambda - \lambda_1)^{k+1}$ is not a factor of the characteristic polynomial. For instance, in Example 5 the eigenvalue $\lambda = 2$ has a multiplicity of 3.

Note also that in Example 5 the dimension of the eigenspace of $\lambda = 2$ is 2. In general, the multiplicity of an eigenvalue is greater than or equal to the dimension of its eigenspace.

EXAMPLE 6   *Finding Eigenvalues and Eigenvectors*

Find the eigenvalues of

$$A = \begin{bmatrix} 1 & 0 & 0 & 0 \\ 0 & 1 & 5 & -10 \\ 1 & 0 & 2 & 0 \\ 1 & 0 & 0 & 3 \end{bmatrix}$$

and find a basis for each of the corresponding eigenspaces.

Solution   The characteristic polynomial of A is

$$|\lambda I - A| = \begin{vmatrix} \lambda - 1 & 0 & 0 & 0 \\ 0 & \lambda - 1 & -5 & 10 \\ -1 & 0 & \lambda - 2 & 0 \\ -1 & 0 & 0 & \lambda - 3 \end{vmatrix}$$

$$= (\lambda - 1)^2(\lambda - 2)(\lambda - 3).$$

Thus, the characteristic equation is $(\lambda - 1)^2(\lambda - 2)(\lambda - 3) = 0$ and the eigenvalues are $\lambda_1 = 1$, $\lambda_2 = 2$, and $\lambda_3 = 3$. (Note that $\lambda_1 = 1$ has a multiplicity of 2.)

You can find a basis for the eigenspace of $\lambda_1 = 1$ as follows.

$$(1)I - A = \begin{bmatrix} 0 & 0 & 0 & 0 \\ 0 & 0 & -5 & 10 \\ -1 & 0 & -1 & 0 \\ -1 & 0 & 0 & -2 \end{bmatrix} \longrightarrow \begin{bmatrix} 1 & 0 & 0 & 2 \\ 0 & 0 & 1 & -2 \\ 0 & 0 & 0 & 0 \\ 0 & 0 & 0 & 0 \end{bmatrix}$$

Letting $s = x_2$ and $t = x_4$ produces

$$\mathbf{x} = \begin{bmatrix} x_1 \\ x_2 \\ x_3 \\ x_4 \end{bmatrix} = \begin{bmatrix} 0s - 2t \\ s + 0t \\ 0s + 2t \\ 0s + t \end{bmatrix} = s\begin{bmatrix} 0 \\ 1 \\ 0 \\ 0 \end{bmatrix} + t\begin{bmatrix} -2 \\ 0 \\ 2 \\ 1 \end{bmatrix}.$$

Therefore, a basis for the eigenspace corresponding to $\lambda_1 = 1$ is

$$B_1 = \{(0, 1, 0, 0), (-2, 0, 2, 1)\}. \qquad \text{Basis for } \lambda_1 = 1$$

For $\lambda_2 = 2$ and $\lambda_3 = 3$, follow the same pattern to obtain the eigenspace bases

$$B_2 = \{(0, 5, 1, 0)\} \qquad \text{Basis for } \lambda_2 = 2$$

$$B_3 = \{(0, -5, 0, 1)\}. \qquad \text{Basis for } \lambda_3 = 3$$

Finding eigenvalues and eigenvectors for matrices of order $n \geq 4$ can be tedious. Moreover, the procedure followed in Example 6 is generally inefficient when used on a computer, because finding roots on a computer is both time consuming and subject to roundoff error. Consequently, numerical methods of approximating the eigenvalues of large matrices are required. These numerical methods can be found in texts on advanced linear algebra and numerical analysis.

Technology Note

Many computers and graphing calculators have built-in programs to approximate the eigen-values and eigenvectors of an $n \times n$ matrix. If you enter the matrix A from Example 6, you should obtain the four eigenvalues

$$\{1 \quad 1 \quad 2 \quad 3\}.$$

Your computer or graphing calculator should also be able to produce a matrix in which the columns are the corresponding eigenvectors, which are sometimes scalar multiples of those you would obtain by hand calculations.

There are a few types of matrices for which eigenvalues are easy to find. The following theorem states that the eigenvalues of an $n \times n$ triangular matrix are the entries on the main diagonal. Its proof follows from the fact that the determinant of a triangular matrix is the product of its diagonal elements.

| Theorem 7.3 | *Eigenvalues for Triangular Matrices* |

If A is an $n \times n$ triangular matrix, then its eigenvalues are the entries on its main diagonal.

EXAMPLE 7   *Finding Eigenvalues for Diagonal and Triangular Matrices*

Find the eigenvalues for each matrix.

(a) $A = \begin{bmatrix} 2 & 0 & 0 \\ -1 & 1 & 0 \\ 5 & 3 & -3 \end{bmatrix}$ (b) $A = \begin{bmatrix} -1 & 0 & 0 & 0 & 0 \\ 0 & 2 & 0 & 0 & 0 \\ 0 & 0 & 0 & 0 & 0 \\ 0 & 0 & 0 & -4 & 0 \\ 0 & 0 & 0 & 0 & 3 \end{bmatrix}$

Solution   (a) Without using Theorem 7.3, you can find that

$$|\lambda I - A| = \begin{vmatrix} \lambda - 2 & 0 & 0 \\ 1 & \lambda - 1 & 0 \\ -5 & -3 & \lambda + 3 \end{vmatrix}$$

$$= (\lambda - 2)(\lambda - 1)(\lambda + 3).$$

Thus, the eigenvalues are $\lambda_1 = 2$, $\lambda_2 = 1$, and $\lambda_3 = -3$, which are simply the main diagonal entries of A.

(b) In this case, use Theorem 7.3 to conclude that the eigenvalues are the main diagonal entries $\lambda_1 = -1$, $\lambda_2 = 2$, $\lambda_3 = 0$, $\lambda_4 = -4$, and $\lambda_5 = 3$.

Eigenvalues and Eigenvectors of Linear Transformations

This section began with definitions of eigenvalues and eigenvectors in terms of matrices. However, they can also be defined in terms of linear transformations. A number λ is called an **eigenvalue** of a linear transformation $T: V \rightarrow V$ if there is a nonzero vector $\mathbf{x}$ such that $T(\mathbf{x}) = \lambda \mathbf{x}$. The vector $\mathbf{x}$ is called an **eigenvector** of T corresponding to λ, and the set of all eigenvectors of λ (with the zero vector) is called the **eigenspace** of λ.

Consider the linear transformation $T: R^3 \rightarrow R^3$, whose matrix relative to the standard basis is

$$A = \begin{bmatrix} 1 & 3 & 0 \\ 3 & 1 & 0 \\ 0 & 0 & -2 \end{bmatrix}.$$

Standard basis:
$B = \{(1, 0, 0), (0, 1, 0), (0, 0, 1)\}$

In Example 5 of Section 6.4, you found that the matrix of T relative to the basis B' is the diagonal matrix

$$A' = \begin{bmatrix} 4 & 0 & 0 \\ 0 & -2 & 0 \\ 0 & 0 & -2 \end{bmatrix}.$$

Nonstandard basis:
$B' = \{(1, 1, 0), (1, -1, 0), (0, 0, 1)\}$

The question now is this: "For a given transformation T, can you find a basis B' whose corresponding matrix is diagonal?" The next example gives an indication of the answer.

EXAMPLE 8   *Finding Eigenvalues and Eigenspaces*

Find the eigenvalues and corresponding eigenspaces of

$$A = \begin{bmatrix} 1 & 3 & 0 \\ 3 & 1 & 0 \\ 0 & 0 & -2 \end{bmatrix}.$$

Solution   Because

$$|\lambda I - A| = \begin{vmatrix} \lambda - 1 & -3 & 0 \\ -3 & \lambda - 1 & 0 \\ 0 & 0 & \lambda + 2 \end{vmatrix}$$

$$= (\lambda + 2)[(\lambda - 1)^2 - 9]$$

$$= (\lambda + 2)(\lambda^2 - 2\lambda - 8) = (\lambda + 2)^2(\lambda - 4),$$

the eigenvalues of A are $\lambda_1 = 4$ and $\lambda_2 = -2$. The eigenspaces for these two eigenvalues are as follows.

$B_1 = \{(1, 1, 0)\}$ Basis for $\lambda_1 = 4$

$B_2 = \{(1, -1, 0), (0, 0, 1)\}$ Basis for $\lambda_2 = -2$

Example 8 illustrates two important and perhaps surprising results.

1. Let $T: R^3 \to R^3$ be the linear transformation whose standard matrix is A, and let B' be the basis of R^3 made up of the three linearly independent eigenvectors found in Example 8. Then A', the matrix of T relative to the basis B', is diagonal.

$$A' = \begin{bmatrix} 4 & 0 & 0 \\ 0 & -2 & 0 \\ 0 & 0 & -2 \end{bmatrix}$$

Nonstandard basis:
$B' = \{(1, 1, 0), (1, -1, 0), (0, 0, 1)\}$

Eigenvalues of A   *Eigenvectors of A*

2. The main diagonal entries of the matrix A' are the eigenvalues of A.

The next section formalizes these two results and also characterizes linear transformations that can be represented by diagonal matrices.

In Exercises 1–6, verify that λ_i is an eigenvalue of A and that $\mathbf{x}_i$ is a corresponding eigenvector.

1. $A = \begin{bmatrix} 1 & 0 \\ 0 & -1 \end{bmatrix}$, $\begin{array}{l} \lambda_1 = 1, \mathbf{x}_1 = (1, 0) \\ \lambda_2 = -1, \mathbf{x}_2 = (0, 1) \end{array}$

2. $A = \begin{bmatrix} 4 & -5 \\ 2 & -3 \end{bmatrix}$, $\begin{array}{l} \lambda_1 = -1, \mathbf{x}_1 = (1, 1) \\ \lambda_2 = 2, \mathbf{x}_2 = (5, 2) \end{array}$

3. $A = \begin{bmatrix} 1 & 1 \\ 1 & 1 \end{bmatrix}$, $\begin{array}{l} \lambda_1 = 0, \mathbf{x}_1 = (1, -1) \\ \lambda_2 = 2, \mathbf{x}_2 = (1, 1) \end{array}$

4. $A = \begin{bmatrix} 2 & 3 & 1 \\ 0 & -1 & 2 \\ 0 & 0 & 3 \end{bmatrix}$, $\begin{array}{l} \lambda_1 = 2, \mathbf{x}_1 = (1, 0, 0) \\ \lambda_2 = -1, \mathbf{x}_2 = (1, -1, 0) \\ \lambda_3 = 3, \mathbf{x}_3 = (5, 1, 2) \end{array}$

5. $A = \begin{bmatrix} -2 & 2 & -3 \\ 2 & 1 & -6 \\ -1 & -2 & 0 \end{bmatrix}$, $\begin{array}{l} \lambda_1 = 5, \mathbf{x}_1 = (1, 2, -1) \\ \lambda_2 = -3, \mathbf{x}_2 = (-2, 1, 0) \\ \lambda_3 = -3, \mathbf{x}_3 = (3, 0, 1) \end{array}$

6. $A = \begin{bmatrix} 0 & 1 & 0 \\ 0 & 0 & 1 \\ 1 & 0 & 0 \end{bmatrix}$, $\lambda_1 = 1, \mathbf{x}_1 = (1, 1, 1)$

7. Use A, λ_i, and $\mathbf{x}_i$ from Exercise 3 for the following.

    (a) Show that $A(c\mathbf{x}_1) = 0(c\mathbf{x}_1)$ for any real number c.

    (b) Show that $A(c\mathbf{x}_2) = 2(c\mathbf{x}_2)$ for any real number c.

8. Use A, λ_i, and $\mathbf{x}_i$ from Exercise 4 for the following.

    (a) Show that $A(c\mathbf{x}_1) = 2(c\mathbf{x}_1)$ for any real number c.

    (b) Show that $A(c\mathbf{x}_2) = -(c\mathbf{x}_2)$ for any real number c.

    (c) Show that $A(c\mathbf{x}_3) = 3(c\mathbf{x}_3)$ for any real number c.

In Exercises 9–12, determine whether $\mathbf{x}$ is an eigenvector of A.

9. $A = \begin{bmatrix} 7 & 2 \\ 2 & 4 \end{bmatrix}$

    (a) $\mathbf{x} = (1, 2)$

    (b) $\mathbf{x} = (2, 1)$

    (c) $\mathbf{x} = (1, -2)$

    (d) $\mathbf{x} = (-1, 0)$

10. $A = \begin{bmatrix} -3 & 10 \\ 5 & 2 \end{bmatrix}$

    (a) $\mathbf{x} = (4, 4)$

    (b) $\mathbf{x} = (-8, 4)$

    (c) $\mathbf{x} = (-4, 8)$

    (d) $\mathbf{x} = (5, -3)$

11. $A = \begin{bmatrix} -1 & -1 & 1 \\ -2 & 0 & -2 \\ 3 & -3 & 1 \end{bmatrix}$

    (a) $\mathbf{x} = (2, -4, 6)$

    (b) $\mathbf{x} = (2, 0, 6)$

    (c) $\mathbf{x} = (2, 2, 0)$

    (d) $\mathbf{x} = (-1, 0, 1)$

12. $A = \begin{bmatrix} 1 & 0 & 5 \\ 0 & -2 & 4 \\ 1 & -2 & 9 \end{bmatrix}$

    (a) $\mathbf{x} = (1, 1, 0)$

    (b) $\mathbf{x} = (-5, 2, 1)$

    (c) $\mathbf{x} = (0, 0, 0)$

    (d) $\mathbf{x} = \left(2\sqrt{6} - 3, -2\sqrt{6} + 6, 3\right)$

In Exercises 13–24, find (a) the characteristic equation and (b) the eigenvalues (and corresponding eigenvectors) of the matrix.

13. $\begin{bmatrix} 6 & -3 \\ -2 & 1 \end{bmatrix}$

14. $\begin{bmatrix} 1 & -4 \\ -2 & 8 \end{bmatrix}$

15. $\begin{bmatrix} 1 & -\frac{3}{2} \\ \frac{1}{2} & -1 \end{bmatrix}$

16. $\begin{bmatrix} \frac{1}{4} & \frac{1}{4} \\ \frac{1}{2} & 0 \end{bmatrix}$

17. $\begin{bmatrix} 2 & 0 & 1 \\ 0 & 3 & 4 \\ 0 & 0 & 1 \end{bmatrix}$

18. $\begin{bmatrix} -5 & 0 & 0 \\ 3 & 7 & 0 \\ 4 & -2 & 3 \end{bmatrix}$

19. $\begin{bmatrix} 1 & 2 & -2 \\ -2 & 5 & -2 \\ -6 & 6 & -3 \end{bmatrix}$

20. $\begin{bmatrix} 3 & 2 & -3 \\ -3 & -4 & 9 \\ -1 & -2 & 5 \end{bmatrix}$

21. $\begin{bmatrix} 0 & -3 & 5 \\ -4 & 4 & -10 \\ 0 & 0 & 4 \end{bmatrix}$

22. $\begin{bmatrix} 1 & -\frac{3}{2} & \frac{5}{2} \\ -2 & \frac{13}{2} & -10 \\ \frac{3}{2} & -\frac{9}{2} & 8 \end{bmatrix}$

23. $\begin{bmatrix} 2 & 0 & 0 & 0 \\ 0 & 2 & 0 & 0 \\ 0 & 0 & 3 & 0 \\ 0 & 0 & 4 & 0 \end{bmatrix}$

24. $\begin{bmatrix} 3 & 0 & 0 & 0 \\ 4 & 1 & 0 & 0 \\ 0 & 0 & 2 & 1 \\ 0 & 0 & 0 & 2 \end{bmatrix}$

In Exercises 25–32, use a graphing utility with matrix capabilities or computer software to find the eigenvalues of the matrix.

25. $\begin{bmatrix} -4 & 5 \\ -2 & 3 \end{bmatrix}$

26. $\begin{bmatrix} 2 & 3 \\ 3 & -6 \end{bmatrix}$

27. $\begin{bmatrix} 7 & 2 \\ -2 & 3 \end{bmatrix}$

28. $\begin{bmatrix} -6 & 2 \\ 3 & -1 \end{bmatrix}$

29. $\begin{bmatrix} 0 & -\frac{1}{2} & 5 \\ -\frac{1}{3} & -\frac{1}{6} & -\frac{1}{4} \\ 0 & 0 & 4 \end{bmatrix}$

30. $\begin{bmatrix} \frac{1}{2} & 0 & 5 \\ -2 & \frac{1}{5} & \frac{1}{4} \\ 0 & 0 & 3 \end{bmatrix}$

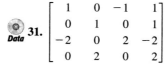

31. $\begin{bmatrix} 1 & 0 & -1 & 1 \\ 0 & 1 & 0 & 1 \\ -2 & 0 & 2 & -2 \\ 0 & 2 & 0 & 2 \end{bmatrix}$

32. $\begin{bmatrix} 1 & -3 & 3 & 3 \\ -1 & 4 & -3 & -3 \\ -2 & 0 & 1 & 1 \\ 1 & 0 & 0 & 0 \end{bmatrix}$

In Exercises 33–36, demonstrate the Cayley-Hamilton Theorem for the given matrix. The **Cayley-Hamilton Theorem** states that a matrix satisfies its characteristic equation. For example, the characteristic equation of

$$A = \begin{bmatrix} 1 & -3 \\ 2 & 5 \end{bmatrix}$$

is $\lambda^2 - 6\lambda + 11 = 0$, and therefore, by the theorem, you have $A^2 - 6A + 11I_2 = O$.

33. $\begin{bmatrix} 4 & 0 \\ -3 & 2 \end{bmatrix}$ **34.** $\begin{bmatrix} 6 & -1 \\ 1 & 5 \end{bmatrix}$

35. $\begin{bmatrix} 1 & 0 & -4 \\ 0 & 3 & 1 \\ 2 & 0 & 1 \end{bmatrix}$ **36.** $\begin{bmatrix} -3 & 1 & 0 \\ -1 & 3 & 2 \\ 0 & 4 & 3 \end{bmatrix}$

37. Perform the following computational checks on the eigenvalues found in Exercises 13–24.

  (a) The sum of the n eigenvalues equals the sum of the diagonal entries of the matrix. (This sum is called the **trace** of A.)
  (b) The product of the n eigenvalues equals $|A|$.

(If λ is an eigenvalue of multiplicity k, remember to enter it k times in the sum or product of these checks.)

38. Prove that $\lambda = 0$ is an eigenvalue of A if and only if A is singular.

39. *Writing* For an invertible matrix A, prove that A and A^{-1} have the same eigenvectors. How are the eigenvalues of A related to the eigenvalues of A^{-1}?

40. *Writing* Prove that A and A^T have the same eigenvalues. Are the eigenspaces the same?

41. Prove that the constant term of the characteristic polynomial is $\pm|A|$.

42. Let $T: R^2 \to R^2$ be given by $T(\mathbf{v}) = \text{proj}_{\mathbf{u}}\mathbf{v}$, where $\mathbf{u}$ is a fixed vector in R^2. Show that the eigenvalues of A (the standard matrix of T) are 0 and 1.

43. *Guided Proof* Prove that a triangular matrix is nonsingular if and only if its eigenvalues are real and nonzero.

  Getting Started: Because this is an "if and only if" statement, you must prove that the statement is true in both directions. Review Theorems 3.2 and 3.7.

  (i) To prove the statement in one direction, assume that the triangular matrix A is nonsingular. Use your knowledge of nonsingular and triangular matrices and determinants to conclude that the entries on the main diagonal of A are nonzero.

  (ii) Because A is triangular, you can use Theorem 7.3 and part (i) to conclude that the eigenvalues are real and nonzero.
  (iii) To prove the statement in the other direction, assume that the eigenvalues of the triangular matrix A are real and nonzero. Repeat parts (i) and (ii) in reverse order to prove that A is nonsingular.

44. *Guided Proof* Prove that if $A^2 = O$, then 0 is the only eigenvalue of A.

  Getting Started: You need to show that if there exists a nonzero vector $\mathbf{x}$ and a real number λ such that $A\mathbf{x} = \lambda\mathbf{x}$, then if $A^2 = O$, λ must be zero.

  (i) Because $A^2 = A \cdot A$, you can write $A^2\mathbf{x}$ as $A(A\mathbf{x})$.
  (ii) Use the fact that $A\mathbf{x} = \lambda\mathbf{x}$ and the properties of matrix multiplication to conclude that $A^2\mathbf{x} = \lambda^2\mathbf{x}$.
  (iii) Because A^2 is a zero matrix, you can conclude that λ must be zero.

45. If the eigenvalues of

$$A = \begin{bmatrix} a & b \\ 0 & d \end{bmatrix}$$

are $\lambda_1 = 0$ and $\lambda_2 = 1$, what are the possible values of a and d?

46. Show that

$$A = \begin{bmatrix} 0 & 1 \\ -1 & 0 \end{bmatrix}$$

has no real eigenvalues.

True or False? In Exercises 47 and 48, determine whether each statement is true or false. If a statement is true, give a reason or cite an appropriate statement in the text. If a statement is false, provide an example that shows that the statement is not true in all cases or cite an appropriate statement in the text.

47. (a) The scalar λ is an eigenvalue of an $n \times n$ matrix A if there exists a vector $\mathbf{x}$ such that $A\mathbf{x} = \lambda\mathbf{x}$.
  (b) If A is an $n \times n$ matrix with eigenvalue λ and corresponding eigenvector $\mathbf{x}$, then every nonzero scalar multiple of $\mathbf{x}$ is also an eigenvector of A.
  (c) To find the eigenvalue(s) of an $n \times n$ matrix A, you can solve the characteristic equation, $\det(\lambda I - A) = 0$.

48. (a) Geometrically, if λ is an eigenvalue of a matrix A and $\mathbf{x}$ is an eigenvector of A corresponding to λ, then multiplying $\mathbf{x}$ by A produces a vector $\lambda\mathbf{x}$ that is parallel to $\mathbf{x}$.
  (b) An $n \times n$ matrix A can have only one eigenvalue.
  (c) If A is an $n \times n$ matrix with an eigenvalue λ, then the set of all eigenvectors of λ is a subspace of R^n.

In Exercises 49–52, find the dimension of the eigenspace corresponding to the eigenvalue $\lambda = 3$.

49. $A = \begin{bmatrix} 3 & 0 & 0 \\ 0 & 3 & 0 \\ 0 & 0 & 3 \end{bmatrix}$ **50.** $A = \begin{bmatrix} 3 & 1 & 0 \\ 0 & 3 & 0 \\ 0 & 0 & 3 \end{bmatrix}$

51. $A = \begin{bmatrix} 3 & 1 & 0 \\ 0 & 3 & 1 \\ 0 & 0 & 3 \end{bmatrix}$ **52.** $A = \begin{bmatrix} 3 & 1 & 1 \\ 0 & 3 & 1 \\ 0 & 0 & 3 \end{bmatrix}$

53. *Calculus*   Let $T: C'[0, 1] \to C[0, 1]$ be given by $T(f) = f'$. Show that $\lambda = 1$ is an eigenvalue of T with corresponding eigenvector $f(x) = e^x$.

54. *Calculus*   For the linear transformation given in Exercise 53, find the eigenvalue corresponding to the eigenvector $f(x) = e^{-2x}$.

55. Let $T: P_2 \to P_2$ be given by

$$T(a_0 + a_1 x + a_2 x^2) = (-3a_1 + 5a_2) + \\ (-4a_0 + 4a_1 - 10a_2)x + 4a_2 x^2.$$

Find the eigenvalues and the eigenvectors of T relative to the standard basis $\{1, x, x^2\}$.

56. Let $T: P_2 \to P_2$ be given by

$$T(a_0 + a_1 x + a_2 x^2) = (2a_0 + a_1 - a_2) + \\ (-a_1 + 2a_2)x - a_2 x^2.$$

Find the eigenvalues and eigenvectors of T relative to the standard basis $\{1, x, x^2\}$.

57. Let $T: M_{2,2} \to M_{2,2}$ be given by

$$T\left(\begin{bmatrix} a & b \\ c & d \end{bmatrix}\right) = \begin{bmatrix} a - c + d & b + d \\ -2a + 2c - 2d & 2b + 2d \end{bmatrix}.$$

Find the eigenvalues and eigenvectors of T relative to the standard basis

$$B = \left\{ \begin{bmatrix} 1 & 0 \\ 0 & 0 \end{bmatrix}, \begin{bmatrix} 0 & 1 \\ 0 & 0 \end{bmatrix}, \begin{bmatrix} 0 & 0 \\ 1 & 0 \end{bmatrix}, \begin{bmatrix} 0 & 0 \\ 0 & 1 \end{bmatrix} \right\}.$$

58. A square matrix A is called **idempotent** if $A^2 = A$. What are the possible eigenvalues of an idempotent matrix?

59. A square matrix A is called **nilpotent** if there exists a positive integer k such that $A^k = 0$. What are the possible eigenvalues of a nilpotent matrix?

60. Find all values of the angle θ for which the matrix

$$A = \begin{bmatrix} \cos \theta & -\sin \theta \\ \sin \theta & \cos \theta \end{bmatrix}$$

has real eigenvalues. Interpret your answer geometrically.

61. Let A be an $n \times n$ matrix such that the sum of the entries in each row is a fixed constant r. Prove that r is an eigenvalue of A. Illustrate this result with a specific example.

7.2 | Diagonalization

The preceding section discussed the eigenvalue problem. In this section, you will look at another classic problem in linear algebra called the **diagonalization problem.** Expressed in terms of matrices,[*] the problem is this: For a square matrix A, does there exist an invertible matrix P such that $P^{-1}AP$ is diagonal?

Recall from Section 6.4 that two square matrices A and B are called **similar** if there exists an invertible matrix P such that

$$B = P^{-1}AP.$$

Matrices that are similar to diagonal matrices are called **diagonalizable.**

Definition of a Diagonalizable Matrix

An $n \times n$ matrix A is **diagonalizable** if A is similar to a diagonal matrix. That is, A is diagonalizable if there exists an invertible matrix P such that

$$P^{-1}AP$$

is a diagonal matrix.

───────────

[*]At the end of this section, the diagonalization problem will be expressed in terms of linear transformations.

Given this definition, the diagonalization problem can be stated as follows. Which square matrices are diagonalizable? Clearly, every diagonal matrix D is diagonalizable, because the identity matrix I can play the role of P to give $D = I^{-1}DI$. Example 1 shows another example of a diagonalizable matrix.

EXAMPLE 1   *A Diagonalizable Matrix*

The matrix from Example 5 of Section 6.4,

$$A = \begin{bmatrix} 1 & 3 & 0 \\ 3 & 1 & 0 \\ 0 & 0 & -2 \end{bmatrix},$$

is diagonalizable because

$$P = \begin{bmatrix} 1 & 1 & 0 \\ 1 & -1 & 0 \\ 0 & 0 & 1 \end{bmatrix}$$

has the property that

$$P^{-1}AP = \begin{bmatrix} 4 & 0 & 0 \\ 0 & -2 & 0 \\ 0 & 0 & -2 \end{bmatrix}.$$

As indicated in Example 8 of the preceding section, the eigenvalue problem is related closely to the diagonalization problem. The next two theorems shed more light on this relationship. The first theorem tells you that similar matrices must have the same eigenvalues.

Theorem 7.4	Similar Matrices Have the Same Eigenvalues

If A and B are similar $n \times n$ matrices, then they have the same eigenvalues.

Proof   Because A and B are similar, there exists an invertible matrix P such that $B = P^{-1}AP$. By the properties of determinants, it follows that

$$\begin{aligned} |\lambda I - B| = |\lambda I - P^{-1}AP| &= |P^{-1}\lambda IP - P^{-1}AP| \\ &= |P^{-1}(\lambda I - A)P| \\ &= |P^{-1}||\lambda I - A||P| \\ &= |P^{-1}||P||\lambda I - A| \\ &= |P^{-1}P||\lambda I - A| \\ &= |\lambda I - A|. \end{aligned}$$

But this means that A and B have the same characteristic polynomial. Hence, they must have the same eigenvalues.

EXAMPLE 2   *Finding Eigenvalues of Similar Matrices*

The matrices A and D are similar.

$$A = \begin{bmatrix} 1 & 0 & 0 \\ -1 & 1 & 1 \\ -1 & -2 & 4 \end{bmatrix} \quad \text{and} \quad D = \begin{bmatrix} 1 & 0 & 0 \\ 0 & 2 & 0 \\ 0 & 0 & 3 \end{bmatrix}$$

Use Theorem 7.4 to find the eigenvalues of A and D.

Solution   Because D is a diagonal matrix, its eigenvalues are simply the entries on its main diagonal—that is,

$$\lambda_1 = 1,$$
$$\lambda_2 = 2, \text{ and}$$
$$\lambda_3 = 3.$$

Moreover, because A is given to be similar to D, you know from Theorem 7.4 that A has the same eigenvalues. Check this by showing that the characteristic polynomial of A is

$$|\lambda I - A| = (\lambda - 1)(\lambda - 2)(\lambda - 3).$$

REMARK: Example 2 simply states that the matrices A and D are similar. Try checking that $D = P^{-1}AP$ using the matrices

$$P = \begin{bmatrix} 1 & 0 & 0 \\ 1 & 1 & 1 \\ 1 & 1 & 2 \end{bmatrix} \quad \text{and} \quad P^{-1} = \begin{bmatrix} 1 & 0 & 0 \\ -1 & 2 & -1 \\ 0 & -1 & 1 \end{bmatrix}.$$

In fact, the columns of P are precisely the eigenvectors of A corresponding to the eigenvalues 1, 2, and 3.

The two diagonalizable matrices given in Examples 1 and 2 provide a clue to the diagonalization problem. Both matrices possess a set of three linearly independent eigenvectors. (See Example 3.) This is characteristic of diagonalizable matrices, as stated in Theorem 7.5.

Theorem 7.5   *Condition for Diagonalization*

An $n \times n$ matrix A is diagonalizable if and only if it has n linearly independent eigenvectors.

Proof   First, assume that A is diagonalizable. Then there exists an invertible matrix P such that $P^{-1}AP = D$ is diagonal. Letting the main entries of D be $\lambda_1, \lambda_2, \ldots, \lambda_n$ and the column vectors of P be $\mathbf{p}_1, \mathbf{p}_2, \ldots, \mathbf{p}_n$ produces

$$PD = [\mathbf{p}_1 \vdots \mathbf{p}_2 \vdots \cdots \vdots \mathbf{p}_n] \begin{bmatrix} \lambda_1 & 0 & \cdots & 0 \\ 0 & \lambda_2 & \cdots & 0 \\ \vdots & \vdots & & \vdots \\ 0 & 0 & \cdots & \lambda_n \end{bmatrix}$$

$$= [\lambda_1\mathbf{p}_1 \vdots \lambda_2\mathbf{p}_2 \vdots \cdots \vdots \lambda_n\mathbf{p}_n].$$

Because $P^{-1}AP = D$, then $AP = PD$, which implies that

$$[A\mathbf{p}_1 \vdots A\mathbf{p}_2 \vdots \cdots \vdots A\mathbf{p}_n] = [\lambda_1\mathbf{p}_1 \vdots \lambda_2\mathbf{p}_2 \vdots \cdots \vdots \lambda_n\mathbf{p}_n].$$

In other words, $A\mathbf{p}_i = \lambda_i\mathbf{p}_i$ for each column vector $\mathbf{p}_i$. This means that the column vectors $\mathbf{p}_i$ of P are eigenvectors of A. Moreover, because P is invertible, its column vectors are linearly independent. Thus A has n linearly independent eigenvectors.

Conversely, assume that A has n linearly independent eigenvectors $\mathbf{p}_1, \mathbf{p}_2, \ldots, \mathbf{p}_n$ with corresponding eigenvalues $\lambda_1, \lambda_2, \ldots, \lambda_n$. Let P be the matrix whose columns are these n eigenvectors. That is, $P = [\mathbf{p}_1 \vdots \mathbf{p}_2 \vdots \cdots \vdots \mathbf{p}_n]$. Because each $\mathbf{p}_i$ is an eigenvector of A, you have $A\mathbf{p}_i = \lambda_i\mathbf{p}_i$ and

$$AP = A[\mathbf{p}_1 \vdots \mathbf{p}_2 \vdots \cdots \vdots \mathbf{p}_n] = [\lambda_1\mathbf{p}_1 \vdots \lambda_2\mathbf{p}_2 \vdots \cdots \vdots \lambda_n\mathbf{p}_n].$$

The right-hand matrix in this equation can be written as the matrix product below.

$$AP = [\mathbf{p}_1 \vdots \mathbf{p}_2 \vdots \cdots \vdots \mathbf{p}_n] \begin{bmatrix} \lambda_1 & 0 & \cdots & 0 \\ 0 & \lambda_2 & \cdots & 0 \\ \vdots & \vdots & & \vdots \\ 0 & 0 & \cdots & \lambda_n \end{bmatrix} = PD$$

Finally, because the vectors $\mathbf{p}_1, \mathbf{p}_2, \ldots, \mathbf{p}_n$ are linearly independent, P is invertible and you can write the equation $AP = PD$ as $P^{-1}AP = D$, which means that A is diagonalizable.

A key result of this proof is the fact that for diagonalizable matrices, *the columns of P consist of the n linearly independent eigenvectors.* Example 3 verifies this important property for the matrices in Examples 1 and 2.

EXAMPLE 3 *Diagonalizable Matrices*

(a) The matrix in Example 1 has the following eigenvalues and corresponding eigenvectors.

$$\lambda_1 = 4, \mathbf{p}_1 = \begin{bmatrix} 1 \\ 1 \\ 0 \end{bmatrix}; \quad \lambda_2 = -2, \mathbf{p}_2 = \begin{bmatrix} 1 \\ -1 \\ 0 \end{bmatrix}; \quad \lambda_3 = -2, \mathbf{p}_3 = \begin{bmatrix} 0 \\ 0 \\ 1 \end{bmatrix}$$

The matrix P whose columns correspond to these eigenvectors is

$$P = \begin{bmatrix} 1 & 1 & 0 \\ 1 & -1 & 0 \\ 0 & 0 & 1 \end{bmatrix}.$$

Moreover, because P is row-equivalent to the identity matrix, the eigenvectors $\mathbf{p}_1, \mathbf{p}_2$, and $\mathbf{p}_3$ are linearly independent.

(b) The matrix in Example 2 has the following eigenvalues and corresponding eigenvectors.

$$\lambda_1 = 1, \mathbf{p}_1 = \begin{bmatrix} 1 \\ 1 \\ 1 \end{bmatrix}; \quad \lambda_2 = 2, \mathbf{p}_2 = \begin{bmatrix} 0 \\ 1 \\ 1 \end{bmatrix}; \quad \lambda_3 = 3, \mathbf{p}_3 = \begin{bmatrix} 0 \\ 1 \\ 2 \end{bmatrix}$$

The matrix P whose columns correspond to these eigenvectors is

$$P = \begin{bmatrix} 1 & 0 & 0 \\ 1 & 1 & 1 \\ 1 & 1 & 2 \end{bmatrix}.$$

Again, because P is row-equivalent to the identity matrix, the eigenvectors $\mathbf{p}_1, \mathbf{p}_2$, and $\mathbf{p}_3$ are linearly independent.

The second part of the proof of Theorem 7.5 and Example 3 suggest the following steps for diagonalizing a matrix.

Steps for Diagonalizing an n × n Square Matrix

Let A be an $n \times n$ matrix.
1. Find n linearly independent eigenvectors $\mathbf{p}_1, \mathbf{p}_2, \ldots, \mathbf{p}_n$ for A with corresponding eigenvalues $\lambda_1, \lambda_2, \ldots, \lambda_n$. If n linearly independent eigenvectors do not exist, then A is not diagonalizable.
2. If A has n linearly independent eigenvectors, let P be the $n \times n$ matrix whose columns consist of these eigenvectors. That is,

$$P = [\mathbf{p}_1 \vdots \mathbf{p}_2 \vdots \cdots \vdots \mathbf{p}_n].$$

3. The diagonal matrix $D = P^{-1}AP$ will have the eigenvalues $\lambda_1, \lambda_2, \ldots, \lambda_n$ on its main diagonal (and zeros elsewhere). Note that the order of the eigenvectors used to form P will determine the order in which the eigenvalues appear on the main diagonal of D.

EXAMPLE 4   *A Matrix That Is Not Diagonalizable*

Show that the matrix A is not diagonalizable.

$$A = \begin{bmatrix} 1 & 2 \\ 0 & 1 \end{bmatrix}$$

Solution   Because A is triangular, the eigenvalues are simply the entries on the main diagonal. Thus the only eigenvalue is $\lambda = 1$. The matrix $(I - A)$ has the following reduced row-echelon form.

$$I - A = \begin{bmatrix} 0 & -2 \\ 0 & 0 \end{bmatrix} \longrightarrow \begin{bmatrix} 0 & 1 \\ 0 & 0 \end{bmatrix}$$

This implies that $x_2 = 0$, and letting $x_1 = t$, you can find that every eigenvector of A has the form

$$\mathbf{x} = \begin{bmatrix} x_1 \\ x_2 \end{bmatrix} = \begin{bmatrix} t \\ 0 \end{bmatrix} = t \begin{bmatrix} 1 \\ 0 \end{bmatrix}.$$

Hence, A does not have two linearly independent eigenvectors, and you can conclude that A is not diagonalizable.

EXAMPLE 5   *Diagonalizing a Matrix*

Show that the matrix A is diagonalizable.

$$A = \begin{bmatrix} 1 & -1 & -1 \\ 1 & 3 & 1 \\ -3 & 1 & -1 \end{bmatrix}$$

Then find a matrix P such that $P^{-1}AP$ is diagonal.

Solution   The characteristic polynomial of A is

$$|\lambda I - A| = \begin{vmatrix} \lambda - 1 & 1 & 1 \\ -1 & \lambda - 3 & -1 \\ 3 & -1 & \lambda + 1 \end{vmatrix} = (\lambda - 2)(\lambda + 2)(\lambda - 3).$$

Thus, the eigenvalues of A are $\lambda_1 = 2$, $\lambda_2 = -2$, and $\lambda_3 = 3$. From these eigenvalues you obtain the following reduced row-echelon forms and corresponding eigenvectors.

Eigenvector

$$2I - A = \begin{bmatrix} 1 & 1 & 1 \\ -1 & -1 & -1 \\ 3 & -1 & 3 \end{bmatrix} \longrightarrow \begin{bmatrix} 1 & 0 & 1 \\ 0 & 1 & 0 \\ 0 & 0 & 0 \end{bmatrix} \qquad \begin{bmatrix} -1 \\ 0 \\ 1 \end{bmatrix}$$

$$-2I - A = \begin{bmatrix} -3 & 1 & 1 \\ -1 & -5 & -1 \\ 3 & -1 & -1 \end{bmatrix} \longrightarrow \begin{bmatrix} 1 & 0 & -\frac{1}{4} \\ 0 & 1 & \frac{1}{4} \\ 0 & 0 & 0 \end{bmatrix} \qquad \begin{bmatrix} 1 \\ -1 \\ 4 \end{bmatrix}$$

$$3I - A = \begin{bmatrix} 2 & 1 & 1 \\ -1 & 0 & -1 \\ 3 & -1 & 4 \end{bmatrix} \longrightarrow \begin{bmatrix} 1 & 0 & 1 \\ 0 & 1 & -1 \\ 0 & 0 & 0 \end{bmatrix} \qquad \begin{bmatrix} -1 \\ 1 \\ 1 \end{bmatrix}$$

Form the matrix P whose columns are the eigenvectors just obtained.

$$P = \begin{bmatrix} -1 & 1 & -1 \\ 0 & -1 & 1 \\ 1 & 4 & 1 \end{bmatrix}$$

This matrix is nonsingular, which implies that the eigenvectors are linearly independent and A is diagonalizable. The inverse of P is

$$P^{-1} = \begin{bmatrix} -1 & -1 & 0 \\ \frac{1}{5} & 0 & \frac{1}{5} \\ \frac{1}{5} & 1 & \frac{1}{5} \end{bmatrix},$$

and it follows that

$$P^{-1}AP = \begin{bmatrix} 2 & 0 & 0 \\ 0 & -2 & 0 \\ 0 & 0 & 3 \end{bmatrix}.$$

EXAMPLE 6   *Diagonalizing a Matrix*

Show that the matrix A is diagonalizable.

$$A = \begin{bmatrix} 1 & 0 & 0 & 0 \\ 0 & 1 & 5 & -10 \\ 1 & 0 & 2 & 0 \\ 1 & 0 & 0 & 3 \end{bmatrix}$$

Then find a matrix P such that $P^{-1}AP$ is diagonal.

Solution   In Example 6 in Section 7.1, you found that the three eigenvalues $\lambda_1 = 1$, $\lambda_2 = 2$, and $\lambda_3 = 3$ have the following eigenvectors.

$$\lambda_1 : \begin{bmatrix} 0 \\ 1 \\ 0 \\ 0 \end{bmatrix}, \begin{bmatrix} -2 \\ 0 \\ 2 \\ 1 \end{bmatrix} \qquad \lambda_2 : \begin{bmatrix} 0 \\ 5 \\ 1 \\ 0 \end{bmatrix} \qquad \lambda_3 : \begin{bmatrix} 0 \\ -5 \\ 0 \\ 1 \end{bmatrix}$$

The matrix whose columns consist of these eigenvectors is

$$P = \begin{bmatrix} 0 & -2 & 0 & 0 \\ 1 & 0 & 5 & -5 \\ 0 & 2 & 1 & 0 \\ 0 & 1 & 0 & 1 \end{bmatrix}.$$

Because P is invertible (check this), its column vectors form a linearly independent set. Hence, A is diagonalizable, and you have

$$P^{-1}AP = \begin{bmatrix} 1 & 0 & 0 & 0 \\ 0 & 1 & 0 & 0 \\ 0 & 0 & 2 & 0 \\ 0 & 0 & 0 & 3 \end{bmatrix}.$$

For a square matrix A of order n to be diagonalizable, the sum of the dimensions of the eigenspaces must be equal to n. One way this can happen is if A has n distinct eigenvalues. Thus, you have the next theorem.

Theorem 7.6

Sufficient Condition for Diagonalization

If an $n \times n$ matrix A has n *distinct* eigenvalues, then the corresponding eigenvectors are linearly independent and A is diagonalizable.

Proof Let $\lambda_1, \lambda_2, \ldots, \lambda_n$ be n distinct eigenvalues of A corresponding to the eigenvectors $\mathbf{x}_1, \mathbf{x}_2, \ldots, \mathbf{x}_n$. To begin, assume that the set of eigenvectors is linearly dependent. Moreover, consider the eigenvectors to be ordered so that the first m eigenvectors are linearly independent, but the first $m + 1$ are dependent, where $m < n$. Then $\mathbf{x}_{m+1}$ can be written as a linear combination of the first m eigenvectors:

$$\mathbf{x}_{m+1} = c_1\mathbf{x}_1 + c_2\mathbf{x}_2 + \cdots + c_m\mathbf{x}_m, \qquad \text{Equation 1}$$

where the c_i's are not all zero. Multiplication of both sides of Equation 1 by A yields

$$A\mathbf{x}_{m+1} = Ac_1\mathbf{x}_1 + Ac_2\mathbf{x}_2 + \cdots + Ac_m\mathbf{x}_m$$
$$\lambda_{m+1}\mathbf{x}_{m+1} = c_1\lambda_1\mathbf{x}_1 + c_2\lambda_2\mathbf{x}_2 + \cdots + c_m\lambda_m\mathbf{x}_m, \qquad \text{Equation 2}$$

whereas multiplication of Equation 1 by λ_{m+1} yields

$$\lambda_{m+1}\mathbf{x}_{m+1} = c_1\lambda_{m+1}\mathbf{x}_1 + c_2\lambda_{m+1}\mathbf{x}_2 + \cdots + c_m\lambda_{m+1}\mathbf{x}_m. \qquad \text{Equation 3}$$

Now, subtracting Equation 2 from Equation 3 produces

$$c_1(\lambda_{m+1} - \lambda_1)\mathbf{x}_1 + c_2(\lambda_{m+1} - \lambda_2)\mathbf{x}_2 + \cdots + c_m(\lambda_{m+1} - \lambda_m)\mathbf{x}_m = \mathbf{0},$$

and, using the fact that the first m eigenvectors are linearly independent, you can conclude that all coefficients of this equation must be zero. That is,

$$c_1(\lambda_{m+1} - \lambda_1) = c_2(\lambda_{m+1} - \lambda_2) = \cdots = c_m(\lambda_{m+1} - \lambda_m) = 0.$$

Because all the eigenvalues are distinct, it follows that $c_i = 0$, $i = 1, 2, \ldots, m$. But this result contradicts our assumption that $\mathbf{x}_{m+1}$ can be written as a linear combination of the first m eigenvectors. Hence, the set of eigenvectors is linearly independent, and from Theorem 7.5 you can conclude that A is diagonalizable.

EXAMPLE 7 *Determining Whether a Matrix Is Diagonalizable*

Determine whether the matrix A is diagonalizable.

$$A = \begin{bmatrix} 1 & -2 & 1 \\ 0 & 0 & 1 \\ 0 & 0 & -3 \end{bmatrix}$$

Solution   Because A is a triangular matrix, its eigenvalues are the main diagonal entries

$$\lambda_1 = 1, \qquad \lambda_2 = 0, \qquad \lambda_3 = -3.$$

Moreover, because these three values are distinct, you can conclude from Theorem 7.6 that A is diagonalizable.

REMARK: Remember that the condition in Theorem 7.6 is sufficient but not necessary for diagonalization, as demonstrated in Example 6. In other words, a diagonalizable matrix need not have distinct eigenvalues.

Diagonalization and Linear Transformations

So far in this section, the diagonalization problem has been considered in terms of matrices. In terms of linear transformations, the diagonalization problem can be stated as follows. For a linear transformation

$$T: V \to V,$$

does there exist a basis B for V such that the matrix for T relative to B is diagonal? The answer is "yes," provided the standard matrix for T is diagonalizable.

EXAMPLE 8   *Finding a Diagonal Matrix for a Linear Transformation*

Let $T: R^3 \to R^3$ be the linear transformation given by

$$T(x_1, x_2, x_3) = (x_1 - x_2 - x_3, x_1 + 3x_2 + x_3, -3x_1 + x_2 - x_3).$$

If possible, find a basis B for R^3 such that the matrix for T relative to B is diagonal.

Solution   The standard matrix for T is given by

$$A = \begin{bmatrix} 1 & -1 & -1 \\ 1 & 3 & 1 \\ -3 & 1 & -1 \end{bmatrix}.$$

From Example 5 you know that A is diagonalizable. Thus, the three linearly independent eigenvectors found in Example 5 can be used to form the basis B. That is,

$$B = \{(-1, 0, 1), (1, -1, 4), (-1, 1, 1)\}.$$

The matrix for T relative to this basis is

$$D = \begin{bmatrix} 2 & 0 & 0 \\ 0 & -2 & 0 \\ 0 & 0 & 3 \end{bmatrix}.$$

In Exercises 1–4, verify that A is diagonalizable by computing $P^{-1}AP$.

1. $A = \begin{bmatrix} -11 & 36 \\ -3 & 10 \end{bmatrix}$, $P = \begin{bmatrix} -3 & -4 \\ -1 & -1 \end{bmatrix}$

2. $A = \begin{bmatrix} 1 & 3 \\ -1 & 5 \end{bmatrix}$, $P = \begin{bmatrix} 3 & 1 \\ 1 & 1 \end{bmatrix}$

3. $A = \begin{bmatrix} -1 & 1 & 0 \\ 0 & 3 & 0 \\ 4 & -2 & 5 \end{bmatrix}$, $P = \begin{bmatrix} 0 & 1 & -3 \\ 0 & 4 & 0 \\ 1 & 2 & 2 \end{bmatrix}$

4. $A = \begin{bmatrix} 0.80 & 0.10 & 0.05 & 0.05 \\ 0.10 & 0.80 & 0.05 & 0.05 \\ 0.05 & 0.05 & 0.80 & 0.10 \\ 0.05 & 0.05 & 0.10 & 0.80 \end{bmatrix}$, $P = \begin{bmatrix} 1 & -1 & 0 & 1 \\ 1 & -1 & 0 & -1 \\ 1 & 1 & 1 & 0 \\ 1 & 1 & -1 & 0 \end{bmatrix}$

In Exercises 5–12, show that the matrix is not diagonalizable.

5. $\begin{bmatrix} 0 & 0 \\ 5 & 0 \end{bmatrix}$

6. $\begin{bmatrix} 1 & \frac{1}{2} \\ -2 & -1 \end{bmatrix}$

7. $\begin{bmatrix} 3 & k \\ 0 & 3 \end{bmatrix}$, $k \neq 0$

8. $\begin{bmatrix} k & 0 \\ 0 & k \end{bmatrix}$

9. $\begin{bmatrix} 1 & -2 & 1 \\ 0 & 1 & 4 \\ 0 & 0 & 2 \end{bmatrix}$

10. $\begin{bmatrix} 2 & 1 & -1 \\ 0 & -1 & 2 \\ 0 & 0 & -1 \end{bmatrix}$

11. $\begin{bmatrix} 1 & 0 & -1 & 1 \\ 0 & 1 & 0 & 1 \\ -2 & 0 & 2 & -2 \\ 0 & 2 & 0 & 2 \end{bmatrix}$

(See Exercise 31, Section 7.1.)

12. $\begin{bmatrix} 1 & -3 & 3 & 3 \\ -1 & 4 & -3 & -3 \\ -2 & 0 & 1 & 1 \\ 1 & 0 & 0 & 0 \end{bmatrix}$

(See Exercise 32, Section 7.1.)

In Exercises 13–16, find the eigenvalues of the matrix and determine whether there are a sufficient number to guarantee that the matrix is diagonalizable. (Recall that the matrix may be diagonalizable even though it is not guaranteed diagonalizable by Theorem 7.6.)

13. $\begin{bmatrix} 1 & 1 \\ 1 & 1 \end{bmatrix}$

14. $\begin{bmatrix} 2 & 0 \\ 5 & 2 \end{bmatrix}$

15. $\begin{bmatrix} 3 & 2 & -3 \\ -3 & -4 & 9 \\ -1 & -2 & 5 \end{bmatrix}$

16. $\begin{bmatrix} 4 & 3 & -2 \\ 0 & 1 & 1 \\ 0 & 0 & -2 \end{bmatrix}$

In Exercises 17–26, for each matrix A, find (if possible) a nonsingular matrix P such that $P^{-1}AP$ is diagonal. Verify that $P^{-1}AP$ is a diagonal matrix with the eigenvalues on the diagonal.

17. $A = \begin{bmatrix} 1 & -\frac{3}{2} \\ \frac{1}{2} & -1 \end{bmatrix}$

(See Exercise 15, Section 7.1.)

18. $A = \begin{bmatrix} \frac{1}{4} & \frac{1}{4} \\ \frac{1}{2} & 0 \end{bmatrix}$

(See Exercise 16, Section 7.1.)

19. $A = \begin{bmatrix} 1 & 2 & -2 \\ -2 & 5 & -2 \\ -6 & 6 & -3 \end{bmatrix}$

(See Exercise 19, Section 7.1.)

20. $A = \begin{bmatrix} 3 & 2 & -3 \\ -3 & -4 & 9 \\ -1 & -2 & 5 \end{bmatrix}$

(See Exercise 20, Section 7.1.)

21. $A = \begin{bmatrix} 0 & -3 & 5 \\ -4 & 4 & -10 \\ 0 & 0 & 4 \end{bmatrix}$

(See Exercise 21, Section 7.1.)

22. $A = \begin{bmatrix} 1 & -\frac{3}{2} & \frac{5}{2} \\ -2 & \frac{13}{2} & -10 \\ \frac{3}{2} & -\frac{9}{2} & 8 \end{bmatrix}$

(See Exercise 22, Section 7.1.)

23. $A = \begin{bmatrix} 1 & 0 & 0 \\ 1 & 2 & 1 \\ 1 & 0 & 2 \end{bmatrix}$

24. $A = \begin{bmatrix} 4 & 0 & 0 \\ 2 & 2 & 0 \\ 0 & 2 & 2 \end{bmatrix}$

25. $A = \begin{bmatrix} 2 & 0 & 0 & 0 \\ 3 & -1 & 0 & 0 \\ 0 & 1 & 1 & 0 \\ 0 & 0 & 1 & -2 \end{bmatrix}$

26. $A = \begin{bmatrix} 1 & 0 & 0 & 0 \\ 1 & 0 & 1 & 0 \\ 0 & 0 & 1 & 0 \\ 0 & 1 & 0 & 1 \end{bmatrix}$

In Exercises 27–30, find a basis B for the domain of T such that the matrix of T relative to B is diagonal.

27. $T: R^2 \rightarrow R^2: T(x, y) = (x + y, x + y)$

28. $T: R^3 \rightarrow R^3: T(x, y, z) = (-2x + 2y - 3z, 2x + y - 6z, -x - 2y)$

29. $T: P_1 \rightarrow P_1: T(a + bx) = a + (a + 2b)x$

30. $T: P_2 \rightarrow P_2: T(a_0 + a_1x + a_2x^2) = (2a_0 + a_2) + (3a_1 + 4a_2)x + a_2x^2$

31. Let A be a diagonalizable $n \times n$ matrix and P an invertible $n \times n$ matrix such that $B = P^{-1}AP$ is the diagonal form of A. Prove the following.

(a) $B^k = P^{-1}A^kP$, where k is a positive integer.

(b) $A^k = PB^kP^{-1}$, where k is a positive integer.

32. Let $\lambda_1, \lambda_2, \ldots, \lambda_n$ be n distinct eigenvalues of the $n \times n$ matrix A. Use the result of Exercise 31 to find the eigenvalues of A^k.

In Exercises 33–36, use the result of Exercise 31 to find the indicated power of A.

33. $A = \begin{bmatrix} 10 & 18 \\ -6 & -11 \end{bmatrix}, A^6$ **34.** $A = \begin{bmatrix} 1 & 3 \\ 2 & 0 \end{bmatrix}, A^7$

35. $A = \begin{bmatrix} 3 & 2 & -3 \\ -3 & -4 & 9 \\ -1 & -2 & 5 \end{bmatrix}, A^8$ **36.** $A = \begin{bmatrix} 2 & 0 & -2 \\ 0 & 2 & -2 \\ 3 & 0 & -3 \end{bmatrix}, A^5$

True or False?   In Exercises 37 and 38, determine whether each statement is true or false. If a statement is true, give a reason or cite an appropriate statement in the text. If a statement is false, provide an example that shows that the statement is not true in all cases or cite an appropriate statement in the text.

37. (a) If A and B are similar $n \times n$ matrices, then they always have the same characteristic polynomial equation.

(b) The fact that an $n \times n$ matrix A has n distinct eigenvalues does not guarantee that A is diagonalizable.

38. (a) If A is a diagonalizable matrix, then it has n linearly independent eigenvectors.

(b) If an $n \times n$ matrix A is diagonalizable, then it must have n distinct eigenvalues.

39. ***Writing***   Can a matrix be similar to two different diagonal matrices? Explain your answer.

40. Are the two matrices similar? If so, find a matrix P such that $B = P^{-1}AP$.

(a) $A = \begin{bmatrix} 1 & 0 & 0 \\ 0 & 2 & 0 \\ 0 & 0 & 3 \end{bmatrix}$

(b) $B = \begin{bmatrix} 3 & 0 & 0 \\ 0 & 2 & 0 \\ 0 & 0 & 1 \end{bmatrix}$

41. Prove that if A is diagonalizable, then A^T is diagonalizable.

42. Prove that the matrix

$$A = \begin{bmatrix} a & b \\ c & d \end{bmatrix}$$

is diagonalizable if $-4bc < (a - d)^2$ and is not diagonalizable if $-4bc > (a - d)^2$.

43. Prove that if A is diagonalizable with n real eigenvalues $\lambda_1, \lambda_2, \ldots,$ and $\lambda_n,$ then

$$|A| = \lambda_1 \lambda_2 \cdots \lambda_n.$$

44. ***Calculus***   If x is a real number, then e^x can be defined by the series

$$e^x = 1 + x + \frac{x^2}{2!} + \frac{x^3}{3!} + \frac{x^4}{4!} + \cdots .$$

In a similar way, if X is a square matrix, you can define e^X by the series

$$e^X = I + X + \frac{1}{2!}X^2 + \frac{1}{3!}X^3 + \frac{1}{4!}X^4 + \cdots .$$

Evaluate e^X, where X is the indicated square matrix.

(a) $X = \begin{bmatrix} 1 & 0 \\ 0 & 1 \end{bmatrix}$ (d) $X = \begin{bmatrix} 0 & 1 \\ 1 & 0 \end{bmatrix}$

(b) $X = \begin{bmatrix} 0 & 0 \\ 0 & 0 \end{bmatrix}$ (e) $X = \begin{bmatrix} 2 & 0 \\ 0 & -2 \end{bmatrix}$

(c) $X = \begin{bmatrix} 1 & 0 \\ 1 & 0 \end{bmatrix}$

45. ***Guided Proof***   Prove that if the eigenvalues of a diagonalizable matrix A are all ± 1, then the matrix is equal to its inverse.

Getting Started: To show that the matrix is equal to its inverse, use the fact that there exists an invertible matrix P such that $D = P^{-1}AP$, where D is a diagonal matrix with ± 1 along its main diagonal.

(i) Let $D = P^{-1}AP$, where D is a diagonal matrix with ± 1 along its main diagonal.

(ii) Find A in terms of P, P^{-1}, and D.

(iii) Use the fact that D is diagonal and the properties of the inverse of a product of matrices to expand to find A^{-1}.

(iv) Conclude that $A^{-1} = A$.

46. ***Guided Proof***   Prove that nonzero nilpotent matrices are not diagonalizable.

Getting Started: From Exercise 59 in Section 7.1, you know that 0 is the only eigenvalue of the nilpotent matrix A. Show that it is impossible for A to be diagonalizable.

(i) Assume that A is diagonalizable, so there exists an invertible matrix P such that $P^{-1}AP = D$, where D is the zero matrix.

(ii) Find A in terms of P, P^{-1}, and D.

(iii) Find a contradiction and conclude that nonzero nilpotent matrices are not diagonalizable.

47. Prove that if A is a nonsingular diagonalizable matrix, then A^{-1} is also diagonalizable.

7.3 | Symmetric Matrices and Orthogonal Diagonalization

For most matrices you must go through much of the diagonalization process before you can finally determine whether diagonalization is possible. One exception is a triangular matrix with distinct entries on the main diagonal. Such a matrix can be recognized as diagonalizable by simple inspection. In this section you will study another type of matrix that is guaranteed to be diagonalizable: a **symmetric** matrix.

Definition of Symmetric Matrix	A square matrix A is **symmetric** if it is equal to its transpose: $$A = A^T.$$

You can determine easily whether a matrix is symmetric by checking whether it is symmetric with respect to its main diagonal.

EXAMPLE 1 *Symmetric Matrices and Nonsymmetric Matrices*

The matrices A and B *are* symmetric, but the matrix C is *not*.

$$A = \begin{bmatrix} 0 & 1 & -2 \\ 1 & 3 & 0 \\ -2 & 0 & 5 \end{bmatrix} \qquad \text{Symmetric}$$

$$B = \begin{bmatrix} 4 & 3 \\ 3 & 1 \end{bmatrix} \qquad \text{Symmetric}$$

$$C = \begin{bmatrix} 3 & 2 & 1 \\ 1 & -4 & 0 \\ 1 & 0 & 5 \end{bmatrix} \qquad \text{Nonsymmetric}$$

DISCOVERY

If you have access to computer software or a graphing calculator that can find eigenvalues, try the following experiment. Pick an arbitrary square matrix and calculate its eigenvalues. Can you find a matrix for which the eigenvalues are not real? Now pick an arbitrary symmetric matrix and calculate its eigenvalues. Can you find a symmetric matrix for which the eigenvalues are not real? What can you conclude about the eigenvalues of a symmetric matrix?

Symmetric matrices have some special properties that are not generally exhibited by nonsymmetric matrices. To see this, consider the following.
1. A nonsymmetric matrix may not be diagonalizable.
2. A nonsymmetric matrix can have eigenvalues that are not real. For instance, the matrix

$$A = \begin{bmatrix} 0 & -1 \\ 1 & 0 \end{bmatrix}$$

has a characteristic equation of $\lambda^2 + 1 = 0$. Hence, its eigenvalues are the imaginary numbers $\lambda_1 = i$ and $\lambda_2 = -i$.
3. For a nonsymmetric matrix, the number of linearly independent eigenvectors corresponding to an eigenvalue can be less than the multiplicity of the eigenvalue. (See Example 4, Section 7.2.)

None of these three properties is possible with a symmetric matrix.

Theorem 7.7	Eigenvalues of Symmetric Matrices

Eigenvalues of Symmetric Matrices

If A is an $n \times n$ symmetric matrix, then the following properties are true.
1. A is diagonalizable.
2. All eigenvalues of A are real.
3. If λ is an eigenvalue of A with multiplicity k, then λ has k linearly independent eigenvectors. That is, the eigenspace of λ has dimension k.

REMARK: Theorem 7.7 is called the **Real Spectral Theorem,** and the set of eigenvalues of A is called the **spectrum** of A.

A general proof of Theorem 7.7 is beyond the scope of this text. The following example verifies that every 2×2 symmetric matrix is diagonalizable.

EXAMPLE 2   *The Eigenvalues and Eigenvectors of a 2 × 2 Symmetric Matrix*

Prove that a symmetric matrix

$$A = \begin{bmatrix} a & c \\ c & b \end{bmatrix}$$

is diagonalizable.

Solution   The characteristic polynomial of A is

$$|\lambda I - A| = \begin{vmatrix} \lambda - a & -c \\ -c & \lambda - b \end{vmatrix} = \lambda^2 - (a + b)\lambda + ab - c^2.$$

As a quadratic in λ, this polynomial has a discriminant of

$$(a + b)^2 - 4(ab - c^2) = a^2 + 2ab + b^2 - 4ab + 4c^2$$
$$= a^2 - 2ab + b^2 + 4c^2$$
$$= (a - b)^2 + 4c^2.$$

Because this discriminant is the sum of two squares, it must be either zero or positive. If $(a - b)^2 + 4c^2 = 0$, then $a = b$ and $c = 0$, which implies that A is already diagonal. That is,

$$A = \begin{bmatrix} a & 0 \\ 0 & a \end{bmatrix}.$$

On the other hand, if $(a - b)^2 + 4c^2 > 0$, then by the Quadratic Formula the characteristic polynomial of A has two distinct real roots, which implies that A has two distinct real eigenvalues. Thus, A is diagonalizable in this case also.

EXAMPLE 3 *Dimensions of the Eigenspaces of a Symmetric Matrix*

Find the eigenvalues of the symmetric matrix

$$A = \begin{bmatrix} 1 & -2 & 0 & 0 \\ -2 & 1 & 0 & 0 \\ 0 & 0 & 1 & -2 \\ 0 & 0 & -2 & 1 \end{bmatrix}$$

and determine the dimensions of the corresponding eigenspaces.

Solution   The characteristic polynomial of A is given by

$$|\lambda I - A| = \begin{vmatrix} \lambda - 1 & 2 & 0 & 0 \\ 2 & \lambda - 1 & 0 & 0 \\ 0 & 0 & \lambda - 1 & 2 \\ 0 & 0 & 2 & \lambda - 1 \end{vmatrix} = (\lambda + 1)^2(\lambda - 3)^2.$$

Thus, the eigenvalues of A are $\lambda_1 = -1$ and $\lambda_2 = 3$. Because each of these eigenvalues has a multiplicity of 2, you know from Theorem 7.7 that the corresponding eigenspaces also have dimension 2. Specifically, the eigenspace of $\lambda_1 = -1$ has a basis of $B_1 = \{(1, 1, 0, 0), (0, 0, 1, 1)\}$ and the eigenspace of $\lambda_2 = 3$ has a basis of $B_2 = \{(1, -1, 0, 0), (0, 0, 1, -1)\}$.

Orthogonal Matrices

To diagonalize a square matrix A, you need to find an *invertible* matrix P such that $P^{-1}AP$ is diagonal. For symmetric matrices, you will see that the matrix P can be chosen to have the special property that $P^{-1} = P^T$. This unusual matrix property is defined as follows.

| *Definition of an Orthogonal Matrix* | A square matrix P is called **orthogonal** if it is invertible and $$P^{-1} = P^T.$$ |

EXAMPLE 4 *Orthogonal Matrices*

(a) The matrix

$$P = \begin{bmatrix} 0 & 1 \\ -1 & 0 \end{bmatrix}$$

is orthogonal because

$$P^{-1} = P^T = \begin{bmatrix} 0 & -1 \\ 1 & 0 \end{bmatrix}.$$

(b) The matrix

$$P = \begin{bmatrix} \frac{3}{5} & 0 & -\frac{4}{5} \\ 0 & 1 & 0 \\ \frac{4}{5} & 0 & \frac{3}{5} \end{bmatrix}$$

is orthogonal because

$$P^{-1} = P^T = \begin{bmatrix} \frac{3}{5} & 0 & \frac{4}{5} \\ 0 & 1 & 0 \\ -\frac{4}{5} & 0 & \frac{3}{5} \end{bmatrix}.$$

In parts (a) and (b) of Example 4, the columns of the matrices P form orthonormal sets in R^2 and R^3, respectively. This result suggests the next theorem.

Theorem 7.8

Property of Orthogonal Matrices

An $n \times n$ matrix P is orthogonal if and only if its column vectors form an orthonormal set.

Proof Suppose that the column vectors of P form an orthonormal set:

$$P = [\mathbf{p}_1 \vdots \mathbf{p}_2 \vdots \cdots \vdots \mathbf{p}_n]$$
$$= \begin{bmatrix} p_{11} & p_{12} & \cdots & p_{1n} \\ p_{21} & p_{22} & \cdots & p_{2n} \\ \vdots & \vdots & & \vdots \\ p_{n1} & p_{n2} & \cdots & p_{nn} \end{bmatrix}.$$

Then the product $P^T P$ has the form

$$P^T P = \begin{bmatrix} p_{11} & p_{21} & \cdots & p_{n1} \\ p_{12} & p_{22} & \cdots & p_{n2} \\ \vdots & \vdots & & \vdots \\ p_{1n} & p_{2n} & \cdots & p_{nn} \end{bmatrix} \begin{bmatrix} p_{11} & p_{12} & \cdots & p_{1n} \\ p_{21} & p_{22} & \cdots & p_{2n} \\ \vdots & \vdots & & \vdots \\ p_{n1} & p_{n2} & \cdots & p_{nn} \end{bmatrix}$$

$$P^T P = \begin{bmatrix} \mathbf{p}_1 \cdot \mathbf{p}_1 & \mathbf{p}_1 \cdot \mathbf{p}_2 & \cdots & \mathbf{p}_1 \cdot \mathbf{p}_n \\ \mathbf{p}_2 \cdot \mathbf{p}_1 & \mathbf{p}_2 \cdot \mathbf{p}_2 & \cdots & \mathbf{p}_2 \cdot \mathbf{p}_n \\ \vdots & \vdots & & \vdots \\ \mathbf{p}_n \cdot \mathbf{p}_1 & \mathbf{p}_n \cdot \mathbf{p}_2 & \cdots & \mathbf{p}_n \cdot \mathbf{p}_n \end{bmatrix}.$$

Because the set $\{\mathbf{p}_1, \mathbf{p}_2, \ldots, \mathbf{p}_n\}$ is orthonormal, you have

$$\mathbf{p}_i \cdot \mathbf{p}_j = 0, \, i \neq j \quad \text{and} \quad \mathbf{p}_i \cdot \mathbf{p}_i = \|\mathbf{p}_i\|^2 = 1.$$

Thus, the matrix composed of dot products has the form

$$P^T P = \begin{bmatrix} 1 & 0 & \cdots & 0 \\ 0 & 1 & \cdots & 0 \\ \vdots & \vdots & & \vdots \\ 0 & 0 & \cdots & 1 \end{bmatrix} = I_n.$$

This implies that $P^T = P^{-1}$, and you can conclude that P is orthogonal.

Conversely, if P is orthogonal, you can reverse the steps above to verify that the column vectors of P form an orthonormal set.

EXAMPLE 5   *An Orthogonal Matrix*

Show that

$$P = \begin{bmatrix} \dfrac{1}{3} & \dfrac{2}{3} & \dfrac{2}{3} \\[2mm] \dfrac{-2}{\sqrt{5}} & \dfrac{1}{\sqrt{5}} & 0 \\[2mm] \dfrac{-2}{3\sqrt{5}} & \dfrac{-4}{3\sqrt{5}} & \dfrac{5}{3\sqrt{5}} \end{bmatrix}$$

is orthogonal by showing that $PP^T = I$. Then show that the column vectors of P form an orthonormal set.

Solution   Because

$$PP^T = \begin{bmatrix} \dfrac{1}{3} & \dfrac{2}{3} & \dfrac{2}{3} \\[2mm] \dfrac{-2}{\sqrt{5}} & \dfrac{1}{\sqrt{5}} & 0 \\[2mm] \dfrac{-2}{3\sqrt{5}} & \dfrac{-4}{3\sqrt{5}} & \dfrac{5}{3\sqrt{5}} \end{bmatrix} \begin{bmatrix} \dfrac{1}{3} & \dfrac{-2}{\sqrt{5}} & \dfrac{-2}{3\sqrt{5}} \\[2mm] \dfrac{2}{3} & \dfrac{1}{\sqrt{5}} & \dfrac{-4}{3\sqrt{5}} \\[2mm] \dfrac{2}{3} & 0 & \dfrac{5}{3\sqrt{5}} \end{bmatrix}$$

$$= \begin{bmatrix} 1 & 0 & 0 \\ 0 & 1 & 0 \\ 0 & 0 & 1 \end{bmatrix} = I_3,$$

it follows that $P^T = P^{-1}$, and you can conclude that P is orthogonal. Moreover, letting

$$\mathbf{p}_1 = \begin{bmatrix} \dfrac{1}{3} \\[2mm] \dfrac{-2}{\sqrt{5}} \\[2mm] \dfrac{-2}{3\sqrt{5}} \end{bmatrix}, \mathbf{p}_2 = \begin{bmatrix} \dfrac{2}{3} \\[2mm] \dfrac{1}{\sqrt{5}} \\[2mm] \dfrac{-4}{3\sqrt{5}} \end{bmatrix}, \text{ and } \mathbf{p}_3 = \begin{bmatrix} \dfrac{2}{3} \\[2mm] 0 \\[2mm] \dfrac{5}{3\sqrt{5}} \end{bmatrix}$$

produces

$$\mathbf{p}_1 \cdot \mathbf{p}_2 = \mathbf{p}_1 \cdot \mathbf{p}_3 = \mathbf{p}_2 \cdot \mathbf{p}_3 = 0$$

and

$$\|\mathbf{p}_1\| = \|\mathbf{p}_2\| = \|\mathbf{p}_3\| = 1.$$

Therefore, $\{\mathbf{p}_1, \mathbf{p}_2, \mathbf{p}_3\}$ is an orthonormal set, as guaranteed by Theorem 7.8.

It can be shown that for a symmetric matrix, the eigenvectors corresponding to distinct eigenvalues are orthogonal. This result is stated in the next theorem.

Theorem 7.9	*Property of Symmetric Matrices*

Let A be an $n \times n$ symmetric matrix. If λ_1 and λ_2 are distinct eigenvalues of A, then their corresponding eigenvectors $\mathbf{x}_1$ and $\mathbf{x}_2$ are orthogonal.

Proof   Let λ_1 and λ_2 be distinct eigenvalues of A with corresponding eigenvectors $\mathbf{x}_1$ and $\mathbf{x}_2$. Thus,

$$A\mathbf{x}_1 = \lambda_1\mathbf{x}_1 \quad \text{and} \quad A\mathbf{x}_2 = \lambda_2\mathbf{x}_2.$$

To prove the theorem, use the following matrix form of the dot product.

$$\mathbf{x}_1 \cdot \mathbf{x}_2 = \begin{bmatrix} x_{11} & x_{12} & \cdots & x_{1n} \end{bmatrix} \begin{bmatrix} x_{21} \\ x_{22} \\ \vdots \\ x_{2n} \end{bmatrix} = \mathbf{x}_1^T\mathbf{x}_2$$

Now you can write

$$
\begin{aligned}
\lambda_1(\mathbf{x}_1 \cdot \mathbf{x}_2) &= (\lambda_1\mathbf{x}_1) \cdot \mathbf{x}_2 \\
&= (A\mathbf{x}_1) \cdot \mathbf{x}_2 \\
&= (A\mathbf{x}_1)^T\mathbf{x}_2 \\
&= (\mathbf{x}_1^T A^T)\mathbf{x}_2 \\
&= (\mathbf{x}_1^T A)\mathbf{x}_2 \qquad \text{Because } A \text{ is symmetric, } A = A^T. \\
&= \mathbf{x}_1^T(A\mathbf{x}_2) \\
&= \mathbf{x}_1^T(\lambda_2\mathbf{x}_2) \\
&= \mathbf{x}_1 \cdot (\lambda_2\mathbf{x}_2) \\
&= \lambda_2(\mathbf{x}_1 \cdot \mathbf{x}_2).
\end{aligned}
$$

This implies that $(\lambda_1 - \lambda_2)(\mathbf{x}_1 \cdot \mathbf{x}_2) = 0$, and because $\lambda_1 \neq \lambda_2$ it follows that $\mathbf{x}_1 \cdot \mathbf{x}_2 = 0$. Therefore, $\mathbf{x}_1$ and $\mathbf{x}_2$ are orthogonal.

EXAMPLE 6   *Eigenvectors of a Symmetric Matrix*

Show that any two eigenvectors of

$$A = \begin{bmatrix} 3 & 1 \\ 1 & 3 \end{bmatrix}$$

corresponding to distinct eigenvalues are orthogonal.

Solution   The characteristic polynomial of A is

$$|\lambda I - A| = \begin{vmatrix} \lambda - 3 & -1 \\ -1 & \lambda - 3 \end{vmatrix} = (\lambda - 2)(\lambda - 4),$$

which implies that the eigenvalues of A are $\lambda_1 = 2$ and $\lambda_2 = 4$. Every eigenvector corresponding to $\lambda_1 = 2$ is of the form

$$\mathbf{x}_1 = \begin{bmatrix} s \\ -s \end{bmatrix}, \ s \neq 0$$

and every eigenvector corresponding to $\lambda_2 = 4$ is of the form

$$\mathbf{x}_2 = \begin{bmatrix} t \\ t \end{bmatrix}, \ t \neq 0.$$

Therefore,

$$\mathbf{x}_1 \cdot \mathbf{x}_2 = \begin{bmatrix} s \\ -s \end{bmatrix} \cdot \begin{bmatrix} t \\ t \end{bmatrix} = st - st = 0,$$

and you can conclude that $\mathbf{x}_1$ and $\mathbf{x}_2$ are orthogonal.

Orthogonal Diagonalization

A matrix A is **orthogonally diagonalizable** if there exists an orthogonal matrix P such that $P^{-1}AP = D$ is diagonal. The following important theorem states that the set of orthogonally diagonalizable matrices is precisely the set of symmetric matrices.

Theorem 7.10   *Fundamental Theorem of Symmetric Matrices*

Let A be an $n \times n$ matrix. Then A is orthogonally diagonalizable and has real eigenvalues if and only if A is symmetric.

Proof   The proof of the theorem in one direction is fairly straightforward. That is, if you assume that A is orthogonally diagonalizable, then there exists an orthogonal matrix P such that $D = P^{-1}AP$ is diagonal. Moreover, because $P^{-1} = P^T$, you have

$$A = PDP^{-1} = PDP^T,$$

which implies that

$$A^T = (PDP^T)^T = (P^T)^T D^T P^T = PDP^T = A.$$

Therefore, A is symmetric.

The proof of the theorem in the other direction is more involved, but it is important because it is constructive. Assume that A is symmetric. If A has an eigenvalue λ of multiplicity k, then, by Theorem 7.7, λ has k linearly independent eigenvectors. Through the Gram-Schmidt orthonormalization process, this set of k vectors can be used to form an orthonormal basis of eigenvectors for the eigenspace corresponding to λ. This procedure is repeated for each eigenvalue of A. The collection of all resulting eigenvectors is orthogonal by Theorem 7.9, and you know from the normalization process that the collection is also orthonormal. Now let P be the matrix whose columns consist of these n orthonormal eigenvectors. By Theorem 7.8, P is an orthogonal matrix. Finally, by Theorem 7.5, you can conclude that $P^{-1}AP$ is diagonal. Hence, A is orthogonally diagonalizable.

EXAMPLE 7   *Determining Whether a Matrix Is Orthogonally Diagonalizable*

Which matrices are orthogonally diagonalizable?

$$A_1 = \begin{bmatrix} 1 & 1 & 1 \\ 1 & 0 & 1 \\ 1 & 1 & 1 \end{bmatrix} \qquad A_2 = \begin{bmatrix} 5 & 2 & 1 \\ 2 & 1 & 8 \\ -1 & 8 & 0 \end{bmatrix}$$

$$A_3 = \begin{bmatrix} 3 & 2 & 0 \\ 2 & 0 & 1 \end{bmatrix} \qquad A_4 = \begin{bmatrix} 0 & 0 \\ 0 & -2 \end{bmatrix}$$

Solution   By Theorem 7.10, the orthogonally diagonalizable matrices are the symmetric ones: A_1 and A_4.

It was mentioned that the second part of the proof of Theorem 7.10 is *constructive.* That is, it gives you steps to follow to diagonalize a symmetric matrix orthogonally. These steps are summarized as follows.

Orthogonal Diagonalization of a Symmetric Matrix

Let A be an $n \times n$ symmetric matrix.
1. Find all eigenvalues of A and determine the multiplicity of each.
2. For *each* eigenvalue of multiplicity 1, choose a unit eigenvector. (Choose any eigenvector and then normalize it.)
3. For each eigenvalue of multiplicity $k \geq 2$, find a set of k linearly independent eigenvectors. (You know from Theorem 7.7 that this is possible.) If this set is not orthonormal, apply the Gram-Schmidt orthonormalization process.
4. The composite of steps 2 and 3 produces an orthonormal set of n eigenvectors. Use these eigenvectors to form the columns of P. The matrix $P^{-1}AP = P^TAP = D$ will be diagonal. (The main diagonal entries of D are the eigenvalues of A.)

> **EXAMPLE 8** *Orthogonal Diagonalization*
>
> Find an orthogonal matrix P that orthogonally diagonalizes
>
> $$A = \begin{bmatrix} -2 & 2 \\ 2 & 1 \end{bmatrix}.$$

Solution

1. The characteristic polynomial of A is

$$|\lambda I - A| = \begin{vmatrix} \lambda + 2 & -2 \\ -2 & \lambda - 1 \end{vmatrix} = (\lambda + 3)(\lambda - 2).$$

So the eigenvalues are $\lambda_1 = -3$ and $\lambda_2 = 2$.

2. For each eigenvalue, find an eigenvector by converting the matrix $\lambda I - A$ to reduced row-echelon form.

Eigenvector

$$-3I - A = \begin{bmatrix} -1 & -2 \\ -2 & -4 \end{bmatrix} \longrightarrow \begin{bmatrix} 1 & 2 \\ 0 & 0 \end{bmatrix} \longrightarrow \begin{bmatrix} -2 \\ 1 \end{bmatrix}$$

$$2I - A = \begin{bmatrix} 4 & -2 \\ -2 & 1 \end{bmatrix} \longrightarrow \begin{bmatrix} 1 & -\frac{1}{2} \\ 0 & 0 \end{bmatrix} \longrightarrow \begin{bmatrix} 1 \\ 2 \end{bmatrix}$$

The eigenvectors $(-2, 1)$ and $(1, 2)$ form an *orthogonal* basis for R^2. Each of these eigenvectors is normalized to produce an *orthonormal* basis.

$$\mathbf{p}_1 = \frac{(-2, 1)}{\|(-2, 1)\|} = \frac{1}{\sqrt{5}}(-2, 1) = \left(\frac{-2}{\sqrt{5}}, \frac{1}{\sqrt{5}} \right)$$

$$\mathbf{p}_2 = \frac{(1, 2)}{\|(1, 2)\|} = \frac{1}{\sqrt{5}}(1, 2) = \left(\frac{1}{\sqrt{5}}, \frac{2}{\sqrt{5}} \right)$$

3. Because each eigenvalue has a multiplicity of 1, go directly to step 4.
4. Using $\mathbf{p}_1$ and $\mathbf{p}_2$ as column vectors, construct the matrix P.

$$P = \begin{bmatrix} \dfrac{-2}{\sqrt{5}} & \dfrac{1}{\sqrt{5}} \\ \dfrac{1}{\sqrt{5}} & \dfrac{2}{\sqrt{5}} \end{bmatrix}$$

Verify that P is correct by computing $P^{-1}AP = P^TAP$.

$$P^TAP = \begin{bmatrix} \dfrac{-2}{\sqrt{5}} & \dfrac{1}{\sqrt{5}} \\ \dfrac{1}{\sqrt{5}} & \dfrac{2}{\sqrt{5}} \end{bmatrix} \begin{bmatrix} -2 & 2 \\ 2 & 1 \end{bmatrix} \begin{bmatrix} \dfrac{-2}{\sqrt{5}} & \dfrac{1}{\sqrt{5}} \\ \dfrac{1}{\sqrt{5}} & \dfrac{2}{\sqrt{5}} \end{bmatrix} = \begin{bmatrix} -3 & 0 \\ 0 & 2 \end{bmatrix}$$

EXAMPLE 9   *Orthogonal Diagonalization*

Find an orthogonal matrix P that diagonalizes

$$A = \begin{bmatrix} 2 & 2 & -2 \\ 2 & -1 & 4 \\ -2 & 4 & -1 \end{bmatrix}.$$

Solution   1. The characteristic polynomial of A,

$$|\lambda I - A| = (\lambda - 3)^2(\lambda + 6),$$

yields the eigenvalues $\lambda_1 = -6$ and $\lambda_2 = 3$. λ_1 has a multiplicity of 1 and λ_2 has a multiplicity of 2.

2. An eigenvector for λ_1 is $\mathbf{v}_1 = (1, -2, 2)$, which normalizes to

$$\mathbf{u}_1 = \frac{\mathbf{v}_1}{\|\mathbf{v}_1\|} = \left(\frac{1}{3}, -\frac{2}{3}, \frac{2}{3}\right).$$

3. Two eigenvectors for λ_2 are $\mathbf{v}_2 = (2, 1, 0)$ and $\mathbf{v}_3 = (-2, 0, 1)$.
   Note that $\mathbf{v}_1$ is orthogonal to $\mathbf{v}_2$ and $\mathbf{v}_3$, as guaranteed by Theorem 7.9. The eigenvectors $\mathbf{v}_2$ and $\mathbf{v}_3$, however, are not orthogonal to each other. To find two orthonormal eigenvectors for λ_2, use the Gram-Schmidt process as follows.

$$\mathbf{w}_2 = \mathbf{v}_2 = (2, 1, 0)$$

$$\mathbf{w}_3 = \mathbf{v}_3 - \left(\frac{\mathbf{v}_3 \cdot \mathbf{w}_2}{\mathbf{w}_2 \cdot \mathbf{w}_2}\right)\mathbf{w}_2 = \left(-\frac{2}{5}, \frac{4}{5}, 1\right)$$

These vectors normalize to

$$\mathbf{u}_2 = \frac{\mathbf{w}_2}{\|\mathbf{w}_2\|} = \left(\frac{2}{\sqrt{5}}, \frac{1}{\sqrt{5}}, 0\right)$$

$$\mathbf{u}_3 = \frac{\mathbf{w}_3}{\|\mathbf{w}_3\|} = \left(\frac{-2}{3\sqrt{5}}, \frac{4}{3\sqrt{5}}, \frac{5}{3\sqrt{5}}\right).$$

4. The matrix P has $\mathbf{u}_1$, $\mathbf{u}_2$, and $\mathbf{u}_3$ as its column vectors.

$$P = \begin{bmatrix} \dfrac{1}{3} & \dfrac{2}{\sqrt{5}} & \dfrac{-2}{3\sqrt{5}} \\[2mm] \dfrac{-2}{3} & \dfrac{1}{\sqrt{5}} & \dfrac{4}{3\sqrt{5}} \\[2mm] \dfrac{2}{3} & 0 & \dfrac{5}{3\sqrt{5}} \end{bmatrix}$$

A check shows that $P^{-1}AP = P^TAP = \begin{bmatrix} -6 & 0 & 0 \\ 0 & 3 & 0 \\ 0 & 0 & 3 \end{bmatrix}.$

In Exercises 1–6, determine whether the matrix is symmetric.

1. $\begin{bmatrix} 1 & 3 \\ 3 & -1 \end{bmatrix}$

2. $\begin{bmatrix} 6 & -2 \\ -2 & 1 \end{bmatrix}$

3. $\begin{bmatrix} 4 & -2 & 1 \\ 3 & 1 & 2 \\ 1 & 2 & 1 \end{bmatrix}$

4. $\begin{bmatrix} 1 & -5 & 4 \\ -5 & 3 & 6 \\ -4 & 6 & 2 \end{bmatrix}$

5. $\begin{bmatrix} 0 & 1 & 2 & -1 \\ 1 & 0 & -3 & 2 \\ 2 & -3 & 0 & 1 \\ -1 & 2 & 1 & -2 \end{bmatrix}$

6. $\begin{bmatrix} 2 & 0 & 3 & 5 \\ 0 & 11 & 0 & -2 \\ 3 & 0 & 5 & 0 \\ 5 & -2 & 0 & 1 \end{bmatrix}$

In Exercises 7–12, find the eigenvalues of the symmetric matrix. For each eigenvalue, find the dimension of the corresponding eigenspace.

7. $\begin{bmatrix} 3 & 1 \\ 1 & 3 \end{bmatrix}$

8. $\begin{bmatrix} 2 & 0 \\ 0 & 2 \end{bmatrix}$

9. $\begin{bmatrix} 3 & 0 & 0 \\ 0 & 2 & 0 \\ 0 & 0 & 2 \end{bmatrix}$

10. $\begin{bmatrix} 2 & 1 & 1 \\ 1 & 2 & 1 \\ 1 & 1 & 2 \end{bmatrix}$

11. $\begin{bmatrix} 0 & 2 & 2 \\ 2 & 0 & 2 \\ 2 & 2 & 0 \end{bmatrix}$

12. $\begin{bmatrix} 0 & 4 & 4 \\ 4 & 2 & 0 \\ 4 & 0 & -2 \end{bmatrix}$

In Exercises 13–20, determine whether the matrix is orthogonal.

13. $\begin{bmatrix} \frac{\sqrt{2}}{2} & \frac{\sqrt{2}}{2} \\ -\frac{\sqrt{2}}{2} & \frac{\sqrt{2}}{2} \end{bmatrix}$

14. $\begin{bmatrix} \frac{2}{3} & -\frac{2}{3} \\ \frac{2}{3} & \frac{1}{3} \end{bmatrix}$

15. $\begin{bmatrix} -4 & 0 & 3 \\ 0 & 1 & 0 \\ 3 & 0 & 4 \end{bmatrix}$

16. $\begin{bmatrix} -\frac{4}{5} & 0 & \frac{3}{5} \\ 0 & 1 & 0 \\ \frac{3}{5} & 0 & \frac{4}{5} \end{bmatrix}$

17. $\begin{bmatrix} \frac{\sqrt{2}}{2} & -\frac{\sqrt{6}}{6} & \frac{\sqrt{3}}{3} \\ 0 & \frac{\sqrt{6}}{3} & \frac{\sqrt{3}}{3} \\ \frac{\sqrt{2}}{2} & \frac{\sqrt{6}}{6} & -\frac{\sqrt{3}}{3} \end{bmatrix}$

18. $\begin{bmatrix} \frac{\sqrt{2}}{3} & 0 & \frac{\sqrt{5}}{2} \\ 0 & \frac{2\sqrt{5}}{5} & 0 \\ -\frac{\sqrt{2}}{6} & -\frac{\sqrt{5}}{5} & \frac{1}{2} \end{bmatrix}$

19. $\begin{bmatrix} \frac{1}{10}\sqrt{10} & 0 & 0 & -\frac{3}{10}\sqrt{10} \\ 0 & 0 & 1 & 0 \\ 0 & 1 & 0 & 0 \\ \frac{3}{10}\sqrt{10} & 0 & 0 & \frac{1}{10}\sqrt{10} \end{bmatrix}$

20. $\begin{bmatrix} 4 & -1 & -4 & 2 \\ -1 & 0 & -17 & -1 \\ 1 & 4 & -1 & 1 \\ 2 & -1 & 1 & 0 \end{bmatrix}$

In Exercises 21–28, find an orthogonal matrix P such that P^TAP diagonalizes A. Verify that P^TAP gives the proper diagonal form.

21. $A = \begin{bmatrix} 1 & 1 \\ 1 & 1 \end{bmatrix}$

22. $A = \begin{bmatrix} 4 & 2 \\ 2 & 4 \end{bmatrix}$

23. $A = \begin{bmatrix} 2 & \sqrt{2} \\ \sqrt{2} & 1 \end{bmatrix}$

24. $A = \begin{bmatrix} 0 & 1 & 1 \\ 1 & 0 & 1 \\ 1 & 1 & 0 \end{bmatrix}$

25. $A = \begin{bmatrix} 0 & 10 & 10 \\ 10 & 5 & 0 \\ 10 & 0 & -5 \end{bmatrix}$

26. $A = \begin{bmatrix} 0 & 3 & 0 \\ 3 & 0 & 4 \\ 0 & 4 & 0 \end{bmatrix}$

27. $A = \begin{bmatrix} 4 & 2 & 0 & 0 \\ 2 & 4 & 0 & 0 \\ 0 & 0 & 4 & 2 \\ 0 & 0 & 2 & 4 \end{bmatrix}$

28. $A = \begin{bmatrix} 1 & 1 & 0 & 0 \\ 1 & 1 & 0 & 0 \\ 0 & 0 & 1 & 1 \\ 0 & 0 & 1 & 1 \end{bmatrix}$

True or False? In Exercises 29 and 30, determine whether each statement is true or false. If a statement is true, give a reason or cite an appropriate statement in the text. If a statement is false, provide an example that shows that the statement is not true in all cases or cite an appropriate statement in the text.

29. (a) Let A be an $n \times n$ matrix. Then A is symmetric if and only if A is orthogonally diagonalizable.
(b) The eigenvectors corresponding to distinct eigenvalues are orthogonal for symmetric matrices.

30. (a) A square matrix P is orthogonal if it is invertible—that is, if $P^{-1} = P^T$.

   (b) If A is an $n \times n$ symmetric matrix, then A is orthogonally diagonalizable and has real eigenvalues.

31. Prove that if A is an $m \times n$ matrix, then A^TA and AA^T are symmetric.

32. Find A^TA and AA^T for the following matrix.
$$A = \begin{bmatrix} 1 & -3 & 2 \\ 4 & -6 & 1 \end{bmatrix}$$

33. Prove that if A is an orthogonal matrix, then $|A| = \pm 1$.

34. Prove that if A and B are $n \times n$ orthogonal matrices, then AB and BA are orthogonal.

35. Show that the following matrix is orthogonal for any value of θ.
$$A = \begin{bmatrix} \cos \theta & -\sin \theta \\ \sin \theta & \cos \theta \end{bmatrix}$$

36. Prove that if a symmetric matrix A has only one eigenvalue λ, then $A = \lambda I$.

37. Prove that if A is an orthogonal matrix, then so are A^T and A^{-1}.

7.4 | Applications of Eigenvalues and Eigenvectors

Population Growth

Matrices can be used to form models for population growth. The first step in this process is to group the population into age classes of equal duration. For instance, if the maximum life span of a member is L years, then the age classes are represented by the following n intervals.

First age class   Second age class nth age class

$$\left[0, \frac{L}{n}\right), \quad \left[\frac{L}{n}, \frac{2L}{n}\right), \ldots, \left[\frac{(n-1)L}{n}, L\right]$$

The number of population members in each age class is then represented by the **age distribution vector**

$$\mathbf{x} = \begin{bmatrix} x_1 \\ x_2 \\ \vdots \\ x_n \end{bmatrix}.$$

Number in first age class
Number in second age class

Number in nth age class

Over a period of L/n years, the *probability* that a member of the ith age class will survive to become a member of the $(i + 1)$th age class is given by p_i, where

$$0 \leq p_i \leq 1, \ i = 1, 2, \ldots, n - 1.$$

The *average number* of offspring produced by a member of the ith age class is given by b_i, where

$$0 \leq b_i, \ i = 1, 2, \ldots, n.$$

These numbers can be written in matrix form, as shown on the following page.

$$A = \begin{bmatrix} b_1 & b_2 & b_3 & \cdots & b_{n-1} & b_n \\ p_1 & 0 & 0 & \cdots & 0 & 0 \\ 0 & p_2 & 0 & \cdots & 0 & 0 \\ \vdots & \vdots & \vdots & & \vdots & \vdots \\ 0 & 0 & 0 & \cdots & p_{n-1} & 0 \end{bmatrix}$$

Multiplying this **age transition matrix** by the age distribution vector for a given time period produces the age distribution vector for the next time period. That is,

$$A\mathbf{x}_i = \mathbf{x}_{i+1}.$$

This procedure is illustrated in Example 1.

EXAMPLE 1   *A Population Growth Model*

A population of rabbits raised in a research laboratory has the following characteristics.

(a) Half of the rabbits survive their first year. Of those, half survive their second year. The maximum life span is 3 years.

(b) During the first year the rabbits produce no offspring. The average number of offspring is 6 during the second year and 8 during the third year.

The laboratory population now consists of 24 rabbits in the first age class, 24 in the second, and 20 in the third. How many rabbits will be in each age class in 1 year?

Solution   The current age distribution vector is

$$\mathbf{x}_1 = \begin{bmatrix} 24 \\ 24 \\ 20 \end{bmatrix} \qquad \begin{matrix} 0 \le \text{age} < 1 \\ 1 \le \text{age} < 2 \\ 2 \le \text{age} \le 3 \end{matrix}$$

and the age transition matrix is

$$A = \begin{bmatrix} 0 & 6 & 8 \\ 0.5 & 0 & 0 \\ 0 & 0.5 & 0 \end{bmatrix}.$$

After 1 year the age distribution vector will be

$$\mathbf{x}_2 = A\mathbf{x}_1 = \begin{bmatrix} 0 & 6 & 8 \\ 0.5 & 0 & 0 \\ 0 & 0.5 & 0 \end{bmatrix} \begin{bmatrix} 24 \\ 24 \\ 20 \end{bmatrix} = \begin{bmatrix} 304 \\ 12 \\ 12 \end{bmatrix}. \qquad \begin{matrix} 0 \le \text{age} < 1 \\ 1 \le \text{age} < 2 \\ 2 \le \text{age} \le 3 \end{matrix}$$

If the pattern of growth in Example 1 continued for another year, the rabbit population would be

$$\mathbf{x}_3 = A\mathbf{x}_2 = \begin{bmatrix} 0 & 6 & 8 \\ 0.5 & 0 & 0 \\ 0 & 0.5 & 0 \end{bmatrix} \begin{bmatrix} 304 \\ 12 \\ 12 \end{bmatrix} = \begin{bmatrix} 168 \\ 152 \\ 6 \end{bmatrix}. \qquad \begin{matrix} 0 \le \text{age} < 1 \\ 1 \le \text{age} < 2 \\ 2 \le \text{age} \le 3 \end{matrix}$$

From the age distribution vectors x_1, x_2, and x_3, you can see that the percent of rabbits in the three age classes changes each year. Suppose that the laboratory prefers a stable growth pattern, one in which the percent in each age class remains the same each year. For this stable growth pattern to be achieved, the $(n + 1)$th age distribution vector must be a scalar multiple of the nth age distribution vector. That is, $Ax_n = x_{n+1} = \lambda x_n$. Hence, the laboratory can obtain a growth pattern in which the percent in each age class remains constant each year by finding an eigenvector of A. Example 2 shows how to solve this problem.

EXAMPLE 2   *Finding a Stable Age Distribution*

Find a stable age distribution for the population given in Example 1.

Solution   To solve this problem, find an eigenvalue λ and corresponding eigenvector x such that

$$Ax = \lambda x.$$

Simulation

Explore this concept further with an electronic simulation available on the *Elementary Linear Algebra* Learning Tools Student CD-ROM.

The characteristic polynomial of A is

$$|\lambda I - A| = \begin{vmatrix} \lambda & -6 & -8 \\ -0.5 & \lambda & 0 \\ 0 & -0.5 & \lambda \end{vmatrix} = \lambda^3 - 3\lambda - 2 = (\lambda + 1)^2(\lambda - 2),$$

which implies that the eigenvalues are -1 and 2. Choosing the positive value, let $\lambda = 2$. To find a corresponding eigenvector, row reduce the matrix $2I - A$ to obtain

$$\begin{bmatrix} 2 & -6 & -8 \\ -0.5 & 2 & 0 \\ 0 & -0.5 & 2 \end{bmatrix} \longrightarrow \begin{bmatrix} 1 & 0 & -16 \\ 0 & 1 & -4 \\ 0 & 0 & 0 \end{bmatrix}.$$

So the eigenvectors of $\lambda = 2$ are of the form

$$x = \begin{bmatrix} x_1 \\ x_2 \\ x_3 \end{bmatrix} = \begin{bmatrix} 16t \\ 4t \\ t \end{bmatrix} = t \begin{bmatrix} 16 \\ 4 \\ 1 \end{bmatrix}.$$

For instance, if $t = 2$, then the initial age distribution vector would be

$$x_1 = \begin{bmatrix} 32 \\ 8 \\ 2 \end{bmatrix} \qquad \begin{array}{l} 0 \le \text{age} < 1 \\ 1 \le \text{age} < 2 \\ 2 \le \text{age} \le 3 \end{array}$$

and the age distribution vector for the next year would be

$$x_2 = Ax_1 = \begin{bmatrix} 0 & 6 & 8 \\ 0.5 & 0 & 0 \\ 0 & 0.5 & 0 \end{bmatrix} \begin{bmatrix} 32 \\ 8 \\ 2 \end{bmatrix} = \begin{bmatrix} 64 \\ 16 \\ 4 \end{bmatrix}. \qquad \begin{array}{l} 0 \le \text{age} < 1 \\ 1 \le \text{age} < 2 \\ 2 \le \text{age} \le 3 \end{array}$$

Notice that the ratio of the three age classes is still 16:4:1, and so the percent of the population in each age class remains the same.

Systems of Linear Differential Equations (Calculus)

A **system of first-order linear differential equations** has the form

$$y_1' = a_{11}y_1 + a_{12}y_2 + \cdots + a_{1n}y_n$$
$$y_2' = a_{21}y_1 + a_{22}y_2 + \cdots + a_{2n}y_n$$
$$\vdots$$
$$y_n' = a_{n1}y_1 + a_{n2}y_2 + \cdots + a_{nn}y_n,$$

where each y_i is a function of t and $y_i' = dy_i/dt$. If you let

$$\mathbf{y} = \begin{bmatrix} y_1 \\ y_2 \\ \vdots \\ y_n \end{bmatrix} \quad \text{and} \quad \mathbf{y}' = \begin{bmatrix} y_1' \\ y_2' \\ \vdots \\ y_n' \end{bmatrix},$$

then the system can be written in matrix form as

$$\mathbf{y}' = A\mathbf{y}.$$

EXAMPLE 3   *Solving a System of Linear Differential Equations*

Solve the system of linear differential equations.

$$y_1' = 4y_1$$
$$y_2' = -y_2$$
$$y_3' = 2y_3$$

Solution   From calculus you know that the solution of the differential equation $y' = ky$ is

$$y = Ce^{kt}.$$

Therefore, the solution of the system is

$$y_1 = C_1e^{4t}$$
$$y_2 = C_2e^{-t}$$
$$y_3 = C_3e^{2t}.$$

The matrix form of the system of linear differential equations in Example 3 is $\mathbf{y}' = A\mathbf{y}$, or

$$\begin{bmatrix} y_1' \\ y_2' \\ y_3' \end{bmatrix} = \begin{bmatrix} 4 & 0 & 0 \\ 0 & -1 & 0 \\ 0 & 0 & 2 \end{bmatrix} \begin{bmatrix} y_1 \\ y_2 \\ y_3 \end{bmatrix}.$$

Thus, the coefficients of t in the solutions $y_i = C_i e^{\lambda_i t}$ are given by the *eigenvalues* of the matrix A.

If A is a *diagonal* matrix, then the solution of

$$\mathbf{y}' = A\mathbf{y}$$

can be obtained immediately, as in Example 3. If A is *not* diagonal, then the solution requires a little more work. First attempt to find a matrix P that diagonalizes A. Then the change of variables given by $\mathbf{y} = P\mathbf{w}$ and $\mathbf{y}' = P\mathbf{w}'$ produces

$$P\mathbf{w}' = \mathbf{y}' = A\mathbf{y} = AP\mathbf{w} \quad \Longrightarrow \quad \mathbf{w}' = P^{-1}AP\mathbf{w},$$

where $P^{-1}AP$ is a diagonal matrix. This procedure is demonstrated in Example 4.

EXAMPLE 4   *Solving a System of Linear Differential Equations*

Solve the system of linear differential equations.

$$y_1{}' = 3y_1 + 2y_2$$
$$y_2{}' = 6y_1 - y_2$$

Solution   First find a matrix P that diagonalizes $A = \begin{bmatrix} 3 & 2 \\ 6 & -1 \end{bmatrix}$. The eigenvalues of A are $\lambda_1 = -3$ and $\lambda_2 = 5$, with corresponding eigenvectors

$$\mathbf{p}_1 = \begin{bmatrix} 1 \\ -3 \end{bmatrix} \quad \text{and} \quad \mathbf{p}_2 = \begin{bmatrix} 1 \\ 1 \end{bmatrix}.$$

Diagonalize A using the matrix P whose columns consist of $\mathbf{p}_1$ and $\mathbf{p}_2$ to obtain

$$P = \begin{bmatrix} 1 & 1 \\ -3 & 1 \end{bmatrix} \quad \text{and} \quad P^{-1}AP = \begin{bmatrix} -3 & 0 \\ 0 & 5 \end{bmatrix}.$$

The system represented by $\mathbf{w}' = P^{-1}AP\mathbf{w}$ has the following form.

$$\begin{bmatrix} w_1{}' \\ w_2{}' \end{bmatrix} = \begin{bmatrix} -3 & 0 \\ 0 & 5 \end{bmatrix} \begin{bmatrix} w_1 \\ w_2 \end{bmatrix} \quad \longrightarrow \quad \begin{matrix} w_1{}' = -3w_1 \\ w_2{}' = 5w_2 \end{matrix}$$

The solution of this system of equations is

$$w_1 = C_1 e^{-3t}$$
$$w_2 = C_2 e^{5t}.$$

To return to the original variables y_1 and y_2, use the substitution $\mathbf{y} = P\mathbf{w}$ and write

$$\begin{bmatrix} y_1 \\ y_2 \end{bmatrix} = \begin{bmatrix} 1 & 1 \\ -3 & 1 \end{bmatrix} \begin{bmatrix} w_1 \\ w_2 \end{bmatrix},$$

which implies that the solution is

$$y_1 = \quad w_1 + w_2 = \quad C_1 e^{-3t} + C_2 e^{5t}$$
$$y_2 = -3w_1 + w_2 = -3C_1 e^{-3t} + C_2 e^{5t}.$$

For the systems of linear differential equations in Examples 3 and 4, you found that each y_i can be written as a linear combination of $e^{\lambda_1 t}, e^{\lambda_2 t}, \ldots, e^{\lambda_n t}$, where $\lambda_1, \lambda_2, \ldots, \lambda_n$ are *distinct real eigenvalues* of the $n \times n$ matrix A. If A has eigenvalues with multiplicity greater than 1 or if A has complex eigenvalues, then the technique for solving the system must be modified. If you take a course on differential equations you will cover these two cases. For now, you can get an idea of the type of modification that is required from the following two systems of linear differential equations.

1. *Eigenvalues with multiplicity greater than* 1: The coefficient matrix of the system

$$\begin{aligned} y_1' &= y_2 \\ y_2' &= -4y_1 + 4y_2 \end{aligned} \quad \text{is} \quad A = \begin{bmatrix} 0 & 1 \\ -4 & 4 \end{bmatrix}.$$

The only eigenvalue of A is $\lambda = 2$, and the solution of the system of linear differential equations is

$$\begin{aligned} y_1 &= C_1 e^{2t} + C_2 t e^{2t} \\ y_2 &= (2C_1 + C_2)e^{2t} + 2C_2 t e^{2t}. \end{aligned}$$

2. *Complex eigenvalues*: The coefficient matrix of the system

$$\begin{aligned} y_1' &= -y_2 \\ y_2' &= y_1 \end{aligned} \quad \text{is} \quad A = \begin{bmatrix} 0 & -1 \\ 1 & 0 \end{bmatrix}.$$

The eigenvalues of A are $\lambda_1 = i$ and $\lambda_2 = -i$, and the solution of the system of linear differential equations is

$$\begin{aligned} y_1 &= C_1 \cos t + C_2 \sin t \\ y_2 &= -C_2 \cos t + C_1 \sin t. \end{aligned}$$

Try checking these solutions by differentiating and substituting into the original systems of equations.

Quadratic Forms

Eigenvalues and eigenvectors can be used to solve the rotation of axes problem introduced in Section 4.8. (You may want to review the discussion presented there.) Recall that classifying the graph of the quadratic equation

$$ax^2 + bxy + cy^2 + dx + ey + f = 0 \qquad \text{Quadratic equation}$$

is fairly straightforward as long as the equation has no *xy*-term (that is, $b = 0$). If the equation has an *xy*-term, however, then the classification is most easily accomplished by first performing a rotation of axes that eliminates the *xy*-term. The resulting equation (relative to the new $x'y'$-axes) will then be of the form

$$a'(x')^2 + c'(y')^2 + d'x' + e'y' + f' = 0.$$

You will see that the coefficients a' and c' are eigenvalues of the matrix

$$A = \begin{bmatrix} a & b/2 \\ b/2 & c \end{bmatrix}.$$

The expression

$$ax^2 + bxy + cy^2 \qquad \textbf{Quadratic form}$$

is called the **quadratic form** associated with the quadratic equation $ax^2 + bxy + cy^2 + dx + ey + f = 0$, and the matrix A is called the **matrix of the quadratic form.** Note that the matrix A is *symmetric* by definition. Moreover, the matrix A will be diagonal if and only if its corresponding quadratic form has no xy-term, as illustrated in Example 5.

EXAMPLE 5   *Finding the Matrix of a Quadratic Form*

Find the matrix of the quadratic form associated with each quadratic equation.
(a) $4x^2 + 9y^2 - 36 = 0$
(b) $13x^2 - 10xy + 13y^2 - 72 = 0$

Solution   (a) Because $a = 4$, $b = 0$, and $c = 9$, the matrix is

$$A = \begin{bmatrix} 4 & 0 \\ 0 & 9 \end{bmatrix}. \qquad \textbf{Diagonal matrix (no } xy\textbf{-term)}$$

(b) Because $a = 13$, $b = -10$, and $c = 13$, the matrix is

$$A = \begin{bmatrix} 13 & -5 \\ -5 & 13 \end{bmatrix}. \qquad \textbf{Nondiagonal matrix (} xy\textbf{-term)}$$

In standard form, the equation $4x^2 + 9y^2 - 36 = 0$ is

$$\frac{x^2}{3^2} + \frac{y^2}{2^2} = 1,$$

which is the equation of the ellipse shown in Figure 7.3. Although it is not apparent by simple inspection, the graph of the equation $13x^2 - 10xy + 13y^2 - 72 = 0$ is similar. In fact, if you rotate the x- and y-axes counterclockwise $45°$ to form a new $x'y'$ coordinate system, this equation takes the form

$$\frac{(x')^2}{3^2} + \frac{(y')^2}{2^2} = 1,$$

which is the equation of the ellipse shown in Figure 7.4.

To see how the matrix of a quadratic form can be used to perform a rotation of axes, let

$$X = \begin{bmatrix} x \\ y \end{bmatrix}.$$

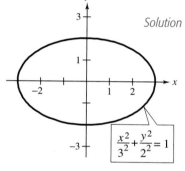

$$\frac{x^2}{3^2} + \frac{y^2}{2^2} = 1$$

Figure 7.3

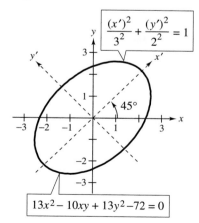

$$\frac{(x')^2}{3^2} + \frac{(y')^2}{2^2} = 1$$

$$13x^2 - 10xy + 13y^2 - 72 = 0$$

Figure 7.4

Then the quadratic expression $ax^2 + bxy + cy^2 + dx + ey + f$ can be written in matrix form as follows.

$$X^T A X + [d \quad e]X + f = [x \quad y]\begin{bmatrix} a & b/2 \\ b/2 & c \end{bmatrix}\begin{bmatrix} x \\ y \end{bmatrix} + [d \quad e]\begin{bmatrix} x \\ y \end{bmatrix} + f$$

$$= ax^2 + bxy + cy^2 + dx + ey + f$$

If $b = 0$, no rotation is necessary. But if $b \neq 0$, then, because A is symmetric, you can apply Theorem 7.10 to conclude that there exists an orthogonal matrix P such that

$$P^T A P = D$$

is diagonal. Thus, if you let

$$P^T X = X' = \begin{bmatrix} x' \\ y' \end{bmatrix},$$

it follows that $X = PX'$, and you have

$$X^T A X = (PX')^T A(PX')$$
$$= (X')^T P^T A P X'$$
$$= (X')^T D X'.$$

The choice of the matrix P must be made with care. Because P is orthogonal, its determinant will be ± 1. It can be shown (see Exercise 49) that if P is chosen so that $|P| = 1$, then P will be of the form

$$P = \begin{bmatrix} \cos \theta & -\sin \theta \\ \sin \theta & \cos \theta \end{bmatrix},$$

where θ gives the angle of rotation of the conic measured from the positive x-axis to the positive x'-axis. This brings you to the following theorem, the **Principal Axes Theorem.**

Principal Axes Theorem

For a conic whose equation is $ax^2 + bxy + cy^2 + dx + ey + f = 0$, the rotation given by $X = PX'$ eliminates the xy-term if P is an orthogonal matrix, with $|P| = 1$, that diagonalizes A. That is,

$$P^T A P = \begin{bmatrix} \lambda_1 & 0 \\ 0 & \lambda_2 \end{bmatrix},$$

where λ_1 and λ_2 are eigenvalues of A. The equation of the rotated conic is given by

$$\lambda_1 (x')^2 + \lambda_2 (y')^2 + [d \quad e]PX' + f = 0.$$

REMARK: Note that the matrix product $[d \quad e]PX'$ has the form

$$[d \quad e]PX' = (d \cos \theta + e \sin \theta)x' + (-d \sin \theta + e \cos \theta)y'.$$

EXAMPLE 6   *Rotation of a Conic*

Perform a rotation of axes to eliminate the xy-term in the quadratic equation $13x^2 - 10xy + 13y^2 - 72 = 0$.

Solution   The matrix of the quadratic form associated with this equation is

$$A = \begin{bmatrix} 13 & -5 \\ -5 & 13 \end{bmatrix}.$$

Because the characteristic polynomial of A is

$$\begin{vmatrix} \lambda - 13 & 5 \\ 5 & \lambda - 13 \end{vmatrix} = (\lambda - 13)^2 - 25 = (\lambda - 8)(\lambda - 18),$$

it follows that the eigenvalues of A are $\lambda_1 = 8$ and $\lambda_2 = 18$. Thus, the equation of the rotated conic is

$$8(x')^2 + 18(y')^2 - 72 = 0,$$

which, when written in the standard form

$$\frac{(x')^2}{3^2} + \frac{(y')^2}{2^2} = 1,$$

is the equation of an ellipse. (See Figure 7.4.)

In Example 6, the eigenvectors of the matrix A are

$$\mathbf{x}_1 = \begin{bmatrix} 1 \\ 1 \end{bmatrix} \quad \text{and} \quad \mathbf{x}_2 = \begin{bmatrix} -1 \\ 1 \end{bmatrix},$$

which you can normalize to form the columns of P as follows.

$$P = \begin{bmatrix} \dfrac{1}{\sqrt{2}} & -\dfrac{1}{\sqrt{2}} \\ \dfrac{1}{\sqrt{2}} & \dfrac{1}{\sqrt{2}} \end{bmatrix}$$

$$= \begin{bmatrix} \cos \theta & -\sin \theta \\ \sin \theta & \cos \theta \end{bmatrix}$$

Note first that $|P| = 1$, which implies that P is a rotation. Moreover, because $\cos 45° = 1/\sqrt{2} = \sin 45°$, you can conclude that $\theta = 45°$, as shown in Figure 7.4.

The orthogonal matrix P specified in the Principal Axes Theorem is not unique. Its entries depend on the ordering of the eigenvalues λ_1 and λ_2 *and* on the subsequent choice of eigenvectors $\mathbf{x}_1$ and $\mathbf{x}_2$. For instance, in the solution of Example 6, any of the following choices of P (see the next page) would have worked.

$$
\begin{array}{cc} \mathbf{x}_1 & \mathbf{x}_2 \end{array}
$$

$$
\begin{bmatrix} -\dfrac{1}{\sqrt{2}} & \dfrac{1}{\sqrt{2}} \\[2mm] -\dfrac{1}{\sqrt{2}} & -\dfrac{1}{\sqrt{2}} \end{bmatrix}
\qquad
\begin{bmatrix} -\dfrac{1}{\sqrt{2}} & -\dfrac{1}{\sqrt{2}} \\[2mm] \dfrac{1}{\sqrt{2}} & -\dfrac{1}{\sqrt{2}} \end{bmatrix}
\qquad
\begin{bmatrix} \dfrac{1}{\sqrt{2}} & \dfrac{1}{\sqrt{2}} \\[2mm] -\dfrac{1}{\sqrt{2}} & \dfrac{1}{\sqrt{2}} \end{bmatrix}
$$

$$
\lambda_1 = 8, \lambda_2 = 18 \qquad \lambda_1 = 18, \lambda_2 = 8 \qquad \lambda_1 = 18, \lambda_2 = 8
$$

$$
\theta = 225° \qquad\qquad \theta = 135° \qquad\qquad \theta = 315°
$$

For any of these choices of P, the graph of the rotated conic will, of course, be the same. (See Figure 7.5.)

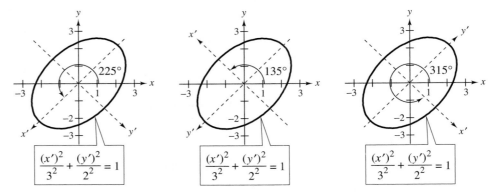

Figure 7.5

The steps used to apply the Principal Axes Theorem are summarized as follows.

1. Form the matrix A and find its eigenvalues λ_1 and λ_2.
2. Find eigenvectors corresponding to λ_1 and λ_2. Normalize these eigenvectors to form the columns of P.
3. If $|P| = -1$, then multiply one of the columns of P by -1 to obtain a matrix of the form

$$
P = \begin{bmatrix} \cos\theta & -\sin\theta \\ \sin\theta & \cos\theta \end{bmatrix}.
$$

4. The angle θ represents the angle of rotation of the conic.
5. The equation of the rotated conic is

$$
\lambda_1(x')^2 + \lambda_2(y')^2 + [d \quad e]PX' + f = 0.
$$

Example 7 shows how to apply the Principal Axes Theorem to rotate a conic whose center has been translated away from the origin.

EXAMPLE 7   *Rotation of a Conic*

Perform a rotation of axes to eliminate the xy-term in the quadratic equation

$$3x^2 - 10xy + 3y^2 + 16\sqrt{2}x - 32 = 0.$$

Solution   The matrix of the quadratic form associated with this equation is

$$A = \begin{bmatrix} 3 & -5 \\ -5 & 3 \end{bmatrix}.$$

The eigenvalues of A are $\lambda_1 = 8$ and $\lambda_2 = -2$, with corresponding eigenvectors of $\mathbf{x}_1 = (-1, 1)$ and $\mathbf{x}_2 = (-1, -1)$. This implies that the matrix P is

$$P = \begin{bmatrix} -\dfrac{1}{\sqrt{2}} & -\dfrac{1}{\sqrt{2}} \\ \dfrac{1}{\sqrt{2}} & -\dfrac{1}{\sqrt{2}} \end{bmatrix} = \begin{bmatrix} \cos\theta & -\sin\theta \\ \sin\theta & \cos\theta \end{bmatrix}, \text{ where } |P| = 1.$$

Because $\cos 135° = -1/\sqrt{2}$ and $\sin 135° = 1/\sqrt{2}$, you can conclude that the angle of rotation is $135°$. Finally, from the matrix product

$$[d \quad e]PX' = [16\sqrt{2} \quad 0]\begin{bmatrix} -\dfrac{1}{\sqrt{2}} & -\dfrac{1}{\sqrt{2}} \\ \dfrac{1}{\sqrt{2}} & -\dfrac{1}{\sqrt{2}} \end{bmatrix}\begin{bmatrix} x' \\ y' \end{bmatrix}$$

$$= -16x' - 16y',$$

you can conclude that the equation of the rotated conic is

$$8(x')^2 - 2(y')^2 - 16x' - 16y' - 32 = 0.$$

In standard form, the equation

$$\frac{(x'-1)^2}{1^2} - \frac{(y'+4)^2}{2^2} = 1$$

is the equation of a hyperbola. Its graph is shown in Figure 7.6.

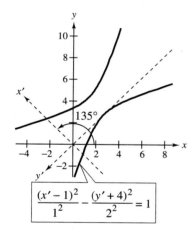

$$\boxed{\dfrac{(x'-1)^2}{1^2} - \dfrac{(y'+4)^2}{2^2} = 1}$$

Figure 7.6

Quadratic forms can also be used to analyze equations of quadric surfaces in space, which are the three-dimensional analogues of conic sections. The equation of a quadric surface in space is a second-degree polynomial of the form

$$ax^2 + by^2 + cz^2 + dxy + exz + fyz + gx + hy + iz + j = 0.$$

There are six basic types of quadric surfaces: ellipsoids, hyperboloids of one sheet, hyperboloids of two sheets, elliptic cones, elliptic paraboloids, and hyperbolic paraboloids. The intersection of a surface with a plane, called the **trace** of the surface in the plane, is useful to help visualize the graph of the surface in space. The six basic types of quadric surfaces, together with their traces, are shown on the following pages.

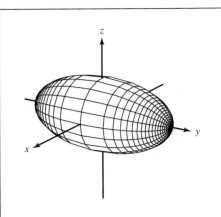

Ellipsoid

$$\frac{x^2}{a^2} + \frac{y^2}{b^2} + \frac{z^2}{c^2} = 1$$

Trace	Plane
Ellipse	Parallel to xy-plane
Ellipse	Parallel to xz-plane
Ellipse	Parallel to yz-plane

The surface is a sphere if $a = b = c \neq 0$.

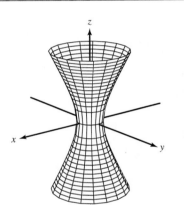

Hyperboloid of One Sheet

$$\frac{x^2}{a^2} + \frac{y^2}{b^2} - \frac{z^2}{c^2} = 1$$

Trace	Plane
Ellipse	Parallel to xy-plane
Hyperbola	Parallel to xz-plane
Hyperbola	Parallel to yz-plane

The axis of the hyperboloid corresponds to the variable whose coefficient is negative.

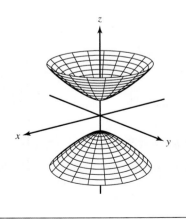

Hyperboloid of Two Sheets

$$\frac{z^2}{c^2} - \frac{x^2}{a^2} - \frac{y^2}{b^2} = 1$$

Trace	Plane
Ellipse	Parallel to xy-plane
Hyperbola	Parallel to xz-plane
Hyperbola	Parallel to yz-plane

The axis of the hyperboloid corresponds to the variable whose coefficient is positive. There is no trace in the coordinate plane perpendicular to this axis.

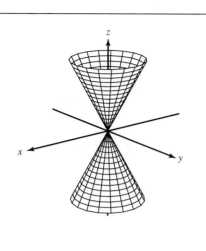

Elliptic Cone

$$\frac{x^2}{a^2} + \frac{y^2}{b^2} - \frac{z^2}{c^2} = 0$$

Trace	Plane
Ellipse	Parallel to xy-plane
Hyperbola	Parallel to xz-plane
Hyperbola	Parallel to yz-plane

The axis of the cone corresponds to the variable whose coefficient is negative. The traces in the coordinate planes parallel to this axis are intersecting lines.

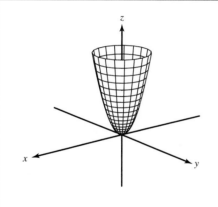

Elliptic Paraboloid

$$z = \frac{x^2}{a^2} + \frac{y^2}{b^2}$$

Trace	Plane
Ellipse	Parallel to xy-plane
Parabola	Parallel to xz-plane
Parabola	Parallel to yz-plane

The axis of the paraboloid corresponds to the variable raised to the first power.

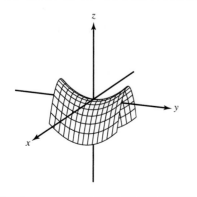

Hyperbolic Paraboloid

$$z = \frac{y^2}{b^2} - \frac{x^2}{a^2}$$

Trace	Plane
Hyperbola	Parallel to xy-plane
Parabola	Parallel to xz-plane
Parabola	Parallel to yz-plane

The axis of the paraboloid corresponds to the variable raised to the first power.

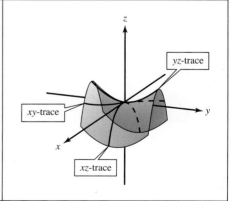

The quadratic form of the equation

$$ax^2 + by^2 + cz^2 + dxy + exz + fyz + gx + hy + iz + j = 0 \qquad \text{Quadric surface}$$

is defined to be

$$ax^2 + by^2 + cz^2 + dxy + exz + fyz. \qquad \text{Quadratic form}$$

The corresponding matrix is

$$A = \begin{bmatrix} a & \dfrac{d}{2} & \dfrac{e}{2} \\ \dfrac{d}{2} & b & \dfrac{f}{2} \\ \dfrac{e}{2} & \dfrac{f}{2} & c \end{bmatrix}.$$

In its three-dimensional version, the Principal Axes Theorem relates the eigenvalues and eigenvectors of A to the equation of the rotated surface, as shown in Example 8.

EXAMPLE 8   *Rotation of a Quadric Surface*

Perform a rotation of axes to eliminate the xz-term in the quadratic equation

$$5x^2 + 4y^2 + 5z^2 + 8xz - 36 = 0.$$

Solution   The matrix A associated with this quadratic equation is

$$A = \begin{bmatrix} 5 & 0 & 4 \\ 0 & 4 & 0 \\ 4 & 0 & 5 \end{bmatrix},$$

which has eigenvalues of $\lambda_1 = 1$, $\lambda_2 = 4$, and $\lambda_3 = 9$. Thus, in the rotated $x'y'z'$-system, the quadratic equation is $(x')^2 + 4(y')^2 + 9(z')^2 - 36 = 0$, which in standard form is

$$\frac{(x')^2}{6^2} + \frac{(y')^2}{3^2} + \frac{(z')^2}{2^2} = 1.$$

The graph of this equation is an ellipsoid. As shown in Figure 7.7, the $x'y'z'$-axes represent a counterclockwise rotation of 45° about the y-axis. Moreover, the orthogonal matrix

$$P = \begin{bmatrix} \dfrac{1}{\sqrt{2}} & 0 & \dfrac{1}{\sqrt{2}} \\ 0 & 1 & 0 \\ -\dfrac{1}{\sqrt{2}} & 0 & \dfrac{1}{\sqrt{2}} \end{bmatrix},$$

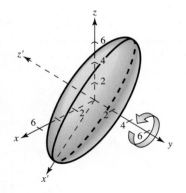

Figure 7.7

whose columns are the eigenvectors of A, has the property that $P^T A P$ is diagonal.

SECTION 7.4   Exercises

Population Growth

In Exercises 1–4, use the age transition matrix A and age distribution vector $\mathbf{x}_1$ to find the age distribution vectors $\mathbf{x}_2$ and $\mathbf{x}_3$.

1. $A = \begin{bmatrix} 0 & 2 \\ \frac{1}{2} & 0 \end{bmatrix}$, $\mathbf{x}_1 = \begin{bmatrix} 10 \\ 10 \end{bmatrix}$

2. $A = \begin{bmatrix} 0 & 4 \\ \frac{1}{16} & 0 \end{bmatrix}$, $\mathbf{x}_1 = \begin{bmatrix} 160 \\ 160 \end{bmatrix}$

3. $A = \begin{bmatrix} 0 & 3 & 4 \\ 1 & 0 & 0 \\ 0 & \frac{1}{2} & 0 \end{bmatrix}$, $\mathbf{x}_1 = \begin{bmatrix} 12 \\ 12 \\ 12 \end{bmatrix}$

4. $A = \begin{bmatrix} 0 & 2 & 2 & 0 \\ \frac{1}{4} & 0 & 0 & 0 \\ 0 & 1 & 0 & 0 \\ 0 & 0 & \frac{1}{2} & 0 \end{bmatrix}$, $\mathbf{x}_1 = \begin{bmatrix} 100 \\ 100 \\ 100 \\ 100 \end{bmatrix}$

5. Find a stable age distribution for the age transition matrix in Exercise 1.

6. Find a stable age distribution for the age transition matrix in Exercise 2.

7. Find a stable age distribution for the age transition matrix in Exercise 3.

8. Find a stable age distribution for the age transition matrix in Exercise 4.

9. A population has the following characteristics.

   (a) A total of 75% of the population survives its first year. Of that 75%, 25% survives the second year. The maximum life span is 3 years.

   (b) The average number of offspring for each member of the population is 2 the first year, 4 the second year, and 2 the third year.

   The population now consists of 120 members in each of the three age classes. How many members will there be in each age class in 1 year? In 2 years?

10. Find the limit (if it exists) of $A^n\mathbf{x}_1$ as n approaches infinity for the matrix.

$$A = \begin{bmatrix} 0 & 2 \\ \frac{1}{2} & 0 \end{bmatrix} \quad \text{and} \quad \mathbf{x}_1 = \begin{bmatrix} a \\ a \end{bmatrix}$$

Systems of Linear Differential Equations (Calculus)

In Exercises 11–14, solve the system of first-order linear differential equations.

11. $\begin{aligned} y_1' &= 2y_1 \\ y_2' &= y_2 \end{aligned}$ **12.** $\begin{aligned} y_1' &= -3y_1 \\ y_2' &= 4y_2 \end{aligned}$

13. $\begin{aligned} y_1' &= -y_1 \\ y_2' &= 6y_2 \\ y_3' &= y_3 \end{aligned}$ **14.** $\begin{aligned} y_1' &= 5y_1 \\ y_2' &= -2y_2 \\ y_3' &= -3y_3 \end{aligned}$

In Exercises 15–22, solve the system of first-order linear differential equations.

15. $\begin{aligned} y_1' &= y_1 - 4y_2 \\ y_2' &= 2y_2 \end{aligned}$ **16.** $\begin{aligned} y_1' &= y_1 - 4y_2 \\ y_2' &= -2y_1 + 8y_2 \end{aligned}$

17. $\begin{aligned} y_1' &= y_1 + 2y_2 \\ y_2' &= 2y_1 + y_2 \end{aligned}$ **18.** $\begin{aligned} y_1' &= y_1 - y_2 \\ y_2' &= 2y_1 + 4y_2 \end{aligned}$

19. $\begin{aligned} y_1' &= -3y_2 + 5y_3 \\ y_2' &= -4y_1 + 4y_2 - 10y_3 \\ y_3' &= 4y_3 \end{aligned}$ **20.** $\begin{aligned} y_1' &= -2y_1 + y_3 \\ y_2' &= 3y_2 + 4y_3 \\ y_3' &= y_3 \end{aligned}$

 21. $\begin{aligned} y_1' &= y_1 - 2y_2 + y_3 \\ y_2' &= 2y_2 + 4y_3 \\ y_3' &= 3y_3 \end{aligned}$ **22.** $\begin{aligned} y_1' &= 2y_1 + y_2 + y_3 \\ y_2' &= y_1 + y_2 \\ y_3' &= y_1 + y_3 \end{aligned}$

In Exercises 23–26, write out the system of first-order linear differential equations represented by the matrix equation $\mathbf{y}' = A\mathbf{y}$. Then verify the indicated general solution.

23. $A = \begin{bmatrix} 1 & 1 \\ 0 & 1 \end{bmatrix}$, $\begin{aligned} y_1 &= C_1 e^t + C_2 t e^t \\ y_2 &= C_2 e^t \end{aligned}$

24. $A = \begin{bmatrix} 1 & -1 \\ 1 & 1 \end{bmatrix}$, $\begin{aligned} y_1 &= C_1 e^t \cos t + C_2 e^t \sin t \\ y_2 &= -C_2 e^t \cos t + C_1 e^t \sin t \end{aligned}$

25. $A = \begin{bmatrix} 0 & 1 & 0 \\ 0 & 0 & 1 \\ 0 & -4 & 0 \end{bmatrix}$, $\begin{aligned} y_1 &= C_1 + C_2 \cos 2t + C_3 \sin 2t \\ y_2 &= 2C_3 \cos 2t - 2C_2 \sin 2t \\ y_3 &= -4C_2 \cos 2t - 4C_3 \sin 2t \end{aligned}$

26. $A = \begin{bmatrix} 0 & 1 & 0 \\ 0 & 0 & 1 \\ 1 & -3 & 3 \end{bmatrix}$,

$\begin{aligned} y_1 &= C_1 e^t + C_2 t e^t + C_3 t^2 e^t \\ y_2 &= (C_1 + C_2)e^t + (C_2 + 2C_3)t e^t + C_3 t^2 e^t \\ y_3 &= (C_1 + 2C_2 + 2C_3)e^t + (C_2 + 4C_3)t e^t + C_3 t^2 e^t \end{aligned}$

Quadratic Forms

In Exercises 27–32, find the matrix of the quadratic form associated with the equation.

27. $x^2 + y^2 - 4 = 0$

28. $x^2 - 4xy + y^2 - 4 = 0$

29. $9x^2 + 10xy - 4y^2 - 36 = 0$

30. $12x^2 - 5xy - x + 2y - 20 = 0$

31. $10xy - 10y^2 + 4x - 48 = 0$

32. $16x^2 - 4xy + 20y^2 - 72 = 0$

In Exercises 33–38, find the matrix A of the quadratic form associated with the equation. In each case, find the eigenvalues of A and an orthogonal matrix P such that $P^T A P$ is diagonal.

33. $2x^2 - 3xy - 2y^2 + 10 = 0$

34. $5x^2 - 2xy + 5y^2 + 10x - 17 = 0$

35. $13x^2 + 6\sqrt{3}\,xy + 7y^2 - 16 = 0$

36. $3x^2 - 2\sqrt{3}\,xy + y^2 + 2x + 2\sqrt{3}\,y = 0$

37. $16x^2 - 24xy + 9y^2 - 60x - 80y + 100 = 0$

38. $17x^2 + 32xy - 7y^2 - 75 = 0$

In Exercises 39–46, use the Principal Axes Theorem to perform a rotation of axes to eliminate the xy-term in the quadratic equation. Identify the resulting rotated conic and give its equation in the new coordinate system.

39. $13x^2 - 8xy + 7y^2 - 45 = 0$

40. $x^2 + 4xy + y^2 - 9 = 0$

41. $7x^2 + 32xy - 17y^2 - 50 = 0$

42. $2x^2 - 4xy + 5y^2 - 36 = 0$

43. $2x^2 + 4xy + 2y^2 + 6\sqrt{2}\,x + 2\sqrt{2}\,y + 4 = 0$

44. $8x^2 + 8xy + 8y^2 + 10\sqrt{2}\,x + 26\sqrt{2}\,y + 31 = 0$

45. $xy + x - 2y + 3 = 0$

46. $5x^2 - 2xy + 5y^2 + 10\sqrt{2}\,x = 0$

In Exercises 47 and 48, find the matrix A of the quadratic form associated with the equation. Then find the equation of the rotated quadric surface in which the xy-, xz-, and yz-terms have been eliminated.

47. $3x^2 - 2xy + 3y^2 + 8z^2 - 16 = 0$

48. $2x^2 + 2y^2 + 2z^2 + 2xy + 2xz + 2yz - 1 = 0$

49. Let P be a 2×2 orthogonal matrix such that $|P| = 1$. Show that there exists a number θ, $0 \leq \theta < 2\pi$, such that

$$P = \begin{bmatrix} \cos\theta & -\sin\theta \\ \sin\theta & \cos\theta \end{bmatrix}.$$

CHAPTER 7 Summary

After studying this chapter, you should have acquired the following skills.

■ Find the characteristic equation and the eigenvalues and corresponding eigenvectors of a given matrix A.

■ Determine whether a given matrix is diagonalizable.

■ Find (if possible) a nonsingular matrix P for a given matrix A such that $P^{-1}AP$ is diagonal.

■ Find a basis B (if possible) for the domain of a linear transformation T such that the matrix of T relative to B is diagonal.

■ Determine whether a given matrix is symmetric.

■ Find the eigenvalues of a given symmetric matrix and determine the dimension of the corresponding eigenspace.

■ Determine whether a given matrix is orthogonal.

■ Find an orthogonal matrix P that diagonalizes A.

■ Use an age transition matrix and an age distribution vector to form a population model and find a stable age distribution for the population.

■ Find a stable age distribution for a given age transition matrix.

■ Solve a system of first-order linear differential equations.

■ Find a matrix of the quadratic form associated with a quadratic equation.

■ Use the Principal Axes Theorem to perform a rotation of axes to eliminate the xy-term in a given quadratic equation.

■ Find the equation of the rotated quadric surface in which the xy-, xz-, and yz-terms have been eliminated.

CHAPTER 7   Review Exercises

In Exercises 1–6, find (a) the characteristic equation of A, (b) the real eigenvalues of A, and (c) a basis for the eigenspace corresponding to each eigenvalue.

1. $A = \begin{bmatrix} 2 & 1 \\ 5 & -2 \end{bmatrix}$
2. $A = \begin{bmatrix} 2 & 1 \\ -4 & -2 \end{bmatrix}$

3. $A = \begin{bmatrix} 9 & 4 & -3 \\ -2 & 0 & 6 \\ -1 & -4 & 11 \end{bmatrix}$
4. $A = \begin{bmatrix} -4 & 1 & 2 \\ 0 & 1 & 1 \\ 0 & 0 & 3 \end{bmatrix}$

5. $A = \begin{bmatrix} 2 & 0 & 1 \\ 0 & 3 & 4 \\ 0 & 0 & 1 \end{bmatrix}$
6. $A = \begin{bmatrix} 2 & 1 & 0 & 0 \\ 1 & 2 & 0 & 0 \\ 0 & 0 & 2 & 1 \\ 0 & 0 & 1 & 2 \end{bmatrix}$

In Exercises 7–10, determine whether A is diagonalizable. If it is, find a nonsingular matrix P such that $P^{-1}AP$ is diagonal.

7. $A = \begin{bmatrix} -2 & -1 & 3 \\ 0 & 1 & 2 \\ 0 & 0 & 1 \end{bmatrix}$
8. $A = \begin{bmatrix} 3 & -2 & 2 \\ -2 & 0 & -1 \\ 2 & -1 & 0 \end{bmatrix}$

9. $A = \begin{bmatrix} 1 & 0 & 2 \\ 0 & 1 & 0 \\ 2 & 0 & 1 \end{bmatrix}$
10. $A = \begin{bmatrix} 2 & -1 & 1 \\ -2 & 3 & -2 \\ -1 & 1 & 0 \end{bmatrix}$

11. Show that if $0 < \theta < \pi$, then the transformation for a counterclockwise rotation through an angle θ has no real eigenvalues.

12. For what value(s) of a does the matrix

$$A = \begin{bmatrix} 0 & 1 \\ a & 1 \end{bmatrix}$$

have the following characteristics?

   (a) A has an eigenvalue of multiplicity 2.
   (b) A has -1 and 2 as eigenvalues.
   (c) A has real eigenvalues.

Writing   In Exercises 13–16, explain why the matrix is not diagonalizable.

13. $A = \begin{bmatrix} 0 & 2 \\ 0 & 0 \end{bmatrix}$
14. $A = \begin{bmatrix} -1 & 2 \\ 0 & -1 \end{bmatrix}$

15. $A = \begin{bmatrix} 3 & 0 & 0 \\ 1 & 3 & 0 \\ 0 & 0 & 3 \end{bmatrix}$
16. $A = \begin{bmatrix} -2 & 3 & 1 \\ 0 & 4 & 3 \\ 0 & 0 & -2 \end{bmatrix}$

In Exercises 17 and 18, determine whether the matrices are similar. If they are, find a matrix P such that $A = P^{-1}BP$.

17. $A = \begin{bmatrix} 1 & 0 \\ 0 & 2 \end{bmatrix}$, $B = \begin{bmatrix} 2 & 0 \\ 0 & 1 \end{bmatrix}$

18. $A = \begin{bmatrix} 1 & 1 & 0 \\ 0 & 1 & 1 \\ 0 & 0 & 1 \end{bmatrix}$, $B = \begin{bmatrix} 1 & 1 & 0 \\ 0 & 1 & 0 \\ 0 & 0 & 1 \end{bmatrix}$

In Exercises 19–22, determine whether the matrix is symmetric, orthogonal, both, or neither.

19. $A = \begin{bmatrix} -\dfrac{\sqrt{2}}{2} & \dfrac{\sqrt{2}}{2} \\ \dfrac{\sqrt{2}}{2} & \dfrac{\sqrt{2}}{2} \end{bmatrix}$
20. $A = \begin{bmatrix} -\dfrac{2}{3} & \dfrac{1}{3} & -\dfrac{2}{3} \\ \dfrac{2}{3} & \dfrac{2}{3} & -\dfrac{1}{3} \\ \dfrac{1}{3} & -\dfrac{2}{3} & \dfrac{2}{3} \end{bmatrix}$

21. $A = \begin{bmatrix} 0 & 0 & 1 \\ 0 & 1 & 0 \\ 1 & 0 & 1 \end{bmatrix}$
22. $A = \begin{bmatrix} \dfrac{4}{5} & 0 & \dfrac{3}{5} \\ 0 & 1 & 0 \\ -\dfrac{3}{5} & 0 & \dfrac{4}{5} \end{bmatrix}$

In Exercises 23–26, find an orthogonal matrix P that diagonalizes A.

23. $A = \begin{bmatrix} 3 & 4 \\ 4 & -3 \end{bmatrix}$
24. $A = \begin{bmatrix} 8 & 15 \\ 15 & -8 \end{bmatrix}$

25. $A = \begin{bmatrix} 2 & 0 & -1 \\ 0 & 1 & 0 \\ -1 & 0 & 2 \end{bmatrix}$
26. $A = \begin{bmatrix} 1 & 2 & 0 \\ 2 & 1 & 0 \\ 0 & 0 & 5 \end{bmatrix}$

In Exercises 27–32, find the steady state probability vector (if it exists) for the matrix. An eigenvector **v** of an $n \times n$ matrix A is called a **steady state probability vector** if $A\mathbf{v} = \mathbf{v}$ and the components of **v** add up to 1.

27. $A = \begin{bmatrix} \frac{2}{3} & \frac{1}{2} \\ \frac{1}{3} & \frac{1}{2} \end{bmatrix}$

28. $A = \begin{bmatrix} \frac{1}{2} & 1 \\ \frac{1}{2} & 0 \end{bmatrix}$

29. $A = \begin{bmatrix} \frac{1}{2} & \frac{1}{4} & 0 \\ \frac{1}{2} & \frac{1}{2} & \frac{1}{2} \\ 0 & \frac{1}{4} & \frac{1}{2} \end{bmatrix}$

30. $A = \begin{bmatrix} \frac{1}{3} & \frac{2}{3} & \frac{1}{3} \\ \frac{1}{3} & \frac{1}{3} & 0 \\ \frac{1}{3} & 0 & \frac{2}{3} \end{bmatrix}$

31. $A = \begin{bmatrix} 0.7 & 0.1 & 0.1 \\ 0.2 & 0.7 & 0.1 \\ 0.1 & 0.2 & 0.8 \end{bmatrix}$

32. $A = \begin{bmatrix} 0.8 & 0.3 \\ 0.2 & 0.7 \end{bmatrix}$

33. Prove that if A is an $n \times n$ symmetric matrix, then P^TAP is symmetric for any $n \times n$ matrix P.

34. Show that the characteristic equation of

$$A = \begin{bmatrix} 0 & 1 & 0 & 0 & \cdots & 0 \\ 0 & 0 & 1 & 0 & \cdots & 0 \\ \vdots & \vdots & \vdots & \vdots & & \vdots \\ 0 & 0 & 0 & 0 & \cdots & 1 \\ -a_0/a_n & -a_1/a_n & -a_2/a_n & -a_3/a_n & \cdots & -a_{n-1}/a_n \end{bmatrix},$$

$a_n \neq 0$, is $p(\lambda) = a_n\lambda^n + a_{n-1}\lambda^{n-1} + \cdots + a_3\lambda^3 + a_2\lambda^2 + a_1\lambda + a_0 = 0$. A is called the **companion matrix** of the polynomial p.

In Exercises 35 and 36, use the result of Exercise 34 to find the companion matrix A of the polynomial and find the eigenvalues of A.

35. $p(\lambda) = -9\lambda + 4\lambda^2$

36. $p(\lambda) = 189 - 120\lambda - 7\lambda^2 + 2\lambda^3$

37. The characteristic equation of the matrix

$$A = \begin{bmatrix} 8 & -4 \\ 2 & 2 \end{bmatrix}$$

is $\lambda^2 - 10\lambda + 24 = 0$. Because $A^2 - 10A + 24I_2 = O$, we can find powers of A by the following process.

$$A^2 = 10A - 24I_2, \quad A^3 = 10A^2 - 24A,$$
$$A^4 = 10A^3 - 24A^2, \ldots$$

Use this process to find the matrices A^2 and A^3.

38. Repeat Exercise 37 for the matrix

$$A = \begin{bmatrix} 9 & 4 & -3 \\ -2 & 0 & 6 \\ -1 & -4 & 11 \end{bmatrix}.$$

39. Let A be an $n \times n$ matrix.

    (a) Prove or disprove that an eigenvector of A is also an eigenvector of A^2.

    (b) Prove or disprove that an eigenvector of A^2 is also an eigenvector of A.

40. Let A be an $n \times n$ matrix. Prove that if $A\mathbf{x} = \lambda\mathbf{x}$, then $\mathbf{x}$ is an eigenvector of $(A + cI)$. What is the corresponding eigenvalue?

41. Let A and B be $n \times n$ matrices. Prove that if A is nonsingular, then AB is similar to BA.

42. (a) Find a symmetric matrix B such that $B^2 = A$ for the matrix

$$A = \begin{bmatrix} 2 & 1 \\ 1 & 2 \end{bmatrix}.$$

    (b) Generalize the result of part (a) by proving that if A is an $n \times n$ symmetric matrix with positive eigenvalues, then there exists a symmetric matrix B such that $B^2 = A$.

43. Find an orthogonal matrix P such that $P^{-1}AP$ is diagonal for the matrix

$$A = \begin{bmatrix} a & b \\ b & a \end{bmatrix}.$$

44. *Writing* Let A be an $n \times n$ idempotent matrix (that is, $A^2 = A$). Describe the eigenvalues of A.

45. *Writing* The following matrix has an eigenvalue $\lambda = 2$ of multiplicity 4.

$$A = \begin{bmatrix} 2 & a & 0 & 0 \\ 0 & 2 & b & 0 \\ 0 & 0 & 2 & c \\ 0 & 0 & 0 & 2 \end{bmatrix}$$

    (a) Under what conditions is A diagonalizable?

    (b) Under what conditions does the eigenspace of $\lambda = 2$ have dimension 1? 2? 3?

46. Determine all $n \times n$ symmetric matrices that have 0 as their only eigenvalue.

True or False? In Exercises 47 and 48, determine whether each statement is true or false. If a statement is true, give a reason or cite an appropriate statement in the text. If a statement is false, provide an example that shows that the statement is not true in all cases or cite an appropriate statement in the text.

47. (a) An eigenvector of an $n \times n$ matrix A is a nonzero vector $\mathbf{x}$ in R^n such that $A\mathbf{x}$ is a scalar multiple of $\mathbf{x}$.
  (b) Similar matrices may or may not have the same eigenvalues.
  (c) To diagonalize a square matrix A, you need to find an invertible matrix P such that $P^{-1}AP$ is diagonal.

48. (a) An eigenvalue of a matrix A is a scalar λ such that $\det(\lambda I - A) = 0$.
  (b) An eigenvector may be the zero vector $\mathbf{0}$.
  (c) A matrix A is orthogonally diagonalizable if there exists an orthogonal matrix P such that $P^{-1}AP = D$ is diagonal.

Population Growth

In Exercises 49–52, use the age transition matrix A and the age distribution vector $\mathbf{x}_1$ to find the age distribution vectors $\mathbf{x}_2$ and $\mathbf{x}_3$. Then find a stable age distribution for the population.

49. $A = \begin{bmatrix} 0 & 1 \\ \frac{1}{4} & 0 \end{bmatrix}, \mathbf{x}_1 = \begin{bmatrix} 100 \\ 100 \end{bmatrix}$

50. $A = \begin{bmatrix} 0 & 1 \\ \frac{3}{4} & 0 \end{bmatrix}, \mathbf{x}_1 = \begin{bmatrix} 32 \\ 32 \end{bmatrix}$

51. $A = \begin{bmatrix} 0 & 3 & 12 \\ 1 & 0 & 0 \\ 0 & \frac{1}{6} & 0 \end{bmatrix}, \mathbf{x}_1 = \begin{bmatrix} 300 \\ 300 \\ 300 \end{bmatrix}$

52. $A = \begin{bmatrix} 0 & 2 & 2 \\ \frac{1}{2} & 0 & 0 \\ 0 & 0 & 0 \end{bmatrix}, \mathbf{x}_1 = \begin{bmatrix} 240 \\ 240 \\ 240 \end{bmatrix}$

Systems of Linear Differential Equations (Calculus)

In Exercises 53–56, solve the system of first-order linear differential equations.

53. $y_1' = y_1 + 2y_2$
   $y_2' = 0$

54. $y_1' = 3y_1$
   $y_2' = y_1 - y_2$

55. $y_1' = y_2$
   $y_2' = y_1$

56. $y_1' = 6y_1 - y_2 + 2y_3$
   $y_2' = 3y_2 - y_3$
   $y_3' = y_3$

Quadratic Forms

In Exercises 57–60, find the matrix A of the quadratic form associated with the equation. In each case, find an orthogonal matrix P such that P^TAP is diagonal. Sketch the graph of each equation.

57. $x^2 + 3xy + y^2 - 3 = 0$
58. $x^2 - \sqrt{3}\,xy + 2y^2 - 10 = 0$
59. $xy - 2 = 0$
60. $9x^2 - 24xy + 16y^2 - 400x - 300y = 0$

CHAPTER 7	**Projects**

1 POPULATION GROWTH AND DYNAMICAL SYSTEMS

Systems of differential equations often arise in biological applications of population growth of various species of animals. These equations are called **dynamical systems** because they describe the changes in a system as functions of time. Suppose that over time t you are studying the populations of predator sharks $y_1(t)$ and their small fish prey $y_2(t)$. One simple model for the relative growths of these populations is

$$y_1'(t) = ay_1(t) + by_2(t) \qquad \text{Predator}$$

$$y_2'(t) = cy_1(t) + dy_2(t) \qquad \text{Prey}$$

where a, b, c, and d are constants that depend on the particular species being studied and on factors such as environment, available food, other competing species, and so on. Generally, the constants a and d are positive, reflecting the growth rates of the individual species. If the species are in a predator–prey relationship, then $b > 0$ and $c < 0$ indicate that an increase in prey fish y_2 would cause an increase in y_1, whereas an increase in the predator y_1 would cause a decrease in y_2.

Suppose that the following system of linear differential equations models the populations of sharks $y_1(t)$ and prey fish $y_2(t)$ with the given initial populations at time $t = 0$.

$$y_1{'}(t) = \quad 0.5y_1(t) + 0.6y_2(t) \qquad y_1(0) = 36$$
$$y_2{'}(t) = -0.4y_1(t) + 3.0y_2(t) \qquad y_2(0) = 121$$

1. Use the diagonalization techniques of this chapter to find the populations $y_1(t)$ and $y_2(t)$ at any time $t > 0$.

2. Interpret the solution in terms of the long-term population trends for the two species. Does one species ultimately disappear? Why or why not?

3. If you have access to a computer software program or graphing calculator, graph the solutions $y_1(t)$ and $y_2(t)$ over the domain $0 \leq t \leq 3$.

4. Use the explicit solution found in part (1) to explain why the quotient $y_2(t)/y_1(t)$ approaches a limit as t increases.

5. If you have access to a computer software program or graphing calculator that can numerically solve differential equations, use it to graph the solution of the original system of equations. Does this numerical approximation appear to be accurate?

For Further Reference You can learn more about dynamical systems and population modeling in most books on differential equations. For example, *Differential Equations and Their Applications*, fourth edition, by Martin Braun, Springer-Verlag, 1993, discusses the theory and applications of systems of linear differential equations. An especially interesting application of *nonlinear* differential equations is given in Section 4.10: "Predator–prey problems; or why the percentage of sharks caught in the Mediterranean Sea rose dramatically during World War I."

2 THE FIBONACCI SEQUENCE

The **Fibonacci sequence** is named after the Italian mathematician Leonard Fibonacci of Pisa (1170–1250). The simplest way to form this sequence is to define the first two terms to be $x_1 = 1$ and $x_2 = 1$, and then define the nth term as the sum of its two immediate predecessors. That is,

$$x_n = x_{n-1} + x_{n-2}.$$

Therefore, the third term is $2 = 1 + 1$, the fourth term is $3 = 2 + 1$, and so on. The formula $x_n = x_{n-1} + x_{n-2}$ is called *recursive* because the first $n - 1$ terms must be calculated before the nth term can be calculated. Is it possible to find an *explicit* formula for the nth term of the Fibonacci sequence? In this project, you will use eigenvalues and diagonalization to derive such a formula.

1. Use the formula $x_n = x_{n-1} + x_{n-2}$ to calculate the first 12 terms in the Fibonacci sequence.

2. Explain how the matrix identity

$$\begin{bmatrix} 1 & 1 \\ 1 & 0 \end{bmatrix} \begin{bmatrix} x_{n-1} \\ x_{n-2} \end{bmatrix} = \begin{bmatrix} x_{n-1} + x_{n-2} \\ x_{n-1} \end{bmatrix}$$

can be used to generate recursively the Fibonacci sequence.

3. Starting with $\begin{bmatrix} x_1 \\ x_2 \end{bmatrix} = \begin{bmatrix} 1 \\ 1 \end{bmatrix}$, show that $A^{n-2} \begin{bmatrix} 1 \\ 1 \end{bmatrix} = \begin{bmatrix} x_n \\ x_{n-1} \end{bmatrix}$, where $A = \begin{bmatrix} 1 & 1 \\ 1 & 0 \end{bmatrix}$.

4. Find a matrix P that diagonalizes A.

5. Derive an explicit formula for the nth term of the Fibonacci sequence. Use this formula to calculate x_1, x_2, and x_3.

6. Use the explicit formula for the nth term of the Fibonacci sequence together with a computer or calculator to find x_{10} and x_{20}.

7. Calculate the quotient x_n/x_{n-1} for various large values of n. Does the quotient appear to be approaching a fixed number as n tends to infinity?

8. Determine the limit of x_n/x_{n-1} as n approaches infinity. Do you recognize this number?

For Further Reference You can learn more about Fibonacci numbers in most books on number theory. You might find it interesting to look at the *Fibonacci Quarterly*, the official journal of the Fibonacci Association.

CHAPTERS 6 & 7 **Cumulative Test**

Take this test as you would take a test in class. After you are done, check your work against the answers given in the back of the book.

1. Determine whether the function $T:R^3 \to R^2$, $T(x, y, z) = (2x, y - x)$ is a linear transformation.

2. Determine whether the function $T: M_{2,2} \to R$, $T(A) = |A + A^T|$ is a linear transformation.

3. Let $T: R^2 \to R^3$ be the linear transformation defined by $T(\mathbf{v}) = A\mathbf{v}$, where

$$A = \begin{bmatrix} 1 & 0 \\ -1 & 0 \\ 0 & 0 \end{bmatrix}.$$

Find (a) $T(1, -2)$ and (b) the preimage of $(1, 3, 0)$.

4. Find the kernel of the linear transformation $T: R^4 \rightarrow R^4$, $T(x_1, x_2, x_3, x_4) = (x_1 - x_2, x_2 - x_1, 0, x_3 + x_4)$.

5. Let $T: R^4 \rightarrow R^2$ be the linear transformation given by $T(\mathbf{v}) = A\mathbf{v}$, where

$$A = \begin{bmatrix} 1 & 0 & 1 & 0 \\ 0 & -1 & 0 & -1 \end{bmatrix}.$$

Find a basis for (a) the kernel of T and (b) the range of T. (c) Determine the rank and nullity of T.

6. Find the standard matrix for the linear transformation given by
$$T(x, y, z) = (y - x, z - y, x - z).$$

7. Find the standard matrix A of the linear transformation $\text{proj}_v\mathbf{u}: R^2 \rightarrow R^2$ that projects an arbitrary vector $\mathbf{u}$ onto the vector

$$\mathbf{v} = \begin{bmatrix} 1 \\ -1 \end{bmatrix}$$

as shown in Figure 7.8. Use this matrix to find the images of the vectors $(1, 1)$ and $(-2, 2)$.

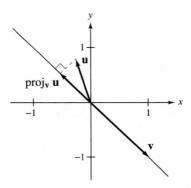

Figure 7.8

8. Find the inverse of the linear transformation $T: R^2 \rightarrow R^2$ given by $T(x, y) = (x - y, 2x + y)$. Verify that $(T^{-1} \circ T)(3, -2) = (3, -2)$.

9. Find the matrix of the linear transformation $T(x, y) = (y, 2x, x + y)$ relative to the bases $B = \{(1, 1), (1, 0)\}$ for R^2 and $B' = \{(1, 0, 0), (1, 1, 0), (1, 1, 1)\}$ for R^3. Use this matrix to find the image of the vector $(0, 1)$.

10. Let $B = \{(1, 0), (0, 1)\}$ and $B' = \{(1, 1), (1, 2)\}$ be bases for R^2.

(a) Find the matrix A of $T: R^2 \rightarrow R^2$, $T(x, y) = (x - 2y, x + 4y)$ relative to the basis B.

(b) Find the transition matrix P from B' to B.

(c) Find the matrix A' of T relative to the basis B'.

(d) Find $[T(\mathbf{v})]_{B'}$ if $[\mathbf{v}]_{B'} = \begin{bmatrix} 3 \\ -2 \end{bmatrix}$.

(e) Verify your answer in part (d) by finding $[\mathbf{v}]_B$ and $[T(\mathbf{v})]_B$.

11. Find the eigenvalues and the corresponding eigenvectors of the matrix
$$A = \begin{bmatrix} 1 & 2 & 1 \\ 0 & 3 & 1 \\ 0 & -3 & -1 \end{bmatrix}.$$

12. Find the eigenvalues and the corresponding eigenvectors of the matrix
$$A = \begin{bmatrix} 1 & -1 & 1 \\ 0 & 1 & 2 \\ 0 & 0 & 1 \end{bmatrix}.$$

13. Find a nonsingular matrix P such that $P^{-1}AP$ is diagonal if
$$A = \begin{bmatrix} 1 & 3 \\ -1 & 5 \end{bmatrix}.$$

14. Find a basis B for R^3 such that the matrix for $T: R^3 \to R^3$, $T(x, y, z) = (2x - 2z, 2y - 2z, 3x - 3z)$ relative to B is diagonal.

15. Find an orthogonal matrix P such that P^TAP diagonalizes the symmetric matrix
$$A = \begin{bmatrix} 1 & 3 \\ 3 & 1 \end{bmatrix}.$$

16. Use the Gram-Schmidt orthonormalization process to find an orthogonal matrix P such that P^TAP diagonalizes the symmetric matrix
$$A = \begin{bmatrix} 0 & 1 & 1 \\ 1 & 0 & 1 \\ 1 & 1 & 0 \end{bmatrix}.$$

17. Let A be an $n \times n$ matrix. Define the terms *eigenvalue* and *eigenvector* of A. How many eigenvalues can A have?

18. Define an *orthogonal matrix* and determine the possible values of its determinant.

19. Prove that if A is similar to B and A is diagonalizable, then B is diagonalizable.

20. Prove that if 0 is an eigenvalue of A, then A is singular.

21. Prove that the eigenvectors corresponding to distinct eigenvalues of a symmetric matrix are orthogonal.

22. Prove that the range of a linear transformation $T: V \to W$ is a subspace of W.

23. Prove that a linear transformation is one-to-one if and only if its kernel is $\{\mathbf{0}\}$.

24. Find all eigenvalues of A if $A^2 = O$.

APPENDIX A

Mathematical Induction and Other Forms of Proofs

In this appendix you will study some basic strategies for writing mathematical proofs—mathematical induction, proof by contradiction, and the use of counterexamples.

Mathematical Induction

To see the need for using mathematical induction, study the problem situation in the following example.

EXAMPLE 1   *Sum of Odd Integers*

Use the pattern to conjecture a formula for the sum of the first n odd integers.

$$1 = 1$$
$$1 + 3 = 4$$
$$1 + 3 + 5 = 9$$
$$1 + 3 + 5 + 7 = 16$$
$$1 + 3 + 5 + 7 + 9 = 25$$

Solution   Notice that the sums on the right are equal to the squares 1^2, 2^2, 3^2, 4^2, and 5^2. Judging from this pattern, you can *conjecture* that the sum of the first n odd integers is n^2,

$$1 + 3 + 5 + 7 + \cdots + (2n - 1) = n^2.$$

Although this formula is valid, it is important to see that recognizing a pattern and then simply *jumping to the conclusion* that the pattern must be true for all values of n is not a logically valid method of proof. There are many examples in which a pattern appears to be developing for small values of n and then at some point the pattern fails. One of the most famous cases of this was the conjecture by the French mathematician Pierre de Fermat (1601–1665), who speculated that all numbers of the form

$$F_n = 2^{2^n} + 1, \quad n = 0, 1, 2, \ldots$$

are prime. For $n = 0, 1, 2, 3,$ and 4, the conjecture is true.

$$F_0 = 3$$
$$F_1 = 5$$
$$F_2 = 17$$
$$F_3 = 257$$
$$F_4 = 65,537$$

The size of the next Fermat number ($F_5 = 4{,}294{,}967{,}297$) is so great that it was difficult for Fermat to determine whether it was prime or not. However, another well-known mathematician, Leonhard Euler (1707–1783), later found the factorization

$$F_5 = 4{,}294{,}967{,}297 = (641)(6{,}700{,}417),$$

which proved that F_5 is *not* prime, and therefore Fermat's conjecture was false.

Just because a rule, pattern, or formula seems to work for several values of n, you cannot simply decide that it is valid for all values of n without proving it so. One legitimate method of proof for such conjectures is the **Principle of Mathematical Induction.**

Principle of Mathematical Induction	Let P_n be a statement involving the positive integer n. If 1. P_1 is true, and 2. the truth of P_k implies the truth of P_{k+1} for every positive integer k, then P_n must be true for all positive integers n.

In the next example, the Principal of Mathematical Induction is used to prove the conjecture from Example 1.

EXAMPLE 2   *Using Mathematical Induction*

Use mathematical induction to prove that the following formula is valid for all positive integers n.

$$S_n = 1 + 3 + 5 + 7 + \cdots + (2n - 1) = n^2$$

Solution   Mathematical induction consists of two distinct parts. First, you must show that the formula is true when $n = 1$.
1. When $n = 1$, the formula is valid, because

$$S_1 = 1 = 1^2.$$

The second part of mathematical induction has two steps. The first step is to assume that the formula is valid for some integer k (the **induction hypothesis**). The second step is to use this assumption to prove that the formula is valid for the next integer, $k + 1$.
2. Assuming that the formula

$$S_k = 1 + 3 + 5 + 7 + \cdots + (2k - 1) = k^2$$

is true, you must show that the formula

$$S_{k+1} = (k + 1)^2$$

is true. Notice in the following steps that because $(2k - 1)$ is the kth term of the sum, $[2(k + 1) - 1]$ is the $(k + 1)$st term.

$$\begin{aligned}
S_{k+1} &= 1 + 3 + 5 + 7 + \cdots + (2k - 1) + [2(k + 1) - 1] \\
&= [1 + 3 + 5 + 7 + \cdots + (2k - 1)] + [2k + 1] \\
&= S_k + (2k + 1) \\
&= k^2 + 2k + 1 \qquad S_k = k^2 \text{ from induction hypothesis.} \\
&= (k + 1)^2
\end{aligned}$$

Combining the results of parts (1) and (2), you can conclude by mathematical induction that the formula is valid for *all* positive integers *n*.

A well-known illustration used to explain why mathematical induction works is the unending line of dominoes shown in the figure at the left. If the line actually contains infinitely many dominoes, then it is clear that you could not knock the entire line down by knocking down only *one domino* at a time. However, suppose it were true that each domino would knock down the next one as it fell. Then you could knock them all down simply by pushing the first one and starting a chain reaction.

Mathematical induction works in the same way. If the truth of P_k implies the truth of P_{k+1} and if P_1 is true, then the chain reaction proceeds as follows:

P_1 implies P_2

P_2 implies P_3

P_3 implies P_4, and so on.

In the next example you will see the proof of a formula that is often used in calculus.

EXAMPLE 3   *Using Mathematical Induction*

Use mathematical induction to prove the formula for the sum of the first *n* squares.

$$S_n = 1^2 + 2^2 + 3^2 + 4^2 + \cdots + n^2 = \frac{n(n+1)(2n+1)}{6}$$

Solution   1. The formula is valid when $n = 1$ because

$$S_1 = 1^2 = \frac{1(1+1)[2(1)+1]}{6} = \frac{1(2)(3)}{6} = 1.$$

2. Assuming that the formula is true for *k*,

$$S_k = 1^2 + 2^2 + 3^2 + 4^2 + \cdots + k^2 = \frac{k(k+1)(2k+1)}{6},$$

you must show that it is true for $k + 1$,

$$S_{k+1} = \frac{(k+1)[(k+1)+1][2(k+1)+1]}{6} = \frac{(k+1)(k+2)(2k+3)}{6}.$$

To do this, write S_{k+1} as the sum of S_k and the $(k+1)$st term, $(k+1)^2$, as follows.

$$
\begin{aligned}
S_{k+1} &= 1^2 + 2^2 + 3^2 + 4^2 + \cdots + k^2 + (k+1)^2 \\
&= [1^2 + 2^2 + 3^2 + 4^2 + \cdots + k^2] + (k+1)^2 \\
&= \frac{k(k+1)(2k+1)}{6} + (k+1)^2 \qquad \text{Induction hypothesis} \\
&= \frac{k(k+1)(2k+1) + 6(k+1)^2}{6}
\end{aligned}
$$

$$= \frac{(k + 1)[k(2k + 1) + 6(k + 1)]}{6}$$

$$= \frac{(k + 1)[2k^2 + 7k + 6]}{6}$$

$$= \frac{(k + 1)(k + 2)(2k + 3)}{6}$$

Combining the results of parts (1) and (2), you can conclude by mathematical induction that the formula is valid for *all* positive integers *n*.

Many of the proofs in linear algebra use mathematical induction. Here is an example from Chapter 2.

EXAMPLE 4   *Using Mathematical Induction in Linear Algebra*

If $A_1, A_2, \ldots, A_n$ are invertible matrices, prove the generalization of Theorem 2.9.

$$(A_1 A_2 A_3 \cdots A_n)^{-1} = A_n^{-1} \cdots A_3^{-1} A_2^{-1} A_1^{-1}$$

Solution   1. The formula is valid trivially when $n = 1$ because $A_1^{-1} = A_1^{-1}$.

2. Assuming that the formula is valid for k,

$$(A_1 A_2 A_3 \cdots A_k)^{-1} = A_k^{-1} \cdots A_3^{-1} A_2^{-1} A_1^{-1},$$

you must show that it is valid for $k + 1$. To do this, use Theorem 2.9, which says that the inverse of a product of two invertible matrices is the product of their inverses in reverse order.

$$
\begin{aligned}
(A_1 A_2 A_3 \cdots A_k A_{k+1})^{-1} &= [(A_1 A_2 A_3 \cdots A_k) A_{k+1}]^{-1} \\
&= A_{k+1}^{-1} (A_1 A_2 A_3 \cdots A_k)^{-1} & \text{\textbf{Theorem 2.9}} \\
&= A_{k+1}^{-1} (A_k^{-1} \cdots A_3^{-1} A_2^{-1} A_1^{-1}) & \text{\textbf{Induction hypothesis}} \\
&= A_{k+1}^{-1} A_k^{-1} \cdots A_3^{-1} A_2^{-1} A_1^{-1}
\end{aligned}
$$

Hence, the formula is valid for $k + 1$.

Combining the results of parts (1) and (2), you can conclude by mathematical induction that the formula is valid for *all* positive integers *n*.

Proof by Contradiction

A second basic strategy for writing a proof is *proof by contradiction*. In mathematical logic, proof by contradiction is described by the following equivalence.

p implies *q*   if and only if   not *q* implies not *p*.

Thus, one way to prove that *q* is a true statement is to assume that *q* is not true. If this leads you to a statement that you know is false, then you have proved that *q* must be true.

The next example shows how to use proof by contradiction to prove that $\sqrt{2}$ is irrational.

EXAMPLE 5   *Using Proof by Contradiction*

Prove that $\sqrt{2}$ is an irrational number.

Solution   Begin by assuming that $\sqrt{2}$ is not an irrational number. Thus, $\sqrt{2}$ is rational and can be written as the quotient of two integers a and b that have no common factor.

$$\sqrt{2} = \frac{a}{b} \qquad \text{Assume that } \sqrt{2} \text{ is a rational number.}$$

$$2 = \frac{a^2}{b^2} \qquad \text{Square each side.}$$

$$2b^2 = a^2 \qquad \text{Multiply each side by } b^2.$$

This implies that 2 is a factor of a^2. Therefore, 2 is also a factor of a, and a can be written as $2c$.

$$2b^2 = (2c)^2 \qquad \text{Substitute } 2c \text{ for } a.$$

$$2b^2 = 4c^2 \qquad \text{Simplify.}$$

$$b^2 = 2c^2 \qquad \text{Divide each side by 2.}$$

So, 2 is a factor of b^2, and is also a factor of b. Thus, 2 is a factor of both a and b. But this is impossible because a and b have no common factor. Therefore, it must be impossible that $\sqrt{2}$ is a rational number. So, you can conclude that $\sqrt{2}$ must be an irrational number.

Proof by contradiction is not a new technique. The proof in the next example was given by Euclid around 300 B.C.

EXAMPLE 6   *Using Proof by Contradiction*

A positive integer greater than 1 is a *prime* if its only positive factors are 1 and itself. Prove that there are infinitely many prime numbers.

Solution   Assume that there are only finitely many primes, $p_1, p_2, \ldots, p_n$. Consider the number $N = p_1 p_2 \cdots p_n + 1$. This number is either prime or composite. If it is composite, then it can be factored as the product of primes. But, none of the primes $(p_1, p_2, \ldots, p_n)$ divides evenly into N. Therefore, N is itself a prime. So, you have found a new prime number, which contradicts the assumption that there are only n prime numbers.

It follows that no list of prime numbers is complete. Hence, there are infinitely many prime numbers.

You can use proof by contradiction to prove many theorems in linear algebra. On the following page is an example from Chapter 3.

EXAMPLE 7   *Using Proof by Contradiction in Linear Algebra*

Let A and B be $n \times n$ matrices such that AB is singular. Prove that either A or B is singular.

Solution   Assume that neither A nor B is singular. Because you know that a matrix is singular if and only if its determinant is zero, $\det(A)$ and $\det(B)$ are both nonzero real numbers. By Theorem 3.5, $\det(AB) = \det(A)\det(B)$. Thus, $\det(AB)$ is not zero because it is a product of two nonzero real numbers. But this contradicts that AB is a singular matrix. So, you can conclude that the assumption was wrong and that either A or B must be singular.

Using Counterexamples

Often you can disprove a statement using a *counterexample*. For instance, when Euler disproved Fermat's conjecture about prime numbers of the form $F_n = 2^{2^n} + 1$, $n = 0$, $1, 2, \ldots$, he used the counterexample $F_5 = 4{,}294{,}967{,}297$, which is not prime.

EXAMPLE 8   *Using a Counterexample*

Use a counterexample to show that the statement is false.
   Every odd number is a prime.

Solution   Certainly, you can list many odd numbers that are prime $(3, 5, 7, 11)$, but the statement above is not true, because 9 and 15 are odd but they are not prime numbers. The numbers 9 and 15 are counterexamples.

Counterexamples can be used to disprove statements in linear algebra, as shown in the next example.

EXAMPLE 9   *Using a Counterexample*

Use a counterexample to show that the statement is false.
   If A and B are square singular matrices of order n, then $A + B$ is a singular matrix of order n.

Solution   Let $A = \begin{bmatrix} 1 & 0 \\ 0 & 0 \end{bmatrix}$ and $B = \begin{bmatrix} 0 & 0 \\ 0 & 1 \end{bmatrix}$. Both A and B are singular of order 2, but

$$A + B = \begin{bmatrix} 1 & 0 \\ 0 & 1 \end{bmatrix}$$

is the identity matrix of order 2, which is not singular.

Exercises

In Exercises 1–4, use mathematical induction to prove that the formula is valid for all positive integers n.

1. $1 + 2 + 3 + \cdots + n = \dfrac{n(n + 1)}{2}$

2. $1^3 + 2^3 + 3^3 + \cdots + n^3 = \dfrac{n^2(n + 1)^2}{4}$

3. $3 + 7 + 11 + 15 + \cdots + (4n - 1) = n(2n + 1)$

4. $\left(1 + \dfrac{1}{1}\right)\left(1 + \dfrac{1}{2}\right)\left(1 + \dfrac{1}{3}\right) \cdots \left(1 + \dfrac{1}{n}\right) = n + 1$

In Exercises 5 and 6, conjecture a formula for the sum of the first n terms of the sequence. Then use mathematical induction to prove that the formula is valid.

5. $2^1,\ 2^2,\ 2^3,\ 2^4, \ldots$

6. $\dfrac{1}{1 \cdot 2},\ \dfrac{1}{2 \cdot 3},\ \dfrac{1}{3 \cdot 4},\ \dfrac{1}{4 \cdot 5}, \ldots$

In Exercises 7 and 8, use mathematical induction to prove the inequality for the indicated integer values of n.

7. $n! > 2^n,\quad n \geq 4$

8. $\dfrac{1}{\sqrt{1}} + \dfrac{1}{\sqrt{2}} + \dfrac{1}{\sqrt{3}} + \cdots + \dfrac{1}{\sqrt{n}} > \sqrt{n},\quad n \geq 2$

9. Prove that for all integers $n > 0$,

$$a^0 + a^1 + a^2 + a^3 + \cdots + a^n = \dfrac{1 - a^{n+1}}{1 - a}, \quad a \neq 1.$$

10. (From Chapter 2) Use mathematical induction to prove that

$$(A_1 A_2 A_3 \cdots A_n)^T = A_n^T \cdots A_3^T A_2^T A_1^T,$$

assuming that $A_1,\ A_2,\ A_3, \ldots, A_n$ are matrices with sizes such that the multiplications are defined.

11. (From Chapter 3) Use mathematical induction to prove that

$$|A_1 A_2 A_3 \cdots A_n| = |A_1||A_2||A_3| \cdots |A_n|,$$

where $A_1,\ A_2,\ A_3, \ldots, A_n$ are square matrices of the same size.

In Exercises 12–17, use proof by contradiction to prove the statement.

12. If p is an integer and p^2 is odd, then p is odd. (*Hint:* An odd number can be written as $2n + 1$, where n is an integer.)

13. If a and b are real numbers and $a \leq b$, then $a + c \leq b + c$.

14. If a, b, and c are real numbers such that $ac \geq bc$ and $c > 0$, then $a \geq b$.

15. If a and b are real numbers and $1 < a < b$, then $\dfrac{1}{a} > \dfrac{1}{b}$.

16. If a and b are real numbers and $(a + b)^2 = a^2 + b^2$, then $a = 0$ or $b = 0$ or $a = b = 0$.

17. If a and b are real numbers and $0 < a < 1$, then $a^2 < a$.

18. Use proof by contradiction to prove that the sum of a rational number and an irrational number is irrational.

19. (From Chapter 4) Use proof by contradiction to prove that in a given vector space, the zero vector is unique.

20. (From Chapter 4) Let $S = \{\mathbf{u}, \mathbf{v}\}$ be a linearly independent set. Use proof by contradiction to prove that the set $\{\mathbf{u} - \mathbf{v}, \mathbf{u} + \mathbf{v}\}$ is linearly independent.

In Exercises 21–27, use a counterexample to show that the statement is false.

21. If a and b are real numbers and $a < b$, then $a^2 < b^2$.

22. The product of two irrational numbers is irrational.

23. If a and b are real numbers such that $a \neq 0$ and $b \neq 0$, then $(a + b)^3 = a^3 + b^3$.

24. If f is a polynomial function and $f(a) = f(b)$, then $a = b$.

25. If f and g are differentiable functions and $y = f(x)g(x)$, then $\dfrac{dy}{dx} = f'(x)g'(x)$.

26. (From Chapter 2) If A, B, and C are matrices and $AC = BC$, then $A = B$.

27. (From Chapter 3) If A is a matrix, then $\det(A^{-1}) = \dfrac{1}{\det A}$.

APPENDIX B

Computer Algebra Systems and Graphing Calculators

To help you become acquainted with the rudiments of using a Computer Algebra System (CAS) or graphing calculator, the following pages contain certain introductory instructions for the following popular systems: MATLAB, Maple, *Mathematica*, and Derive, as well as information about various graphing calculator models.

Introduction to MATLAB

The purpose of this introduction is to illustrate some basic MATLAB commands for linear algebra. It is suggested that you complete the following series of simple examples while actually working at the computer terminal. Once you have signed on to MATLAB, you should type in your commands to the right of the prompt >>.

1. Enter the matrix

$$a = \begin{bmatrix} 1 & 2 & 3 \\ 4 & 5 & 6 \\ 7 & 8 & 0 \end{bmatrix}$$

by typing **a = [1 2 3; 4 5 6; 7 8 0]** and hitting the **return** or **enter** key. Be sure to separate the entries in each row by spaces and to terminate the rows with semicolons. MATLAB should return the following matrix.

$$a = \begin{matrix} 1 & 2 & 3 \\ 4 & 5 & 6 \\ 7 & 8 & 0 \end{matrix}$$

2. Perform the following elementary commands on the matrix *a*. You should obtain the indicated results.

(a) **inv(a)**

$$\text{ans} = \begin{matrix} -1.7778 & 0.8889 & -0.1111 \\ 1.5556 & -0.7778 & 0.2222 \\ -0.1111 & 0.2222 & -0.1111 \end{matrix}$$

(b) **a′**

$$\text{ans} = \begin{matrix} 1 & 4 & 7 \\ 2 & 5 & 8 \\ 3 & 6 & 0 \end{matrix}$$

(c) **a*inv(a)**

$$\text{ans} = \begin{matrix} 1.0000 & 0 & -0.0000 \\ -0.0000 & 1.0000 & 0 \\ 0.0000 & -0.0000 & 1.0000 \end{matrix}$$

(d) **2*a**

$$\text{ans} = \begin{matrix} 2 & 4 & 6 \\ 8 & 10 & 12 \\ 14 & 16 & 0 \end{matrix}$$

(e) **rref(a)**

$$\text{ans} = \begin{matrix} 1 & 0 & 0 \\ 0 & 1 & 0 \\ 0 & 0 & 1 \end{matrix}$$

(f) **det(a)**

$$\text{ans} = 27$$

(g) **rank(a)**

$$\text{ans} = 3$$

(h) **diag(a)**

$$\text{ans} = \begin{matrix} 1 \\ 5 \\ 0 \end{matrix}$$

3. Enter the matrix **b = [7; 16; 7].** You can solve the system of linear equations $a\mathbf{x} = \mathbf{b}$ in different ways:

(a) **a\b**

$$\text{ans} = \begin{matrix} 1 \\ 0 \\ 2 \end{matrix}$$

(b) **c = [a b]** and then **rref(c)**

$$c = \begin{matrix} 1 & 2 & 3 & 7 \\ 4 & 5 & 6 & 16 \\ 7 & 8 & 0 & 7 \end{matrix}$$

$$>> \textbf{rref(c)}$$

$$\text{ans} = \begin{matrix} 1 & 0 & 0 & 1 \\ 0 & 1 & 0 & 0 \\ 0 & 0 & 1 & 2 \end{matrix}$$

4. Enter the matrices **A = [1 2; 3 4]** and **B = [2 0; −3 5]** and perform the following matrix operations. Notice that MATLAB distinguishes between upper- and lowercase letters; using the uppercase letter B to label the matrix above is not the same as using the lowercase letter b.

(a) **A + B**

$$\text{ans} = \begin{matrix} 3 & 2 \\ 0 & 9 \end{matrix}$$

(b) **A − B**

$$\text{ans} = \begin{matrix} -1 & 2 \\ 6 & -1 \end{matrix}$$

(c) **A * B**

$$\text{ans} = \begin{matrix} -4 & 10 \\ -6 & 20 \end{matrix}$$

5. What do the following commands do?

(a) **zeros(3, 5)**
(b) **eye(5)**
(c) **rand(6)**
(d) **hilb(6)**

6. Describe the effects of the following commands on the matrix **a = [1 2 3; 4 5 6; 7 8 0].**

(a) **a(2, 3)**
(b) **a(2, :)**
(c) **a(:, 3)**

7. You can obtain more information about MATLAB by typing **help.**
8. You can exit MATLAB by typing **exit.**

Introduction to Maple

The purpose of this introduction is to illustrate some basic Maple commands for linear algebra. It is suggested that you complete the following series of simple examples while actually working at the computer terminal. Once you have signed on to Maple, you should type in your commands to the right of the prompt >.

Begin by loading the linear algebra package that comes with Maple by typing **with(linalg);** and hitting the **return** or **enter** key. Notice that all commands in Maple terminate with a semicolon ;.

1. Enter the matrix

$$a = \begin{bmatrix} 1 & 2 & 3 \\ 4 & 5 & 6 \\ 7 & 8 & 0 \end{bmatrix}$$

by typing **a: = matrix ([[1, 2, 3], [4, 5, 6], [7, 8, 0]]);** and hitting the **return** or **enter** key. Be sure to separate the entries in each row by commas and place each row of the matrix within brackets. Maple should return the following matrix.

$$a: = \begin{bmatrix} 1 & 2 & 3 \\ 4 & 5 & 6 \\ 7 & 8 & 0 \end{bmatrix}$$

2. Perform the following elementary commands on the matrix a. You should obtain the indicated results.

(a) **inverse(a);**

$$\begin{bmatrix} -\frac{16}{9} & \frac{8}{9} & -\frac{1}{9} \\ \frac{14}{9} & -\frac{7}{9} & \frac{2}{9} \\ -\frac{1}{9} & \frac{2}{9} & -\frac{1}{9} \end{bmatrix}$$

(b) **transpose(a);**

$$\begin{bmatrix} 1 & 4 & 7 \\ 2 & 5 & 8 \\ 3 & 6 & 0 \end{bmatrix}$$

(c) **multiply(a, inverse(a));**

$$\begin{bmatrix} 1 & 0 & 0 \\ 0 & 1 & 0 \\ 0 & 0 & 1 \end{bmatrix}$$

(d) **scalarmul(a, 2);**

$$\begin{bmatrix} 2 & 4 & 6 \\ 8 & 10 & 12 \\ 14 & 16 & 0 \end{bmatrix}$$

(e) **rref(a);**

$$\begin{bmatrix} 1 & 0 & 0 \\ 0 & 1 & 0 \\ 0 & 0 & 1 \end{bmatrix}$$

(f) **det(a);**

27

(g) **rank(a);**

3

(h) **diag(a, inverse(a));**

$$\begin{bmatrix} 1 & 2 & 3 & 0 & 0 & 0 \\ 4 & 5 & 6 & 0 & 0 & 0 \\ 7 & 8 & 0 & 0 & 0 & 0 \\ 0 & 0 & 0 & -\frac{16}{9} & \frac{8}{9} & -\frac{1}{9} \\ 0 & 0 & 0 & \frac{14}{9} & -\frac{7}{9} & \frac{2}{9} \\ 0 & 0 & 0 & -\frac{1}{9} & \frac{2}{9} & -\frac{1}{9} \end{bmatrix}$$

3. Enter the column matrix **b: = transpose(matrix([[7, 16, 7]]));.** You can solve the system of linear equations $a\mathbf{x} = \mathbf{b}$ in different ways:

   (a) **linsolve(a, b);**

   $$\begin{bmatrix} 1 \\ 0 \\ 2 \end{bmatrix}$$

   (b) **c: = augment(a, b);** and then **rref(c);**

   $$c: = \begin{bmatrix} 1 & 2 & 3 & 7 \\ 4 & 5 & 6 & 16 \\ 7 & 8 & 0 & 7 \end{bmatrix}$$

   $$\begin{bmatrix} 1 & 0 & 0 & 1 \\ 0 & 1 & 0 & 0 \\ 0 & 0 & 1 & 2 \end{bmatrix}$$

4. Enter the matrices **A: = matrix([[1, 2], [3, 4]]);** and **B: = matrix([[2, 0], [−3, 5]]);** and perform the following matrix operations. Notice that Maple distinguishes between upper- and lowercase letters; using the uppercase letter B to label the matrix above is not the same as using the lowercase letter b.

   (a) **matadd(A, B);**

   $$\begin{bmatrix} 3 & 2 \\ 0 & 9 \end{bmatrix}$$

   (b) **matadd(A, B, 1, −1);**

   $$\begin{bmatrix} -1 & 2 \\ 6 & -1 \end{bmatrix}$$

   (c) **multiply(A, B);**

   $$\begin{bmatrix} -4 & 10 \\ -6 & 20 \end{bmatrix}$$

5. What do the following commands do?
   (a) **matrix(3, 5, 0);**
   (b) **band([1], 5);**
   (c) **randmatrix(4, 6);**
   (d) **hilbert(6);**

6. Describe the effects of the following commands on the matrix **a: = matrix([[1, 2, 3], [4, 5, 6], [7, 8, 0]]);.**
   (a) **submatrix(a, 2..2, 3..3);**
   (b) **submatrix(a, 2..2, 1..3);**
   (c) **submatrix(a, 1..3, 3..3);**

7. You can obtain more information about a particular topic, such as eigenvalues, by typing **?eigenvalues.**

8. You can exit Maple by choosing **Exit** from the file menu.

Introduction to Mathematica

The purpose of this introduction is to illustrate some basic *Mathematica* commands for linear algebra. It is suggested that you complete the following series of simple examples while actually working at the computer terminal. Once you have signed on to *Mathematica*, you should type in your commands to the right of the prompt.

1. Enter the matrix

$$a = \begin{bmatrix} 1 & 2 & 3 \\ 4 & 5 & 6 \\ 7 & 8 & 0 \end{bmatrix}$$

by typing **a = {{1, 2, 3}, {4, 5, 6}, {7, 8, 0}}** and evaluating the expression (cell). Be sure to separate the entries in each row by commas and place each row of the matrix within braces. *Mathematica* should return the following matrix. You can obtain a more natural matrix display by typing **%//MatrixForm.**

> *Out[1]* =
>
> $$\{\{1, 2, 3\}, \{4, 5, 6\}, \{7, 8, 0\}\}$$
>
> *In[2]* =
>
> %//MatrixForm
>
> *Out[2]//MatrixForm* =
>
> $$\begin{pmatrix} 1 & 2 & 3 \\ 4 & 5 & 6 \\ 7 & 8 & 0 \end{pmatrix}$$

2. Perform the following elementary commands on the matrix *a*. You should obtain the indicated results.

   (a) **Inverse[a] //MatrixForm**

   > *Out[3]//MatrixForm* =
   >
   > $$\begin{pmatrix} -\frac{16}{9} & \frac{8}{9} & -\frac{1}{9} \\ \frac{14}{9} & -\frac{7}{9} & \frac{2}{9} \\ -\frac{1}{9} & \frac{2}{9} & -\frac{1}{9} \end{pmatrix}$$

   (b) **Transpose[a]**

   > *Out[4]* =
   >
   > $$\{\{1, 4, 7\}, \{2, 5, 8\}, \{3, 6, 0\}\}$$

   (c) **a.Inverse[a]**

   > *Out[5]* =
   >
   > $$\{\{1, 0, 0\}, \{0, 1, 0\}, \{0, 0, 1\}\}$$

(d) **2∗a**

$Out[6] =$

$$\{\{2, 4, 6\}, \{8, 10, 12\}, \{14, 16, 0\}\}$$

(e) **RowReduce[a]**

$Out[7] =$

$$\{\{1, 0, 0\}, \{0, 1, 0\}, \{0, 0, 1\}\}$$

(f) **Det[a]**

$Out[8] =$

$$27$$

3. Enter the matrix **b** = {**7, 16, 7**}. You can solve the system of linear equations $a\mathbf{x} = \mathbf{b}$ in different ways:

(a) **LinearSolve[a, b]**

$Out[10] =$

$$\{1, 0, 2\}$$

(b) **c = Table[Append[a[[i]], b[[i]]], {i, 1, 3}]** and then **RowReduce[c]**

$Out[11] =$

$$\{\{1, 2, 3, 7\}, \{4, 5, 6, 16\}, \{7, 8, 0, 7\}\}$$

$In[12] =$

RowReduce[c]//MatrixForm

$Out[12]//MatrixForm=$

$$\begin{pmatrix} 1 & 0 & 0 & 1 \\ 0 & 1 & 0 & 0 \\ 0 & 0 & 1 & 2 \end{pmatrix}$$

4. Enter the matrices **A** = {{**1, 2**}, {**3, 4**}} and **B** = {{**2, 0**}, {**−3, 5**}} and perform the following matrix operations. Notice that *Mathematica* distinguishes between upper- and lowercase letters; using the uppercase letter B to label the matrix above is not the same as using the lowercase letter b.

(a) **A + B**

$Out[15] =$

$$\{\{3, 2\}, \{0, 9\}\}$$

(b) **A − B**

$Out[16] =$

$$\{\{-1, 2\}, \{6, -1\}\}$$

(c) **A.B** (matrix multiplication is not A ∗ B)

$Out[17] =$

$$\{\{-4, 10\}, \{-6, 20\}\}$$

5. What do the following commands do?
   (a) **Table[0, {i, 1, 3}, {j, 1, 5}]**
   (b) **IdentityMatrix[5]**
   (c) **Table[Random[Integer,{− 100, 100}], {4}, {6}]//MatrixForm**
   (d) **Table[1/(i + j − 1), {i, 1, 6}, {j, 1, 6}]**
6. Describe the effects of the following commands on the matrix **a = {{1, 2, 3}, {4, 5, 6}, {7, 8, 0}}**.
   (a) **a[[2, 3]]**
   (b) **a[[2]]**
   (c) **Transpose[a] [[3]]**
7. You can obtain more information about a particular topic, such as determinants, by typing **?Det.**
8. You can exit *Mathematica* by choosing **Exit** from the file menu.

Introduction to Derive

The purpose of this introduction is to illustrate some basic Derive commands for linear algebra. It is suggested that you complete the following series of simple examples while actually working at the computer terminal.

1. Enter the matrix

$$a = \begin{bmatrix} 1 & 2 & 3 \\ 4 & 5 & 6 \\ 7 & 8 & 0 \end{bmatrix}$$

by using the **Author/Expression** feature—i.e., **a: = [[1, 2, 3], [4, 5, 6], [7, 8, 0]]** and hitting the **return** or **enter** key. Be sure to separate the entries in each row by commas and place each row of the matrix within brackets. Derive should return the following matrix.

$$a: = \begin{bmatrix} 1 & 2 & 3 \\ 4 & 5 & 6 \\ 7 & 8 & 0 \end{bmatrix}$$

2. Perform the following elementary operations on the matrix *a* by typing the given expression and **Simplifying** it. You should obtain the indicated results.
   (a) **a^{-1}**

$$\begin{bmatrix} -\frac{16}{9} & \frac{8}{9} & -\frac{1}{9} \\ \frac{14}{9} & -\frac{7}{9} & \frac{2}{9} \\ -\frac{1}{9} & \frac{2}{9} & -\frac{1}{9} \end{bmatrix}$$

(b) **a`**

$$\begin{bmatrix} 1 & 4 & 7 \\ 2 & 5 & 8 \\ 3 & 6 & 0 \end{bmatrix}$$

(c) **a ⋆ a⁻¹**

$$\begin{bmatrix} 1 & 0 & 0 \\ 0 & 1 & 0 \\ 0 & 0 & 1 \end{bmatrix}$$

(d) **a ⋆ 2**

$$\begin{bmatrix} 2 & 4 & 6 \\ 8 & 10 & 12 \\ 14 & 16 & 0 \end{bmatrix}$$

(e) **Row_Reduce(a)**

$$\begin{bmatrix} 1 & 0 & 0 \\ 0 & 1 & 0 \\ 0 & 0 & 1 \end{bmatrix}$$

(f) **det(a)**

27

3. Enter the column matrix **b: =** $[[7], [16], [7]]$. You can solve the system of linear equations $a\mathbf{x} = \mathbf{b}$ in several ways:

(a) You can row reduce the augmented matrix $[a \vdots b]$.
   Row_Reduce(a, b)

$$\begin{bmatrix} 1 & 0 & 0 & 1 \\ 0 & 1 & 0 & 0 \\ 0 & 0 & 1 & 2 \end{bmatrix}$$

(b) You can **Simplify** the expression $\begin{bmatrix} x \\ y \\ z \end{bmatrix} = a^{-1} \star b.$

$$\begin{bmatrix} x = 1 \\ y = 0 \\ z = 2 \end{bmatrix}$$

(c) You can **Solve** the equation $a \begin{bmatrix} x \\ y \\ z \end{bmatrix} = b.$

$$[x = 1, \ y = 0, \ z = 2]$$

4. Enter the matrices **"A"**: = [[1, 2], [3, 4]] and **"B"**: = [[2, 0], [−3, 5]] and perform the following matrix operations.
   (a) **"A"** + **"B"**

   $$\begin{bmatrix} 3 & 2 \\ 0 & 9 \end{bmatrix}$$

   (b) **"A"** − **"B"**

   $$\begin{bmatrix} -1 & 2 \\ 6 & -1 \end{bmatrix}$$

   (c) **"A"** ⋆ **"B"**

   $$\begin{bmatrix} -4 & 10 \\ -6 & 20 \end{bmatrix}$$

5. What do the following commands do?
   (a) **Vector(Vector(0, j, 1, 5), k, 1, 3)**
   (b) **Identity_Matrix(5)**
   (c) **Vector(Vector(Random(100), j, 1, 6), k, 1, 4)**
   (d) **Vector(Vector(1/(j + k − 1), j, 1, 6), k, 1, 6)**

6. Describe the effects of the following commands on the matrix **a**: = [[1, 2, 3], [4, 5, 6], [7, 8, 0]].
   (a) **Element(Element(a, 2), 3)**
   (b) **Element(a, 2)**
   (c) **Element(a`, 3)`**

7. You can obtain more information about a particular topic by choosing **Contents** from the help menu.

8. You can exit Derive by choosing **Exit** from the file menu.

Graphing Calculators

The following chart lists various graphing calculators[*] and their built-in matrix capabilities. For example, the first row indicates that all the calculators can perform matrix addition.

Matrix operation	TI–92	TI–89	TI–86	TI–83 Plus	TI–83
Addition	*x*	*x*	*x*	*x*	*x*
Scalar multiplication	*x*	*x*	*x*	*x*	*x*
Multiplication	*x*	*x*	*x*	*x*	*x*
Elementary row operations	*x*	*x*	*x*	*x*	*x*
Reduced row-echelon form	*x*	*x*	*x*	*x*	*x*
Determinant	*x*	*x*	*x*	*x*	*x*
Inverse	*x*	*x*	*x*	*x*	*x*
Transpose	*x*	*x*	*x*	*x*	*x*

[*]For other calculator models, see your user's manual.

Matrix operation	TI–92	TI–89	TI–86	TI–83 Plus	TI–83
Norm	x	x	x		
Power	x	x	x	x	x
Eigenvalues	x	x	x		
LU decomposition	x	x	x		
Identity matrix	x	x	x	x	x
Random matrix	x	x	x	x	x
Concatenate matrices	x	x	x	x	x
Condition number			x		
Complex matrices	x	x	x		
Vectors	x	x	x		

Technology Pitfalls

Many matrix computations in linear algebra can be performed by modern graphing calculators and computer software. However, you should be aware that these devices are not perfect because of roundoff error—that is, the calculator or computer must approximate fractions by decimals either by rounding off the last digit or by truncating.

For example, evaluate $\frac{1}{6}$ using your calculator. On the TI-86 you will get 0.166666666667, which is not correct, because the calculator has rounded up the last digit. To analyze this more closely, evaluate $\left[\left(\frac{1}{6}\right) \cdot 3\right] - 0.5$. Do you get the expected answer of 0? Normally this kind of rounding will not affect your answers to problems in linear algebra, but it can happen. Here are a couple of simple examples.

1. Calculate the determinant of the matrix $A = \begin{bmatrix} 3 & 11 \\ 2 & 6 \end{bmatrix}$ on the TI-86. Then calculate the greatest integer of the determinant, int det A. Again, roundoff error has produced an error.

2. The Hilbert matrix of order 6 is

$$H = \begin{bmatrix} 1 & \frac{1}{2} & \frac{1}{3} & \frac{1}{4} & \frac{1}{5} & \frac{1}{6} \\ \frac{1}{2} & \frac{1}{3} & \frac{1}{4} & \frac{1}{5} & \frac{1}{6} & \frac{1}{7} \\ \frac{1}{3} & \frac{1}{4} & \frac{1}{5} & \frac{1}{6} & \frac{1}{7} & \frac{1}{8} \\ \frac{1}{4} & \frac{1}{5} & \frac{1}{6} & \frac{1}{7} & \frac{1}{8} & \frac{1}{9} \\ \frac{1}{5} & \frac{1}{6} & \frac{1}{7} & \frac{1}{8} & \frac{1}{9} & \frac{1}{10} \\ \frac{1}{6} & \frac{1}{7} & \frac{1}{8} & \frac{1}{9} & \frac{1}{10} & \frac{1}{11} \end{bmatrix}.$$

This matrix is notorious for causing poor results in matrix computations. Try calculating $\text{inv}(H) \cdot H$ to see if you obtain the identity matrix.

You might find it interesting to discover other examples for which your calculator gives erroneous results. In general, you should always be aware that such errors are possible, and accept your calculator output with some skepticism.

ANSWER KEY

Chapter 1

SECTION 1.1   (*page 11*)

1. Linear   **3.** Not linear   **5.** Not linear

7. $x = 2t$
$y = t$

9. $x = 1 - s - t$
$y = s$
$z = t$

11. $x_1 = 5$
$x_2 = 3$

13. $x = \frac{3}{2}$
$y = \frac{3}{2}$
$z = 0$

15. $x_1 = -t$
$x_2 = 2t$
$x_3 = t$

17. $x = 2$
$y = 0$

19. No solution

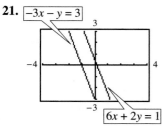

21. $\boxed{-3x - y = 3}$ $\boxed{6x + 2y = 1}$

Inconsistent

23. $\boxed{2x - 8y = 3}$ $\boxed{\frac{1}{2}x + y = 0}$

$x = \frac{1}{2}$
$y = -\frac{1}{4}$

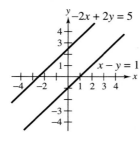

25. $\boxed{4x - 8y = 9}$ $\boxed{0.8x - 1.6y = 1.8}$

$x = \frac{9}{4} + 2t$
$y = t$

27. $x_1 = -1$
$x_2 = -1$

29. $u = 40$
$v = 40$

33. $x = 7$
$y = 1$

35. $x_1 = 8$
$x_2 = 7$

37. $x = 1$
$y = 2$
$z = 3$

39. No solution

41. $x_1 = \frac{5}{2} - \frac{1}{2}t$
$x_2 = 4t - 1$
$x_3 = t$

43. No solution

45. $x = 1$
$y = 0$
$z = 3$
$w = 2$

47. $x_1 = -15$
$x_2 = 40$
$x_3 = 45$
$x_4 = -75$

49. $x = -1.2$
$y = -0.6$
$z = 2.4$

51. $x_1 = \frac{1}{5}$
$x_2 = -\frac{4}{5}$
$x_3 = \frac{1}{2}$

53. $x = 6.8813$
$y = -163.3111$
$z = -210.2915$
$w = -59.2913$

55. $x = 0$
$y = 0$
$z = 0$

57. $x = -\frac{3}{5}t$
$y = \frac{4}{5}t$
$z = t$

59. (a) True, you can describe the entire solution set using parametric representation.

$ax + by = c$

Choosing $y = t$ as the free variable, the solution is

$x = \dfrac{c}{a} - \dfrac{b}{a}t$, $y = t$, where t is any real number.

(b) False, for example consider the system

$x_1 + x_2 + x_3 = 1$
$x_1 + x_2 + x_3 = 2$

which is an inconsistent system.

(c) False, a consistent system may have only one solution.

61. $3x_1 - x_2 = 4$
$-3x_1 + x_2 = -4$

(The answer is not unique.)

63. $x = 3$
$y = -4$

65. $x = \cos\theta$
$y = \sin\theta$

67. $k = -2$

69. All $k \neq \pm 1$

71. $k = \frac{8}{3}$

73. $k = 1, -2$

, Three lines intersecting at one point

(b) Three coincident lines

(c) Three lines having no common point

77. Answers will vary. (*Hint:* Choose three different values of x and solve the resulting system of linear equations in the variables a, b, and c.)

79. $x - 4y = -3$
$5x - 6y = 13$

$x - 4y = -3$
$14y = 28$

$x - 4y = -3$
$y = 2$

$x = 5$
$y = 2$

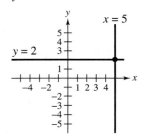

Intersection points are all the same.

SECTION 1.2 (*page 25*)

1. 3×3 **3.** 1×5 **5.** 4×5

7. Reduced row-echelon form

9. Not in row-echelon form

11. Not in row-echelon form

13. $x_1 = 0$
$x_2 = 2$

15. $x_1 = 2$
$x_2 = -1$
$x_3 = -1$

17. $x_1 = -26$
$x_2 = 13$
$x_3 = -7$
$x_4 = 4$

19. $x = 3$
$y = 2$

21. No solution

23. $x = 4$
$y = -2$

25. $x_1 = 4$
$x_2 = -3$
$x_3 = 2$

27. $x_1 = 1 + 2t$
$x_2 = 2 + 3t$
$x_3 = t$

29. $x = 100 + 96t - 3s$
$y = s$
$z = 54 + 52t$
$w = t$

31. $x = 0$
$y = 2 - 4t$
$z = t$

33. $x_1 = 23.5361 + 0.5278t$
$x_2 = 18.5444 + 4.1111t$
$x_3 = 7.4306 + 2.1389t$
$x_4 = t$

35. $x_1 = 2$
$x_2 = -2$
$x_3 = 3$
$x_4 = -5$
$x_5 = 1$

37. $x_1 = 1$
$x_2 = -2$
$x_3 = 6$
$x_4 = -3$
$x_5 = -4$
$x_6 = 3$

39. $x_1 = 0$
$x_2 = -t$
$x_3 = t$

41. $x_1 = -t$
$x_2 = s$
$x_3 = 0$
$x_4 = t$

43. (a) Two equations in two unknowns

(b) All real $k \neq -\frac{4}{3}$

(c) Two equations in three unknowns

(d) All real k

45. (a) $a + b + c = 0$

(b) $a + b + c \neq 0$

(c) Not possible

47. (a) $x = \frac{8}{3} - \frac{5}{6}t$
$y = -\frac{8}{3} + \frac{5}{6}t$
$z = t$

(b) $x = \frac{18}{7} - \frac{11}{14}t$
$y = -\frac{20}{7} + \frac{13}{14}t$
$z = t$

(c) $x = 3 - t$
$y = -3 + t$
$z = t$

(d) Each system has an infinite number of solutions.

49. $\begin{bmatrix} 1 & 0 \\ 0 & 1 \end{bmatrix}$

51. $\begin{bmatrix} 1 & 0 \\ 0 & 1 \end{bmatrix}, \begin{bmatrix} 1 & k \\ 0 & 0 \end{bmatrix}, \begin{bmatrix} 0 & 1 \\ 0 & 0 \end{bmatrix}, \begin{bmatrix} 0 & 0 \\ 0 & 0 \end{bmatrix}$

53. (a) True, in the notation $m \times n$, m is the number of rows of the matrix. Thus, a 6×3 matrix has 6 rows.

(b) True, on page 18, after Example 4, the sentence reads, "It can be shown that every matrix is row-equivalent to a matrix in row-echelon form."

(c) False, consider the row-echelon form.

$$\begin{bmatrix} 1 & 0 & 0 & 0 & 0 \\ 0 & 1 & 0 & 0 & 1 \\ 0 & 0 & 1 & 0 & 2 \\ 0 & 0 & 0 & 1 & 3 \end{bmatrix}$$

which gives the solution $x_1 = 0$, $x_2 = 0$, $x_3 = 2$, and $x_4 = 3$.

(d) True, Theorem 1.1 states that if a homogenous system has fewer equations than variables, then it must have an infinite number of solutions.

55. $ad - bc \neq 0$　　　　**57.** $\lambda = 1, 3$

59. Yes, it is possible:
$$x_1 + x_2 + x_3 = 0$$
$$x_1 + x_2 + x_3 = 1$$

61. $\alpha = \pi/3,\, 5\pi/3$; $\beta = \pi/2$ (*Hint:* Let $u = \cos \alpha$ and $v = \sin \beta$.)

63. The rows have been interchanged.

SECTION 1.3 (*page 37*)

1. (a) $p(x) = 29 - 18x + 3x^2$

(b)

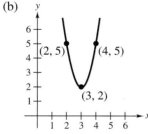

3. (a) $p(x) = 2x$

(b)

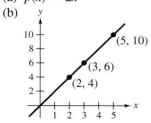

5. (a) $p(z) = 7 + \frac{7}{2}z + \frac{3}{2}z^2$

$p(x) = 7 + \frac{7}{2}(x - 2002) + \frac{3}{2}(x - 2002)^2$

(b)

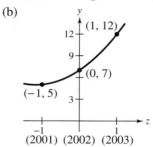

7. y is not a function of x because the x-value of 3 is repeated.

9. $p(x) = 1 + x$

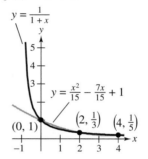

11. $p(x) = -3x + x^3$

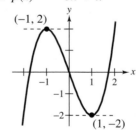

13. $p(z) = 249 + 2.7z + 0.05z^2$ (where $z = x - 1990$)

Year 2005: $p = 300.75$ million

Year 2010: $p = 323$ million

15. (a)
$$\begin{aligned}
a_0 - a_1 + a_2 - a_3 &= 7625.0 \\
a_0 &= 9421.0 \\
a_0 + a_1 + a_2 + a_3 &= 10{,}003.0 \\
a_0 + 2a_1 + 4a_2 + 8a_3 &= 10{,}530.0
\end{aligned}$$

(b) $p(z) = 9421.0 + 995.83z - 607z^2 + 193.1667z^3$

(where $z = x - 2000$)

17. $p(x) = -\dfrac{4}{\pi^2}x^2 + \dfrac{4}{\pi}x$

$\sin\dfrac{\pi}{3} \approx \dfrac{8}{9} \approx 0.889$

(Actual value is $\sqrt{3}/2 \approx 0.866$.)

19. Solve the system:

$\begin{aligned} p(-1) &= a_0 - a_1 + a_2 = 0 \\ p(0) &= a_0 = 0 \\ p(1) &= a_0 + a_1 + a_2 = 0 \\ & a_0 = a_1 = a_2 = 0 \end{aligned}$

21. (a) $\begin{aligned} x_1 &= s \\ x_2 &= t \\ x_3 &= 600 - s \\ x_4 &= s - t \\ x_5 &= 500 - t \\ x_6 &= s \\ x_7 &= t \end{aligned}$ (b) $\begin{aligned} x_1 &= 0 \\ x_2 &= 0 \\ x_3 &= 600 \\ x_4 &= 0 \\ x_5 &= 500 \\ x_6 &= 0 \\ x_7 &= 0 \end{aligned}$ (c) $\begin{aligned} x_1 &= 0 \\ x_2 &= -500 \\ x_3 &= 600 \\ x_4 &= 500 \\ x_5 &= 1000 \\ x_6 &= 0 \\ x_7 &= -500 \end{aligned}$

23. (a) $\begin{aligned} x_1 &= 100 + t \\ x_2 &= -100 + t \\ x_3 &= 200 + t \\ x_4 &= t \end{aligned}$ (b) $\begin{aligned} x_1 &= 100 \\ x_2 &= -100 \\ x_3 &= 200 \\ x_4 &= 0 \end{aligned}$ (c) $\begin{aligned} x_1 &= 200 \\ x_2 &= 0 \\ x_3 &= 300 \\ x_4 &= 100 \end{aligned}$

25. $\begin{aligned} I_1 &= 0 \\ I_2 &= 1 \\ I_3 &= 1 \end{aligned}$ **27.** (a) $\begin{aligned} I_1 &= 1 \\ I_2 &= 2 \\ I_3 &= 1 \end{aligned}$ (b) $\begin{aligned} I_1 &= 0 \\ I_2 &= 1 \\ I_3 &= 1 \end{aligned}$

REVIEW EXERCISES – CHAPTER 1   *(page 40)*

1. $\begin{aligned} x &= -\tfrac{1}{4} + \tfrac{1}{2}s - \tfrac{3}{2}t \\ y &= s \\ z &= t \end{aligned}$

3. Row-echelon form (not reduced)

5. Not in row-echelon form

7. $\begin{aligned} x_1 &= -2t \\ x_2 &= t \\ x_3 &= 0 \end{aligned}$ **9.** $\begin{aligned} x &= \tfrac{1}{2} \\ y &= \tfrac{3}{2} \end{aligned}$ **11.** $\begin{aligned} x &= -12 \\ y &= -8 \end{aligned}$

13. $\begin{aligned} x &= 0 \\ y &= 0 \end{aligned}$ **15.** No solution   **17.** $\begin{aligned} x_1 &= -\tfrac{1}{2} \\ x_2 &= \tfrac{4}{5} \end{aligned}$

19. $\begin{aligned} x &= 0 \\ y &= 0 \end{aligned}$ **21.** $\begin{aligned} x &= 2 \\ y &= -3 \\ z &= 3 \end{aligned}$ **23.** $\begin{aligned} x &= \tfrac{1}{2} \\ y &= -\tfrac{1}{3} \\ z &= 1 \end{aligned}$

25. $\begin{aligned} x &= 4 + 3t \\ y &= 5 + 2t \\ z &= t \end{aligned}$ **27.** $\begin{aligned} x &= \tfrac{3}{2} - 2t \\ y &= 1 + 2t \\ z &= t \end{aligned}$ **29.** $\begin{aligned} x_1 &= 1 \\ x_2 &= 4 \\ x_3 &= -3 \\ x_4 &= -2 \end{aligned}$

31. $\begin{aligned} x_1 &= 0 \\ x_2 &= 0 \\ x_3 &= 0 \end{aligned}$ **33.** $\begin{aligned} x_1 &= -4t \\ x_2 &= -\tfrac{1}{2}t \\ x_3 &= t \end{aligned}$ **35.** $k = \pm 1$

37. (a) $b = 2a$ and $a \neq -3$

(b) $b \neq 2a$

(c) $a = -3$ and $b = -6$

39. The two matrices are row-equivalent because each is row-equivalent to

$\begin{bmatrix} 1 & 0 & 0 \\ 0 & 1 & 0 \\ 0 & 0 & 1 \end{bmatrix}.$

41. $\begin{bmatrix} 1 & 0 & -1 & -2 & \cdots & 2 - n \\ 0 & 1 & 2 & 3 & \cdots & n - 1 \\ 0 & 0 & 0 & 0 & \cdots & 0 \\ \vdots & & & & & \vdots \\ 0 & 0 & 0 & 0 & \cdots & 0 \end{bmatrix}$

43. (a) False. See page 3, following Example 2.

(b) True. See page 5, Example 4(b).

45. (a) $\begin{aligned} 6x + y + 3z &= 62 \\ x + y + z &= 18 \\ x - y &= 0 \end{aligned}$

where x = number of touchdowns,

y = number of extra points,

z = number of field goals

(b) $\begin{aligned} x &= 8 \\ y &= 8 \\ z &= 2 \end{aligned}$

47. (a) $p(x) = 90 - \frac{135}{2}x + \frac{25}{2}x^2$

(b)

49. $p(x) = 50 + \frac{15}{2}x + \frac{5}{2}x^2$

(First year is represented by $x = 0$.)

Fourth-year sales: $p(3) = 95$

51. (a) $a_0 \qquad\qquad\qquad = 80$

$a_0 + 4a_1 + 16a_2 = 68$

$a_0 + 80a_1 + 6400a_2 = 30$

(b) and (c) $a_0 = 80$

$a_1 = -\frac{25}{8}$

$a_2 = \frac{1}{32}$

(e) There is precisely one polynomial function of degree $n - 1$ (or less) that fits n distinct points.

53. (a) $x_1 = 100 - r + t$ (b) $x_1 = 50$

$x_2 = 300 - r + s \qquad\quad x_2 = 250$

$x_3 = r \qquad\qquad\qquad\quad x_3 = 100$

$x_4 = -s + t \qquad\qquad x_4 = 0$

$x_5 = s \qquad\qquad\qquad\quad x_5 = 50$

$x_6 = t \qquad\qquad\qquad\quad x_6 = 50$

Chapter 2

SECTION 2.1 *(page 55)*

1. (a) $\begin{bmatrix} 3 & -2 \\ 1 & 7 \end{bmatrix}$ (b) $\begin{bmatrix} -1 & 0 \\ 3 & -9 \end{bmatrix}$

(c) $\begin{bmatrix} 2 & -2 \\ 4 & -2 \end{bmatrix}$ (d) $\begin{bmatrix} 0 & -1 \\ 5 & -10 \end{bmatrix}$

3. (a) $\begin{bmatrix} 7 & 3 \\ 1 & 9 \\ -2 & 15 \end{bmatrix}$ (b) $\begin{bmatrix} 5 & -5 \\ 3 & -1 \\ -4 & -5 \end{bmatrix}$

(c) $\begin{bmatrix} 12 & -2 \\ 4 & 8 \\ -6 & 10 \end{bmatrix}$ (d) $\begin{bmatrix} 11 & -6 \\ 5 & 3 \\ -7 & 0 \end{bmatrix}$

5. (a) $c_{21} = -6$ (b) $c_{13} = 29$

7. $x = 3$, $y = 2$, $z = 1$

9. (a) $\begin{bmatrix} 0 & 15 \\ 6 & 12 \end{bmatrix}$ (b) $\begin{bmatrix} -2 & 2 \\ 31 & 14 \end{bmatrix}$

11. (a) $\begin{bmatrix} 6 & -21 & 15 \\ 8 & -23 & 19 \\ 4 & 7 & 5 \end{bmatrix}$ (b) $\begin{bmatrix} 9 & 0 & 13 \\ 7 & -2 & 21 \\ 1 & 4 & -19 \end{bmatrix}$

13. (a) Not defined   (b) $\begin{bmatrix} 3 & -4 \\ 10 & 16 \\ 26 & 46 \end{bmatrix}$

15. (a) $\begin{bmatrix} -1 & 19 \\ 4 & -27 \\ 0 & 14 \end{bmatrix}$ (b) Not defined

17. (a) $\begin{bmatrix} 60 & 72 \\ -20 & -24 \\ 10 & 12 \\ 60 & 72 \end{bmatrix}$ (b) Not defined

19. (a) $\begin{bmatrix} 5 & -2 & 5 & 6 & 1 & 8 \\ 0 & 5 & 5 & -1 & -3 & 8 \\ 6 & -8 & -1 & 4 & 7 & -9 \\ 5 & 0 & 9 & 2 & 3 & 3 \\ 12 & 1 & -7 & -8 & 6 & 8 \\ 5 & 2 & 10 & -10 & -6 & -5 \end{bmatrix}$

(b) $\begin{bmatrix} 1 & 8 & -13 & 11 & 3 & 3 \\ 7 & -13 & 1 & 11 & -2 & 3 \\ -3 & -3 & -10 & -2 & 0 & 1 \\ 1 & 7 & 6 & 6 & 2 & -5 \\ 1 & -4 & -7 & 4 & 11 & 3 \\ 1 & -8 & 9 & -2 & -11 & -1 \end{bmatrix}$

(c) $\begin{bmatrix} 2 & -2 & -5 & -2 & -3 & -8 \\ 6 & -27 & 10 & -2 & -21 & 1 \\ 1 & 22 & -34 & 15 & 37 & 7 \\ 4 & -4 & -11 & 1 & 2 & -1 \\ 8 & -4 & -11 & 9 & -10 & 17 \\ -2 & -11 & -30 & 10 & 3 & 9 \end{bmatrix}$

19. –CONTINUED–

(d) $\begin{bmatrix} 8 & 16 & 21 & -21 & -8 & 28 \\ 15 & -19 & 20 & 6 & 6 & -8 \\ -4 & 0 & -12 & -2 & -5 & 11 \\ 16 & -6 & 12 & 3 & 11 & 6 \\ 20 & 9 & 1 & -26 & -6 & 23 \\ -10 & -26 & 3 & 33 & 9 & -33 \end{bmatrix}$

21. $\begin{bmatrix} 1 & 3 \\ 2 & -1 \end{bmatrix}\begin{bmatrix} x_1 \\ x_2 \end{bmatrix} = \begin{bmatrix} -1 \\ 3 \end{bmatrix}$

$\begin{bmatrix} x_1 \\ x_2 \end{bmatrix} = \begin{bmatrix} \frac{8}{7} \\ -\frac{5}{7} \end{bmatrix}$

23. $\begin{bmatrix} 8 & -8 \\ -3 & 2 \end{bmatrix}\begin{bmatrix} x_1 \\ x_2 \end{bmatrix} = \begin{bmatrix} 0 \\ 0 \end{bmatrix}$

$\begin{bmatrix} x_1 \\ x_2 \end{bmatrix} = \begin{bmatrix} 0 \\ 0 \end{bmatrix}$

25. $\begin{bmatrix} 1 & 0 & 2 \\ 3 & -2 & 1 \\ -2 & 2 & -1 \end{bmatrix}\begin{bmatrix} x_1 \\ x_2 \\ x_3 \end{bmatrix} = \begin{bmatrix} 5 \\ 8 \\ -3 \end{bmatrix}$

$\begin{bmatrix} x_1 \\ x_2 \\ x_3 \end{bmatrix} = \begin{bmatrix} 5 \\ \frac{7}{2} \\ 0 \end{bmatrix}$

27. $\begin{bmatrix} -5 & 2 \\ 3 & -1 \end{bmatrix}$

29. $a = 7, b = -4, c = -\frac{1}{2}, d = \frac{7}{2}$

31. $w = z, x = -y$

33. $\begin{bmatrix} 1 & 0 & 0 \\ 0 & 4 & 0 \\ 0 & 0 & 9 \end{bmatrix}$

35. $AB = \begin{bmatrix} -10 & 0 \\ 0 & -12 \end{bmatrix}$

$BA = \begin{bmatrix} -10 & 0 \\ 0 & -12 \end{bmatrix}$

37. Proof   **39.** 2   **41.** Proof

43. Let $A = \begin{bmatrix} a_{11} & a_{12} \\ a_{21} & a_{22} \end{bmatrix}$.

Then the given matrix equation expands to

$\begin{bmatrix} a_{11} + a_{21} & a_{12} + a_{22} \\ a_{11} + a_{21} & a_{12} + a_{22} \end{bmatrix} = \begin{bmatrix} 1 & 0 \\ 0 & 1 \end{bmatrix}$.

Because $a_{11} + a_{21} = 1$ and $a_{11} + a_{21} = 0$ cannot both be true, you can conclude that there is no solution.

45. (a) $A^2 = \begin{bmatrix} i^2 & 0 \\ 0 & i^2 \end{bmatrix} = \begin{bmatrix} -1 & 0 \\ 0 & -1 \end{bmatrix}$

$A^3 = \begin{bmatrix} i^3 & 0 \\ 0 & i^3 \end{bmatrix} = \begin{bmatrix} -i & 0 \\ 0 & -i \end{bmatrix}$

$A^4 = \begin{bmatrix} i^4 & 0 \\ 0 & i^4 \end{bmatrix} = \begin{bmatrix} 1 & 0 \\ 0 & 1 \end{bmatrix}$

(b) $B^2 = \begin{bmatrix} -i^2 & 0 \\ 0 & -i^2 \end{bmatrix} = \begin{bmatrix} 1 & 0 \\ 0 & 1 \end{bmatrix} = I$

47. Proof

49. Proof

51. (a) True. On page 50, ". . . for the product of two matrices to be defined, the number of columns of the first matrix must equal the number of rows of the second matrix."

(b) True. On page 54, ". . . the system $A\mathbf{x} = \mathbf{b}$ is consistent if and only if $\mathbf{b}$ can be expressed as . . . a linear combination, where the coefficients of the linear combination are a solution of the system."

53. $\begin{bmatrix} 110 & 99 & 77 & 33 \\ 44 & 22 & 66 & 66 \end{bmatrix}$

55. $[1250 \quad 1331.25 \quad 981.25]$

This row matrix represents the profits at each of the three outlets.

57. (a)

2000

	0–17	18–64	65+
Northeast	4.64	11.79	2.62
Midwest	5.92	14.00	2.94
South	9.09	22.12	4.42
Mountain	1.75	3.98	0.72
Pacific	4.30	9.97	1.74

2015

	0–17	18–64	65+
Northeast	4.06	10.99	2.63
Midwest	5.12	13.23	3.26
South	8.36	22.25	5.63
Mountain	1.69	4.07	1.05
Pacific	4.81	10.74	2.12

(b)

	0–17	18–64	65+
Northeast	−0.58	−0.80	0.01
Midwest	−0.80	−0.77	0.32
South	−0.73	0.13	1.21
Mountain	−0.06	0.09	0.33
Pacific	0.51	0.77	0.38

(c) The Pacific region and the 65+ age group show relative growth.

59. $\begin{bmatrix} -1 & 4 & | & 0 \\ -1 & 1 & | & 0 \\ 0 & 0 & | & 5 \end{bmatrix}$

61. $\mathbf{b} = 3\begin{bmatrix} 1 \\ 3 \end{bmatrix} + 0\begin{bmatrix} -1 \\ -3 \end{bmatrix} - 2\begin{bmatrix} 2 \\ 1 \end{bmatrix} = \begin{bmatrix} -1 \\ 7 \end{bmatrix}$

(The answer is not unique.)

63. $\mathbf{b} = 1\begin{bmatrix} 1 \\ 1 \\ 2 \end{bmatrix} + 2\begin{bmatrix} 1 \\ 0 \\ -1 \end{bmatrix} + 0\begin{bmatrix} -5 \\ -1 \\ -1 \end{bmatrix} = \begin{bmatrix} 3 \\ 1 \\ 0 \end{bmatrix}$

SECTION 2.2 *(page 69)*

1. $\begin{bmatrix} 3 & 2 \\ 13 & 4 \end{bmatrix}$ **3.** $\begin{bmatrix} 0 & -12 \\ 12 & -24 \end{bmatrix}$ **5.** $\begin{bmatrix} 7 & 7 \\ 28 & 14 \end{bmatrix}$

7. $\begin{bmatrix} 3 & \frac{2}{3} \\ -\frac{4}{3} & \frac{11}{3} \\ \frac{10}{3} & 0 \end{bmatrix}$ **9.** $\begin{bmatrix} -14 & -4 \\ 7 & -17 \\ -17 & -2 \end{bmatrix}$

11. $\begin{bmatrix} -3 & -5 & -10 \\ -2 & -5 & -5 \end{bmatrix}$ **13.** $\begin{bmatrix} 1 & 6 & -1 \\ -2 & -2 & -8 \end{bmatrix}$

15. $\begin{bmatrix} 12 & -4 \\ 8 & 4 \end{bmatrix}$ **17.** $AC = BC = \begin{bmatrix} 12 & -6 & 9 \\ 16 & -8 & 12 \\ 4 & -2 & 3 \end{bmatrix}$

19. $\begin{bmatrix} 1 & 0 \\ 0 & 1 \end{bmatrix}$ **21.** $\begin{bmatrix} 2 & 2 \\ 0 & 0 \end{bmatrix}$

23. (a) $\begin{bmatrix} 2 & 4 \\ -1 & 3 \\ 3 & -5 \end{bmatrix}$ (b) $\begin{bmatrix} 20 & 10 & -14 \\ 10 & 10 & -18 \\ -14 & -18 & 34 \end{bmatrix}$

(c) $\begin{bmatrix} 14 & -10 \\ -10 & 50 \end{bmatrix}$

25. (a) $\begin{bmatrix} 6 & 0 & 7 \\ 0 & -4 & 5 \end{bmatrix}$ (b) $\begin{bmatrix} 85 & 35 \\ 35 & 41 \end{bmatrix}$

(c) $\begin{bmatrix} 36 & 0 & 42 \\ 0 & 16 & -20 \\ 42 & -20 & 74 \end{bmatrix}$

27. (a) $\begin{bmatrix} 0 & 8 & -2 & 0 \\ -4 & 4 & 3 & 0 \\ 3 & 0 & 5 & -3 \\ 2 & 1 & 1 & 2 \end{bmatrix}$

(b) $\begin{bmatrix} 68 & 26 & -10 & 6 \\ 26 & 41 & 3 & -1 \\ -10 & 3 & 43 & 5 \\ 6 & -1 & 5 & 10 \end{bmatrix}$

(c) $\begin{bmatrix} 29 & -14 & 5 & -5 \\ -14 & 81 & -3 & 2 \\ 5 & -3 & 39 & -13 \\ -5 & 2 & -13 & 13 \end{bmatrix}$

29. $(A + B)(A - B) = A^2 + BA - AB - B^2$, which is not necessarily equal to $A^2 - B^2$ because AB is not necessarily equal to BA.

31. $(AB)^T = B^T A^T = \begin{bmatrix} 2 & -5 \\ 4 & -1 \end{bmatrix}$

33. (a) True. See Theorem 2.1, part 1.
(b) True. See Theorem 2.3, part 1.
(c) False. See Theorem 2.6, part 4, or Example 9.
(d) True. See Example 10.

35. (a) $a = 3$ and $b = -1$
(b) $a + b = 1$
$\quad\quad\quad b = 1$
$\quad a \quad\quad = 1$
No solution
(c) $a + b + c = 0; a = -c \to b = 0 \to c = 0 \to a = 0$
$\quad\quad b + c = 0$
$\quad a \quad + c = 0$
(d) $a = -3t;$ let $t = 1$: $a = -3, b = 1, c = 1$
$\quad b = t$
$\quad c = t$

37. $\begin{bmatrix} 1 & 0 & 0 \\ 0 & -1 & 0 \\ 0 & 0 & 1 \end{bmatrix}$

39. $\begin{bmatrix} \pm 3 & 0 \\ 0 & \pm 2 \end{bmatrix}$

41. $\begin{bmatrix} -4 & 0 \\ 8 & 2 \end{bmatrix}$

43. $\begin{bmatrix} 0 & 0 & 0 \\ 0 & 0 & 0 \\ 0 & 0 & 0 \end{bmatrix}$

45–53. Proof

55. Skew-symmetric

57. Symmetric

59. Proof

61. (a) $\frac{1}{2}(A + A^T)$

$$= \frac{1}{2}\left(\begin{bmatrix} a_{11} & a_{12} & \cdots & a_{1n} \\ a_{21} & a_{22} & \cdots & a_{2n} \\ \vdots & \vdots & & \vdots \\ a_{n1} & a_{n2} & \cdots & a_{nn} \end{bmatrix} + \begin{bmatrix} a_{11} & a_{21} & \cdots & a_{n1} \\ a_{12} & a_{22} & \cdots & a_{n2} \\ \vdots & \vdots & & \vdots \\ a_{1n} & a_{2n} & \cdots & a_{nn} \end{bmatrix} \right)$$

$$= \frac{1}{2}\begin{bmatrix} 2a_{11} & a_{12}+a_{21} & \cdots & a_{1n}+a_{n1} \\ a_{21}+a_{12} & 2a_{22} & \cdots & a_{2n}+a_{n2} \\ \vdots & & \vdots & \vdots \\ a_{n1}+a_{1n} & a_{n2}+a_{2n} & \cdots & 2a_{nn} \end{bmatrix}$$

(b) $\frac{1}{2}(A - A^T)$

$$= \frac{1}{2}\left(\begin{bmatrix} a_{11} & a_{12} & \cdots & a_{1n} \\ a_{21} & a_{22} & \cdots & a_{2n} \\ \vdots & \vdots & & \vdots \\ a_{n1} & a_{n2} & \cdots & a_{nn} \end{bmatrix} - \begin{bmatrix} a_{11} & a_{21} & \cdots & a_{n1} \\ a_{12} & a_{22} & \cdots & a_{n2} \\ \vdots & \vdots & & \vdots \\ a_{1n} & a_{2n} & \cdots & a_{nn} \end{bmatrix} \right)$$

$$= \frac{1}{2}\begin{bmatrix} 0 & a_{12}-a_{21} & \cdots & a_{1n}-a_{n1} \\ a_{21}-a_{12} & 0 & \cdots & a_{2n}-a_{n2} \\ \vdots & & \vdots & \vdots \\ a_{n1}-a_{1n} & a_{n2}-a_{2n} & \cdots & 0 \end{bmatrix}$$

(c) Proof

(d) $A = \frac{1}{2}(A - A^T) + \frac{1}{2}(A + A^T) =$

$$\underbrace{\begin{bmatrix} 0 & 4 & -\frac{1}{2} \\ -4 & 0 & -\frac{1}{2} \\ \frac{1}{2} & \frac{1}{2} & 0 \end{bmatrix}}_{\text{Skew-symmetric}} + \underbrace{\begin{bmatrix} 2 & 1 & \frac{7}{2} \\ 1 & 6 & \frac{1}{2} \\ \frac{7}{2} & \frac{1}{2} & 1 \end{bmatrix}}_{\text{Symmetric}}$$

63. (a) $A = \begin{bmatrix} 0 & 1 \\ 1 & 0 \end{bmatrix},\ B = \begin{bmatrix} -1 & 1 \\ 1 & 0 \end{bmatrix}$
(The answer is not unique.)

(b) Proof

SECTION 2.3 *(page 83)*

1. $AB = \begin{bmatrix} 1 & 0 \\ 0 & 1 \end{bmatrix} = BA$

3. $AB = \begin{bmatrix} 1 & 0 & 0 \\ 0 & 1 & 0 \\ 0 & 0 & 1 \end{bmatrix} = BA$

5. $\begin{bmatrix} 7 & -2 \\ -3 & 1 \end{bmatrix}$

7. $\begin{bmatrix} -19 & -33 \\ -4 & -7 \end{bmatrix}$

9. $\begin{bmatrix} 1 & 1 & -1 \\ -3 & 2 & -1 \\ 3 & -3 & 2 \end{bmatrix}$

11. Singular

13. $\begin{bmatrix} -\frac{3}{2} & \frac{3}{2} & 1 \\ \frac{9}{2} & -\frac{7}{2} & -3 \\ -1 & 1 & 1 \end{bmatrix}$

15. $\begin{bmatrix} 0 & -2 & 0.8 \\ -10 & 4 & 4.4 \\ 10 & -2 & -3.2 \end{bmatrix}$

17. $\begin{bmatrix} 1 & 0 & 0 \\ -\frac{3}{4} & \frac{1}{4} & 0 \\ \frac{7}{20} & -\frac{1}{4} & \frac{1}{5} \end{bmatrix}$

19. Singular

21. $\begin{bmatrix} -24 & 7 & 1 & -2 \\ -10 & 3 & 0 & -1 \\ -29 & 7 & 3 & -2 \\ 12 & -3 & -1 & 1 \end{bmatrix}$

23. Singular

25. (a) $x = 4$ (b) $x = -8$ (c) $x = 0$
      $y = 8$ $y = -11$ $y = 0$

27. (a) $x_1 = -5$
      $x_2 = -40.5$
      $x_3 = 48$
   (b) $x_1 = 0$
      $x_2 = 0$
      $x_3 = 0$

29. $x_1 = 1$
     $x_2 = -2$
     $x_3 = 3$
     $x_4 = 0$
     $x_5 = 1$
     $x_6 = -2$

31. (a) $\begin{bmatrix} 35 & 17 \\ 4 & 10 \end{bmatrix}$
   (b) $\begin{bmatrix} 2 & -7 \\ 5 & 6 \end{bmatrix}$

   (c) $\begin{bmatrix} -31 & 40 \\ -56 & 1 \end{bmatrix}$
   (d) $\begin{bmatrix} 1 & \frac{5}{2} \\ -\frac{7}{2} & 3 \end{bmatrix}$

33. (a) $\frac{1}{16}\begin{bmatrix} 138 & 56 & -84 \\ 37 & 26 & -71 \\ 24 & 34 & 3 \end{bmatrix}$ (b) $\frac{1}{4}\begin{bmatrix} 4 & 6 & 1 \\ -2 & 2 & 4 \\ 3 & -8 & 2 \end{bmatrix}$

(c) $\frac{1}{16}\begin{bmatrix} 7 & 0 & 34 \\ 28 & -40 & -14 \\ 30 & 14 & -25 \end{bmatrix}$ (d) $\frac{1}{8}\begin{bmatrix} 4 & -2 & 3 \\ 6 & 2 & -8 \\ 1 & 4 & 2 \end{bmatrix}$

35. $x = 4$ **37.** $\begin{bmatrix} -1 & \frac{1}{2} \\ \frac{3}{4} & -\frac{1}{4} \end{bmatrix}$

39. (a) True. See Theorem 2.7.
(b) True. See Theorem 2.10, part 1.
(c) False. See Theorem 2.9.
(d) True. See "Finding the Inverse of a Matrix by Gauss-Jordan Elimination," part 2, on page 74.

41–47. Proof

49. The sum of two invertible matrices is not necessarily invertible. For example, let
$$A = \begin{bmatrix} 1 & 0 \\ 0 & 1 \end{bmatrix} \quad \text{and} \quad B = \begin{bmatrix} -1 & 0 \\ 0 & -1 \end{bmatrix}.$$

51. (a) $\begin{bmatrix} -1 & 0 & 0 \\ 0 & \frac{1}{3} & 0 \\ 0 & 0 & \frac{1}{2} \end{bmatrix}$ (b) $\begin{bmatrix} 2 & 0 & 0 \\ 0 & 3 & 0 \\ 0 & 0 & 4 \end{bmatrix}$

53. (a) Proof (b) $H = \begin{bmatrix} 0 & -1 & 0 \\ -1 & 0 & 0 \\ 0 & 0 & 1 \end{bmatrix}$

55. $A = PDP^{-1}$ **57.** $\begin{bmatrix} 1 & 0 \\ -1 & 0 \end{bmatrix}$
No, A is not necessarily equal to D.

SECTION 2.4 *(page 95)*

1. Elementary, multiply Row 2 by 2

3. Elementary, add 2 times Row 1 to Row 2

5. Not elementary

7. Elementary, add -5 times Row 2 to Row 3

9. $\begin{bmatrix} 0 & 0 & 1 \\ 0 & 1 & 0 \\ 1 & 0 & 0 \end{bmatrix}$ **11.** $\begin{bmatrix} 0 & 0 & 1 \\ 0 & 1 & 0 \\ 1 & 0 & 0 \end{bmatrix}$

13. $\begin{bmatrix} 0 & 1 \\ 1 & 0 \end{bmatrix}$ **15.** $\begin{bmatrix} 0 & 0 & 1 \\ 0 & 1 & 0 \\ 1 & 0 & 0 \end{bmatrix}$

17. $\begin{bmatrix} \frac{1}{k} & 0 & 0 \\ 0 & 1 & 0 \\ 0 & 0 & 1 \end{bmatrix}$ **19.** $\begin{bmatrix} 1 & 0 & 0 & 0 \\ 0 & 1 & -k & 0 \\ 0 & 0 & 1 & 0 \\ 0 & 0 & 0 & 1 \end{bmatrix}$

21. $\begin{bmatrix} 1 & 0 \\ 1 & 1 \end{bmatrix}\begin{bmatrix} 1 & -1 \\ 0 & 1 \end{bmatrix}\begin{bmatrix} 1 & 0 \\ 0 & -2 \end{bmatrix}$
(The answer is not unique.)

23. $\begin{bmatrix} 1 & 1 \\ 0 & 1 \end{bmatrix}\begin{bmatrix} 1 & 0 \\ 3 & 1 \end{bmatrix}\begin{bmatrix} 1 & 0 \\ 0 & -1 \end{bmatrix}$
(The answer is not unique.)

25. $\begin{bmatrix} 1 & 0 & 0 \\ -1 & 1 & 0 \\ 0 & 0 & 1 \end{bmatrix}\begin{bmatrix} 1 & -2 & 0 \\ 0 & 1 & 0 \\ 0 & 0 & 1 \end{bmatrix}$
(The answer is not unique.)

27. $\begin{bmatrix} 1 & 0 & 0 \\ 2 & 1 & 0 \\ 0 & 0 & 1 \end{bmatrix}\begin{bmatrix} 2 & 0 & 0 \\ 0 & 1 & 0 \\ 0 & 0 & 1 \end{bmatrix}\begin{bmatrix} 1 & 0 & 0 \\ 0 & 2 & 0 \\ 0 & 0 & 1 \end{bmatrix}$
$\begin{bmatrix} 1 & 0 & 0 \\ 0 & 1 & 0 \\ 0 & 2 & 1 \end{bmatrix}\begin{bmatrix} 1 & 0 & 0 \\ 0 & 1 & 0 \\ 0 & 0 & -3 \end{bmatrix}\begin{bmatrix} 1 & 0 & -4 \\ 0 & 1 & 0 \\ 0 & 0 & 1 \end{bmatrix}$
$\begin{bmatrix} 1 & 0 & 0 \\ 0 & 1 & 5 \\ 0 & 0 & 1 \end{bmatrix}$ (The answer is not unique.)

29. (a) True. See "Remark" following the definition of an elementary matrix on page 86.
(b) False. Multiplication of a matrix by a scalar is not a single elementary row operation so it cannot be represented by a corresponding elementary matrix.
(c) True. See "Definition of Row Equivalence" on page 89.
(d) True. See Theorem 2.13.

31. (a) EA will have two rows interchanged. (The same rows are interchanged in E.)
(b) $E^2 = I_n$

33. $A^{-1} = \begin{bmatrix} 1 & -a & 0 \\ -b & ab+1 & 0 \\ 0 & 0 & \frac{1}{c} \end{bmatrix}$

35. No. For example, $\begin{bmatrix} 1 & 0 \\ 2 & 1 \end{bmatrix}\begin{bmatrix} 1 & 1 \\ 0 & 1 \end{bmatrix} = \begin{bmatrix} 1 & 1 \\ 2 & 3 \end{bmatrix}$.

37. $\begin{bmatrix} 1 & 0 \\ -2 & 1 \end{bmatrix}\begin{bmatrix} 1 & 0 \\ 0 & 1 \end{bmatrix}$

(The answer is not unique.)

39. $\begin{bmatrix} 1 & 0 & 0 \\ 2 & 1 & 0 \\ -1 & 1 & 1 \end{bmatrix}\begin{bmatrix} 3 & 0 & 1 \\ 0 & 1 & -1 \\ 0 & 0 & 2 \end{bmatrix}$

(The answer is not unique.)

41. (a) $\begin{bmatrix} 1 & 0 & 0 \\ 0 & 1 & 0 \\ -1 & 2 & 1 \end{bmatrix}\begin{bmatrix} 2 & 1 & 0 \\ 0 & 1 & -1 \\ 0 & 0 & 3 \end{bmatrix}$

(The answer is not unique.)

(b) $\mathbf{y} = \begin{bmatrix} 1 \\ 2 \\ -5 \end{bmatrix}$ (c) $\mathbf{x} = \begin{bmatrix} \frac{1}{3} \\ \frac{1}{3} \\ -\frac{5}{3} \end{bmatrix}$

43. First factor the matrix $A = LU$. Then, for each right-hand side $\mathbf{b}_i$, solve $L\mathbf{y} = \mathbf{b}_i$ and $U\mathbf{x} = \mathbf{y}$.

45. Idempotent   **47.** Not idempotent

49. Not idempotent

51. *Case 1.* $b = 1, a = 0$
Case 2. $b = 0, a =$ any real number

53–57. Proof

SECTION 2.5 *(page 110)*

1. Not stochastic   **3.** Stochastic

5. Stochastic   **7.** Next month: 350 people
In 2 months: 475 people

9.

	In 1 month	In 2 months
Nonsmokers	5025	5047
Smokers of less than 1 pack/day	2500	2499
Smokers of more than 1 pack/day	2475	2454

11. Tomorrow: 25 students
In 2 days: 44 students
In 30 days: 40 students

13. Proof

15. Uncoded: $[19\ \ 5\ \ 12], [12\ \ 0\ \ 3], [15\ \ 14\ \ 19],$
$[15\ \ 12\ \ 9], [4\ \ 1\ \ 20], [5\ \ 4\ \ 0]$
Encoded: $-48, 5, 31,\ -6, -6, 9,\ -85, 23, 43,$
$-27, 3, 15,\ -115, 36, 59,\ 9, -5, -4$

17. Uncoded: $[3\ \ 15], [13\ \ 5], [0\ \ 8], [15\ \ 13], [5\ \ 0],$
$[19\ \ 15], [15\ \ 14]$
Encoded: 48, 81, 28, 51, 24, 40, 54, 95, 5, 10,
64, 113, 57, 100

19. HAPPY_NEW_YEAR   **21.** FILM_AT_ELEVEN

23. MEET_ME_TONIGHT_RON

25. $A^{-1} = \begin{bmatrix} -3 & 2 & 2 \\ -4 & 2 & 3 \\ 2 & -1 & -1 \end{bmatrix}$;

_SEPTEMBER_THE_ELEVENTH_WE_WILL
_ALWAYS_REMEMBER

27. $D = \begin{matrix} & \text{Coal} & \text{Steel} \\ \begin{bmatrix} 0.1 & 0.2 \\ 0.8 & 0.1 \end{bmatrix} & \begin{matrix} \text{Coal} \\ \text{Steel} \end{matrix} \end{matrix}$ $X = \begin{bmatrix} 20{,}000 \\ 40{,}000 \end{bmatrix}\begin{matrix} \text{Coal} \\ \text{Steel} \end{matrix}$

29. $X = \begin{bmatrix} 8622.0 \\ 4685.0 \\ 3661.4 \end{bmatrix}\begin{matrix} \text{Farmer} \\ \text{Baker} \\ \text{Grocer} \end{matrix}$

31. (a)

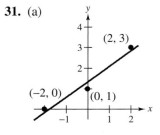

(b) $y = \frac{4}{3} + \frac{3}{4}x$

(c) $\frac{1}{6}$

33. (a)

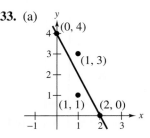

(b) $y = 4 - 2x$

(c) 2

35. $y = -\frac{1}{3} + 2x$ **37.** $y = 1.3 + 0.6x$

39. $y = 0.412x + 3$ **41.** $y = -0.5x + 7.5$

43. (a) $y = 685 - 240x$ (b) 349

45. (a) $y = 3338.1t - 20{,}774.5$
(b) $y = 3338.1t - 20{,}774.5$

Review Exercises – Chapter 2 *(page 114)*

1. $\begin{bmatrix} -13 & -8 & 18 \\ 0 & 11 & -19 \end{bmatrix}$

3. $\begin{bmatrix} 14 & -2 & 8 \\ 14 & -10 & 40 \\ 36 & -12 & 48 \end{bmatrix}$

5. $\begin{bmatrix} 4 & 6 & 3 \\ 0 & 6 & -10 \\ 0 & 0 & 6 \end{bmatrix}$

7. $5x + 4y = 2$
$-x + y = -22$

9. $\begin{bmatrix} 2 & 3 & 1 \\ 2 & -3 & -3 \\ 4 & -2 & 3 \end{bmatrix} \begin{bmatrix} x_1 \\ x_2 \\ x_3 \end{bmatrix} = \begin{bmatrix} 10 \\ 22 \\ -2 \end{bmatrix}$

11. $A^T = \begin{bmatrix} 1 & 0 \\ 2 & 1 \\ -3 & 2 \end{bmatrix}$, $A^T A = \begin{bmatrix} 1 & 2 & -3 \\ 2 & 5 & -4 \\ -3 & -4 & 13 \end{bmatrix}$,

$AA^T = \begin{bmatrix} 14 & -4 \\ -4 & 5 \end{bmatrix}$

13. $\begin{bmatrix} 1 & -1 \\ 2 & -3 \end{bmatrix}$

15. $\begin{bmatrix} \frac{3}{20} & \frac{3}{20} & \frac{1}{10} \\ \frac{3}{10} & -\frac{1}{30} & -\frac{2}{15} \\ -\frac{1}{5} & -\frac{1}{5} & \frac{1}{5} \end{bmatrix}$

17. $\begin{bmatrix} 5 & 4 \\ -1 & 1 \end{bmatrix} \begin{bmatrix} x_1 \\ x_2 \end{bmatrix} = \begin{bmatrix} 2 \\ -22 \end{bmatrix}$

$A^{-1} = \begin{bmatrix} \frac{1}{9} & -\frac{4}{9} \\ \frac{1}{9} & \frac{5}{9} \end{bmatrix}$

$\begin{bmatrix} x_1 \\ x_2 \end{bmatrix} = \begin{bmatrix} 10 \\ -12 \end{bmatrix}$

19. $\begin{bmatrix} \frac{1}{14} & \frac{1}{42} \\ -\frac{1}{21} & \frac{2}{21} \end{bmatrix}$

21. $\begin{bmatrix} 1 & 0 & -4 \\ 0 & 1 & 0 \\ 0 & 0 & 1 \end{bmatrix}$

23. $\begin{bmatrix} 1 & 3 \\ 0 & 1 \end{bmatrix} \begin{bmatrix} 2 & 0 \\ 0 & 1 \end{bmatrix}$
(The answer is not unique.)

25. $\begin{bmatrix} 1 & 0 & 0 \\ 0 & 1 & 0 \\ 0 & 0 & 4 \end{bmatrix} \begin{bmatrix} 1 & 0 & 0 \\ 0 & 1 & -2 \\ 0 & 0 & 1 \end{bmatrix} \begin{bmatrix} 1 & 0 & 1 \\ 0 & 1 & 0 \\ 0 & 0 & 1 \end{bmatrix}$
(The answer is not unique.)

27. $\begin{bmatrix} -1 & 0 \\ 0 & -1 \end{bmatrix}$ and $\begin{bmatrix} 1 & 0 \\ 0 & 1 \end{bmatrix}$
(The answer is not unique.)

29. $\begin{bmatrix} 0 & 0 \\ 0 & 0 \end{bmatrix}, \begin{bmatrix} 1 & 0 \\ 0 & 1 \end{bmatrix}$, and $\begin{bmatrix} 1 & 0 \\ 0 & 0 \end{bmatrix}$
(The answer is not unique.)

31. (a) $a = -1$
$b = -1$
$c = 1$
(b) Proof

33. Proof

35. $\begin{bmatrix} 1 & 0 \\ 3 & 1 \end{bmatrix} \begin{bmatrix} 2 & 5 \\ 0 & -1 \end{bmatrix}$
(The answer is not unique.)

37. $x = 4$, $y = 1$, $z = -1$

39. (a) False. See Theorem 2.1, part 1, page 60.
(b) True. See Theorem 2.6, part 2, page 67.

41. (a) False. The matrix $\begin{bmatrix} 1 & 0 \\ 0 & 0 \end{bmatrix}$ is not invertible.
(b) False. See Exercise 53, page 71.

43. Not stochastic

45. $PX = \begin{bmatrix} 80 \\ 112 \end{bmatrix}$, $P^2X = \begin{bmatrix} 68 \\ 124 \end{bmatrix}$, $P^3X = \begin{bmatrix} 65 \\ 127 \end{bmatrix}$

47. (a) $\begin{bmatrix} 110,000 \\ 100,000 \\ 90,000 \end{bmatrix}$ Region 1
Region 2
Region 3

(b) $\begin{bmatrix} 123,125 \\ 100,000 \\ 76,875 \end{bmatrix}$ Region 1
Region 2
Region 3

49. Uncoded: $[15 \quad 14], [5 \quad 0], [9 \quad 6], [0 \quad 2], [25 \quad 0],$
$[12 \quad 1], [14 \quad 4]$

Encoded: 103, 44, 25, 10, 57, 24, 4, 2, 125, 50, 62, 25, 78, 32

51. $A^{-1} = \begin{bmatrix} 3 & 2 \\ 4 & 3 \end{bmatrix}$; ALL_SYSTEMS_GO

53. $A^{-1} = \begin{bmatrix} -2 & -1 & 0 \\ 0 & 1 & 1 \\ -5 & -3 & -3 \end{bmatrix}$; INVASION_AT_DAWN

55. $D = \begin{bmatrix} 0.20 & 0.50 \\ 0.30 & 0.10 \end{bmatrix}$, $X = \begin{bmatrix} 133,333 \\ 133,333 \end{bmatrix}$

57. $y = \frac{20}{3} - \frac{3}{2}x$

59. $y = \frac{2}{5} - \frac{9}{5}x$

61. (a) $y = 19 + 14x$
(b) 41.4 kilograms per square kilometer

63. (a) $y = 3.594x + 49.664$
(b) $y = 3.594x + 49.664$
(c)

x	Actual	Estimated
6	\$70.93	\$71.23
7	\$75.31	\$74.82
8	\$78.62	\$78.42
9	\$81.33	\$82.01
10	\$85.89	\$85.60

(d) \$103.57   (e) $x \approx 14$ (Year 2004)

Chapter 3

SECTION 3.1 *(page 127)*

1. 1   **3.** 5   **5.** 27   **7.** -24
9. 6   **11.** $\lambda^2 - 4\lambda - 5$
13. (a) $M_{11} = 4$ (b) $C_{11} = 4$
        $M_{12} = 3$ $C_{12} = -3$
        $M_{21} = 2$ $C_{21} = -2$
        $M_{22} = 1$ $C_{22} = 1$
15. (a) $M_{11} = 23$ $M_{12} = -8$ $M_{13} = -22$
        $M_{21} = 5$ $M_{22} = -5$ $M_{23} = 5$
        $M_{31} = 7$ $M_{32} = -22$ $M_{33} = -23$
    (b) $C_{11} = 23$ $C_{12} = 8$ $C_{13} = -22$
        $C_{21} = -5$ $C_{22} = -5$ $C_{23} = -5$
        $C_{31} = 7$ $C_{32} = 22$ $C_{33} = -23$
17. (a) $4(-5) + 5(-5) + 6(-5) = -75$
    (b) $2(8) + 5(-5) - 3(22) = -75$
19. -58 **21.** -30 **23.** -0.022
25. $4x - 2y - 2$ **27.** -168 **29.** 0
31. $65{,}644w + 62{,}256x + 12{,}294y - 24{,}672z$
33. -100 **35.** -43.5 **37.** -1098
39. 329   **41.** -24 **43.** 0   **45.** -30
47. (a) False. See "Definition of the Determinant of a 2×2 Matrix," page 120.
    (b) True. See the first line after "Remark," page 121.
    (c) False. See "Definition of Minors and Cofactors of a Matrix," page 121.

49. $\lambda = \pm 3$ **51.** $\lambda = -1, -8,$ or 3
53. Expanding along the first row, the determinant of a 4×4 matrix involves four 3×3 determinants. Each of these 3×3 determinants requires six triple products. Hence, there are $4(6) = 24$ quadruple products.

55. $\begin{vmatrix} 1 & 1 & 1 \\ a & b & c \\ a^2 & b^2 & c^2 \end{vmatrix} = (bc^2 - cb^2) - (ac^2 - ca^2) + (ab^2 - a^2b)$
$$= bc^2 + ca^2 + ab^2 - ba^2 - ac^2 - cb^2.$$
The right side of the original equation yields
$(a - b)(b - c)(c - a) = (ab - b^2 - ac + bc)(c - a)$
$$= abc - cb^2 - ac^2 + bc^2 - ba^2 + ab^2 + ca^2 - abc$$
$$= bc^2 + ca^2 + ab^2 - ba^2 - ac^2 - cb^2.$$

57. $\begin{vmatrix} x & 0 & c \\ -1 & x & b \\ 0 & -1 & a \end{vmatrix} = x \begin{vmatrix} x & b \\ -1 & a \end{vmatrix} + c \begin{vmatrix} -1 & x \\ 0 & -1 \end{vmatrix}$
$$= x(ax + b) + c(1)$$
$$= ax^2 + bx + c.$$

59. Expand along the row of zeros; then the determinant is zero.

SECTION 3.2 *(page 137)*

1. The first row is 2 times the second row. If one row of a matrix is a multiple of another row, then the determinant of the matrix is zero.

3. The second row consists entirely of zeros. If one row of a matrix consists entirely of zeros, then the determinant of the matrix is zero.

5. The second and third columns are interchanged. If two columns of a matrix are interchanged, then the determinant of the matrix changes sign.

7. The first row of the matrix is multiplied by 5. If a row in a matrix is multiplied by a scalar, then the determinant of the matrix is multiplied by that scalar.

9. The matrix is multiplied by 5. If an $n \times n$ matrix is multiplied by a scalar c, then the determinant of the matrix is multiplied by c^n.

11. -4 times the first row has been added to the second row. If a scalar multiple of one row of a matrix is added to another row, then the determinant of the matrix is unchanged.

13. -5 times the second column has been added to the third column. If a scalar multiple of one column of a matrix is added to another column, then the determinant of the matrix is unchanged.

15. The sixth column is 2 times the first column. If one column of a matrix is a multiple of another column, then the determinant of the matrix is zero.

17. 28 **19.** 17 **21.** -60

23. 223 **25.** -1344 **27.** 136

29. -1100

31. (a) True. See Theorem 3.3, part 1, page 130.

(b) True. See Theorem 3.3, part 3, page 130.

(c) True. See Theorem 3.4, part 2, page 133.

33. k **35.** -1 **37.** 1

39. Proof

41. $\cos^2 \theta + \sin^2 \theta = 1$

43. Not possible. The determinant is equal to $\cos^2 x + \sin^2 x$, which cannot equal zero because $\cos^2 x + \sin^2 x = 1$.

45. Proof

SECTION 3.3 *(page 145)*

1. (a) 0

(b) -1

(c) $\begin{bmatrix} -2 & -3 \\ 4 & 6 \end{bmatrix}$

(d) 0

3. (a) 2

(b) -6

(c) $\begin{bmatrix} 1 & 4 & 3 \\ -1 & 0 & 3 \\ 0 & 2 & 0 \end{bmatrix}$

(d) -12

5. -44 **7.** 54

9. (a) -2 (b) -2 (c) 0

11. (a) 14
(b) 196
(c) 196
(d) 56
(e) $\frac{1}{14}$

13. (a) 29
(b) 841
(c) 841
(d) 232
(e) $\frac{1}{29}$

15. (a) 22
(b) 22
(c) 484
(d) 88
(e) $\frac{1}{22}$

17. (a) -115
(b) -115
(c) 13,225
(d) -1840
(e) $-\frac{1}{115}$

19. (a) -15
(b) -125
(c) 243
(d) -15
(e) $-\frac{1}{5}$

21. (a) 8
(b) 4
(c) 64
(d) 8
(e) $\frac{1}{2}$

23. Singular **25.** Nonsingular

27. Nonsingular **29.** Singular

31. The solution is not unique because the determinant of the coefficient matrix is zero.

33. The solution is unique because the determinant of the coefficient matrix is nonzero.

35. $k = -1, 4$ **37.** $k = 24$ **39.** Proof

41. $\begin{bmatrix} 0 & 1 \\ 0 & 0 \end{bmatrix}$ and $\begin{bmatrix} 1 & 0 \\ 0 & 0 \end{bmatrix}$ **43.** 0
(The answer is not unique.)

45. Proof

47. No; in general, $P^{-1}AP \neq A$. For example, let
$P = \begin{bmatrix} 1 & 2 \\ 3 & 5 \end{bmatrix}$, $P^{-1} = \begin{bmatrix} -5 & 2 \\ 3 & -1 \end{bmatrix}$,
and $A = \begin{bmatrix} 2 & 1 \\ -1 & 0 \end{bmatrix}$.
Then you have
$P^{-1}AP = \begin{bmatrix} -27 & -49 \\ 16 & 29 \end{bmatrix} \neq A$.
However, the equation $|P^{-1}AP| = |A|$ is true in general because
$$|P^{-1}AP| = |P^{-1}|\,|A|\,|P|$$
$$= |P^{-1}|\,|P|\,|A| = \frac{1}{|P|}\,|P|\,|A| = |A|.$$

49. (a) False. See Theorem 3.6, page 140.

(b) True. See Theorem 3.8, page 143.

(c) True. See "Equivalent Conditions for a Nonsingular Matrix," parts 1 and 2, page 144.

51. Proof **53.** Orthogonal

55. Not orthogonal **57.** Orthogonal

59. Proof

61. (a) $\begin{bmatrix} \frac{2}{3} & \frac{2}{3} & \frac{1}{3} \\ -\frac{2}{3} & \frac{1}{3} & \frac{2}{3} \\ \frac{1}{3} & -\frac{2}{3} & \frac{2}{3} \end{bmatrix}$ (b) $\begin{bmatrix} \frac{2}{3} & \frac{2}{3} & \frac{1}{3} \\ -\frac{2}{3} & \frac{1}{3} & \frac{2}{3} \\ \frac{1}{3} & -\frac{2}{3} & \frac{2}{3} \end{bmatrix}$ (c) 1

A is orthogonal.

63. Proof

Section 3.4 (*page 152*)

1. $\begin{bmatrix} 1 & 2 \\ 0 & -3 \end{bmatrix}\begin{bmatrix} 1 \\ 0 \end{bmatrix} = 1\begin{bmatrix} 1 \\ 0 \end{bmatrix}$;

$\begin{bmatrix} 1 & 2 \\ 0 & -3 \end{bmatrix}\begin{bmatrix} -1 \\ 2 \end{bmatrix} = \begin{bmatrix} 3 \\ -6 \end{bmatrix} = -3\begin{bmatrix} -1 \\ 2 \end{bmatrix}$

3. $\begin{bmatrix} 1 & 1 & 1 \\ 0 & 1 & 0 \\ 1 & 1 & 1 \end{bmatrix}\begin{bmatrix} 1 \\ 0 \\ 1 \end{bmatrix} = \begin{bmatrix} 2 \\ 0 \\ 2 \end{bmatrix} = 2\begin{bmatrix} 1 \\ 0 \\ 1 \end{bmatrix}$

$\begin{bmatrix} 1 & 1 & 1 \\ 0 & 1 & 0 \\ 1 & 1 & 1 \end{bmatrix}\begin{bmatrix} -1 \\ 0 \\ 1 \end{bmatrix} = \begin{bmatrix} 0 \\ 0 \\ 0 \end{bmatrix} = 0\begin{bmatrix} -1 \\ 0 \\ 1 \end{bmatrix}$

$\begin{bmatrix} 1 & 1 & 1 \\ 0 & 1 & 0 \\ 1 & 1 & 1 \end{bmatrix}\begin{bmatrix} -1 \\ 1 \\ -1 \end{bmatrix} = \begin{bmatrix} -1 \\ 1 \\ -1 \end{bmatrix} = 1\begin{bmatrix} -1 \\ 1 \\ -1 \end{bmatrix}$

5. (a) $\lambda^2 - \lambda - 2 = 0$ (b) $2, -1$

(c) $\lambda = 2$: $\mathbf{x} = \begin{bmatrix} 5 \\ 2 \end{bmatrix}$; $\lambda = -1$: $\mathbf{x} = \begin{bmatrix} 1 \\ 1 \end{bmatrix}$

7. (a) $\lambda^2 - 2\lambda - 3 = 0$ (b) $3, -1$

(c) $\lambda = 3$: $\mathbf{x} = \begin{bmatrix} 1 \\ 1 \end{bmatrix}$; $\lambda = -1$: $\mathbf{x} = \begin{bmatrix} -1 \\ 3 \end{bmatrix}$

9. (a) $\lambda^2 - 3\lambda - 18 = 0$ (b) $6, -3$

(c) $\lambda = 6$: $\mathbf{x} = \begin{bmatrix} 1 \\ 2 \end{bmatrix}$; $\lambda = -3$: $\mathbf{x} = \begin{bmatrix} -4 \\ 1 \end{bmatrix}$

11. (a) $\lambda^3 - 3\lambda^2 - 4\lambda + 12 = 0$ (b) $2, 3, -2$

(c) $\lambda = 2$: $\mathbf{x} = \begin{bmatrix} -1 \\ 0 \\ 1 \end{bmatrix}$; $\lambda = 3$: $\mathbf{x} = \begin{bmatrix} -1 \\ 1 \\ 1 \end{bmatrix}$;

$\lambda = -2$: $\mathbf{x} = \begin{bmatrix} 1 \\ -1 \\ 4 \end{bmatrix}$

13. Eigenvalues: $\lambda = -3, 1$

Eigenvectors: $\lambda = -3$: $\mathbf{x} = \begin{bmatrix} -1 \\ 1 \end{bmatrix}$

$\lambda = 1$: $\mathbf{x} = \begin{bmatrix} -5 \\ 1 \end{bmatrix}$

15. Eigenvalues: $\lambda = 1, 2, 3$

Eigenvectors: $\lambda = 1$: $\mathbf{x} = \begin{bmatrix} 0 \\ -1 \\ 1 \end{bmatrix}$

$\lambda = 2$: $\mathbf{x} = \begin{bmatrix} 1 \\ 0 \\ 1 \end{bmatrix}$

$\lambda = 3$: $\mathbf{x} = \begin{bmatrix} 2 \\ 0 \\ 1 \end{bmatrix}$

17. Eigenvalues: $\lambda = -3, -1, 3, 5$

Eigenvectors: $\lambda = -3$: $x = \begin{bmatrix} 0 \\ 0 \\ 1 \\ -1 \end{bmatrix}$

$\lambda = -1$: $x = \begin{bmatrix} 0 \\ 1 \\ 0 \\ 0 \end{bmatrix}$

$\lambda = 3$: $x = \begin{bmatrix} 1 \\ 0 \\ 0 \\ 0 \end{bmatrix}$

$\lambda = 5$: $x = \begin{bmatrix} 0 \\ 0 \\ 5 \\ 3 \end{bmatrix}$

19. (a) False. If x is an eigenvector corresponding to λ, then any *non-zero* multiple of x is also an eigenvector corresponding to λ. See page 149, first paragraph.

(b) False. If $\lambda = a$ is an eigenvalue of the matrix, then $\lambda = a$ is a solution of the characteristic equation $|\lambda I - A| = 0$. See page 149, second paragraph.

Section 3.5 (*page 164*)

1. $\text{adj}(A) = \begin{bmatrix} 4 & -2 \\ -3 & 1 \end{bmatrix}$, $A^{-1} = \begin{bmatrix} -2 & 1 \\ \frac{3}{2} & -\frac{1}{2} \end{bmatrix}$

3. $\mathrm{adj}(A) = \begin{bmatrix} 0 & 0 & 0 \\ 0 & -12 & -6 \\ 0 & 4 & 2 \end{bmatrix}$, A^{-1} does not exist.

5. $\mathrm{adj}(A) = \begin{bmatrix} -7 & -12 & 13 \\ 2 & 3 & -5 \\ 2 & 3 & -2 \end{bmatrix}$, $A^{-1} = \begin{bmatrix} \frac{7}{3} & 4 & -\frac{13}{3} \\ -\frac{2}{3} & -1 & \frac{5}{3} \\ -\frac{2}{3} & -1 & \frac{2}{3} \end{bmatrix}$

7. $\mathrm{adj}(A) = \begin{bmatrix} 7 & 1 & 9 & -13 \\ 7 & 1 & 0 & -4 \\ -4 & 2 & -9 & 10 \\ 2 & -1 & 9 & -5 \end{bmatrix}$,

$A^{-1} = \begin{bmatrix} \frac{7}{9} & \frac{1}{9} & 1 & -\frac{13}{9} \\ \frac{7}{9} & \frac{1}{9} & 0 & -\frac{4}{9} \\ -\frac{4}{9} & \frac{2}{9} & -1 & \frac{10}{9} \\ \frac{2}{9} & -\frac{1}{9} & 1 & -\frac{5}{9} \end{bmatrix}$

9. Proof   **11.** Proof

13. $|\mathrm{adj}(A)| = \begin{vmatrix} -2 & 0 \\ -1 & 1 \end{vmatrix} = -2$,

$|A|^{2-1} = \begin{vmatrix} 1 & 0 \\ 1 & -2 \end{vmatrix}^{2-1} = -2$

15. Proof

17. $x_1 = 1$ **19.** $x_1 = 2$ **21.** $x_1 = \frac{3}{4}$
$x_2 = 2$ $x_2 = -2$ $x_2 = -\frac{1}{2}$

23. Cramer's Rule does not apply because the coefficient matrix has a determinant of zero.

25. Cramer's Rule does not apply because the coefficient matrix has a determinant of zero.

27. $x_1 = 1$ **29.** $x_1 = 1$ **31.** $x_1 = 0$
$x_2 = 1$ $x_2 = \frac{1}{2}$ $x_2 = -\frac{1}{2}$
$x_3 = 2$ $x_3 = \frac{3}{2}$ $x_3 = \frac{1}{2}$

33. 5   **35.** 1

37. $x = \dfrac{4k-3}{2k-1}$, $y = \dfrac{4k-1}{2k-1}$

The system will be inconsistent if $k = \frac{1}{2}$.

39. 3   **41.** 3

43. Collinear   **45.** Not collinear

47. $3y - 4x = 0$ **49.** $x = -2$

51. $\frac{1}{3}$ **53.** 2

55. Not coplanar   **57.** Coplanar

59. $4x - 10y + 3z = 27$ **61.** $x + y + z = 0$

63. Incorrect. The numerator and denominator should be interchanged.

65. Correct

REVIEW EXERCISES – CHAPTER 3   (*page 166*)

1. 10   **3.** 0   **5.** 0

7. -6 **9.** 1620   **11.** 82

13. -64 **15.** -1 **17.** -1

19. (a) -1 (b) -5 (c) $\begin{bmatrix} 1 & -2 \\ 2 & 1 \end{bmatrix}$ (d) 5

21. (a) -12 (b) -1728 (c) 144 (d) -300

23. (a) -20 (b) $-\frac{1}{20}$

25. $x_1 = 1$ **27.** $x_1 = -\frac{3}{2}$
$x_2 = -1$ $x_2 = 2$
$x_3 = 2$ $x_3 = \frac{5}{2}$

29. Unique solution   **31.** Unique solution

33. Not a unique solution

35. (a) False. See "Definition of Minors and Cofactors of a Matrix," page 121.

(b) False. See Theorem 3.3, part 1, page 130.

(c) False. See Theorem 3.9, page 145.

37. 128   **39.** Proof   **41.** 0

43. $\lambda = 7, \begin{bmatrix} 1 \\ 1 \end{bmatrix}$ and $\lambda = -8, \begin{bmatrix} -2 \\ 1 \end{bmatrix}$

45. $\lambda = 1, \begin{bmatrix} 1 \\ 1 \\ 0 \end{bmatrix}$; $\lambda = 3, \begin{bmatrix} 0 \\ 1 \\ 0 \end{bmatrix}$; and $\lambda = 4, \begin{bmatrix} 0 \\ 0 \\ 1 \end{bmatrix}$

47. $-\frac{1}{2}$ **49.** $-uv$

51. Row reduction is generally preferred for matrices with few zeros. For a matrix with many zeros, it is often easier to expand along a row or column having many zeros.

53. $\begin{bmatrix} 1 & 0 & 0 \\ -2 & 1 & 0 \\ -4 & 1 & 1 \end{bmatrix}$

55. Unique solution: $x = 0.6$
$ y = 0.5$

57. Unique solution: $x_1 = \frac{1}{2}$
$ x_2 = -\frac{1}{3}$
$ x_3 = 1$

59. 16 **61.** $x - 2y = -4$

63. $9x + 4y - 3z = 0$

65. (a) False. See Theorem 3.11, page 159.

(b) False. See "Test for Collinear Points in the *xy*-Plane," page 161.

CUMULATIVE TEST CHAPTERS 1–3 *(page 172)*

1. $x_1 = 2,\ x_2 = -3,\ x_3 = -2$

2. $x_1 = s - 2t,\ x_2 = 2 + t,\ x_3 = t,\ x_4 = s$

3. $x_1 = -2s,\ x_2 = s,\ x_3 = 2t,\ x_4 = t$ **4.** $k = 12$

5. $BA = \begin{bmatrix} 13{,}275.00 & 15{,}500.00 \end{bmatrix}$
The entries represent the total values (in dollars) of the models sent to each of the warehouses.

6. $x = -3,\ y = 4$

7. $A^T A = \begin{bmatrix} 17 & 22 & 27 \\ 22 & 29 & 36 \\ 27 & 36 & 45 \end{bmatrix}$

8. (a) $\begin{bmatrix} -\frac{1}{4} & \frac{1}{8} \\ \frac{1}{6} & \frac{1}{12} \end{bmatrix}$ **9.** $\begin{bmatrix} 1 & 0 & -1 \\ 0 & 0 & 1 \\ \frac{3}{5} & \frac{1}{5} & -\frac{9}{5} \end{bmatrix}$

(b) $\begin{bmatrix} -\frac{2}{7} & \frac{1}{7} \\ \frac{1}{7} & \frac{2}{21} \end{bmatrix}$

10. $\begin{bmatrix} 0 & 1 \\ 1 & 0 \end{bmatrix}\begin{bmatrix} 1 & 0 \\ 2 & 1 \end{bmatrix}\begin{bmatrix} 1 & 0 \\ 0 & -4 \end{bmatrix}$ **11.** -34
(The answer is not unique.)

12. (a) 14 (b) -10 (c) -140 (d) $\frac{1}{14}$

13. (a) 567 (b) 7 (c) $\frac{1}{7}$ (d) 343

14. $y = \frac{7}{6}x^2 + \frac{1}{6}x + 1$

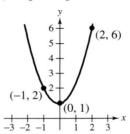

15. $a = 1,\ b = 0,\ c = 2$

16. No; proof **17.** Proof **18.** Proof

19. (a) A is row-equivalent to B if there exist elementary matrices $E_1, \ldots, E_k$ such that $A = E_k \ldots E_1 B$.

(b) Proof

20. Proof

21. (a) $\begin{bmatrix} 0 & 0 \\ 0 & 0 \end{bmatrix}, \begin{bmatrix} 1 & 0 \\ 0 & 1 \end{bmatrix}, \begin{bmatrix} 1 & 0 \\ 0 & 0 \end{bmatrix}$ (b) 0 or 1
(The answer is not unique.)

22. (a) Not necessarily; proof
(b) Yes; proof

23. $\begin{bmatrix} 0 & 0 \\ 0 & 0 \end{bmatrix}, \begin{bmatrix} 0 & 1 \\ 0 & 0 \end{bmatrix}, \begin{bmatrix} 1 & a \\ 0 & 0 \end{bmatrix}, \begin{bmatrix} 1 & 0 \\ 0 & 1 \end{bmatrix}$

Chapter 4

SECTION 4.1 *(page 183)*

1. $\mathbf{v} = \left(-3, \frac{9}{2}\right)$ **3.** $\mathbf{v} = (-8, -1)$

  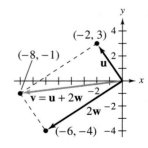

5. $\mathbf{v} = \left(-\frac{9}{2}, \frac{7}{2}\right)$

7. (a)

(b)

(c)

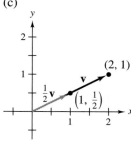

9. $\mathbf{u} - \mathbf{v} = (-1, 0, 4)$
$\mathbf{v} - \mathbf{u} = (1, 0, -4)$

11. $(6, 12, 6)$

13. $\left(\frac{7}{2}, 3, \frac{5}{2}\right)$

15. (a)

(b)

(c)

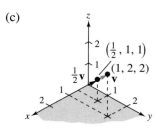

17. (a) and (b)

19. (a) $(4, -2, -8, 1)$ (b) $(8, 12, 24, 34)$

21. (a) $(-9, 3, 2, -3, 6)$
    (b) $(-2, -18, -12, 18, 36)$

23. (a) $(1, 6, -5, -3)$
    (b) $(-1, -8, 10, 0)$
    (c) $\left(-\frac{3}{2}, 11, -\frac{13}{2}, -\frac{21}{2}\right)$
    (d) $\left(\frac{1}{4}, 3, -3, -1\right)$

25. $\left(\frac{1}{2}, -\frac{7}{2}, -\frac{9}{2}, 2\right)$ **27.** $\mathbf{v} = \mathbf{u} + \mathbf{w}$

29. $\mathbf{v} = \mathbf{u} + 2\mathbf{w}$ **31.** $\mathbf{v} = -\mathbf{u}$

33. $\left(-1, \frac{5}{3}, 6, \frac{2}{3}\right)$ **35.** $\mathbf{v} = \mathbf{u}_1 + 2\mathbf{u}_2 - 3\mathbf{u}_3$

37. It is not possible to write $\mathbf{v}$ as a linear combination of $\mathbf{u}_1, \mathbf{u}_2,$ and $\mathbf{u}_3$.

39. $\mathbf{v} = 5\mathbf{u}_1 - \mathbf{u}_2 + \mathbf{u}_3 + 2\mathbf{u}_4 - 5\mathbf{u}_5 + 3\mathbf{u}_6$

41. (a) True. Two vectors in R^n are equal if and only if their corresponding components are equal, that is $\mathbf{u} = \mathbf{v}$ if $\mathbf{u}_1 = \mathbf{v}_1, \mathbf{u}_2 = \mathbf{v}_2, \dots, \mathbf{u}_u = \mathbf{v}_u$.

    (b) False. The vector $c\mathbf{v}$ is $|c|$ times as long as $\mathbf{v}$ and has the same direction as $\mathbf{v}$ if c is positive and the opposite direction as $\mathbf{v}$ if c is negative.

43. No

45. Answers will vary.   **47.** Proof

49. (a) Add $-\mathbf{v}$ to both sides.
    (b) Associative property and additive identity
    (c) Additive inverse
    (d) Additive identity

51. (a) Additive identity
    (b) Distributive property
    (c) Add $-c\mathbf{0}$ to both sides.
    (d) Additive inverse and associative property
    (e) Additive inverse
    (f) Additive identity

53. (a) Additive inverse
   (b) Transitive property
   (c) Add **v** to both sides.
   (d) Associative property
   (e) Additive inverse
   (f) Additive identity

55. If $\mathbf{b} = x_1\mathbf{a}_1 + \cdots + x_n\mathbf{a}_n$ is a linear combination of the columns of A, then a solution of $A\mathbf{x} = \mathbf{b}$ is

$$\mathbf{x} = \begin{bmatrix} x_1 \\ \vdots \\ x_n \end{bmatrix}.$$

The system $A\mathbf{x} = \mathbf{b}$ is inconsistent if $\mathbf{b}$ is not a linear combination of the columns of A.

SECTION 4.2   *(page 191)*

1. $(0, 0, 0, 0)$

3. $\begin{bmatrix} 0 & 0 & 0 \\ 0 & 0 & 0 \end{bmatrix}$

5. $0 + 0x + 0x^2 + 0x^3$

7. $-(v_1, v_2, v_3, v_4) = (-v_1, -v_2, -v_3, -v_4)$

9. $-\begin{bmatrix} a_{11} & a_{12} & a_{13} \\ a_{21} & a_{22} & a_{23} \end{bmatrix} = \begin{bmatrix} -a_{11} & -a_{12} & -a_{13} \\ -a_{21} & -a_{22} & -a_{23} \end{bmatrix}$

11. $-(a_0 + a_1x + a_2x^2 + a_3x^3) = -a_0 - a_1x - a_2x^2 - a_3x^3$

13. The set is a vector space.

15. The set is not a vector space. Axiom 1 fails because $x^5 + (-x^5 + x) = x$, which is not a fifth-degree polynomial. (Axioms 4, 5, and 6 also fail.)

17. The set is a vector space.

19. The set is a vector space.

21. The set is not a vector space. Axiom 1 fails because

$$\begin{bmatrix} 1 & 0 \\ 0 & 0 \end{bmatrix} + \begin{bmatrix} 0 & 0 \\ 0 & 1 \end{bmatrix} = \begin{bmatrix} 1 & 0 \\ 0 & 1 \end{bmatrix},$$

which is not singular.

23. The set is a vector space.

25. (a) The set is not a vector space because Axiom 8 fails.
$$(1 + 2)(1, 1) = \quad 3(1, 1) \quad = (3, 1)$$
$$1(1, 1) + 2(1, 1) = (1, 1) + (2, 1) = (3, 2)$$

(b) The set is not a vector space because Axiom 2 fails.
$$(1, 2) + (2, 1) = (1, 0)$$
$$(2, 1) + (1, 2) = (2, 0)$$
(Axioms 4, 5, and 8 also fail.)

(c) The set is not a vector space. Axiom 6 fails because $(-1)(1, 1) = (\sqrt{-1}, \sqrt{-1})$, which is not in R^2. (Axioms 8 and 9 also fail.)

27. Proof

29. The set is not a vector space because Axiom 5 fails. Because $(1, 1)$ is the additive identity, $(0, 0)$ has no additive inverse. (Axioms 7 and 8 also fail.)

31. (a) True. See page 186.
   (b) False. See Example 6, page 190.
   (c) False. With standard operations on R^2, the additive inverse axiom is not satisfied.

33. (a) Add $-\mathbf{w}$ to both sides.
   (b) Associative property
   (c) Additive inverse
   (d) Additive identity

35. Proof   **37.** Proof

SECTION 4.3   *(page 200)*

1. Because W is nonempty and $W \subset R^4$, you need only check that W is closed under addition and scalar multiplication. Given
$$(x_1, x_2, x_3, 0) \in W \quad \text{and} \quad (y_1, y_2, y_3, 0) \in W,$$
it follows that
$$(x_1, x_2, x_3, 0) + (y_1, y_2, y_3, 0)$$
$$= (x_1 + y_1, x_2 + y_2, x_3 + y_3, 0) \in W.$$
Furthermore, for any real number c and $(x_1, x_2, x_3, 0) \in W$, it follows that
$$c(x_1, x_2, x_3, 0) = (cx_1, cx_2, cx_3, 0) \in W.$$

3. Because W is nonempty and $W \subset M_{2,2}$, you need only check that W is closed under addition and scalar multiplication. Given

$$\begin{bmatrix} 0 & a_1 \\ b_1 & 0 \end{bmatrix} \in W \quad \text{and} \quad \begin{bmatrix} 0 & a_2 \\ b_2 & 0 \end{bmatrix} \in W$$

it follows that

$$\begin{bmatrix} 0 & a_1 \\ b_1 & 0 \end{bmatrix} + \begin{bmatrix} 0 & a_2 \\ b_2 & 0 \end{bmatrix} = \begin{bmatrix} 0 & a_1 + a_2 \\ b_1 + b_2 & 0 \end{bmatrix} \in W.$$

Furthermore, for any real number c and

$$\begin{bmatrix} 0 & a \\ b & 0 \end{bmatrix} \in W$$

it follows that

$$c\begin{bmatrix} 0 & a \\ b & 0 \end{bmatrix} = \begin{bmatrix} 0 & ca \\ cb & 0 \end{bmatrix} \in W.$$

5. Recall from calculus that continuity implies integrability; thus $W \subset V$. Furthermore, because W is nonempty, you need only check that W is closed under addition and scalar multiplication. Given continuous functions f, $g \in W$, it follows that $f + g$ is continuous, and thus $f + g \in W$. Also, for any real number c and for a continuous function $f \in W$, cf is continuous. Thus, $cf \in W$.

7. Not closed under addition:

$(0, 0, -1) + (0, 0, -1) = (0, 0, -2)$

Not closed under scalar multiplication:

$2(0, 0, -1) = (0, 0, -2)$

9. Not closed under scalar multiplication: $(-1)\, e^x = -e^x$

11. Not closed under addition:

$$\begin{bmatrix} 1 & 0 \\ 0 & 0 \end{bmatrix} + \begin{bmatrix} 0 & 0 \\ 0 & 1 \end{bmatrix} = \begin{bmatrix} 1 & 0 \\ 0 & 1 \end{bmatrix}$$

13. (b), (c), (d), and (e)

15. W is a subspace of R^3. (W is nonempty and closed under addition and scalar multiplication.)

17. W is a subspace of R^3. (W is nonempty and closed under addition and scalar multiplication.)

19. W is not a subspace of R^3.

Not closed under addition:

$(1, 1, 1) + (1, 1, 1) = (2, 2, 2)$, which is not in W.

Not closed under scalar multiplication:

$2(1, 1, 1) = (2, 2, 2)$, which is not in W.

21. (a) True. See "Remark," page 194.

(b) True. See Theorem 4.6, page 196-197.

(c) False. There may be elements of W which are not elements of U or vice-versa.

23–31. Proof

SECTION 4.4 *(page 213)*

1. (a) **u** cannot be written as a linear combination of the given vectors.

(b) $\mathbf{v} = \frac{1}{4}(2, -1, 3) + \frac{3}{2}(5, 0, 4)$

(c) $\mathbf{w} = 8(2, -1, 3) - 3(5, 0, 4)$

3. (a) $\mathbf{u} = -\frac{7}{4}(2, 0, 7) + \frac{5}{4}(2, 4, 5) + 0(2, -12, 13)$

(b) **v** cannot be written as a linear combination of the given vectors.

(c) $\mathbf{w} = -\frac{1}{6}(2, 0, 7) + \frac{1}{3}(2, 4, 5) + 0(2, -12, 13)$

5. S spans R^2.

7. S does not span R^2. (It spans a line in R^2.)

9. S does not span R^2. (It spans a line in R^2.)

11. S spans R^3.

13. S does not span R^3. (It spans a plane in R^3.)

15. S does not span R^3. (It spans a plane in R^3.)

17. Linearly independent

19. Linearly dependent

21. Linearly independent

23. Linearly dependent

25. Linearly independent

27. Linearly independent

29. $(3, 4) - 4(-1, 1) - \frac{7}{2}(2, 0) = (0, 0)$,

$(3, 4) = 4(-1, 1) + \frac{7}{2}(2, 0)$

(The answer is not unique.)

31. $(1, 1, 1) - (1, 1, 0) - (0, 0, 1) - 0(0, 1, 1) = (0, 0, 0)$

$(1, 1, 1) = (1, 1, 0) + (0, 0, 1) - 0(0, 1, 1)$

(The answer is not unique.)

33. (a) All $t \neq 1, -2$ (b) All $t \neq \frac{1}{2}$

35. (a) $\begin{bmatrix} 6 & -19 \\ 10 & 7 \end{bmatrix} = 3A - 2B$

(b) Not a linear combination of A and B

(c) $\begin{bmatrix} -2 & 28 \\ 1 & -11 \end{bmatrix} = -A + 5B$

(d) $\begin{bmatrix} 0 & 0 \\ 0 & 0 \end{bmatrix} = 0A + 0B$

37. (a) Linearly dependent
(b) Linearly independent
(c) Linearly independent

39. (a) Any set of three vectors in R^2 must be linearly dependent.
(b) The second vector is a scalar multiple of the first vector.
(c) The first vector is the zero vector.

41. S_1 and S_2 span the same subspace.

43. (a) False. See "Definition of Linear Dependence and Linear Independence," page 207.
(b) True. See Theorem 4.8, page 211.

45–51. Proof

53. The theorem requires that only one of the vectors be a linear combination of the others. Because $(-1, 0, 2) = 0(1, 2, 3) - (1, 0, -2)$, there is no contradiction.

55. Proof

57. On $[0, 1]$,
$$f_2(x) = |x| = x = \tfrac{1}{3}(3x)$$
$$= \tfrac{1}{3}f_1(x) \implies \{f_1, f_2\} \text{ dependent}$$

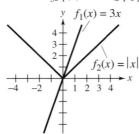

On $[-1, 1]$, f_1 and f_2 are not multiples of each other. For if they were, $cf_1(x) = f_2(x)$, then $c(3x) = |x|$. But if $x = 1, c = \tfrac{1}{3}$, whereas if $x = -1, c = -\tfrac{1}{3}$.

59. Proof

SECTION 4.5 *(page 224)*

1. R^6: $\{(1, 0, 0, 0, 0, 0), (0, 1, 0, 0, 0, 0), (0, 0, 1, 0, 0, 0),$
$(0, 0, 0, 1, 0, 0), (0, 0, 0, 0, 1, 0), (0, 0, 0, 0, 0, 1)\}$

3. $M_{2,4}$: $\left\{ \begin{bmatrix} 1 & 0 & 0 & 0 \\ 0 & 0 & 0 & 0 \end{bmatrix}, \begin{bmatrix} 0 & 1 & 0 & 0 \\ 0 & 0 & 0 & 0 \end{bmatrix}, \right.$
$\begin{bmatrix} 0 & 0 & 1 & 0 \\ 0 & 0 & 0 & 0 \end{bmatrix}, \begin{bmatrix} 0 & 0 & 0 & 1 \\ 0 & 0 & 0 & 0 \end{bmatrix},$
$\begin{bmatrix} 0 & 0 & 0 & 0 \\ 1 & 0 & 0 & 0 \end{bmatrix}, \begin{bmatrix} 0 & 0 & 0 & 0 \\ 0 & 1 & 0 & 0 \end{bmatrix},$
$\left. \begin{bmatrix} 0 & 0 & 0 & 0 \\ 0 & 0 & 1 & 0 \end{bmatrix}, \begin{bmatrix} 0 & 0 & 0 & 0 \\ 0 & 0 & 0 & 1 \end{bmatrix} \right\}$

5. S is linearly dependent. (S does span R^2.)

7. S is linearly dependent and does not span R^2.

9. S is linearly dependent and does not span R^3.

11. S is linearly dependent and does not span R^3.

13. S is linearly dependent. (S does span P_2.)

15. S is linearly dependent.

17. S does not span $M_{2,2}$. (S is linearly independent.)

19. The set is a basis for R^2.

21. The set is not a basis for R^2.

23. S is a basis for R^2.   **25.** S is not a basis for R^3.

27. S is a basis for R^4.   **29.** S is a basis for $M_{2,2}$.

31. S is a basis for P_3.

33. S is not a basis for P_3.

35. S is a basis for R^3.
$(8, 3, 8) = 2(4, 3, 2) - (0, 3, 2) + 3(0, 0, 2)$

37. S is not a basis for R^3.

39. 6   **41.** 8

43. $\begin{bmatrix} 1 & 0 & 0 \\ 0 & 0 & 0 \\ 0 & 0 & 0 \end{bmatrix}, \begin{bmatrix} 0 & 0 & 0 \\ 0 & 1 & 0 \\ 0 & 0 & 0 \end{bmatrix}, \begin{bmatrix} 0 & 0 & 0 \\ 0 & 0 & 0 \\ 0 & 0 & 1 \end{bmatrix},$
$\dim(D_{3,3}) = 3$

45. $\{(1, 0), (0, 1)\}, \{(1, 0), (1, 1)\}, \{(0, 1), (1, 1)\}$

47. $\{(1, 1), (1, 0)\}$ (The answer is not unique.)

49. (a) Line through the origin
(b) $\{(2, 1)\}$
(c) 1

51. (a) Line through the origin
(b) $\{(2, 1, -1)\}$
(c) 1

53. (a) $\{(2, 1, 0, 1), (-1, 0, 1, 0)\}$
(b) 2

55. (a) $\{(0, 6, 1, -1)\}$
(b) 1

57. (a) False. If the dimension of V is n, then every spanning set of V must have at least n vectors.
(b) True. Find a set of n basis vectors in V that will span V and add any other vector.

59. Proof   **61.** Proof

63. (a) Basis for S_1: $\{(1, 0, 0), (1, 1, 0)\}$, dimension $= 2$
Basis for S_2: $\{(0, 0, 1), (0, 1, 0)\}$, dimension $= 2$
Basis for $S_1 \cap S_2$: $\{(0, 1, 0)\}$, dimension $= 1$
Basis for $S_1 + S_2$: $\{(1, 0, 0), (0, 1, 0), (0, 0, 1)\}$, dimension $= 3$
(b) No, it is not possible.

65. Proof

SECTION 4.6 *(page 239)*

1. (a) 2
(b) $\{(1, 0), (0, 1)\}$
(c) $\left\{ \begin{bmatrix} 1 \\ 0 \end{bmatrix}, \begin{bmatrix} 0 \\ 1 \end{bmatrix} \right\}$

3. (a) 2
(b) $\{(1, 2), (0, 1)\}$
(c) $\left\{ \begin{bmatrix} 1 \\ \frac{1}{2} \end{bmatrix}, \begin{bmatrix} 0 \\ 1 \end{bmatrix} \right\}$

5. (a) 2
(b) $\left\{ \left(1, 0, \frac{1}{4}\right), \left(0, 1, \frac{3}{2}\right) \right\}$
(c) $\left\{ \begin{bmatrix} 1 \\ 0 \\ -\frac{2}{5} \end{bmatrix}, \begin{bmatrix} 0 \\ 1 \\ \frac{3}{5} \end{bmatrix} \right\}$

7. (a) 2
(b) $\{(1, 2, -2, 0), (0, 0, 0, 1)\}$
(c) $\left\{ \begin{bmatrix} 1 \\ 0 \\ \frac{19}{7} \end{bmatrix}, \begin{bmatrix} 0 \\ 1 \\ \frac{8}{7} \end{bmatrix} \right\}$

9. $\{(1, 0, -1, 0), (0, 1, 0, 0), (0, 0, 0, 1)\}$

11. $\{(1, 0, 0, 0), (0, 1, 0, 0), (0, 0, 1, 0), (0, 0, 0, 1)\}$

13. Solution space is $\{(0, 0)\}$, dim $= 0$

15. $\{(-2, 1, 0), (-3, 0, 1)\}$, dim $= 2$

17. $\{(-1, 2, 1)\}$, dim $= 1$

19. $\{(2, -2, 0, 1), (-1, 1, 1, 0)\}$, dim $= 2$

21. (a) $\{(-1, -3, 2)\}$
(b) 1

23. (a) $\{(-3, 0, 1), (2, 1, 0)\}$
(b) 2

25. (a) $\{(8, -9, -6, 6)\}$
(b) 1

27. (a) Consistent
(b) $\mathbf{x} = t(2, -4, 1) + (3, 5, 0)$

29. (a) Inconsistent   (b) Not applicable

31. (a) Consistent
(b) $\mathbf{x} = t(5, 0, -6, -4, 1) + s(-2, 1, 0, 0, 0) + (1, 0, 2, -3, 0)$

33. $\begin{bmatrix} -1 \\ 4 \end{bmatrix} + 2\begin{bmatrix} 2 \\ 0 \end{bmatrix} = \begin{bmatrix} 3 \\ 4 \end{bmatrix}$

35. $-\frac{5}{4}\begin{bmatrix} 1 \\ -1 \\ 2 \end{bmatrix} + \frac{3}{4}\begin{bmatrix} 3 \\ 1 \\ 0 \end{bmatrix} - \frac{1}{2}\begin{bmatrix} 0 \\ 0 \\ 1 \end{bmatrix} = \begin{bmatrix} 1 \\ 2 \\ -3 \end{bmatrix}$

37. Four vectors in R^3 must be linearly dependent.

39. Proof

41. (a) $\begin{bmatrix} 1 & 0 \\ 0 & 1 \end{bmatrix}, \begin{bmatrix} 0 & 1 \\ 1 & 0 \end{bmatrix}$
(b) $\begin{bmatrix} 1 & 0 \\ 0 & 0 \end{bmatrix}, \begin{bmatrix} 0 & 1 \\ 0 & 0 \end{bmatrix}$
(c) $\begin{bmatrix} 1 & 0 \\ 0 & 0 \end{bmatrix}, \begin{bmatrix} 0 & 0 \\ 0 & 1 \end{bmatrix}$

43. (a) m
(b) r
(c) r
(d) R^n
(e) R^m

45. Answers will vary.   **47.** Proof

49. (a) False. See "Remark," page 227.
(b) False. See Theorem 4.19, page 237–238.

51. (a) rank(A) = rank(B) = 3

nullity(A) = $n - r = 5 - 3 = 2$

(b) Choosing x_3 and x_5 as the free variables, you have

$x_1 = -3s + 4t$

$x_2 = s - 2t$

$x_3 = s$

$x_4 = 2t$

$x_5 = t$

A basis for the nullspace is $\left\{ \begin{bmatrix} -3 \\ 1 \\ 1 \\ 0 \\ 0 \end{bmatrix}, \begin{bmatrix} 4 \\ -2 \\ 0 \\ 2 \\ 1 \end{bmatrix} \right\}$.

(c) $\{(1, 0, 3, 0, -4), (0, 1, -1, 0, 2), (0, 0, 0, 1, -2)\}$

(d) $\{(1, 2, 3, 4), (2, 5, 7, 9), (0, 1, 2, -1)\}$

(e) Linearly dependent

(f) (i) and (iii) are linearly independent.

53. Proof   **55.** Proof

SECTION 4.7 *(page 253)*

1. $\begin{bmatrix} 8 \\ -3 \end{bmatrix}$ **3.** $\begin{bmatrix} 5 \\ 4 \\ 3 \end{bmatrix}$ **5.** $\begin{bmatrix} -1 \\ 2 \\ 0 \\ 1 \end{bmatrix}$ **7.** $\begin{bmatrix} 3 \\ 2 \end{bmatrix}$

9. $\begin{bmatrix} 1 \\ -1 \\ 2 \end{bmatrix}$ **11.** $\begin{bmatrix} 0 \\ -1 \\ 2 \end{bmatrix}$ **13.** $\begin{bmatrix} \frac{3}{2} & -\frac{1}{2} \\ -2 & 1 \end{bmatrix}$

15. $\begin{bmatrix} 2 & -1 \\ 4 & 3 \end{bmatrix}$ **17.** $\begin{bmatrix} 1 & 2 & -\frac{1}{2} \\ 0 & \frac{1}{2} & 0 \\ 0 & -\frac{1}{3} & \frac{1}{12} \end{bmatrix}$

19. $\begin{bmatrix} 1 & 1 & 1 \\ 3 & 5 & 4 \\ 3 & 6 & 5 \end{bmatrix}$ **21.** $\begin{bmatrix} -24 & 7 & 1 & -2 \\ -10 & 3 & 0 & -1 \\ -29 & 7 & 3 & -2 \\ 12 & -3 & -1 & 1 \end{bmatrix}$

23. (a) $\begin{bmatrix} -\frac{1}{3} & \frac{1}{3} \\ \frac{3}{4} & -\frac{1}{2} \end{bmatrix}$ (b) $\begin{bmatrix} 6 & 4 \\ 9 & 4 \end{bmatrix}$ (d) $\begin{bmatrix} 6 \\ 3 \end{bmatrix}$

25. (a) $\begin{bmatrix} 4 & 5 & 1 \\ -7 & -10 & -1 \\ -2 & -2 & 0 \end{bmatrix}$

(b) $\begin{bmatrix} \frac{1}{2} & \frac{1}{2} & -\frac{5}{4} \\ -\frac{1}{2} & -\frac{1}{2} & \frac{3}{4} \\ \frac{3}{2} & \frac{1}{2} & \frac{5}{4} \end{bmatrix}$ (d) $\begin{bmatrix} \frac{11}{4} \\ -\frac{9}{4} \\ \frac{5}{4} \end{bmatrix}$

27. (a) $\begin{bmatrix} -\frac{48}{5} & -24 & \frac{4}{5} \\ 4 & 10 & \frac{1}{2} \\ -\frac{6}{5} & -5 & -\frac{2}{5} \end{bmatrix}$

(b) $\begin{bmatrix} \frac{3}{32} & \frac{17}{20} & \frac{5}{4} \\ -\frac{1}{16} & -\frac{3}{10} & -\frac{1}{2} \\ \frac{1}{2} & \frac{6}{5} & 0 \end{bmatrix}$ (d) $\begin{bmatrix} \frac{279}{160} \\ -\frac{61}{80} \\ -\frac{7}{10} \end{bmatrix}$

29. $\begin{bmatrix} 4 \\ 11 \\ 1 \end{bmatrix}$ **31.** $\begin{bmatrix} 0 \\ 3 \\ 2 \end{bmatrix}$

33. (a) False. See Theorem 4.20, page 246–247.

(b) True. See the discussion following Example 4, page 249.

35. QP

37. If B is the standard basis, then

$[B' \vdots B] = [B' \vdots I] \implies [I \vdots (B')^{-1}]$

shows that P^{-1}, the transition matrix from B to B', is $(B')^{-1}$.

If B' is the standard basis, then

$[B' \vdots B] = [I \vdots B]$

shows that P^{-1}, the transition matrix from B to B', is B.

SECTION 4.8 *(page 263)*

1. (b), (c), and (d)   **3.** (a), (b), and (d)

5. -2 **7.** 0

9. Linearly independent   **11.** Linearly dependent

13. Linearly dependent   **15.** Linearly dependent

17. $y = C_1 \sin x + C_2 \cos x$

19. $y = C_1 + C_2 \sin x + C_3 \cos x$

21. Proof   **23.** Proof

25. No. For instance, consider $y'' = 1$. Two solutions are $y = \dfrac{x^2}{2}$ and $y = \dfrac{x^2}{2} + 1$. Their sum is not a solution.

27. Parabola   **29.** Hyperbola

$y^2 + x = 0$

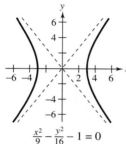

$\dfrac{x^2}{9} - \dfrac{y^2}{16} - 1 = 0$

31. Point

$9x^2 + 25y^2 - 36x - 50y + 61 = 0$

33. Ellipse   **35.** Parabola

$x^2 + 4y^2 + 4x + 32y + 64 = 0$

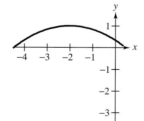

$x^2 + 4x + 6y - 2 = 0$

37. $\dfrac{(y')^2}{2} - \dfrac{(x')^2}{2} = 1$ **39.** $\dfrac{(x')^2}{6} + \dfrac{(y')^2}{4} = 1$

41. $(x')^2 + \dfrac{(y')^2}{4} = 1$ **43.** $y' + 4 = -(x')^2$

45. $y' = 0$

47. Proof   **49.** Proof

REVIEW EXERCISES – CHAPTER 4   *(page 265)*

1. (a) $(0, 2, 5)$ **3.** (a) $(3, 1, 4, 4)$
     (b) $(2, 0, 4)$ (b) $(0, 4, 4, 2)$
     (c) $(-2, 2, 1)$ (c) $(3, -3, 0, 2)$
     (d) $(-5, 6, 5)$ (d) $(9, -7, 2, 7)$

5. $\left(\dfrac{1}{2}, -4, -4\right)$ **7.** $\mathbf{v} = \dfrac{9}{8}\mathbf{u}_1 + \dfrac{1}{8}\mathbf{u}_2 + 0\mathbf{u}_3$

9. $O_{3,4} = \begin{bmatrix} 0 & 0 & 0 & 0 \\ 0 & 0 & 0 & 0 \\ 0 & 0 & 0 & 0 \end{bmatrix}$,

$-A = \begin{bmatrix} -a_{11} & -a_{12} & -a_{13} & -a_{14} \\ -a_{21} & -a_{22} & -a_{23} & -a_{24} \\ -a_{31} & -a_{32} & -a_{33} & -a_{34} \end{bmatrix}$

11. W is a subspace of R^2.

13. W is a subspace of R^3.

15. W is not a subspace of $C[-1, 1]$.

17. (a) W is a subspace of R^3.
(b) W is not a subspace of R^3.

	Does S span R^3?	Is S linearly independent?	Is S a basis for R^3?
19.	(a) Yes	(b) Yes	(c) Yes
21.	(a) No	(b) No	(c) No

23. S is a basis for P_3.

25. (a) $\{(-3, 0, 4, 1), (-2, 1, 0, 0)\}$ (b) 2

27. (a) $\{(2, 3, 7, 0), (-1, 0, 0, 1)\}$ (b) 2

29. $\{(8, 5)\}$
Rank(A) and nullity(A) = $1 + 1 = 2$

31. $\{(3, 0, 1, 0), (-1, -2, 0, 1)\}$
Rank(A) and nullity(A) = $2 + 2 = 4$

33. (a) 2   (b) $\{(1, 0), (0, 1)\}$

35. (a) 3   (b) $\{(1, 0, 0), (0, 1, 0), (0, 0, 1)\}$

37. $\begin{bmatrix} -2 \\ 8 \end{bmatrix}$ **39.** $\begin{bmatrix} 2 \\ -1 \\ -1 \end{bmatrix}$ **41.** $\begin{bmatrix} \frac{2}{5} \\ -\frac{1}{4} \end{bmatrix}$

43. $\begin{bmatrix} 3 \\ 1 \\ 0 \\ 1 \end{bmatrix}$ **45.** $\begin{bmatrix} -12 \\ 6 \end{bmatrix}$ **47.** $\begin{bmatrix} 1 & 3 \\ -1 & 1 \end{bmatrix}$

49. $\begin{bmatrix} -\frac{1}{2} & \frac{1}{2} \\ -\frac{3}{2} & -\frac{5}{2} \end{bmatrix}$

51. Basis for W: $\{x, x^2, x^3\}$
Basis for U: $\{(x - 1), (x - 1)^2, (x - 1)^3\}$
Basis for $W \cap U$: $\{x(x - 1), x^2(x - 1)\}$

53. No. For example, the set
$\{x^2 + x, x^2 - x, 1\}$
is a basis for P_2.

55. Yes, W is a subspace of V.

57. Proof   **59.** Answers will vary.

61. (a) True. See "Definition of Vector Addition and Scalar Multiplication in R^n," page 179.
(b) False. See Theorem 4.3, part 2, page 181.
(c) True. See "Definition of Vector Space" and the discussion following, page 186.

63. (a) True. See Example 2, page 186.
(b) False. See "Definition of Vector Space," part 4, page 186.
(c) True. See discussion following "Summary of Important Vector Spaces," page 189.

65. (a) and (d) are solutions.

67. e^x

69. Linearly independent

71. Linearly dependent

73. Circle

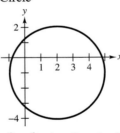

$x^2 + y^2 - 4x + 2y - 4 = 0$

75. Parabola

$(5, -4)$
$2x^2 - 20x - y + 46 = 0$

77. $\dfrac{(x')^2}{6} - \dfrac{(y')^2}{6} = 1$

Chapter 5

SECTION 5.1 *(page 283)*

1. 5

3. 3

5. $3\sqrt{6}$

7. (a) 5
   (b) $\sqrt{6}$
   (c) $\sqrt{21}$

9. (a) $\sqrt{6}$
   (b) $\sqrt{11}$
   (c) $\sqrt{13}$

11. (a) $\left(-\frac{5}{13}, \frac{12}{13}\right)$
   (b) $\left(\frac{5}{13}, -\frac{12}{13}\right)$

13. (a) $\left(\frac{3}{\sqrt{38}}, \frac{2}{\sqrt{38}}, -\frac{5}{\sqrt{38}}\right)$
   (b) $\left(-\frac{3}{\sqrt{38}}, -\frac{2}{\sqrt{38}}, \frac{5}{\sqrt{38}}\right)$

15. $\pm\dfrac{1}{\sqrt{14}}$

17. $\left(2\sqrt{2}, 2\sqrt{2}\right)$

19. $\left(1, \sqrt{3}, 0\right)$

21. (a) $(4, 4, 3)$ (b) $\left(-2, -2, -\frac{3}{2}\right)$

23. $2\sqrt{2}$

25. $2\sqrt{3}$

27. (a) -6
   (b) 25
   (c) 25
   (d) $(-12, 18)$
   (e) -30

29. (a) 5
   (b) 50
   (c) 50
   (d) $(0, 10, 25, 20)$
   (e) 25

31. -7

33. (a) $\|\mathbf{u}\| = 13$,
      $\|\mathbf{v}\| = 17$
   (b) $\left(-\frac{8}{17}, -\frac{15}{17}\right)$
   (c) $\left(-\frac{5}{13}, \frac{12}{13}\right)$
   (d) 140
   (e) 169
   (f) 289

35. (a) $\|\mathbf{u}\| = 13$,
      $\|\mathbf{v}\| = 13$
   (b) $\left(-\frac{12}{13}, \frac{5}{13}\right)$
   (c) $\left(-\frac{5}{13}, -\frac{12}{13}\right)$
   (d) 0
   (e) 169
   (f) 169

37. (a) $\|\mathbf{u}\| = 26$,
      $\|\mathbf{v}\| = 13$
   (b) $\left(-\frac{5}{13}, -\frac{12}{13}\right)$
   (c) $\left(-\frac{10}{26}, \frac{24}{26}\right) = \left(-\frac{5}{13}, \frac{12}{13}\right)$
   (d) 238
   (e) 676
   (f) 169

39. (a) $\|\mathbf{u}\| = 13$, $\|\mathbf{v}\| = 13$
   (b) $\left(0, -\frac{5}{13}, -\frac{12}{13}\right)$
   (c) $\left(0, -\frac{5}{13}, -\frac{12}{13}\right)$
   (d) -169
   (e) 169
   (f) 169

41. (a) $\|\mathbf{u}\| = \sqrt{14}$, $\|\mathbf{v}\| = \sqrt{14}$
   (b) $\left(\frac{2}{\sqrt{14}}, 0, \frac{1}{\sqrt{14}}, \frac{1}{\sqrt{14}}, \frac{2}{\sqrt{14}}, -\frac{2}{\sqrt{14}}\right)$
   (c) $\left(0, -\frac{2}{\sqrt{14}}, -\frac{2}{\sqrt{14}}, \frac{1}{\sqrt{14}}, -\frac{1}{\sqrt{14}}, \frac{2}{\sqrt{14}}\right)$
   (d) 7
   (e) 14
   (f) 14

43. (a) $\|\mathbf{u}\| = \sqrt{14}$, $\|\mathbf{v}\| = 4$
   (b) $\left(-\frac{1}{4}, 0, \frac{1}{4}, \frac{1}{2}, -\frac{1}{2}, \frac{1}{4}, \frac{1}{4}, -\frac{1}{2}\right)$
   (c) $\left(\frac{1}{\sqrt{14}}, -\frac{1}{\sqrt{14}}, -\frac{2}{\sqrt{14}}, \frac{1}{\sqrt{14}}, -\frac{1}{\sqrt{14}}, -\frac{1}{\sqrt{14}}, \frac{2}{\sqrt{14}}, -\frac{1}{\sqrt{14}}\right)$
   (d) -4
   (e) 14
   (f) 16

45. $|(3, 4) \cdot (2, -3)| \leq \|(3, 4)\| \, \|(2, -3)\|$
      $6 \leq 5\sqrt{13}$

47. 1.713 radians
      $(98.13°)$

49. 1.080 radians
      $(61.87°)$

51. $\dfrac{\pi}{4}$

53. $\mathbf{v} = (t, 0)$

55. $\mathbf{v} = (2t, 3t)$

57. $\mathbf{v} = (t, 4t, s)$

59. Orthogonal

61. Parallel

63. Orthogonal

65. Orthogonal

67. Neither

69. (a) False. See "Definition of Length of a Vector in R^n," page 271.
   (b) False. See "Definition of Dot Product in R^n," page 275.

71. (a) $\|\mathbf{u} \cdot \mathbf{v}\|$ is meaningless because $\mathbf{u} \cdot \mathbf{v}$ is a scalar.
   (b) $\mathbf{u} + (\mathbf{u} \cdot \mathbf{v})$ is meaningless because $\mathbf{u}$ is a vector and $\mathbf{u} \cdot \mathbf{v}$ is a scalar.

73. $\|(5, 1)\| \le \|(4, 0)\| + \|(1, 1)\|$
   $\sqrt{26} \le 4 + \sqrt{2}$

75. $\|(2, 0)\|^2 = \|(1, -1)\|^2 + \|(1, 1)\|^2$
   $4 = \left(\sqrt{2}\right)^2 + \left(\sqrt{2}\right)^2$

77. (a) $\theta = \dfrac{\pi}{2}$ provided $\mathbf{u} \ne \mathbf{0}$ and $\mathbf{v} \ne \mathbf{0}$

   (b) $0 \le \theta < \dfrac{\pi}{2}$

   (c) $\dfrac{\pi}{2} < \theta \le \pi$

79. $\theta = \arccos\left(\dfrac{1}{\sqrt{3}}\right) \approx 54.7°$

81. Proof   **83.** Proof

85. Proof

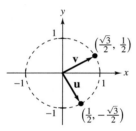

87. $\mathbf{u} \cdot \mathbf{v} = 17{,}139.05$, which represents the total value of the three crops.

89. Proof

SECTION 5.2 *(page 295)*

1. (a) -33 **3.** (a) 15
   (b) 5 (b) $\sqrt{57}$
   (c) 13 (c) 5
   (d) $2\sqrt{65}$ (d) $2\sqrt{13}$

5. (a) -34 **7.** (a) 14
   (b) $\sqrt{97}$ (b) 2
   (c) $\sqrt{101}$ (c) $\sqrt{58}$
   (d) $\sqrt{266}$ (d) $\sqrt{34}$

9. (a) $\dfrac{16}{15}$

   (b) $\dfrac{\sqrt{10}}{5}$

   (c) $\dfrac{2\sqrt{210}}{15}$

   (d) $\sqrt{2}$

11. (a) $\dfrac{2}{e} \approx 0.736$

   (b) $\dfrac{\sqrt{6}}{3} \approx 0.816$

   (c) $\sqrt{\dfrac{e^2}{2} - \dfrac{1}{2e^2}} \approx 1.904$

   (d) $\sqrt{\dfrac{e^2}{2} + \dfrac{2}{3} - \dfrac{1}{2e^2} - \dfrac{4}{e}} \approx 1.680$

13. (a) -6 (b) $\sqrt{35}$ (c) $\sqrt{7}$ (d) $3\sqrt{6}$
15. (a) -4 (b) $\sqrt{11}$ (c) $\sqrt{2}$ (d) $\sqrt{21}$
17. Proof   **19.** Proof
21. Axiom 4 fails.
   $\langle(0, 1), (0, 1)\rangle = 0$, but $(0, 1) \ne 0$.
23. Axiom 2 fails.
   $\langle(1, 0), (1, 0) + (1, 0)\rangle = \langle(1, 0), (2, 0)\rangle = 4$
   $\langle(1, 0), (1, 0)\rangle + \langle(1, 0), (1, 0)\rangle = 1 + 1 \qquad = 2$

25. 2.103 radians   **27.** $\dfrac{\pi}{2}$ **29.** $\dfrac{\pi}{2}$
   $(120.5°)$

31. (a) $|\langle(5, 12), (3, 4)\rangle| \le \|(5, 12)\|\,\|(3, 4)\|$
   $63 \le (13)(5)$
   (b) $\|(5, 12) + (3, 4)\| \le \|(5, 12)\| + \|(3, 4)\|$
   $8\sqrt{5} \le 13 + 5$

33. (a) $|\langle 2x, 3x^2 + 1\rangle| \le \|2x\|\,\|3x^2 + 1\|$
   $0 \le (2)\left(\sqrt{10}\right)$
   (b) $\|2x + 3x^2 + 1\| \le \|2x\| + \|3x^2 + 1\|$
   $\sqrt{14} \le 2 + \sqrt{10}$

35. (a) $|\langle \sin x, \cos x\rangle| \le \|\sin x\|\,\|\cos x\|$
   $0 \le \left(\sqrt{\pi}\right)\left(\sqrt{\pi}\right)$
   (b) $\|\sin x + \cos x\| \le \|\sin x\| + \|\cos x\|$
   $\sqrt{2\pi} \le \sqrt{\pi} + \sqrt{\pi}$

37. Because

$$\langle f, g \rangle = \int_{-\pi}^{\pi} \cos x \sin x \, dx$$

$$= \frac{1}{2} \sin^2 x \Big]_{-\pi}^{\pi} = 0,$$

f and g are orthogonal.

39. The functions $f(x) = x$ and $g(x) = \frac{1}{2}(5x^3 - 3x)$ are orthogonal because

$$\langle f, g \rangle = \int_{-1}^{1} x \frac{1}{2}(5x^3 - 3x) \, dx$$

$$= \frac{1}{2} \int_{-1}^{1} (5x^4 - 3x^2) \, dx = \frac{1}{2}(x^5 - x^3) \Big]_{-1}^{1} = 0.$$

41. (a) $\left(\frac{8}{5}, \frac{4}{5}\right)$ (b) $\left(\frac{4}{5}, \frac{8}{5}\right)$

(c)

43. (a) $\left(0, \frac{5}{2}, -\frac{5}{2}\right)$ (b) $\left(-\frac{5}{14}, -\frac{15}{14}, \frac{10}{14}\right)$

45. $\text{proj}_g f = 0$ **47.** $\text{proj}_g f = \dfrac{2e^x}{e^2 - 1}$

49. $\text{proj}_g f = 0$

51. (a) False. See the introduction to this section, page 285.
(b) True. See paragraph after "Remark," page 294.

53. (a) $\langle \mathbf{u}, \mathbf{v} \rangle = 4(2) + 2(2)(-2) = 0 \implies \mathbf{u}$ and $\mathbf{v}$ are orthogonal.
(b) Not orthogonal in the Euclidean sense

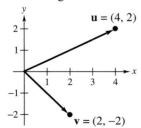

55–61. Proof

SECTION 5.3   *(page 310)*

1. Neither   **3.** Orthonormal   **5.** Orthogonal
7. Neither   **9.** Orthonormal

11. The set $\{1, x, x^2, x^3\}$ is orthogonal because
$\langle 1, x \rangle = 0, \langle 1, x^2 \rangle = 0, \langle 1, x^3 \rangle = 0,$
$\langle x, x^2 \rangle = 0, \langle x, x^3 \rangle = 0, \langle x^2, x^3 \rangle = 0.$
Furthermore, the set is orthonormal because
$\|1\| = 1, \|x\| = 1, \|x^2\| = 1,$ and $\|x^3\| = 1.$
Hence, $\{1, x, x^2, x^3\}$ is an orthonormal basis for P_3.

13. $\begin{bmatrix} \dfrac{4\sqrt{13}}{13} \\ \dfrac{7\sqrt{13}}{13} \end{bmatrix}$ **15.** $\begin{bmatrix} \dfrac{\sqrt{10}}{2} \\ -2 \\ -\dfrac{\sqrt{10}}{2} \end{bmatrix}$ **17.** $\begin{bmatrix} 11 \\ 2 \\ 15 \end{bmatrix}$

19. $\left\{\left(\frac{3}{5}, \frac{4}{5}\right), \left(\frac{4}{5}, -\frac{3}{5}\right)\right\}$

21. $\left\{\left(\frac{1}{3}, -\frac{2}{3}, \frac{2}{3}\right), \left(\frac{2}{3}, \frac{2}{3}, \frac{1}{3}\right), \left(\frac{2}{3}, -\frac{1}{3}, -\frac{2}{3}\right)\right\}$

23. $\left\{\left(\frac{4}{5}, -\frac{3}{5}, 0\right), \left(\frac{3}{5}, \frac{4}{5}, 0\right), (0, 0, 1)\right\}$

25. $\left\{\left(0, \frac{\sqrt{2}}{2}, \frac{\sqrt{2}}{2}\right), \left(\frac{\sqrt{6}}{3}, \frac{\sqrt{6}}{6}, -\frac{\sqrt{6}}{6}\right), \left(\frac{\sqrt{3}}{3}, -\frac{\sqrt{3}}{3}, \frac{\sqrt{3}}{3}\right)\right\}$

27. $\left\{\left(-\frac{4\sqrt{2}}{7}, \frac{3\sqrt{2}}{14}, \frac{5\sqrt{2}}{14}\right)\right\}$

29. $\left\{\left(\frac{3}{5}, \frac{4}{5}, 0\right), \left(\frac{4}{5}, -\frac{3}{5}, 0\right)\right\}$

31. $\left\{\left(\frac{\sqrt{6}}{6}, \frac{\sqrt{6}}{3}, -\frac{\sqrt{6}}{6}, 0\right), \left(\frac{\sqrt{3}}{3}, 0, \frac{\sqrt{3}}{3}, \frac{\sqrt{3}}{3}\right)\right\}$

33. $\langle x, 1 \rangle = \int_{-1}^{1} x \, dx = \dfrac{x^2}{2} \Big]_{-1}^{1} = 0$

35. $\langle x^2, 1 \rangle = \int_{-1}^{1} x^2 \, dx = \dfrac{x^3}{3} \Big]_{-1}^{1} = \dfrac{2}{3}$

37. (a) True. See "Definition of Orthogonal and Orthonormal Sets," page 298.
(b) True. See Corollary to Theorem 5.10, page 301.
(c) False. See "Remark," page 306.

39. $\left\{\left(\frac{3\sqrt{10}}{10}, 0, \frac{\sqrt{10}}{10}, 0\right), \left(0, -\frac{2\sqrt{5}}{5}, 0, \frac{\sqrt{5}}{5}\right)\right\}$

41. $\left\{\left(\frac{\sqrt{2}}{2}, 0, \frac{\sqrt{2}}{2}, 0\right), \left(\frac{\sqrt{6}}{6}, 0, -\frac{\sqrt{6}}{6}, \frac{\sqrt{6}}{3}\right)\right\}$

43. Orthonormal

45. $\{x^2, x, 1\}$

47. $\left\{ \dfrac{1}{\sqrt{2}}(x^2 - 1), -\dfrac{1}{\sqrt{6}}(x^2 - 2x + 1) \right\}$

49. $\left\{ \left(\dfrac{2}{3}, -\dfrac{1}{3}\right), \left(\dfrac{\sqrt{2}}{6}, \dfrac{2\sqrt{2}}{3}\right) \right\}$

51. Proof

53. Proof

55. $\left\{ \left(\dfrac{1}{\sqrt{2}}, 0, \dfrac{1}{\sqrt{2}}, 0\right), \left(0, -\dfrac{1}{\sqrt{2}}, 0, \dfrac{1}{\sqrt{2}}\right),\right.$

$\left.\left(\dfrac{1}{\sqrt{2}}, 0, -\dfrac{1}{\sqrt{2}}, 0\right), \left(0, \dfrac{1}{\sqrt{2}}, 0, \dfrac{1}{\sqrt{2}}\right) \right\}$

57. $N(A) = \text{span}\{(3, -1, 2)\}$
$N(A^T) = \text{span}\{(-1, -1, 1)\}$
$R(A) = \text{span}\{(1, 0, 1), (0, 1, 1)\}$
$R(A^T) = \text{span}\{(1, 1, -1), (0, 2, 1)\}$

59. $N(A) = \text{span}\{(1, 0, -1)\}$
$N(A^T) = \{(0, 0)\}$
$R(A) = \mathbf{R}^2$
$R(A^T) = \text{span}\{(1, 0, 1), (0, 1, 0)\}$

61. Proof

SECTION 5.4 *(page 325)*

1. Not orthogonal

3. Orthogonal

5. $\text{span}\left\{ \begin{bmatrix} 0 \\ 1 \\ 0 \end{bmatrix} \right\}$

7. $\text{span}\left\{ \begin{bmatrix} 0 \\ 0 \\ 1 \\ 0 \end{bmatrix}, \begin{bmatrix} 2 \\ -1 \\ 0 \\ 1 \end{bmatrix} \right\}$

9. $\text{span}\left\{ \begin{bmatrix} 1 \\ 2 \\ 0 \\ 0 \end{bmatrix}, \begin{bmatrix} 0 \\ 1 \\ 0 \\ 1 \end{bmatrix} \right\}$

11. $\begin{bmatrix} 0 \\ \frac{2}{3} \\ \frac{2}{3} \\ \frac{2}{3} \end{bmatrix}$

13. $\begin{bmatrix} \frac{5}{3} \\ \frac{8}{3} \\ \frac{13}{3} \end{bmatrix}$

15. $\begin{bmatrix} 1 \\ -1 \\ 2 \end{bmatrix}$

17. $R(A) = \text{span}\left\{ \begin{bmatrix} 1 \\ 0 \end{bmatrix}, \begin{bmatrix} 0 \\ 1 \end{bmatrix} \right\} = \mathbf{R}^2$

$N(A^T) = \left\{ \begin{bmatrix} 0 \\ 0 \end{bmatrix} \right\}$

$N(A) = \text{span}\left\{ \begin{bmatrix} -3 \\ 0 \\ 1 \end{bmatrix} \right\}$

$R(A^T) = \text{span}\left\{ \begin{bmatrix} 1 \\ 0 \\ 3 \end{bmatrix}, \begin{bmatrix} 0 \\ 1 \\ 0 \end{bmatrix} \right\}$

19. $R(A) = \text{span}\left\{ \begin{bmatrix} 1 \\ 0 \\ 1 \\ 1 \end{bmatrix}, \begin{bmatrix} 0 \\ 1 \\ 1 \\ 2 \end{bmatrix} \right\}$

$N(A^T) = \text{span}\left\{ \begin{bmatrix} -1 \\ -1 \\ 1 \\ 0 \end{bmatrix}, \begin{bmatrix} -1 \\ -2 \\ 0 \\ 1 \end{bmatrix} \right\}$

$N(A) = \text{span}\left\{ \begin{bmatrix} -1 \\ -1 \\ 0 \\ 1 \end{bmatrix}, \begin{bmatrix} 0 \\ -1 \\ 1 \\ 0 \end{bmatrix} \right\}$

$R(A^T) = \text{span}\left\{ \begin{bmatrix} 1 \\ 0 \\ 0 \\ 1 \end{bmatrix}, \begin{bmatrix} 0 \\ 1 \\ 1 \\ 1 \end{bmatrix} \right\}$

21. $\mathbf{x} = \begin{bmatrix} 1 \\ -1 \end{bmatrix}$

23. $\mathbf{x} = \begin{bmatrix} 2 \\ -2 \\ 1 \end{bmatrix}$

25. $y = -x + \dfrac{1}{3}$

27. $y = 1.2x + 0.45$

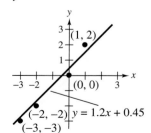

29. $y = \frac{3}{7}x^2 + \frac{6}{5}x + \frac{26}{35}$

31. Possible models are:
$S = 125.03t^2 + 866.0t + 15,704$
(Target, $t = 2$ is 1992)
$S = 237.673t^3 - 3106.98t^2 + 24,993.8t + 15,604$
(Wal-Mart, $t = 2$ is 1992)
For 2005, $t = 15$ yields
$S \approx \$56,826$ million (Target) and
$S \approx \$493,587$ million (Wal-Mart)

33. (a) False. See discussion after Example 2, pages 313 and 314.
(b) True. See "Definition of Direct Sum," page 315.
(c) True. See discussion preceding Example 7, page 320.

35. Proof.

37. Proof.

39. If A has orthonormal columns, then $A^T A = I$ and the normal equations become
$A^T A \mathbf{x} = A^T \mathbf{b}$
$\mathbf{x} = A^T \mathbf{b}.$

SECTION 5.5 *(page 341)*

1. $(0, 0, 1)$

3. $(1, 0, 0)$

5. $(-2, 3, -1)$

7. $(-1, -1, -1)$

9. $(5, -4, -3)$

11. $(1, -1, -3)$

13. 1

15. $6\sqrt{5}$

17. 1

19. 6

21. 3

23–31. Proof

33. (a) $g(x) = -\frac{1}{6} + x$
(b)

35. (a) $g(x) \approx 0.1945 + 6x$
(b)

37. (a) $g(x) \approx 0.1148 + 0.6644x$
(b)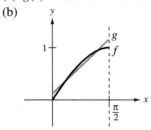

39. (a) $g(x) = 1.5x^2 - 0.6x + 0.05$
(b)

41. (a) $g(x) \approx -0.0505 + 1.3122x - 0.4177x^2$
(b)

43. $g(x) = 2 \sin x + \sin 2x + \frac{2}{3} \sin 3x$

45. $g(x) = \frac{\pi^2}{3} + 4 \cos x + \cos 2x + \frac{4}{9} \cos 3x$

47. $g(x) = \dfrac{1}{2\pi}(1 - e^{-2\pi})(1 + \cos x + \sin x)$

49. $g(x) = (1 + \pi) - 2\sin x - \sin 2x - \frac{2}{3}\sin 3x$

51. $g(x) = \sin 2x$

53. $g(x) = 2\left(\sin x + \dfrac{\sin 2x}{2} + \dfrac{\sin 3x}{3} + \cdots + \dfrac{\sin nx}{n}\right)$

REVIEW EXERCISES – CHAPTER 5 *(page 343)*

1. (a) $\sqrt{5}$
  (b) $\sqrt{17}$
  (c) 6
  (d) $\sqrt{10}$

3. (a) 3
  (b) 3
  (c) 4
  (d) $\sqrt{10}$

5. $\|\mathbf{v}\| = \sqrt{38}$

$\mathbf{u} = \left(\dfrac{5}{\sqrt{38}}, \dfrac{3}{\sqrt{38}}, -\dfrac{2}{\sqrt{38}}\right)$

7. $\dfrac{\pi}{2}$

9. π

11. $\left(-\frac{9}{13}, \frac{45}{13}\right)$

13. $\left(\frac{18}{29}, \frac{12}{29}, \frac{24}{29}\right)$

15. (a) -2 (b) $\dfrac{3\sqrt{11}}{2}$

17. Triangle Inequality:

$$\left\|(2, -\tfrac{1}{2}, 1) + (\tfrac{3}{2}, 2, -1)\right\|$$
$$\leq \left\|(2, -\tfrac{1}{2}, 1)\right\| + \left\|(\tfrac{3}{2}, 2, -1)\right\|$$
$$\dfrac{\sqrt{67}}{2} \leq \sqrt{\dfrac{15}{2}} + \sqrt{\dfrac{53}{4}}$$
$$4.093 \leq 6.379$$

Cauchy-Schwarz Inequality:

$$\left|\langle(2, -\tfrac{1}{2}, 1), (\tfrac{3}{2}, 2, -1)\rangle\right|$$
$$\leq \left\|(2, -\tfrac{1}{2}, 1)\right\| \left\|(\tfrac{3}{2}, 2, -1)\right\|$$
$$2 \leq \sqrt{\tfrac{15}{2}}\sqrt{\tfrac{53}{4}} \approx 9.969$$

19. $(s, 3t, 4t)$

21. $\left\{\left(\dfrac{1}{\sqrt{2}}, \dfrac{1}{\sqrt{2}}\right), \left(-\dfrac{1}{\sqrt{2}}, \dfrac{1}{\sqrt{2}}\right)\right\}$

23. $\left\{\left(0, \frac{3}{5}, \frac{4}{5}\right), (1, 0, 0), \left(0, \frac{4}{5}, -\frac{3}{5}\right)\right\}$

25. (a) $(-1, 4, -2) = 2(0, 2, -2) - (1, 0, -2)$

  (b) $\left\{\left(0, \dfrac{1}{\sqrt{2}}, -\dfrac{1}{\sqrt{2}}\right), \left(\dfrac{1}{\sqrt{3}}, -\dfrac{1}{\sqrt{3}}, -\dfrac{1}{\sqrt{3}}\right)\right\}$

  (c) $(-1, 4, -2) = 3\sqrt{2}\left(0, \dfrac{1}{\sqrt{2}}, -\dfrac{1}{\sqrt{2}}\right) - \sqrt{3}\left(\dfrac{1}{\sqrt{3}}, -\dfrac{1}{\sqrt{3}}, -\dfrac{1}{\sqrt{3}}\right)$

27. (a) $\dfrac{1}{4}$ (b) $\dfrac{1}{\sqrt{5}}$ (c) $\dfrac{1}{\sqrt{30}}$ (d) $\left\{\sqrt{3}x, \sqrt{5}(4x^2 - 3x)\right\}$

29. Let $f(x) = \sqrt{1 - x^2}$ and $g(x) = 2x\sqrt{1 - x^2}$.

$$\langle f, g\rangle = \int_{-1}^{1}\sqrt{1 - x^2}\, 2x\sqrt{1 - x^2}\, dx$$
$$= \int_{-1}^{1} 2x(1 - x^2)\, dx = 2\int_{-1}^{1}(x - x^3)\, dx$$
$$= 2\left[\dfrac{x^2}{2} - \dfrac{x^4}{4}\right]_{-1}^{1} = 0$$

31. $\left\{\left(-\dfrac{1}{\sqrt{2}}, 0, \dfrac{1}{\sqrt{2}}\right), \left(-\dfrac{1}{\sqrt{6}}, \dfrac{2}{\sqrt{6}}, -\dfrac{1}{\sqrt{6}}\right)\right\}$
(The answer is not unique.)

33. (a) 0   (b) Orthogonal
  (c) Because $\langle f, g\rangle = 0$, it follows that $|\langle f, g\rangle| \leq \|f\|\|g\|$.

35–41. Proof

43. $\text{span}\left\{\begin{bmatrix} 2 \\ -1 \\ 3 \end{bmatrix}\right\}$

45. $R(A) = \text{span}\left\{\begin{bmatrix} 0 \\ 0 \\ 1 \end{bmatrix}, \begin{bmatrix} 1 \\ -3 \\ 0 \end{bmatrix}\right\}$

$N(A^T) = \text{span}\left\{\begin{bmatrix} 3 \\ 1 \\ 0 \end{bmatrix}\right\}$

$N(A) = \text{span}\left\{\begin{bmatrix} 1 \\ 0 \\ -1 \end{bmatrix}\right\}$

$R(A^T) = \text{span}\left\{\begin{bmatrix} 0 \\ 1 \\ 0 \end{bmatrix}, \begin{bmatrix} 1 \\ 0 \\ 1 \end{bmatrix}\right\}$

47. Possible models are:

$y_I = -0.0025x^3 + 0.0608x^2 - 0.292x + 2.98$
(ice cream, $x = 3$ is 1993)

$y_T = -0.004527x^4 + 0.11552x^3 - 1.0232x^2$
$+ 3.810x + 0.79$
(T-bone steak, $x = 3$ is 1993)

$y_G = 0.0205x^2 - 0.257x + 2.84$
(ground beef, $x = 3$ is 1993)

For 2005, $x = 15$ yields

$y_I \approx 3.84$ dollars per pound (ice cream)
y_T is not applicable to 2005. The price is negative.
$y_G \approx 3.60$ dollars per pound (ground beef)

49. $(0, 1, -1)$ **51.** 7   **53.** 1

55. $g(x) = \dfrac{3}{5}x$ **57.** $g(x) = \dfrac{2}{\pi}$

 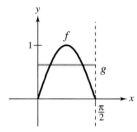

59. $g(x) = \frac{2}{35}(-10x^2 + 24x + 3)$

61. $g(x) = \dfrac{\pi^2}{3} - 4\cos x$

63. (a) True. See Theorem 5.18, part 1, page 330.
    (b) False. See Theorem 5.17, part 1, page 329.
    (c) True. See discussion before Theorem 5.19, page 335.

CUMULATIVE TEST — CHAPTERS 4 AND 5 (*page 349*)

1. (a) $(2, 1)$

    (b) $(-3, 6)$

    (c) $(-14, 8)$

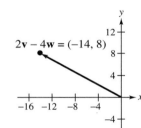

2. $\mathbf{w} = 4\mathbf{v}_1 + \mathbf{v}_2 - 2\mathbf{v}_3$

3. Proof

4. Yes   **5.** No   **6.** Yes

7. (a) A set of vectors $\{\mathbf{v}_1, \ldots, \mathbf{v}_n\}$ is linearly independent if the vector equation $c_1\mathbf{v}_1 + \cdots + c_n\mathbf{v}_n = \mathbf{0}$ has only the trivial solution.
    (b) Linearly dependent

8. The dimension is 6. One basis is

$$\left\{ \begin{bmatrix} 1 & 0 & 0 \\ 0 & 0 & 0 \\ 0 & 0 & 0 \end{bmatrix}, \begin{bmatrix} 0 & 1 & 0 \\ 1 & 0 & 0 \\ 0 & 0 & 0 \end{bmatrix}, \begin{bmatrix} 0 & 0 & 1 \\ 0 & 0 & 0 \\ 1 & 0 & 0 \end{bmatrix}, \right.$$
$$\left. \begin{bmatrix} 0 & 0 & 0 \\ 0 & 1 & 0 \\ 0 & 0 & 0 \end{bmatrix}, \begin{bmatrix} 0 & 0 & 0 \\ 0 & 0 & 1 \\ 0 & 1 & 0 \end{bmatrix}, \begin{bmatrix} 0 & 0 & 0 \\ 0 & 0 & 0 \\ 0 & 0 & 1 \end{bmatrix} \right\}$$

9. (a) A set of vectors $\{\mathbf{v}_1, \ldots, \mathbf{v}_n\}$ in a vector space V is a basis for V if the set is linearly independent and spans V.
(b) Yes

10. $\left\{\begin{bmatrix} -1 \\ 1 \\ 0 \\ 0 \end{bmatrix}, \begin{bmatrix} 0 \\ 0 \\ -1 \\ 1 \end{bmatrix}\right\}$
11. $\begin{bmatrix} -1 \\ 3 \\ -2 \end{bmatrix}$

12. $\begin{bmatrix} \frac{1}{4} & \frac{1}{2} & -\frac{1}{2} \\ \frac{7}{4} & \frac{1}{2} & \frac{1}{2} \\ -\frac{3}{4} & -\frac{1}{2} & \frac{1}{2} \end{bmatrix}$

13. (a) $\sqrt{5}$ (b) $\sqrt{11}$ (c) 4

14. $-\frac{5}{12}$

15. $\left\{(1, 0, 0), \left(0, \frac{\sqrt{2}}{2}, \frac{\sqrt{2}}{2}\right), \left(0, -\frac{\sqrt{2}}{2}, \frac{\sqrt{2}}{2}\right)\right\}$

16. $\frac{1}{13}(-3, 2)$

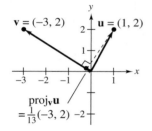

17. $R(A) = \mathbf{R}^3$

$N(A^T) = \left\{\begin{bmatrix} 0 \\ 0 \\ 0 \end{bmatrix}\right\}$

$N(A) = \text{span}\left\{\begin{bmatrix} 0 \\ 1 \\ -1 \\ 0 \end{bmatrix}\right\}$

$R(A^T) = \left\{\begin{bmatrix} 0 \\ 1 \\ 1 \\ 0 \end{bmatrix}, \begin{bmatrix} -1 \\ 0 \\ 0 \\ 1 \end{bmatrix}, \begin{bmatrix} 1 \\ 1 \\ 1 \\ 1 \end{bmatrix}\right\}$

18. $\text{span}\left\{\begin{bmatrix} -1 \\ -1 \\ 1 \end{bmatrix}\right\}$

19–21. Proof
22. $y = \frac{7}{3} - x$

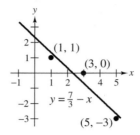

23. (a) 3
(b) One basis consists of the first three rows of A.
(c) One basis consists of columns 1, 3, and 4 of A.
(d) $\left\{\begin{bmatrix} 2 \\ 1 \\ 0 \\ 0 \\ 0 \\ 0 \end{bmatrix}, \begin{bmatrix} -3 \\ 0 \\ 5 \\ -1 \\ 1 \\ 0 \end{bmatrix}, \begin{bmatrix} -2 \\ 0 \\ 3 \\ -7 \\ 0 \\ 1 \end{bmatrix}\right\}$ (e) No
(f) No (g) Yes (h) No

24. Proof

Chapter 6

SECTION 6.1 (*page 362*)

1. (a) $(-1, 7)$ (b) $(11, -8)$
3. (a) $(1, 5, 4)$ (b) $(5, -6, t)$
5. (a) $(-14, -7)$ (b) $(1, 1, t)$
7. Not linear **9.** Not linear **11.** Linear
13. Linear **15.** $(3, 11, -8)$ **17.** $(0, -6, 8)$
19. $T: R^4 \rightarrow R^3$ **21.** $T: R^5 \rightarrow R^2$ **23.** $T: R^2 \rightarrow R^2$
25. (a) $(10, 12, 4)$ (b) $(-1, 0)$
(c) The system represented by

$$\begin{bmatrix} 1 & 2 \\ -2 & 4 \\ -2 & 2 \end{bmatrix} \begin{bmatrix} x_1 \\ x_2 \end{bmatrix} = \begin{bmatrix} 1 \\ 1 \\ 1 \end{bmatrix}$$

is inconsistent.

27. (a) $(-1, 1, 2, 1)$ (b) $(-1, 1, \frac{1}{2}, 1)$

29. (a) $\left(0, 4\sqrt{2}\right)$ (b) $\left(2\sqrt{3} - 2, 2\sqrt{3} + 2\right)$

    (c) $\left(-\dfrac{5}{2}, \dfrac{5\sqrt{3}}{2}\right)$

31. (a) True. D_x is a linear transformation and thus preserves addition and scalar multiplication.

    (b) True. D_x is a linear transformation and thus preserves addition and scalar multiplication.

    (c) False because $\sin 2x \neq 2 \sin x$ for all x

33. (a) -1 (b) $\frac{1}{12}$ (c) -4

35. $T(1, 0) = \left(\frac{1}{2}, \frac{1}{2}\right)$ **37.** $x^2 - 3x - 5$

    $T(0, 2) = (1, -1)$

39. (a) True. See discussion before "Definition of a Linear Transformation," page 353.

    (b) False because $\cos(x_1 + x_2) \neq \cos x_1 + \cos x_2$

    (c) True. See Example 10, page 361.

41. (a) $(x, 0)$ (b) Projection onto the x-axis

43. $\left(\frac{1}{2}(x + y), \frac{1}{2}(x + y)\right)$

45. Proof

47. $T(3, 4) = \left(\frac{7}{2}, \frac{7}{2}\right)$

    $T(T(3, 4)) = \left(\frac{7}{2}, \frac{7}{2}\right)$

    T is a projection onto the line $y = x$.

49. Proof   **51.** $(t, 0)$

53. Assume that h and k are not both zero. Because $T(0, 0) = (0 - h, 0 - k) = (-h, -k) \neq (0, 0)$, it follows that T is not a linear transformation.

55. If $T(x, y) = (x - h, y - k) = (x, y)$, then $h = 0$ and $k = 0$, which contradicts the assumption that h and k are not both zero. Therefore, T has no fixed points.

57. Let $T: R^3 \to R^3$ be given by $T(x, y, z) = (0, 0, 0)$. Then, if $\{\mathbf{v}_1, \mathbf{v}_2, \mathbf{v}_3\}$ is any set of vectors in R^3, the set $\{T(\mathbf{v}_1), T(\mathbf{v}_2), T(\mathbf{v}_3)\} = \{\mathbf{0}, \mathbf{0}, \mathbf{0}\}$ is linearly dependent.

59–63. Proof

Section 6.2   *(page 376)*

1. R^3

3. $\{a_1x + a_2x^2 + a_3x^3: a_1, a_2, a_3 \text{ are real}\}$

5. $\{(0, 0)\}$

7. (a) $\{(0, 0)\}$ **9.** (a) $\{(-4, -2, 1)\}$

    (b) $\{(1, 0), (0, 1)\}$ (b) $\{(1, 0), (0, 1)\}$

11. (a) $\{(-1, 1, 1, 0)\}$

    (b) $\{(1, 0, -1, 0), (0, 1, -1, 0), (0, 0, 0, 1)\}$

13. (a) $\{(0, 0)\}$ (b) 0

    (c) R^2 (d) 2

15. (a) $\{(0, 0)\}$

    (b) 0

    (c) $\{(4s, 4t, s - t): s \text{ and } t \text{ are real}\}$

    (d) 2

17. (a) $\{(-11t, 6t, 4t): t \text{ is real}\}$

    (b) 1

    (c) R^2

    (d) 2

19. (a) $\{(t, -3t): t \text{ is real}\}$

    (b) 1

    (c) $\{(3t, t): t \text{ is real}\}$

    (d) 1

21. (a) $\{(2s - t, t, 4s, -5s, s): s \text{ and } t \text{ are real}\}$

    (b) 2

    (c) $\{(7r, 7s, 7t, 8r + 20s + 2t): r, s, \text{ and } t \text{ are real}\}$

    (d) 3

23. (a) $\{(s + t, s, -2t): s \text{ and } t \text{ are real}\}$

    (b) 2

    (c) $\{(2t, -2t, t): t \text{ is real}\}$

    (d) 1

25. Nullity $= 1$ **27.** Nullity $= 3$

    Kernel: a line   Kernel: R^3

    Range: a plane   Range: $\{(0, 0, 0)\}$

29. Nullity $= 0$

    Kernel: $\{(0, 0, 0)\}$

    Range: R^3

31. Nullity $= 2$

    Kernel: $\{(x, y, z): x + 2y + 2z = 0\}$ (plane)

    Range: $\{(t, 2t, 2t), t \text{ is real}\}$ (line)

33. 2

35. 4

Zero   Standard Basis

37. (a) $(0, 0, 0, 0)$ $\{(1, 0, 0, 0), (0, 1, 0, 0),$
$(0, 0, 1, 0), (0, 0, 0, 1)\}$

(b) $\begin{bmatrix} 0 \\ 0 \\ 0 \\ 0 \end{bmatrix}$ $\left\{ \begin{bmatrix} 1 \\ 0 \\ 0 \\ 0 \end{bmatrix}, \begin{bmatrix} 0 \\ 1 \\ 0 \\ 0 \end{bmatrix}, \begin{bmatrix} 0 \\ 0 \\ 1 \\ 0 \end{bmatrix}, \begin{bmatrix} 0 \\ 0 \\ 0 \\ 1 \end{bmatrix} \right\}$

(c) $\begin{bmatrix} 0 & 0 \\ 0 & 0 \end{bmatrix}$ $\left\{ \begin{bmatrix} 1 & 0 \\ 0 & 0 \end{bmatrix}, \begin{bmatrix} 0 & 1 \\ 0 & 0 \end{bmatrix}, \begin{bmatrix} 0 & 0 \\ 1 & 0 \end{bmatrix}, \begin{bmatrix} 0 & 0 \\ 0 & 1 \end{bmatrix} \right\}$

(d) $p(x) = 0$ $\{1, x, x^2, x^3\}$

(e) $(0, 0, 0, 0, 0)$ $\{(1, 0, 0, 0, 0), (0, 1, 0, 0, 0),$
$(0, 0, 1, 0, 0), (0, 0, 0, 1, 0)\}$

39. The set of constant functions: $p(x) = a_0$

41. (a) Rank $= 1$, nullity $= 2$

(b) $\{(1, 0, -2), (1, 2, 0)\}$

43. Because $|A| = -1 \neq 0$, the homogenous equation $A\mathbf{x} = \mathbf{0}$ has only the trivial solution. Thus, $\ker(T) = \{(0, 0)\}$ and T is one-to-one (by Theorem 6.6). Furthermore, because

rank$(T) = \dim(R^2) - $ nullity(T)
$= 2 - 0 = 2 = \dim(R^2)$

T is onto (by Theorem 6.7).

45. Because $|A| = -1 \neq 0$, the homogenous equation $A\mathbf{x} = \mathbf{0}$ has only the trivial solution. Thus, $\ker(T) = \{(0, 0, 0)\}$ and T is one-to-one (by Theorem 6.6). Furthermore, because

rank$(T) = \dim(R^3) - $ nullity(T)
$= 3 - 0 = 3 = \dim(R^3)$

T is onto (by Theorem 6.7).

47. (a) False. See "Definition of Kernel of a Linear Transformation," page 365.

(b) False. See Theorem 6.4.

(c) True. See discussion before Theorem 6.6, page 372.

(d) True. See discussion before "Definition of Isomorphism," page 375.

49. (a) Rank $= n$

(b) Rank $< n$

51. $mn = jk$

53. Proof

55. Although they are not the same, they have the same dimension (4) and are isomorphic.

SECTION 6.3 *(page 387)*

1. $\begin{bmatrix} 1 & 2 \\ 1 & -2 \end{bmatrix}$

3. $\begin{bmatrix} 2 & -3 \\ 1 & -1 \\ -4 & 1 \end{bmatrix}$

5. $\begin{bmatrix} 1 & 1 & 0 \\ 1 & -1 & 0 \\ -1 & 0 & 1 \end{bmatrix}$

7. $\begin{bmatrix} 0 & -2 & 3 \\ 4 & 0 & 11 \end{bmatrix}$

9. $\begin{bmatrix} 35 \\ -7 \end{bmatrix}$

11. $\begin{bmatrix} 0 \\ 0 \end{bmatrix}$

13. (a) $\begin{bmatrix} -1 & 0 \\ 0 & -1 \end{bmatrix}$

(b) $(-3, -4)$

(c)

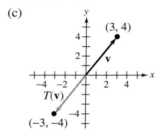

15. (a) $\begin{bmatrix} -\dfrac{\sqrt{2}}{2} & -\dfrac{\sqrt{2}}{2} \\ \dfrac{\sqrt{2}}{2} & -\dfrac{\sqrt{2}}{2} \end{bmatrix}$

(b) $\left(-4\sqrt{2}, 0\right)$

(c)

17. (a) $\begin{bmatrix} 1 & 0 & 0 \\ 0 & 1 & 0 \\ 0 & 0 & -1 \end{bmatrix}$

(b) $(3, 2, -2)$

(c)

19. (a) $\begin{bmatrix} -1 & 0 \\ 0 & -1 \end{bmatrix}$

(b) $(-1, -2)$

(c)

21. (a) $\begin{bmatrix} \frac{9}{10} & \frac{3}{10} \\ \frac{3}{10} & \frac{1}{10} \end{bmatrix}$

(b) $\left(\frac{21}{10}, \frac{7}{10}\right)$

(c)

23. (a) $\begin{bmatrix} \frac{4}{5} & \frac{3}{5} \\ \frac{3}{5} & -\frac{4}{5} \end{bmatrix}$

(b) $\left(\frac{16}{5}, -\frac{13}{5}\right)$

(c)

25. (a) $\begin{bmatrix} 2 & 3 & -1 \\ 3 & 0 & -2 \\ 2 & -1 & 1 \end{bmatrix}$ (b) $\begin{bmatrix} 9 \\ 5 \\ -1 \end{bmatrix}$

27. (a) $\begin{bmatrix} 1 & -1 & 0 & 0 \\ 0 & 0 & 1 & 0 \\ 1 & 2 & 0 & -1 \\ 0 & 0 & 0 & 1 \end{bmatrix}$ (b) $\begin{bmatrix} 1 \\ 1 \\ 2 \\ -1 \end{bmatrix}$

29. $A = \begin{bmatrix} 2 & -4 \\ -1 & -5 \end{bmatrix}, A' = \begin{bmatrix} 0 & 2 \\ 7 & -3 \end{bmatrix}$

31. $A = \begin{bmatrix} 0 & 0 & 0 \\ 1 & 0 & 0 \\ 0 & 0 & 0 \end{bmatrix}, A' = \begin{bmatrix} 0 & 0 & 0 \\ 1 & 0 & 0 \\ 0 & 0 & 0 \end{bmatrix}$

33. $A = \begin{bmatrix} -4 & -1 \\ -2 & 5 \end{bmatrix}, A' = \begin{bmatrix} 5 & 3 & 2 \\ 4 & -3 & 1 \\ -2 & -3 & -1 \end{bmatrix}$

35. $T^{-1}(x, y) = \left(\dfrac{x + y}{2}, \dfrac{x - y}{2}\right)$

37. T is not invertible.   **39.** $T^{-1}(x, y) = \left(\dfrac{x}{5}, \dfrac{y}{5}\right)$

41. $(9, 5, 4)$ **43.** $(-1, 5)$

45. $(2, -4, -3, 3)$ **47.** $(9, 16, -20)$

49. $\begin{bmatrix} 0 & 0 & 0 \\ 1 & 0 & 0 \\ 0 & 1 & 0 \\ 0 & 0 & 1 \end{bmatrix}$ **51.** $\begin{bmatrix} 0 & 1 & 0 & 0 \\ 0 & 0 & 0 & 0 \\ 0 & 0 & 1 & 1 \\ 0 & 0 & 0 & 1 \end{bmatrix}$

53. $3 - 2e^x - 2xe^x$

55. (a) $\begin{bmatrix} 0 & 0 & 0 & 0 \\ 1 & 0 & 0 & 0 \\ 0 & \frac{1}{2} & 0 & 0 \\ 0 & 0 & \frac{1}{3} & 0 \\ 0 & 0 & 0 & \frac{1}{4} \end{bmatrix}$ (b) $6x - x^2 + \frac{3}{4}x^4$

57. (a) True. See discussion before Theorem 6.11, page 381.

   (b) False. See Example 3, page 382.

   (c) False. See Theorem 6.12, page 383.

59. $\begin{bmatrix} 1 & 0 & 0 & 0 & 0 & 0 \\ 0 & 0 & 0 & 1 & 0 & 0 \\ 0 & 1 & 0 & 0 & 0 & 0 \\ 0 & 0 & 0 & 0 & 1 & 0 \\ 0 & 0 & 1 & 0 & 0 & 0 \\ 0 & 0 & 0 & 0 & 0 & 1 \end{bmatrix}$

61. Proof

63. Sometimes it is preferable to use a nonstandard basis. For example, some linear transformations have diagonal matrix representations relative to a nonstandard basis.

SECTION 6.4 *(page 395)*

1. (a) $A' = \begin{bmatrix} 4 & -3 \\ \frac{5}{3} & -1 \end{bmatrix}$

   (b) $A' = \begin{bmatrix} 1 & 0 \\ \frac{2}{3} & \frac{1}{3} \end{bmatrix} \begin{bmatrix} 2 & -1 \\ -1 & 1 \end{bmatrix} \begin{bmatrix} 1 & 0 \\ -2 & 3 \end{bmatrix}$

3. (a) $A' = \begin{bmatrix} 1 & 0 & 0 \\ 0 & 1 & 0 \\ 0 & 0 & 1 \end{bmatrix}$

   (b) $A' = \begin{bmatrix} \frac{1}{2} & \frac{1}{2} & -\frac{1}{2} \\ \frac{1}{2} & -\frac{1}{2} & \frac{1}{2} \\ -\frac{1}{2} & \frac{1}{2} & \frac{1}{2} \end{bmatrix} \begin{bmatrix} 1 & 0 & 0 \\ 0 & 1 & 0 \\ 0 & 0 & 1 \end{bmatrix} \begin{bmatrix} 1 & 1 & 0 \\ 1 & 0 & 1 \\ 0 & 1 & 1 \end{bmatrix}$

5. (a) $A' = \begin{bmatrix} \frac{7}{3} & \frac{10}{3} & -\frac{1}{3} \\ -\frac{1}{6} & \frac{4}{3} & \frac{8}{3} \\ \frac{2}{3} & -\frac{4}{3} & -\frac{2}{3} \end{bmatrix}$

   (b) $A' = \begin{bmatrix} \frac{2}{3} & -\frac{1}{3} & \frac{1}{3} \\ -\frac{1}{3} & \frac{1}{6} & \frac{1}{3} \\ \frac{1}{3} & \frac{1}{3} & -\frac{1}{3} \end{bmatrix} \begin{bmatrix} 1 & -1 & 2 \\ 2 & 1 & -1 \\ 1 & 2 & 1 \end{bmatrix} \begin{bmatrix} 1 & 0 & 1 \\ 0 & 2 & 2 \\ 1 & 2 & 0 \end{bmatrix}$

7. (a) $\begin{bmatrix} 6 & 4 \\ 9 & 4 \end{bmatrix}$

   (b) $[\mathbf{v}]_B = \begin{bmatrix} 2 \\ -1 \end{bmatrix}$, $[T(\mathbf{v})]_B = \begin{bmatrix} 4 \\ -4 \end{bmatrix}$

   (c) $A' = \begin{bmatrix} 0 & -\frac{4}{3} \\ 9 & 7 \end{bmatrix}$, $P^{-1} = \begin{bmatrix} -\frac{1}{3} & \frac{1}{3} \\ \frac{3}{4} & -\frac{1}{2} \end{bmatrix}$

   (d) $\begin{bmatrix} -\frac{8}{3} \\ 5 \end{bmatrix}$

9. (a) $\begin{bmatrix} \frac{1}{2} & \frac{1}{2} & -\frac{1}{2} \\ \frac{1}{2} & -\frac{1}{2} & \frac{1}{2} \\ -\frac{1}{2} & \frac{1}{2} & \frac{1}{2} \end{bmatrix}$

   (b) $[\mathbf{v}]_B = \begin{bmatrix} 1 \\ 0 \\ -1 \end{bmatrix}$, $[T(\mathbf{v})]_B = \begin{bmatrix} 2 \\ -1 \\ -2 \end{bmatrix}$

   (c) $A' = \begin{bmatrix} 1 & 0 & 0 \\ 0 & 2 & 0 \\ 0 & 0 & 3 \end{bmatrix}$, $P^{-1} = \begin{bmatrix} 1 & 1 & 0 \\ 1 & 0 & 1 \\ 0 & 1 & 1 \end{bmatrix}$

   (d) $\begin{bmatrix} 1 \\ 0 \\ -3 \end{bmatrix}$

11. Proof   **13.** Proof   **15.** I_n

17–23. Proof

25. The matrix of I relative to B and B' is the square matrix whose columns are the coordinates of $\mathbf{v}_1, \ldots, \mathbf{v}_n$ relative to the standard basis. The matrix of I relative to B, or relative to B', is the identity matrix.

27. (a) True. See discussion, page 390, and note that $A' = P^{-1}AP \implies PA'P^{-1} = PP^{-1}APP^{-1} = A$.

   (b) False. Unless it is a diagonal matrix; see Example 5, pages 393–94.

SECTION 6.5 *(page 403)*

1. (a) $(3, -5)$ (b) $(2, 1)$

   (c) $(a, 0)$ (d) $(0, -b)$

   (e) $(-c, -d)$ (f) (f, g)

3. (a) (y, x) (b) Reflection in the line $y = x$

5. (a) Vertical contraction

   (b)

7. (a) Horizontal shear

   (b)
   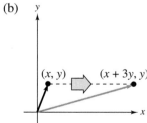

9. $\{(0, t): t \text{ is real}\}$ **11.** $\{(t, 0): t \text{ is real}\}$

13.

15.

17.

19.

21.

23.

25. $T(1, 0) = (2, 0)$, $T(0, 1) = (0, 3)$, $T(2, 2) = (4, 6)$

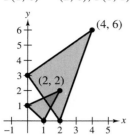

27. Horizontal expansion

29. Reflection in the line $y = x$

31. Reflection in the x-axis followed by a vertical expansion

33. Vertical shear followed by a horizontal expansion

35. $\begin{bmatrix} \dfrac{\sqrt{3}}{2} & -\dfrac{1}{2} & 0 \\ \dfrac{1}{2} & \dfrac{\sqrt{3}}{2} & 0 \\ 0 & 0 & 1 \end{bmatrix}$ **37.** $\begin{bmatrix} \dfrac{1}{2} & 0 & \dfrac{\sqrt{3}}{2} \\ 0 & 1 & 0 \\ -\dfrac{\sqrt{3}}{2} & 0 & \dfrac{1}{2} \end{bmatrix}$

39. $\left(\dfrac{\sqrt{3} - 1}{2}, \dfrac{\sqrt{3} + 1}{2}, 1 \right)$

41. $\left(\dfrac{1 + \sqrt{3}}{2}, 1, \dfrac{1 - \sqrt{3}}{2} \right)$

43. $90°$ about the x-axis   **45.** $180°$ about the y-axis

47. $90°$ about the z-axis

49. $\begin{bmatrix} 0 & 1 & 0 \\ 0 & 0 & -1 \\ -1 & 0 & 0 \end{bmatrix}$

Line segment from $(0, 0, 0)$ to $(1, -1, -1)$

51. $\begin{bmatrix} \dfrac{\sqrt{3}}{4} & -\dfrac{1}{4} & \dfrac{\sqrt{3}}{2} \\ \dfrac{1}{2} & \dfrac{\sqrt{3}}{2} & 0 \\ -\dfrac{3}{4} & \dfrac{\sqrt{3}}{4} & \dfrac{1}{2} \end{bmatrix}$

Line segment from $(0, 0, 0)$ to

$\left(\dfrac{3\sqrt{3} - 1}{4}, \dfrac{1 + \sqrt{3}}{2}, \dfrac{\sqrt{3} - 1}{4} \right)$

REVIEW EXERCISES – CHAPTER 6   (*page 405*)

1. (a) $(2, -4)$ (b) $(4, 4)$

3. Linear, $\begin{bmatrix} 1 & 2 \\ -1 & -1 \end{bmatrix}$ **5.** Not linear

7. Linear, $\begin{bmatrix} 1 & -1 & 0 \\ 0 & 1 & -1 \\ -1 & 0 & 1 \end{bmatrix}$

9. $T(1, 1) = \left(\frac{3}{2}, \frac{3}{2} \right)$ **11.** $A^2 = I$

$T(0, 1) = (1, 1)$

13. $A^3 = \begin{bmatrix} \cos 3\theta & -\sin 3\theta \\ \sin 3\theta & \cos 3\theta \end{bmatrix}$

15. (a) $T: R^3 \to R^2$

(b) $(3, -12)$

(c) $\left\{ \left(-\frac{5}{2}, 3 - 2t, t \right): t \text{ is real} \right\}$

17. (a) $T: R^3 \to R^3$

(b) $(-2, -4, -5)$

(c) $(2, 2, 2)$

19. (a) $\{(2, -1, 0, 0), (2, 0, 1, -2)\}$

(b) $\{(5, 0, 4), (0, 5, 8)\}$

21. (a) $\{(0, 0)\}$

(b) $\left\{ \left(1, 0, \frac{1}{2} \right), \left(0, 1, -\frac{1}{2} \right) \right\}$

(c) Rank $= 2$

(d) Nullity $= 0$

23. 3   **25.** 2

27. $A = \begin{bmatrix} 2 & 0 \\ 0 & 1 \end{bmatrix}$, $A^{-1} = \begin{bmatrix} \frac{1}{2} & 0 \\ 0 & 1 \end{bmatrix}$

29. T has no inverse.

31. $A = \begin{bmatrix} 0 & 0 & 0 \\ 0 & 1 & 0 \\ 0 & 1 & 0 \end{bmatrix}$, $A' = \begin{bmatrix} 0 & 0 \\ 1 & 1 \end{bmatrix}$

33.

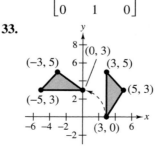

35. (a) One-to-one
(b) Onto
(c) Invertible

37. (a) One-to-one
(b) Onto
(c) Invertible

39. (a) $(0, 1, 1)$
(b) $(0, 1, 1)$

41. $A' = \begin{bmatrix} 3 & -1 \\ 1 & -1 \end{bmatrix}$,

$A' = P^{-1}AP = \begin{bmatrix} \frac{1}{2} & -\frac{1}{2} \\ \frac{1}{2} & \frac{1}{2} \end{bmatrix} \begin{bmatrix} 1 & -3 \\ -1 & 1 \end{bmatrix} \begin{bmatrix} 1 & 1 \\ -1 & 1 \end{bmatrix}$

43. (a) $\begin{bmatrix} 0 & 0 & 0 \\ 0 & \frac{1}{5} & \frac{2}{5} \\ 0 & \frac{2}{5} & \frac{4}{5} \end{bmatrix}$
(b) Answers will vary.
(c) Answers will vary.

45. Answers will vary. **47.** Proof

49. (a) Proof
(b) Nullity $= 3$, rank $= 1$
(c) $\{1 - x, 1 - x^2, 1 - x^3\}$

51. $\text{Ker}(T) = \{\mathbf{v}: \langle \mathbf{v}, \mathbf{v}_0 \rangle = 0\}$
Range $= R$
Rank $= 1$
Nullity $= \dim(V) - 1$

53. $mn = pq$

55. (a) Vertical expansion
(b)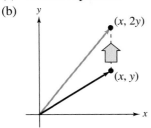

57. (a) Vertical shear
(b)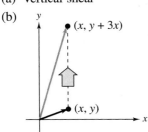

59.

61.

63. Reflection in the line $y = x$ followed by a horizontal expansion

65. $\begin{bmatrix} \frac{\sqrt{2}}{2} & -\frac{\sqrt{2}}{2} & 0 \\ \frac{\sqrt{2}}{2} & \frac{\sqrt{2}}{2} & 0 \\ 0 & 0 & 1 \end{bmatrix}, (\sqrt{2}, 0, 1)$

67. $\begin{bmatrix} \frac{\sqrt{3}}{2} & -\frac{1}{4} & \frac{\sqrt{3}}{4} \\ \frac{1}{2} & \frac{\sqrt{3}}{4} & -\frac{3}{4} \\ 0 & \frac{\sqrt{3}}{2} & \frac{1}{2} \end{bmatrix}$

69. $(0, 0, 0), \left(\frac{\sqrt{2}}{2}, \frac{\sqrt{2}}{2}, 0\right), (0, \sqrt{2}, 0),$
$\left(-\frac{\sqrt{2}}{2}, \frac{\sqrt{2}}{2}, 0\right), (0, 0, 1), \left(\frac{\sqrt{2}}{2}, \frac{\sqrt{2}}{2}, 1\right),$
$\left(0, \sqrt{2}, 1\right), \left(-\frac{\sqrt{2}}{2}, \frac{\sqrt{2}}{2}, 1\right)$

71. (a) False. See "Elementary Matrices for Linear Transformations in the Plane," page 396.

(b) True. See "Elementary Matrices for Linear Transformations in the Plane," page 396.

(c) True. See discussion following Example 4, page 401.

73. (a) False. See "Remark," page 355.

(b) False. See Theorem 6.7, page 373.

(c) True. See discussion following Example 5, page 394.

Chapter 7

SECTION 7.1 *(page 421)*

1. $\begin{bmatrix} 1 & 0 \\ 0 & -1 \end{bmatrix}\begin{bmatrix} 1 \\ 0 \end{bmatrix} = 1\begin{bmatrix} 1 \\ 0 \end{bmatrix}, \begin{bmatrix} 1 & 0 \\ 0 & -1 \end{bmatrix}\begin{bmatrix} 0 \\ 1 \end{bmatrix} = -1\begin{bmatrix} 0 \\ 1 \end{bmatrix}$

3. $\begin{bmatrix} 1 & 1 \\ 1 & 1 \end{bmatrix}\begin{bmatrix} 1 \\ -1 \end{bmatrix} = 0\begin{bmatrix} 1 \\ -1 \end{bmatrix}, \begin{bmatrix} 1 & 1 \\ 1 & 1 \end{bmatrix}\begin{bmatrix} 1 \\ 1 \end{bmatrix} = 2\begin{bmatrix} 1 \\ 1 \end{bmatrix}$

5. $\begin{bmatrix} -2 & 2 & -3 \\ 2 & 1 & -6 \\ -1 & -2 & 0 \end{bmatrix}\begin{bmatrix} 1 \\ 2 \\ -1 \end{bmatrix} = 5\begin{bmatrix} 1 \\ 2 \\ -1 \end{bmatrix},$

$\begin{bmatrix} -2 & 2 & -3 \\ 2 & 1 & -6 \\ -1 & -2 & 0 \end{bmatrix}\begin{bmatrix} -2 \\ 1 \\ 0 \end{bmatrix} = -3\begin{bmatrix} -2 \\ 1 \\ 0 \end{bmatrix},$

$\begin{bmatrix} -2 & 2 & -3 \\ 2 & 1 & -6 \\ -1 & -2 & 0 \end{bmatrix}\begin{bmatrix} 3 \\ 0 \\ 1 \end{bmatrix} = -3\begin{bmatrix} 3 \\ 0 \\ 1 \end{bmatrix}$

7. (a) $\begin{bmatrix} 1 & 1 \\ 1 & 1 \end{bmatrix}\begin{bmatrix} c \\ -c \end{bmatrix} = 0\begin{bmatrix} c \\ -c \end{bmatrix}$

(b) $\begin{bmatrix} 1 & 1 \\ 1 & 1 \end{bmatrix}\begin{bmatrix} c \\ c \end{bmatrix} = \begin{bmatrix} 2c \\ 2c \end{bmatrix} = 2\begin{bmatrix} c \\ c \end{bmatrix}$

9. (a) No   (b) Yes   (c) Yes   (d) No

11. (a) Yes   (b) No   (c) Yes   (d) Yes

13. (a) $\lambda^2 - 7\lambda = 0$

(b) $\lambda = 0, (1, 2)$

$\lambda = 7, (3, -1)$

15. (a) $\lambda^2 - \frac{1}{4} = 0$

(b) $\lambda = -\frac{1}{2}, (1, 1)$

$\lambda = \frac{1}{2}, (3, 1)$

17. (a) $\lambda^3 - 6\lambda^2 + 11\lambda - 6 = 0$

(b) $\lambda = 1, (1, 2, -1)$

$\lambda = 2, (1, 0, 0)$

$\lambda = 3, (0, 1, 0)$

19. (a) $\lambda^3 - 3\lambda^2 - 9\lambda + 27 = 0$

(b) $\lambda = -3, (1, 1, 3)$

$\lambda = 3, (1, 0, -1), (1, 1, 0)$

21. (a) $\lambda^3 - 8\lambda^2 + 4\lambda + 48 = 0$

(b) $\lambda = -2, (3, 2, 0)$

$\lambda = 4, (5, -10, -2)$

$\lambda = 6, (1, -2, 0)$

23. (a) $\lambda^4 - 7\lambda^3 + 16\lambda^2 - 12\lambda = 0$

(b) $\lambda = 2, (1, 0, 0, 0), (0, 1, 0, 0)$

$\lambda = 3, (0, 0, 3, 4)$

$\lambda = 0, (0, 0, 0, 1)$

25. $\lambda = -2, 1$ **27.** $\lambda = 5, 5$

29. $\lambda = 4, -\frac{1}{2}, \frac{1}{3}$ **31.** $\lambda = 0, 3$

33. $p(\lambda) = \lambda^2 - 6\lambda + 8, p(A) = \begin{bmatrix} 0 & 0 \\ 0 & 0 \end{bmatrix}$

35. $p(\lambda) = \lambda^3 - 5\lambda^2 + 15\lambda - 27, p(A) = \begin{bmatrix} 0 & 0 & 0 \\ 0 & 0 & 0 \\ 0 & 0 & 0 \end{bmatrix}$

37.

Exercise	Trace of A	Determinant of A
13	7	0
14	9	0
15	0	$-\frac{1}{4}$
16	$\frac{1}{4}$	$-\frac{1}{8}$
17	6	6
18	5	-105
19	3	-27
20	4	0
21	8	-48
22	$\frac{31}{2}$	$\frac{29}{8}$
23	7	0
24	8	12

39. Assume that λ is an eigenvalue of A, with corresponding eigenvector $\mathbf{x}$. Because A is invertible (from Exercise 38), $\lambda \neq 0$. Then, $A\mathbf{x} = \lambda\mathbf{x}$ implies that $\mathbf{x} = A^{-1}A\mathbf{x} = A^{-1}\lambda\mathbf{x} = \lambda A^{-1}\mathbf{x}$, which in turn implies that $(1/\lambda)\mathbf{x} = A^{-1}\mathbf{x}$. Thus $\mathbf{x}$ is an eigenvector of A^{-1}, and its corresponding eigenvalue is $1/\lambda$.

41. Proof   **43.** Proof

45. $a = 0,\ d = 1$ or $a = 1,\ d = 0$

47. (a) True. See "Definition of Eigenvalue and Eigenvector," page 411.

(b) True. See discussion before Theorem 7.1, page 413.

(c) True. See Theorem 7.2, page 414.

49. Dim $= 3$

51. Dim $= 1$

53. $T(e^x) = \dfrac{d}{dx}[e^x] = e^x = 1(e^x)$

55. $\lambda = -2,\ 3 + 2x$

$\lambda = 4,\ -5 + 10x + 2x^2$

$\lambda = 6,\ -1 + 2x$

57. $\lambda = 0,\ \begin{bmatrix} 1 & 0 \\ 1 & 0 \end{bmatrix}, \begin{bmatrix} 1 & 1 \\ 0 & -1 \end{bmatrix}$

$\lambda = 3,\ \begin{bmatrix} 1 & 0 \\ -2 & 0 \end{bmatrix}$

59. The only possible eigenvalue is 0.

61. Proof

SECTION 7.2 *(page 432)*

1. $P^{-1} = \begin{bmatrix} 1 & -4 \\ -1 & 3 \end{bmatrix}$, $P^{-1}AP = \begin{bmatrix} 1 & 0 \\ 0 & -2 \end{bmatrix}$

3. $P^{-1} = \begin{bmatrix} \frac{2}{3} & -\frac{2}{3} & 1 \\ 0 & \frac{1}{4} & 0 \\ -\frac{1}{3} & \frac{1}{12} & 0 \end{bmatrix}$, $P^{-1}AP = \begin{bmatrix} 5 & 0 & 0 \\ 0 & 3 & 0 \\ 0 & 0 & -1 \end{bmatrix}$

5. There is only one eigenvalue, $\lambda = 0$, and the dimension of its eigenspace is 1. Therefore, the matrix is not diagonalizable.

7. There is only one eigenvalue, $\lambda = 3$, and the dimension of its eigenspace is 1. Therefore, the matrix is not diagonalizable.

9. There are two eigenvalues, 1 and 2. The dimension of the eigenspace for the repeated eigenvalue 1 is 1. Therefore, the matrix is not diagonalizable.

11. There are two repeated eigenvalues, 0 and 3. The eigenspace associated with 3 is of dimension 1. Therefore, the matrix is not diagonalizable.

13. $\lambda = 0,\ 2$ Matrix is diagonalizable.

15. $\lambda = 0,\ 2$ Insufficient number of eigenvalues to guarantee diagonalizability

17. $P = \begin{bmatrix} 3 & 1 \\ 1 & 1 \end{bmatrix}$

(The answer is not unique.)

19. $P = \begin{bmatrix} 1 & -1 & 1 \\ 1 & 0 & 1 \\ 3 & 1 & 0 \end{bmatrix}$

(The answer is not unique.)

21. $P = \begin{bmatrix} 3 & -1 & -5 \\ 2 & 2 & 10 \\ 0 & 0 & 2 \end{bmatrix}$

(The answer is not unique.)

23. A is not diagonalizable.

25. $P = \begin{bmatrix} 4 & 0 & 0 & 0 \\ 4 & 0 & -2 & 0 \\ 4 & 3 & 1 & 0 \\ 1 & 1 & 1 & 1 \end{bmatrix}$

(The answer is not unique.)

27. $\{(1, -1), (1, 1)\}$ **29.** $\{(-1 + x), x\}$

31. Proof

33. $\begin{bmatrix} -188 & -378 \\ 126 & 253 \end{bmatrix}$

35. $\begin{bmatrix} 384 & 256 & -384 \\ -384 & -512 & 1152 \\ -128 & -256 & 640 \end{bmatrix}$

37. (a) True. See the proof of Theorem 7.4, page 424.

(b) False. See Theorem 7.6, page 430.

39. Yes, the order of elements on the main diagonal may change.

41–47. Proof

SECTION 7.3 *(page 444)*

1. Symmetric   **3.** Not symmetric

5. Symmetric   **7.** $\lambda = 2,\ \dim = 1$

$\qquad\qquad\qquad\qquad\quad\ \lambda = 4,\ \dim = 1$

9. $\lambda = 2,\ \dim = 2$ **11.** $\lambda = -2,\ \dim = 2$

$\ \ \lambda = 3,\ \dim = 1$ $\qquad\ \lambda = \ \ 4,\ \dim = 1$

13. Orthogonal

15. Not orthogonal

17. Orthogonal

19. Orthogonal

21. $P = \begin{bmatrix} \dfrac{\sqrt{2}}{2} & \dfrac{\sqrt{2}}{2} \\ -\dfrac{\sqrt{2}}{2} & \dfrac{\sqrt{2}}{2} \end{bmatrix}$

(The answer is not unique.)

23. $P = \begin{bmatrix} \dfrac{\sqrt{3}}{3} & \dfrac{\sqrt{6}}{3} \\ -\dfrac{\sqrt{6}}{3} & \dfrac{\sqrt{3}}{3} \end{bmatrix}$

(The answer is not unique.)

25. $P = \begin{bmatrix} -\dfrac{2}{3} & -\dfrac{1}{3} & \dfrac{2}{3} \\ \dfrac{1}{3} & \dfrac{2}{3} & \dfrac{2}{3} \\ \dfrac{2}{3} & -\dfrac{2}{3} & \dfrac{1}{3} \end{bmatrix}$

(The answer is not unique.)

27. $P = \begin{bmatrix} \dfrac{\sqrt{2}}{2} & 0 & \dfrac{\sqrt{2}}{2} & 0 \\ -\dfrac{\sqrt{2}}{2} & 0 & \dfrac{\sqrt{2}}{2} & 0 \\ 0 & \dfrac{\sqrt{2}}{2} & 0 & \dfrac{\sqrt{2}}{2} \\ 0 & -\dfrac{\sqrt{2}}{2} & 0 & \dfrac{\sqrt{2}}{2} \end{bmatrix}$

(The answer is not unique.)

29. (a) True. See Theorem 7.10, page 440.

(b) True. See Theorem 7.9, page 439.

31. Proof   **33.** Proof

35. $A^{-1} = \left(\dfrac{1}{\cos^2 \theta + \sin^2 \theta} \right) \begin{bmatrix} \cos \theta & \sin \theta \\ -\sin \theta & \cos \theta \end{bmatrix}$

$\qquad = \begin{bmatrix} \cos \theta & \sin \theta \\ -\sin \theta & \cos \theta \end{bmatrix} = A^T$

37. Proof

SECTION 7.4 *(page 459)*

1. $\mathbf{x}_2 = \begin{bmatrix} 20 \\ 5 \end{bmatrix},\ \mathbf{x}_3 = \begin{bmatrix} 10 \\ 10 \end{bmatrix}$

3. $\mathbf{x}_2 = \begin{bmatrix} 84 \\ 12 \\ 6 \end{bmatrix},\ \mathbf{x}_3 = \begin{bmatrix} 60 \\ 84 \\ 6 \end{bmatrix}$ **5.** $\mathbf{x} = t\begin{bmatrix} 2 \\ 1 \end{bmatrix}$

7. $\mathbf{x} = t\begin{bmatrix} 8 \\ 4 \\ 1 \end{bmatrix}$ **9.** $\mathbf{x}_2 = \begin{bmatrix} 960 \\ 90 \\ 30 \end{bmatrix},\ \mathbf{x}_3 = \begin{bmatrix} 2340 \\ 720 \\ \frac{45}{2} \end{bmatrix}$

11. $y_1 = C_1 e^{2t}$
$\quad\ y_2 = C_2 e^t$

13. $y_1 = C_1 e^{-t}$
$\quad\ y_2 = C_2 e^{6t}$
$\quad\ y_3 = C_3 e^t$

15. $y_1 = C_1 e^t - 4C_2 e^{2t}$
$\quad\ y_2 = C_2 e^{2t}$

17. $y_1 = \ \ C_1 e^{-t} + C_2 e^{3t}$
$\quad\ y_2 = -C_1 e^{-t} + C_2 e^{3t}$

19. $y_1 = 3C_1 e^{-2t} - \ \ 5C_2 e^{4t} - \ C_3 e^{6t}$
$\quad\ y_2 = 2C_1 e^{-2t} + 10C_2 e^{4t} + 2C_3 e^{6t}$
$\quad\ y_3 = \qquad\qquad\quad\ 2C_2 e^{4t}$

21. $y_1 = C_1 e^t - 2C_2 e^{2t} - 7C_3 e^{3t}$
$\quad\ y_2 = \qquad\quad\ C_2 e^{2t} + 8C_3 e^{3t}$
$\quad\ y_3 = \qquad\qquad\qquad\ 2C_3 e^{3t}$

23. $y_1' = y_1 + y_2$
$\quad\ y_2' = \qquad y_2$

25. $y_1' = y_2$
$\quad\ y_2' = y_3$
$\quad\ y_3' = -4y_2$

27. $\begin{bmatrix} 1 & 0 \\ 0 & 1 \end{bmatrix}$ **29.** $\begin{bmatrix} 9 & 5 \\ 5 & -4 \end{bmatrix}$ **31.** $\begin{bmatrix} 0 & 5 \\ 5 & -10 \end{bmatrix}$

33. $A = \begin{bmatrix} 2 & -\frac{3}{2} \\ -\frac{3}{2} & -2 \end{bmatrix},$

$P = \begin{bmatrix} \dfrac{1}{\sqrt{10}} & -\dfrac{3}{\sqrt{10}} \\ \dfrac{3}{\sqrt{10}} & \dfrac{1}{\sqrt{10}} \end{bmatrix},$

$\lambda_1 = -\frac{5}{2},\ \lambda_2 = \frac{5}{2}$

35. $A = \begin{bmatrix} 13 & 3\sqrt{3} \\ 3\sqrt{3} & 7 \end{bmatrix},$

$P = \begin{bmatrix} \dfrac{1}{2} & \dfrac{\sqrt{3}}{2} \\ -\dfrac{\sqrt{3}}{2} & \dfrac{1}{2} \end{bmatrix},$

$\lambda_1 = 4,\ \lambda_2 = 16$

37. $A = \begin{bmatrix} 16 & -12 \\ -12 & 9 \end{bmatrix},$

$P = \begin{bmatrix} \frac{3}{5} & -\frac{4}{5} \\ \frac{4}{5} & \frac{3}{5} \end{bmatrix},$

$\lambda_1 = 0,\ \lambda_2 = 25$

39. Ellipse, $5(x')^2 + 15(y')^2 - 45 = 0$

41. Hyperbola, $-25(x')^2 + 15(y')^2 - 50 = 0$

43. Parabola, $4(y')^2 + 4x' + 8y' + 4 = 0$

45. Hyperbola,
$$\frac{1}{2}\left[-(x')^2 + (y')^2 - 3\sqrt{2}x' - \sqrt{2}y' + 6\right] = 0$$

47. $A = \begin{bmatrix} 3 & -1 & 0 \\ -1 & 3 & 0 \\ 0 & 0 & 8 \end{bmatrix}$,

$2(x')^2 + 4(y')^2 + 8(z')^2 - 16 = 0$

49. Let $P = \begin{bmatrix} a & b \\ c & d \end{bmatrix}$ be a 2×2 orthogonal matrix such that $|P| = 1$. Define $\theta \in (0, 2\pi)$ as follows.

(i) If $a = 1$, then $c = 0$, $b = 0$, and $d = 1$, so let $\theta = 0$.

(ii) If $a = -1$, then $c = 0$, $b = 0$, and $d = -1$, so let $\theta = \pi$.

(iii) If $a \geq 0$ and $c > 0$, let $\theta = \arccos(a)$, $0 < \theta \leq \pi/2$.

(iv) If $a \geq 0$ and $c < 0$, let $\theta = 2\pi - \arccos(a)$, $3\pi/2 \leq \theta < 2\pi$.

(v) If $a \leq 0$ and $c > 0$, let $\theta = \arccos(a)$, $\pi/2 \leq \theta < \pi$.

(vi) If $a \leq 0$ and $c < 0$, let $\theta = 2\pi - \arccos(a)$, $\pi < \theta \leq 3\pi/2$.

In each of these cases, check that
$$P = \begin{bmatrix} a & b \\ c & d \end{bmatrix} = \begin{bmatrix} \cos\theta & -\sin\theta \\ \sin\theta & \cos\theta \end{bmatrix}.$$

REVIEW EXERCISES – CHAPTER 7 *(page 461)*

1. (a) $\lambda^2 - 9 = 0$
(b) $\lambda = -3$, $\lambda = 3$
(c) A basis for $\lambda = -3$ is $(1, -5)$ and a basis for $\lambda = 3$ is $(1, 1)$.

3. (a) $\lambda^3 - 20\lambda^2 + 128\lambda - 256 = 0$
(b) $\lambda = 4$, $\lambda = 8$
(c) A basis for $\lambda = 4$ is $(1, -2, -1)$ and a basis for $\lambda = 8$ is $(4, -1, 0)$, $(3, 0, 1)$.

5. (a) $\lambda^3 - 6\lambda^2 + 11\lambda - 6 = 0$
(b) $\lambda = 1$, $\lambda = 2$, $\lambda = 3$
(c) A basis for $\lambda = 1$ is $(1, 2, -1)$, a basis for $\lambda = 2$ is $(1, 0, 0)$, and a basis for $\lambda = 3$ is $(0, 1, 0)$.

7. Not diagonalizable

9. $P = \begin{bmatrix} 1 & 0 & 1 \\ 0 & 1 & 0 \\ 1 & 0 & -1 \end{bmatrix}$
(The answer is not unique.)

11. The characteristic equation of
$$A = \begin{bmatrix} \cos\theta & -\sin\theta \\ \sin\theta & \cos\theta \end{bmatrix}$$
is $\lambda^2 - (2\cos\theta)\lambda + 1 = 0$. The roots of this equation are $\lambda = \cos\theta \pm \sqrt{\cos^2\theta - 1}$. If $0 < \theta < \pi$, then $-1 < \cos\theta < 1$, which implies that $\sqrt{\cos^2\theta - 1}$ is imaginary.

13. A has only one eigenvalue, $\lambda = 0$, and the dimension of its eigenspace is 1. Therefore, the matrix is not diagonalizable.

15. A has only one eigenvalue, $\lambda = 3$, and the dimension of its eigenspace is 2. Therefore, the matrix is not diagonalizable.

17. $P = \begin{bmatrix} 0 & 1 \\ 1 & 0 \end{bmatrix}$

19. Both orthogonal and symmetric

21. Symmetric

23. $P = \begin{bmatrix} \dfrac{2}{\sqrt{5}} & -\dfrac{1}{\sqrt{5}} \\ \dfrac{1}{\sqrt{5}} & \dfrac{2}{\sqrt{5}} \end{bmatrix}$
(The answer is not unique.)

25. $P = \begin{bmatrix} \dfrac{1}{\sqrt{2}} & 0 & \dfrac{1}{\sqrt{2}} \\ 0 & 1 & 0 \\ -\dfrac{1}{\sqrt{2}} & 0 & \dfrac{1}{\sqrt{2}} \end{bmatrix}$
(The answer is not unique.)

27. $\left(\frac{3}{5}, \frac{2}{5}\right)$ **29.** $\left(\frac{1}{4}, \frac{1}{2}, \frac{1}{4}\right)$ **31.** $\left(\frac{4}{16}, \frac{5}{16}, \frac{7}{16}\right)$

33. Proof

35. $A = \begin{bmatrix} 0 & 1 \\ 0 & \frac{9}{4} \end{bmatrix}$, $\lambda_1 = 0$, $\lambda_2 = \frac{9}{4}$

37. $A^2 = \begin{bmatrix} 56 & -40 \\ 20 & -4 \end{bmatrix}$, $A^3 = \begin{bmatrix} 368 & -304 \\ 152 & -88 \end{bmatrix}$

39. Proof **41.** Proof

43. $P = \begin{bmatrix} \dfrac{1}{\sqrt{2}} & -\dfrac{1}{\sqrt{2}} \\ \dfrac{1}{\sqrt{2}} & \dfrac{1}{\sqrt{2}} \end{bmatrix}$

45. (a) $a = b = c = 0$
  (b) Dim $= 1$ if $a \neq 0$, $b \neq 0$, $c \neq 0$.
  Dim $= 2$ if exactly one is 0.
  Dim $= 3$ if exactly two are 0.

47. (a) True. See "Definition of Eigenvalue and Eigenvector," page 411.
  (b) False. See Theorem 7.4, page 424.
  (c) True. See "Definition of a Diagonalizable Matrix," page 423.

49. $\mathbf{x}_2 = \begin{bmatrix} 100 \\ 25 \end{bmatrix}$, $\mathbf{x}_3 = \begin{bmatrix} 25 \\ 25 \end{bmatrix}$, $\mathbf{x} = t \begin{bmatrix} 2 \\ 1 \end{bmatrix}$

51. $\mathbf{x}_2 = \begin{bmatrix} 4500 \\ 300 \\ 50 \end{bmatrix}$, $\mathbf{x}_3 = \begin{bmatrix} 1500 \\ 4500 \\ 50 \end{bmatrix}$, $\mathbf{x} = t \begin{bmatrix} 24 \\ 12 \\ 1 \end{bmatrix}$

53. $y_1 = -2C_1 + C_2 e^t$
  $y_2 = C_1$

55. $y_1 = C_1 e^t + C_2 e^{-t}$
  $y_2 = C_1 e^t - C_2 e^{-t}$

57. $A = \begin{bmatrix} 1 & \frac{3}{2} \\ \frac{3}{2} & 1 \end{bmatrix}$

$P = \begin{bmatrix} \dfrac{1}{\sqrt{2}} & -\dfrac{1}{\sqrt{2}} \\ \dfrac{1}{\sqrt{2}} & \dfrac{1}{\sqrt{2}} \end{bmatrix}$

$5(x')^2 - (y')^2 = 6$

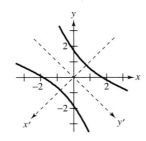

59. $A = \begin{bmatrix} 0 & \frac{1}{2} \\ \frac{1}{2} & 0 \end{bmatrix}$

$P = \begin{bmatrix} \dfrac{1}{\sqrt{2}} & -\dfrac{1}{\sqrt{2}} \\ \dfrac{1}{\sqrt{2}} & \dfrac{1}{\sqrt{2}} \end{bmatrix}$

$(x')^2 - (y')^2 = 4$

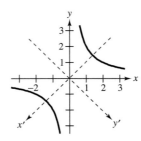

CUMULATIVE TEST FOR CHAPTERS 6 AND 7 (*page 465*)

1. Yes, T is a linear transformation.

2. No, T is not a linear transformation.

3. (a) $(1, -1, 0)$ (b) $(1, -3)$

4. Span$\{(1, 1, 0, 0), (0, 0, -1, 1)\}$

5. (a) Span$\{(0, -1, 0, 1), (1, 0, -1, 0)\}$
  (b) Span$\{(1, 0), (0, 1)\}$
  (c) Rank $= 2$, nullity $= 2$

6. $\begin{bmatrix} -1 & 1 & 0 \\ 0 & -1 & 1 \\ 1 & 0 & -1 \end{bmatrix}$

7. $\begin{bmatrix} \frac{1}{2} & -\frac{1}{2} \\ -\frac{1}{2} & \frac{1}{2} \end{bmatrix}$, $T(1, 1) = (0, 0)$, $T(-2, 2) = (-2, 2)$

8. $T^{-1}(x, y) = \left(\frac{1}{3}x + \frac{1}{3}y, -\frac{2}{3}x + \frac{1}{3}y\right)$

9. $\begin{bmatrix} -1 & -2 \\ 0 & 1 \\ 2 & 1 \end{bmatrix}$, $T(0, 1) = (1, 0, 1)$

10. (a) $A = \begin{bmatrix} 1 & -2 \\ 1 & 4 \end{bmatrix}$ (b) $P = \begin{bmatrix} 1 & 1 \\ 1 & 2 \end{bmatrix}$
  (c) $A' = \begin{bmatrix} -7 & -15 \\ 6 & 12 \end{bmatrix}$ (d) $\begin{bmatrix} 9 \\ -6 \end{bmatrix}$
  (e) $[\mathbf{v}]_B = \begin{bmatrix} 1 \\ -1 \end{bmatrix}$, $[T(\mathbf{v})]_B = \begin{bmatrix} 3 \\ -3 \end{bmatrix}$

11. $\lambda = 1, \begin{bmatrix} 1 \\ 0 \\ 0 \end{bmatrix}$; $\lambda = 0, \begin{bmatrix} -1 \\ -1 \\ 3 \end{bmatrix}$; $\lambda = 2, \begin{bmatrix} 1 \\ 1 \\ -1 \end{bmatrix}$

12. $\lambda = 1$ (three times), $\begin{bmatrix} 1 \\ 0 \\ 0 \end{bmatrix}$

13. $P = \begin{bmatrix} 3 & 1 \\ 1 & 1 \end{bmatrix}$ (The answer is not unique.)

14. $\{(0, 1, 0), (1, 1, 1), (2, 2, 3)\}$

15. $\begin{bmatrix} \dfrac{1}{\sqrt{2}} & \dfrac{1}{\sqrt{2}} \\ -\dfrac{1}{\sqrt{2}} & \dfrac{1}{\sqrt{2}} \end{bmatrix}$

16. $\begin{bmatrix} \dfrac{1}{\sqrt{3}} & \dfrac{1}{\sqrt{2}} & \dfrac{1}{\sqrt{6}} \\ \dfrac{1}{\sqrt{3}} & 0 & -\dfrac{2}{\sqrt{6}} \\ \dfrac{1}{\sqrt{3}} & -\dfrac{1}{\sqrt{2}} & \dfrac{1}{\sqrt{6}} \end{bmatrix}$

17. λ is an eigenvalue of A if there exists a nonzero vector **x** such that $A\mathbf{x} = \lambda\mathbf{x}$. **x** is called an eigenvector of A. If A is an $n \times n$ matrix, then A can have n eigenvalues, possibly complex and possibly repeated.

18. P is orthogonal if $P^{-1} = P^T$. The possible eigenvalues of an orthogonal matrix are 1 and -1.

19–23. Proof

24. 0 is the only eigenvalue.

INDEX

A

A and *A'* are similar, 392
Abstraction, 185
Addition
    of matrices, 47
    of vectors, 175, 177, 179, 186
Additive identity
    of a matrix, 61
    of a vector, 181, 186
        properties of, 181
Additive inverse
    of a matrix, 61
    of a vector, 171, 181, 186
        properties of, 181
Adjoining two matrices, 74
Adjoint of a matrix, 153, 154
Age distribution vector, 445
Age transition matrix, 446
Algebra of matrices, 60
Algebraic properties of the cross product, 329
Angle between two vectors, 279, 288
Approximation
    Fourier, 337, 339
    least squares, 333, 336
    nth-order, 338
Area of a triangle in the xy-plane, 160
Associative property
    of matrix addition, 60
    of scalar multiplication, 177, 186
    of vector addition, 177, 186
Augmented matrix, 15
Axioms, for vector space, 185

B

Back-substitution, 6, 18
Bases and linear dependence, 219
Basis, 215
    change of, 245
    coordinates relative to, 242
    orthogonal, 298
    orthonormal, 298
    standard, 215
    tests for, 223
Basis for the row space of a matrix, 227
Bessel's Inequality, 344
Block multiply two matrices, 59

C

Cancellation properties, 81
Cauchy, Augustin-Louis (1789–1857), 120, 278

Cauchy-Schwarz Inequality, 278, 291
Cayley, Arthur (1821–1895), 45
Cayley-Hamilton Theorem, 170, 422
Change of basis, 245
Characteristic equation, 149, 415
Characteristic polynomial, 170, 415
Circle, 259
Closed economic system, 104
Closure
    under scalar multiplication, 177, 186
    under vector addition, 177, 186
Coded row matrix, 101
Codomain, 352
Coefficient, 2, 54
    leading, 2
Coefficient matrix, 15
Cofactor, 121
    expansion by, 123
    matrix of, 153
    sign patterns for, 121
Column
    matrix, 46
    of a matrix, 14
    space, 226
    subscript, 14
    vector, 46, 226
Column-equivalent, 132
Commutative property
    of matrix addition, 60
    of vector addition, 177, 186
Companion matrix, 462
Component of a vector, 175
Composition, 382
Computer graphics, 399
Condition for diagonalization, 425
Conditions that yield a zero determinant, 133
Conic, 258, 259
    rotation, 261
Consistent system of linear equations, 5
Constant term, 2
Contraction, 397
Coordinate matrix of a vector, 242
Coordinates
    relative to an orthonormal basis, 302
    of a vector, relative to a basis, 242
Counterexample, A6
Cramer's Rule, 158, 159
Cross product of two vectors, 327
    algebraic properties of, 329
    geometric properties of, 330
Cryptogram, 100
Cryptography, 100

D

Derive, A16
Determinant, 119, 120, 122
    evaluation by elementary column operations, 132
    evaluation by elementary row operations, 130
    expansion by cofactors, 123
    of an inverse matrix, 143
    of an invertible matrix, 142
    of a matrix product, 139
    properties of, 139
    of a scalar multiple of a matrix, 140
    of a transpose, 145
    of a triangular matrix, 126
    of a 2×2 matrix, 120
    zero, 133
Diagonal matrix, 125
Diagonalizable, 423
Diagonalization
    condition for, 425
    problem, 423
    sufficient condition for, 430
Difference of two vectors, 163, 164, 176, 177, 179
Differential operator, 361
Dimension
    of the solution space, 235
    of a vector space, 221
Direct sum, 268, 315
Directed line segment, 175
Distance between two vectors, 275, 288
Distributive property, 60
    of scalar multiplication, 177, 186
Domain, 352
Dot product of two vectors, 275
    properties of, 276
Dynamical systems, 463

E

Eigenspace, 413, 419
Eigenvalue, 148
    of a linear transformation, 419
    of a matrix, 410, 411, 414
    multiplicity of, 417
    of a symmetric matrix, 435
    for triangular matrix, 419
Eigenvalue problem, 148, 410
Eigenvector, 148
    of a linear transformation, 419
    of a matrix, 410, 411, 414
Electrical network, 34